PE STRUCTURAL
REFERENCE MANUAL

TENTH EDITION

Alan Williams, PhD, SE, FICE, C Eng

PPI®

PPI2PASS.COM
A KAPLAN COMPANY

Report Errors for This Book

PPI is grateful to every reader who notifies us of a possible error. Your feedback allows us to improve the quality and accuracy of our products. Report errata at **ppi2pass.com**.

PE STRUCTURAL REFERENCE MANUAL
Tenth Edition

Current release of this edition: 5

Release History

date	edition number	revision number	update
Jan 2022	10	3	Minor corrections.
May 2022	10	4	Minor corrections.
Dec 2022	10	5	Minor corrections.

PPI
ppi2pass.com

ISBN: 978-1-59126-846-8

Table of Contents

Appendices
Table of Contents

Preface and Acknowledgments

I wrote the *PE Structural Reference Manual* to be a comprehensive resource to help you prepare for the National Council of Examiners for Engineering and Surveying (NCEES) 16-hour PE Structural exam. In each of this book's chapters, I have presented the most useful equations in the exam-adopted codes and standards and provided guidelines for selecting and applying these equations.

This, the tenth edition of the book, is updated to conform to the 2018 *International Building Code* (IBC). Several of the codes and specifications adopted by reference in the 2018 IBC have been revised since the ninth edition of this book was published. The tenth edition has extensive revisions and additions to Chapters 1, 2, and 5 through 9. Additionally, the text has been updated to be consistent with NCEES design standards and specifications, and all examples are worked using both allowable stress and strength design methods. The more significant changes to the chapters are as follows.

Chapter 1, Vertical Forces, Incidental Lateral Forces, and Other Variable Forces, is updated to conform to the 2016 edition of ASCE/SEI7, *Minimum Design Loads and Associated Criteria for Buildings and Other Structures*. Changes have been made to the calculation of ice loads, wind load on ice-covered members, tsunami loads, reduction of live loads, ground snow load maps and tables, inclusion of snow importance factor in the calculation of drift loads, drifts caused by roof projections, and the height and width of drifts.

Chapter 2, Reinforced Concrete Design, is expanded to include discussions of slender walls and special structural walls.

Chapter 5, Structural Steel Design, is updated to conform to the fifteenth edition of the AISC *Steel Construction Manual* and AISC 360-15. The new concepts introduced include design for tear-out strength at bolt holes, provision in the *Steel Construction Manual* of new tables for the direct design of members subject to combined forces, new tables for the design of composite columns, changes to plate girder shear and stiffener requirements, revised shear lag factors, and an additional high-strength bolt category.

Chapter 6, Wood Design, is updated to conform to the 2018 edition of the *National Design Specification* (NDS) *for Wood Construction*. The major changes are to the adjustment factors for load duration, stability, flat use,

incising, buckling stiffness, and format conversion. The effective bending stiffness of cross-laminated timber has been modified.

Chapter 7, Reinforced Masonry Design, is updated to conform to the 2016 edition of TMS 402/602, *Building Code Requirements and Specification for Masonry Structures*. The changes introduced include revised requirements for reinforcement placement and splicing, deep beam requirements, shear friction and boundary element requirements for shear walls, revised anchor bolt provisions, and quality assurance provisions.

Chapter 8, Lateral Forces, is updated to conform to ASCE/SEI7-2016. New concepts include revised wind speed maps, calculation of wind velocity pressure, definition of enclosed buildings, revised zone widths for external pressures, and elimination of the IBC alternate wind design procedure. Updated seismic requirements include new seismic design maps, changes to site coefficients, seismic response coefficients, redundancy factors, accidental torsion, collector design, diaphragm classification, and the simplified lateral force procedure.

Chapter 9, Bridge Design, is updated to conform to the eighth edition of *AASHTO LRFD Bridge Design Specifications*. For reinforced concrete construction, the determination of skin reinforcement, cracking moment, and fatigue limits have been modified. For prestressed concrete construction, allowable stresses and the determination of flexural strength have been modified.

I would like to thank the staff at PPI, including Michael Wordelman, lead editor; Indira Prabhu Kumar, editor; Grace Wong, editorial operations director; Scott Marley, editorial manager; Damon Larson, production editor; Beth Christmas, production manager; Jeri Jump, project manager; Richard Iriye and Stan Info Solutions, typesetters; Kim Burton-Weisman, proofreader; Tom Bergstrom, cover design and technical drawings; Louis Eleazar, technical drawings; Anna Howland, content manager; and Sam Webster, publishing systems manager.

Finally, if you find an error in this book, please let me know by using the error reporting form on the PPI website at **ppi2pass.com**. Valid submitted errors will be posted to the errata page and incorporated into future printings of this book.

Alan Williams, PhD, SE, FICE, C Eng

Introduction

PART 1: HOW TO USE THIS BOOK

This book, *PE Structural Reference Manual*, is designed to help you prepare for the 16-hour PE Structural exam administered by the National Council of Examiners for Engineering and Surveying (NCEES). The NCEES PE Structural exam will test your knowledge of structural principles by presenting problems that cover the design of an entire structure or a portion of a structure.

The exam is given in four modules—two concerning vertical forces and two concerning lateral forces. Chapter 1 covers vertical forces, incidental lateral forces, and other variable forces. Chapter 2 through Chap. 9 are organized around the eight areas in which these forces are applied, which are

- reinforced concrete design
- foundations and retaining structures
- prestressed concrete design
- structural steel design
- wood design
- reinforced masonry design
- lateral forces (wind and seismic)
- bridge design

The structural design principles presented in each chapter build on those in earlier chapters, so you should read the chapters in the order in which they are presented. The examples within each chapter should also be read in sequence. Taken together in this way, they constitute the solution to a complete design problem similar to that on the exam.

The last nine chapters of this book contain practice problems that make use of the same concepts discussed in the first nine chapters. The practice problems in Chap. 10 correspond to the topics covered in Chap. 1, those in Chap. 11 correspond to Chap. 2, and so on.

Your solutions to the PE Structural exam problems must be based on the NCEES-adopted codes and design standards. Therefore, you should carefully review the appropriate sections of the exam-adopted design standards and codes that are presented, analyzed, and explained in each chapter of this book. Each example in this book focuses on one specific code principle and offers a clear interpretation of that principle.

The "Codes and References" section lists the PE Structural design standards that code-based problems on the exam will reference. You will not receive credit for solutions based on other editions or standards. All problems are in customary U.S. (English) units, and you will not receive credit for solutions using SI units.

Abbreviations are used throughout this book to refer to the design standards and codes referenced by the PE Structural exam. This book's "Codes and References" section lists these abbreviations, each followed by the full name of its appropriate design standard or code. This book also cites other publications that discuss pertinent structural design procedures, and these publications are also listed in the "Codes and References" section. Text references to all other publications are numbered as endnotes in each chapter, and the publications are cited in the "References" section at the end of each chapter. These references are provided for your additional review.

As you prepare for the PE Structural exam, the following suggestions may also help.

- Become intimately familiar with this book. This means knowing the order of the chapters, the approximate locations of important figures and tables, and so on.

- Use the subject title tabs along the side of each page.

- Skim through a chapter to familiarize yourself with the subjects before starting the practice problems.

- To minimize time spent searching for often-used formulas and data, prepare a one-page summary of all the important formulas and information in each subject area. You can then refer to this summary during the exam instead of searching in this book.

- Use the index extensively. Every significant term, law, theorem, concept, and code (including sections, equations, and tables) has been indexed. If you don't recognize a term used, look for it in the index. Some subjects appear in more than one chapter. Use the index to learn all there is to know about a particular subject.

PART 2: EVERYTHING YOU EVER WANTED TO KNOW ABOUT THE PE STRUCTURAL EXAM

ABOUT THE EXAM

The PE Structural exam is offered in two components. The first component—vertical forces (gravity/other) and incidental lateral forces—takes place on a Friday. The second component—lateral forces (wind/earthquake)—takes place on a Saturday. Each component comprises a morning breadth module and an afternoon depth module, as outlined in Table 1.

Each morning breadth module lasts four hours and contains 40 multiple-choice problems that cover a range of structural engineering topics specific to vertical and lateral forces. Each afternoon depth module also lasts four hours, but instead of multiple-choice problems, each module contains constructed-response (essay) problems. You may choose either the bridges or the buildings depth module, but you must choose the same depth module within both exam components. That is, if you choose to work buildings for the lateral forces component, you must also work buildings for the vertical forces component.

According to NCEES, the vertical forces (gravity/other) and incidental lateral depth module in buildings covers loads, lateral earth pressures, analysis methods, general structural considerations (e.g., element design), structural systems integration (e.g., connections), and foundations and retaining structures. The same depth module in bridges covers gravity loads, superstructures, substructures, and lateral loads other than wind and seismic. It may also require pedestrian bridge and/or vehicular bridge knowledge.

The lateral forces (wind/earthquake) depth module in buildings covers lateral forces, lateral force distribution, analysis methods, general structural considerations (e.g., element design), structural systems integration (e.g., connections), and foundations and retaining structures. The same depth module in bridges covers gravity loads, superstructures, substructures, and lateral forces. It may also require pedestrian bridge and/or vehicular bridge knowledge.

WHAT DOES "MOST NEARLY" REALLY MEAN?

One of the more disquieting aspects of the exam's multiple-choice questions is that the available answer choices are seldom exact. Answer choices generally have only two or three significant digits. Exam questions ask, "Which answer choice is most nearly the correct value?" or they instruct you to complete the sentence, "The value is approximately ..." A lot of self-confidence is required to move on to the next question when you

don't find an exact match for the answer you calculated, or if you have had to split the difference because no available answer choice is close.

NCEES has described it like this.

> Many of the questions on NCEES exams require calculations to arrive at a numerical answer. Depending on the method of calculation used, it is very possible that examinees working correctly will arrive at a range of answers. The phrase "most nearly" is used to accommodate answers that have been derived correctly but that may be slightly different from the correct answer choice given on the exam. You should use good engineering judgment when selecting your choice of answer. For example, if the question asks you to determine the load on a beam, you should literally select the answer option that is most nearly what you calculated, regardless of whether it is more or less than your calculated value. However, if the question asks you to size a beam to carry a load, you should select an answer option that will safely carry the load. Typically, this requires selecting a value that is closest to but larger than the load.

The difference is significant. Suppose you were asked to calculate "most nearly" the volumetric pure water flow required to dilute a contaminated stream to an acceptable concentration. Suppose, also, that you calculated 823 gpm. If the answer choices were (A) 600 gpm, (B) 800 gpm, (C) 1000 gpm, and (D) 1200 gpm, you would go with answer choice (B), because it is most nearly what you calculated. If, however, you were asked to select a pump or pipe with the same rated capacities, you would have to go with choice (C). Got it?

HOW MUCH MATHEMATICS IS NEEDED FOR THE EXAM?

There are no pure mathematics questions (algebra, geometry, trigonometry, etc.) on the PE Structural exam. However, you will need to apply your knowledge of these subjects to the exam questions.

Generally, only simple algebra, trigonometry, and geometry are needed on the PE Structural exam. You will need to use trigonometric, logarithm, square root, exponentiation, and similar buttons on your calculator. There is no need to use any other method for these functions.

Except for simple quadratic equations, you will probably not need to find the roots of higher-order equations. Occasionally, it will be convenient to use the equation-solving capability of an advanced calculator. However, other solution methods will always exist. For second-order (quadratic) equations, it does not matter if you find roots by factoring, completing the square, using the quadratic equation, or using your calculator's root finder.

Table 1 NCEES PE Structural Exam Component/Module Specifications

Friday: vertical forces (gravity/other) and incidental lateral forces

morning breadth module	analysis of structures (13)
4 hours	generation of loads (5)
40 multiple-choice problems	load distribution and analysis methods (8)
	design and details of structures (27)
	general structural considerations (3)
	structural systems integration (2)
	structural steel (5)
	cold-formed steel (1)
	concrete (5)
	wood (4)
	masonry (3)
	foundations and retaining structures (4)
afternoon depth module[a]	buildings[b]
4 hours	steel structure (25% of your score)
essay problems	concrete structure (25% of your score)
	wood structure (25% of your score)
	masonry structure (25% of your score)
	bridges
	concrete superstructure (25% of your score)
	other elements of bridges (e.g., culverts, abutments, and retaining walls) (25% of your score)
	steel superstructure (50% of your score)

Saturday: lateral forces (wind/earthquake)

morning breadth module	analysis of structures (15)
4 hours	generation of loads (7)
40 multiple-choice problems	load distribution and analysis methods (8)
	design and details of structures (25)
	general structural considerations (3)
	structural systems integration (2)
	structural steel (5)
	cold-formed steel (1)
	concrete (5)
	wood (3)
	masonry (3)
	foundations and retaining structures (3)
afternoon depth module[a]	buildings[c]
4 hours	steel structure (25% of your score)
essay problems	concrete structure (25% of your score)
	wood and/or masonry structure (25% of your score)
	general analysis (e.g., existing structures, secondary structures, nonbuilding structures, and/or computer verification) (25% of your score)
	bridges
	piers or abutments (25% of your score)
	foundations (25% of your score)
	general analysis of seismic forces (50% of your score)

[a]Afternoon sessions focus on a single area of practice. You must choose *either* the buildings or bridges depth module, and you must work the same depth module across both exam components.

[b]At least one problem will contain a multistory building, and at least one problem will contain a foundation.

[c]At least two problems will include seismic content with a seismic design category of D or above. At least one problem will include wind content of at least 140 mph. Problems may include a multistory building and/or a foundation.

There is little or no use of calculus on the exam. Rarely, you may need to take a simple derivative to find a maximum or minimum of some function. Even rarer is the need to integrate to find an average, moment of inertia, statical moment, or shear flow.

Basic statistical analysis of observed data may be necessary. Statistical calculations are generally limited to finding means, medians, standard deviations, variances, percentiles, and confidence limits. Usually, the only population distribution you need to be familiar with is the normal curve. Probability, reliability, hypothesis testing, and statistical quality control are not explicit exam subjects, though their concepts may appear peripherally in some problems.

The PE Structural exam is concerned with numerical answers, not with proofs or derivations. You will not be asked to prove or derive formulas.

Occasionally, a calculation may require an iterative solution method. Generally, there is no need to complete more than two iterations. You will not need to program your calculator to obtain an "exact" answer. Nor will you generally need to use complex numerical methods.

IS THE EXAM TRICKY?

Other than providing superfluous data, the PE Structural exam is not a "tricky exam." The exam questions are difficult in their own right. NCEES does not provide misleading or conflicting statements to try to get you to fail. However, commonly made mistakes are represented in the available answer choices. Thus, the alternative answers (known as distractors) will be logical.

Questions are generally practical, dealing with common and plausible situations that you might experience in your job. You will not be asked to design a structure for reduced gravity on the moon, to design a mud-brick road, to analyze the effects of a nuclear bomb blast on a structure, or to use bamboo for tension reinforcement.

WHAT MAKES THE QUESTIONS DIFFICULT?

Some questions are difficult because the pertinent theory is not obvious. There may be only one acceptable procedure, and it may be heuristic (or defined by a code) such that nothing else will be acceptable.

Some questions are difficult because the data needed is hard to find. Some data just isn't available unless you happen to have brought the right reference book. Many of the structural questions are of this nature. There is no way to solve most structural steel questions without the *Steel Construction Manual*. Designing an eccentrically loaded concrete column without published interaction diagrams is nearly impossible in six minutes.

Some questions are difficult because they defy the imagination. Three-dimensional structural questions fit this description. If you cannot visualize the question, you probably cannot solve it.

Some questions are difficult because the computational burden is high, and they just take a long time.

Some questions are difficult because the terminology is obscure, and you may not know what the terms mean. This can happen in almost any subject.

WHAT REFERENCE MATERIAL IS PERMITTED IN THE EXAM?

The PE Structural exam is an open-book exam. Check your state's exam requirements and restrictions, as some states restrict which books and materials can be used for the exam. (The PPI website has a listing of state boards at **ppi2pass.com**.)

Personal notes in a three-ring binder and other semipermanent covers can usually be used. Some states use a "shake test" to eliminate loose papers from binders. Make sure that nothing escapes from your binders when they are inverted and shaken.

The references you bring into the exam room in the morning do not have to be the same as the references you use in the afternoon. However, you cannot share books with other examinees during the exam.

A few states do not permit collections of solved problems such as *Schaum's Outline Series*, sample exams, and solutions manuals. A few states maintain a formal list of banned books.

Strictly speaking, loose paper and scratch pads are not permitted in the exam. Certain types of preprinted graphs and logarithmically scaled graph papers (which are almost never needed) should be three-hole punched and brought in a three-ring binder. An exception to this restriction may be made for laminated and oversize charts, graphs, and tables that are commonly needed for particular types of questions. However, there probably aren't any such items for the PE Structural exam.

MAY TABS BE PLACED ON PAGES?

It is common to tab pages in your books in an effort to reduce the time required to locate useful sections. Inasmuch as some states consider Post-it notes to be "loose paper," your tabs should be of the more permanent variety. Although you can purchase tabs with gummed attachment points, it is also possible to use transparent tape to attach the Post-its you have already placed in your books.

CAN YOU WRITE AND MARK IN YOUR BOOKS?

During your preparation, you may write anything you want, anywhere in your books, including this one. You can use pencil, pen, or highlighter in order to further your understanding of the content. However, during the exam, you must avoid the appearance of taking notes about the exam. This means that you should write only on the scratch paper that is provided. During the exam, other than drawing a line across a wide table of numbers, you should not write in your books.

WHAT ABOUT CALCULATORS?

The PE Structural exam requires the use of a scientific calculator. It is a good idea to bring a spare calculator with you to the exam.

NCEES has banned communicating and text-editing calculators from the exam site. Only select types of calculators are permitted. Check the current list of permissible devices at the PPI website (**ppi2pass.com**). Contact your state board to determine if nomographs and specialty slide rules are permitted.

The exam has not been optimized for any particular brand or type of calculator. In fact, for most calculations, a $15 scientific calculator will produce results as satisfactory as those from a $200 calculator. There are definite benefits to having built-in statistical functions, graphing, unit-conversion, and equation-solving capabilities. However, these benefits are not so great as to give anyone an unfair advantage.

It is essential that a calculator used for the PE Structural exam have the following functions.

- trigonometric and inverse trigonometric functions
- hyperbolic and inverse hyperbolic functions
- π
- $\sqrt{x}$ and x^2
- both common and natural logarithms
- y^x and e^x

You may not share calculators with other examinees.

Laptops, tablet computers, and electronic readers are not permitted in the exam. Their use has been considered, but no states actually permit them.

You may not use a walkie-talkie, cell phone, or other communications or text-messaging device during the exam.

Be sure to take your calculator with you whenever you leave the exam room for any length of time.

HOW ARE THE EXAM COMPONENTS GRADED AND SCORED?

For the morning multiple-choice problems, answers are recorded on an answer sheet that is machine graded. The minimum number of points for passing (referred to by NCEES as the "cut score") varies from administration to administration. The cut score is determined through a rational procedure, without the benefit of knowing examinees' performance on the exam. That is, the exam is not graded on a curve. The cut score is selected based on what you are expected to know, not on allowing a certain percentage of engineers "through."

The grading of multiple-choice problems is straightforward, since a computer grades your score sheet. Either you get the problem right or you don't. There is no deduction for incorrect answers, so guessing is encouraged. However, if you mark two or more answers, no credit is given for the problem.

Solutions for the afternoon essay problems are evaluated for overall compliance with established scoring criteria and for general quality. The scores from each of the morning and afternoon modules are combined for a component's final score.

Exam results are given a pass/fail grade approximately 10–12 weeks after the exam date. You will receive the results of your exam from either your state board by mail or online through your MyNCEES account. You will receive a pass or fail notice only and will not receive a numerical score. Diagnostic reports that outline areas of strength and weakness are provided to those who do not pass.

HOW YOU SHOULD GUESS

NCEES produces defensible licensing exams. As a result, there is no pattern to the placement of correct responses. Therefore, it most likely will not help you to guess all "A," "B," "C," or "D."

The proper way to guess is as an engineer. You should use your knowledge of the subject to eliminate illogical answer choices. Illogical answer choices are those that violate good engineering principles, that are outside normal operating ranges, or that require extraordinary assumptions. Of course, this requires you to have some basic understanding of the subject in the first place. Otherwise, it's back to random guessing. That's the reason that the minimum passing score is higher than 25%.

You won't get any points using the "test-taking skills" that helped with tests prepared by amateurs. You won't be able to eliminate any [verb] answer choices from "Which [noun] ..." questions. You won't find problems with options of the "more than 50" and "less than 50" variety. You won't find one answer choice among the four that has a different number of significant digits, or has a verb in a different tense, or has some singular/plural discrepancy with the stem. The distractors will always match the stem, and they will be logical.

CHEATING AND EXAM SUBVERSION

There aren't very many ways to cheat on an open-book exam. The proctors are well trained in spotting the few ways that do exist. It goes without saying that you should not talk to other examinees in the room, nor should you pass notes back and forth. You should not write anything into your books or take notes on the contents of the exam. The number of people who are released to use the restroom may be limited to prevent discussions.

NCEES regularly reuses good problems that have appeared on previous exams. Therefore, exam integrity is a serious issue with NCEES, which goes to great lengths to make sure nobody copies the questions. You may not keep your exam booklet, enter text from questions into your calculator, or copy problems into your own material.

The proctors are concerned about exam subversion, which generally means activity that might invalidate the exam or the exam process. The most common form of exam subversion involves trying to copy exam problems for future use.

NCEES has become increasingly unforgiving about the loss of its intellectual property. NCEES routinely prosecutes violators and seeks financial redress for loss of its exam problems, as well as invalidating any engineering license you may have earned by taking one of its exams while engaging in prohibited activities. Your state board may impose additional restrictions on your right to retake any exam if you are convicted of such activities. In addition to tracking down the sources of any exam problem compilations that it becomes aware of, NCEES is also aggressive in pursuing and prosecuting examinees who disclose the contents of the exam in internet forum and "chat" environments. Your constitutional rights to free speech and expression will not protect you from civil prosecution for violating the nondisclosure agreement that NCEES requires you to sign before taking the exam. If you wish to participate in a dialogue about a particular exam subject, you must do so in such a manner that does not violate the essence of your nondisclosure agreement. This requires decoupling your discussion from the exam and reframing the question to avoid any exam particulars.

PART 3: HOW TO PREPARE FOR AND PASS THE PE STRUCTURAL EXAM

WHAT SHOULD YOU STUDY?

The exam covers many diverse subjects. Strictly speaking, you don't have to study every subject on the exam in order to pass. However, the more subjects you study, the more you'll improve your chances of passing. You should decide early in the preparation process which subjects you are going to study. The strategy you select will depend on your background. The four most common strategies are as follows.

- A broad approach is the key to success for examinees who have recently completed their academic studies. This strategy is to review the fundamentals of a broad range of undergraduate subjects (which means studying all or most of the chapters in this book). The exam includes enough fundamental problems to make this strategy worthwhile. Overall, it's the best approach.

- Engineers who have little time for preparation tend to concentrate on the subject areas in which they hope to find the most problems. By studying the list of exam subjects, some have been able to focus on those subjects that will give them the highest probability of finding enough problems that they can answer. This strategy works as long as the exam has enough of the types of questions they need. Too often, though, examinees who pick and choose subjects to review can't find enough problems to complete the exam.

- Engineers who have been away from classroom work for a long time tend to concentrate on the subjects in which they have had extensive experience, in the hope that the exam will feature lots of problems in those subjects. This method is seldom successful.

- Some engineers plan on modeling their solutions from similar problems they have found in textbooks, collections of solutions, and old exams. These engineers often spend a lot of time compiling and indexing the example and sample problem types in all of their books. This is not a legitimate preparation method, and it is almost never successful.

HOW LONG SHOULD YOU STUDY?

We've all heard stories of the person who didn't crack a book until the week before the exam and still passed it with flying colors. Yes, these people really exist. However, I'm not one of them, and you probably aren't either. In fact, I'm convinced that these people are as rare as the ones who have taken the exam five times and still can't pass it.

A thorough review takes approximately 300 hours. Most of this time is spent solving problems. Some of it may be spent in class; some is spent at home. Some examinees spread this time over a year. Others cram it all into two months. Most classroom review courses last for three or four months. The best time to start studying will depend on how much time you can spend per week.

DO YOU NEED A REVIEW SCHEDULE?

It is important that you develop and adhere to a review outline and schedule. Once you have decided which subjects you are going to study, you can allocate the available time to those subjects in a manner that makes sense to you. If you are not taking a classroom review course (where the order of preparation is determined by the lectures), you should make an outline of subjects for self-study to use for scheduling your preparation. A fill-in-the-dates schedule is provided in Table 2 at the end of this introduction.

A SIMPLE PLANNING SUGGESTION

Designate some location (a drawer, a corner, a cardboard box, or even a paper shopping bag left on the floor) as your "exam catch-all." Use your catch-all during the months before the exam when you have revelations about things you should bring with you. For example, you might realize that the plastic ruler marked off in tenths of an inch that is normally kept in the kitchen junk drawer can help you with some soil pressure questions. Or you might decide that a certain book is particularly valuable, or that it would be nice to have dental floss after lunch, or that large rubber bands and clips are useful for holding books open.

It isn't actually necessary to put these treasured items in the catch-all during your preparation. You can, of course, if it's convenient. But if these items will have other functions during the time before the exam, at least write yourself a note and put the note into the catch-all. When you go to pack your exam kit a few days before the exam, you can transfer some items immediately, and the notes will be your reminders for the other items that are back in the kitchen drawer.

HOW YOU CAN MAKE YOUR REVIEW REALISTIC

During the exam, you must be able to recall solution procedures, formulas, and important data quickly. You must remain sharp for eight hours or more. If you played a sport back in school, your coach tried to put you in game-related situations. Preparing for the PE Structural exam isn't much different than preparing for a big game. Some part of your preparation should be realistic and representative of the exam environment.

There are several things you can do to make your review more representative. For example, if you gather most of your review resources (i.e., books) in advance and try to use them exclusively during your review, you will become more familiar with them. (Of course, you can also add to or change your references if you find inadequacies.)

Learning to use your time wisely is one of the most important lessons you can learn during your review. You will undoubtedly encounter questions that end up taking much longer than you expected. In some instances, you will cause your own delays by spending too much time looking through books for things you need (or just by looking for the books themselves!). Other times, the questions will entail too much work. It is important that a portion of your review involves solving problems so that you learn to recognize these situations and so that you can make intelligent decisions about skipping such questions during the exam.

WHAT TO DO A FEW DAYS BEFORE THE EXAM

There are a few things you should do a week or so before the exam.

You should arrange for childcare and transportation. Since the exam does not always start or end at the designated time, make sure that your childcare and transportation arrangements are flexible.

Check PPI's website for last-minute updates and errata to any PPI books you might have and are bringing to the exam.

Obtain a separate copy of this book's index (e.g., by photocopying it from this book).

If it's convenient, visit the exam location in order to find the building, parking areas, exam room, and restrooms. If it's not convenient, you can find driving directions and/or site maps online.

Take the battery cover off your calculator and check to make sure you are bringing the correct size replacement batteries. Some calculators require a different kind of battery for their "permanent" memories. Put the cover back on and secure it with a piece of masking tape. Write your name on the tape to identify your calculator.

If your spare calculator is not the same as your primary calculator, spend a few minutes familiarizing yourself with how it works. In particular, you should verify that your spare calculator is functional.

PREPARE YOUR CAR

- [ ] Gather snow chains, shovel, and tarp to kneel on while installing chains.
- [ ] Check tire pressures.
- [ ] Check your car's spare tire.
- [ ] Check for tire installation tools.
- [ ] Verify that you have the vehicle manual.
- [ ] Check fluid levels (oil, gas, water, brake fluid, transmission fluid, window-washing solution).
- [ ] Fill up car with gas.
- [ ] Check battery and charge if necessary.
- [ ] Know something about your fuse system (where they are, how to replace them, etc.).
- [ ] Assemble all required maps.
- [ ] Fix anything that might slow you down (missing wiper blades, etc.).
- [ ] Check your car's taillights.
- [ ] Affix the current DMV registration sticker.
- [ ] Fix anything that might get you pulled over on the way to the exam (burned-out taillight or headlight, broken lenses, bald tires, missing license plate, noisy muffler, etc.).
- [ ] Treat the inside windows with anti-fog solution.
- [ ] Put a roll of paper towels in the back seat.
- [ ] Gather exact change for any bridge tolls or toll roads.
- [ ] Put $20 in your car's glove box.
- [ ] Check for current registration and proof of insurance.
- [ ] Locate a spare door and ignition key.
- [ ] Find your roadside-assistance cards and phone numbers.
- [ ] Plan alternate routes.

PREPARE YOUR EXAM KITS

Second in importance to your scholastic preparation is the preparation of your two exam kits. The first kit consists of a bag, box (plastic milk crates hold up better than cardboard in the rain), or wheeled travel suitcase containing items to be brought with you into the exam room.

- [ ] your exam authorization notice
- [ ] government-issued photo identification (e.g., driver's license)
- [ ] this book
- [ ] other textbooks and reference books

- [ ] regular dictionary
- [ ] scientific/engineering dictionary
- [ ] review course notes in a three-ring binder
- [ ] cardboard boxes or plastic milk crates to use as bookcases
- [ ] primary calculator
- [ ] spare calculator
- [ ] instruction booklets for your calculators
- [ ] extra calculator batteries
- [ ] straightedge and rulers
- [ ] compass
- [ ] protractor
- [ ] scissors
- [ ] stapler
- [ ] transparent tape
- [ ] magnifying glass
- [ ] small (jeweler's) screwdriver for fixing your glasses or for removing batteries from your calculator
- [ ] unobtrusive (quiet) snacks or candies, already unwrapped
- [ ] two small plastic bottles of water
- [ ] travel pack of tissue (keep in your pocket)
- [ ] handkerchief
- [ ] headache remedy
- [ ] personal medication
- [ ] $5.00 in miscellaneous change
- [ ] back-up reading glasses
- [ ] light, comfortable sweater
- [ ] loose shoes or slippers
- [ ] cushion for your chair
- [ ] earplugs
- [ ] wristwatch
- [ ] several large trash bags ("raincoats" for your boxes of books)
- [ ] roll of paper towels
- [ ] wire coat hanger (to hang up your jacket)
- [ ] extra set of car keys

The second kit consists of the following items and should be left in a separate bag or box in your car in case they are needed.

- [ ] copy of your application
- [ ] proof of delivery
- [ ] light lunch
- [ ] beverage in thermos or cans
- [ ] sunglasses
- [ ] extra pair of prescription glasses
- [ ] raincoat, boots, gloves, hat, and umbrella

[ ] street map of the exam area
[ ] parking permit
[ ] battery-powered desk lamp
[ ] your cell phone
[ ] piece of rope

PREPARE FOR THE WORST

All of the occurrences listed in this section have happened to examinees. Granted, you cannot prepare for every eventuality. But even though each occurrence is a low-probability event, taken together these occurrences are worth considering in advance.

- Imagine getting a flat tire, getting stuck in traffic, or running out of gas on the way to the exam.

- Imagine rain and snow as you are carrying your cardboard boxes of books into the exam room. Would plastic trash bags be helpful?

- Imagine arriving late. Can you get into the exam without having to make two trips from your car?

- Imagine having to park two blocks from the exam site. How are you going to get everything to the exam room? Can you actually carry everything that far? Could you use a furniture dolly, a supermarket basket, or perhaps a helpmate?

- Imagine a Star Trek convention, a square-dancing contest, construction, or an auction taking place in the next room.

- Imagine a site without any heat, with poor lighting, or with sunlight streaming directly into your eyes.

- Imagine a hard folding chair and a table with one short leg.

- Imagine a site next to an airport with frequent take-offs, or next to a construction site with a pile driver, or next to the NHRA State Championship.

- Imagine a seat where someone nearby chews gum with an open mouth; taps his pencil or drums her fingers; or wheezes, coughs, and sneezes for eight hours.

- Imagine the distraction of someone crying or of proctors evicting yelling and screaming examinees who have been found cheating. Imagine the tragedy of another examinee's serious medical emergency.

- Imagine a delay of an hour while they find someone to unlock the building, turn on the heat, or wait for the head proctor to bring instructions.

- Imagine a power outage occurring sometime during the exam.

- Imagine a proctor who (a) tells you that one of your favorite books can't be used during the exam,

(b) accuses you of cheating, or (c) calls "time's up" without giving you any warning.

- Imagine not being able to get your lunch out of your car or find a restaurant.

- Imagine getting sick or nervous during the exam.

- Imagine someone stealing your calculator during lunch.

WHAT TO DO THE DAY BEFORE THE EXAM

Take the day before the exam off from work to relax. Do not cram. A good night's sleep is the best way to start the exam. If you live a considerable distance from the exam site, consider getting a hotel room in which to spend the night.

Practice setting up your exam work environment. Carry your boxes to the kitchen table. Arrange your "bookcases" and supplies. Decide what stays on the floor in boxes and what gets an "honored position" on the tabletop.

Use your checklist to make sure you have everything. Make sure your exam kits are packed and ready to go. Wrap your boxes in plastic bags in case it's raining when you carry them from the car to the exam room.

Calculate your wake-up time and set the alarms on two bedroom clocks. Select and lay out your clothing items. (Dress in layers.) Select and lay out your breakfast items.

If it's going to be hot on exam day, put your (plastic) bottles of water in the freezer.

Make sure you have gas in your car and money in your wallet.

WHAT TO DO THE DAY OF THE EXAM

Turn off the quarterly and hourly alerts on your wristwatch. Leave your cell phone in your car. If you must bring it, set it on silent mode. Bring a morning newspaper.

You should arrive at least 30 minutes before the exam starts. This will allow time for finding a convenient parking place, bringing your materials to the exam room, adjusting to room and seating changes, and calming down. Be prepared, though, to find that the exam room is not open or ready at the designated time.

WHAT TO DO DURING THE EXAM

All of the procedures typically associated with timed, proctored, computer-graded assessment tests will be in effect when you take the PE Structural exam.

The proctors will distribute the exam booklets and answer sheets if they are not already on your tables. You should not open the booklets until instructed to do so. You may read the information on the front and back covers, and you should write your name in the appropriate blank spaces.

Listen carefully to everything the proctors say. Do not ask your proctors any engineering questions. Even if they are knowledgeable in engineering, they will not be permitted to answer your questions.

Answers to questions are recorded on an answer sheet contained in the test booklet. The proctors will guide you through the process of putting your name and other biographical information on this sheet when the time comes, which will take approximately 15 minutes. You will be given the full four hours to answer questions. Time to initialize the answer sheet is not part of your four hours.

The common suggestions to "completely fill the bubbles, and erase completely" apply here. NCEES provides each examinee with a mechanical pencil with HB lead. Use of ballpoint pens and felt-tip markers is prohibited.

If you finish the exam early and there are still more than 30 minutes remaining, you will be permitted to leave the room. If you finish less than 30 minutes before the end of the exam, you may be required to remain until the end. This is done to be considerate of the people who are still working.

When you leave, you must return your exam booklet. You may not keep the exam booklet for later review.

If there are any questions that you think were flawed, in error, or unsolvable, ask a proctor for a "reporting form" on which you can submit your comments. Follow your proctor's advice in preparing this document.

HOW TO SOLVE MULTIPLE-CHOICE QUESTIONS

When you begin each session of the exam, observe the following suggestions.

- Use only the pencil provided.

- Do not spend an inordinate amount of time on any single question. If you have not answered a question in a reasonable amount of time, make a note of it and move on.

- Set your wristwatch alarm for five minutes before the end of each four-hour session, and use that remaining time to guess at all of the remaining questions. Odds are that you will be successful with about 25% of your guesses, and these points will more than make up for the few points that you might earn by working during the last five minutes.

- Make mental notes about any questions for which you cannot find a correct response, that appear to have two correct responses, or that you believe have some technical flaw. Errors in the exam are rare, but they do occur. Such errors are usually discovered during the scoring process and discounted from the exam, so it is not necessary to tell your proctor, but be sure to mark the one best answer before moving on.

- Make sure all of your responses on the answer sheet are dark. Completely fill the bubbles.

SOLVE QUESTIONS CAREFULLY

Many points are lost to carelessness. Keep the following items in mind when you are solving the end-of-chapter questions. Hopefully, these suggestions will be automatic during the exam.

[ ]   Did you recheck your mathematical equations?

[ ]   Do the units cancel out in your calculations?

[ ]   Did you convert between radius and diameter?

[ ]   Did you convert between feet and inches?

[ ]   Did you convert from gage to absolute pressures?

[ ]   Did you convert between kPa and Pa?

[ ]   Did you recheck all data obtained from other sources, tables, and figures?

SHOULD YOU TALK TO OTHER EXAMINEES AFTER THE EXAM?

The jury is out on this question. People react quite differently to the exam experience. Some people are energized. Most are exhausted. Some people need to unwind by talking with other examinees, describing every detail of their experience, and dissecting every exam question. Other people need lots of quiet space. Most engineers are in this latter category.

Since everyone who took the exam has seen it, you will not be violating your "oath of silence" if you talk about the details with other examinees immediately after the exam. It's difficult not to ask how someone else approached a question that had you completely stumped. However, keep in mind that it is very disquieting to think you answered a question correctly, only to have someone tell you where you went wrong.

To ensure you do not violate the nondisclosure agreement you signed before taking the exam, make sure you do not discuss any exam particulars with people who have not also taken the exam.

AFTER THE EXAM

Yes, there is something to do after the exam. Most people go home, throw their exam "kits" into the corner, and collapse. A week later, when they can bear to think about the experience again, they start integrating their exam kits back into their normal lives. The calculators go back into the drawer, the books go back on the shelves, the $5.00 in change goes back into the piggy bank, and all of the miscellaneous stuff brought to the exam is put back wherever it came.

Here's what I suggest you do as soon as you get home.

[ ] Thank your partner and children for helping you during your preparation.

[ ] Take any paperwork you received on exam day out of your pocket, purse, or wallet. Put this inside your *PE Structural Reference Manual*.

[ ] Reflect on any statements regarding exam secrecy to which you signed your agreement.

[ ] Call your employer and tell him/her that you need to take a mental health day on Monday.

A few days later, when you can face the world again, do the following.

[ ] Make notes about anything you would do differently if you had to take the exam over again.

[ ] Consolidate all of your application paperwork, correspondence to/from your state, and any paperwork that you received on exam day.

[ ] If you took a live prep course, call or email the instructor (or write a note) to say, "Thanks."

[ ] Return any books you borrowed.

[ ] Write thank-you notes to all of the people who wrote letters of recommendation or reference for you.

[ ] Find and read the chapter in this book that covers ethics. There were no ethics questions on your PE Structural exam, but it doesn't make any difference. Ethical behavior is expected of an SE in any case. Spend a few minutes reflecting on how your performance (obligations, attitude, presentation, behavior, appearance, etc.) might be about to change once you are licensed. Consider how you are going to be a role model for others around you.

[ ] Put all of your review books, binders, and notes someplace where they will be out of sight.

FINALLY

By the time you've "undone" all of your preparations, you might have thought of a few things that could help future examinees. If you have any sage comments about how to prepare, any suggestions about what to do during or bring to the exam, any comments on how to improve this book, or any funny anecdotes about your experience, I hope you will share these with me.

AND THEN, THERE'S THE WAIT ...

Waiting for the exam results is its own form of mental torture.

Although the actual machine grading "only takes seconds," consider the following facts: (a) NCEES prepares multiple exams for each administration, in case one becomes unusable (i.e., is inappropriately released) before the exam date. (b) Since the actual version of the exam used is not known until after it is finally given, the cut score determination occurs after the exam date.

I wouldn't be surprised to hear that NCEES receives dozens, if not hundreds, of claims from well-meaning examinees who were 100% certain that the exams they took were fatally flawed to some degree—that there wasn't a correct answer for such-and-such question— that there were two answers for such-and-such question —or even, perhaps, that such-and-such question was missing from their exam booklet altogether. Each of these claims must be considered as a potential adjustment to the cut score.

Then the exams must actually be graded. Since grading nearly 50,000 exams (counting all the FE and PE exams) requires specialized equipment, software, and training not normally possessed by the average employee, as well as time to do the work (also not normally possessed by the average employee), grading is invariably outsourced.

Outsourced grading cannot begin until all of the states have returned their score sheets to NCEES and NCEES has sorted, separated, organized, and consolidated the score sheets into whatever sequence is best. During grading, some of the score sheets "pop out" with any number of abnormalities that demand manual scoring.

After the individual exams are scored, the results are analyzed in a variety of ways. Some of the analysis looks at passing rates by such delineators as degree, major, university, site, and state. Part of the analysis looks for similarities between physically adjacent examinees (to look for cheating). Part of the analysis looks for exam sites that have statistically abnormal group performance. And some of the analysis looks for exam questions that have a disproportionate fraction of successful or unsuccessful examinees. Anyway, you get the idea: Grading is not merely putting your exam sheet in an electronic reader. All of these steps have to be completed for 100% of the examinees before any results can go out.

Once NCEES has graded your test and notified your state, when you hear about it depends on when the work is done by your state. Some states have to approve the results at a board meeting; others prepare the certificates before sending out notifications. Some states are more computerized than others. Some states have 50 examinees, while others have 10,000. Some states are shut down by blizzards and hurricanes; others are administratively challenged—understaffed, inadequately trained, or over budget.

There is no pattern to the public release of results. None. The exam results are not released to all states simultaneously. (The states with the fewest examinees often receive their results soonest.) They are not released alphabetically by state or examinee name. The people who failed are not notified first (or last). Your coworker might receive his or her notification today, and you might be waiting another three weeks for yours.

Some states post the names of the successful examinees, unsuccessful examinees, or both on their official state websites before the results go out. Others update their websites after the results go out. Some states don't list much of anything on their websites.

Remember, too, that the size or thickness of the envelope you receive from your state does not mean anything. Some states send a big congratulations package and certificate. Others send a big package with a new application to repeat the exam. Some states send a postcard. Some send a one-page letter. Some states send you an invoice for your license fees. (Ahh, what a welcome bill!) You just have to open it to find out.

AND WHEN YOU PASS ...

[ ] Celebrate.

[ ] Notify the people who wrote letters of recommendation or reference for you.

[ ] Ask your employer for a raise.

[ ] Tell the folks at PPI (who have been rootin' for you all along) the good news.

Table 2 *Schedule for Self-Study*

chapter number	subject	date to start	date to finish
1, 10	Vertical Forces, Incidental Lateral Forces, and Other Variable Forces		
2, 11	Reinforced Concrete Design		
3, 12	Foundations and Retaining Structures		
4, 13	Prestressed Concrete Design		
5, 14	Structural Steel Design		
6, 15	Wood Design		
7, 16	Reinforced Masonry Design		
8, 17	Lateral Forces		
9, 18	Bridge Design		

Codes and References

NCEES CODES

AASHTO: *AASHTO LRFD Bridge Design Specifications*, eighth ed., 2017. American Association of State Highway and Transportation Officials, Washington, DC.

ACI 318: *Building Code Requirements for Structural Concrete*, 2014. American Concrete Institute, Farmington Hills, MI.

AISC: *Seismic Design Manual*, third ed., 2018. American Institute of Steel Construction, Inc., Chicago, IL.

AISC: *Steel Construction Manual*, fifteenth ed., 2017. American Institute of Steel Construction, Inc., Chicago, IL.

AISI S100: *North American Specification for the Design of Cold-Formed Steel Structural Members*, 2016 ed. American Iron and Steel Institute, Washington, DC.

ASCE/SEI7: *Minimum Design Loads and Associated Criteria for Buildings and Other Structures*, 2016 ed. American Society of Civil Engineers, Reston, VA.

IBC: *International Building Code*, 2018 ed. International Code Council, Falls Church, VA.

NDS: *National Design Specification for Wood Construction*, 2018 ed., and *National Design Specification Supplement, Design Values for Wood Construction*, 2018 ed., American Wood Council, Leesburg, VA.

NDS (SDPWS): *Special Design Provisions for Wind and Seismic*, 2015 ed. American Wood Council, Leesburg, VA.

TMS 402/602: *Building Code Requirements and Specification for Masonry Structures*, 2016. The Masonry Society, Boulder, CO.

ADDITIONAL RECOMMENDED REFERENCES

American Institute of Steel Construction. *AISC Basic Design Values Cards*, 2017. Chicago, IL: American Institute of Steel Construction. (This resource is available online as a downloadable PDF.)

American Iron and Steel Institute. AISI S213: *North American Standard for Cold-Formed Steel Framing—Lateral Design, 2015 Edition with Supplement No. 1*, 2016. Washington, DC: American Iron and Steel Institute.

Building Seismic Safety Council of the National Institute of Building Sciences. *NEHRP Recommended Seismic Provisions for New Buildings and Other Structures*. Washington, DC: Building Seismic Safety Council; the Earthquake Hazards Reduction Program; the Federal Emergency Management Agency; and the U.S. Department of Homeland Security.

Precast/Prestressed Concrete Institute. *PCI Design Handbook: Precast and Prestressed Concrete*, seventh ed., 2010. Chicago, IL: Precast/Prestressed Concrete Institute.

Simpson Strong-Tie Company. *Wood Construction Connectors Catalog*. Pleasanton, CA: Simpson Strong-Tie Company. (This resource is available online as a downloadable PDF.)

United States Department of Defense. *Seismic Design for Buildings*. Washington, DC: United States Department of Defense.

1 Vertical Forces, Incidental Lateral Forces, and Other Variable Forces

1. DEAD LOADS

Nomenclature

a	beam spacing	ft
A_T	tributary area	ft
B_T	tributary width	ft
l	beam span	ft
V_{DB}	beam dead load reaction	lbf
V_{DC}	total dead load applied to column	lbf
V_{DG}	girder dead load reaction	lbf
w_B	beam self-weight	lbf/ft
w_G	girder self-weight	lbf/ft
w_L	uniform line load	lbf/ft
w_{Lij}	member ij uniform line load	lbf/ft
w_S	slab weight plus superimposed dead load	lbf/ft^2
w_T	total equivalent uniformly distributed dead load	lbf/ft^2
W_{Dij}	total dead load on member ij	lbf
W_{DB}	total dead load on beam	lbf
W_{DG}	total dead load on girder	lbf

General Principles

A *dead load* is defined in the *International Building Code*[1] (IBC) Sec. 202 and in *Minimum Design Loads for Buildings and Other Structures*[2] ASCE/SEI7 Sec. 3.1.1 as the weight of construction materials and fixed service equipment incorporated into the building. These consist of permanent walls and partitions, floors, roofs, ceilings, stairways, finishes, cladding, architectural features, and fixed equipment such as cranes, HVAC (heating, ventilating, and air conditioning) systems, automatic sprinkler systems, and utility services.

To provide a uniform approach in design, a list of customary weights of building materials is given in ASCE/SEI7 Table C31-1A, and a list of the customary densities of building materials is given in ASCE/SEI7 Table C3.1-2. In using these tables, however, allowance must be made for possible variations in the actual materials used. Allowance is also required for the possible future addition of permanent equipment loads and for additional wearing surfaces.

Over the life of a structure, partition walls not permanently attached may be rearranged many times in order to divide large floor areas into smaller offices and cubicles. IBC Sec. 1607.5 specifies a value of 15 lbf/ft^2 for this partition load; moreover, because the load is not permanently fixed in position, it is considered a nonreducible *live* load. This partition load must be added to the specified floor live load whether or not partitions are shown on the construction documents; however, a partition load need not be applied when the floor live load exceeds 80 lbf/ft^2.

Tributary Areas for a Slab with an Aspect Ratio Exceeding Two

Figure 1.1 shows the floor layout of a reinforced concrete building. The floor slab is supported on beams, which run east-west and are supported by either columns or girders. Girders run north-south and are supported by columns. The aspect ratio of the slab panels exceeds two, and the slab is designed as a one-way slab spanning north-south between beams. Design of the slab may be by elastic analysis or by the approximate method given in the American Concrete Institute's *Building Code Requirements for Structural Concrete and Commentary*[3] ACI Sec. 6.5.2.

Dead loads are usually uniformly distributed, and the loading on a specific member of the structure may be determined using the tributary area principle. The *tributary area* for a specific member is the loaded area that is considered to directly load that member.

Tributary Area of a Beam

The *tributary area* for a typical interior beam, beam 78, is shown shaded in Fig. 1.1. The tributary area is the product of the tributary width and the span of beam 78.

Figure 1.1 *Tributary Areas*

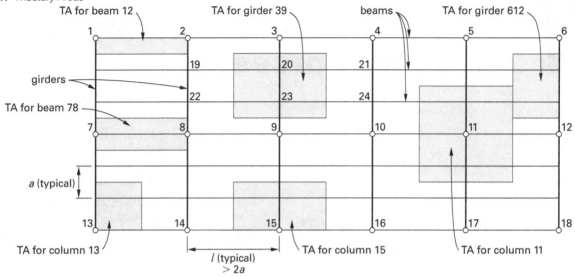

TA = tributary area
O = column (typical)

plan view
(not to scale)

For a beam, the tributary width is measured from the midpoint of the slab panel on one side of the beam to the midpoint of the slab panel on the opposite side of the beam. For the floor layout shown in Fig. 1.1, the tributary width equals the spacing between beams. For a beam spacing of a and a beam span of l, the tributary width is

$$B_T = a$$

The tributary area is

$$A_T = B_T l = al$$

Applied Dead Load on a Beam

The tributary area method assumes that all members have pin connections.

For a slab weight plus superimposed dead load of w_S (in lbf/ft^2) and a beam self-weight of w_B (in lbf/ft), the uniform line load on beam 78 is

$$w_{L78} = w_S B_T + w_B$$
$$= w_S a + w_B$$

The total dead load on beam 78 is

$$W_{D78} = l w_{L78}$$
$$= l(w_S a + w_B)$$
$$= w_S A_T + w_B l$$

The dead load reaction at end 7 of beam 78 is

$$V_{D78} = \frac{l(w_S a + w_B)}{2}$$
$$= V_{D87}$$

The dead loading acting on beam 78 is shown in Fig. 1.2.

Figure 1.2 *Dead Load on Beam 78*

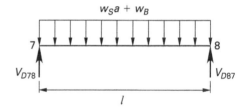

For edge beam 12, the corresponding values are

$$w_{L12} = w_S B_T + w_B$$
$$= \frac{w_S a}{2} + w_B$$
$$W_{D12} = l\left(\frac{w_S a}{2} + w_B\right)$$
$$V_{D12} = \frac{l\left(\dfrac{w_S a}{2} + w_B\right)}{2}$$
$$= V_{D21}$$

Example 1.1

The detail shown is of a typical interior beam in an office building whose floor layout is as shown in Fig. 1.1. The beams span 30 ft and are spaced at 10 ft centers. The construction is of normal weight concrete with a weight of 150 lbf/ft^3. A 1.5 in terrazzo finish and a suspended metal lath and gypsum plaster ceiling are provided. Provision is required for mechanical ducts. Determine the dead load acting on a typical interior beam and the support reactions.

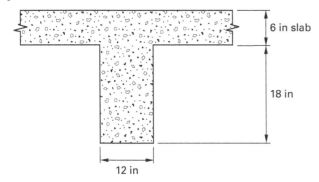

Solution

The customary weights of building materials are given in ASCE/SEI7 Table C3.1-1A. From this table, the weight of a 1.5 in terrazzo finish is 19 lbf/ft^2, the weight of a suspended metal lath and gypsum plaster ceiling is 10 lbf/ft^2, and the allowance for mechanical ducts is 4 lbf/ft^2. The superimposed dead load is

$$
\begin{aligned}
w_{\text{superimposed}} &= w_{\text{terrazzo}} + w_{\text{ceiling}} + w_{\text{ducts}} \\
&= 19 \ \frac{\text{lbf}}{\text{ft}^2} + 10 \ \frac{\text{lbf}}{\text{ft}^2} + 4 \ \frac{\text{lbf}}{\text{ft}^2} \\
&= 33 \ \text{lbf/ft}^2
\end{aligned}
$$

The slab self-weight is

$$
\begin{aligned}
w_{\text{slab}} = t w_{\text{concrete}} &= \frac{(6 \ \text{in})\left(150 \ \dfrac{\text{lbf}}{\text{ft}^3}\right)}{12 \dfrac{\text{in}}{\text{ft}}} \\
&= 75 \ \text{lbf/ft}^2
\end{aligned}
$$

The total distributed dead load is

$$
\begin{aligned}
w_S &= w_{\text{superimposed}} + w_{\text{slab}} \\
&= 33 \ \frac{\text{lbf}}{\text{ft}^2} + 75 \ \frac{\text{lbf}}{\text{ft}^2} \\
&= 108 \ \text{lbf/ft}^2
\end{aligned}
$$

The self-weight of the beam per foot run is

$$
\begin{aligned}
w_B &= \frac{(1 \ \text{ft})(18 \ \text{in})\left(150 \ \dfrac{\text{lbf}}{\text{ft}^3}\right)}{12 \dfrac{\text{in}}{\text{ft}}} \\
&= 225 \ \text{lbf/ft}
\end{aligned}
$$

The tributary area of the beam is

$$
\begin{aligned}
A_T &= al \\
&= (10 \ \text{ft})(30 \ \text{ft}) \\
&= 300 \ \text{ft}^2
\end{aligned}
$$

The total dead load on the beam is

$$
\begin{aligned}
W_{DB} &= w_S A_T + w_B l \\
&= \frac{\left(108 \ \dfrac{\text{lbf}}{\text{ft}^2}\right)(300 \ \text{ft}^2) + \left(225 \ \dfrac{\text{lbf}}{\text{ft}}\right)(30 \ \text{ft})}{1000 \ \dfrac{\text{lbf}}{\text{kip}}} \\
&= 39.15 \ \text{kips}
\end{aligned}
$$

The dead load support reactions are

$$
\begin{aligned}
V_{DB} &= \frac{W_{DB}}{2} \\
&= \frac{39.15 \ \text{kips}}{2} \\
&= 19.575 \ \text{kips}
\end{aligned}
$$

Tributary Area of a Girder

The tributary area for a typical interior girder, girder 39, is shown shaded in Fig. 1.1. The tributary area is given by the product of the tributary width and a length of $2a$ located centrally on the span.

For a girder, the tributary width is measured from the midpoint between girders on one side of the girder to the midpoint between girders on the opposite side of the girder. For the floor layout shown in Fig. 1.1, the tributary width equals the spacing between girders.

For a girder spacing of l and a girder span of $3a$, the tributary width is

$$
B_T = l
$$

The tributary area is

$$
\begin{aligned}
A_T &= 2aB_T \\
&= 2al
\end{aligned}
$$

Applied Dead Load on a Girder

For the floor layout shown in Fig. 1.1, the floor slab plus ancillary superimposed dead load is assumed to be supported by the beams. Hence, a girder supports its own weight, w_G, as well as the end reactions of the beams framing into each side at the third points of the span. As shown in Fig. 1.3, beams 2019 and 2021 frame into node 20 on girder 39, and beams 2322 and 2324 frame into node 23 on the girder.

Figure 1.3 *Dead Load on Girder 39*

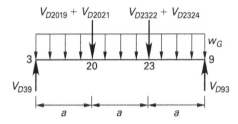

Since all beams are identical, the total dead load on girder 39 is

$$W_{D39} = 2l(w_S a + w_B) + 3aw_G$$

The dead load reaction at end 3 of girder 39 is

$$V_{D39} = l(w_S a + w_B) + \frac{3aw_G}{2}$$
$$= V_{D93}$$

For edge girder 612, beams frame into only one side of the girder, and the dead load reactions are

$$V_{D\,612} = \frac{l(w_S a + w_B)}{2} + \frac{3aw_G}{2}$$
$$= V_{D\,126}$$

Example 1.2

The detail shown is of a typical interior girder in an office building whose floor layout is as shown in Fig. 1.1. The beams span 30 ft and are spaced at 10 ft centers. The construction is of normal weight concrete with a weight of 150 lbf/ft³. A 1.5 in terrazzo finish and a suspended metal lath and gypsum plaster ceiling are provided. Provision is required for mechanical ducts. Determine the dead load acting on a typical interior girder and the support reactions.

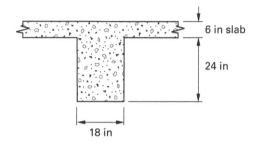

Solution

Because of the symmetry of the structure, the support reactions of all interior beams are equal and are derived in Ex. 1.1 as

$$V_{DB} = 19.575 \text{ kips}$$

The magnitude of each concentrated load acting at the third points of the girder span is

$$2V_{DB} = (2)(19.575 \text{ kips})$$
$$= 39.15 \text{ kips}$$

The self-weight of the girder per foot run is

$$w_G = \frac{(18 \text{ in})(24 \text{ in})\left(150 \ \dfrac{\text{lbf}}{\text{ft}^3}\right)}{\left(12 \ \dfrac{\text{in}}{\text{ft}}\right)^2}$$
$$= 450 \text{ lbf/ft}$$

The loads acting on a girder are shown in the illustration. The support reactions are

$$V_{DG} = 2V_{DB} + \frac{w_G l}{2}$$
$$= 39.15 \text{ kips} + \frac{\left(450 \ \dfrac{\text{lbf}}{\text{ft}}\right)(30 \text{ ft})}{(2)\left(1000 \ \dfrac{\text{lbf}}{\text{kip}}\right)}$$
$$= 45.90 \text{ kips}$$

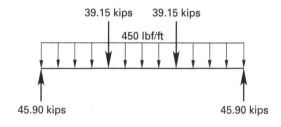

Tributary Area of a Column

The tributary area for a typical interior column, column 11, is shown shaded in Fig. 1.1. The tributary area is measured from the midpoints between adjacent columns on each side of column 11. For the floor layout shown in Fig. 1.1, with a beam span of l and a girder span of $3a$, the tributary area is

$$A_T = 3al$$

Applied Dead Load on a Column

The dead load applied to a column may be determined by summing the end reactions of the beams and girders framing into the column. Framing into interior column 11 are girders 115 and 1117 and beams 1110 and 1112, and the total dead load applied to column 11 is

$$V_{D\,11} = V_{D\,115} + V_{D\,1117} + V_{D\,1110} + V_{D\,1112}$$

Alternatively, the dead load on a column may be determined using the tributary area method. To use this method, convert the weight of the beams and girders to loads uniformly distributed over the area of the slab. Thus, the equivalent total uniformly distributed dead load on the floor is

$$w_T = w_S + \frac{w_B}{a} + \frac{w_G}{l}$$

The dead load applied to column 11 is

$$\begin{aligned} V_{D\,11} &= w_T A_T \\ &= w_T(3al) \end{aligned}$$

The corner column 13 has one beam and one girder framing into it. The tributary area of column 13 is the shaded area shown in Fig. 1.1. The members framing into the column are edge beam 1314 and edge girder 137, and the column supports the end reactions from these members. Hence, the total dead load applied to column 13 is

$$\begin{aligned} V_{D\,13} &= V_{D\,1314} + V_{D\,137} \\ &= \frac{l\left(\dfrac{w_S a}{2} + w_B\right)}{2} + \frac{l(w_S a + w_B)}{2} \\ &\quad + \frac{3aw_G}{2} \end{aligned}$$

Alternatively, using the tributary area method,

$$\begin{aligned} A_T &= \left(\frac{3a}{2}\right)\!\left(\frac{l}{2}\right) \\ &= \frac{3al}{4} \end{aligned}$$

The dead load applied to column 13 is

$$\begin{aligned} V_{D\,13} &= A_T w_T \\ &= \left(\frac{3al}{4}\right)\!\left(w_S + \frac{w_B}{a} + \frac{w_G}{l}\right) \end{aligned}$$

Example 1.3

The typical floor layout of an office building is shown in Fig. 1.1. The beams span 30 ft and are spaced at 10 ft centers. The construction is of normal weight concrete with a weight of 150 lbf/ft³. A 1.5 in terrazzo finish and a suspended metal lath and gypsum plaster ceiling are provided. Provision is required for mechanical ducts. Determine the dead load acting on a typical interior column, column 11. Check the solution using the tributary area method.

Solution

The members framing into interior column 11 are girders 115 and 1117 and beams 1110 and 1112, and the total dead load applied to column 11 is

$$V_{D\,11} = V_{D\,115} + V_{D\,1117} + V_{D\,1110} + V_{D\,1112}$$

From Ex. 1.2, the end reaction from girder 115 is

$$\begin{aligned} V_{D\,115} &= 45.90 \text{ kips} \\ &= V_{D\,1117} \end{aligned}$$

From Ex. 1.1, the end reaction from beam 1110 is

$$\begin{aligned} V_{D\,1110} &= 19.575 \text{ kips} \\ &= V_{D\,1112} \end{aligned}$$

Hence, the dead load acting on column 11 is

$$\begin{aligned} V_{D\,11} &= 2V_{D\,115} + 2V_{D\,1110} \\ &= (2)(45.90 \text{ kips} + 19.575 \text{ kips}) \\ &= 130.95 \text{ kips} \end{aligned}$$

Alternatively, for the floor layout shown in Fig. 1.1, with a beam span of l and a girder span of $3a$, the tributary area of column 11 is

$$\begin{aligned} A_T &= 3al \\ &= (3)(10 \text{ ft})(30 \text{ ft}) \\ &= 900 \text{ ft}^2 \end{aligned}$$

Figure 1.4 *Slab Panels Supported on Four Sides with l ≤ 2a*

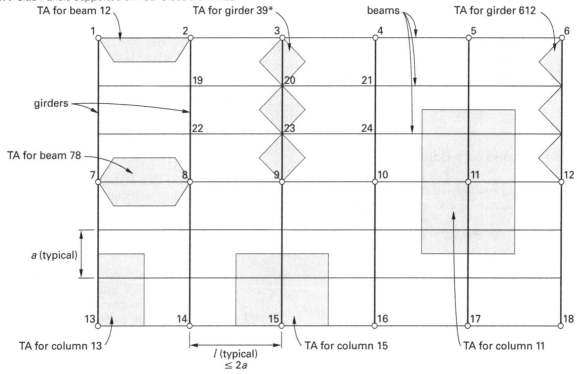

TA = tributary area
O = column (typical)
* In addition, the girder carries beam reactions at 20 and 23.

plan view

The total distributed dead load on the floor is

$$w_T = w_S + \frac{w_B}{a} + \frac{w_G}{l}$$

$$= 108 \ \frac{\text{lbf}}{\text{ft}^2} + \frac{225 \ \frac{\text{lbf}}{\text{ft}}}{10 \ \text{ft}} + \frac{450 \ \frac{\text{lbf}}{\text{ft}}}{30 \ \text{ft}}$$

$$= 145.5 \ \text{lbf/ft}^2$$

The dead load applied to column 11 is

$$V_{D\,11} = w_T A_T$$

$$= \frac{\left(145.5 \ \frac{\text{lbf}}{\text{ft}^2}\right)(900 \ \text{ft}^2)}{1000 \ \frac{\text{lbf}}{\text{kip}}}$$

$$= 130.95 \ \text{kips}$$

Tributary Areas for a Slab with an Aspect Ratio of Two or Less

If a slab panel has an aspect ratio greater than two, the slab is assumed to act as a one-way slab with the slab

weight supported by the two longer boundary members. In Fig. 1.1, the tributary area for a typical interior beam, beam 78, is the rectangular area shown shaded. The slab weight is assumed to be supported entirely by the beams, with none of the slab weight supported by the girders.

If the aspect ratio of a slab panel is two or less, the slab is considered a two-way slab with all four boundary members sharing the load from the slab.

Tributary Area of a Beam

As shown in Fig. 1.4, the tributary area for a typical interior beam, beam 78, is bounded by 45° lines drawn from the corners of adjacent panels to a line midway between adjacent beams. The tributary area consists of the double trapezoid shown shaded.

For a beam spacing of a and a beam span of l, the tributary area is

$$A_T = al - \frac{4a^2}{8}$$

$$= al - \frac{a^2}{2}$$

Applied Dead Load on a Beam

For a slab weight plus superimposed dead load of w_S lbf/ft^2 and a beam self weight of w_B lbf/ft, the total dead load on beam 78 is

$$W_{D\,78} = w_S A_T + w_B l$$
$$= w_S\left(al - \frac{a^2}{2}\right) + w_B l$$

The dead load reaction at end 7 of beam 78 is

$$V_{D\,78} = \frac{W_{D\,78}}{2}$$
$$= V_{D\,87}$$

The dead load acting on beam 78 is shown in Fig. 1.5.

Figure 1.5 *Dead Load on Beam 78*

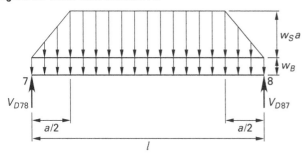

Tributary Area of a Girder

The tributary area for a typical interior girder, girder 39, is shown shaded in Fig. 1.4. The corresponding load on the girder due to the slab weight plus the superimposed dead load of w_S consists of three isosceles triangles as shown in Fig. 1.6. For a beam spacing of a and a beam span of l, the tributary area is

$$A_T = 6\left(\frac{a^2}{4}\right)$$
$$= \frac{3a^2}{2}$$

Figure 1.6 *Dead Load on Girder 39*

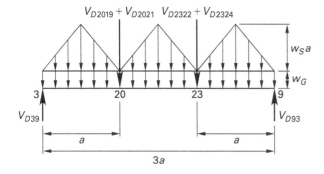

Applied Dead Load on a Girder

The slab weight plus superimposed dead load supported by the girder is

$$w_S A_T = \frac{3 w_S a^2}{2}$$

In addition, the girder supports its own weight as well as the end reactions of the beams that frame into each side at the third points of the span. As shown in Fig. 1.6, beams 2019 and 2021 frame into node 20 on girder 39, and beams 2322 and 2324 frame into node 23 on the same girder.

Since all beams are identical, the concentrated loads applied to girder 39 at the third points of the span are each

$$W_{DB} = w_S\left(al - \frac{a^2}{2}\right) + w_B l$$

For a girder self-weight of w_G (in lbf/ft), the total self-weight of girder 39 is $3aw_G$.

Summing these three components gives the total dead load on girder 39 as

$$W_{D\,39} = w_S A_T + 2 W_{DB} + 3 a w_G$$
$$= \frac{3 w_S a^2}{2} + 2 w_S\left(al - \frac{a^2}{2}\right)$$
$$\quad + 2 w_B l + 3 a w_G$$
$$= \frac{w_S a^2}{2} + 2 w_S a l + 2 w_B l + 3 a w_G$$

The dead load reaction at end 3 of girder 39 is

$$V_{D\,39} = \frac{W_{D\,39}}{2}$$
$$= V_{D\,93}$$

For edge girder 612, beams frame into one side only of the girder, and the total dead load on girder 612 is

$$W_{D\,612} = \frac{3 w_S a^2}{4} + w_S\left(al - \frac{a^2}{2}\right)$$
$$\quad + w_B l + 3 a w_G$$
$$= \frac{w_S a^2}{4} + w_S a l + w_B l + 3 a w_G$$

Tributary Area of a Column

The tributary area for a typical interior column, column 11, is shown shaded in Fig. 1.4. The tributary area is measured from the midpoint between adjacent columns on each side of column 11. For the floor layout shown in Fig. 1.1, with a beam span of l and a girder span of $3a$, the tributary area is

$$A_T = 3al$$

Applied Dead Load on a Column

Dead loads applied to a column may be determined by summing the end reactions of the beams and girders that frame into the column. The members framing into interior column 11 are girders 115 and 1117 and beams 1110 and 1112, and the total dead load applied to column 11 is

$$V_{D\,11} = V_{D\,115} + V_{D\,1117} + V_{D\,1110} + V_{D\,1112}$$

Alternatively, the dead load on a column may be determined using the tributary area method. To use this method, convert the weight of the beams and girders to uniformly distributed loads over the area of the slab. The equivalent total uniformly distributed dead load on the floor is

$$w_T = w_S + \frac{w_B}{a} + \frac{w_G}{l}$$

The dead load applied to column 11 is

$$\begin{aligned} V_{D\,11} &= w_T A_T \\ &= w_T(3al) \end{aligned}$$

The corner column 13 has one beam and one girder framing into it. The tributary area of column 13 is the shaded area shown in Fig. 1.4. The members framing into the column are edge beam 1314 and edge girder 137, and the column supports the end reactions from these members. Hence, the total dead load applied to column 13 is

$$V_{D\,13} = V_{D\,1314} + V_{D137}$$

Alternatively, using the tributary area method,

$$\begin{aligned} A_T &= \left(\frac{3a}{2}\right)\left(\frac{l}{2}\right) \\ &= \frac{3al}{4} \end{aligned}$$

The dead load applied to column 13 is

$$\begin{aligned} V_{D\,13} &= A_T w_T \\ &= \left(\frac{3al}{4}\right)\left(w_S + \frac{w_B}{a} + \frac{w_G}{l}\right) \end{aligned}$$

Example 1.4

The typical floor layout of an office building is shown in Fig. 1.4. The beams span 20 ft and are spaced at 10 ft centers. The construction is of normal weight concrete with a weight of 150 lbf/ft^3. A 1.5 in terrazzo finish and a suspended metal lath and gypsum plaster ceiling are provided. Provision is required for mechanical ducts. Determine the dead load acting on a typical interior column, column 11. Check the solution using the tributary area method.

Solution

From previous examples, the total distributed dead load, w_S, is 108 lbf/ft^2; the self-weight of the beam per foot run, w_B, is 225 lbf/ft; and the self-weight of the girder per foot run, w_G, is 450 lbf/ft.

The members framing into interior column 11 are girders 115 and 1117 and beams 1110 and 1112, and the total dead load applied to column 11 is

$$V_{D\,11} = V_{D\,115} + V_{D\,1117} + V_{D\,1110} + V_{D\,1112}$$

The sum of the reactions from the two girders is

$$\begin{aligned} V_{D\,115} + V_{D\,1117} &= \frac{w_S a^2}{2} + 2w_S al + 2w_B l + 3aw_G \\[6pt] &= \frac{\left(108\ \dfrac{\text{lbf}}{\text{ft}^2}\right)(10\ \text{ft})^2}{2} \\[6pt] &\quad + (2)\left(108\ \frac{\text{lbf}}{\text{ft}^2}\right)(10\ \text{ft})(20\ \text{ft}) \\[6pt] &\quad + (2)\left(225\ \frac{\text{lbf}}{\text{ft}}\right)(20\ \text{ft}) \\[6pt] &\quad + (3)(10\ \text{ft})\left(450\ \frac{\text{lbf}}{\text{ft}}\right) \\[6pt] &= 71{,}100\ \text{lbf} \end{aligned}$$

The sum of the reactions from the two beams is

$$V_{D\,1110} + V_{D\,1112} = w_S\left(al - \frac{a^2}{2}\right) + w_B l$$

$$= \left(108\ \frac{\text{lbf}}{\text{ft}^2}\right)$$

$$\times (10\ \text{ft})(20\ \text{ft}) - \frac{(10\ \text{ft})^2}{2}$$

$$+ \left(225\ \frac{\text{lbf}}{\text{ft}}\right)(20\ \text{ft})$$

$$= 20{,}700\ \text{lbf}$$

The dead load acting on column 11 is

$$V_{D\,11} = V_{D\,115} + V_{D\,1117} + V_{D\,1110} + V_{D\,1112}$$

$$= \frac{71{,}100\ \text{lbf} + 20{,}700\ \text{lbf}}{1000\ \dfrac{\text{lbf}}{\text{kip}}}$$

$$= 91.80\ \text{kips}$$

Alternatively, for the floor layout shown in Fig. 1.4, with a beam span of $l = 20$ ft and a girder span of $3a = 30$ ft, the tributary area of column 11 is

$$A_T = 3al$$

$$= (3)(10\ \text{ft})(20\ \text{ft})$$

$$= 600\ \text{ft}^2$$

The total distributed dead load on the floor is

$$w_T = w_S + \frac{w_B}{a} + \frac{w_G}{l}$$

$$= 108\ \frac{\text{lbf}}{\text{ft}^2} + \frac{225\ \dfrac{\text{lbf}}{\text{ft}}}{10\ \text{ft}}$$

$$+ \frac{450\ \dfrac{\text{lbf}}{\text{ft}}}{20\ \text{ft}}$$

$$= 153\ \text{lbf/ft}^2$$

The dead load applied to column 11 is

$$V_{D\,11} = w_T A_T$$

$$= \frac{\left(153\ \dfrac{\text{lbf}}{\text{ft}^2}\right)(600\ \text{ft}^2)}{1000\ \dfrac{\text{lbf}}{\text{kip}}}$$

$$= 91.8\ \text{kips}$$

2. LIVE LOADS

Nomenclature

A_I	influence area	ft^2
F	number of inches of rise per foot of pitched roof	in/ft
K_{LL}	live load element factor	–
L	reduced distributed design live load of area supported by member	lbf/ft^2
L_o	unreduced distributed design live load of area supported by member	lbf/ft^2
L_r	reduced roof live load	lbf/ft^2
R_1	reduction factor	–
R_2	reduction factor	–
W	total load supported by member	kips

General Principles

A *live load* is defined in IBC Sec. 202 as a load produced by the use and occupancy of the structure that does not include construction or environmental loads such as wind load, snow load, rain load, earthquake load, flood load, or dead load. The live loads used in the design of the structure must be the maximum loads anticipated for the intended use of the structure, but they must not be less than the uniformly distributed live loads given in IBC Table 1607.1.

In addition to the uniformly distributed live load given in the tables, a concentrated live load is also specified. This concentrated load is assumed to be uniformly distributed over an area of 2.5 ft by 2.5 ft, and it must be located so as to produce the maximum load effect in the structural members. These two loads—the uniformly distributed load and the concentrated load—are applied separately (not simultaneously), and the structure must be designed for whichever load produces the greater load effect in the structural members.

Floor Design Live Loads

Non-load-bearing walls that are supported on and permanently attached to the floor slab apply line loads on the slab and are considered as dead load. Large floor areas are frequently divided into smaller offices and cubicles by means of moveable partitions. These partitions may be relocated during the life of the structure, and they are conservatively considered to be a nonreducible live load. The partition live load is specified in IBC Sec. 1607.5 as a uniformly distributed live load of 15 lbf/ft^2. This partition load must be added to the specified floor live load whether or not partitions are shown on the construction documents; however, a partition load need not be applied when the floor live load exceeds 80 lbf/ft^2.

Forces

Influence Area

ASCE/SEI7 Sec. C4.7 introduces the concept of the *influence area* to provide a rational method for determining permissible reductions in floor live load. The influence area is defined in ASCE/SEI7 Sec. C4.7.1 as that floor area over which the influence surface for structural effects is significantly different from zero. The influence area is derived as the product of the tributary area, A_T, and the live load element factor, K_{LL}.

$$A_I = K_{LL}A_T$$

Values of K_{LL} are given in IBC Table 1607.11.1 and are shown in Table 1.1.

Table 1.1 *Values for Live Load Element Factor, K_{LL}*

element	K_{LL}
interior columns	4
exterior columns without cantilever slabs	4
edge columns with cantilever slabs	3
corner columns with cantilever slabs	2
edge beams without cantilever slabs	2
interior beams	2
all other members	1

Values of K_{LL} may also be determined by reference to Fig. 1.7. For a typical interior beam, beam 78, the hatched area is the influence area, and the live load element factor is

$$K_{LL} = \frac{A_I}{A_T} = \frac{(2a)l}{al}$$
$$= 2$$

For a typical interior column, column 11, the cross-hatched area is the influence area, and the live load element factor is

$$K_{LL} = \frac{A_I}{A_T} = \frac{(6a)(2l)}{(3a)(l)}$$
$$= 4$$

Reduction in Floor Live Load

The actual live load on a member that is supporting a large floor area is likely to be less than that calculated using the full nominal uniform live load specified in IBC Table 1607.1. A large floor area is unlikely to be completely loaded with the full uniform load, and this reduces the design load on a member supporting a large floor. For members that have a value of $A_I \geq 400$ ft^2, floor live loads may be reduced in accordance with IBC Eq. 16-23. The permitted reduction increases with the influence area supported by the member. Similarly, for members supporting two or more floors, the calculation of A_I includes the floors above the floor containing the member.

The reduced design uniform live load of area supported by the member is given by IBC Eq. 16-23 as

$$L = L_o\left(0.25 + \frac{15}{\sqrt{K_{LL}A_T}}\right)$$

$$\geq 0.5L_o \quad \text{[for members supporting only one floor]}$$
$$\geq 0.4L_o \quad \text{[for members supporting two or more floors]}$$

The total design live load acting on the member is

$$W = LA_T$$

The following limitations are imposed on the application of IBC Eq. 16-23 in IBC Sec. 1607.11.1.1 through IBC Sec. 1607.11.1.3.

- The *tributary area of a one-way slab* is restricted to the product of the slab span and a tributary width of $B_T = 1.5$ times the slab span.

- A *heavy live load*—a live load exceeding 100 lbf/ft^2— may not be reduced for a member supporting only one floor. For a member that supports two or more floors, the heavy live load may be reduced by a maximum of 20%, but the live load must not be less than the value as calculated by IBC Eq. 16-23.

- The *live load for a passenger vehicle garage* may not be reduced for a member supporting only one floor. For a member that supports two or more floors, the live load may be reduced by a maximum of 20%, but the live load must not be less than the value as calculated by IBC Eq. 16-23.

In addition, IBC Table 1607.1 states that live loads may not be reduced in facilities used as

- armories and drill rooms, except for a 20% reduction permitted for members supporting two or more floors

- assembly areas

- dining rooms and restaurants

- stack rooms, except for a 20% reduction permitted for members supporting two or more floors

- dance halls and ballrooms

- gymnasiums and other recreational facilities

- sidewalks, vehicular driveways, and yards when subject to trucking, except for a 20% reduction permitted for members supporting two or more floors

- wholesale stores, except for a 20% reduction permitted for members supporting two or more floors

- pedestrian yards and terraces

Figure 1.7 *Influence Areas*

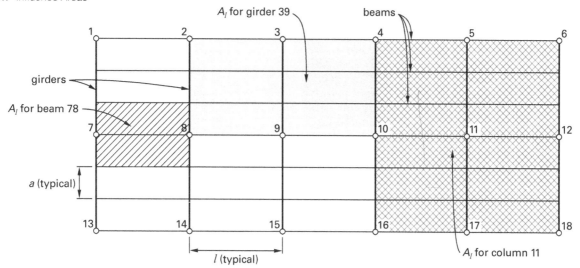

A_I = influence area
O = column (typical)

plan view
(not to scale)

Example 1.5

The typical floor layout of an office facility is shown in Fig. 1.1. The beams span 30 ft and are spaced at 10 ft centers. Determine the live load acting on a typical interior beam, beam 78.

Solution

From IBC Table 1607.1, the unreduced live load for an office building is

$$L_o = 50 \ \text{lbf/ft}^2$$

From Table 1.1, the live load element factor for an interior beam is

$$K_{LL} = 2$$

From Ex. 1.1, the area tributary to beam 78 is

$$A_T = 300 \ \text{ft}^2$$

The influence area is

$$A_I = K_{LL}A_T$$
$$= (2)(300 \ \text{ft}^2)$$
$$= 600 \ \text{ft}^2$$
$$> 400 \ \text{ft}^2 \quad \text{[IBC Eq. 16-23 is applicable]}$$

The reduced design live load for beam 78 is

$$L = L_o \left(0.25 + \frac{15}{\sqrt{K_{LL}A_T}} \right)$$
$$= \left(50 \ \frac{\text{lbf}}{\text{ft}^2} \right) \left(0.25 + \frac{15}{\sqrt{600 \ \text{ft}^2}} \right)$$
$$= 43.12 \ \text{lbf/ft}^2 \quad \text{[satisfactory]}$$
$$> 0.5L_o \quad \text{[minimum for member supporting one floor]}$$

In accordance with IBC Sec. 1607.5, an additional 15 lbf/ft² must be added to allow for the nonreducible weight of moveable partitions. Hence, the total live load intensity is

$$L = 43.12 \ \frac{\text{lbf}}{\text{ft}^2} + 15 \ \frac{\text{lbf}}{\text{ft}^2}$$
$$= 58.12 \ \text{lbf/ft}^2$$

The total design live load is

$$W = LA_T$$
$$= \frac{\left(58.12 \ \frac{\text{lbf}}{\text{ft}^2} \right) (300 \ \text{ft}^2)}{1000 \ \frac{\text{lbf}}{\text{kip}}}$$
$$= 17.44 \ \text{kips}$$

Example 1.6

The typical floor layout of a four-story multiuse building is shown in Fig. 1.7, and the building elevation is shown in the illustration. The beams span 30 ft and are spaced at 10 ft centers. Floor 2 is a public assembly area with moveable seats, and floors 3 and 4 are used for general office occupancy. Public access to the roof is not permitted, and rain, snow, and corridor loads may be neglected. Determine the floor live load produced on a typical interior column, column 11, at each story. Use live load reductions in accordance with IBC Sec. 1607.11.1.

```
               5@30 ft = 150 ft
        ┌─────────────────────────────────────┐
roof    ├──────┬──────┬──────┬──────┬──────────┤
        │office│      │      │      │          │
        │occup-│      │      │      │          │
4th floor│ancy │      │      │      │          │
        ├──────┼──────┼──────┼──────┼──────────┤
        │office│      │      │      │          │
        │occup-│      │      │      │          │
3rd floor│ancy │      │      │      │          │
        ├──────┼──────┼──────┼──────┼──────────┤
        │public│      │      │      │          │
        │assem-│      │      │      │          │
2nd floor│bly  │      │      │      │          │
        ├──────┼──────┼──────┼──────┼──────────┤
        │      │      │      │      │          │
1st floor│     │      │      │      │          │
 //////////////////////////////////////////////
```

Solution

From IBC Table 1607.1, the reducible live load for office occupancy on floors 3 and 4 is

$$L_o = 50 \text{ lbf/ft}^2$$

Also from IBC Table 1607.1, the nonreducible live load for an assembly area with moveable seats on floor 2 is

$$L_o = 100 \text{ lbf/ft}^2$$

From Table 1.1, the live load element factor for an interior column is

$$K_{LL} = 4$$

From Ex. 1.3, the area tributary to column 11 at each floor level is

$$A_T = 900 \text{ ft}^2$$

At each floor level, the live load on the column must account for the tributary floor area supported by the column above that floor level.

Fourth floor to roof

There is no floor live load acting on the column in the top story.

Third floor to fourth floor

$$
\begin{aligned}
A_I &= K_{LL} A_T \\
&= (4)(900 \text{ ft}^2) \\
&= 3600 \text{ ft}^2 \\
&> 400 \text{ ft}^2 \quad \text{[IBC Eq. 16-23 is applicable]}
\end{aligned}
$$

The reduced live load at the fourth floor is

$$
\begin{aligned}
L &= L_o\left(0.25 + \frac{15}{\sqrt{K_{LL}A_T}}\right) \\
&= \left(50 \ \frac{\text{lbf}}{\text{ft}^2}\right)\left(0.25 + \frac{15}{\sqrt{3600 \text{ ft}^2}}\right) \\
&= 25 \text{ lbf/ft}^2 \quad \text{[satisfactory]} \\
&= 0.5L_o \quad \text{[minimum for member supporting one floor]}
\end{aligned}
$$

In accordance with IBC Sec. 1607.5, an additional 15 lbf/ft^2 must be added to allow for the nonreducible weight of moveable partitions. Hence, the total live load intensity is

$$
\begin{aligned}
L &= 25 \ \frac{\text{lbf}}{\text{ft}^2} + 15 \ \frac{\text{lbf}}{\text{ft}^2} \\
&= 40 \text{ lbf/ft}^2
\end{aligned}
$$

The live load above the third floor is

$$
\begin{aligned}
W_3 &= L A_T \\
&= \frac{\left(40 \ \dfrac{\text{lbf}}{\text{ft}^2}\right)(900 \text{ ft}^2)}{1000 \ \dfrac{\text{lbf}}{\text{kip}}} \\
&= 36 \text{ kips}
\end{aligned}
$$

Second floor to third floor

The column supports the floor live load from the third and fourth floor. The influence area is the sum of the influence areas at the third and fourth floors.

$$
\begin{aligned}
A_I &= 2K_{LL}A_T \\
&= (2)(4)(900 \text{ ft}^2) \\
&= 7200 \text{ ft}^2 \\
&> 400 \text{ ft}^2 \quad \text{[IBC Eq. 16-23 is applicable]}
\end{aligned}
$$

The reduced live load at the third floor is

$$
\begin{aligned}
L &= L_o\left(0.25 + \frac{15}{\sqrt{K_{LL}A_T}}\right) \\
&= \left(50\ \frac{\text{lbf}}{\text{ft}^2}\right)\left(0.25 + \frac{15}{\sqrt{7200\ \text{ft}^2}}\right) \\
&= 21.34\ \text{lbf/ft}^2 \quad [\text{satisfactory}] \\
&> 0.4L_o \quad [\text{minimum for member supporting two floors}]
\end{aligned}
$$

In accordance with IBC Sec. 1607.5, an additional 15 lbf/ft^2 must be added to allow for the nonreducible weight of moveable partitions. Hence, the total live load intensity on both the third and fourth floors is

$$
\begin{aligned}
L &= 21.34\ \frac{\text{lbf}}{\text{ft}^2} + 15\ \frac{\text{lbf}}{\text{ft}^2} \\
&= 36.34\ \text{lbf/ft}^2
\end{aligned}
$$

The total live load above the second floor is

$$
\begin{aligned}
W_2 = LA_T &= \frac{\left(36.34\ \dfrac{\text{lbf}}{\text{ft}^2}\right)(900\ \text{ft}^2)(2\ \text{floors})}{1000\ \dfrac{\text{lbf}}{\text{kip}}} \\
&= 65.41\ \text{kips}
\end{aligned}
$$

First floor to second floor

The column supports the floor live load from the second, third, and fourth floor. From IBC Table 1607.1, the distributed live load on the second floor may not be reduced, and

$$
L = L_o = 100\ \text{lbf/ft}^2
$$

Since L is greater than 80 lbf/ft^2, the partition load of 15 lbf/ft^2 is not added to the floor load in accordance with IBC Sec. 1607.5.

The additional load applied to the column by the second floor distributed live load is

$$
\begin{aligned}
W = LA_T &= \frac{\left(100\ \dfrac{\text{lbf}}{\text{ft}^2}\right)(900\ \text{ft}^2)}{1000\ \dfrac{\text{lbf}}{\text{kip}}} \\
&= 90\ \text{kips}
\end{aligned}
$$

The total live load above the first floor is

$$
\begin{aligned}
W_1 = W + W_2 &= 90\ \text{kips} + 65.41\ \text{kips} \\
&= 155.41\ \text{kips}
\end{aligned}
$$

Distribution of Floor Loads

For continuous floor members, IBC Sec. 1607.12 requires that the design force on a member be determined using the most adverse distribution of the live load. For continuous beams, this requires analysis of the loading arrangements shown in Fig. 1.8 using skip or checkerboard loading to determine maximum span and support moments.

Figure 1.8 *Partial Loading Conditions*

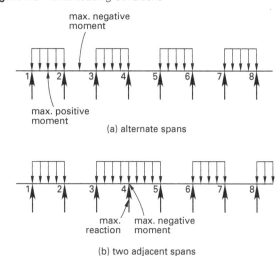

In Fig. 1.8(a), alternate spans are loaded. This produces maximum positive moments at the center of the loaded spans 12, 34, 56, and 78. This also produces maximum negative moments at the center of the unloaded spans 23, 45, and 67.

In Fig. 1.8(b), two adjacent spans are loaded with alternate spans loaded beyond these. This produces maximum negative moment and maximum shear at support 4.

To simplify the procedure, ACI Sec. 6.4.2 allows consideration of the following modified arrangements.

- Live load applied to alternate spans. This produces the maximum positive span moment.

- Live load applied to any two adjacent spans, with all other spans unloaded. This produces the maximum negative support moment.

Roof Design Live Loads

In accordance with IBC Sec. 202, a live load on a roof is typically produced

- during maintenance, by workers, equipment, and materials

- during the life of the structure, by moveable objects such as planters or other similar small decorative appurtenances that are not occupancy related

- by the use and occupancy of the roof, such as for roof gardens or assembly areas

A roof load is assumed to act on the horizontal projection of the roof surface. For continuous roof construction, IBC Sec. 1607.13.1 requires that the design force on a member be determined using the most adverse distribution of the live load. For a continuous beam, the loading arrangements shown in Fig. 1.8 must be analyzed using skip loading to determine maximum span and support moments.

Snow, rain, and wind loads must also be considered. Roof live loads are specified in IBC Table 1607.1, and the normal value is $L_o = 20$ lbf/ft^2. A roof used for a roof garden or for assembly purposes has a specified live load of 100 lbf/ft^2.

Reduction in Roof Live Load

In accordance with IBC Sec. 1607.13.3, if an area of a roof is used for a roof garden or other similar purposes, its uniformly distributed live load may be reduced in the same manner as a floor live load.

In accordance with IBC Table 1607.1, Part 26, footnote m, assembly area live load may not be reduced.

For a normal roof live load, the reduced load is determined in accordance with IBC Sec. 1607.13.2.1. The reduced load depends on the tributary area supported by the member and on the roof slope. The reduced roof live load is given by IBC Eq. 16-26 as

$$L_r = L_o R_1 R_2$$
$$\geq 12 \; \frac{\text{lbf}}{\text{ft}^2}$$
$$\leq 20 \; \frac{\text{lbf}}{\text{ft}^2}$$

In accordance with IBC Table 1607.1 part 26, footnote m, assembly area live load must not be reduced.

From IBC Eq. 16-27, Eq. 16-28, and Eq. 16-29,

$$R_1 = 1.0 \quad \text{[for } A_T \leq 200 \text{ ft}^2]$$
$$= 1.2 - 0.001 A_T \quad \text{[for } 200 \text{ ft}^2 < A_T < 600 \text{ ft}^2]$$
$$= 0.6 \quad \text{[for } A_T \geq 600 \text{ ft}^2]$$

From IBC Eq. 16-30, Eq. 16-31, and Eq. 16-32,

$$R_2 = 1.0 \quad \text{[for } F \leq 4]$$
$$= 1.2 - 0.05 F \quad \text{[for } 4 < F < 12]$$
$$= 0.6 \quad \text{[for } F \geq 12]$$

F is the number of inches of rise per foot of a pitched roof. The reduced live load, L_r, is applied per square foot of horizontal projection of a pitched roof.

Example 1.7

The roof framing layout of an office facility is shown in Fig. 1.1. The beams span 30 ft and are spaced at 10 ft centers. The roof is nominally flat. Public access to the roof is not permitted, and rain and snow loads may be neglected. Determine the roof live load acting on a typical interior column, column 11.

Solution

For a roof without public access, the unreduced roof live load is obtained from IBC Table 1607.1 as

$$L_o = 20 \text{ lbf/ft}^2$$

The area tributary to column 11 is obtained from Ex. 1.3 as

$$A_T = 900 \text{ ft}^2$$

This is greater than 600 ft^2, so from IBC Eq. 16-29, the reduction factor is

$$R_1 = 0.6$$

For a flat roof, the rise per foot is

$$F = 0$$

This is less than 4, so from IBC Eq. 16-30, the reduction factor is

$$R_2 = 1.0$$

The reduced roof live load is given by IBC Eq. 16-26 as

$$L_r = L_o R_1 R_2$$
$$= \left(20 \; \frac{\text{lbf}}{\text{ft}^2} \right)(0.6)(1.0)$$
$$= 12 \text{ lbf/ft}^2 \quad \text{[satisfactory]}$$

The minimum permissible value of L_r is 12 lbf/ft^2, so this is satisfactory. The total roof live load is

$$W = L_r A_T$$
$$= \frac{\left(12 \; \dfrac{\text{lbf}}{\text{ft}^2} \right)(900 \text{ ft}^2)}{1000 \; \dfrac{\text{lbf}}{\text{kip}}}$$
$$= 10.80 \text{ kips}$$

3. SNOW LOADS

Nomenclature

C_e	exposure factor	–
C_s	roof slope factor	–
C_t	thermal factor	–
d_S	horizontal extent of surcharge	ft
h	vertical separation between edge of upper roof and edge of lower adjacent roof	ft
h_b	height of balanced snow load	ft
h_c	clear height from top of balanced snow load to (1) closest point on adjacent upper roof, (2) top of parapet, or (3) top of projection on roof	ft
h_d	height of snow drift	ft
h_o	height of obstruction above roof level	ft
h_T	total depth of snow	ft
I_s	snow importance factor	–
l_u	length of roof upwind of drift	ft
L_{lower}	length of lower roof	–
L_{upper}	length of upper roof	–
p_d	maximum intensity of drift surcharge load	lbf/ft²
p_f	snow load on flat roof	lbf/ft²
p_g	ground snow load	lbf/ft²
p_{LD}	distributed snow load on leeward slope	lbf/ft²
p_{LS}	surcharge load on leeward slope	lbf/ft²
p_m	minimum snow load for low-slope roof	lbf/ft²
p_s	balanced sloped roof snow load	lbf/ft²
$p_{sliding}$	sliding snow surcharge load	lbf/ft²
p_T	total snow load	lbf/ft²
p_W	snow load on windward slope	lbf/ft²
R	thermal resistance	ft²-hr-°F/ Btu
s	horizontal separation between two buildings	ft
S	roof slope run per rise of one unit	–
w	snow drift width	ft
W	horizontal distance from eaves to ridge	ft

Symbols

θ	roof slope	deg
γ	snow density	lbf/ft³

General Principles

In accordance with IBC Sec. 1608.1, design snow load is determined by ASCE/SEI7 Chap. 7. Snow load is independent of tributary area and (as noted in ASCE/SEI7 Sec. C7.3) is not reducible. The design provisions of ASCE/SEI7 include consideration of roof slope, thermal resistance of roof sheathing materials, rain-on-snow surcharge, exposure to wind, the effects of sliding snow, ice dams, snow drifts, unbalanced loads, and partial loading.

Ground Snow Loads[4]

Determining the design snow load requires an accurate estimate of anticipated snowfall. The ground snow load, p_g, is obtained by interpolation from the contours of ASCE/SEI7 Fig. 7.2-1. This map, which shows snow loads in the United States, is based on over 40 years of snow depth records. The values given are the result of a statistical analysis of the records using a 2% annual probability of being exceeded. This is equivalent to a 50-year mean recurrence interval.

In some areas, the ground snow load is too variable to allow accurate mapping, and a site-specific case study is required. Values for Alaska are separately tabulated in ASCE/SEI7 Table 7.2-1, and values recorded at National Weather Service locations are provided in ASCE/SEI7 Table C7.2-1.

Design Snow Loads on Roofs

The principal factor affecting snow load on a roof is the slope of the roof. Roof slope is divided into three categories: flat roof, low-slope roof, and sloped roof.

Snow Loads for Flat Roofs

A *flat roof* is generally considered a roof with a slope of not more than 5°, which is approximately 1 in/ft. The snow load on a flat roof is calculated from ASCE/SEI7 Eq. 7.3-1 as

$$p_f = 0.7 C_e C_t I_s p_g$$

In this equation, four factors are applied to the ground snow load, p_g, to obtain the snow load on a flat roof, p_f. The first, the factor of 0.7, accounts for the fact that less snow accumulates on a roof than on the ground.

Exposure Factor, C_e

Wind tends to blow snow off a roof, so a roof in an exposed, windswept location accumulates less snow than a roof in a sheltered area. The exposure factor, C_e, accounts for differences in roof exposure and the surrounding terrain. Values for C_e are given in ASCE/SEI7 Table 7.3-1 and in the abbreviated Table 1.2. Of the factors included in ASCE/SEI7 Eq. 7.3-1, the exposure factor has the greatest effect.

The three categories of exposure are described as fully exposed, partially exposed, and sheltered. A *sheltered* roof is one surrounded by conifers that qualify as obstructions. A *fully exposed* roof is one without any shelter provided by higher structures or trees. A building or other obstruction within a distance of $10h_o$ is considered to provide shelter, where h_o is the height of the

Table 1.2 Exposure Factor, C_e

| | exposure of roof | | |
terrain category	fully exposed	partially exposed	sheltered
B: urban, suburban, and wooded areas with numerous closely spaced obstructions the size of single-family dwellings or larger	0.9	1.0	1.2
C: open terrain, including flat open country and grasslands, with scattered obstructions having heights generally less than 30 feet	0.9	1.0	1.1
D: flat, unobstructed areas and water surfaces, including smooth mud flats, salt flats, and unbroken ice	0.8	0.9	1.0
windswept, mountainous locations above the treeline	0.7	0.8	n/a
areas in Alaska where trees do not exist within a 2-mile radius	0.7	0.8	n/a

obstruction above the roof level. A *partially exposed* roof is one that does not fall into the category of either fully exposed or sheltered.

Thermal Factor, C_t

More snow accumulates on a cold roof than on a warm roof. This effect is reflected in the values of the *thermal factor* given in ASCE/SEI7 Table 7.3-2 and in the abbreviated Table 1.3.

An uninsulated roof permits heat from within the building to melt roof snow, consequently reducing the snow load. For a continuously heated greenhouse whose roof has a thermal resistance less than 2 ft²-hr-°F/Btu, snow melt is considerable, and a thermal factor of 0.85 applies. For unheated garages and agricultural buildings where the temperature in winter can be below zero, a thermal factor of 1.2 applies. For structures intentionally kept below freezing, snow melt is negligible, and a thermal factor of 1.3 applies. For a building that is kept heated throughout winter, a thermal factor of 1.0 applies.

As specified in ASCE/SEI7 Sec. 7.4.1, roofs with a thermal factor of $C_t \le 1.0$ are designated as warm roofs, and roofs with a thermal factor of $C_t > 1.0$ are designated as cold roofs. A gable roof, with air vents that allow air to

Table 1.3 Thermal Factor, C_t

thermal condition	C_t
all structures except as indicated below	1.0
structures kept just above freezing and others with cold, ventilated roofs in which the thermal resistance, R, between the ventilated space and the heated space exceeds 25 ft²-hr-°F/Btu	1.1
unheated and open-air structures	1.2
structures intentionally kept below zero	1.3
continuously heated greenhouses with roofs having a thermal resistance, R, less than 2 ft²-hr-°F/Btu	0.85

flow in at the eaves and out at the ridge, is designated a cold roof with a thermal factor of $C_t = 1.1$. An unventilated gable roof over a heated building is designated a warm roof with a thermal factor of $C_t = 1.0$.

Snow Importance Factor, I_s

The *snow importance factor* takes into account the consequences of roof failure. Essential facilities are given higher importance factors, which increases their design snow loads and reduces the possibility that overload will cause collapse.

In ASCE/SEI7 Table 1.5-1, risk categories are assigned to buildings based on the nature of their occupancies, and the importance factors corresponding to the risk categories are given in ASCE/SEI7 Table 1.5-2. Risk categories, nature of occupancy, and snow importance factors are combined in Table 1.4.

Example 1.8

A three-story residential facility with a nominally flat roof is located in a suburban area of Anchorage, Alaska. The roof of the facility is considered partially exposed. Determine the design roof snow load.

Solution

From ASCE/SEI7 Table 7.2-1, the ground snow load, p_g, for Anchorage, Alaska, is 50 lbf/ft². From Table 1.2, the exposure factor, C_e, is 1.0 for terrain category B, partially exposed. From Table 1.3, the thermal factor, C_t, is 1.0 for a residential structure with a warm roof. From Table 1.4, the snow importance factor, I_s, is 1.0 for risk category II.

Table 1.4 Risk Category and Snow Importance Factor, I_s

risk category	snow importance factor
I: low-hazard structures such as agricultural facilities, minor storage buildings, and temporary facilities	0.80
II: buildings other than those listed in risk categories I, III, and IV such as residential, commercial, and office buildings	1.00
III: facilities, other than those included in risk category IV, that could cause a substantial economic impact and/or mass disruption of day-to-day civilian life in the event of failure, including	1.10
• buildings with high occupant loads, such as buildings where more than 300 people congregate	
• schools with capacities of more than 250	
• colleges with capacities of more than 500	
• health care facilities with capacities of 50 or more	
• jails	
• power stations	
• buildings housing toxic materials that will endanger the safety of the public if released (if the authority having jurisdiction is satisfied that the risk involved is appropriate to this category)	
IV: essential facilities such as	1.20
• hospitals, fire and police stations, post-earthquake recovery centers, and buildings housing equipment for these facilities	
• buildings posing a substantial hazard to the community if they were to collapse	
• buildings housing toxic materials that will endanger the safety of the public if released	

The flat roof snow load is

$$p_f = 0.7 C_e C_t I_s p_g$$
$$= (0.7)(1.0)(1.0)(1.0)\left(50 \; \frac{\text{lbf}}{\text{ft}^2}\right)$$
$$= 35 \; \text{lbf/ft}^2$$

Minimum Snow Loads for Low-Slope Roofs

A low-slope roof is defined in ASCE/SEI7 Sec. 7.3.4 as a monoslope, hip, or gable roof with a slope less than 15°. The minimum value for the roof snow load for a low-slope roof is specified in ASCE/SEI7 Sec. 7.3.4 as

$$p_m = I_s p_g \qquad [\text{for } p_g \leq 20 \; \text{lbf/ft}^2]$$
$$p_m = 20 I_s \qquad [\text{for } p_g > 20 \; \text{lbf/ft}^2]$$

The minimum snow load is also applied to curved roofs where the vertical angle from the eaves to the crown is less than 10°.

The minimum roof snow load for a low-slope roof is a separate uniform load case; it is not applied in combination with drift, sliding, unbalanced, or partial loads.

Rain-on-Snow Surcharge Loads

Where the ground snow load is 20 lbf/ft² or less, the temporary roof load contributed by a heavy rain may be significant. In accordance with ASCE/SEI7 Sec. 7.10, in locations where the ground snow load is 20 lbf/ft² or less, a roof with a slope in degrees of less than $W/50$ (with W in feet) must have a 5 lbf/ft² rain-on-snow surcharge imposed. This surcharge is not applied in combination with drift, sliding, unbalanced, or partial snow loads.

Example 1.9

A fire station with a nominally flat roof is located in a suburban area and is considered partially exposed. The ground snow load is 10 lbf/ft². Determine the design roof snow load.

Solution

From Table 1.2, the exposure factor, C_e, is 1.0 for terrain category B, partially exposed. From Table 1.3, the thermal factor, C_t, is 1.0 for a heated structure. From Table 1.4, the snow importance factor, I_s, is 1.2 for risk category IV.

From ASCE/SEI7 Eq. 7.3-1, the flat roof snow load is

$$p_f = 0.7 C_e C_t I_s p_g$$
$$= (0.7)(1.0)(1.0)(1.2)\left(10 \; \frac{\text{lbf}}{\text{ft}^2}\right)$$
$$= 8.4 \; \text{lbf/ft}^2$$

Since the ground snow load does not exceed 20 lbf/ft² and the slope of the roof in degrees is less than $W/50$, a

5 lbf/ft² rain-on-snow surcharge must be imposed in accordance with ASCE/SEI7 Sec. 7.10. The total flat roof snow load is

$$p_f = 8.4 \ \frac{\text{lbf}}{\text{ft}^2} + 5 \ \frac{\text{lbf}}{\text{ft}^2}$$

$$= 13.4 \ \text{lbf/ft}^2$$

Since the ground snow load does not exceed 20 lbf/ft² and the slope of the roof is less than 15°, ASCE/SEI7 Sec. 7.3.4 specifies a minimum flat roof snow load of

$$p_m = I_s p_g$$

$$= (1.2)\left(10 \ \frac{\text{lbf}}{\text{ft}^2}\right)$$

$$= 12 \ \text{lbf/ft}^2$$

The design flat roof snow load is 13.4 lbf/ft², so the minimum does not govern.

Balanced Snow Loads for Sloped Roofs

A snow load acting on a sloping surface is assumed to act on the horizontal projection of that surface. A balanced snow load on a sloped roof is assumed to be uniformly distributed on the horizontal projection as shown in Fig. 1.9.

Because of the combined effects of wind and snow sliding, less snow accumulates on a sloped roof than on a flat roof. The balanced snow load for a sloped roof, p_s, is calculated from ASCE/SEI7 Eq. 7.4-1 by multiplying the flat roof snow load by the roof slope factor, C_s, to give

$$p_s = C_s p_f$$

Figure 1.9 *Balanced Snow Load for Gable Roof*

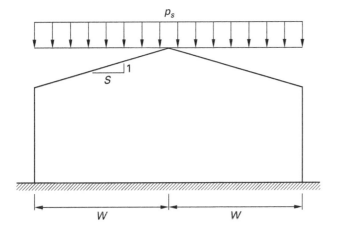

The value of the roof slope factor depends on the following factors.

- roof slope
- roof temperature
- slipperiness of the roof surface
- presence of obstructions
- presence of sufficient space below the eaves to accept sliding snow

Snow load decreases as the roof slope increases, and for slopes of 70° or more the snow load is assumed to be zero. Roof materials that are considered to be slippery are defined in ASCE/SEI7 Sec. 7.4 as metal, slate, glass, and bituminous, rubber, and plastic membranes with smooth surfaces. A membrane with an imbedded aggregate or mineral granule surface is not considered smooth. Asphalt shingles, wood shingles and shakes are not considered smooth.

Warm Roof Slope Factor ($C_t \le 1.0$)

For a *warm roof*—one with a thermal factor, C_t, equal to 1.0 or less—the roof slope factor is determined using ASCE/SEI7 Fig. 7.4-1a. There are two lines on this graph, one solid and one dashed. The dashed line is used if the roof has an unobstructed, slippery surface and

- the roof is unventilated and its thermal resistance, R, is at least 30 ft²-hr-°F/Btu
- the roof is ventilated and its thermal resistance, R, is at least 20 ft²-hr-°F/Btu

For warm roofs that do not meet these conditions, the roof slope factor is determined using the solid line in ASCE/SEI7 Fig. 7.4-1a.

Cold Roof Slope Factor ($C_t \ge 1.1$)

For a *cold roof*—one with a thermal factor, C_t, equal to 1.1 or greater—the roof slope factor is determined using

- ASCE/SEI7 Fig. 7.4-1b if the thermal factor is 1.1
- ASCE/SEI7 Fig. 7.4-1c if the thermal factor is 1.2 or greater

Each of these graphs has two lines, one solid and one dashed. The dashed line is used if the roof has an unobstructed, slippery surface, and the solid line is used for all other surfaces.

Example 1.10

An office building with a gable roof is located in a suburban area and is considered partially exposed. The building has an unobstructed slate roof with a slope of 6 on 12. The ground snow load, p_g, is 30 lbf/ft². Determine the balanced snow load on the roof.

Solution

The ground snow load is

$$p_g = 30 \ \frac{\text{lbf}}{\text{ft}^2}$$

This is greater than 20 lbf/ft^2, so the rain-on-snow surcharge does not apply. From Table 1.2, for terrain category B, partially exposed, the exposure factor, C_e, is 1.0. From Table 1.3, for a heated structure, the thermal factor, C_t, is 1.0. From Table 1.4, for risk category II, the snow importance factor, I_s, is 1.0.

From ASCE/SEI7 Eq. 7.3-1, the flat roof snow load is

$$p_f = 0.7 C_e C_t I_s p_g$$
$$= (0.7)(1.0)(1.0)(1.0)\left(30 \ \frac{\text{lbf}}{\text{ft}^2}\right)$$
$$= 21 \ \text{lbf/ft}^2$$

The roof slope is 6 on 12, or

$$S = \frac{6}{12}$$
$$= 26.6°$$

This is greater than 15°, so in accordance with ASCE/SEI7 Sec. 7.3.4, the minimum flat roof load does not apply. The flat roof snow load is 21 lbf/ft^2.

The office building is heated in winter, so it is considered to have a warm roof, and its slate roof is considered a slippery surface. For a warm roof with a slope of 26.6° and a slippery surface, the slope factor is determined from the dashed line in ASCE/SEI7 Fig. 7.4-1a.

$$C_s = 0.7$$

The balanced sloped roof snow load is given by ASCE/SEI7 Eq. 7.4-1.

$$p_s = C_s p_f$$
$$= (0.7)\left(21 \ \frac{\text{lbf}}{\text{ft}^2}\right)$$
$$= 14.7 \ \text{lbf/ft}^2$$

Balanced Snow Loads for Curved Roofs

Those portions of a curved roof where the slope exceeds 70° are considered to be free of snow and to have a slope factor, C_s, of zero. A balanced load for a curved roof is determined from the diagrams in ASCE/SEI7 Fig. 7.4-2, with the roof slope factor, C_s, determined from ASCE/SEI7 Fig. 7.4-1 as before.

Balanced Snow Loads for Sawtooth Roofs

As specified in ASCE/SEI7 Sec. 7.4.5 and shown in Fig. 1.10, a sawtooth roof is designed for a roof slope factor, C_s of 1.0, with no reduction in snow load because of slope. This gives a value for the balanced snow load of $p_s = p_f$.

Figure 1.10 *Balanced Snow Load for a Sawtooth Roof*

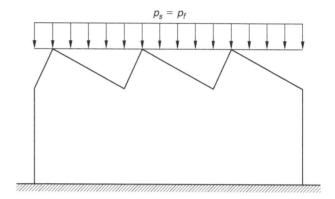

Ice Dam Formation Along Eaves

An uninsulated attic below a warm, sloped roof will increase the tendency of snow to melt. When the melt water reaches the eaves overhang, which is at ambient temperature, it may freeze and form an ice dam, as shown in Fig. 1.11.

Figure 1.11 *Formation of Ice Dam on a Sloped Roof*

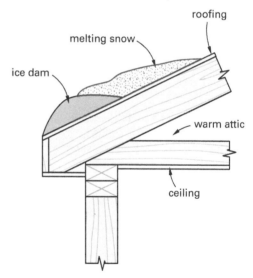

As specified in ASCE/SEI7 Sec. 7.4.5, this situation can occur with

- unventilated warm roofs with R-values less than 30 ft^2-hr-°F/Btu

- ventilated warm roofs with R-values less than 20 ft^2-hr-°F/Btu

For these conditions, all overhanging portions of the roof must be designed for a uniformly distributed load of two times the flat roof snow load. This surcharge is applied only to the overhanging portion of the structure, not to the entire building. In designing the overhang, snow load on the remainder of the roof should be ignored, as shown in Fig. 1.12.

Figure 1.12 *Ice Dam Loads on Eave Overhangs*

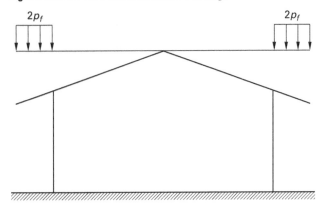

Unbalanced Snow Loads for Hip and Gable Roofs

When wind blows over a gable roof, it removes snow from the windward side of the gable and deposits it on the leeward side, producing an unbalanced load. Moreover, more snow is deposited near the leeward side of the ridge than is deposited near the leeward eaves.[4] This complex situation is idealized in ASCE/SEI7 Sec. 7.6.1, as illustrated in Fig. 1.13(a).

From ASCE/SEI7 Sec. 7.6.1, the unbalanced load on the windward slope consists of a uniform load of

$$p_W = 0.3p_s$$

The unbalanced load on the leeward slope consists of two superimposed loads: a uniform load over the full length of the leeward slope, plus a surcharge load, representing the drift, that extends from the ridge a horizontal distance of

$$d_S = \frac{8(\sqrt{S})\,h_d}{3}$$

The height of the leeward drift, h_d, is obtained from ASCE/SEI7 Fig. 7.6-1 and is

$$h_d/\sqrt{I_s} = 0.43 l_u^{1/3}(p_g + 10)^{1/4} - 1.5$$

The value of the uniform load on the leeward slope is

$$p_{LD} = p_s$$

Figure 1.13 *Unbalanced Snow Load on Gable Roof*

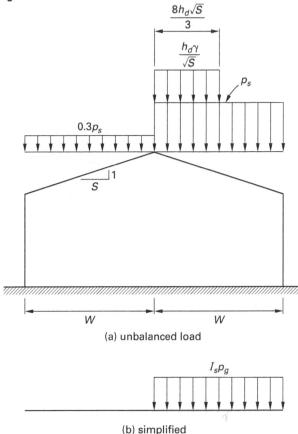

(a) unbalanced load

(b) simplified

The value of the surcharge load on the leeward slope is

$$p_{LS} = \frac{h_d\gamma}{\sqrt{S}}$$

The density of the snow is given by ASCE/SEI7 Eq. 7.7-1 as

$$\gamma = 0.13p_g + 14$$
$$\leq 30.0 \text{ lb/ft}^3$$

A further simplification, as shown in Fig. 1.13(b), is made for simple roof rafter systems with a supporting ridge beam and an eave-to-ridge distance, W, of less than 20 ft.

When determining unbalanced loads, winds from all directions must be accounted for. In accordance with ASCE/SEI7 Sec. 7.6, balanced and unbalanced loading conditions must be analyzed separately to determine the most critical situation.

As specified in ASCE/SEI7 Fig. 7.6-2, analysis for unbalanced snow loads is not required for a gable roof slope greater than 30.2° or less than 2.38°.

Example 1.11

An office building with a gable roof is located in a suburban area and is considered partially exposed. The building has an unobstructed slate roof with a slope of 6 on 12 and a horizontal distance from eaves to ridge of 25 ft. The ground snow load is 30 lbf/ft^2. Determine the unbalanced snow load on the roof.

Solution

From Ex. 1.10, the balanced sloped roof snow load, p_s, is 14.7 lbf/ft^2, and the roof slope, θ, is 26.6°. This is between 2.38° and 30.2°, so analysis for the unbalanced snow load is required.

From ASCE/SEI7 Sec. 7.6.1, the load on the windward slope is

$$p_W = 0.3p_s$$
$$= (0.3)\left(14.7 \ \frac{\text{lbf}}{\text{ft}^2}\right)$$
$$= 4.4 \ \text{lbf/ft}^2$$

From ASCE/SEI7 Sec. 7.6.1, the distributed load on the leeward slope is

$$p_{LD} = p_s = 14.7 \ \text{lbf/ft}^2$$

The length of the upwind fetch is

$$l_u = W$$
$$= 25 \ \text{ft}$$

I_s is 1.0, and from ASCE/SEI7 Fig. 7.6-1, the height of the leeward snow drift is

$$h_d = 0.43l_u^{1/3}(p_g + 10)^{1/4} - 1.5$$
$$\text{[from ASCE/SEI7 Fig. 7.6-1]}$$
$$= (0.43)(25 \ \text{ft})^{1/3}\left(30 \ \frac{\text{lbf}}{\text{ft}^2} + 10\right)^{1/4} - 1.5$$
$$= 1.66 \ \text{ft}$$

The roof slope run for a rise of one is

$$S = \frac{12}{6} = 2.0$$

From ASCE/SEI7 Eq. 7.7-1, the density of the snow is

$$\gamma = 0.13p_g + 14$$
$$= (0.13)\left(30 \ \frac{\text{lbf}}{\text{ft}^2}\right) + 14$$
$$= 17.9 \ \text{lbf/ft}^3$$
$$< 30.0 \ \text{lbf/ft}^3 \qquad \text{[satisfactory]}$$

The surcharge load on the leeward slope is

$$p_{LS} = \frac{h_d\gamma}{\sqrt{S}}$$
$$= \frac{(1.66 \ \text{ft})\left(17.9 \ \frac{\text{lbf}}{\text{ft}^3}\right)}{\sqrt{2.0}}$$
$$= 21.0 \ \text{lbf/ft}^2$$

The horizontal extent of the surcharge is

$$d_S = \frac{8\sqrt{S}\,h_d}{3}$$
$$= \frac{8\sqrt{2.0}\,(1.66 \ \text{ft})}{3}$$
$$= 6.3 \ \text{ft}$$

The unbalanced loading condition is shown.

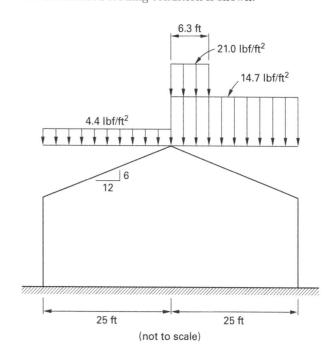

(not to scale)

Unbalanced Snow Loads for Curved Roofs

Those portions of a curved roof where the slope exceeds 70° are considered free of snow with a slope factor, C_s, of zero. If the slope of a straight line from the eaves (or the 70° point, if present) to the crown is less than 10° or greater than 60°, unbalanced snow loads are neglected. Unbalanced loads are determined from the unbalanced load diagrams in ASCE/SEI7 Fig. 7.4-2, with the roof slope factor, C_s, determined from the appropriate curve in ASCE/SEI7 Fig. 7.4-1.

Unbalanced Snow Loads for Sawtooth Roofs

On a sawtooth roof, snow can drift and slide into the roof's valleys, leading to an unbalanced snow load. In accordance with ASCE/SEI7 Sec. 7.6.3, a sawtooth roof with a slope greater than 1.79° ($\frac{3}{8}$ in/ft) must be analyzed for unbalanced snow load. As specified in ASCE/SEI7 Sec. 7.4.4, the roof slope factor, C_s, for a sawtooth roof is 1.0, and balanced snow load is $p_s = p_f$.

As shown in Fig. 1.14, the unbalanced snow load increases from a minimum value of $p_f/2$ at the ridge to a value of $2p_f/C_e$ at the valley, with a maximum value equivalent to the snow level at the valley no higher than the snow level at the ridge. Snow levels are determined by dividing snow load by snow density as defined by ASCE/SEI7 Eq. 7.7-1.

Figure 1.14 *Unbalanced Snow Load on Sawtooth Roof*

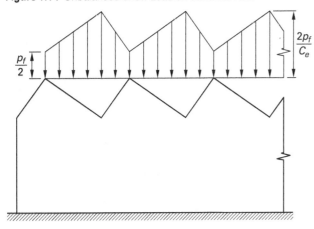

Sliding Snow Loads

Snow may slide from a sloped upper roof onto a lower roof, and this must be considered in design of the lower roof. The amount of snow deposited on the lower roof depends on the size of both roofs and their juxtaposition. The effects of rain-on-snow loads and of partial, unbalanced, and drift loads are neglected. Sliding need not be considered for an upper roof that is

- slippery with a slope no greater than $\frac{1}{4}$ in/ft
- nonslippery with a slope no greater than 2 in/ft

Sliding Loads for Adjoining Buildings

When two buildings adjoin, as shown in Fig. 1.15, a sliding snow load is calculated for the upper roof and superimposed on the balanced snow load of the lower roof.

Figure 1.15 *Sliding Snow Load for Adjoining Buildings*

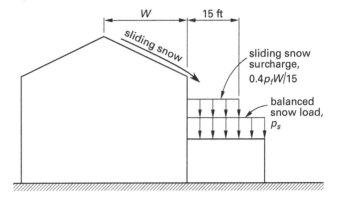

As specified by ASCE/SEI7 Sec. 7.9, the sliding load consists of 40% of the flat roof snow load for the upper roof, p_f, extended over the horizontal distance from the upper roof's eave to its ridge, W. This sliding load is uniformly distributed on the lower roof over a distance of 15 ft. This gives a value for the surcharge load of

$$p_{\text{sliding}} = \frac{0.4p_f W}{15}$$

Where the lower roof is less than 15 ft wide, the surcharge extends over the full width of the lower roof. The surcharge load is reduced if a portion of the sliding snow from the upper roof is blocked from sliding onto the lower roof by snow already on the lower roof.

Example 1.12

An office building with a gable roof is located in a suburban area and is considered partially exposed. The building has an unobstructed slate roof with a slope of 6 on 12 and a horizontal distance from eaves to ridge of 25 ft. The ground snow load is 30 lbf/ft². An unheated garage with a roof slope of 1 on 48 adjoins the office building, as shown in Fig. 1.15. The vertical separation, h, between the edge of the upper roof and the edge of the garage roof is 12 ft. Determine the snow load on the garage roof.

Solution

Slate is considered a slippery surface, and the roof slope is greater than $\frac{1}{4}$ on 12, so in accordance with ASCE/SEI7 Sec. 7.9, sliding snow must be considered.

As determined in Ex. 1.10, the flat roof snow load for the upper roof, p_f, is 21 lbf/ft^2. The surcharge imposed on the lower roof is

$$p_{\text{sliding}} = \frac{0.4 p_f W}{15}$$

$$= \frac{(0.4)\left(21 \ \dfrac{\text{lbf}}{\text{ft}^2}\right)(25 \ \text{ft})}{15 \ \text{ft}}$$

$$= 14 \ \text{lbf/ft}^2$$

For the unheated garage roof, the roof slope is

$$S = \frac{1}{48}$$

$$= 1.19°$$

This is less than 15°, so the garage roof qualifies as a low-slope roof in accordance with ASCE/SEI7 Sec. 7.3.4. The ground snow load is

$$p_g = 30 \ \text{lbf/ft}^2$$

This is greater than 20 lbf/ft^2, so the rain-on-snow surcharge does not apply. From Table 1.2, the exposure factor for terrain category B, partially exposed, is

$$C_e = 1.0$$

From Table 1.3, the thermal factor for an unheated structure is

$$C_t = 1.2$$

From Table 1.4, the importance factor for risk category I is

$$I_s = 0.8$$

The flat roof snow load for the garage roof is calculated from ASCE/SEI7 Eq. 7.3-1 as

$$p_f = 0.7 C_e C_t I_s p_g$$

$$= (0.7)(1.0)(1.2)(0.8)\left(30 \ \dfrac{\text{lbf}}{\text{ft}^2}\right)$$

$$= 20.2 \ \text{lbf/ft}^2$$

From ASCE/SEI7 Fig. 7.4-1c, the slope factor for a low slope roof is

$$C_s = 1.0$$

The balanced snow load on the garage roof is given by ASCE/SEI7 Eq. 7.4-1 as

$$p_s = C_s p_f$$

$$= (1.0)\left(20.2 \ \dfrac{\text{lbf}}{\text{ft}^2}\right)$$

$$= 20.2 \ \text{lbf/ft}^2$$

The total snow load on the lower roof is

$$p_T = p_{\text{sliding}} + p_s$$

$$= 14 \ \frac{\text{lbf}}{\text{ft}^2} + 20.2 \ \frac{\text{lbf}}{\text{ft}^2}$$

$$= 34.2 \ \text{lbf/ft}^2$$

From ASCE/SEI7 Eq. 7.7-1, the density of the snow is

$$\gamma = 0.13 p_g + 14$$

$$= \left(0.13 \ \frac{1}{\text{ft}}\right)\left(30 \ \frac{\text{lbf}}{\text{ft}^2}\right) + 14 \ \frac{\text{lbf}}{\text{ft}^3}$$

$$= 17.9 \ \text{lbf/ft}^3$$

$$< 30.0 \ \text{lbf/ft}^3 \qquad \text{[satisfactory]}$$

The total depth of snow on the lower roof is

$$h_T = \frac{p_T}{\gamma}$$

$$= \frac{34.2 \ \dfrac{\text{lbf}}{\text{ft}^2}}{17.9 \ \dfrac{\text{lbf}}{\text{ft}^3}}$$

$$= 1.9 \ \text{ft}$$

This is less than the difference in height, h, of 12 ft, so the sliding snow from the upper roof is not blocked from sliding onto the lower roof by snow already on the lower roof. The loading condition on the garage roof is as shown.

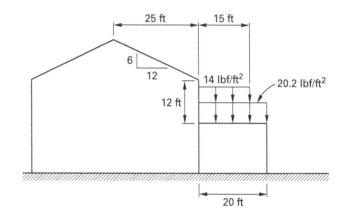

Sliding Loads for Separated Buildings

When two buildings are separated, as shown in Fig. 1.16, sliding snow from the upper roof onto the lower roof must be considered in design of the lower roof when both of these are true

- the horizontal separation of the two roofs is less than 15 ft

- the vertical separation of the two roofs is greater than the horizontal separation

Figure 1.16 Sliding Snow Load for Separated Buildings

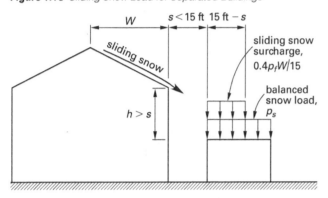

From ASCE/SEI7 Sec. 7.9:

1. The total sliding load per unit length of eaves is $p_{\text{sliding}} = 0.4p_f W(15-s)/15$ lbf/ft.

2. This is distributed on the lower roof over a distance of $(15-s)$ ft.

3. The surcharge on the lower roof is then, $p_{\text{sliding}}/(15-s) = 0.4p_f W/15$ lbf/ft^2.

Example 1.13

An office building with a gable roof is located in a suburban area and is considered partially exposed. The building has an unobstructed slate roof with a slope of 6 on 12 and a horizontal distance from eaves to ridge of 25 ft. The ground snow load is 30 lbf/ft^2. An unheated garage with a roof slope of 1 on 48 is located 8 ft from the office building. The vertical separation between the edge of the upper roof and the edge of the garage roof is 12 ft. Determine the snow load on the garage roof.

Solution

The horizontal separation between the two roofs, s, is 8 ft, which is less than 15 ft. The vertical separation of the two roofs, h, is 12 ft, which is greater than the horizontal separation. Because both conditions are met, in accordance with ASCE/SEI7 Sec. 7.9, a sliding snow surcharge must be considered on the lower roof.

The surcharge imposed on the lower roof is the same as in Ex. 1.12.

$$p_{\text{sliding}} = 14 \text{ lbf/ft}^2$$

The horizontal extent of the sliding snow surcharge on the lower roof is

$$15 \text{ ft} - s = 15 \text{ ft} - 8 \text{ ft} = 7 \text{ ft}$$

The balanced snow load on the lower roof is the same as in Ex. 1.12.

$$p_s = 20.2 \text{ lbf/ft}^2$$

The loading condition on the garage roof is as shown.

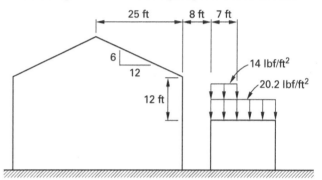

Snow Drift Loads

A *snow drift* is formed when snow that has already been deposited once is picked up by wind and carried in suspension. When the wind velocity is reduced, the snow is deposited again, forming a drift.

Wind velocity is reduced when the wind encounters obstructions such as on a stepped roof or at a parapet. The size of the drift depends on the amount of snow available, which in turn depends on both the length of the upwind fetch and the ground snow load. Snow drifts will build up on both the lee side and, to a lesser extent, on the windward side of an obstruction.

Figure 1.17(a) shows a lower roof that is subjected to a leeward drift. When the wind changes direction by 180°, as shown in Fig. 1.17(b), the lower roof is now subject to a windward drift. When the upwind fetch, l_u, is longer for the windward drift than for the leeward drift, the windward drift may govern the design of the lower roof. Hence, ASCE/SEI7 Sec. 7.7.1 requires that, at a step, the larger of the windward and leeward drifts must be used in design. The drift is in addition to the balanced snow load of the lower roof.

To simplify the design process, ASCE/SEI7 Fig. 7.7-2 approximates the shape of a drift with a triangle, and the drift surcharge is superimposed on the balanced snow load of the lower roof.

As specified in ASCE/SEI7 Sec. 7.7.2, drift loads may also be caused by adjacent structures or higher terrain and must be considered in design.

Snow Drifts for Leeward Roof Steps

A leeward step drift is formed by wind blowing off a higher roof upwind of a lower roof, as shown in Fig. 1.17(a). In accordance with ASCE/SEI7 Sec. 7.7.1, the configuration of the surcharge load is as shown in Fig. 1.18.

Figure 1.17 *Snow Drift Formation*

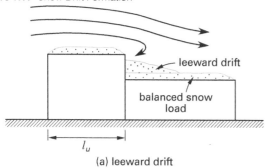

(a) leeward drift

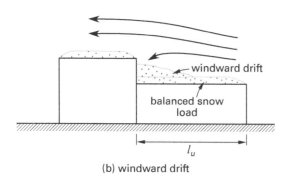

(b) windward drift

Figure 1.18 *Configuration of Snow Drift*

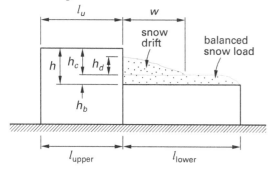

The balanced snow load on the lower roof is calculated from ASCE/SEI7 Eq. 7.4-1. The flat roof snow load, p_f, is multiplied by the roof slope factor, C_s, to give

$$p_s = C_s p_f$$

Snow density is calculated from ASCE/SEI7 Eq. 7.7-1 as

$$\gamma = 0.13 p_g + 14 \ \le 30 \text{ pcf}$$

The depth of the balanced snow load on the lower roof is

$$h_b = \frac{p_s}{\gamma}$$

The clear height from the top of the balanced snow load to the upper roof level, as shown in Fig. 1.18, is

$$h_c = h - h_b$$

Drift loads need not be considered when the clear height, h_c, is

$$h_c < 0.2 h_b$$

The height of the drift, h_d, is obtained from ASCE/SEI7 Fig. 7.6-1; the length of the upwind fetch, l_u, is taken as equal to the length of the upper roof.

$$h_d / \sqrt{I_s} = 0.43 l_u^{1/3} (p_g + 10)^{1/4} - 1.5$$
$$\le h_c$$
$$\le 0.6 l_{\text{lower}}$$

The maximum intensity of the drift surcharge load at the step is

$$p_d = h_d \gamma$$

If $h_d \le h_c$, then the drift has no height constraints, and the width of the drift is

$$w = 4 h_d$$
$$\le 8 h_c$$

If $h_d > h_c$, then the height of the drift is constrained and is taken as h_c, and the width of the drift is

$$w = \frac{4 h_d^2}{h_c}$$
$$\le 8 h_c$$

If the width of the drift, w, exceeds the length of the lower roof, the triangle representing the drift should taper to zero.

Snow Drifts for Windward Roof Steps

A windward step drift is formed by wind blowing off a lower roof upwind of a higher roof, as shown in Fig. 1.17(b). The configuration of the surcharge load is shown in Fig. 1.18. The height of a windward drift, h_d, is taken as 75% of the value for a leeward drift (from ASCE/SEI7 Fig. 7.6-1), or

$$h_d = 0.75 \left(\left(0.43 l_u^{1/3} (p_g + 10)^{1/4} - 1.5 \right) \sqrt{I_s} \right)$$

The length of the upwind fetch, l_u, is taken as equal to the length of the lower roof.

Example 1.14

A commercial building with a warm roof sloped at 1 on 48 is located in a suburban area and is considered partially exposed. The building has a stepped roof with an elevation difference of 8 ft. The length of the upper roof is 100 ft, and the length of the lower roof is 150 ft. The ground snow load is 30 lbf/ft². Determine the design snow load on the lower roof.

Solution

For the lower roof, the roof slope is

$$S = \frac{1}{48} = 1.19°$$
$$< 15°$$

This is less than 15°, so the lower roof is a low-slope roof in accordance with ASCE/SEI7 Sec. 7.3.4.

The ground snow load, p_g, is 30 lbf/ft². This is greater than 20 lbf/ft², so the rain-on-snow surcharge does not apply (ASCE/SEI7 Sec. 7.10). From Table 1.2, the exposure factor, C_e, for terrain category B, partially exposed, is 1.0. From Table 1.3, the thermal factor, C_t, for a warm roof is 1.0. From Table 1.4, the importance factor, I_s, for risk category II is 1.0.

From ASCE/SEI7 Eq. 7.3-1, the flat roof snow load is

$$p_f = 0.7 C_e C_t I_s p_g = (0.7)(1.0)(1.0)(1.0)\left(30 \ \frac{\text{lbf}}{\text{ft}^2}\right)$$
$$= 21 \ \text{lbf/ft}^2$$

This is greater than $20I_s = 20$ lbf/ft², so the minimum roof snow load of ASCE/SEI7 Sec. 7.3.4 does not govern. From ASCE/SEI7 Fig. 7.4-1a, the slope factor, C_s, for a low slope roof is 1.0. From ASCE/SEI7 Eq. 7.4-1, the balanced snow load on the lower roof is

$$p_s = C_s p_f = (1.0)\left(21 \ \frac{\text{lbf}}{\text{ft}^2}\right) = 21 \ \text{lbf/ft}^2$$

From ASCE/SEI7 Eq. 7.7-1, snow density is

$$\gamma = 0.13 p_g + 14 = (0.13 \ \text{ft}^{-1})\left(30 \ \frac{\text{lbf}}{\text{ft}^2}\right) + 14 \ \frac{\text{lbf}}{\text{ft}^3}$$
$$= 17.9 \ \text{lbf/ft}^3$$
$$< 30.0 \ \text{lbf/ft}^3 \qquad \text{[satisfactory]}$$

The depth of the balanced snow load on the lower roof is

$$h_b = \frac{p_s}{\gamma} = \frac{21 \ \dfrac{\text{lbf}}{\text{ft}^2}}{17.9 \ \dfrac{\text{lbf}}{\text{ft}^3}} = 1.2 \ \text{ft}$$

From Fig. 1.18, the clear height from the top of the balanced snow load to the upper roof level is

$$h_c = h - h_b = 8 \ \text{ft} - 1.2 \ \text{ft} = 6.8 \ \text{ft}$$

This is greater than $0.2h_b = 0.24$ ft, so in accordance with ASCE/SEI7 Sec. 7.7.1, drift loads must be considered. The length of the upwind fetch producing a leeward drift at the roof step is

$$l_u = L_{\text{upper}} = 100 \ \text{ft}$$

I_s is 1.0, and from ASCE/SEI7 Fig. 7.6-1, the height of the leeward snow drift is

$$h_d = 0.43(l_u)^{1/3}(p_g + 10)^{1/4} - 1.5$$
$$= (0.43)(100 \ \text{ft})^{1/3}\left(30 \ \frac{\text{lbf}}{\text{ft}^2} + 10 \ \frac{\text{lbf}}{\text{ft}^2}\right)^{1/4} - 1.5 \ \text{ft}$$
$$= 3.5 \ \text{ft}$$

This is less than $h_c = 6.8$ ft, so the drift height is not constrained. The length of the upwind fetch producing a windward drift at the roof step is

$$l_u = L_{\text{lower}} = 150 \ \text{ft}$$

$I_s = 1.0$, and from ASCE/SEI7 Fig. 7.6-1, the height of the windward snow drift is

$$h_d = 0.75(0.43 l_u^{1/3}(p_g + 10)^{1/4} - 1.5)$$
$$= (0.75)\left((0.43)(150 \ \text{ft})^{1/3}\left(30 \ \frac{\text{lbf}}{\text{ft}^2} + 10 \ \frac{\text{lbf}}{\text{ft}^2}\right)^{1/4} - 1.5 \ \text{ft}\right)$$
$$= 3.2 \ \text{ft}$$

This is less than the height of the leeward drift, so the leeword drift governs, and the critical drift height is

$$h_d = 3.5 \ \text{ft}$$

This is less than $h_c = 6.8$ ft, so the drift height is not constrained, and from ASCE/SEI7 Sec. 7.7.1, the width of the drift is

$$w = 4h_d = (4)(3.5 \ \text{ft})$$
$$= 14 \ \text{ft}$$
$$< 8h_c \qquad \text{[satisfactory]}$$

The maximum intensity of the drift surcharge load at the roof step is

$$p_d = h_d \gamma = (3.5 \text{ ft}) \left(17.9 \ \frac{\text{lbf}}{\text{ft}^3} \right) = 62.7 \ \text{lbf/ft}^2$$

The loading diagram for the low roof is shown.

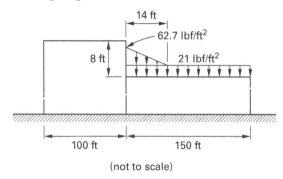

(not to scale)

Drifts at Roof Projections and Parapets

For drift loads caused by roof projections and parapet walls, a windward drift is assumed, and the height of the drift is taken as three-quarters the drift height determined from ASCE/SEI7 Fig. 7.6-1, which gives

$$h_d/\sqrt{I_s} = 0.75 \left(0.43 l_u^{1/3} (p_g + 10)^{1/4} - 1.5 \right)$$

For parapet walls, the upwind fetch is the length of roof upwind of the parapet, as shown in Fig. 1.19.

Figure 1.19 Snow Drift at Parapet Wall

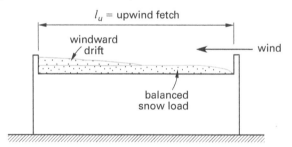

For projections, the fetch is taken as the length of roof either upwind or downwind of the projection, whichever is greater. A drift load is not required if

- the crosswind width of the projection is less than 15 ft

- the clear distance between the top surface of the balanced snow load and the bottom face of the projection is at least 2 ft

Snow Drift Loads for Separated Buildings

As shown in Fig. 1.20, drift loads must also be considered on the roof of a building that lies within the wind shadow of an adjacent building with a higher roof. In accordance with ASCE/SEI7 Sec. 7.7.2, a *leeward drift* will form on the lower roof if the horizontal separation distance from the upper roof, s, is less than 20 ft and also less than $6h$, where h is the vertical separation distance.

Figure 1.20 Leeward Drift Load for Separated Buildings

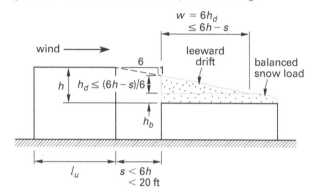

These restrictions ensure that the lower roof is designed for a leeward drift when the lower roof is in the wind shadow of the upper roof. As shown in ASCE/SEI7 Fig. C7.7-2, the trajectory of the wind shadow extends from the top of the upper roof downward at a slope of 1 on 6. The drift is deposited on the balanced snow load of the lower roof, with the top of the drift sloped at 1 on 6 to match the wind shadow boundary.

The height of the drift, h_d, is obtained from ASCE/SEI7 Fig. 7.6-1, with the length of the upwind fetch, l_u, equal to the length of the upper roof.

$$h_d/\sqrt{I_s} = 0.43 l_u^{1/3} (p_g + 10)^{1/4} - 1.5$$

The horizontal extent of the drift is obtained by tapering the top of the drift at a slope of 1 on 6 to match the wind shadow boundary, which gives

$$w = 6h_d$$

However, when the drift height is constrained by the top boundary of the wind shadow region, these values are reduced. To simplify the design process, ASCE/SEI7 Sec. 7.7.2 specifies that in this case the horizontal extent of the drift is

$$w = 6h - s$$

For a drift with a slope of 1 on 6, the resulting drift height is

$$h_d = \frac{6h - s}{6}$$

Windward drifts are designed in accordance with ASCE/SEI7 Sec. 7.7.1, with the height of a drift taken as 75% of the drift height determined from ASCE/SEI7 Fig. 7.6-1, which gives

$$h_d/\sqrt{I_s} = 0.75 \left(0.43 l_u^{1/3}(p_g + 10)^{1/4} - 1.5 \right)$$

As shown in Fig. 1.21, the drift is truncated by neglecting the front portion of the drift that lies between the two buildings.

Figure 1.21 *Windward Drift Load for Separated Buildings*

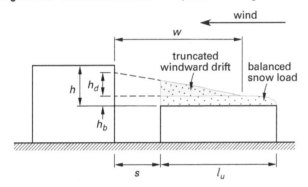

Example 1.15

Two commercial buildings with warm roofs sloped at 1 on 48 are located in a suburban area that is considered partially exposed. The buildings are separated a distance of 8 ft. The ground snow load is 30 lbf/ft², and the difference in elevation between the roofs of the two buildings is 8 ft. The length of the upper roof is 100 ft, and the length of the lower roof is 150 ft. Determine the design snow load on the lower roof.

Solution

From Ex. 1.14, the balanced snow load on the lower roof, p_s, is 21 lbf/ft². The heights of the windward and leeward snow drifts, h_d, are 3.2 ft and 3.5 ft, respectively; the height of the leeward snow drift governs.

From ASCE/SEI7 Sec. 7.7.2, the horizontal extent of the drift is the smaller of

$$w = 6h_d = (6)(3.5 \text{ ft})$$
$$= 21 \text{ ft} \quad [\text{governs}]$$

And,

$$w = 6h - s = (6)(8 \text{ ft}) - 8 \text{ ft}$$
$$= 40 \text{ ft}$$

The loading diagram for the low roof is shown in the illustration.

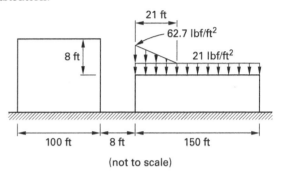

(not to scale)

Partial Loading

Wind scour, snow melt, and snow removal operations can reduce the snow load on some portions of the roof system. This can reduce stress in some members of the roof system, but can also produce stress in other members that is higher than the stress caused when the whole roof is fully loaded.

Consideration is not required for multiple checkerboard loading arrangements, as described in Sec. 2 and specified in IBC Sec. 1607.11. Instead, ASCE/SEI7 Sec. 7.5.1 requires analysis of the loading arrangements shown in Fig. 1.22.

The three partial loading conditions that require consideration are

- full balanced snow load on either exterior span and half the balanced snow load on all other spans (see Fig. 1.22(a))

- half balanced snow load on either exterior span and full balanced snow load on all other spans (see Fig. 1.22(b))

- full balanced snow load on any two adjacent spans and half the balanced snow load on all other spans (see Fig. 1.22(c))

In all the loading conditions shown in Fig. 1.22, the end span may be replaced by a cantilever.

In a gable roof whose slope is 2.38° (½ on 12) or greater, partial loading conditions need not be considered for members spanning perpendicular to the ridge, because the governing load condition in this situation is an unbalanced snow drift load.

Figure 1.22 *Partial Loading Conditions for Snow Loads*

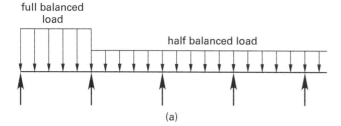

(a)

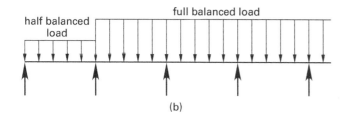

(b)

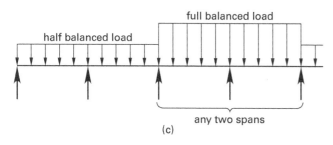

(c)

4. MOVING LOADS

Nomenclature

A	area under the influence line	ft-kips/kip, ft²-kips/kip
c	length of distributed load	ft
F	axial force in truss member	kips
h	distance between top and bottom chords of truss	ft
l	span length	ft
M	bending moment	ft-kips
M_{max}	maximum bending	ft-kips
R	support reaction	kips
V	shear force	kips
w	uniformly distributed moving load	kips/ft
W	concentrated load	kips
y	influence line ordinate	kips/kip, ft-kips/kip

General Principles

A *moving load* is a load that moves continuously over a structure. Examples of moving loads include a train or other vehicle crossing a bridge and an overhead traveling crane moving along a runway girder.

Figure 1.23 *Static and Moving Loads*

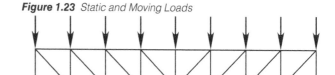

(a) static loads

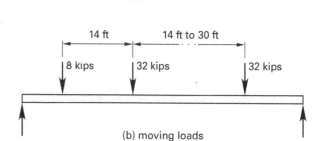

(b) moving loads

The design force in a member of a structure subjected to static loads is readily determined. In Fig. 1.23(a), the static loads are applied to the top chord of the roof joist, and the member forces may be calculated.

The design force in a member of a structure subjected to moving loads may be more difficult to determine. In Fig. 1.23(b), the design truck load specified in AASHTO[5] Sec. 3.6.1.2 is applied to a simply supported bridge deck. Because this is a moving load, the bending moment at any section of the deck depends on the location of the design truck. To find the maximum bending moment at a particular section of the deck, trial-and-error positioning of the design truck on the deck can be used to determine the critical location. Alternatively, an influence line can be utilized to determine the location of the design truck that produces the maximum moment at a specific section of the deck.

Train of Wheel Loads on a Simply Supported Beam

When the moving loads consist of a train of wheel loads on a simply supported beam, the location of the loads that will produce the maximum possible bending moment can be determined from expressions in *AISC Manual*[6] Table 3-23, cases 43, 44, and 45. The maximum moment will occur under one of the wheel loads, and it will occur when this wheel and the centroid of the train are equal distances from and on opposite sides of the center of the span. The maximum moment usually occurs under one of the wheels adjacent to the centroid of the train.

For bridge decks spanning not less than 40.28 ft, the AASHTO design truck produces the maximum moment under the central axle when this axle is located 2.33 ft from the center of the span, as shown in Fig. 1.24.

Figure 1.24 *Maximum Moment Due to AASHTO Design Truck Load*

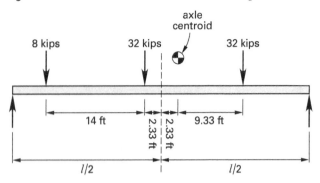

For bridge decks spanning from 10 ft to 40.28 ft, the AASHTO design tandem load governs, as shown in Fig. 1.25. For bridge decks spanning less than 10 ft, the AASHTO design truck governs when a single 32 kip axle is located at midspan.

Figure 1.25 *Maximum Moment Due to AASHTO Design Tandem Load*

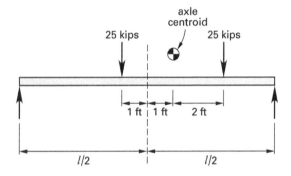

Example 1.16

Determine the maximum moment produced by the AASHTO design tandem in a deck slab spanning 10 ft. The distance between the tandem axles is 4 ft.

Solution

The design tandem is positioned as shown in Fig. 1.25, and the maximum moment is produced under the lead axle. The maximum moment is given by *AISC Manual* Table 3-23, case 44, as

$$M_{\max} = \frac{W\left(l - \dfrac{a}{2}\right)^2}{2l}$$

$$= \frac{(25 \text{ kips})\left(10 \text{ ft} - \dfrac{4 \text{ ft}}{2}\right)^2}{(2)(10 \text{ ft})}$$

$$= 80 \text{ ft-kips}$$

Influence Lines

The *influence line* for a member in a structure is a line whose ordinate at any given location represents the force in the member when a unit load is placed at that location.

An influence line may be constructed using *Müller-Breslau's principle*.[7] According to this principle, when a restraint in a structure is replaced with a unit virtual displacement, an elastic curve is produced which is the influence line for the restraint. The virtual displacement corresponding to a moment is a rotation, and that corresponding to a force is a linear deflection. The displacement is applied in the same direction as the restraint.

Figure 1.26 shows the application of Müller-Breslau's principle to obtain influence lines for the simply supported beam 12. The influence line for the support reaction at end 2 of the beam is obtained by applying a unit virtual displacement in the line of action of R_2. This results in the influence line for R_2 shown at (b). Similarly, as shown at (c), the influence line for shear at point 3 is obtained by cutting the beam at point 3 and displacing the cut ends a unit distance apart. The maximum shear produced at point 3 by a concentrated load W is obtained by locating W at point 3; this maximum shear is

$$V_3 = W(\text{maximum influence line ordinate at point 3})$$

$$= \frac{Wb}{l}$$

The influence line for bending moment at point 3 is produced by cutting the beam at point 3 and imposing a unit virtual rotation, as shown in Fig. 1.26(d). The maximum moment produced at point 3 by a concentrated load W is obtained by locating W at point 3; this maximum moment is

$$M_3 = W(\text{influence line ordinate at point 3})$$

$$= \frac{Wab}{l}$$

Beam with Partial Distributed Load

The influence line for bending moment at point 3 of a simply supported beam of length l is shown in Fig. 1.27(a), and a uniformly distributed moving load w of length $c < l$ is applied to the beam as shown in Fig. 1.27(b). As shown in Fig. 1.27(a), the maximum moment at point 3 is produced when point 3 divides the distributed load in the same ratio as it divides the span. This occurs when

$$\frac{c_1}{c_2} = \frac{a}{b}$$

Figure 1.26 *Application of Müller-Breslau's Principle*

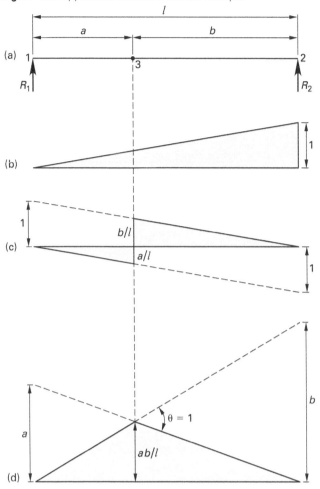

Figure 1.27 *Beam with Partial Distributed Load*

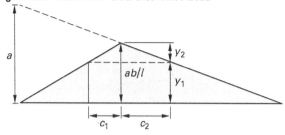

(a) moment influence line

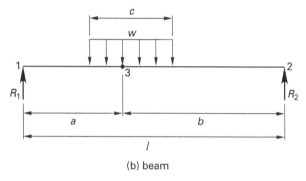

(b) beam

The maximum moment at point 3 is

$$M_{max} = w(\text{area under the influence line})$$

$$= w\left(cy_1 + \frac{cy_2}{2}\right)$$

$$= wc\left(y_1 + \frac{y_2}{2}\right)$$

From the geometric relationships established in the figure, the indicated dimensions are given by

$$y_1 = \frac{a(b - c_2)}{l}$$

$$y_2 = \frac{ac_2}{l}$$

Example 1.17

Determine the maximum bending moment at point 3 in the simply supported beam shown in Fig. 1.27 due to a distributed load of 4 kips/ft over a length of 12 ft. The beam spans a distance of 48 ft, and point 3 is located 16 ft from the support.

Solution

The maximum moment at point 3 is produced when point 3 divides the distributed load in the same ratio as it divides the span. As shown in Fig. 1.27, this occurs when

$$\frac{c_1}{c_2} = \frac{a}{b} = \frac{16 \text{ ft}}{32 \text{ ft}}$$

Find c_1 and c_2.

$$c = c_1 + c_2 = 12 \text{ ft}$$
$$c_1 = 4 \text{ ft}$$
$$c_2 = 8 \text{ ft}$$

From the geometric relationships in Fig. 1.27, the dimensions y_1 and y_2 are

$$y_1 = \frac{a(b - c_2)}{l} = \left(\frac{(16 \text{ ft})(32 \text{ ft} - 8 \text{ ft})}{48 \text{ ft}}\right)\left(\frac{1 \text{ kip}}{1 \text{ kip}}\right)$$

$$= 8 \text{ ft-kips/kip}$$

$$y_2 = \frac{ac_2}{l} = \left(\frac{(16 \text{ ft})(8 \text{ ft})}{48 \text{ ft}}\right)\left(\frac{1 \text{ kip}}{1 \text{ kip}}\right)$$

$$= 2.67 \text{ ft-kips/kip}$$

The area under the influence line is

$$A = c\left(y_1 + \frac{y_2}{2}\right)$$

$$= (12 \text{ ft})\left(8 \frac{\text{ft-kips}}{\text{kip}} + \frac{2.67 \frac{\text{ft-kips}}{\text{kip}}}{2}\right)$$

$$= 112 \text{ ft}^2\text{-kips/kip}$$

The maximum moment at point 3 is

$$M_{\text{max}} = wA = \left(4 \frac{\text{kips}}{\text{ft}}\right)\left(112 \frac{\text{ft}^2\text{-kips}}{\text{kip}}\right)$$

$$= 448 \text{ ft-kips}$$

Beam with Internal Hinges

Figure 1.28 shows the application of Müller-Breslau's principle to obtain influence lines for a three-span beam with a central drop-in span. The influence line for the support reaction R_2 is obtained by applying a unit virtual displacement in the line of action of R_2. This results in the influence line for R_2 shown at Fig. 1.28(b). The maximum reaction produced at point 2 by a concentrated load W is obtained by locating W at point 6; this maximum reaction is

$$R_2 = W(\text{influence line ordinate at point 6})$$

$$= W\left(1 + \frac{b}{a}\right)$$

The maximum reaction produced at point 2 by a distributed load w of unlimited length is obtained by locating w over length 17; this maximum reaction is

$$R_2 = wA = w\left(\frac{1 + \dfrac{b}{a}}{2}\right)l_{17}$$

Similarly, as shown in Fig. 1.28(c), the influence line for shear at point 5 in the end span is obtained by cutting the beam at 5 and displacing the cut ends a unit dis-

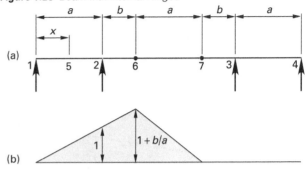

Figure 1.28 *Beam with Internal Hinges*

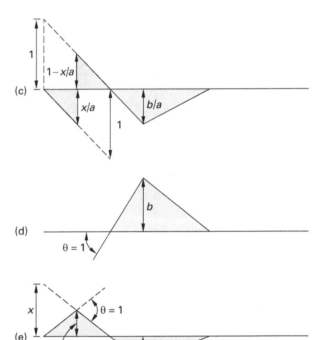

tance apart. The maximum shear produced in the beam at point 5 by a concentrated load W is usually obtained by locating W at point 5. Locating W at point 6 will govern only if $a < b$.

The maximum shear produced in the beam at point 5 by a distributed load w of unlimited length is obtained by locating w over the appropriate portions of the influence line and determining the corresponding areas. With the distributed load located over length 52, the shear is

$$V_5 = wA = w\left(\frac{1 - \dfrac{x}{a}}{2}\right)l_{52}$$

With the distributed load located simultaneously over lengths 15 and 27, the shear is

$$V_5 = w(\text{sum of areas under the influence line})$$

$$= w\left(\left(\frac{\dfrac{x}{a}}{2}\right)l_{15} + \left(\frac{\dfrac{b}{a}}{2}\right)l_{27}\right)$$

The influence line for bending moment at point 2 is produced by cutting the beam at point 2 and imposing a unit virtual rotation, as shown at Fig. 1.28(d). The maximum moment produced in the beam at point 2 by a concentrated load W is obtained by locating W at point 6; this maximum moment, which produces tension in the top fiber of the beam, is

$$M_2 = W(\text{influence line ordinate at point 6})$$
$$= Wb$$

The maximum moment produced in the beam at point 2 by a distributed load w of unlimited length is obtained by locating w over length 27; this maximum moment, which produces tension in the top fiber of the beam, is

$$M_2 = wA$$
$$= w\left(\frac{b}{2}\right)l_{27}$$

Similarly, as shown at Fig. 1.28(e), the influence line for bending moment at point 5 is produced by cutting the beam at point 5 and imposing a unit virtual rotation. By locating the concentrated load W at point 6, the maximum moment producing tension in the top fiber of the beam at point 5 is obtained as

$$M_2 = W(\text{influence line ordinate at point 6})$$
$$= \frac{Wbx}{a}$$

By locating the concentrated load W at point 5, the maximum moment producing tension in the bottom fiber of the beam at point 5 is obtained as

$$M_2 = W(\text{influence line ordinate at point 5})$$
$$= \frac{W(a-x)x}{a}$$

Example 1.18

A three-span bridge with a central drop-in span is shown. Determine the maximum shear produced by the AASHTO lane load of 0.64 kips/ft at the midpoint, 5, of the end span.

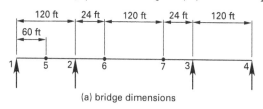

(a) bridge dimensions

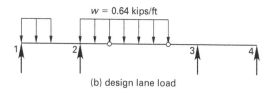

(b) design lane load

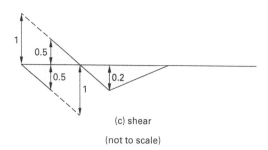

(c) shear

(not to scale)

Solution

The influence line for shear is shown in the illustration, and placing the design lane load as shown produces the maximum shear at point 5. With the distributed load located simultaneously over lengths 15 and 27, the shear is

$$V_5 = w(\text{sum of areas under the influence line})$$

$$= w\left(\left(\frac{0.5\,\dfrac{\text{kip}}{\text{kip}}}{2}\right)l_{15} + \left(\frac{0.2\,\dfrac{\text{kip}}{\text{kip}}}{2}\right)l_{27}\right)$$

$$= \left(0.64\,\frac{\text{kips}}{\text{ft}}\right)$$

$$\times \left(\left(\frac{0.5\,\dfrac{\text{kip}}{\text{kip}}}{2}\right)(60\text{ ft}) + \left(\frac{0.2\,\dfrac{\text{kip}}{\text{kip}}}{2}\right)(144\text{ ft})\right)$$

$$= 18.82\text{ kips}$$

Continuous Beams

Müller-Breslau's principle may be applied to multispan continuous beams to obtain the general shape of the influence line without calculating the ordinates of the curve. This provides a visual indication of where the live load should be placed in order to obtain the maximum effect.

For the four-span continuous beam shown in Fig. 1.29, the influence line for the support reaction R_1 is obtained by applying a unit virtual displacement in the line of action of R_1. This results in the influence line for R_1 shown at Fig. 1.29(b). The maximum reaction produced at the support by a distributed load is obtained by simultaneously loading spans 12 and 34 with the distributed load. The maximum reaction produced at the support by a concentrated load is obtained by placing the concentrated load at point 1.

Figure 1.29 *Qualitative Influence Lines*

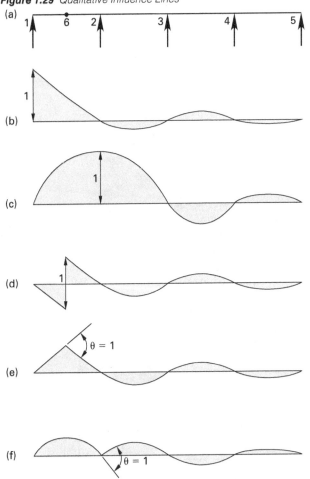

The influence line for the support reaction R_2 is obtained by applying a unit virtual displacement in the line of action of R_2. This results in the influence line for R_2 shown at Fig. 1.29(c). The maximum reaction produced at the support by a distributed load is obtained by simultaneously loading spans 12, 23, and 45 with the

distributed load. The maximum reaction produced at the support by a concentrated load is obtained by placing the concentrated load at point 2.

The influence line for shear at point 6 is obtained by cutting the beam at point 6 and displacing the cut ends a unit distance apart. This results in the influence line for V_6 shown at Fig. 1.29(d). The maximum shear produced at point 6 by a distributed load is obtained either by simultaneously loading length 62 and span 34 or by simultaneously loading length 16 and spans 23 and 45 with the distributed load. The maximum reaction produced at point 6 by a concentrated load is obtained by placing the concentrated load at point 6.

The influence line for moment at point 6 is obtained by cutting the beam at point 6 and imposing a unit virtual rotation on the cut ends. This results in the influence line for M_6 shown at Fig. 1.29(e). The maximum moment produced at point 6 by a distributed load is obtained by simultaneously loading spans 12 and 34 with the distributed load. The maximum moment produced at point 6 by a concentrated load is obtained by placing the concentrated load at point 6.

The influence line for moment at support 2 is obtained by cutting the beam at point 2 and imposing a unit virtual rotation on the cut ends. This results in the influence line for M_2 shown at Fig. 1.29(f). The maximum moment produced at support 2 by a distributed load is obtained by simultaneously loading spans 12, 23, and 45 with the distributed load. The maximum moment produced at support 2 by a concentrated load is obtained by placing the concentrated load at a distance of approximately 0.58 times the span length from support 1.

In order to accurately draw the influence lines, influence line coefficients may be used. These are available for a number of standard cases and allow a rapid determination of the maximum forces produced by wheel loads.[8,9] For nonstandard situations, several methods may be used to determine the required influence lines.[10,11]

Pin-Jointed Trusses

Figure 1.30 shows a detail of a through-truss bridge. The deck slab is supported on longitudinal stringer beams, which are in turn supported on cross girders. The cross girders are located at the bottom chord panel points of the trusses, so that vehicle loads are transferred directly to the truss bottom chord panel points.

The longitudinal stringers are assumed to be simply supported on the cross girders, and as a moving load moves over the deck slab, the load is transferred linearly from one panel point to the next. Hence, in constructing the influence line for axial force in a truss member, the influence line ordinates at the panel points on either side of a panel are connected with a straight line.

Figure 1.30 *Through-Truss Details*

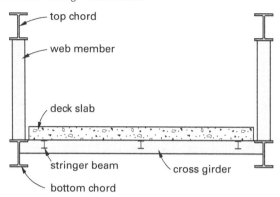

Figure 1.31 *Influence Lines for a Warren Truss*

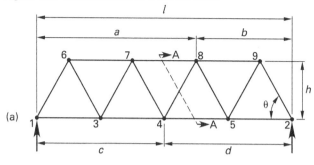

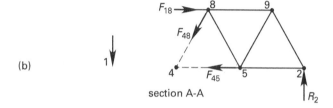

section A-A

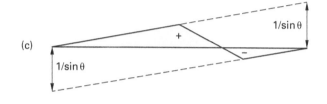

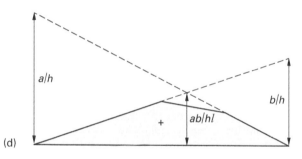

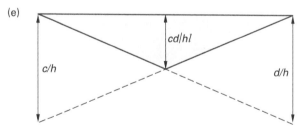

The construction of influence lines for axial force in a truss member is shown in Fig. 1.31 for a Warren truss with the load from cross girders applied directly to the bottom chord panel points.

The influence lines for axial force in members 78, 48, and 45 are obtained by considering the free-body diagram shown at Fig. 1.31(b), which is formed by cutting the truss at section A-A. With a unit virtual load to the left of panel point 4, the tensile force in web member 48 is obtained by resolving vertically and is given by

$$F_{48} = \frac{R_2}{\sin\theta}$$

Similarly, with a unit virtual load to the right of panel point 5, the compressive force in member 48 is

$$F_{48} = \frac{R_1}{\sin\theta}$$

The influence line for F_{48} is shown at Fig. 1.31(c) and is obtained by multiplying the influence lines for end reactions of the truss by $1/\sin\theta$ and connecting the influence ordinates at panel points 4 and 5 with a straight line. Positive sense of the influence line indicates tension in member 48.

With a unit virtual load to the left of panel point 4, the tensile force in the bottom chord member 45 is obtained by taking moments about panel point 8; this force is

$$F_{45} = \frac{bR_2}{h}$$

Similarly, with a unit virtual load to the right of panel point 5, the tensile force in member 45 is

$$F_{45} = \frac{aR_1}{h}$$

The influence line for F_{45} is shown at Fig. 1.31(d). This line is obtained by multiplying the influence line for moment at panel point 8 by $1/h$ and connecting the

influence ordinates at panel points 4 and 5 with a straight line. Positive sense of the influence line indicates tension in member 45.

With a unit virtual load to the left of panel point 4, the compressive force in the top chord member 78 is obtained by taking moments about panel point 4. This force is

$$F_{78} = \frac{dR_2}{h}$$

Similarly, with a unit virtual load to the right of panel point 4, the compressive force in member 78 is

$$F_{78} = \frac{cR_1}{h}$$

The influence line for F_{78} is shown at Fig. 1.31(e) and is obtained by multiplying the influence line for moment at panel point 4 by $1/h$. Negative sense of the influence line indicates compression in member 78.

5. THERMAL LOADS

Nomenclature

D	dead load	kips
h	height	ft
l	span length	ft
L	live load	kips
T	self-straining load	kips
T	temperature	°F

Symbols

α	coefficient of thermal expansion	in/in-°F
δ	displacement	in

General Principles

Thermal stresses occur when a material is unable to freely expand or contract in response to temperature differential. The resulting force produced in a structure is known as the *self-straining force*, and this is defined as a force resulting from temperature change, shrinkage, creep, moisture change, or diffcrential settlement.

When thermal effects produce a compressive force in a member, the stiffness of the member is reduced, and the stability of the structure is thus compromised. Expansion joints may be required in the structure to prevent this.

Load Combinations

Self-straining load effects should be calculated based on an assessment of the most probable value that can be expected at any arbitrary point in time. When using strength design, ASCE/SEI7 Sec. 2.3.4 requires the load factor for T to be established with consideration of whether the maximum value of the self-straining load occurs simultaneously with other applied loads. A load factor of 1.2 is usually used, but the load factor should not be taken as less than 1.0. ASCE/SEI7 Sec. C2.3.4 requires consideration of the load combinations

$$1.2D + 1.2T + 0.5L$$
$$1.2D + 1.6L + 1.0T$$

When using allowable stress design, ASCE/SEI7 Sec. 2.4.4 requires that the load factor for T be not less than 0.75. ASCE/SEI7 Sec. C2.4.4 requires consideration of the load combinations

$$1.0D + 1.0T$$
$$1.0D + 0.75(L + T)$$

Example 1.19

The columns in the rigid steel bent shown may be considered fixed at top and bottom. The bent is subjected to a temperature rise of 50°F. The coefficient of thermal expansion of steel is 6.5×10^{-6} in/in-°F, and the columns are W14 × 90 sections. Determine the bending moment produced at the bottom of each column.

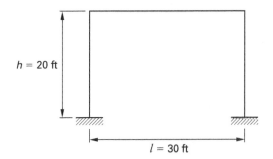

Solution

Because of the symmetry of the structure, the lateral displacement at the top of each column is half the thermal expansion of the beam and is

$$\delta = \frac{\alpha T l}{2}$$
$$= \frac{\left(6.5 \times 10^{-6} \, \dfrac{\text{in}}{\text{in-°F}}\right)(50°\text{F})(30 \text{ ft})\left(12 \, \dfrac{\text{in}}{\text{ft}}\right)}{2}$$
$$= 0.059 \text{ in}$$

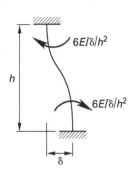

As shown, a lateral displacement of δ to a fixed-ended member produces an end moment of

$$M = \frac{6EI\delta}{h^2}$$

From the *AISC Manual*, the moment of inertia of the W14 × 90 beam is

$$I = 999 \text{ in}^4$$

The moment produced at the bottom of each column is

$$M = \frac{6EI\delta}{h^2}$$
$$= \frac{(6)\left(29{,}000 \dfrac{\text{kips}}{\text{in}^2}\right)(999 \text{ in}^4)(0.059 \text{ in})}{\left((20 \text{ ft})\left(12 \dfrac{\text{in}}{\text{ft}}\right)\right)^2}$$
$$= 178 \text{ in-kips}$$

6. SHRINKAGE AND CREEP

Nomenclature

D	dead load	kips
h	height	ft
l	span length	ft
L	live load	kips
T	self-straining load	kips
T	temperature	°F

Symbols

α	coefficient of thermal expansion	in/in-°F
δ	displacement	in
ϵ_{sh}	shrinkage strain	–

Shrinkage

The *shrinkage* of a concrete member is due to the gradual loss of moisture from the concrete. Approximately 40% of the total shrinkage occurs within 28 days of pouring the concrete and approximately 90% within one year. The total amount of shrinkage increases when

- the water-cement ratio of the concrete mix is increased

- the ratio of surface area to volume of a member is increased

- the ambient relative humidity decreases

Cracks due to shrinkage can occur in a restrained concrete member. These cracks may be controlled by introducing control joints in slabs and by providing shrinkage reinforcement as specified in ACI Sec. 9.7.2.3 for deep beams and ACI Sec. 7.6.4 for slabs.

Forces produced in a restrained member are treated as self-straining forces in the same manner as thermal forces. For concrete with a shrinkage strain of $\epsilon_{sh} = 300 \times 10^{-6}$ in/in and a coefficient of thermal expansion of $\alpha = 6 \times 10^{-6}$ in/in-°F, the shrinkage is equivalent to a fall in temperature of

$$T = \frac{\epsilon_{sh}}{\alpha}$$
$$= \frac{300 \times 10^{-6} \dfrac{\text{in}}{\text{in}}}{6 \times 10^{-6} \dfrac{\text{in}}{\text{in-°F}}}$$
$$= 50°F$$

In a prestressed concrete member, shrinkage is one of the factors contributing to loss of prestress, causing approximately 30% of the total losses. Calculation of prestress losses is specified in ACI Sec. R20.3.2.6.

Creep

When a concrete member is subject to a sustained compressive force, the compressive strain in the member will gradually increase. This gradual increase is known as *creep*.

As long as the member is loaded, creep will continue, but the rate of creep decreases. When the member is unloaded, the elastic strain is recovered immediately, and some of the creep strain is recovered over time. Approximately 50% of the total creep occurs within 28 days of loading the member and approximately 75% within six months.

The total amount of creep increases when

- the applied force increases

- the compressive strength of the member decreases

- the member is loaded at an earlier age

Creep and shrinkage cause long-term deflections in reinforced concrete beams that are in addition to the initial elastic deflection. In accordance with ACI Table 24.2.4.1.3, the increase in deflection after five years is approximately twice the initial elastic deflection.

In a prestressed concrete member, creep is one of the factors contributing to loss of prestress, causing approximately 30% of the total losses. Calculation of prestress losses is specified in ACI Sec. R20.3.2.6.

7. IMPACT LOADS

Nomenclature

I	dynamic factor	–
IM	dynamic load allowance	–
M	bending moment	ft-kips
M_{DT}	bending moment produced by the design truck load	ft-kips
M_{LL}	bending moment produced by the design lane load	ft-kips

General Principles

Impact loads are defined in IBC Sec. 202 as loads that result from moving machinery, elevators, craneways, vehicles, and other similar forces, as well as kinetic loads, pressure, and possible surcharge from fixed or moving loads. Impact loads are applied to the structure instantaneously, as distinct from the normal static live loads tabulated in IBC Table 1607.1.

As specified in IBC Sec. 1607.10, the normal live loads are assumed to include adequate allowance for ordinary impact conditions. Additional provisions are made for loads that involve unusual impact or vibration forces. To approximate the dynamic effects of these loads, an equivalent static load, equal to the dynamic load effects, is applied.

In IBC Sec. 1607.10.2 the weights of machinery are increased as follows.

- shaft and motor driven light machinery: 20%

- reciprocating machinery and power driven units: 50%

In accordance with IBC Sec. 1607.10.1, structural members subject to dynamic loads from elevators must be designed for the impact loads prescribed by ASME A17.1.[12]

Vehicle Impact Loads on a Bridge Deck

A *vertical impact load* is produced when a moving vehicle crosses a rut or a projection in the road surface. To account for this, AASHTO Sec. 3.6.2 stipulates that the static axle loads of the design truck and the design tandem must be increased by the dynamic factor. The dynamic load allowance, *IM*, is given by AASHTO Table 3.6.2.1-1 as

- 75% for deck joints for all limit states

- 15% for all other components for fatigue and fracture limit states

- 33% for all other components for all other limit states

The dynamic factor to be applied to the static load is

$$I = 1 + \frac{IM}{100}$$

The dynamic factor is not applied to the design lane load or pedestrian loads.

Example 1.20

The span bending moment produced by the design truck load on a simply supported bridge deck is 120 ft-kips, and the moment produced by the design lane load is 50 ft-kips. Determine the total design moment after applying the dynamic load allowance.

Solution

The dynamic load allowance for the span moment is given by AASHTO Table 3.6.2.1-1 as

$$IM = 33\%$$

The dynamic factor is

$$\begin{aligned} I &= 1 + \frac{IM}{100} \\ &= 1 + \frac{33}{100} \\ &= 1.33 \end{aligned}$$

The span moment caused by the design truck load, including the dynamic load allowance, is

$$\begin{aligned} M_{DLA} &= IM_{DT} = 1.33 M_{DT} \\ &= (1.33)(120 \text{ ft-kips}) \\ &= 160 \text{ ft-kips} \end{aligned}$$

The total moment produced by the design lane load and the design truck load, including the dynamic load allowance, is

$$\begin{aligned} M_{\text{total}} &= M_{DLA} + M_{LL} \\ &= 160 \text{ ft-kips} + 50 \text{ ft-kips} \\ &= 210 \text{ ft-kips} \end{aligned}$$

Impact Loads on a Vehicle Barrier

The impact load on a vehicle barrier is specified in ASCE/SEI7 Sec. 4.5.3 and shown in Fig. 1.32.

Figure 1.32 *Vehicle Barrier Load*

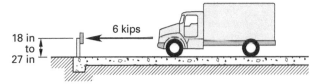

The barrier is designed to resist a single load of 6 kips applied horizontally in any direction. The load is assumed to act at heights between 18 in and 27 in above the floor surface.

Crane Loads

The design of crane systems is covered in IBC Sec. 1607.14. The supporting structure for the crane is designed for the maximum wheel load, vertical impact force, lateral force, and longitudinal force, all applied as a simultaneous load combination. For a typical bridge crane, the maximum wheel load is the sum of the applicable weights of the bridge and trucks, the rated capacity of the crane, and the weight of the hoist and trolley.

Impact forces are generated by the lifting and depositing of the load and by the movement of the trucks on the rails. To account for these effects, IBC Sec. 1607.14.2 requires that maximum wheel loads be increased by these percentages.

- 25% for power-operated monorail cranes

- 25% for cab-operated or remotely operated powered bridge cranes

- 10% for pendant-operated powered bridge cranes

- 0% for hand-geared bridge cranes or monorail cranes

The acceleration and deceleration of the trolley on the bridge beam causes lateral forces to be applied to the runway girders. For electrically powered trolleys, IBC Sec. 1607.14.3 specifies that the lateral force is 20% of the sum of the rated capacity of the crane, the weight of the hoist, and the weight of the trolley.

The acceleration and deceleration of the trucks on the runway girders causes longitudinal forces to be applied to the runway girders. For electrically powered bridges, IBC Sec. 1607.14.4 specifies that the longitudinal force is 10% of the maximum wheel loads.

8. SETTLEMENT

Nomenclature

d_{ij}	distribution factor at end i of member ij	–
EI	flexural rigidity	ft^2-kips
l	span length	ft
M_F	fixed-end moment	in-kips, ft-kips
R	support reaction	kips
s_{ij}	stiffness factor at end i of member ij	ft-kips

Symbols

δ	displacement	in

General Principles

Settlement is the downward movement of a building during and after construction. Settlement occurs due to the consolidation of the subsoil under the foundations as the subsoil supports the building weight.

When settlement occurs uniformly over the whole building, it does not cause any structural problems. However, unless the subsoil is homogeneous and the load distribution on the foundations is uniform, the settlement may not be uniform. The unequal downward movement of a building's foundations is called *differential settlement*, and it causes cracking and other damage to the building.

Damage can be particularly severe to brittle materials and finishes such as brick, concrete blocks, and ceramic tiles. Settlement continues for several years after construction and may be characterized into three phases.

- short-term settlement

- consolidation settlement

- creep

When building on unsatisfactory subsoil that is liable to cause differential settlement, several techniques are available to prevent future structural problems. These are

- excavate the unsatisfactory subsoil and replace it with suitable material

- construct the building on an adequately designed raft foundation

- construct the building on a piled foundation bearing on suitable material below the unsatisfactory subsoil

- stabilize the subsoil by injecting a cement slurry

- stabilize the subsoil by using a vibroflotation technique

- preconsolidate the subsoil using temporary surcharge

Settlement of Supports

The moment distribution technique[13] provides a convenient solution for the effects of known settlements in continuous beams and simple framed structures.[13] When the initial fixed-end moments due to the settlements have been established, the distribution of moments can proceed and the final moments be obtained. As shown in Fig 1.33(a), the fixed-end moment induced at end 1 by a displacement δ when end 2 is fixed is

$$M_{F12} = \frac{6EI\delta}{l^2}$$

As shown in Fig 1.33(b), the fixed-end moment induced at end 1 by a displacement δ when end 2 is simply supported is

$$M_{F12} = \frac{3EI\delta}{l^2}$$

Figure 1.33 *Fixed-End Moments Due to Displacements*

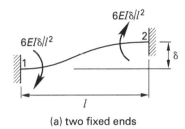

(a) two fixed ends

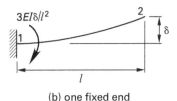

(b) one fixed end

Example 1.21

The continuous three-span beam shown in the illustration has a uniform flexural rigidity, EI, of 40,000 ft^2-kips. Determine the support reaction produced at end support 1 by a settlement of 1 in at support 2.

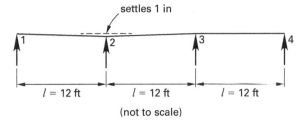

(not to scale)

Solution

Allowing for the hinged ends at 1 and 4, the modified stiffness of the members is

$$s_{21} = s_{34} = \frac{3EI}{l}$$

$$s_{23} = s_{32} = \frac{4EI}{l}$$

The modified distribution factors are

$$d_{21} = \frac{s_{21}}{(s_{21} + s_{23})} = \frac{3}{7}$$
$$= d_{34}$$
$$d_{23} = \frac{s_{23}}{(s_{23} + s_{21})} = \frac{4}{7}$$
$$= d_{32}$$

Allowing for the hinged end at 1, the initial fixed-end moments are

$$M_{F\,21} = \frac{-3EI\delta}{l^2}$$
$$= \frac{(-3)(40{,}000 \ \text{ft}^2\text{-kips})(1 \ \text{in})}{(12 \ \text{ft})^2}$$
$$= -833 \ \text{in-kips}$$
$$M_{F\,23} = \frac{6EI\delta}{l^2}$$
$$= \frac{(6)(40{,}000 \ \text{ft}^2\text{-kips})(1 \ \text{in})}{(12 \ \text{ft})^2}$$
$$= 1666 \ \text{in-kips}$$
$$= M_{F\,32}$$

The distribution of moments is given in the table shown (all values in in-kips).

member	21	23	32	34
distribution factor	3/7	4/7	4/7	3/7
fixed-end moments	−833	1666	1666	
distribution	−357	−476	−952	−714
carry-over		−476	−238	
distribution	204	272	136	102
carry-over		68	136	
distribution	−29	−39	−78	−58
carry-over		−39	−20	
distribution	17	22	11	9
carry-over		6	11	
distribution	−3	−3	−6	−5
carry-over		−3	−1	
distribution	1	2		1
final moments	−1000	1000	665	−665

The required support reaction at support 1 is

$$R_1 = \frac{M_{21}}{l_{12}} = \frac{1000 \ \text{in-kips}}{(12 \ \text{ft})\left(12 \ \dfrac{\text{in}}{\text{ft}}\right)}$$
$$= 7 \ \text{kips}$$

9. PONDING AND RAIN LOADS

Nomenclature

A	area of roof being drained	ft^2
d_h	additional depth of water, or hydraulic head, above the inlet of the overflow drain at its design flow	in
d_s	depth of water up to the inlet of the overflow drain when the primary system is blocked (static head)	in
i	design rainfall intensity as given in IBC Fig. 1611.1 or as specified by the local jurisdiction	in/hr
Q	required flow rate	gal/min
R	rain load on the undeflected roof	lbf/ft^2

Symbols

ρ_w	density of rainwater, 62.4	lbf/ft^3

General Principles

Rain-induced roof collapses are frequently reported. These failures are generally related to the poor design, poor execution, or poor maintenance of the roof drainage system.[14]

The design for rain loads is covered in IBC Sec. 1611 and ASCE/SEI7 Chap. 8. Industrial buildings with low-slope roofs and parapet walls are particularly susceptible to rain overloads. These types of building are usually provided with both a primary drainage system and a secondary emergency drainage system as shown in Fig. 1.34. The primary drainage system is designed to discharge the *design rainfall*, which is the 1-hour duration, 100-year rainfall intensity rate provided in the rainfall intensity maps given in IBC Fig. 1611.1. The secondary drainage system is designed with the same capacity in case the primary system becomes blocked by debris or ice. Typically, the secondary system consists of scuppers in the parapet walls to allow rainwater to discharge from the roof.

Figure 1.34 *Primary and Secondary Drainage Systems*

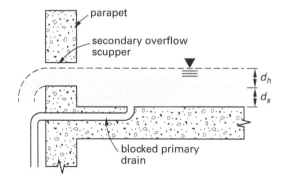

The roof must be designed to support the maximum depth of rainwater that can build up. This maximum depth is the sum of two distances. The first is the *static head*, d_s, which is the distance from the inlet of the primary system to the inlet of the secondary system, and the second is the *hydraulic head*, d_h, which is the height that develops above the secondary inlet at its design flow.

Some roofs are provided with a system of controlled drainage. This limits the drainage from the roof to a rate less than the rainfall rate, intentionally allowing rainwater to build up to a level that is controlled by a secondary drainage system. Such a roof must be designed to support the controlled depth of rainwater that can build up.

Design Rain Loads

To determine the rain load on a roof, find the rainfall intensity at the location of the building. This value is obtained from the rainfall intensity maps given in IBC Fig. 1611.1 or is specified by the local jurisdiction. Use the rainfall intensity to determine the required capacity of the drainage system. The required flow rate is given by ASCE/SEI7 Eq. C8.3-1 as

$$Q = 0.0104Ai$$

A is the area in square feet of the roof being drained, i is the rainfall intensity in inches per hour, and 0.0104 is the conversion factor needed to give the flow rate, Q, in gallons per minute.

When the type of overflow drain to be used for the secondary drainage system has been determined, the hydraulic head, d_h, can be obtained from ASCE/SEI7 Table C8.3-3. This table relates hydraulic head to flow rate, Q, for several types of overflow drain.

With both the static head and the hydraulic head known, the rain load on the roof is given by ASCE/SEI7 Eq. 8.3-1 as

$$R = 5.2(d_s + d_h)$$

The coefficient 5.2 has units of lbf/ft^2-in. This is the weight of rainwater per inch of depth, and it is derived from the density of rainwater.

$$5.2 = \frac{\rho_w}{12\,\frac{in}{ft}} = \frac{62.4\,\frac{lbf}{ft^3}}{12\,\frac{in}{ft}}$$
$$= 5.2\ lbf/ft^2\text{-}in$$

Example 1.22

A 3000 ft^2 roof with a pitch of $\frac{1}{4}$ in/ft is enclosed by parapet walls and is provided with a single secondary overflow scupper in the parapet wall. The closed top scupper is 6 in wide and 4 in high and is set 2 in above the roof surface. The specified design rainfall intensity is 1.5 in/hr. Determine the design rain load.

Solution

From ASCE/SEI7 Eq.C8.3-1, the required flow rate for the overflow scupper is

$$Q = 0.0104Ai$$
$$= \left(0.0104 \; \frac{\text{gal-hr}}{\text{in-min-ft}^2}\right)(3000 \text{ ft}^2)\left(1.5 \; \frac{\text{in}}{\text{hr}}\right)$$
$$= 46.8 \text{ gal/min}$$

The hydraulic head is obtained by interpolation from ASCE 7 Table C8.3-3 as

$$d_h = 1 + \frac{46.8 \; \dfrac{\text{gal}}{\text{min}} - 18 \; \dfrac{\text{gal}}{\text{min}}}{50 \; \dfrac{\text{gal}}{\text{min}} - 18 \; \dfrac{\text{gal}}{\text{min}}}$$
$$= 1.90 \text{ in}$$

The height of the secondary overflow scupper above the roof surface, d_s, is given as 2 in. From ASCE/SEI7 Eq. 8.3-1, the rain load on the roof is

$$R = 5.2(d_s + d_h) = \left(5.2 \; \frac{\text{lbf}}{\text{ft}^2\text{-in}}\right)(2 \text{ in} + 1.90 \text{ in})$$
$$= 20.28 \text{ lbf/ft}^2$$

Ponding Instability

IBC Sec. 1611.2 requires that susceptible bays of roofs be evaluated for ponding instability in accordance with ASCE/SEI7 Sec. 8.4. A *susceptible bay* is defined in IBC Sec. 202 as a roof or a portion thereof that meets two conditions.

- The slope is less than $\frac{1}{4}$ inch per foot.

- The secondary drainage system is functional, but the primary drainage system is blocked.

A roof surface with a slope of $\frac{1}{4}$ inch per foot or greater toward points of free drainage is not considered a susceptible bay.

A susceptible bay must be analyzed to ensure that the structural members that are supporting the roof possess sufficient stiffness to prevent progressive deflection, or instability, as rainwater and melting snow build up. On a roof with a slope of less than $\frac{1}{4}$ in/ft, rainwater may accumulate and form a pond. Under the weight of the pond, the roof may deflect, which causes the pond to deepen and allows more water to collect, leading to still larger deflections. If the roof does not possess adequate stiffness to resist this progression, overloading and failure may result.

The design method to prevent instability consists of an iterative structural analysis. The incremental deflection is determined, and the increased roof load that results from this deflection is added to the original roof load. Alternatively, AISC 360 App. 2 presents a direct design method to preclude ponding instability in steel structures.[15]

10. FLUID LOADS

Nomenclature

D	dead load	kips
E	earthquake load	kips
L	live load	kips
F	load due to fluids with well-defined pressures and maximum heights	lbf/ft^2
H	load due to lateral pressure	kips
W	wind load	kips

General Principles

Fluid pressure is the pressure exerted by a fluid at any point within the fluid. *Fluid load*, F, is the structural effect produced by the weight of a storage tank on its supporting structure due to the fluid stored within. A fluid load is similar to a dead load insofar as the maximum weight of the contents of a storage tank can be calculated based on the tank's capacity. However, a fluid load is not a permanent load, and emptying and filling the tank causes fluctuations in the force applied to the structure.

In calculating seismic base shear, ASCE/SEI7 Sec. 15.4.3 requires that the maximum weight of the storage tank's contents be included when determining the effective seismic weight, W. Hence, it is logical to use the maximum weight of the storage tank's contents in calculating the resistance to uplift, using the same load factor as dead load, D. When using strength design, IBC Sec. 1605.2 requires consideration of the load combination

$$0.9(D+F) + 1.0E + 1.6H \qquad \text{[IBC 16-7]}$$

When using allowable stress design, IBC Sec. 1605.3.1 requires consideration of the load combination

$$0.6(D+F) + 0.7E + H \qquad \text{[IBC 16-16]}$$

However, when considering wind load, the most critical load combination is with the storage tank empty. For strength design, IBC Sec. 1605.2 requires consideration of the load combination

$$0.9D + 1.0W + 1.6H \qquad \text{[IBC 16-6]}$$

When using allowable stress design, IBC Sec. 1605.3.1 requires consideration of the load combination

$$0.6D + 0.6W + H \qquad \text{[IBC 16-15]}$$

11. ICE LOADS

Nomenclature

d_i	weight of ice	kips
D	dead load	kips
D_i	weight of ice	kips
I_i	importance factor for ice thickness	–
I_w	importance factor for concurrent wind pressure	–
L	live load	kips
S	snow load	kips
W_i	wind load	kips

General Principles

An *ice-sensitive structure* is defined in IBC Sec. 202 as a structure for which the effect of an atmospheric ice load governs the design of a structure or a part of a structure. Examples of ice-sensitive structures include lattice structures, guyed masts, overhead lines, light suspension and cable-stayed bridges, aerial cable systems, amusement rides, open catwalks and platforms, flagpoles, and signs.

IBC Sec. 1614.1 requires that ice-sensitive structures be designed for the effects of atmospheric ice loads in accordance with ASCE/SEI7 Chap. 10. The adverse effects of ice accretion are the increase in the dead load imposed on the structure and the increase in the projected area exposed to wind.

Ice Formation

There are four types of atmospheric ice formation.

- *Freezing rain* occurs when warm, moist air encounters a layer of subfreezing air, causing raindrops to freeze on contact with a structure. Freezing rain is the cause of the most severe ice loads.

- *In-cloud icing* occurs when fog or a cloud comes into contact with a structural member that is at or below freezing temperature. In-cloud icing can cause significant loadings on structures that are very tall or that are at high elevations.

- *Snow accretions* may occur anywhere snow falls, due to capillary forces and interparticle freezing. On circular members such as wires, cables, and guys, a cylindrical sleeve may form with a density much higher than the density of the snowfall on the ground.

- *Hoarfrost* is the accumulation of ice crystals formed when water vapor is deposited directly on an exposed member. Hoarfrost has very little weight and does not constitute a significant loading problem.

Importance Factor

In ASCE/SEI7 Table 1.5-2, importance factors are provided relating to the design for ice loads. These are shown in Table 1.5. An increase in the importance factor results in an increase in the design ice load for a building. Use of the importance factors ensures that essential facilities are designed for higher loads, reducing the possibility of collapse due to overload.

Table 1.5 *Risk Category and Importance Factors I_i and I_w*

risk category	ice importance factor for thickness, I_i	ice importance factor for wind, I_w
I	0.80	1.00
II	1.00	1.00
III	1.15	1.00
IV	1.25	1.00

These importance factors correspond to the risk categories from ASCE/SEI7 Table 1.5-1, which are assigned to buildings based on the nature of their occupancies (see Table 1.4).

Load Combinations

Load combinations for strength design that include atmospheric ice and wind-on-ice loads are given in ASCE/SEI7 Sec. 2.3.3. The following load combinations are applicable.

- The governing loading condition when a structure is subjected to maximum occupancy live load is

$$1.2D + 1.6L + 0.2D_i + 0.5S$$

- The governing loading condition when a structure is subjected to maximum wind load increasing the effects of dead load is

$$1.2D + W_i + D_i + 0.5L + 0.5S$$

In this load combination, $0.5L$ is replaced with L for garages, places of public assembly, and any areas where L is greater than 100 lbf/ft^2.

- When dead load governs, the applicable combination is

$$1.2D + D_i$$

- The governing loading condition when a structure is subjected to maximum wind load opposing the effects of dead load is

$$0.9D + W_i + D_i$$

Load combinations for allowable stress design that include atmospheric ice and wind-on-ice loads are given in ASCE/SEI7 Sec. 2.4.3. The following load combinations are applicable.

- The governing loading condition when a structure is subjected to maximum occupancy live load is

$$D + L + 0.7D_i$$

- The governing loading condition when a structure is subjected to maximum snow load is

$$D + 0.7W_i + 0.7D_i + S$$

- The governing loading condition when a structure is subjected to maximum wind load opposing the effects of dead load is

$$0.6D + 0.7W_i + 0.7D_i$$

- When dead load governs, the applicable combination is

$$1.2D + D_i$$

Design Procedure

ASCE/SEI7 Sec. 10.4 provides details for calculating the weight of ice due to freezing rain on several structural shapes. The equivalent radial ice thickness due to freezing rain is obtained from the maps given in ASCE/SEI7 Fig. 10.4-2 through Fig. 10.4-6.

ASCE/SEI7 Sec. 10.5 provides details for calculating the projected area of the ice accreted on structural members. The wind speeds concurrent with these increased projected areas are obtained from the maps given in ASCE/SEI7 Fig. 10.4-2 through Fig. 10.4-6.

The following information is needed to determine the weight of ice due to freezing rain on structural members.

- ice importance factor for thickness, I_i, from Table 1.5 (ASCE/SEI7 Tables 1.5-1 and 1.5-2)

- equivalent diameter of the structural shape, D_c, from ASCE/SEI7 Fig. 10.4-1

- nominal ice thickness, t, from ASCE/SEI7 Fig. 10.4-2 through Fig. 10.4-6

- height factor, f_z, defined by ASCE/SEI7 Eq. 10.4-4 as

$$f_z = \left(\frac{z}{33}\right)^{0.10}$$

z is height above the ground. For values of z over 900 ft, f_z is taken as 1.4.

- topographic factor, K_{zt}, defined by ASCE/SEI7 Eq. 26.8-1 as

$$K_{zt} = (1 + K_1 K_2 K_3)^2$$

- design ice thickness, t_d, defined by ASCE/SEI7 Eq. 10.4-5 as

$$t_d = t I_i f_z (K_{zt})^{0.35}$$

- cross-sectional area of ice, A_i, defined by ASCE/SEI7 Eq. 10.4-1 as

$$A_i = \pi t_d (D_c + t_d)$$

- weight of ice, w_i, determined from ASCE/SEI7 Sec. 10.4.1 as

$$w_i = \gamma_i A_i$$

The density of ice, γ_i, is 56 lbf/ft.

The following additional information is needed to determine the wind load on ice-covered structural members.

- design ice thickness, t_d, defined by ASCE/SEI7 Eq. 10.4-5 as

$$t_d = t I_i f_z (K_{zt})^{0.35}$$

- ice importance factor for wind, I_w, from Table 1.5 (ASCE/SEI7 Tables 1.5-1 and 1.5-2)

- concurrent three-second gust speeds, V_c, from ASCE/SEI7 Fig. 10.4-2 through Fig. 10.4-6

- site exposure category, defined in ASCE/SEI7 Sec. 26.7.3

- wind parameter, K_z, from ASCE/SEI7 Table 26.10-1

- wind parameter, K_d, from ASCE/SEI7 Table 26.6-1

- ground elevation factor, K_e, from ASCE/SEI7 Table 26.9-1

- wind velocity pressure, q_z, defined by ASCE/SEI7 Eq. 26.10-1 as

$$q_z = 0.00256 K_z K_{zt} K_d K_e V_c^2 I_w$$

- wind gust effect factor, G, from ASCE/SEI7 Sec. 26.11

- wind force coefficient, C_f, from ASCE/SEI7 Fig. 29.3-1 and 29.4-1 through 29.4-3

- design wind force, F, defined by ASCE/SEI7 Eq. 29.3-1 as

$$F = q_z G C_f A_f$$

A_f is the projection area that includes the ice cover.

Example 1.23

A lattice framework supporting a road sign over a highway is 16 ft above grade and is located on flat terrain in Cleveland, Ohio. The struts in the framework are HSS3 × 0.250. Determine the ice load on the struts.

Solution

For a risk category of I, the ice importance factor for thickness from Table 1.5 is $I_i = 0.80$.

The equivalent diameter of the structural shape from ASCE/SEI7 Fig. 10.4-1 is $D_c = D = 3$ in.

The nominal ice thickness from ASCE/SEI7 Fig. 10.4-2 is $t = 1.5$ in.

The elevation, z, is between 0 ft and 900 ft, so from ASCE/SEI7 Eq. 10.4-4, the height factor is

$$f = \left(\frac{z}{33}\right)^{0.10} = \left(\frac{16 \text{ ft}}{33 \text{ ft}}\right)^{0.10}$$
$$= 0.93$$

The topographic factor for flat terrain is defined by ASCE/SEI7 Sec. 26.8.2 as $K_{zt} = 1.0$.

From ASCE/SEI7 Eq. 10.4-5, the design ice thickness is

$$t_d = t I_i f_z (K_{zt})^{0.35}$$
$$= (1.5 \text{ in})(0.80 \text{ in})(0.93)(1.0)^{0.35}$$
$$= 1.12 \text{ in}$$

From ASCE/SEI7 Eq. 10.4-1, the cross-sectional area of the ice on the struts is

$$A_i = \pi t_d (D_c + t_d)$$
$$= \pi (1.12 \text{ in})(3 \text{ in} + 1.12 \text{ in})$$
$$= 14.49 \text{ in}^2$$

From ASCE/SEI7 Sec. 10.4.1, the weight of the ice on the struts is

$$w_i = \gamma_i A_i$$
$$= \frac{\left(56 \dfrac{\text{lbf}}{\text{ft}^3}\right)(14.49 \text{ in}^2)}{\left(12 \dfrac{\text{in}}{\text{ft}}\right)^2}$$
$$= 5.64 \text{ lbf/ft}$$

12. STATIC EARTH PRESSURE

Nomenclature

h	total height of retaining wall	ft
h_{dry}	distance from surface of backfill to ground water table	ft
h_{eq}	height of backfill equivalent to surcharge load	ft
h_K	height of shear key	ft
h_{sub}	distance from bottom of footing to ground water table	ft
H	load due to lateral earth pressures, ground water pressure or pressure of bulk materials	kips
H_A	load due to active earth pressure behind wall	kips/ft
H_S	load due to pressure behind wall caused by surcharge on surface of backfill	kips/ft
H_P	load due to passive earth pressure in front of wall	kips/ft
K_A	Rankine coefficient of active earth pressure	–
K_P	Rankine coefficient of passive earth pressure	–
p_A	active earth pressure	lbf/ft^2
p_P	passive earth pressure	lbf/ft^2
p_S	uniform earth pressure due to surcharge load	lbf/ft^2
w	surcharge load on backfill	lbf/ft^2
z	depth below backfill surface	ft

Symbols

γ_S	density of backfill soil	lbf/ft^3
γ_{sub}	density of submerged soil	lbf/ft^3
γ_W	density of freshwater	lbf/ft^3
ϕ	angle of internal friction	degrees

General Principles

Lateral loads caused by soil pressure are addressed in ASCE/SEI7 Sec. 3.2 and in IBC Sec. 1610. *Earth pressure* is defined as the pressure exerted on the back of a retaining wall that retains an earth backfill. Figure 1.35 illustrates the three types of earth pressure.

- active earth pressure (Fig. 1.35(a))

- earth pressure at rest (Fig. 1.35(b))

- passive earth pressure (Fig. 1.35(c))

Retaining walls that are free to move and rotate at the top are designed for active earth pressure as the wall is free to deflect away from the earth backfill.

Forces

Figure 1.35 *Earth Pressure on Retaining Walls*

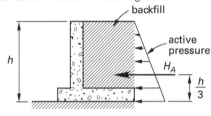

(a) active earth pressure

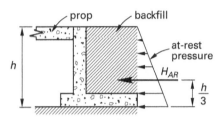

(b) earth pressure at rest

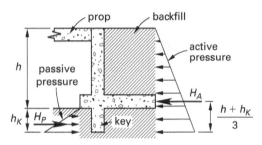

(c) passive earth pressure

Basement walls braced by floors that restrict movement at the top must be designed for earth pressure at rest, which may be as much as twice the value of active earth pressure. In accordance with IBC Sec. 1610.1, basement walls extending not more than 8 ft below grade and supporting flexible floor systems may be designed for active earth pressure.

To increase the factor of safety against sliding, a retaining wall may be designed with a key (see Fig. 1.35(c)). This serves to mobilize passive earth pressure in front of the key. Passive pressure may be as much as 13 times the value of active earth pressure; however, mobilizing the full value of passive earth pressure would require unacceptably large movement of the wall.

Design active pressure and design at-rest pressure are given for a range of soil classifications in IBC Table 1610.1. IBC Sec. 1610.1 requires that these values be used in design unless other values are substantiated by a specific geotechnical investigation.

Design Requirements

Active Pressure

For the retaining wall shown in Fig. 1.35(a), which has a soil backfill with a density of γ_S and an angle of internal friction of ϕ, the active earth pressure behind the wall at a depth z below the fill surface is given by Rankine's theory as

$$p_A = K_A \gamma_S z$$

In this equation, K_A is the Rankine coefficient of active earth pressure and is

$$K_A = \frac{1 - \sin\phi}{1 + \sin\phi}$$

The maximum active pressure occurs at the bottom of the heel and is

$$p_A = K_A \gamma_S h$$

The total horizontal force on the retaining wall is

$$H_A = \frac{p_A h}{2}$$

$$= \frac{K_A \gamma_S h^2}{2}$$

The line of action of the total horizontal force intersects the retaining wall at a distance of $h/3$ above the bottom of the heel (see Fig. 1.35(a)).

Example 1.24

The retaining wall shown in Fig. 1.35(a) retains a soil backfill with a density of 120 lbf/ft³ and an angle of internal friction of 35°. The total height of the retaining wall is 15 ft. Determine the total horizontal force on the retaining wall per foot of wall.

Solution

The Rankine coefficient of active earth pressure is

$$K_A = \frac{1 - \sin\phi}{1 + \sin\phi}$$

$$= \frac{1 - \sin 35°}{1 + \sin 35°}$$

$$= 0.27$$

The maximum active pressure at the bottom of the heel is

$$p_A = K_A \gamma_S h$$
$$= (0.27)\left(120 \ \frac{\text{lbf}}{\text{ft}^3}\right)(15 \ \text{ft})$$
$$= 486 \ \text{lbf/ft}^2$$

For a 1 ft length of wall, the total horizontal force on the retaining wall is

$$H_A = \frac{p_A h}{2}$$
$$= \frac{\left(486 \ \dfrac{\text{lbf}}{\text{ft}^2}\right)(15 \ \text{ft})}{(2)\left(1000 \ \dfrac{\text{lbf}}{\text{kip}}\right)}$$
$$= 3.65 \ \text{kips/ft}$$

Passive Pressure

For the retaining wall shown in Fig. 1.35(c), which is constructed on a soil with a density of γ_S and an angle of internal friction of ϕ, the passive earth pressure that can be developed in front of the key at a depth z below ground surface is given by Rankine's theory as

$$p_P = K_P \gamma_S z$$

In this equation, K_P is the Rankine coefficient of passive earth pressure and is

$$K_P = \frac{1 + \sin \phi}{1 - \sin \phi}$$

The maximum passive pressure occurs at the bottom of the key and is

$$p_P = K_P \gamma_S h_K$$

In this equation, h_K is the height of the key. The total horizontal force on the key due to passive pressure is

$$H_P = \frac{p_P h_K}{2} = \frac{K_P \gamma_S h_K{}^2}{2}$$

The line of action of the total horizontal force due to passive pressure is located $h_K/3$ above the bottom of the key (see Fig. 1.35(c)).

Example 1.25

The retaining wall shown in Fig. 1.35(c) is constructed on a soil with a density of 120 lbf/ft^3 and an angle of internal friction of 35°. The total height of the key is 5 ft. Determine the total horizontal force per foot of wall due to passive pressure on the key.

Solution

The Rankine coefficient of passive earth pressure is

$$K_P = \frac{1 + \sin \phi}{1 - \sin \phi}$$
$$= \frac{1 + \sin 35°}{1 - \sin 35°}$$
$$= 3.69$$

The maximum passive pressure that can be developed at the bottom of the key is

$$p_P = K_P \gamma_S h_K$$
$$= (3.69)\left(120 \ \frac{\text{lbf}}{\text{ft}^3}\right)(5 \ \text{ft})$$
$$= 2214 \ \text{lbf/ft}^2$$

For a 1 ft length of wall, the total force due to passive pressure that can be developed on the key is

$$H_P = \frac{p_P h_K}{2}$$
$$= \frac{\left(2214 \ \dfrac{\text{lbf}}{\text{ft}^2}\right)(5 \ \text{ft})}{(2)\left(1000 \ \dfrac{\text{lbf}}{\text{kip}}\right)}$$
$$= 5.54 \ \text{kips/ft}$$

Hydrostatic Pressure and Soil Lateral Pressure

A retaining wall is usually provided with a drainage system behind the wall. Where this is not possible and adjacent soil is below the free-water table, design must include the effects of submerged soil pressure plus hydrostatic pressure as shown in Fig. 1.36.

For the retaining wall shown in Fig. 1.36, which has a soil backfill with a dry density of γ_S, the active earth pressure behind the wall at a depth of h_{dry} below the fill surface is given by Rankine's theory as

$$p_A = K_A \gamma_S h_{\text{dry}}$$

Figure 1.36 *Effect of Water Table*

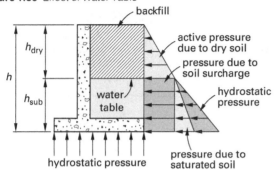

The soil backfill above the water table applies a surcharge load to the portion of the retaining wall below the water table with a uniform pressure of

$$p_A = K_A \gamma_S h_{\text{dry}}$$

Below the water table, the soil density is reduced by hydrostatic pressure. The submerged soil density is

$$\gamma_{\text{sub}} = \gamma_S - \gamma_W$$

Due to the submerged soil, the active pressure at the bottom of the heel is

$$p_A = K_A \gamma_{\text{sub}} h_{\text{sub}}$$

The hydrostatic pressure at the bottom of the heel is

$$p_W = \gamma_W h_{\text{sub}}$$

The hydrostatic pressure under the foundation is also

$$p_W = \gamma_W h_{\text{sub}}$$

In accordance with ASCE/SEI7 Table C3-2, the density of freshwater is

$$\gamma_W = 62 \ \text{lbf/ft}^3$$

In accordance with ASCE/SEI7 Table C3-2, the density of saltwater is

$$\gamma_W = 64 \ \text{lbf/ft}^3$$

Example 1.26

The retaining wall shown in Fig. 1.36 retains a soil backfill with a density of 120 lbf /ft³ and an angle of internal friction of 35°. The total height of the retaining wall is 15 ft, and the water table is 7.5 ft below the surface of the backfill. Determine the total horizontal force on the retaining wall per foot of wall.

Solution

The Rankine coefficient of active earth pressure is

$$\begin{aligned}
K_A &= \frac{1 - \sin \phi}{1 + \sin \phi} \\
&= \frac{1 - \sin 35°}{1 + \sin 35°} \\
&= 0.27
\end{aligned}$$

The active pressure of the dry soil at a depth of h_{dry} is

$$\begin{aligned}
p_A &= K_A \gamma_S h_{\text{dry}} \\
&= (0.27)\left(120 \ \frac{\text{lbf}}{\text{ft}^3}\right)(7.5 \ \text{ft}) \\
&= 243 \ \text{lbf/ft}^2
\end{aligned}$$

For a 1 ft length of wall, the horizontal force of the dry soil on the retaining wall is

$$\begin{aligned}
H_A &= \frac{p_A h_{\text{dry}}}{2} \\
&= \frac{\left(243 \ \dfrac{\text{lbf}}{\text{ft}^2}\right)(7.5 \ \text{ft})}{(2)\left(1000 \ \dfrac{\text{lbf}}{\text{kip}}\right)} \\
&= 0.91 \ \text{kip/ft}
\end{aligned}$$

The soil above the water table applies a surcharge load to the retaining wall below the water table with a uniform pressure of

$$p_A = 243 \ \text{lbf/ft}^2$$

For a 1 ft length of wall, the horizontal force of the surcharge load on the retaining wall is

$$\begin{aligned}
H_A &= p_A h_{\text{sub}} \\
&= \frac{\left(243 \ \dfrac{\text{lbf}}{\text{ft}^2}\right)(7.5 \ \text{ft})}{1000 \ \dfrac{\text{lbf}}{\text{kip}}} \\
&= 1.82 \ \text{kips/ft}
\end{aligned}$$

Below the water table, the soil density is reduced by hydrostatic pressure, and the submerged soil density is

$$\begin{aligned}
\gamma_{\text{sub}} &= \gamma_S - \gamma_W \\
&= 120 \ \frac{\text{lbf}}{\text{ft}^3} - 62 \ \frac{\text{lbf}}{\text{ft}^3} \\
&= 58 \ \text{lbf/ft}^3
\end{aligned}$$

The active pressure of the submerged soil at a depth of h_{sub} is

$$p_A = K_A \gamma_{\text{sub}} h_{\text{sub}}$$
$$= (0.27)\left(58 \ \frac{\text{lbf}}{\text{ft}^3}\right)(7.5 \ \text{ft})$$
$$= 117.45 \ \text{lbf/ft}^2$$

For a 1 ft length of wall, the horizontal force of the submerged soil on the retaining wall is

$$H_A = \frac{p_A h_{\text{sub}}}{2}$$
$$= \frac{\left(117.45 \ \dfrac{\text{lbf}}{\text{ft}^2}\right)(7.5 \ \text{ft})}{(2)\left(1000 \ \dfrac{\text{lbf}}{\text{kip}}\right)}$$
$$= 0.44 \ \text{kip/ft}$$

Below the water table, the hydrostatic pressure at a depth of z is

$$p_W = \gamma_W z$$

The hydrostatic pressure at the bottom of the heel at a depth of h_{sub} is

$$p_W = \gamma_W h_{\text{sub}}$$
$$= \left(62 \ \frac{\text{lbf}}{\text{ft}^3}\right)(7.5 \ \text{ft})$$
$$= 465 \ \text{lbf/ft}^2$$

For a 1 ft length of wall, the hydrostatic force on the retaining wall is

$$H_W = \frac{p_W h_{\text{sub}}}{2}$$
$$= \frac{\left(465 \ \dfrac{\text{lbf}}{\text{ft}^2}\right)(7.5 \ \text{ft})}{(2)\left(1000 \ \dfrac{\text{lbf}}{\text{kip}}\right)}$$
$$= 1.74 \ \text{kips/ft}$$

The total horizontal force on a 1 ft length of the wall is the sum of the individual forces.

$$H_{\text{total}} = 0.91 \ \frac{\text{kip}}{\text{ft}} + 1.82 \ \frac{\text{kips}}{\text{ft}} + 0.44 \ \frac{\text{kip}}{\text{ft}} + 1.74 \ \frac{\text{kips}}{\text{ft}}$$
$$= 4.91 \ \text{kips/ft}$$

Surcharge Loads

A *surcharge load* is a superimposed load applied to the surface of the backfill that produces a lateral load on the rear face of the retaining wall. A surcharge load may consist of live load, such as from vehicles on driveways or highways adjacent to the retaining wall, or of dead load, such as from sloping backfills, road paving, or adjacent building foundations.

The surcharge load may often be represented by an equivalent height of backfill as shown in Fig. 1.37.

Figure 1.37 Surcharge Load

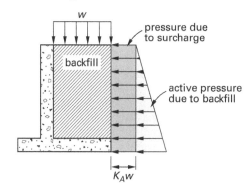

The retaining wall shown in Fig. 1.37 retains a backfill with a density of γ_S and has a uniformly distributed surcharge load of w applied to the surface of the backfill. The height of backfill equivalent to the surcharge is

$$h_{\text{eq}} = \frac{w}{\gamma_S}$$

The uniform surcharge pressure is

$$p_s = K_A \gamma_S h_{\text{eq}} = K_A w$$

The surcharge on walls adjacent to highways is covered by AASHTO Sec. 3.11.6.4 and AASHTO Table 3.11.6.4-2 and is shown in Fig. 1.38.

Figure 1.38 Surcharge Due to Vehicular Load

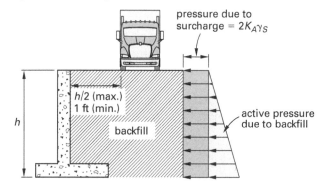

When vehicular traffic acts on the surface of the backfill behind the wall, within a distance from the back of the wall not greater than one-half the wall height but not less than 1 ft, the height of backfill equivalent to the surcharge is

$$h_{eq} = 2 \text{ ft}$$

For the retaining wall shown in Fig. 1.38, which retains a backfill that has a density of γ_S and a Rankine coefficient of active earth pressure of K_A, the uniform surcharge pressure is

$$p_S = K_A \gamma_S h_{eq} = 2K_A \gamma_S$$

Expansive Soils

When soil with expansion potential is present behind basement and retaining walls, large and unpredictable pressures are produced, and severe damage may result. ASCE/SEI7 Sec. C3.2.1 recommends that expansive material be removed from the backfill and replaced with non-expansive, freely draining sand or gravel.

Load Combinations

Strength Design

For strength design or load and resistance factor design, the load combinations given in IBC Sec. 1605.2 apply.

$1.4(D + F)$	[IBC Eq. 16-1]
$1.2(D + F) + 1.6(L + H)$	[IBC Eq. 16-2]
$\quad + 0.5(L_r \text{ or } S \text{ or } R)$	
$1.2(D + F) + 1.6(L_r \text{ or } S \text{ or } R)$	[IBC Eq. 16-3]
$\quad + 1.6H + (f_1 L \text{ or } 0.5W)$	
$1.2(D + F) + 1.0W + f_1 L$	[IBC Eq. 16-4]
$\quad + 1.6H + 0.5(L_r \text{ or } S \text{ or } R)$	
$1.2(D + F) + 1.0E + f_1 L$	[IBC Eq. 16-5]
$\quad + 1.6H + f_2 S$	
$0.9D + 1.0W + 1.6H$	[IBC Eq. 16-6]
$0.9(D + F) + 1.0E + 1.6H$	[IBC Eq. 16-7]

The load factor f_1 is 1.0 for places of public assembly, garages, and live loads greater than 100 lbf/ft^2, and 0.5 otherwise. The load factor f_2 is 0.7 for roof configurations that do not shed snow and 0.2 otherwise. The load factor for H is subject to these qualifications.

- When the effect of H is permanent and counteracts the effects of other loads, the load factor for H is 0.9.

- For all other conditions, the load factor for H is zero.

Allowable Stress Design

For allowable stress design, the load combinations given in IBC Sec. 1605.3.1 apply.

$D + F$	[IBC Eq. 16-8]
$D + H + F + L$	[IBC Eq. 16-9]
$D + H + F + (L_r \text{ or } S \text{ or } R)$	[IBC Eq. 16-10]
$D + H + F + 0.75L$	[IBC Eq. 16-11]
$\quad + 0.75(L_r \text{ or } S \text{ or } R)$	
$D + H + F + (0.6W \text{ or } 0.7E)$	[IBC Eq. 16-12]
$D + H + F + 0.75(0.6W) + 0.75L$	[IBC Eq. 16-13]
$\quad + 0.75(L_r \text{ or } S \text{ or } R)$	
$D + H + F + 0.75(0.7E)$	[IBC Eq. 16-14]
$\quad + 0.75L + 0.75S$	
$0.6D + 0.6W + H$	[IBC Eq. 16-15]
$0.6(D + F) + 0.7E + H$	[IBC Eq. 16-16]

The load factor for H is subject to these qualifications.

- When the effect of H is permanent and counteracts the effects of other loads, the load factor on H is 0.6.

- For all other conditions, the load factor on H is zero.

Stability Requirements

IBC Sec. 1807.2.3 requires that a retaining wall be designed to resist sliding and overturning using a minimum factor of safety of 1.5. The load factors and load combinations of IBC Sec. 1605 are not applicable to this requirement. Design is based on nominal loads, with a load factor of 1.0 for all loads except earthquake loads, which are assigned a load factor of 0.7. Where earthquake loads are included in the design, the minimum factor of safety against sliding and overturning is reduced to 1.1.

13. HYDROSTATIC LOADS

Nomenclature

A_W	dam wall cross-sectional area	ft^2
b	width at top of wall	ft
B	width at bottom of wall	ft
h	total height of wall	ft
H_W	total horizontal force on dam	kips
M_O	overturning moment about toe	ft-kips
M_R	restoring moment about toe	ft-kips
p_W	hydrostatic pressure	lbf/ft^2
V_W	upward hydrostatic force	kips
W_G	weight of wall	kips
x	distance from wall centroid to heel	ft

x' distance from wall centroid to point where   ft
        resultant of H_W and W_G cuts base
z depth below backfill surface   ft

Symbols

γ_G density of concrete   lbf/ft^3
γ_W density of freshwater   lbf/ft^3

General Principles

In accordance with ASCE/SEI7 Sec. 3.2.2, elements that are below grade and roughly horizontal, such as basement floors, must be designed for the upward water pressure when they are to be constructed below the water table.

This upward pressure consists of the full hydrostatic pressure applied over the entire area. This is also a consideration in the design of cofferdams, caissons, and gravity dams.

Design Requirements

Figure 1.39(a) shows a mass concrete gravity dam. The hydrostatic pressure exerted on the dam at a depth z below the water level is

$$p_W = \gamma_W z$$

The maximum hydrostatic pressure occurs at the bottom of the heel at a depth h below the water level. This pressure is

$$p_W = \gamma_W h$$

The total horizontal force on the dam wall per foot of length is

$$H_W = \frac{p_W h}{2}$$
$$= \frac{\gamma_W h^2}{2}$$

The line of action of this force is located $h/3$ above the bottom of the heel.

The total upward hydrostatic force on the base of the dam wall per foot of length is

$$V_W = p_W B$$
$$= \gamma_W h B$$

The weight per foot of length of a dam wall with a cross-sectional area of A_W and a density of γ_G is

$$W_G = A_W \gamma_G$$

Figure 1.39 Mass Concrete Gravity Dam

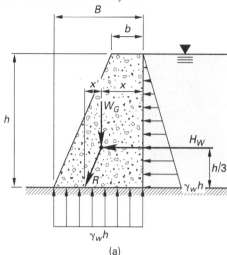

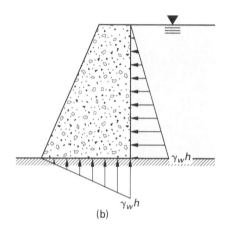

This weight acts at the centroid of the dam wall.

Traditionally, tensile stress is not allowed in the concrete. In order to keep such stress from forming, the line of action of the resultant of H_W and W_G must cross the base within its middle third. The dam dimensions must be adjusted to ensure this.

Should seepage occur under the dam, the hydrostatic pressure under the base reduces to zero at the toe. The pressure distribution under the dam is then triangular as shown in Fig. 1.39(b).

Example 1.27

The mass concrete gravity dam in Fig. 1.39(a) has a total height of 15 ft, a width at the top of the wall of 6 ft, and a width at the bottom of the wall of 12 ft. The concrete in the dam has a density of 144 lbf/ft^3, and the water density is 62 lbf/ft^3. Determine the total horizontal force on the dam wall, whether tension develops in the dam wall, and whether the factor of safety against overturning is satisfactory.

Forces

Solution

The maximum hydrostatic pressure at the bottom of the heel is

$$p_W = \gamma_W h = \left(62 \; \frac{\text{lbf}}{\text{ft}^3}\right)(15 \text{ ft})$$
$$= 930 \text{ lbf/ft}^2$$

The total hydrostatic force on the dam wall per foot of length is

$$H_W = \frac{p_W h}{2} = \frac{\left(930 \; \frac{\text{lbf}}{\text{ft}^2}\right)(15 \text{ ft})}{(2)\left(1000 \; \frac{\text{lbf}}{\text{kip}}\right)}$$
$$= 6.98 \text{ kips/ft}$$

The cross-sectional area of the dam wall is

$$A_W = \frac{h(b+B)}{2}$$
$$= \frac{(15 \text{ ft})(6 \text{ ft} + 12 \text{ ft})}{2}$$
$$= 135 \text{ ft}^2$$

The weight of the dam wall per foot of length is

$$W_G = A_W \gamma_G = \frac{(135 \text{ ft}^2)\left(144 \; \frac{\text{lbf}}{\text{ft}^3}\right)}{1000 \; \frac{\text{lbf}}{\text{kip}}}$$
$$= 19.44 \text{ kips/ft}$$

The distance of the wall centroid from the heel is

$$x = \frac{(b^2 + bB + B^2)}{3(b+B)}$$
$$= \frac{\left((6 \text{ ft})^2 + (6 \text{ ft})(12 \text{ ft}) + (12 \text{ ft})^2\right)}{(3)(6 \text{ ft} + 12 \text{ ft})}$$
$$= 4.67 \text{ ft}$$

The line of action of the resultant of H_W and W_G cuts the base at a distance from the centroid of

$$x' = \left(\frac{h}{3}\right)\left(\frac{H_W}{W_G}\right)$$
$$= \left(\frac{15 \text{ ft}}{3}\right)\left(\frac{6.98 \; \frac{\text{kips}}{\text{ft}}}{19.44 \; \frac{\text{kips}}{\text{ft}}}\right)$$
$$= 1.80 \text{ ft}$$

The line of action of the resultant of H_W and W_G cuts the base at a distance from the heel of

$$x + x' = 4.67 \text{ ft} + 1.80 \text{ ft}$$
$$= 6.47 \text{ ft}$$

To ensure that no tensile stress occurs in the concrete dam wall, the line of action of the resultant of H_W and W_G must cross the base a distance from the heel of not more than

$$\frac{2B}{3} = \frac{(2)(12 \text{ ft})}{3}$$
$$= 8 \text{ ft}$$

This is greater than the actual distance of 6.47 ft, so tensile stress does not occur in the dam wall.

The total upward hydrostatic force on the base of the dam per foot of length is

$$V_W = p_W B$$
$$= \frac{\left(930 \; \frac{\text{lbf}}{\text{ft}^2}\right)(12 \text{ ft})}{1000 \; \frac{\text{lbf}}{\text{kip}}}$$
$$= 11.16 \text{ kips/ft}$$

The overturning moment about the toe per foot of length is

$$M_O = \frac{H_W h}{3} + \frac{V_W B}{2}$$
$$= \frac{\left(6.98 \; \frac{\text{kips}}{\text{ft}}\right)(15 \text{ ft})}{3} + \frac{\left(11.16 \; \frac{\text{kips}}{\text{ft}}\right)(12 \text{ ft})}{2}$$
$$= 101.86 \text{ ft-kips}$$

The restoring moment about the toe per foot of length is

$$M_R = W_G(B - x)$$
$$= \left(19.44 \, \frac{\text{kips}}{\text{ft}}\right)(12 \, \text{ft} - 4.67 \, \text{ft})$$
$$= 142.50 \, \text{ft-kips}$$

The factor of safety against overturning is

$$\frac{M_R}{M_O} = \frac{142.50 \, \text{ft-kips}}{101.86 \, \text{ft-kips}}$$
$$= 1.4$$

This is less than 1.5, so the factor of safety is unsatisfactory.

14. HYDRAULIC LOADS

Nomenclature

a	shape factor or drag coefficient	–
d_h	equivalent surcharge depth	ft
D	dead load	kips
F_a	flood load	kips
g	acceleration due to gravity, 32.2	ft/sec^2
L	live load	kips
L_r	roof live load	kips
R	rain load	kips
S	snow load	kips
v	average velocity of water	ft/sec
W	wind load	kips

Flood Loads

Flooding is a temporary condition of inundation of normally dry land from the

- overflow of inland or tidal waters
- rapid accumulation or runoff of surface waters from any source

IBC Sec. 1612.1 requires that structures in flood hazard areas be designed and constructed to resist the effects of flood hazards and flood loads, and IBC Sec. 16.12.2 requires that this must be done in accordance with ASCE/SEI Chap. 5 and ASCE/SEI 24.[16]

A number of terms pertaining to flood loads are defined in ASCE/SEI7 Sec. 5.2. A *flood hazard area* is an area subject to flooding during the design flood. Flood hazard areas are delineated on the community's legally designated *flood hazard map*.

A *special flood hazard area* is an area within a flood plain that is subject to a 1% or greater chance of flooding in any given year. The special flood hazard area is further subdivided into the

- *coastal V-zone* extending from offshore to the inland limit of a primary frontal dune along an open coast
- *coastal A-zone* extending landward of a V-zone or landward of an open coast without mapped V-zones, subject to astronomical tides, storm surges, seiches, or tsunamis, not riverine flooding, and the potential for breaking wave heights not less than 1.5 ft

The *design flood* is the greater of two potential flood events:

- the flood associated with the flood hazard area (as defined above)
- the *base flood*, which is the flood affecting the special flood hazard areas (as defined above) on the community's flood insurance rate map

The *design flood elevation* is the elevation of the design flood, including wave height. In accordance with ASCE/SEI7 Sec. 5.4.2, the design depth is increased by 1 ft for surfaces exposed to free water.

In accordance with ASCE/SEI7, *flood loads* are the loads or pressures on surfaces of structures caused by the presence of flood waters. These loads are either hydrostatic or hydrodynamic in nature.

Hydrostatic loads are caused by the depth of water to the level of the design flood elevation. The water is either stagnant or moves at less than 5 ft/sec. The resulting pressure acts perpendicular to the surface involved, both above and below ground level. The design load is equal to the product of the water pressure multiplied by the surface area on which it acts.

Hydrodynamic loads are caused by the flow of water moving at moderate to high velocity above ground level. Hydrodynamic loads are lateral loads, and they are caused by the impact of the moving water on the surface area of a structure as well as by drag forces as the water flows around the structure.

Wave loads are a special type of hydrodynamic load. Wave loads result from water waves propagating over the water surface and striking the structure.

Flood waters can carry debris, such as logs and ice floes, that can strike structures and produce impact loads. Impact loads are covered in ASCE/SEI7 Sec. C5.4.5.

Flood Load Combinations

Strength Design

For strength design, the load combinations given in ASCE/SEI7 Sec. 2.3.2, Eq. 1 through Eq. 7, are applicable with the following changes.

In V-zones or coastal A-zones, replace combinations 4 and 5 with

(4) $1.2D + 1.0W + 2.0F_a + L + 0.5(L_r \text{ or } S \text{ or } R)$

(5) $0.9D + 1.0W + 2.0F_a$

In noncoastal A-zones, replace combinations 4 and 5 with

(4) $1.2D + 0.5W + 1.0F_a + L + 0.5(L_r \text{ or } S \text{ or } R)$

(5) $0.9D + 0.5W + 1.0F_a$

Otherwise, the ASCE/SEI7 load combinations for strength design are equivalent to the IBC load combinations for strength design given in Sec. 12.

Allowable Stress Design

For allowable stress design, the load combinations given in ASCE/SEI7 Sec. 2.4.1, Eq. 1 through Eq. 8, are applicable with the following changes.

In V-zones or coastal A-zones, replace combinations 5, 6, and 7 with

(5) $D + 1.5F_a + 0.6W$

(6) $D + 0.75(0.6W) + 0.75L + 1.5F_a$
       $+ 0.75(L_r \text{ or } S \text{ or } R)$

(7) $0.6D + 0.6W + 1.5F_a$

In noncoastal A-zones, replace combinations 5, 6, and 7 with

(5) $D + 0.75F_a + 0.6W$

(6) $D + 0.75(0.6W) + 0.75L + 0.75F_a$
       $+ 0.75(L_r \text{ or } S \text{ or } R)$

(7) $0.6D + 0.6W + 0.75F_a$

Otherwise, the ASCE/SEI7 load combinations for allowable stress design are equivalent to the IBC load combinations for allowable stress design given in Sec. 12.

Stream Flow

If the water velocity is no more than 10 ft/sec, the dynamic effects of stream flow may be represented by an equivalent hydrostatic load. To do this, increase the design flood elevation by an equivalent surcharge depth, which is given by ASCE/SEI7 Eq. 5.4-1 as

$$d_h = \frac{a \mathrm{v}^2}{2g}$$

In this equation, the *shape factor* (or *drag coefficient*), a, must not be less than 1.25.

Wave Action

ASCE/SEI7 Sec. C5.4.4 recommends that buildings should be elevated above the wave crest elevation to avoid serious damage due to wave forces. Design for wave action is covered in ASCE/SEI7 Sec. 5.4.4, and three approaches are approved.

- analytical methods given in ASCE/SEI7 Sec. 5.4.4

- advanced numerical modeling procedures

- laboratory test procedures

The analytical methods in ASCE/SEI7 Sec. 5.4.4 include procedures for determining wave forces on vertical pilings, columns, and vertical and nonvertical walls.

Scour Effects

Scour is defined in FEMA P-550[17] Sec. 3.7 as a localized loss of soil that results in a lowering of the ground surface around a building foundation or a bridge pier.[17] Scour is similar to erosion, which can also lower the ground surface, but erosion occurs over an extensive area and, generally, over a longer time period, while scour can result from a single flood event and typically results in depressions immediately around foundations and piles. Buildings located in flood hazard areas and coastal areas and bridge piers located in stream flows are particularly prone to scour.

Scour is caused by obstructions in a stream flow. When a stream flow encounters an obstruction, the stream must flow around the obstruction, and the flow accelerates. This loosens the soil, which is carried away in the stream flow. Scour effects increase as stream velocity and turbulence increase and as soil density decreases.

Scour reduces the embedment of a building's foundation into the soil, which causes a shallow foundation to collapse or be more susceptible to differential settlement. In the case of pile foundations, scour increases the unbraced length of the piles, which can cause overstressing of the piles and possible differential settlement.

Because of the variable nature of scour action, buildings located in flood hazard areas and coastal areas should be designed with conservative foundation details.

Tsunami Loads

In accordance with IBC Sec. 1615.1, the design and construction of risk category III and IV structures located in a tsunami design zone must be in accordance with ASCE/SEI7 Chap. 6.

Several terms pertaining to tsunami loads are defined in ASCE/SEI7 Sec. 6.2.

A *tsunami* is a series of waves with variable long periods caused by earthquake-induced uplift or subsidence of the sea floor.

A *tsunami design zone* is an area identified on the ASCE tsunami design geodatabase within which structures are analyzed and designed for inundation by the maximum considered tsunami.

Maximum considered tsunami has a 2% probability of being exceeded in a 50-year period.

15. REFERENCES

1. International Code Council. *International Building Code.* Falls Church, VA: International Code Council, 2018.

2. American Society of Civil Engineers. *Minimum Design Loads and Associated Criteria for Buildings and Other Structures.* Reston, VA: American Society of Civil Engineers, 2016.

3. American Concrete Institute. *Building Code Requirements for Structural Concrete and Commentary.* Farmington Hills, MI: American Concrete Institute, 2014.

4. O'Rourke, M. *Snow Loads: Guide to the Snow Load Provisions of ASCE 7-16.* Reston, VA: American Society of Civil Engineers, 2017.

5. *AASHTO LRFD Bridge Design Specifications*, 8th ed. Washington, DC: American Association of State Highway and Transportation Officials, 2017.

6. American Institute of Steel Construction. *Steel Construction Manual*, 15th ed. Chicago, IL: American Institute of Steel Construction, 2017.

7. Williams, A. *Structural Analysis in Theory and Practice.* Burlington, MA: Elsevier/International Code Council, 2009.

8. American Institute of Steel Construction. *Moments, Shears, and Reactions: Continuous Highway Bridge Tables.* Chicago, IL: American Institute of Steel Construction, 1986.

9. Graudenz, H. *Bending Moment Coefficients in Continuous Beams.* London: Pitman, 1964.

10. Portland Cement Association. *Influence Lines Drawn as Deflection Curves.* Skokie, IL: Portland Cement Association, 1948.

11. Williams, A. "The Determination of Influence Lines for Bridge Decks Monolithic with Their Piers." *Structural Engineer* 42 no. 5 (1964): 161-166.

12. American Society of Mechanical Engineers. *Safety Code for Elevators and Escalators.* New York, NY: American Society of Mechanical Engineers, 2007.

13. Lightfoot, E. *Moment Distribution.* London: E. & F. N. Spon Ltd., 1961.

14. Parolini, M., and J. Lawson. "The Making of a Rain-Induced Roof Collapse and the Lessons Learned for our Profession." *Proceedings of the Structural Engineers Association of California 2016 Convention, Maui* (October 2016).

15. American Institute of Steel Construction. *Specification for Structural Steel Buildings.* Chicago, IL: American Institute of Steel Construction, 2016.

16. American Society of Civil Engineers. ASCE 24, *Flood-Resistance Design and Construction Standard.* Reston, VA: American Society of Civil Engineers, 2005.

17. Federal Emergency Management Agency. FEMA P-550, *Recommended Residential Construction for Coastal Areas*, Washington, DC: Federal Emergency Management Agency, 2009.

Reinforced Concrete Design

1. GENERAL REQUIREMENTS

The *International Building Code*[1] (IBC) adopts by reference the American Concrete Institute's[2] *Building Code Requirements for Structural Concrete and Commentary* (ACI). Some sections of the ACI code are modified by the IBC, and these exceptions are given in IBC Sec. 1905.

The strength design method is the only design method presented in the code.

In the *strength design method*, factored loads are applied to the member to determine the required ultimate strength. This required strength must not exceed the design strength, which is calculated as the member's nominal strength multiplied by a resistance factor, ϕ.

2. STRENGTH DESIGN PRINCIPLES

Nomenclature

D	dead load	kips or lbf
E	earthquake load	kips or lbf
F	load due to weight and pressure of fluids	kips or lbf
H	load due to pressure of soil	kips or lbf
L	live load	kips or lbf
L_r	roof live load	kips or lbf
Q	service level force	kips or lbf
R	load due to rainwater	kips or lbf
S	snow load	kips or lbf
T	effect of temperature, shrinkage, creep, differential settlement, and shrinkage-compensating concrete	kips or lbf
U	required strength to resist factored load	kips or lbf
w	distributed load	kips/ft
W	wind load	kips or lbf

Symbols

γ	load factor	–
ϕ	strength-reduction factor	–

Required Strength

The required ultimate strength of a member consists of the most critical combination of factored loads applied to the member. Factored loads consist of working (i.e., service) loads, Q, multiplied by the appropriate load factor, γ. In accordance with ACI Sec. 5.3.1, the required strength, U, is defined by seven combinations as follows.

$$U = 1.4D \qquad \text{[ACI 5.3.1a]}$$
$$U = 1.2D + 1.6L + 0.5(L_r \text{ or } S \text{ or } R) \text{ [ACI 5.3.1b]}$$
$$U = 1.2D + 1.6(L_r \text{ or } S \text{ or } R) \qquad \text{[ACI 5.3.1c]}$$
$$+(1.0L \text{ or } 0.5W)$$
$$U = 1.2D + 1.0W + 1.0L \qquad \text{[ACI 5.3.1d]}$$
$$+0.5(L_r \text{ or } S \text{ or } R)$$
$$U = 1.2D + 1.0E + 1.0L + 0.2S \qquad \text{[ACI 5.3.1e]}$$
$$U = 0.9D + 1.0W \qquad \text{[ACI 5.3.1f]}$$
$$U = 0.9D + 1.0E \qquad \text{[ACI 5.3.1g]}$$

Replace $1.0L$ with $0.5L$ in ACI Eq. 5.3.1c, Eq. 5.3.1d, and Eq. 5.3.1e except for garages, places of public assembly, and all areas where L is greater than 100 lbf/ft².

The cumulative structural effects of temperature, creep, shrinkage, differential settlement, and shrinkage-compensating concrete, T, must be considered where appropriate. The load factor on T should be established by considering its likely magnitude and the probability that the maximum effect of T will occur simultaneously with other loads. The load factor on T must have a value not less than 1.0.

Illustration for Ex. 2.1

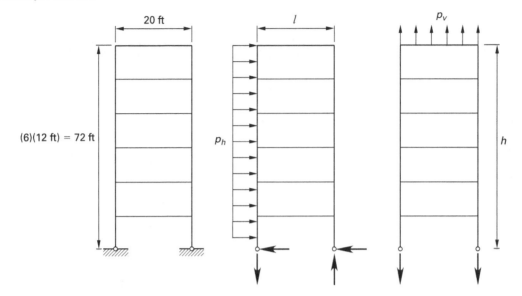

When a fluid load, F, is present, it must be included with the same load factor as the dead load, D, in ACI Eq. 5.3.1a through Eq. 5.3.1e and Eq. 5.3.1g. If the effect of F is not permanent but when present counteracts the primary load, F must not be included in ACI Eq. 5.3.1a through Eq. 5.3.1g. When a lateral soil pressure load, H, is present, it must be included in the load combinations with load factors that depend on how H interacts with other loads. When H acts alone or adds to the effect of other loads, it has a load factor of 1.6. When the effect of H is permanent and counteracts the effects of other loads, it has a load factor of 0.9. When the effect of H is not permanent but counteracts the effects of other loads when present, it should not be included.

Example 2.1

Illustration for Ex. 2.1 shows a typical frame of a six-story office building. The loading on the frame is as follows.

> roof dead load, including
> cladding and columns, $w_{Dr} = 1.2$ kips/ft
> roof live load, w_{Lr} $\quad = 0.4$ kip/ft
> floor dead load, including
> cladding and columns, w_D $= 1.6$ kips/ft
> floor live load, w_L $\quad = 1.25$ kips/ft
> horizontal wind pressure, p_h $= 1.0$ kips/ft
> vertical wind pressure, p_v $\quad = 0.5$ kip/ft

Determine the maximum and minimum required loads on the first-floor columns.

Solution

The axial load on one column due to the dead load is

$$
\begin{aligned}
D &= \frac{l(w_{Dr} + 5w_D)}{2} \\
&= \frac{(20\text{ ft})\left(1.2\ \dfrac{\text{kips}}{\text{ft}} + (5\text{ stories})\left(1.6\ \dfrac{\text{kips}}{\text{ft}}\right)\right)}{2} \\
&= 92\text{ kips}
\end{aligned}
$$

The axial load on one column due to the floor live load is

$$
\begin{aligned}
L &= \frac{l(5w_L)}{2} \\
&= \frac{(20\text{ ft})\left((5\text{ stories})\left(1.25\ \dfrac{\text{kips}}{\text{ft}}\right)\right)}{2} \\
&= 62.5\text{ kips}
\end{aligned}
$$

The axial load on one column due to the roof live load is

$$
\begin{aligned}
L_r &= \frac{lw_{Lr}}{2} \\
&= \frac{(20\text{ ft})\left(0.4\ \dfrac{\text{kip}}{\text{ft}}\right)}{2} \\
&= 4\text{ kips}
\end{aligned}
$$

The axial load on one column due to horizontal wind pressure is obtained by taking moments about the base of the other column and is given by

$$W_h = \pm \frac{p_h h^2}{2l}$$
$$= \pm \frac{\left(1 \dfrac{\text{kip}}{\text{ft}}\right)(72 \text{ ft})^2}{(2)(20 \text{ ft})}$$
$$= \pm 129.6 \text{ kips}$$

The axial load on one column due to the vertical wind pressure is obtained by resolving forces at the column bases and is given by

$$W_v = -\frac{p_v l}{2}$$
$$= -\frac{\left(0.5 \dfrac{\text{kip}}{\text{ft}}\right)(20 \text{ ft})}{2}$$
$$= -5 \text{ kips}$$

The maximum strength level required load on a column is

$$U = 1.2D + 1.6L + 0.5L_r \quad \text{[ACI 5.3.1b]}$$
$$= (1.2)(92 \text{ kips}) + (1.6)(62.5 \text{ kips})$$
$$\quad + (0.5)(4 \text{ kips})$$
$$= 212 \text{ kips} \quad \text{[compression]}$$

$$U = 1.2D + 1.6L_r + 0.5L \quad \text{[ACI 5.3.1c]}$$
$$= (1.2)(92 \text{ kips}) + (1.6)(4 \text{ kips})$$
$$\quad + (0.5)(62.5 \text{ kips})$$
$$= 148 \text{ kips} \quad \text{[compression]}$$

$$U = 1.2D + 1.6L_r + 0.5W \quad \text{[ACI 5.3.1c]}$$
$$= (1.2)(92 \text{ kips}) + (1.6)(4 \text{ kips})$$
$$\quad + (0.5)(129.6 \text{ kips} - 5 \text{ kips})$$
$$= 179 \text{ kips} \quad \text{[compression]}$$

$$U = 1.2D + 1.0W + 0.5L + 0.5L_r \quad \text{[ACI 5.3.1d]}$$
$$= (1.2)(92 \text{ kips})$$
$$\quad + (1.0)(129.6 \text{ kips} - 5 \text{ kips})$$
$$\quad + (0.5)(62.5 \text{ kips}) + (0.5)(4 \text{ kips})$$
$$= 268 \text{ kips} \quad \text{[compression; governs]}$$

The minimum strength level design load on a column is

$$U = 0.9D + 1.0W_h + 1.0W_v \quad \text{[ACI 5.3.1f]}$$
$$= (0.9)(92 \text{ kips}) + (1.0)(-129.6 \text{ kips})$$
$$\quad + (1.0)(-5 \text{ kips})$$
$$= -52 \text{ kips} \quad \text{[tension]}$$

Design Strength

The design strength of a member consists of the nominal, or theoretical ultimate, strength of the member multiplied by the appropriate strength reduction factor, ϕ. The reduction factor is defined in ACI Sec. 21.2.1 as

$\phi = 0.90$ [for flexure of tension-controlled sections]

$\phi = 0.75$ [for shear and torsion]

$\phi = 0.75$ [for compression-controlled sections with spiral reinforcement]

$\phi = 0.65$ [for compression-controlled sections with lateral ties]

$\phi = 0.65$ [for bearing on concrete surfaces]

$\phi = 0.75$ [for strut-and-tie models]

3. STRENGTH DESIGN OF REINFORCED CONCRETE BEAMS

Nomenclature

a	depth of equivalent rectangular stress block	in
A_g	gross area of concrete section	in^2
A_{max}	maximum area of tension reinforcement, $\rho_t bd$	in^2
A_s	area of tension reinforcement	in^2
A_s'	area of compression reinforcement	in^2
A_{sf}	reinforcement area to develop the outstanding flanges	in^2
A_{sw}	reinforcement area to balance the residual moment	in^2
A_t	additional tension reinforcement	in^2
b	width of compression face of member	in
b_w	web width	in
c_f	compressive force in flanges	kips
c	distance from extreme compression fiber to neutral axis	in
C_u	compressive force in the concrete	kips
d	distance from extreme compression fiber to centroid of tension reinforcement	in
d'	distance from extreme compression fiber to centroid of compression reinforcement	in
f_c'	compressive strength of concrete	lbf/in^2
f_s'	stress in compression reinforcement	lbf/in^2
f_y	yield strength of reinforcement	lbf/in^2
h_f	flange depth	in
K_u	design moment factor, M_u/bd^2	lbf/in^2
M_f	design moment strength of the outstanding flanges	in-lbf or ft-kips
M_{max}	maximum design flexural strength for a tension-controlled section	in-lbf or ft-kips

M_n	nominal flexural strength of a member	in-lbf or ft-kips
M_r	residual moment $(M_u - M_{max})$	in-lbf or ft-kips
M_u	factored moment on the member	in-lbf or ft-kips
T_u	tensile force in the reinforcement	kips

Symbols

β_1	compression zone factor	–
ϵ_c	strain at external compression fiber	–
$\epsilon_{c(max)}$	maximum strain at external compression fiber, 0.003	–
ϵ_t	strain in tension reinforcement	–
ϵ_s'	strain in compression reinforcement	–
ϵ_{ty}	strain in tension reinforcement used to define a compression-controlled section	–
ρ	ratio of tension reinforcement, A_s/bd	–
ρ_b	reinforcement ratio producing balanced strain conditions	–
ρ_{max}	maximum tension reinforcement ratio in a rectangular beam with tension reinforcement only	–
ρ_{min}	minimum allowable reinforcement ratio	–
ρ_t	reinforcement ratio producing a tension-controlled section	–
ρ_w	residual reinforcement ratio	–
ω	tension reinforcement index, $\rho f_y/f_c'$	–
ω_{max}	maximum tension reinforcement index in a rectangular beam with tension reinforcement only	–

Beams with Tension Reinforcement Only

In accordance with ACI Sec. 22.2.2.3, a rectangular stress block is assumed in the concrete, as shown in Fig. 2.1, and it is also assumed that the tension reinforcement has yielded.[2,3] The nominal flexural strength of a rectangular beam is derived[3] as

$$M_n = A_s f_y d\left(1 - \frac{0.59\rho f_y}{f_c'}\right)$$

Equating the tensile and compressive forces acting on the section gives the depth of the equivalent rectangular stress block as

$$a = \frac{A_s f_y}{0.85 f_c' b}$$

$$M_n = A_s f_y\left(d - \frac{a}{2}\right)$$

The maximum permissible factored moment on the member, or required moment strength, must not exceed ϕM_n. For a tension-controlled section, where $\phi = 0.9$,

$$M_u = 0.9 M_n$$

The required reinforcement ratio for a given factored moment is then

$$\rho = \frac{\left(0.85 f_c'\right)\left(1 - \sqrt{1 - \dfrac{K_u}{0.383 f_c'}}\right)}{f_y}$$

$$K_u = \frac{M_u}{bd^2}$$

These expressions may be readily programmed using a handheld calculator.[4] Alternatively, design tables may be used,[5,6,7] and rearranging the expression[4] in terms of the tension reinforcement index, ω, gives

$$\frac{M_u}{f_c' bd^2} = \omega(0.9 - 0.5294\omega)$$

$$\omega = \frac{\rho f_y}{f_c'}$$

Appendix 2.A provides a design aid that tabulates the tension reinforcement index, ω, against $M_u/f_c' bd^2$.

Tension-Controlled and Compression-Controlled Sections

As specified in ACI Sec. 22.2.2.1 and shown in Fig. 2.1, the nominal flexural strength of a member is reached when the strain in the extreme compression fiber reaches a value of 0.003. Depending on the strain in the tension steel, the section is classified as either tension-controlled or compression-controlled, and the strength-reduction factor varies from a value of 0.90 to 0.65.

ACI Table 21.2.2 defines a tension-controlled section as one in which the strain in the extreme tension steel $\epsilon_t \geq 0.005$ when the concrete reaches its ultimate strain of $\epsilon_c = 0.003$. From Fig. 2.1, for a value of $\epsilon_t = 0.005$, the neutral axis depth ratio is given by

$$\frac{c}{d} = 0.375$$

For a tension-controlled section, the following relationships are obtained from Fig. 2.1.

$$\phi = 0.90$$
$$\epsilon_t = 0.005$$
$$a = 0.375\beta_1 d$$

Figure 2.1 *Rectangular Stress Block*[3]

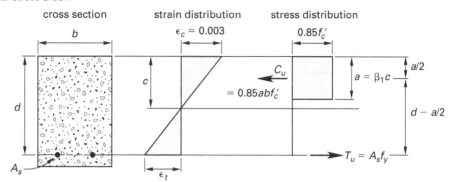

$$C_u = 0.319\beta_1 f_c' bd$$

$$\rho_t = 0.319\beta_1 \frac{f_c'}{f_y}$$

$$K_u = \omega f_c'(0.9 - 0.529\omega)$$

$$\omega = 0.319\beta_1$$

$$\beta_1 = 0.85 \quad [f_c' \le 4000 \text{ lbf/in}^2]$$

$$= 0.85 - \frac{f_c' - 4000 \frac{\text{lbf}}{\text{in}^2}}{20,000 \frac{\text{lbf}}{\text{in}^2}}$$

$$[4000 \text{ lbf/in}^2 < f_c' \le 8000 \text{ lbf/in}^2]$$

$$= 0.65 \text{ minimum} \quad [f_c' > 8000 \text{ lbf/in}^2]$$

The strength reduction factor for a tension-controlled section is given by ACI Sec. 21.2.1 as

$$\phi = 0.90$$

In a tension-controlled section at failure, the strength of the reinforcement is fully used and wide cracks and large deflections are produced, giving adequate warning of impending failure.

Table 2.1 lists various factors and their values for typical concrete strengths in tension-controlled beams where the tension strain, ϵ_t, is 0.005 and the yield strength, f_y, is 60,000 lbf/in^2.

Table 2.1 *Typical Values for Singly Reinforced Concrete Beams with $\epsilon_t = 0.005$, $f_y = 60,000$ lbf/in^2, and $\phi = 0.90$*

f_c' (lbf/in^2)	β_1	ρ_t	$\rho_{\min}$	K_u (lbf/in^2)	ω
3000	0.85	0.0136	0.0033	615	0.271
3500	0.85	0.0158	0.0033	718	0.271
4000	0.85	0.0181	0.0033	820	0.271
4500	0.83	0.0199	0.0034	906	0.265
5000	0.80	0.0213	0.0035	975	0.255

ACI Table 21.2.2 defines a compression-controlled section as that in which the strain in the extreme tension steel, $\epsilon_t \le \epsilon_{ty} = f_y/E_s$ when the concrete reaches its ultimate strain of $\epsilon_c = 0.003$. For grade 60 deformed reinforcement bars, ACI Sec. 21.2.2.1 assumes a strain limit of

$$\epsilon_{ty} = 0.002$$

The neutral axis depth ratio is given by

$$\frac{c}{d} = 0.600$$

The strength reduction factor for this condition, for members with rectangular stirrups, is given by ACI Sec. 21.2.1 as

$$\phi = 0.65$$

For sections that lie in the transition region between the tension-controlled and compression-controlled limits, the strength reduction factor is obtained from ACI Fig. R21.2.2b as

$$\phi = 0.75 + 50\epsilon_t - 0.10 \quad [\text{with spiral reinforcement}]$$

$$= 0.65 + \frac{250\epsilon_t}{3} - \frac{0.5}{3} \quad [\text{without spiral reinforcement}]$$

In accordance with ACI Fig. R21.2.2b, it is permitted to use $\phi = 0.65$ for sections in the transition stage.

Maximum Reinforcement Ratio

To ensure an under-reinforced section with ductile behavior and adequate warning of impending failure, ACI Sec. 9.3.3.1 limits the maximum reinforcement ratio in accordance with a prescribed strain distribution. For members with $P_u < 0.10 f_c' A_g$ the maximum reinforcement ratio is that which produces a tensile strain in the reinforcing bar closest to the edge of the beam, of

$$\epsilon_t = 0.004$$

From ACI Sec. 22.2.2.1, the maximum concrete compressive strain is

$$\epsilon_c = 0.003$$

The depth of the neutral axis is obtained from Fig. 2.1 as

$$c = \left(\frac{\epsilon_c}{\epsilon_c + \epsilon_t}\right)d$$
$$= \left(\frac{0.003}{0.003 + 0.004}\right)d$$
$$= 0.429d$$

The corresponding strength reduction factor is

$$\phi = 0.817$$

The maximum allowable reinforcement ratio is derived from Fig. 2.1 as

$$\rho_{\max} = 0.364\beta_1 \frac{f'_c}{f_y}$$

Minimum Reinforcement Ratio

A minimum reinforcement ratio is required to ensure that the flexural strength of the reinforced section is greater than that of the uncracked concrete section. If this is not the case, sudden failure will occur when the modulus of rupture of the concrete is exceeded and the first flexural crack forms.

In accordance with ACI Sec. 9.6.1.2, the minimum permissible reinforcement ratio is the greater of

$$\rho_{\min} = \frac{3\sqrt{f'_c}}{f_y} \quad \text{or} \quad \frac{200}{f_y}$$

The exception is that the minimum reinforcement need not exceed 33% more than that required by analysis. For slabs and footings, ACI Table 24.4.3.2 requires a minimum reinforcement area for grade 60 deformed bars of

$$A_{s,\min} = 0.0018bh$$

Analysis Procedure for a Singly Reinforced Beam or Slab

Given the section properties b, d, A_s, f_y, and f'_c, the analysis of the section consists of calculating the

- stress block depth using $a = A_s f_y / 0.85 f'_c b$

- nominal strength using $M_n = A_s f_y (d - a/2)$

- design strength using ϕM_n

Example 2.2

A reinforced concrete slab is simply supported over a span of 12 ft. The slab has a concrete compressive strength of 3000 lbf/in², and the reinforcement consists of no. 4 grade 60 bars at 11 in on center with an effective depth of 6 in. The total dead load, including the self-weight of the slab, is 120 lbf/ft².

(a) Consider a slab with a 12 in width. What is the tension reinforcement area provided?

(b) What is the depth of the rectangular stress block?

(c) What is the lever-arm of the internal resisting moment?

(d) What is the nominal flexural strength of a 12 in wide slab?

(e) What is the maximum permissible factored moment on a 12 in wide slab?

(f) What is the applied factored dead load moment in ACI Eq. 5.3.1b on a 12 in wide slab?

(g) What is the maximum permissible strength level live load moment on a 12 in wide slab?

(h) What is the maximum permissible service level live load moment on a 12 in wide slab?

(i) What is the permissible service level live load?

Solution

(a) Consider a 12 in wide slab.

The area of one no. 4 bar is 0.20 in². The reinforcement area provided in a 12 in width is

$$A_s = \frac{(0.20 \text{ in}^2)(12 \text{ in})}{11 \text{ in}} = 0.22 \text{ in}^2$$

(b) Equating the tensile and compressive forces acting on the section gives the depth of the equivalent rectangular stress block as

$$a = \frac{A_s f_y}{0.85 f'_c b} = \frac{(0.22 \text{ in}^2)\left(60{,}000 \dfrac{\text{lbf}}{\text{in}^2}\right)}{(0.85)\left(3000 \dfrac{\text{lbf}}{\text{in}^2}\right)(12 \text{ in})}$$
$$= 0.43 \text{ in}$$

(c) The lever-arm of the internal resisting moment is obtained from Fig. 2.1 as

$$d - \frac{a}{2} = 6 \text{ in} - \frac{0.43 \text{ in}}{2}$$
$$= 5.78 \text{ in}$$

(d) The nominal moment of resistance is

$$M_n = A_s f_y \left(d - \frac{a}{2} \right) = \frac{(0.22 \text{ in}^2)\left(60 \frac{\text{kips}}{\text{in}^2} \right)(5.78 \text{ in})}{12 \frac{\text{in}}{\text{ft}}}$$

$$= 6.36 \text{ ft-kips}$$

(e) The limiting reinforcement ratio for a tension-controlled section is

$$\rho_t = 0.319\beta_1 \frac{f_c'}{f_y} = (0.319)(0.85)\left(\frac{3000 \frac{\text{lbf}}{\text{in}^2}}{60,000 \frac{\text{lbf}}{\text{in}^2}} \right)$$

$$= 0.0136$$

The reinforcement ratio provided is

$$\rho = \frac{A_s}{bd}$$

$$= \frac{0.22 \text{ in}^2}{(12 \text{ in})(6 \text{ in})}$$

$$= 0.003$$

$$< \rho_t$$

The section is tension-controlled, and the strength reduction factor is

$$\phi = 0.9$$

The maximum permissible factored moment is

$$M_u = \phi M_n = (0.9)(6.36 \text{ ft-kips})$$

$$= 5.73 \text{ ft-kips}$$

(f) The applied factored dead load moment is

$$M_{uD} = \frac{1.2 w_D l^2}{8}$$

$$= \frac{(1.2)\left(0.12 \frac{\text{kip}}{\text{ft}} \right)(12 \text{ ft})^2}{8}$$

$$= 2.59 \text{ ft-kips}$$

(g) From ACI Eq. 5.3.1b, the maximum permissible strength level live load moment is

$$M_{uL} = M_u - M_{uD}$$

$$= 5.73 \text{ ft-kips} - 2.59 \text{ ft-kips}$$

$$= 3.14 \text{ ft-kips}$$

(h) The maximum permissible service level live load moment is

$$M_L = \frac{M_{uL}}{1.6} = \frac{3.14 \text{ ft-kips}}{1.6}$$

$$= 1.96 \text{ ft-kips}$$

(i) The permissible service level live load is

$$w_L = \frac{8M_L}{l^2}$$

$$= \frac{(8)(1.96 \text{ ft-kips})\left(1000 \frac{\text{lbf}}{\text{kip}} \right)}{(12 \text{ ft})^2}$$

$$= 109 \text{ lbf/ft}$$

Design Procedure for a Singly Reinforced Beam

The procedure to select a suitable section to resist a given bending moment, M_u, consists of the following steps.

step 1: Assume beam dimensions and concrete strength.

step 2: Calculate the design moment factor from

$$K_u = \frac{M_u}{bd^2}$$

step 3: Calculate the ratio

$$\frac{K_u}{f_c'}$$

step 4: Assume a tension-controlled section, since generally this is the case, and determine the reinforcement index, ω, from App. 2.A.

step 5: Determine the required reinforcement from

$$\rho = \frac{\omega f_c'}{f_y}$$

step 6: Check that the beam complies with the maximum reinforcement requirements of

$$\rho \leq 0.364\beta_1 \frac{f_c'}{f_y}$$

Increase the beam size or f_c' if necessary.

step 7: Check that the beam complies with tension-controlled reinforcement requirements of

$$\rho \leq 0.319\beta_1 \frac{f_c'}{f_y}$$

Increase the beam size or f_c' if necessary.

step 8: Check that the beam complies with minimum reinforcement requirements of ACI Sec. 9.6.1.2.

$$\rho_{\min} = \frac{3\sqrt{f_c'}}{f_y} > \frac{200}{f_y}$$

Increase the beam size or f_c' if necessary.

Example 2.3

A reinforced concrete beam with an effective depth of 16 in and a width of 12 in is reinforced with grade 60 bars and has a concrete compressive strength of 3000 lbf/in². Determine the area of tension reinforcement required if the beam supports a total factored moment of 150 ft-kips.

Solution

The design moment factor is

$$\begin{aligned} K_u = \frac{M_u}{bd^2} &= \frac{(150 \text{ ft-kips})\left(12 \frac{\text{in}}{\text{ft}}\right)\left(1000 \frac{\text{lbf}}{\text{kip}}\right)}{(12 \text{ in})(16 \text{ in})^2} \\ &= 586 \text{ lbf/in}^2 \end{aligned}$$

$$\begin{aligned} \frac{K_u}{f_c'} &= \frac{586 \frac{\text{lbf}}{\text{in}^2}}{3000 \frac{\text{lbf}}{\text{in}^2}} \\ &= 0.195 \end{aligned}$$

From App. 2.A, assuming a tension-controlled section, the corresponding tension reinforcement index is

$$\omega = 0.255$$

The required reinforcement ratio is

$$\begin{aligned} \rho = \frac{\omega f_c'}{f_y} &= \frac{(0.255)\left(3000 \frac{\text{lbf}}{\text{in}^2}\right)}{60,000 \frac{\text{lbf}}{\text{in}^2}} \\ &= 0.0128 \end{aligned}$$

The limiting reinforcement ratio for a tension-controlled section is

$$\begin{aligned} \rho_t = 0.319\beta_1 \frac{f_c'}{f_y} \\ = (0.319)(0.85)\left(\frac{3000 \frac{\text{lbf}}{\text{in}^2}}{60,000 \frac{\text{lbf}}{\text{in}^2}}\right) \\ = 0.0136 \\ > \rho \end{aligned}$$

Therefore, the section is tension-controlled.

The minimum allowable reinforcement ratio is the greater of

$$\begin{aligned} \rho_{\min} &= \frac{200 \frac{\text{lbf}}{\text{in}^2}}{f_y} \\ &= \frac{200 \frac{\text{lbf}}{\text{in}^2}}{60,000 \frac{\text{lbf}}{\text{in}^2}} \\ &= 0.0033 \\ &< \rho \quad \text{[satisfactory]} \end{aligned}$$

$$\begin{aligned} \rho_{\min} = \frac{3\sqrt{f_c'}}{f_y} &= \frac{3\sqrt{3000 \frac{\text{lbf}}{\text{in}^2}}}{60,000 \frac{\text{lbf}}{\text{in}^2}} \\ &= 0.00274 \\ &< \rho \quad \text{[satisfactory]} \end{aligned}$$

The reinforcement area required is

$$\begin{aligned} A_s = \rho bd \\ = (0.0128)(12 \text{ in})(16 \text{ in}) \\ = 2.45 \text{ in}^2 \end{aligned}$$

Beams with Compression Reinforcement

A reinforced concrete beam with compression reinforcement is shown in Fig. 2.2. Compression reinforcement and additional tension reinforcement are required when the factored moment on the member exceeds the design flexural strength of a singly reinforced member with the strain in the tension steel, $\epsilon_t = 0.005$. The residual moment is given by

$$M_r = M_u - M_{\max}$$

Figure 2.2 *Beam with Compression Reinforcement*[2]

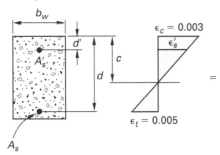 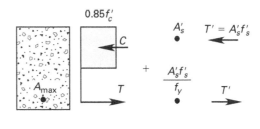

The area of compression reinforcement is

$$A_s' = \frac{M_r}{\phi f_s'(d - d')}$$

The area of additional tension reinforcement is

$$A_t = \frac{A_s' f_s'}{f_y}$$

The stress in the compression reinforcement is

$$f_s' = \left(87{,}000 \ \frac{\text{lbf}}{\text{in}^2}\right)\left(1 - \frac{d'}{c}\right)$$
$$\leq f_y$$

The neutral axis depth is

$$c = 0.375d$$

Analysis Procedure for a Beam with Compression Reinforcement

To analyze a beam with compression reinforcement, an initial estimate of the neutral axis depth is required. Assuming that the compressive strain in the concrete is 0.003, the strain and the stress in the compression and tension reinforcement may be determined. The total compressive force in the concrete and the compression reinforcement is then compared with the tensile force in the tension reinforcement. The initial estimate of the neutral axis depth is adjusted until these two values are equal. The maximum nominal moment provided by the section is obtained by taking moments of the forces in the concrete and in the compression reinforcement about the centroid of the tension reinforcement.

Design Procedure for a Beam with Compression Reinforcement

The procedure to select a suitable section to resist a given bending moment, M_u, consists of the following steps.

step 1: Assume beam dimensions and concrete strength.

step 2: Determine ρ_t for $\epsilon_t = 0.005$ from Table 2.1.

step 3: Calculate the maximum area of tension reinforcement from

$$A_{\max} = \rho_t b d$$

step 4: Determine the design moment factor, K_u, from Table 2.1.

step 5: Calculate the maximum design flexural strength from

$$M_{\max} = K_u b d^2$$

step 6: Calculate the residual moment from
$$M_r = M_u - M_{\max}$$

step 7: Determine the additional tension steel from

$$A_t = \frac{M_r}{\phi f_y(d - d')}$$

step 8: Calculate the required total area of tension reinforcement from

$$A_s = A_{\max} + A_t$$

step 9: Find the neutral axis depth for $\epsilon_t = 0.005$ from

$$c = 0.375d$$

step 10: Calculate the stress in compression reinforcement from

$$f_s' = \left(87{,}000 \ \frac{\text{lbf}}{\text{in}^2}\right)\left(1 - \frac{d'}{c}\right)$$

step 11: Calculate the required area of compression steel from

$$A_s' = \frac{A_t f_y}{f_s'}$$

Increase beam size or f_c' if necessary.

Example 2.4

A reinforced concrete beam with an effective depth of 16 in and a width of 12 in is reinforced with grade 60 bars and has a concrete compressive strength of 3000 lbf/in². The depth to the centroid of the compression reinforcement is 3 in. Determine the areas of tension and compression reinforcement required if the beam supports a total factored moment of 178 ft-kips.

Solution

From Table 2.1, the maximum allowable tension reinforcement ratio in a tension-controlled beam with a concrete strength of 3000 lbf/in² and grade 60 reinforcement bars is

$$\rho_t = 0.0136$$

The corresponding tension reinforcement area is

$$
\begin{aligned}
A_{\max} &= \rho_t b d \\
&= (0.0136)(12 \text{ in})(16 \text{ in}) \\
&= 2.611 \text{ in}^2
\end{aligned}
$$

The corresponding tension reinforcement index is

$$
\begin{aligned}
\omega &= \frac{\rho_t f_y}{f'_c} \\
&= \frac{(0.0136)\left(60 \ \dfrac{\text{kips}}{\text{in}^2}\right)}{3 \ \dfrac{\text{kips}}{\text{in}^2}} \\
&= 0.271
\end{aligned}
$$

From App. 2.A, the corresponding maximum design flexural strength is

$$
\begin{aligned}
M_{\max} &= 0.205 f'_c b d^2 \\
&= \frac{(0.205)\left(3000 \ \dfrac{\text{lbf}}{\text{in}^2}\right)(12 \text{ in})(16 \text{ in})^2}{\left(1000 \ \dfrac{\text{lbf}}{\text{kip}}\right)\left(12 \ \dfrac{\text{in}}{\text{ft}}\right)} \\
&= 157.4 \text{ ft-kips}
\end{aligned}
$$

The residual moment is

$$
\begin{aligned}
M_r &= M_u - M_{\max} \\
&= 178 \text{ ft-kips} - 157.4 \text{ ft-kips} \\
&= 20.6 \text{ ft-kips}
\end{aligned}
$$

The additional area of tension reinforcement required is

$$
\begin{aligned}
A_t &= \frac{M_r}{\phi f_y (d - d')} \\
&= \frac{(20.6 \text{ ft-kips})\left(12 \ \dfrac{\text{in}}{\text{ft}}\right)}{(0.9)\left(60 \ \dfrac{\text{kips}}{\text{in}^2}\right)(16 \text{ in} - 3 \text{ in})} \\
&= 0.352 \text{ in}^2
\end{aligned}
$$

The total required area of tension reinforcement is

$$
\begin{aligned}
A_s &= A_{\max} + A_t = 2.611 \text{ in}^2 + 0.352 \text{ in}^2 \\
&= 2.963 \text{ in}^2
\end{aligned}
$$

The neutral axis depth is

$$
\begin{aligned}
c &= 0.375 d \\
&= (0.375)(16 \text{ in}) \\
&= 6.0 \text{ in}
\end{aligned}
$$

The stress in the compression steel is

$$
\begin{aligned}
f'_s &= \left(87,000 \ \frac{\text{lbf}}{\text{in}^2}\right)\left(1 - \frac{d'}{c}\right) \\
&= \left(87,000 \ \frac{\text{lbf}}{\text{in}^2}\right)\left(1 - \frac{3 \text{ in}}{6.0 \text{ in}}\right) \\
&= 43,500 \text{ lbf/in}^2
\end{aligned}
$$

The required area of compression reinforcement is

$$
\begin{aligned}
A'_s &= \frac{A_t f_y}{f'_s} \\
&= \frac{(0.352 \text{ in}^2)\left(60 \ \dfrac{\text{kips}}{\text{in}^2}\right)}{43.50 \ \dfrac{\text{kips}}{\text{in}^2}} \\
&= 0.486 \text{ in}^2
\end{aligned}
$$

Flanged Section with Tension Reinforcement

When the rectangular stress block is wholly contained in the flange, a flanged section may be designed as a rectangular beam. The effective width of the flange is limited by ACI Table 6.3.2.1.

Figure 2.3 *Flanged Section with Tension Reinforcement*[2]

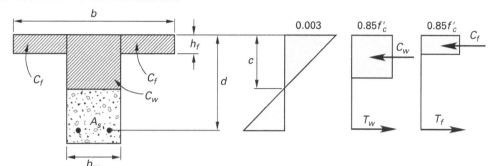

When the depth of the rectangular stress block exceeds the flange thickness, the flanged beam is designed as shown in Fig. 2.3. The area of reinforcement required to balance the compressive force in the outstanding flanges is

$$A_{sf} = \frac{0.85 f'_c h_f (b - b_w)}{f_y}$$

The corresponding design moment strength is

$$M_f = \phi A_{sf} f_y \left(d - \frac{h_f}{2} \right)$$

The beam web must develop the residual moment, which is given by

$$M_r = M_u - M_f$$

The value of $M_r / f'_c b_w d^2$ is determined. The corresponding value of ω is obtained from App. 2.A, and the additional area of reinforcement required to balance the residual moment is

$$A_{sw} = \frac{\omega b_w d f'_c}{f_y}$$

The total area of reinforcement required is

$$A_s = A_{sf} + A_{sw}$$

Example 2.5

A reinforced concrete flanged beam with a flange width of 24 in, a web width of 12 in, a flange depth of 3 in, and an effective depth of 16 in is reinforced with grade 60 reinforcement. If the concrete compressive strength is 3000 lbf/in², determine the area of tension reinforcement required to support an applied factored moment of 250 ft-kips.

Solution

Assume that the depth of the rectangular stress block exceeds the depth of the flange.

The area of tension reinforcement required to balance the compression force in the flange is

$$
\begin{aligned}
A_{sf} &= \frac{0.85 f'_c h_f (b - b_w)}{f_y} \\
&= \frac{(0.85)\left(3 \dfrac{\text{kips}}{\text{in}^2}\right)(3 \text{ in})(24 \text{ in} - 12 \text{ in})}{60 \dfrac{\text{kips}}{\text{in}^2}} \\
&= 1.53 \text{ in}^2
\end{aligned}
$$

Assuming the section is tension-controlled, the corresponding design moment strength is

$$
\begin{aligned}
M_f &= \phi A_{sf} f_y \left(d - \frac{h_f}{2} \right) \\
&= \frac{(0.9)(1.53 \text{ in}^2)\left(60 \dfrac{\text{kips}}{\text{in}^2}\right)\left(16 \text{ in} - \dfrac{3 \text{ in}}{2}\right)}{12 \dfrac{\text{in}}{\text{ft}}} \\
&= 99.83 \text{ ft-kips}
\end{aligned}
$$

The residual moment to be developed by the web is

$$
\begin{aligned}
M_r &= M_u - M_f = 250 \text{ ft-kips} - 99.83 \text{ ft-kips} \\
&= 150.17 \text{ ft-kips}
\end{aligned}
$$

$$
\begin{aligned}
\frac{M_r}{f'_c b_w d^2} &= \frac{(150.17 \text{ ft-kips})\left(12 \dfrac{\text{in}}{\text{ft}}\right)}{\left(3 \dfrac{\text{kips}}{\text{in}^2}\right)(12 \text{ in})(16 \text{ in})^2} \\
&= 0.196
\end{aligned}
$$

From App. 2.A, the corresponding tension reinforcement index is

$$\omega = 0.257$$

The reinforcement required to develop the residual moment is

$$A_{sw} = \frac{\omega b_w d f_c'}{f_y}$$

$$= \frac{(0.257)(12 \text{ in})(16 \text{ in})\left(3 \dfrac{\text{kips}}{\text{in}^2}\right)}{60 \dfrac{\text{kips}}{\text{in}^2}}$$

$$= 2.47 \text{ in}^2$$

The total tension reinforcement area required is

$$A_s = A_{sf} + A_{sw}$$

$$= 1.53 \text{ in}^2 + 2.47 \text{ in}^2$$

$$= 4.00 \text{ in}^2$$

The depth of the equivalent rectangular stress block is given by

$$a = \frac{A_{sw} f_y}{0.85 f_c' b_w}$$

$$= \frac{(2.47 \text{ in}^2)\left(60 \dfrac{\text{kips}}{\text{in}^2}\right)}{(0.85)\left(3 \dfrac{\text{kips}}{\text{in}^2}\right)(12 \text{ in})}$$

$$= 4.84 \text{ in}$$

$$> h_f \quad \text{[as assumed]}$$

For a tension-controlled section, the maximum depth of the equivalent rectangular stress block is given by

$$a_t = 0.375\beta_1 d = (0.375)(0.85)(16 \text{ in}) = 5.10 \text{ in}$$

$$> a \quad \text{[The section is tension-controlled as assumed.]}$$

Analysis of a Flanged Section

The following steps are used to analyze a flanged beam when the depth of the stress block exceeds the flange thickness.

step 1: Calculate the compressive force developed by the outstanding flanges from

$$C_f = 0.85 f_c' h_f (b - b_w)$$

step 2: Calculate the area of tension reinforcement needed to balance the compressive force from

$$A_{sf} = \frac{C_f}{f_y}$$

step 3: Calculate the corresponding design moment strength of the outstanding flanges from

$$M_f = \phi A_{sf} f_y \left(d - \frac{h_f}{2}\right)$$

step 4: Calculate the residual reinforcement area from

$$A_{sw} = A_s - A_{sf}$$

step 5: Calculate the residual reinforcement ratio referred to the web from

$$\rho_w = \frac{A_{sw}}{b_w d}$$

step 6: Calculate the design moment strength of the residual reinforcement from

$$M_w = \phi A_{sw} f_y d (1 - 0.59 \rho_w f_y / f_c')$$

step 7: Calculate the total design moment strength of the section from

$$M_u = M_f + M_w$$

Alternatively, the value of $M_w/(f_c' b_w d^2)$ may be determined from App. 2.A using the calculated value of the reinforcement index, $\omega_w = \rho_w f y / f_c'$. The value of M_w can then be determined from $M_w/(f_c' b_w d^2)$ using the known values f_c', b_w, and d.

4. SERVICEABILITY REQUIREMENTS FOR BEAMS

Nomenclature

A_b	area of individual bar	in^2
A_s	area of tension reinforcement	in^2
A_{sk}	area of skin reinforcement per unit height in one side face	in^2
A_{ts}	area of nonprestressed reinforcement in a tie	in^2
c_c	clear cover to tension reinforcement	in
d_b	diameter of bar	in
E_c	modulus of elasticity of concrete, $33 w_c^{1.5}\sqrt{f_c'}$	lbf/in^2
E_s	modulus of elasticity of reinforcement, 29,000	kips/in^2
f_r	modulus of rupture of concrete, $7.5\sqrt{f_c'}$	lbf/in^2
f_s	calculated stress in reinforcement at service loads	kips/in^2
h	overall dimension of member	in
I_{cr}	moment of inertia of cracked transformed section, $b_w(kd)^3/3 + nA_s(d - kd)^2$	in^4
I_e	effective moment of inertia	in^4

I_g	moment of inertia of gross concrete section, $b_w h^3/12$	in^4
k	neutral axis depth factor at service load, $\sqrt{2\rho n + (\rho n)^2} - \rho n$	–
l	span length of beam or one-way slab, projection of cantilever	ft
M_a	maximum moment in member at stage deflection is required	in-lbf or ft-kips
M_{cr}	cracking moment, $2f_r I_g/h$	in-lbf or ft-kips
n	modular ratio, E_s/E_c	–
s	center-to-center spacing of tension reinforcement	in
w_c	unit weight of concrete	lbf/ft^3

Symbols

δ	deflection	in
ξ	time-dependent factor for sustained load	–
λ_Δ	multiplier for additional long-time deflection	–
ρ'	reinforcement ratio for compression reinforcement, A_s'/bd	–

Control of Crack Widths

Limitations on reinforcement distribution are specified to control cracking and to protect the reinforcement from corrosion. Crack width and reinforcement corrosion increase when reinforcement stress increases, concrete cover increases, or the area of concrete surrounding each bar increases.

In accordance with ACI Sec. 24.3, crack width is controlled by limiting the spacing of tension reinforcement to a value given by ACI Table 24.3.2, where f_s is in units of kips/in^2.

$$s = \frac{600}{f_s} - 2.5c_c$$
$$\leq \frac{480}{f_s}$$

As shown in Fig. 2.4, s is the center-to-center spacing, in inches, of the tension reinforcement nearest to the extreme tension face, and c_c is the clear concrete cover, in inches, from the nearest surface in tension to the surface of the tension reinforcement. Where there is only one bar nearest to the extreme tension face, s is taken as the width of the extreme tension face. Controlling the spacing of tension reinforcement limits the width of surface cracks to an acceptable level.

Figure 2.4 *Tension Reinforcement Details*

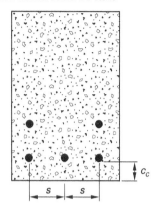

The stress in the reinforcement at service load may be either calculated or assumed to be equal to $\frac{2}{3}f_y$.

When the depth of the beam exceeds 36 in, ACI Sec. 9.7.2.3 requires that skin reinforcement be placed along both side faces of the web, in the lower half of the beam.

Example 2.6

The beam shown is reinforced with eight no. 9 grade 60 bars. Clear cover of $1\frac{1}{2}$ in is provided to the no. 4 stirrups. Determine the skin reinforcement required, and check that the spacing of the reinforcement conforms to ACI Sec. 24.3.

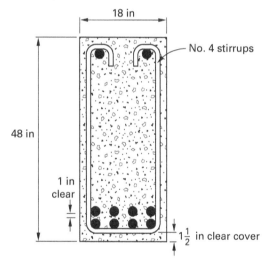

Solution

The clear cover provided to the tension reinforcement is given by

$$c_c = 1.5 \text{ in} + 0.5 \text{ in}$$
$$= 2 \text{ in}$$

The stress in the reinforcement at service load is assumed equal to

$$f_s = \frac{2}{3}f_y = \left(\frac{2}{3}\right)\left(60 \; \frac{\text{kips}}{\text{in}^2}\right)$$
$$= 40 \; \text{kips/in}^2$$

The maximum allowable bar spacing is given by ACI Table 24.3.2 as

$$s = \frac{600 \; \dfrac{\text{kips}}{\text{in}^2}}{f_s} - 2.5c_c$$
$$= \frac{600 \; \dfrac{\text{kips}}{\text{in}^2}}{40 \; \dfrac{\text{kips}}{\text{in}^2}} - (2.5)(2 \; \text{in})$$
$$= 10 \; \text{in}$$

The actual bar spacing is given by

$$s' = \frac{18 \; \text{in} - (2)(1.5 \; \text{in}) - (2)(0.5 \; \text{in}) - 1.128 \; \text{in}}{3}$$
$$= 4.29 \; \text{in}$$

Since 4.29 in is less than 10 in, this bar spacing is satisfactory.

The depth of the beam is

$$h = 48 \; \text{in} > 36 \; \text{in}$$

Therefore, skin reinforcement is required.

Using no. 3 bars, the maximum allowable spacing is

$$s_{sk} = 10 \; \text{in}$$

The bars must extend for a distance, $h/2$, from the tension face.

The reinforcement layout is shown.

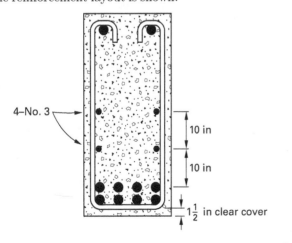

Deflection Limitations

Two methods are given in ACI 318 to control deflections. One method is to provide a minimum overall thickness for beams or slabs. The other method is to calculate the beam and slab deflections and compare them to permissible values.

Calculation of the deflections is not required if the limiting thickness requirements for beams and slabs are met.

The allowable, immediate deflection of flexural members supporting nonsensitive elements is specified in ACI Table 24.2.2 as $l/180$ for flat roofs due to L_r, S, and R and $l/360$ for floors due to L. The total deflection occurring after the attachment of nonsensitive elements is limited to $l/240$. The total deflection occurring after the attachment of deflection sensitive elements is limited to $l/480$.

For normal weight concrete and grade 60 reinforcement, ACI Table 7.3.1.1 and Table 9.3.1.1 provide span/depth ratios applicable to members supporting nonsensitive elements. These ratios are shown in Table 2.2.

Table 2.2 *Span/Depth Ratios*

end conditions	beam	slab
simply supported	$\dfrac{l}{16}$	$\dfrac{l}{20}$
one end continuous	$\dfrac{l}{18.5}$	$\dfrac{l}{24}$
both ends continuous	$\dfrac{l}{21}$	$\dfrac{l}{28}$
cantilever	$\dfrac{l}{8}$	$\dfrac{l}{10}$

For grade 40 reinforcement, the tabulated values are multiplied by the factor 0.8. For lightweight concrete, the tabulated values are multiplied by the greater of

$$R = 1.65 - 0.005w_c$$

Or,

$$R = 1.09$$

Deflection Determination

Short-term deflections may be calculated by using the effective moment of inertia given by ACI Sec. 24.2.3.5 and illustrated in Fig. 2.5 as

$$I_e = \left(\frac{M_{cr}}{M_a}\right)^3 I_g + \left(1 - \left(\frac{M_{cr}}{M_a}\right)^3\right)I_{cr} \quad \text{[ACI 24.2.3.5a]}$$

Figure 2.5 *Service Load Conditions*

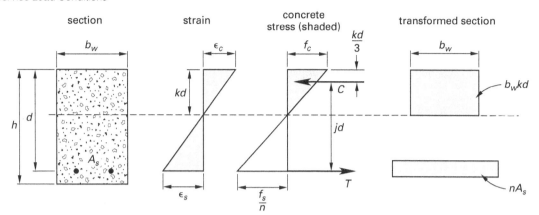

Additional long-term deflection is estimated from ACI Sec. 24.2.4.1.1 by multiplying the short-term deflection by the multiplier

$$\lambda_\Delta = \frac{\xi}{1 + 50\rho'} \quad \text{[ACI 24.2.4.1.1]}$$

ξ is the time-dependent factor for sustained load defined in ACI Sec. 24.2.4.1.3 and shown in Table 2.3.

Table 2.3 *Value of* ξ

time period (mo)	ξ
60	2.0
12	1.4
6	1.2
3	1.0

The deflection is calculated for each loading case using the appropriate value of the effective moment of inertia. Thus, the short-term deflection, δ_D, may be calculated for dead load only and the short-term deflection, $\delta_{(D+L)}$, may be calculated for the total applied load. The live load deflection is then given by

$$\delta_L = \delta_{(D+L)} - \delta_D$$

The final total deflection, including additional long-term deflection, is given by

$$\delta_T = \delta_D(1 + \lambda_\Delta) + \delta_L = \delta_{(D+L)} + \lambda_\Delta\delta_D$$

Example 2.7

A reinforced concrete beam of normal weight concrete and spanning 12 ft has an effective depth of 16 in, an overall depth of 18 in, and a compressive strength of 3000 lbf/in², and it is reinforced with three no. 8 grade 60 bars. The beam is 12 in wide. The bending moment due to sustained dead load is 60 ft-kips. Neglect the weight of the nondeflection sensitive elements, which are attached immediately after removing the falsework.

The transient floor live load moment is 30 ft-kips. Compare the beam deflections with the allowable values and determine the final beam deflection due to long-term effects and transient loads. The concrete weight is 150 lbf/ft³.

Solution

The allowable live load deflection for floors is given by ACI Table 24.2.2 as

$$\delta_L = \frac{l}{360} = \frac{(12 \text{ ft})\left(12 \frac{\text{in}}{\text{ft}}\right)}{360} = 0.40 \text{ in}$$

The allowable deflection after attachment of nonsensitive elements is

$$\delta_{(\lambda D + L)} = \frac{l}{240} = \frac{(12 \text{ ft})\left(12 \frac{\text{in}}{\text{ft}}\right)}{240} = 0.60 \text{ in}$$

From ACI Sec. 19.2.2.1 for $w_c = 150$ lbf/ft³,

$$E_c = 33\sqrt{w_c^3 f_c'}$$

$$= \frac{33\sqrt{\left(150 \frac{\text{lbf}}{\text{ft}^3}\right)^3\left(3000 \frac{\text{lbf}}{\text{in}^2}\right)}}{1000 \frac{\text{lbf}}{\text{kip}}}$$

$$= 3320 \text{ kips/in}^2$$

$$E_s = 29{,}000 \text{ kips/in}^2 \quad \text{[from ACI Sec. 20.2.2.2]}$$

$$n = \frac{E_s}{E_c} = \frac{29{,}000 \frac{\text{kips}}{\text{in}^2}}{3320 \frac{\text{kips}}{\text{in}^2}} = 8.73$$

$$\rho = \frac{A_s}{b_w d} = \frac{2.37 \text{ in}^2}{(12 \text{ in})(16 \text{ in})} = 0.0123$$

$$\rho n = (0.0123)(8.73) = 0.108$$

From App. 2.B, the corresponding neutral axis depth factor is

$$k = 0.3691$$

The moment of inertia of the cracked transformed section is

$$
\begin{aligned}
I_{cr} &= \frac{b_w(kd)^3}{3} + nA_s(d - kd)^2 \\
&= \frac{(12 \text{ in})\big((0.3691)(16 \text{ in})\big)^3}{3} \\
&\quad + (8.73)(2.37 \text{ in}^2)\big(16 \text{ in} - (0.3691)(16 \text{ in})\big)^2 \\
&= 2932 \text{ in}^4
\end{aligned}
$$

$$I_g = \frac{(12 \text{ in})(18 \text{ in})^3}{12} = 5832 \text{ in}^4$$

The modulus of rupture, where $\lambda = 1.0$ for normal weight concrete (ACI Sec. 19.2.4.2), is given by ACI Eq. 19.2.3.1 as

$$f_r = 7.5\lambda\sqrt{f'_c} = (7.5)(1.0)\sqrt{3000 \frac{\text{lbf}}{\text{in}^2}} = 411 \text{ lbf/in}^2$$

The cracking moment is given by ACI Eq. 24.2.3.5b as

$$
\begin{aligned}
M_{cr} &= \frac{2f_r I_g}{h} = \frac{(2)\left(411 \frac{\text{lbf}}{\text{in}^2}\right)(5832 \text{ in}^4)}{(18 \text{ in})\left(12 \frac{\text{in}}{\text{ft}}\right)\left(1000 \frac{\text{lbf}}{\text{kip}}\right)} \\
&= 22.18 \text{ ft-kips} \\
&< 60 \text{ ft-kips} \quad [\text{Section is cracked.}]
\end{aligned}
$$

The effective moment of inertia for the dead load bending moment is given by ACI Eq. 24.2.3.5a as

$$
\begin{aligned}
I_e &= \left(\frac{M_{cr}}{M_D}\right)^3 I_g + \left(1 - \left(\frac{M_{cr}}{M_D}\right)^3\right) I_{cr} \\
&= \left(\frac{22.18 \text{ ft-kips}}{60 \text{ ft-kips}}\right)^3 (5832 \text{ in}^4) \\
&\quad + \left(1 - \left(\frac{22.18 \text{ ft-kips}}{60 \text{ ft-kips}}\right)^3\right)(2932 \text{ in}^4) \\
&= 3078 \text{ in}^4
\end{aligned}
$$

The corresponding short-term deflection due to dead load is

$$
\begin{aligned}
\delta_D &= \frac{180 M_D L^2}{E_c I_e} \\
&= \frac{(180)(60 \text{ ft-kips})(12 \text{ ft})^2}{\left(3320 \frac{\text{kips}}{\text{in}^2}\right)(3078 \text{ in}^4)} \\
&= 0.152 \text{ in}
\end{aligned}
$$

The effective moment of inertia for the dead load plus live load is

$$
\begin{aligned}
I_e &= \left(\frac{M_{cr}}{M_{(D+L)}}\right)^3 I_g + \left(1 - \left(\frac{M_{cr}}{M_{(D+L)}}\right)^3\right) I_{cr} \\
&= \left(\frac{22.18 \text{ ft-kips}}{90 \text{ ft-kips}}\right)^3 (5832 \text{ in}^4) \\
&\quad + \left(1 - \left(\frac{22.18 \text{ ft-kips}}{90 \text{ ft-kips}}\right)^3\right)(2932 \text{ in}^4) \\
&= 2975 \text{ in}^4
\end{aligned}
$$

The corresponding short-term deflection due to the dead load plus live load is

$$\delta_{(D+L)} = \frac{(180)(90 \text{ ft-kips})(12 \text{ ft})^2}{\left(3320 \frac{\text{kips}}{\text{in}^2}\right)(2975 \text{ in}^4)} = 0.236 \text{ in}$$

The short-term deflection due to transient live load is

$$
\begin{aligned}
\delta_L &= \delta_{(D+L)} - \delta_D = 0.236 \text{ in} - 0.152 \text{ in} \\
&= 0.084 \text{ in} \\
&< 0.40 \text{ in} \quad [\text{satisfactory}]
\end{aligned}
$$

The multiplier for additional long-term deflection is given by ACI Eq. 24.2.4.1.1 as

$$\lambda_\Delta = \frac{\xi}{1 + 50\rho'} = \frac{2}{1 + 0} = 2$$

The deflection due to short-term live loads and long-term dead loads is

$$
\begin{aligned}
\delta_{(\lambda D + L)} &= \lambda_\Delta \delta_D + \delta_L \\
&= (2)(0.152 \text{ in}) + 0.084 \text{ in} \\
&= 0.388 \text{ in} \\
&< 0.60 \text{ in} \quad [\text{satisfactory}]
\end{aligned}
$$

The final deflection due to long-term and short-term effects is

$$\begin{aligned}\delta_T &= \delta_D(1 + \lambda_\Delta) + \delta_L \\ &= (0.152 \text{ in})(1 + 2) + 0.084 \text{ in} \\ &= 0.540 \text{ in}\end{aligned}$$

5. ELASTIC DESIGN METHOD

Nomenclature

f_c	actual stress in concrete	lbf/in^2
f_s	actual tensile stress in reinforcement	lbf/in^2
j	lever-arm factor	–
j_{bal}	balanced lever-arm factor	–
k_{bal}	balanced neutral axis depth factor	–
M	service design moment	ft-kips
M_{bal}	balanced service design moment	ft-kips
p_{cb}	permissible concrete stress	lbf/in^2
p_{st}	permissible steel stress	lbf/in^2

Symbols

ρ_{bal}	balanced tension reinforcement ratio, $p_{cb}k_{\text{bal}}/2p_{st}$	–

Determination of Working Stress Values

The elastic design method is the procedure previously used in the ACI code. It may still be used to determine the stresses in a member as required by ACI Sec. 24.3.2. The straight-line theory, illustrated in Fig. 2.5, is used to calculate the stresses in a member under the action of the applied service loads and to ensure that these stresses do not exceed permissible values. The permissible stresses are

$$\begin{aligned}p_{cb} &= \text{maximum permissible stress in the concrete} \\ &= 0.45f_c'\end{aligned}$$

$$\begin{aligned}p_{st} &= \text{maximum permissible stress} \\ &\quad \text{in the reinforcement} \\ &= 20 \text{ kips/in}^2 \quad \text{[grade 40 reinforcement]} \\ &= 24 \text{ kips/in}^2 \quad \text{[grade 60 reinforcement]}\end{aligned}$$

From Fig. 2.5, the neutral axis depth factor is derived as

$$k = \sqrt{2\rho n + (\rho n)^2} - \rho n$$

Appendix 2.B tabulates values of k against ρn. In addition, the lever-arm factor is derived as

$$j = 1 - \frac{k}{3}$$

The stress in the reinforcement due to an applied service moment, M, is

$$f_s = \frac{M}{A_s j d}$$

The stress in the concrete is

$$f_c = \frac{2M}{jkb_w d^2}$$

For a balanced design, the stress in the reinforcement and the maximum stress in the concrete should simultaneously reach their permissible values. Then, the corresponding design values will be

$$k_{\text{bal}} = \frac{np_{cb}}{p_{st} + np_{cb}}$$

$$j_{\text{bal}} = 1 - \frac{k_{\text{bal}}}{3}$$

$$\rho_{\text{bal}} = \frac{p_{cb}k_{\text{bal}}}{2p_{st}}$$

$$M_{\text{bal}} = A_{s(\text{bal})}p_{st}j_{\text{bal}}d$$

Example 2.8

A reinforced concrete beam with an effective depth of 16 in and a width of 12 in is reinforced with grade 60 bars and has a concrete cylinder strength of 3000 lbf/in^2. Using the elastic design method, determine the area of tension reinforcement required if the beam supports a total service moment of 50 ft-kips.

Solution

From Ex. 2.7, the modular ratio is given as

$$n = 8.73$$

The permissible concrete and reinforcement stresses are

$$p_{cb} = 1350 \text{ lbf/in}^2$$

$$p_{st} = 24{,}000 \text{ lbf/in}^2$$

Reinforced Concrete

Sufficient accuracy is obtained by assuming that the neutral axis depth factor equals the balanced value. Then,

$$k_{bal} = \frac{np_{cb}}{p_{st} + np_{cb}}$$

$$= \frac{(8.73)\left(1350 \; \frac{lbf}{in^2}\right)}{24{,}000 \; \frac{lbf}{in^2} + (8.73)\left(1350 \; \frac{lbf}{in^2}\right)}$$

$$= 0.329$$

$$j_{bal} = 1 - \frac{k_{bal}}{3} = 1 - \frac{0.329}{3} = 0.89$$

$$A_{s(bal)} = \frac{M_{bal}}{p_{st}j_{bal}d}$$

$$= \frac{(50 \; \text{ft-kips})\left(12 \; \frac{in}{ft}\right)}{\left(24 \; \frac{kips}{in^2}\right)(0.89)(16 \; in)}$$

$$= 1.76 \; in^2$$

$$f_c = \frac{2A_s p_{st}}{b_w kd}$$

$$= \frac{(2)(1.76 \; in^2)\left(24{,}000 \; \frac{lbf}{in^2}\right)}{(12 \; in)(0.329)(16 \; in)}$$

$$= 1334 \; lbf/in^2$$

$$< p_{cb} \quad \text{[satisfactory]}$$

6. BEAMS IN SHEAR

Nomenclature

a_v	shear span, distance between concentrated load and face of supports	ft
A_{cs}	effective cross-sectional area of a strut in a strut-and-tie model taken perpendicular to the axis of the strut	in^2
A_{cp}	area enclosed by outside perimeter of concrete cross section	in^2
A_f	area of reinforcement in bracket or corbel resisting factored moment	in^2
A_h	area of shear reinforcement parallel to flexural tension reinforcement	in^2
A_l	total area of longitudinal reinforcement to resist torsion	in^2
A_n	area of reinforcement in bracket or corbel resisting tensile force N_{uc}	in^2
A_{nz}	effective cross-sectional area of the face of a nodal zone	in^2

A_o	gross area enclosed by shear flow, $0.85A_{oh}$	in^2
A_{oh}	area enclosed by centerline of the outermost closed transverse torsional reinforcement	in^2
A_s	area of nonprestressed tension reinforcement	in^2
A_t	area of one leg of a closed stirrup resisting torsion within a distance s	in^2
A_v	area of shear reinforcement perpendicular to flexural tension reinforcement	in^2
A_{vf}	area of shear friction reinforcement	in^2
A_{vh}	area of shear reinforcement parallel to flexural tension	in^2
b	width of a deep beam	in
b	width of compression face of member	in
b_w	web width or diameter of circular section	in
C	compressive force acting on a nodal zone	kips
f_{ce}	effective compressive strength of concrete in a strut or node	lbf/in^2
f_y	yield strength of reinforcement	lbf/in^2
f_{yt}	yield strength of transverse reinforcement	lbf/in^2
h	overall thickness of member	in
l_a	anchorage length of a reinforcing bar	in
l_b	width of bearing plate	in
l_n	clear span measured face-to-face of supports	ft
M_u	factored moment at section	ft-kips
N_{uc}	factored tensile force applied at top of corbel	kips
p_{cp}	outside perimeter of the concrete cross section	in
p_h	perimeter of centerline of outermost closed transverse torsional reinforcement	in
R	support reaction acting on a nodal zone	kips
s	spacing of shear or torsion reinforcement in direction parallel to longitudinal reinforcement	in
s_2	spacing of horizontal reinforcement	in
T	tension force acting on a nodal zone	kips
T_n	nominal torsional moment strength	ft-kips
T_u	factored torsional moment at section	ft-kips
V_c	nominal shear strength provided by concrete	kips
V_s	nominal shear strength provided by shear reinforcement	kips
V_u	factored shear force at section	kips
w_s	effective width of strut perpendicular to the axis of the strut	in
w_t	effective width of concrete concentric with a tie	in

Symbols

α angle between inclined stirrups and   degree
longitudinal axis of member

β_n factor to account for the effect of the   –
anchorage of ties on the effective
compressive strength of a nodal zone

β_s factor to account for the effect of   –
cracking and confining reinforcement on
the effective compressive strength of the
concrete in a strut

λ correction factor related to unit weight   –
of concrete, as given in ACI Table
19.2.4.2

μ coefficient of friction   –

ρ reinforcement ratio, $A_s/b_w d$ –

Design for Shear

When the support reaction produces a compressive stress in the member, as shown in Fig. 2.6, the critical section for shear is located at a distance, d, from the support equal to the effective depth. This location is applicable only when loads are applied near or at the top of the beam, and no concentrated load occurs within a distance from the support equal to the effective depth. When the applied factored shear force, V_u, exceeds the shear capacity of the concrete, ϕV_c, shear reinforcement, with a capacity ϕV_s, is added to the section to give a combined shear capacity of

$$\phi V_c + \phi V_s > V_u \quad [\phi = 0.75]$$

When V_u is less than $\phi V_c/2$, the concrete section is adequate to carry the shear without any shear reinforcement. Within the range $\phi V_c/2 \leq V_u \leq \phi V_c$, a minimum area of shear reinforcement is specified by ACI Table 9.6.3.3 as the greater of

$$A_{v(\min)} = \frac{0.75 b_w s \sqrt{f_c'}}{f_{yt}}$$

$$A_{v(\min)} = \frac{50 b_w s}{f_{yt}}$$

When $f_c' > 4.44$ kips/in^2, the first expression governs.

Exceptions to this requirement are given in ACI Table 9.6.3.1.

Shear Capacity of Concrete

The nominal shear capacity of the concrete section is given by ACI Sec. 22.5.5.1 as

$$V_c = 2 b_w d \lambda \sqrt{f_c'} \quad \text{[ACI 22.5.5.1]}$$

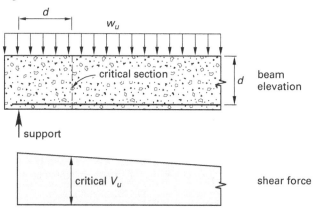

Figure 2.6 *Critical Section for Shear*

This value is conservative and is usually sufficiently accurate. A more precise value is provided by ACI Table 22.5.5.1 as the least of

$$V_c = \left(1.9\lambda\sqrt{f_c'} + \frac{2500\rho_w V_u d}{M_u}\right) b_w d$$

$$= (1.9\lambda\sqrt{f_c'} + 2500\rho_w)\, d b_w$$

$$= 3.5\lambda\sqrt{f_c'}\, d b_w$$

$$\frac{V_u d}{M_u} \leq 1.0$$

$$\sqrt{f_c'} \leq 100 \text{ lbf/in}^2 \begin{bmatrix} \text{unless web reinforcement} \\ \text{provided is not less than that} \\ \text{specified in ACI Table 9.6.3.3} \end{bmatrix}$$

M_u is the factored moment occurring simultaneously with V_u at the section being analyzed.

$\lambda = 1.0$ [for normal weight concrete]

$\lambda = 0.85$ to 1.0 [coarse blend sand lightweight (a)]

$\lambda = 0.75$ to 0.85 [lightweight, fine blend (b)]

Note:

(a)   based on the absolute volume of normal weight coarse aggregate as a fraction of the absolute volume of coarse aggregate.

(b)   based on the absolute volume of normal weight fine aggregate as a fraction of the total absolute volume of fine aggregate.

Shear Capacity of Stirrups

The nominal shear capacity of the inclined stirrups shown in Fig. 2.7 is given by ACI Sec. 22.5.10.5.4 as

$$V_s = \frac{A_v f_{yt}(\sin\alpha + \cos\alpha)d}{s} \quad \text{[ACI 22.5.10.5.4]}$$

In this equation, the minimum value of α is 45°.

Figure 2.7 *Beam with Inclined Stirrups*

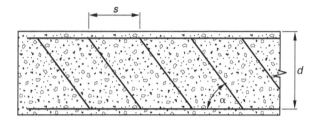

When the shear reinforcement is vertical, ACI Sec. 22.5.10.5.3 gives the nominal shear capacity as

$$V_s = \frac{A_v f_{yt} d}{s} \qquad \text{[ACI 22.5.10.5.3]}$$

The nominal shear strength of the shear reinforcement is limited by ACI Sec. 22.5.1.2 to a value of

$$V_s = 8\sqrt{f_c'}\, b_w d = 4V_c \quad \text{[for } \lambda = 1.0]$$

If additional shear capacity is required, the size of the concrete section must be increased.

The spacing of the stirrups is limited to a maximum value of $d/2$ or 24 in when

$$V_s \leq 4\sqrt{f_c'}\, b_w d = 2V_c \quad \text{[for } \lambda = 1.0]$$

The spacing of the stirrups is limited to a maximum value of $d/4$ or 12 in when

$$V_s > 4\sqrt{f_c'}\, b_w d$$

Example 2.9

A reinforced concrete beam of normal weight concrete with an effective depth of 16 in and a width of 12 in is reinforced with grade 60 bars and has a concrete compressive strength of 3000 lbf/in². Determine the shear reinforcement required when

(a)   the factored shear force $V_u = 9$ kips, the factored moment $M_u = 20$ ft-kips, and the reinforcement ratio $\rho_w = 0.015$

(b)   the factored shear force $V_u = 14$ kips

(c)   the factored shear force $V_u = 44$ kips

(d)   the factored shear force $V_u = 71$ kips

(e)   the factored shear force $V_u = 120$ kips

Solution

(a) The shear strength provided by the concrete is given by ACI Eq. 22.5.5.1 as

$$\phi V_c = 2\phi b_w d\lambda\sqrt{f_c'}$$

$$= \frac{(2)(0.75)(12\text{ in})(16\text{ in})(1.0)\sqrt{3000\, \dfrac{\text{lbf}}{\text{in}^2}}}{1000\, \dfrac{\text{lbf}}{\text{kip}}}$$

$$= 15.8 \text{ kips} < 2V_u$$

Using ACI Sec. 22.5.5.1 to verify that the section is adequate

$$\frac{V_u d}{M_u} = \frac{(9\text{ kips})(16\text{ in})}{(20\text{ ft-kips})\left(12\, \dfrac{\text{in}}{\text{ft}}\right)}$$

$$= 0.60 < 1.0 \quad \text{[satisfactory]}$$

$$\phi V_c = \phi\left(1.9\lambda\sqrt{f_c'} + \frac{2500\rho_w V_u d}{M_u}\right)b_w d$$

$$= \frac{(0.75)\left[\begin{array}{c}(1.9)(1.0)\sqrt{3000\, \dfrac{\text{lbf}}{\text{in}^2}} \\[2mm] + \dfrac{(2500)(0.015)(9\text{ kips})(16\text{ in})}{(20\text{ ft-kips})\left(12\, \dfrac{\text{in}}{\text{ft}}\right)}\end{array}\right] \times (12\text{ in})(16\text{ in})}{1000\, \dfrac{\text{lbf}}{\text{kip}}}$$

$$= 18.2 \text{ kips} > 2V_u$$

In accordance with ACI Sec. 9.6.3.1, shear reinforcement is not required.

(b) Because $\phi V_c/2 < V_u < \phi V_c$, and the conditions of ACI Table 9.6.3.1 are not met, the minimum shear reinforcement specified by ACI Sec. 9.6.3.3 is required, and this is given by

$$\frac{A_{v(\text{min})}}{s} = \frac{50 b_w}{f_y} \quad \text{[governs]}$$

$$= \frac{\left(50\, \dfrac{\text{lbf}}{\text{in}^2}\right)(12\text{ in})\left(12\, \dfrac{\text{in}}{\text{ft}}\right)}{60{,}000\, \dfrac{\text{lbf}}{\text{in}^2}}$$

$$= 0.12 \text{ in}^2/\text{ft}$$

Shear reinforcement consisting of two arms of no. 3 bars at 8 in spacing provides a reinforcement area of

$$\frac{A_v}{s} = 0.33 \text{ in}^2/\text{ft}$$

The spacing of 8 in does not exceed $d/2$ and is satisfactory.

(c) The factored shear force exceeds the shear strength of the concrete, and the shear strength required from shear reinforcement is given by ACI Eq. 22.5.1.1 as

$$\begin{aligned} \phi V_s &= V_u - \phi V_c = 44 \text{ kips} - 15.8 \text{ kips} \\ &= 28.2 \text{ kips} \\ &< 2\phi V_c \end{aligned}$$

In accordance with ACI Sec. 9.7.6.2.2, stirrups are required at a maximum spacing of $d/2 = 8$ in. The area of shear reinforcement required is given by ACI Eq. 22.5.10.5.3 as

$$\begin{aligned} \frac{A_v}{s} = \frac{\phi V_s}{\phi d f_{yt}} &= \frac{(28.2 \text{ kips})\left(12 \dfrac{\text{in}}{\text{ft}}\right)}{(0.75)(16 \text{ in})\left(60 \dfrac{\text{kips}}{\text{in}^2}\right)} \\ &= 0.47 \text{ in}^2/\text{ft} \end{aligned}$$

Shear reinforcement consisting of two arms of no. 4 bars at 8 in spacing provides a reinforcement area of

$$\begin{aligned} \frac{A_v}{s} &= 0.60 \text{ in}^2/\text{ft} \\ &> 0.47 \text{ in}^2/\text{ft} \quad \text{[satisfactory]} \end{aligned}$$

(d) The shear strength required from the shear reinforcement is given by ACI Eq. 22.5.1.1 as

$$\begin{aligned} \phi V_s &= V_u - \phi V_c \\ &= 71 \text{ kips} - 15.8 \text{ kips} \\ &= 55.2 \text{ kips} \\ &> 2\phi V_c \end{aligned}$$

In accordance with ACI Sec. 9.7.6.2.2, stirrups are required at a maximum spacing of $d/4 = 4$ in. The area of shear reinforcement required is given by ACI Eq. 22.5.10.5.3 as

$$\begin{aligned} \frac{A_v}{s} = \frac{\phi V_s}{\phi d f_{yt}} &- \frac{(55.2 \text{ kips})\left(12 \dfrac{\text{in}}{\text{ft}}\right)}{(0.75)(16 \text{ in})\left(60 \dfrac{\text{kips}}{\text{in}^2}\right)} \\ &= 0.92 \text{ in}^2/\text{ft} \end{aligned}$$

Shear reinforcement consisting of two arms of no. 4 bars at 4 in spacing provides a reinforcement area of

$$\begin{aligned} \frac{A_v}{s} &= 1.2 \text{ in}^2/\text{ft} \\ &> 0.92 \quad \text{[satisfactory]} \end{aligned}$$

(e) The shear strength required from the shear reinforcement is given by ACI Eq. 22.5.1.1 as

$$\begin{aligned} \phi V_s &= V_u - \phi V_c \\ &= 120 \text{ kips} - 15.8 \text{ kips} \\ &= 104.2 \text{ kips} \\ &> 4\phi V_c \end{aligned}$$

In accordance with ACI Sec. 22.5.1.2, the section size is inadequate.

Shear Capacity of Inclined Bars

When a single, bent-up bar or group of bars equidistant from the support is used as shear reinforcement, the nominal shear capacity is given by ACI Sec. 22.5.10.6.2a as

$$V_s = A_v f_{yt} \sin \alpha \quad \text{[ACI 22.5.10.6.2a]}$$
$$\le 3b_w d \sqrt{f_c'}$$

When a series of equally spaced bent-up bars is used, as shown in Fig. 2.8, the nominal shear capacity is given by ACI Sec. 22.5.10.5.4 as

$$V_s = \frac{A_v f_y (\sin \alpha + \cos \alpha) d}{s} \quad \text{[ACI 22.5.10.5.4]}$$

Figure 2.8 *Beam with Inclined Bars*

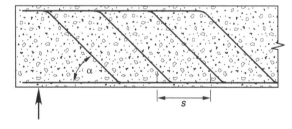

Only the center three-fourths of the inclined bar is considered effective; this limits the spacing, measured in a direction parallel to the longitudinal reinforcement, to a maximum value of

$$s_{\max} = 0.375 d (1 + \cot \alpha)$$

This value is halved, in accordance with ACI Sec. 9.7.6.2.2, when V_s exceeds $4\sqrt{f_c'}\, b_w d$.

Illustration for Ex. 2.10

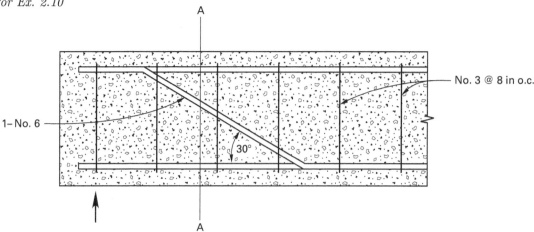

In accordance with ACI Sec. 22.5.10.6.1, the minimum permitted angle of inclination of the inclined bars is 30°. When shear reinforcement consists of both stirrups and inclined bars, the total combined shear resistance is given by the sum of the shear resistances of each type. The nominal combined shear resistance must not exceed $8\sqrt{f_c'}\,b_w d$.

Example 2.10

The reinforced concrete beam of normal weight concrete shown in *Illustration for Ex. 2.10* has an effective depth of 16 in, a width of 12 in, a concrete compressive strength of 3000 lbf/in², and is reinforced with grade 60 bars. Determine the design shear capacity provided at section A-A. Each U-stirrup has two vertical legs.

Solution

The nominal shear strength of the concrete is given by ACI Eq. 22.5.5.1 as

$$V_c = 2b_w d\lambda\sqrt{f_c'}$$

$$= \frac{(2)(12 \text{ in})(16 \text{ in})(1.0)\sqrt{3000 \dfrac{\text{lbf}}{\text{in}^2}}}{1000 \dfrac{\text{lbf}}{\text{kip}}}$$

$$= 21.0 \text{ kips}$$

The nominal shear strength of the vertical stirrups is given by ACI Eq. 22.5.10.5.3 as

$$V_{s(\text{str})} = \frac{A_v f_{yt} d}{s} = \frac{\left(0.33 \dfrac{\text{in}^2}{\text{ft}}\right)\left(60 \dfrac{\text{kips}}{\text{in}^2}\right)(16 \text{ in})}{12 \dfrac{\text{in}}{\text{ft}}}$$

$$= 26.4 \text{ kips}$$

The nominal shear strength of the inclined bar is given by ACI Eq. 22.5.10.6.2b as

$$V_{s(\text{bar})} = A_v f_y \sin\alpha = (0.44 \text{ in}^2)\left(60 \dfrac{\text{kips}}{\text{in}^2}\right)(\sin 30°)$$

$$= 13.2 \text{ kips}$$

The combined shear strength of the shear reinforcement is

$$V_s = V_{s(\text{str})} + V_{s(\text{bar})}$$

$$= 39.6 \text{ kips}$$

$$< 2V_c$$

In accordance with ACI Sec. 9.7.6.2.2, stirrup spacing must not exceed $d/2 = 8$ in, and the spacing provided is satisfactory. The minimum shear reinforcement required is specified by ACI Sec. 9.6.3.3 as

$$\frac{A_v}{s} = \frac{50b_w}{f_{yt}} \quad \text{[governs]}$$

$$= \frac{\left(50 \dfrac{\text{lbf}}{\text{in}^2}\right)(12 \text{ in})\left(12 \dfrac{\text{in}}{\text{ft}}\right)}{60{,}000 \dfrac{\text{lbf}}{\text{in}^2}}$$

$$= 0.12 \text{ in}^2/\text{ft}$$

$$< 0.33 \quad \text{[satisfactory]}$$

The total design shear capacity at section A-A is

$$\phi V_n = \phi(V_c + V_s)$$

$$= (0.75)(21 \text{ kips} + 39.6 \text{ kips})$$

$$= 45.5 \text{ kips}$$

Figure 2.9 *Minimum Shear Reinforcement for a Deep Beam*

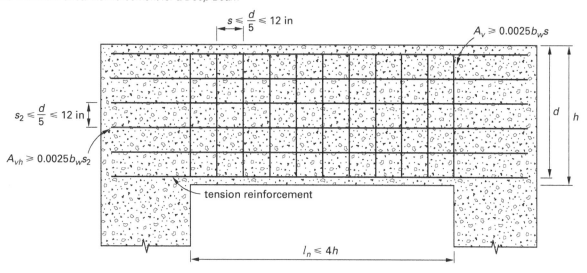

7. DEEP BEAMS

As shown in Fig. 2.9, a *deep beam*, as defined in ACI Sec. 9.9.1.1, is a beam in which the ratio of clear span to overall depth does not exceed 4. Deep beam conditions also apply to regions of beams loaded with concentrated loads within twice the beam depth from a support. The nominal shear strength of a deep beam is limited by ACI Sec. 9.9.2.1 to a maximum of

$$V_n = 10 b_w d \sqrt{f_c'}$$

Along the side faces of deep beams indicated in Fig. 2.9, minimum areas of vertical and horizontal reinforcement are specified in ACI Sec. 9.9.4.3 in order to restrain cracking.

Strut-and-Tie Model[8]

In accordance with ACI Sec. 9.9.1.2 and Sec. 9.9.1.3, deep beams must be designed using either nonlinear analysis or by the *strut-and-tie* method given in ACI Chap. 23. In the strut-and-tie method, a member is divided, as shown in Fig. 2.10, into *discontinuity*, or *D-regions*, in which the beam theory of ACI Sec. 22.2 does not apply, and *B-regions* in which beam theory does apply. In addition, for the strut-and-tie method to apply, the deep beam must be loaded so that compression struts can develop between the loads and the supports.

Figure 2.10 *B- and D-Regions*

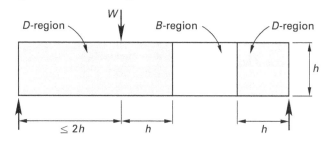

As shown in Fig. 2.11, a strut-and-tie model may be constructed to represent the internal forces in a deep beam. Compression struts are formed in the concrete to resist compressive forces. The strength of these struts is governed by the transverse tension developed by the lateral spread of the applied compression force. Using crack control reinforcement, as specified in ACI Sec. 23.5, to resist the transverse tension increases the strength of the strut.

Figure 2.11 *Strut-and-Tie Model*

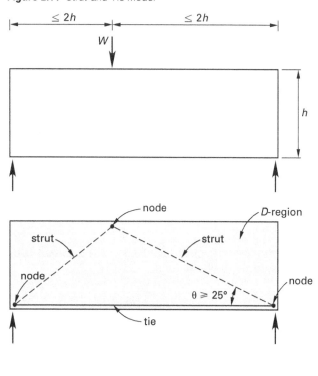

As shown in Fig. 2.12, struts may be either prism shaped or bottle shaped.

Figure 2.12 *Prism and Bottle-Shaped Struts*

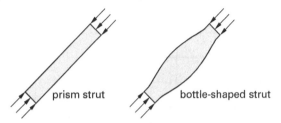

prism strut   bottle-shaped strut

Ties consist of tension reinforcement and the surrounding concrete that is concentric with the axis of the tie. The concrete does not contribute to the strength of the tie.

Nodes occur where the axes of struts, ties, concentrated loads, and support reactions acting on the joint intersect. The angle between the axes of a strut and a tie at a node is limited by ACI Sec. 23.2.7 to a minimum of $\theta = 25°$ in order to mitigate cracking.

Strut Nominal Strength

The effective compressive strength of the concrete in a strut is specified in ACI Sec. 23.4.3 as

$$f_{ce} = 0.85\beta_s f_c' \qquad \text{[ACI 23.4.3]}$$

$\beta_s = 1.0$ $\left[\begin{array}{l}\text{for a strut of uniform cross section,}\\ \text{as in the compression zone of a beam}\end{array}\right]$

$\beta_s = 0.60\lambda$ $\left[\begin{array}{l}\text{for an unreinforced,}\\ \text{bottle-shaped strut}\end{array}\right]$

$\beta_s = 0.75$ $\left[\begin{array}{l}\text{for a bottle-shaped strut with reinforce-}\\ \text{ment as specified in ACI Sec. 23.5}\end{array}\right]$

$\beta_s = 0.40$ $\left[\begin{array}{l}\text{for struts in a tension member}\\ \text{or the tension flange of a member}\end{array}\right]$

$\beta_s = 0.60\lambda$ [for all other cases]

$\lambda = 1.0$ [for normal weight concrete]

$\lambda = 0.85$ [for sand lightweight concrete]

$\lambda = 0.75$ [for all lightweight concrete]

In accordance with ACI Sec. 23.4.1, the nominal compressive strength of a strut is

$$\begin{aligned} F_{ns} &= f_{ce}A_{cs} \qquad \text{[without longitudinal reinforcement]}\\ &= f_{ce}w_s b\\ &= f_{ce}A_{cs} + A'_s f'_s \text{ [with longitudinal reinforcement]}\end{aligned}$$

The strength reduction factor for strut-and-tie models is given by ACI Sec. 21.2.1 as

$$\phi = 0.75$$

Tie Nominal Strength

In a reinforced concrete beam, the nominal strength of a reinforcing bar acting as a tie is given by ACI Sec. 23.7.2 as

$$F_{nt} = A_{ts}f_y$$

If the bars in a tie are in one layer, as shown in Fig. 2.13, the width of the tie may be taken as the diameter of the bars in the tie plus twice the cover to the surface of the bars.

Figure 2.13 *Nodal Zone*

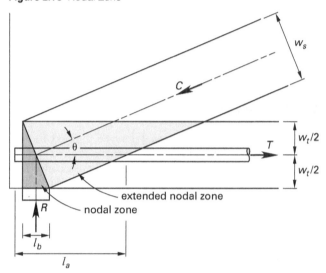

Nodal Zone Nominal Strength

As defined in ACI Sec. 2.3 and shown in Fig. 2.13, a *nodal zone* is the volume of concrete surrounding a node that is assumed to transfer strut-and-tie forces through the node. The node illustrated in Fig. 2.13 is classified as C-C-T, with two of the members acting on the node in compression and the third member in tension. Similarly, when all three members acting on the node are in compression, the node is classified as C-C-C. The effective compressive strength of the concrete in a node is specified in ACI Sec. 23.9.2 as

$$f_{ce} = 0.85\beta_n f_c' \quad \text{[ACI 23.9.2]}$$

$\beta_n = 1.0$ $\left[\begin{array}{l}\text{for a nodal zone bounded on all sides}\\ \text{by struts or bearing areas or both}\end{array}\right]$

$ = 0.80$ $\left[\begin{array}{l}\text{for a nodal zone anchoring}\\ \text{one tie}\end{array}\right]$

$ = 0.60$ $\left[\begin{array}{l}\text{for a nodal zone anchoring}\\ \text{two or more ties}\end{array}\right]$

In accordance with ACI Sec. 23.9.1, the nominal compressive strength of a nodal zone is

$$F_{nn} = f_{ce}A_{nz} = f_{ce}w_s b \quad \text{[ACI 23.9.1]}$$

The faces of the nodal zone shown in Fig. 2.13 are perpendicular to the axes of the strut, tie, and bearing plate, and the lengths of the faces are in direct proportion to the forces acting. The node has equal stresses on all faces and is termed a *hydrostatic nodal zone*. The effective width of the strut shown in Fig. 2.13 is

$$w_s = w_t \cos\theta + l_b \sin\theta$$

The *extended nodal zone* shown in Fig. 2.13 is that portion of the member bounded by the intersection of the effective strut width and the effective tie width. As specified in ACI Sec. 23.8.3, the anchorage length of the reinforcement is measured from the point of intersection of the bar and the extended nodal zone. The reinforcement may be anchored by a plate, by hooks, or by a straight development length. Alternatively, the tie reinforcement may be developed within the nodal zone.

Modeling Procedure

The modeling procedure for struts, ties, and nodes involves the following steps.

step 1: Determine the reactions on the model.

step 2: Select the location of the members by aligning the direction of struts in the direction of the anticipated cracking.

step 3: Determine the areas of struts, ties, and nodes necessary to provide the required strength.

step 4: Provide anchorage for the ties.

step 5: Provide crack control reinforcement.

Example 2.11

A reinforced concrete beam of normal weight concrete with a clear span of 6 ft, an effective depth of 26 in, and a width of 12 in, as shown in *Illustration for Ex. 2.11*, has a concrete compressive strength of 4500 lbf/in². The factored applied force of 80 kips includes an allowance for the self weight of the beam. Determine the number of grade 60, no. 8 bars required for tension reinforcement and check that the equivalent concrete strut and nodal zone at the left support comply with the requirements of ACI Chap. 23.

Solution

The clear span-to-depth ratio is given by

$$\frac{l_n}{h} = \frac{72 \text{ in}}{28 \text{ in}} = 2.6$$
$$< 4 \quad \text{[satisfies ACI Sec. 9.9.1.1]}$$

The idealized strut-and-tie model is shown in the illustration, and the angle between the struts and the tie is

$$\theta = \arctan\frac{26 \text{ in}}{42 \text{ in}} = 31.8°$$
$$> 25° \quad \text{[satisfies ACI Sec. 23.2.7]}$$

The equivalent tie force is determined from the strut-and-tie model as

$$T = \frac{(40 \text{ kips})(42 \text{ in})}{26 \text{ in}} = 64.62 \text{ kips}$$

The strength reduction factor is given by ACI Sec. 21.2.1 as

$$\phi = 0.75$$

The necessary reinforcement area is given by

$$A_{ts} = \frac{T}{\phi f_y} = \frac{64.62 \text{ kips}}{(0.75)\left(60 \ \dfrac{\text{kips}}{\text{in}^2}\right)} = 1.44 \text{ in}^2$$

Use two no. 8 bars, which gives an area of

$$A = 1.58 \text{ in}^2 > A_{ts} \quad \text{[satisfactory]}$$

As shown in the illustration, the dimensions of the nodal zone are

$$w_t = \text{equivalent tie width} = d_b + 2c$$
$$= 1 \text{ in} + (2)(1.5 \text{ in})$$
$$= 4 \text{ in}$$

$$l_b = \text{width of equivalent support strut}$$
$$= w_t \tan\theta$$
$$= (4 \text{ in})\tan 31.8°$$
$$= 2.48 \text{ in}$$

$$w_s = \text{width of equivalent concrete strut}$$
$$= \frac{w_t}{\cos\theta}$$
$$= \frac{4 \text{ in}}{\cos 31.8°}$$
$$= 4.70 \text{ in}$$

The stress in the equivalent tie is

$$f_T = \frac{T}{bw_t} = \frac{64.62 \text{ kips}}{(12 \text{ in})(4 \text{ in})} = 1.35 \text{ kips/in}^2$$

Illustration for Ex. 2.11

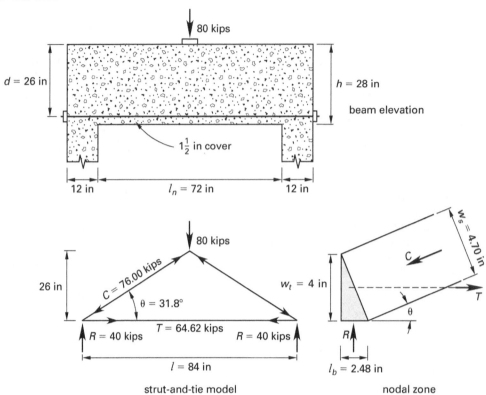

strut-and-tie model   nodal zone

For a hydrostatic nodal zone,

f_C = stress in the equivalent concrete strut = f_T

    = 1.35 kips/in^2

f_R = stress in the equivalent support strut = f_T

    = 1.35 kips/in^2

For an unreinforced, bottle-shaped strut with normal weight concrete ($\lambda = 1.0$), the design compressive strength of the concrete in the strut is given by ACI Eq. 23.4.3 as

$$\phi f_{ce} = 0.85\phi\beta_s f_c'$$
$$= (0.85)(0.75)(0.6)(1.0)\left(4.5\ \frac{\text{kips}}{\text{in}^2}\right)$$
$$= 1.72\ \text{kips/in}^2$$
$$> f_C \quad [\text{satisfactory}]$$

The design compressive strength of a nodal zone anchoring one layer of reinforcing bars without confining reinforcement is given by ACI Eq. 23.9.2 as

$$\phi f_{ce} = \phi 0.85\beta_n f_c'$$
$$= (0.75)(0.85)(0.8)\left(4.5\ \frac{\text{kips}}{\text{in}^2}\right)$$
$$= 2.30\ \text{kips/in}^2$$
$$> f_C \quad [\text{satisfactory}]$$

The anchorage length available for the tie reinforcement, using 2 in end cover, is

$$l_a = \frac{h-d}{\tan\theta} + \frac{l_b}{2} + 6\ \text{in} - 2\ \text{in}$$
$$= \frac{28\ \text{in} - 26\ \text{in}}{\tan 31.8°} + \frac{2.48\ \text{in}}{2} + 6\ \text{in} - 2\ \text{in}$$
$$= 8.47\ \text{in}$$

The development length for a grade 60, no. 8 bar, with 2.5 in side cover and 2 in end cover and with a standard 90° hook, is given by ACI Sec. 25.4.3.1 as

$$l_{dh} = \frac{(0.7)\left(1200 \; \dfrac{\text{lbf}}{\text{in}^2}\right) d_b}{\sqrt{f'_c}}$$

$$= \frac{(0.7)\left(1200 \; \dfrac{\text{lbf}}{\text{in}^2}\right)(1 \text{ in})}{\sqrt{4500 \; \dfrac{\text{lbf}}{\text{in}^2}}}$$

$$= 12.5 \text{ in}$$
$$> l_a \quad \text{[Anchorage length is inadequate.]}$$

Use an end plate to anchor the bars.

8. CORBELS

A corbel is a cantilever bracket supporting a load-bearing member. As shown in Fig. 2.14 and specified in ACI Sec. 16.5.1.1, the shear span-to-depth ratio, a_v/d, and the ratio of horizontal tensile force to vertical force, N_{uc}/V_u, are limited to a maximum value of unity. The depth of the corbel at the outside edge of bearing area must not be less than $d/2$.

Figure 2.14 *Corbel Details*

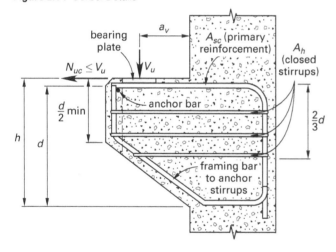

At the face of the support, the forces acting on the corbel are a shear force, V_u, a moment ($V_u a_v + N_{uc}(h - d)$), and a tensile force, N_{uc}. These require reinforcement areas of A_{vf}, A_f, and A_n, respectively. The shear friction reinforcement area is derived from ACI Eq. 22.9.4.2 as

$$A_{vf} = \frac{V_u}{\phi f_y \mu}$$

Also, from ACI Table 22.9.4.4, for normal weight concrete, the factored shear force on the section is

$$V_u \leq 0.2\phi f'_c b_w d$$

$$\leq (480 + 0.08 f'_c)\phi b_w d$$

$$\leq 1600 \phi b_w d$$

$$\mu = \text{coefficient of friction at face of support,}$$
$$\quad \text{as given by ACI 22.9.4.2}$$

$$= 1.4\lambda \quad \text{[for concrete placed monolithically]}$$

The correction factor related to the unit weight of concrete is defined by ACI Sec. 19.2.4.2 as

$$\lambda = 1.0 \quad \text{[for normal weight concrete]}$$
$$= 0.75 \quad \text{[for all lightweight concrete]}$$

In accordance with ACI Sec. 16.5.3.5, the tensile force N_{uc} may not be less than $0.2V_u$, and the corresponding area of reinforcement required is

$$A_n = \frac{N_{uc}}{\phi f_y}$$

The required area of primary tension reinforcement is given by ACI Sec. 16.5.5.1 as the greater of

$$A_{sc} = A_f + A_n$$

$$A_{sc} = \frac{2A_{vf}}{3} + A_n$$

$$\frac{A_{sc}}{bd} = \frac{0.04 f'_c}{f_y}$$

The minimum required area of closed ties distributed over a depth of $2d/3$ is given by ACI Sec. 16.5.5.2 as

$$A_h = \frac{A_{sc} - A_n}{2}$$

ACI Table 21.2.1 gives the value of the strength reduction factor for corbels as $\phi = 0.75$.

Example 2.12

The reinforced concrete corbel shown has a width of 15 in, is reinforced with grade 60 bars, and has a concrete compressive strength of 3000 lbf/in². Determine whether the corbel is adequate for the applied factored loads indicated.

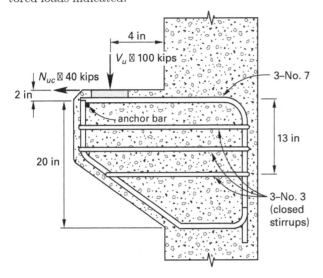

Solution

$$0.2\phi f'_c b_w d = (0.2)(0.75)\left(3\ \frac{\text{kips}}{\text{in}^2}\right)(15\ \text{in})(20\ \text{in})$$
$$= 135\ \text{kips}$$
$$> V_u$$
$$(480 + 0.08f'_c)\phi b_w d = (0.72)(0.75)(15\ \text{in})(20\ \text{in})$$
$$= 162\ \text{kips}$$
$$> V_u$$
$$1.6\phi b_w d = (1.6)(0.75)(15\ \text{in})(20\ \text{in})$$
$$= 360\ \text{kips}$$
$$> V_u$$

The corbel conforms to ACI Table 22.9.4.4.

The shear friction reinforcement area is given by ACI Sec. R22.9.4.2 as

$$A_{vf} = \frac{V_u}{\phi f_y \mu} = \frac{100\ \text{kips}}{(0.75)\left(60\ \dfrac{\text{kips}}{\text{in}^2}\right)(1.4)}$$
$$= 1.59\ \text{in}^2$$

The tension reinforcement area required is

$$A_n = \frac{N_{uc}}{\phi f_y} = \frac{40\ \text{kips}}{(0.75)\left(60\ \dfrac{\text{kips}}{\text{in}^2}\right)}$$
$$= 0.889\ \text{in}^2$$

The factored moment acting on the corbel is

$$M_u = V_u a_v + N_{uc}(h - d)$$
$$= (100\ \text{kips})(4\ \text{in}) + (40\ \text{kips})(2\ \text{in})$$
$$= 480\ \text{in-kips}$$

The area of flexural reinforcement required for $\phi = 0.75$ for corbels as given by ACI Sec. R21.2.1 for corbels is

$$A_f = \frac{0.85 b d f'_c\left(1 - \sqrt{1 - \dfrac{M_u}{0.319 b_w d^2 f'_c}}\right)}{f_y}$$
$$= 0.545\ \text{in}^2$$

The primary reinforcement area required is given by ACI Sec. 16.5.5.1 as

$$A_{sc} = A_f + A_n$$
$$= 0.545\ \text{in}^2 + 0.889\ \text{in}^2$$
$$= 1.434\ \text{in}^2$$

Three no. 7 bars are provided, giving an area of

$$A'_s = 1.80\ \text{in}^2$$
$$> 1.434\ \text{in}^2 \quad \text{[satisfactory]}$$

Also, from ACI Sec. 16.5.5.1, the area of primary reinforcement must not be less than

$$\frac{2A_{vf}}{3} + A_n = \frac{(2)(1.59\ \text{in}^2)}{3} + 0.889\ \text{in}^2$$
$$= 1.95\ \text{in}^2$$
$$> A'_s \quad \text{[unsatisfactory]}$$

The area of closed stirrups required is given by ACI Sec. 16.5.5.2 as

$$A_h = \frac{A_{sc} - A_n}{2}$$
$$= \frac{1.95\ \text{in}^2 - 0.889\ \text{in}^2}{2}$$
$$= 0.53\ \text{in}^2$$

Three no. 3 closed stirrups are provided, giving an area of

$$A_h = 0.66\ \text{in}^2$$
$$> 0.53\ \text{in}^2 \quad \text{[satisfactory]}$$

9. BEAMS IN TORSION

The terminology used in torsion design is illustrated in Fig. 2.15 and Fig. 2.16.

Figure 2.15 *Torsion in Flanged Section*

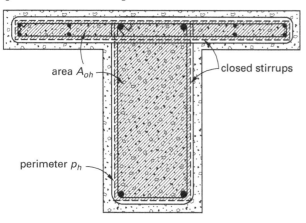

In accordance with ACI Sec. 22.7.1.1, for a statically determinate member the equilibrium torsional effects may be neglected, and closed stirrups and longitudinal torsional reinforcement are not required when the factored torque $T_u < \phi T_{th}$.

$$T_{th} = \lambda \sqrt{f_c'} \left(\frac{A_{cp}^2}{p_{cp}} \right) \quad \text{[ACI Table 22.7.4.1(a)]}$$

When this value is exceeded, reinforcement must be provided to resist the full torsion. When both shear and torsion reinforcements are required, the sum of the individual areas must be provided.

Assuming that the angle between the compression diagonal and the longitudinal axis of the member, θ, is 45°, ACI Sec. 22.7.6.1 specifies that the nominal torsional moment strength is the lesser of

$$T_n = \frac{2A_o A_t f_{yt}}{s} \quad \text{[ACI 22.7.6.1a]}$$

$$T_n = \frac{2A_o A_l f_y}{p_h} \quad \text{[ACI 22.7.6.1b]}$$

In accordance with ACI Sec. 22.7.6.1.1, for these two equations it is permitted to take

$$A_o = 0.85 A_{oh}$$

From ACI Eq. 22.7.6.1a and 22.7.6.1b, it follows that the corresponding longitudinal reinforcement required is

$$A_l = \frac{A_t p_h f_{yt}}{f_y s}$$

The minimum area of longitudinal reinforcement required is specified in ACI Sec. 9.6.4.3 as

$$A_l = \frac{5 A_{cp} \sqrt{f_c'}}{f_y} - \frac{A_t p_h f_{yt}}{f_y s}$$

Where,

$$\frac{A_t}{s} \geq \frac{25 b_w}{f_{yt}}$$

The maximum spacing of longitudinal reinforcement, specified in ACI Sec. 9.7.5.1, is 12 in.

The minimum diameter, specified in ACI Sec. 9.7.5.2, is

$$d_b = 0.042s \geq \text{no. 3 bar}$$

The minimum combined area of stirrups for combined shear and torsion is given by ACI Sec. 9.6.4.2 as

$$\frac{A_v + 2A_t}{s} = \frac{0.75 \sqrt{f_c'} b_w}{f_{yt}} \geq \frac{50 b_w}{f_{yt}}$$

The maximum spacing of closed stirrups is given by ACI Sec. 9.7.6.3 as

$$s = \frac{p_h}{8} \leq 12 \text{ in}$$

In accordance with ACI Sec. 22.7.5.1, when redistribution of internal forces occurs in an indeterminate structure upon cracking, a member may be designed for the factored torsion causing cracking, which is given by

$$T_u = 4\phi \lambda \sqrt{f_c'} \left(\frac{A_{cp}^2}{p_{cp}} \right)$$

Example 2.13

A simply supported reinforced concrete beam of normal weight concrete with an overall depth of 19 in, and a width of 12 in is reinforced with grade 60 bars and has a concrete compressive strength of 3000 lbf/in². Determine the combined shear and torsion reinforcement required when

(a) the factored shear force is 5 kips and the factored torsion is 2 ft-kips

(b) the factored shear force is 15 kips and the factored torsion is 4 ft-kips

Use no. 3 stirrups with 1.5 in cover.

Figure 2.16 *Torsion in Rectangular Section*

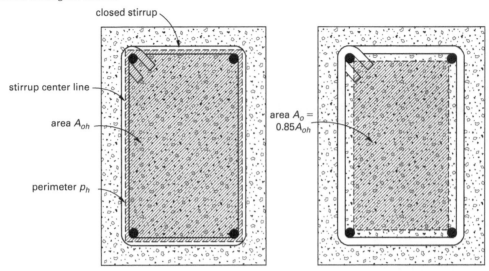

Solution

(a) The area enclosed by the outside perimeter of the beam is

$$A_{cp} = (19 \text{ in})(12 \text{ in})$$
$$= 228 \text{ in}^2$$

The length of the outside perimeter of the beam is

$$p_{cp} = (2)(19 \text{ in} + 12 \text{ in})$$
$$= 62 \text{ in}$$

Torsional reinforcement is not required in accordance with ACI Table 22.2.4.1(a) when the factored torque is less than

$$T_u = \phi \lambda \sqrt{f_c'} \left(\frac{A_{cp}^2}{p_{cp}} \right)$$
$$= (0.75)(1.0)\sqrt{3000 \ \frac{\text{lbf}}{\text{in}^2}}$$
$$\times \left(\frac{(228 \text{ in}^2)^2}{(62 \text{ in})\left(12 \ \frac{\text{in}}{\text{ft}}\right)\left(1000 \ \frac{\text{lbf}}{\text{kip}}\right)} \right)$$
$$= 2.87 \text{ ft-kips}$$
$$> 2.0 \quad \text{[Closed stirrups are not required.]}$$

The shear strength provided by the concrete was determined in Ex. 2.9 as

$$\phi V_c = 15.8 \text{ kips}$$
$$> 2V_u \quad \text{[Shear stirrups are not required.]}$$

(b) Because $\phi V_c/2 < V_u < \phi V_c$, minimum shear reinforcement is required, and because $T_u > 2.87$ ft-kips, closed stirrups are necessary. Using no. 3 stirrups with 1.5 in cover, the area enclosed by the centerline of the stirrups is

$$A_{oh} = (19 \text{ in} - 3 \text{ in} - 0.375 \text{ in})$$
$$\times (12 \text{ in} - 3 \text{ in} - 0.375 \text{ in})$$
$$= 134.77 \text{ in}^2$$

From ACI Eq. 22.7.6.1a, the required area of one arm of a closed stirrup is given by

$$\frac{A_t}{s} = \frac{T_u}{1.7\phi A_{oh} f_{yt}}$$
$$= \frac{(4 \text{ ft-kips})\left(12 \ \frac{\text{in}}{\text{ft}}\right)}{(1.7)(0.75)\left(\frac{134.77 \text{ in}^2}{12 \ \frac{\text{in}}{\text{ft}}}\right)\left(60 \ \frac{\text{kips}}{\text{in}^2}\right)}$$
$$= 0.056 \text{ in}^2/\text{ft} \quad \text{[per arm]}$$

From ACI Sec. 9.6.4.2(b), the governing minimum combined shear and torsion reinforcement area (2 arms) is given by

$$\frac{A_v + 2A_t}{s} = \frac{50 b_w}{f_{yt}} = \frac{(50)(12 \text{ in})\left(12 \ \frac{\text{in}}{\text{ft}}\right)}{60,000 \ \frac{\text{lbf}}{\text{in}^2}}$$
$$= 0.12 \text{ in}^2/\text{ft} \quad \text{[governs]}$$

ACI Sec. 9.6.4.2(a) does not govern.

The perimeter of the centerline of the closed stirrups is

$$p_h = (2)\big(19\text{ in} + 12\text{ in} - (2)(3.375\text{ in})\big)$$
$$= 48.50\text{ in}$$

The governing maximum permissible spacing of the closed stirrups is specified in ACI Sec. 9.7.6.3 as

$$s_{\max} = \frac{p_h}{8}$$
$$= \frac{48.50\text{ in}}{8}$$
$$= 6\text{ in}$$

Closed stirrups consisting of two arms of no. 3 bars at 6 in spacing provides an area of

$$\frac{A}{s} = 0.44\text{ in}^2/\text{ft}$$
$$> 0.12 \quad [\text{satisfactory}]$$

The required area of the longitudinal reinforcement is

$$A_l = \left(\frac{A_t}{s}\right)p_h\left(\frac{f_{yt}}{f_y}\right)$$

$$= \frac{\left(0.056\ \dfrac{\text{in}^2}{\text{ft}}\right)(48.50\text{ in})\left(\dfrac{60\ \dfrac{\text{kips}}{\text{in}^2}}{60\ \dfrac{\text{kips}}{\text{in}^2}}\right)}{12\ \dfrac{\text{in}}{\text{ft}}}$$

$$= 0.23\text{ in}^2$$

Because the required value of $A_t/s = 0.056$ in²/ft per arm is less than $25b_w/f_{yt} = 0.060$ in²/ft, the minimum permissible area of longitudinal reinforcement is given by ACI Sec. 9.6.4.3(b) as

$$A_{l(\min)} = \frac{5A_{cp}\sqrt{f_c'}}{f_y} - \left(\frac{25b_w}{f_{yt}}\right)p_h\left(\frac{f_{yt}}{f_y}\right)$$

$$= \frac{(5)(228\text{ in}^2)\sqrt{3000\ \dfrac{\text{lbf}}{\text{in}^2}}}{60{,}000\ \dfrac{\text{lbf}}{\text{in}^2}}$$

$$- \left(0.0050\ \dfrac{\text{in}^2}{\text{in}}\right)(48.50\text{ in})\left(\dfrac{60\ \dfrac{\text{kips}}{\text{in}^2}}{60\ \dfrac{\text{kips}}{\text{in}^2}}\right)$$

$$= 1.041\text{ in}^2 - 0.242\text{ in}^2$$
$$= 0.799\text{ in}^2 \quad [\text{governs}]$$

Using eight no. 3 bars around the perimeter of the closed stirrups, with four on a side, gives a longitudinal steel area of

$$A_l = 0.88\text{ in}^2 > 0.799\text{ in}^2 \quad [\text{satisfactory}]$$

The maximum spacing of the longitudinal bars occurs at the top of the beam and is

$$s = 12\text{ in} - 2(1.5\text{ in} + 0.375\text{ in}) - 0.375\text{ in}$$
$$= 7.875\text{ in}$$
$$< 12\text{ in} \quad [\text{satisfies ACI Sec. 9.7.5.1}]$$

10. CONCRETE COLUMNS

Nomenclature

A_{ch}	area of core of spirally reinforced compression member measured to outside diameter of spiral	in²
A_g	gross area of concrete section	in²
A_{st}	total area of longitudinal reinforcement	in²
C_m	a factor relating actual moment diagram to an equivalent uniform moment diagram	–
d_{agg}	nominal maximum size of coarse aggregate	in
E_c	modulus of elasticity of concrete	lbf/in²
E_s	modulus of elasticity of the reinforcement	lbf/in²
h	overall thickness of member	in
I_g	moment of inertia of the gross concrete section	–
I_{se}	moment of inertia of reinforcement about the centroid of the member cross section	–
k	effective length factor for compression members	–
l_c	length of a compression member in a frame, measured from center-to-center of the joints in the frame	ft or in
l_u	unsupported length of compression member	ft or in
M_c	factored moment to be used for design of compression member	ft-kips
M_1	smaller factored end moment on a compression member, negative if member is bent in single curvature, positive if bent in double curvature	ft-kips
M_{1ns}	factored end moment on a compression member at the end at which M_1 acts, due to loads that cause no appreciable sidesway, calculated using a first-order elastic frame analysis	ft-kips

M_{1s}	factored end moment on compression members at the end at which M_1 acts, due to loads that cause appreciable sidesway, calculated using a first-order elastic frame analysis	ft-kips
M_2	larger factored end moment on compression member, always positive	ft-kips
M_{2ns}	factored end moment on compression member at the end at which M_2 acts, due to loads that cause no appreciable sidesway, calculated using a first-order elastic frame analysis	ft-kips
M_{2s}	factored end moment on compression member at the end at which M_2 acts, due to loads that cause appreciable sidesway, calculated using a first-order elastic frame analysis	ft-kips
P_c	critical load, $\pi^2 EI/(kl_u)^2$	kips
P_n	nominal axial load strength at given eccentricity	kips
P_o	nominal axial load strength at zero eccentricity	kips
P_u	factored axial load at given eccentricity $\leq \phi P_n$	kips
Q	stability index for a story, $\sum P_u \Delta_o / V_u l_c$	–
r	radius of gyration of cross section of a compression member	in
V_{us}	factored horizontal shear in a story	kips

Symbols

β_{dns}	ratio of sustained load to maximum load ≤ 1.0	–
γ	the ratio of the distance between centroids of the longitudinal reinforcement to the overall diameter of the column	–
δ	moment magnification factor for frames braced against sidesway to reflect effects of member curvature between ends of compression members	–
δ_s	moment magnification factor for frames not braced against sidesway to reflect lateral drift resulting from lateral and gravity loads	–
Δ_o	relative lateral deflection between the top and bottom of a story due to V_{us}, computed using a first-order elastic frame analysis	in
ρ	ratio of A_{st} to A_g	–
ρ_s	ratio of volume of spiral reinforcement to total volume of core (out-to-out of spirals) of a spirally reinforced compression member	–
Ψ	stiffness ratio at the end of a column	–

Reinforcement Requirements

ACI Sec. 10.6.1.1 limits the area of longitudinal reinforcement to not more than 8% and not less than 1% of the gross area of the section. For columns with rectangular or circular ties, a minimum of four bars is required. For columns with spirals, a minimum of six longitudinal bars is required. The minimum ratio of volume of spiral reinforcement to volume of core is given by ACI Sec. 25.7.3.3 as

$$\rho_s = 0.45 f_c' \left(\frac{\dfrac{A_g}{A_{ch}} - 1}{f_{yt}} \right) \qquad \text{[ACI 25.7.3.3]}$$

In accordance with ACI Sec. 25.7.3.1, the clear spacing between spirals cannot exceed 3 in nor be less than the greater of 1 in or $\frac{4}{3} d_{agg}$. The minimum diameter of the spiral is $\frac{3}{8}$ in.

For rectangular columns, the minimum tie size specified by ACI Sec. 25.7.2.2 is no. 3 for longitudinal bars of no. 10 or smaller and no. 4 for longitudinal bars larger than no. 10. The maximum vertical spacing of ties is given by ACI Sec. 25.7.2.1 as

$$s_{max} \leq 16 \times \text{longitudinal bar diameters}$$
$$\leq 48 \times \text{tie bar diameters}$$
$$\leq \text{least dimension of the column}$$

A clear spacing between ties of at least $\frac{4}{3} d_{agg}$ is required.

Ties must be provided, as shown in Fig. 2.17, to support every corner and alternate bar, and no bar should be more than 6 in clear from a supported bar.

Figure 2.17 *Column Ties*

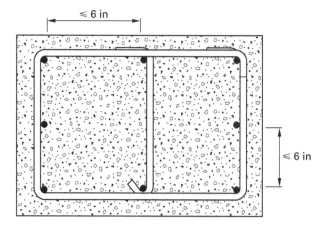

Example 2.14

A 24 in diameter spirally reinforced column with a 1.5 in cover to the spiral is reinforced with grade 60 bars and has a concrete compressive strength of 4500 lbf/in². Determine the required diameter and pitch of the spiral.

Solution

From ACI Sec. 25.7.3.2, the minimum permissible diameter of spiral reinforcement is

$$d_b = 0.375 \text{ in}$$
$$A_b = \text{area of spiral bar} = 0.11 \text{ in}^2$$
$$A_g = \text{gross area of column}$$
$$= \frac{\pi(24 \text{ in})^2}{4}$$
$$= 452 \text{ in}^2$$
$$A_{ch} = \text{area of core}$$
$$= \frac{\pi(21 \text{ in})^2}{4}$$
$$= 346 \text{ in}^2$$
$$d_s = \text{mean diameter of spiral}$$
$$= 21 \text{ in} - 0.375 \text{ in}$$
$$= 20.625 \text{ in}$$

From ACI Eq. 25.7.3.3, the minimum allowable spiral reinforcement ratio is

$$\rho_s = 0.45 f_c' \left(\frac{\dfrac{A_g}{A_{ch}} - 1}{f_{yt}} \right)$$

$$= (0.45)\left(4.5 \ \frac{\text{kips}}{\text{in}^2}\right)\left(\frac{\dfrac{452 \text{ in}^2}{346 \text{ in}^2} - 1}{60 \ \dfrac{\text{kips}}{\text{in}^2}} \right)$$

$$= 0.0103$$
$$= \frac{\text{volume of spiral bar per turn}}{\text{volume of concrete core per turn}}$$
$$= \frac{A_b \pi d_s}{A_{ch} s}$$
$$= \frac{(0.11 \text{ in}^2)\,\pi\,(20.625 \text{ in})}{(346 \text{ in}^2)s}$$
$$s = 2 \text{ in} \quad \text{[satisfactory]}$$

The calculated pitch lies between the maximum of 3 in and the minimum of 1 in specified in ACI Sec. 25.7.3.1.

Effective Length and Slenderness Ratio

The effective column length may be determined from the alignment charts given in ACI Sec. R6.2.5 and shown in Fig. 2.18.[2]

To use the alignment charts, the stiffness ratios at each end of the column must be calculated, and this is given by ACI Fig. R6.2.5 as

$$\Psi = \frac{\displaystyle\sum \frac{E_c I_c}{l_c}}{\displaystyle\sum \frac{E_b I_b}{l_b}}$$

The subscript c refers to the columns meeting at a joint, and the subscript b refers to the beams meeting at a joint.

For a non-sway frame, the effective length factor k may be conservatively taken as unity, as indicated in ACI Sec. 6.6.4.4.3. The slenderness ratio is defined as kl_u/r, and the radius of gyration is given by ACI Sec. 6.2.5.1 as

$$r = 0.25 \times \text{diameter of circular column}$$
$$= 0.30 \times \text{dimension of a rectangular column}$$
$$\qquad \text{in the direction stability is being considered}$$
$$= \sqrt{(I_g \,/\, A_g)}$$

A non-sway column is defined in ACI Sec. 6.6.4.3(a) as one in which the secondary moments due to P-delta effects do not exceed 5% of the primary moments due to lateral loads. ACI Sec. 6.6.4.3(b) specifies a story within a structure as non-sway, provided that the stability index, Q, does not exceed 0.05 where the stability index is given by

$$Q = \frac{\sum P_u \Delta_o}{V_{us} l_c} \qquad \text{[ACI 6.6.4.4.1]}$$

V_{us} is the total factored story shear, and $\sum P_u$ is the total factored vertical load on a story.

Example 2.15

Determine the slenderness ratio of columns 12 and 34 of the sway frame shown. The columns are 18 in square and have an unsupported height of 9 ft. All members of the frame have identical EI values.

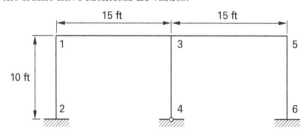

Figure 2.18 *Alignment Charts for k*

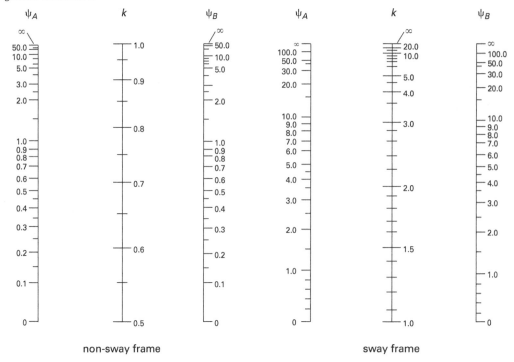

non-sway frame   sway frame

Adapted from *Building Code Requirements for Structural Concrete and Commentary* (ACI 318-14), Fig. R6.25, copyright © 2014, by the American Concrete Institute.

Solution

For column 12, the stiffness ratio of the fixed base is given by AISC 360[9] Comm. App. 7.2 as

$$\Psi_2 = 1.0$$

At joint 1, the relative stiffness value of the beam is

$$\sum \frac{E_b I_b}{l_b} = \frac{1}{15}$$

At joint 1, the relative stiffness value of the column is

$$\sum \frac{E_c I_c}{l_c} = \frac{1}{10}$$

The stiffness ratio at joint 1 is

$$\Psi_1 = \frac{\sum \dfrac{E_c I_c}{l_c}}{\sum \dfrac{E_b I_b}{l_b}} = \frac{15}{10}$$
$$= 1.5$$

From the alignment chart, for a sway frame, the effective length factor for column 12 is

$$k_{12} = 1.38$$

The radius of gyration of the column, in accordance with ACI Sec. 6.2.5.1, is

$$r = 0.30h = (0.30)(18 \text{ in}) = 5.4 \text{ in}$$

The slenderness ratio of column 12 is

$$\frac{kl_u}{r} = \frac{(1.38)(9 \text{ ft})\left(12 \dfrac{\text{in}}{\text{ft}}\right)}{5.4 \text{ in}} = 27.6$$

For column 34, the stiffness ratio of the pinned base is given by AISC 360 Comm. App. 7.2 as

$$\Psi_4 = 10$$

At joint 3, the sum of the relative stiffness values of the two beams is

$$\sum \frac{E_b I_b}{l_b} = \frac{1}{15} + \frac{1}{15} = \frac{2}{15}$$

At joint 3, the relative stiffness value of the column is

$$\sum \frac{E_c I_c}{l_c} = \frac{1}{10}$$

The stiffness ratio at joint 3 is

$$\Psi_3 = \frac{\sum \dfrac{E_c I_c}{l_c}}{\sum \dfrac{E_b I_b}{l_b}} = \frac{15}{20} = 0.75$$

From the alignment chart, for a sway frame, the effective length factor for column 34 is

$$k_{34} = 1.85$$

The slenderness ratio of column 34 is

$$\frac{k l_u}{r} = \frac{(1.85)(9 \text{ ft})\left(12 \dfrac{\text{in}}{\text{ft}}\right)}{5.4 \text{ in}} = 37.0$$

Short Column with Axial Load

In accordance with ACI Sec. 6.2.5(a), a column in a sway frame is classified as a short column, and slenderness effects may be ignored, when the slenderness ratio is

$$\frac{k l_u}{r} \le 22 \quad \text{[ACI 6.2.5a]}$$

For a non-sway frame, a column is classified as a short column, in accordance with ACI Sec. 6.2.5(b), when the slenderness ratio is both

$$\frac{k l_u}{r} \le 34 + \frac{12 M_1}{M_2}$$

$$\frac{k l_u}{r} \le 40$$

The term M_1 / M_2 is negative if the column is bent in single curvature and positive for double curvature. In addition, columns in a story may be considered braced when the bracing elements in the story have a total stiffness of at least 12 times the gross stiffness of the columns within the story.

For a short column with spiral reinforcement, allowing for accidental eccentricity, ACI Sec. 22.4.2.2 gives the design axial load capacity as

$$\phi P_n = 0.85\phi \left(\begin{array}{c} 0.85 f'_c (A_g - A_{st}) \\ + A_{st} f_y \end{array} \right) \quad [\phi = 0.75]$$

For a short column with lateral tie reinforcement, allowing for accidental eccentricity, ACI Sec. 22.4.2.2 gives the design axial load capacity as

$$\phi P_n = 0.80\phi \left(\begin{array}{c} 0.85 f'_c (A_g - A_{st}) \\ + A_{st} f_y \end{array} \right) \quad [\phi = 0.65]$$

Example 2.16

An 18 in square column is reinforced with 12 no. 9 grade 60 bars and has a concrete compressive strength of 4000 lbf/in². The column, which is braced against sidesway, has an unsupported height of 9 ft and supports axial load only without end moments. Determine the lateral ties required and the design axial load capacity.

Solution

The minimum tie size specified by ACI Sec. 25.7.2.2 is no. 3 for longitudinal bars of size no. 9. ACI Sec. 25.7.2.1 specifies a tie spacing not greater than

$$h = 18 \text{ in}$$
$$48 d_t = (48)(0.375 \text{ in}) = 18 \text{ in}$$
$$16 d_b = (16)(1.128 \text{ in}) = 18 \text{ in}$$

From ACI Sec. 6.6.4.4.3, the effective length factor, k, for a column braced against sidesway is

$$k = 1.0$$

The radius of gyration, in accordance with ACI Sec. 6.2.5.1, is

$$r = 0.3h = (0.3)(18 \text{ in}) = 5.4 \text{ in}$$

The slenderness ratio is

$$\frac{k l_u}{r} = \frac{(1.0)(9 \text{ ft})\left(12 \dfrac{\text{in}}{\text{ft}}\right)}{5.4 \text{ in}} = 20.0$$

In accordance with ACI Sec. 6.2.5(b), the column, which is bent in single curvature, may be classified as a short column provided that

$$\frac{k l_u}{r} \le 34 + \frac{12 M_1}{M_2} \le 40$$

$$20 < 34$$

Where,

$$M_1 = 0$$
$$M_2 = M_{\min}$$

The column is a short column, and the design axial load capacity is given by ACI Sec. 22.4.2.2 as

$$\phi P_n = 0.80\phi\left(0.85f_c'(A_g - A_{st}) + A_{st}f_y\right)$$

$$= (0.80)(0.65)\begin{pmatrix} (0.85)\left(4\ \dfrac{\text{kips}}{\text{in}^2}\right) \\ \times(324\ \text{in}^2 - 12\ \text{in}^2) \\ +(12\ \text{in}^2)\left(60\ \dfrac{\text{kips}}{\text{in}^2}\right) \end{pmatrix}$$

$$= 926\ \text{kips}$$

Short Column with End Moments

The axial load carrying capacity of a column decreases as end moments are applied to the column. Design of the column may then be obtained by means of a computer program, such as Structure Point's *spColumn*[10], based on ACI 318. Alternatively, approximate design values may be obtained from the interaction diagrams given in App. 2.C through App. 2.H.

Example 2.17

A 24 in diameter tied column with a 1.5 in cover to the $\frac{3}{8}$ in diameter ties is reinforced with 14 no. 9 grade 60 bars and has a concrete compressive strength of 4000 lbf/in². The column, which is braced against sidesway, has an unsupported height of 9 ft and is bent in single curvature with factored end moments of $M_1 = M_2$ = 400 ft-kips. Determine the maximum factored axial load that the column can carry.

Solution

The column may be classified as a short column, and slenderness effects do not have to be considered. The ratio of the distance between centroids of the longitudinal reinforcement to the overall diameter of the column is

$$\gamma = \frac{24\ \text{in} - (2)(1.5\ \text{in}) - (2)(0.375\ \text{in}) - 1.125\ \text{in}}{24\ \text{in}}$$

$$= 0.80$$

The reinforcement ratio is

$$\rho = \frac{A_{st}}{A_g} = \frac{14\ \text{in}^2}{\pi(12\ \text{in})^2} = 0.031$$

$$\frac{M_u}{A_g h} = \frac{(400\ \text{ft-kips})\left(12\ \dfrac{\text{in}}{\text{ft}}\right)}{\pi(12\ \text{in})^2(24\ \text{in})} = 0.44\ \text{kip/in}^2$$

From App. 2.D, with $\gamma = 0.75$, the design axial stress is

$$\frac{P_u}{A_g} = 1.2\ \text{kips/in}^2$$

From App. 2.E, with $\gamma = 0.90$, the design axial stress is

$$\frac{P_u}{A_g} = 1.7\ \text{kips/in}^2$$

By interpolation, for $\gamma = 0.80$, the design axial stress is

$$\frac{P_u}{A_g} = 1.2\ \frac{\text{kips}}{\text{in}^2} + \left(0.5\ \frac{\text{kip}}{\text{in}^2}\right)\left(\frac{5}{15}\right) = 1.37\ \text{kips/in}^2$$

The maximum factored axial load that the column can carry is

$$P_u = 1.37A_g = 1.37\pi(12\ \text{in})^2 = 620\ \text{kips}$$

Long Column Without Sway[11]

ACI 318 requires that the design of long columns be based on the factored loading from a second-order analysis, which must satisfy one of three potential analysis approaches: nonlinear second-order analysis, elastic second-order analysis, or moment magnification. The nonlinear second-order analysis is too complex to be demonstrated in this chapter, as is the elastic second-order analysis. Both are typically performed using frame analysis software. More commonly, the moments due to second-order effects are estimated by multiplying the first-order solution by appropriately defined moment magnification (or amplification) factors.

ACI 318 contains simplified criteria for determining when slenderness amplification factors do not have to be calculated.

The moment magnification analysis takes the secondary bending stresses caused by *P*-delta effects, which are estimated by amplifying the primary bending moments by a moment magnification factor. The column is then designed for the axial force and the magnified bending moment using the short column design procedure. Nonsway and sway frames are treated separately. For a nonsway column, the magnification factor is given by ACI Sec. 6.6.4.5.2 as

$$\delta = \frac{C_m}{1 - \dfrac{P_u}{0.75P_c}} \quad \text{[ACI 6.6.4.5.2]}$$

$$\geq 1.0$$

P_u is the factored axial load, and the Euler critical load is given by

$$P_c = \frac{\pi^2 EI}{(kl_u)^2} \quad \text{[ACI 6.6.4.4.2]}$$

The flexural rigidity, in accordance with ACI Sec. 6.6.4.4.4, may be taken as either

(a) $\quad (EI)_{eff} = \dfrac{0.4 E_c I_g}{1 + \beta_{dns}}$

(b) $\quad (EI)_{eff} = \dfrac{(0.2 E_c I_g + E_s I_{se})}{1 + \beta_{dns}}$

(c) $\quad (EI)_{eff} = \dfrac{E_c I}{1 + \beta_{dns}}$

In equation (c), I is obtained from ACI Table 6.6.3.1.1(b) and is

$$I = \left[0.80 + \frac{25 A_{st}}{A_g} \right]\left(1 - \frac{M_u}{P_u h} - \frac{0.5 P_u}{P_o} \right) I_g$$

$$\geq 0.35 I_g$$

$$\leq 0.87 I_g$$

For simplification, it may be assumed that $\beta_{dns} = 0.6$. Then, as given by ACI Sec. R6.6.4.4.4, Eq. (a) reduces to

$$EI = 0.25 E_c I_g$$

The factor C_m corrects for a nonuniform bending moment on a column and is defined by ACI Sec. 6.6.4.5.3 as

$$C_m = 0.6 - \frac{0.4 M_1}{M_2} \quad \text{[ACI 6.6.4.5.3a]}$$

$$= 1.0 \text{ for columns with transverse loads between supports}$$

The column is now designed for the magnified moment given by ACI Sec. 6.6.4.5.1 as

$$M_c = \delta M_2 \quad \text{[ACI 6.6.4.5.1]}$$

The design procedure for a slender non-sway column is as follows.

step 1: Calculate forces on the column using first-order analysis.

step 2: Calculate end restraints and effective length.

step 3: Calculate C_m, P_c, δ, and M_c.

step 4: Use interaction diagrams to check the adequacy of the column.

Example 2.18

A 24 in diameter tied column with a 1.5 in cover to the $\frac{3}{8}$ in diameter ties is reinforced with 14 no. 9 grade 60 bars and has a concrete compressive strength of 4000 lbf/in². The column, which is braced against sidesway, has an unsupported height of 12 ft and is bent in

single curvature with factored end moments of $M_1 = M_2 = 400$ ft-kips. Determine whether the column can carry a factored axial load of 700 kips. The concrete is normal weight concrete.

Solution

$k = 1.0$ and $r = 6.0$ in. The slenderness ratio is

$$\frac{k l_u}{r} = \frac{(1.0)(12 \text{ ft})\left(12 \ \dfrac{\text{in}}{\text{ft}} \right)}{6.0 \text{ in}} = 24$$

In accordance with ACI Eq. 6.2.5(b), slenderness effects may be neglected when

$$\frac{k l_u}{r} \leq 34 + \frac{12 M_1}{M_2} \leq 40$$

$$34 - \frac{(12)(400 \text{ ft-kips})}{400 \text{ ft-kips}} = 22$$

$$< 24$$

The column is a long column, and secondary effects must be considered.

From ACI Sec. 19.2.2.1(b), the modulus of elasticity is

$$E_c = 57{,}000 \sqrt{f'_c}$$

$$= 57{,}000 \sqrt{4000 \ \frac{\text{lbf}}{\text{in}^2}}$$

$$= 3.61 \times 10^6 \text{ lbf/in}^2$$

The gross moment of inertia is given by

$$I_g = \frac{\pi (12 \text{ in})^4}{4} = 16{,}288 \text{ in}^4$$

The effective flexural rigidity is given by ACI Sec. R6.6.4.4.4 as

$$EI = 0.25 E_c I_g$$

$$= (0.25)\left(3.61 \times 10^3 \ \frac{\text{kips}}{\text{in}^2} \right)(16{,}288 \text{ in}^4)$$

$$= 14.68 \times 10^6 \text{ in}^2\text{-kips}$$

The critical load is given by ACI Eq. 6.6.4.4.2 as

$$P_c = \frac{\pi^2 EI}{(k l_u)^2} = \frac{\pi^2 (14.68 \times 10^6 \text{ in}^2\text{-kips})}{\left((1.0)(12 \text{ ft})\left(12 \ \dfrac{\text{in}}{\text{ft}} \right) \right)^2}$$

$$= 6987 \text{ kips}$$

The moment correction factor is defined by ACI Eq. 6.6.4.5.3a as

$$C_m = 0.6 - \frac{0.4M_1}{M_2}$$

$$= 0.6 + \frac{(0.4)(400 \text{ ft-kips})}{400 \text{ ft-kips}}$$

$$= 1.0$$

The moment magnification factor is given by ACI Eq. 6.6.4.5.2 as

$$\delta = \frac{C_m}{1 - \dfrac{P_u}{0.75 P_c}} = \frac{1.0}{1 - \dfrac{700 \text{ kips}}{(0.75)(6987 \text{ kips})}}$$

$$= 1.15$$

The magnified end moment is given by ACI Eq. 6.6.4.5.1 as

$$M_c = \delta M_2 = (1.15)(400 \text{ ft-kips}) = 461 \text{ ft-kips}$$

From Ex. 2.17, $\gamma = 0.80$, $\rho = 0.031$, and

$$\frac{M_u}{A_g h} = \frac{(461 \text{ ft-kips})\left(12 \dfrac{\text{in}}{\text{ft}}\right)}{\pi (12 \text{ in})^2 (24 \text{ in})} = 0.51 \text{ kip/in}^2$$

From App. 2.D, with $\gamma = 0.75$, the design axial stress is

$$\frac{P_u}{A_g} = 0.4 \text{ kip/in}^2$$

From App. 2.E, with $\gamma = 0.90$, the design axial stress is

$$\frac{P_u}{A_g} = 1.2 \text{ kips/in}^2$$

By interpolation, for $\gamma = 0.80$, the allowable design axial stress is

$$\frac{P_u}{A_g} = 0.4 \frac{\text{kip}}{\text{in}^2} + \left(0.8 \frac{\text{kip}}{\text{in}^2}\right)\left(\frac{5}{15}\right) = 0.67 \text{ kip/in}^2$$

The maximum factored axial load that the column can carry is

$$P_u = 0.67 A_g$$

$$= (0.67)\pi(12 \text{ in})^2$$

$$= 303 \text{ kips}$$

$$< 700 \text{ kips}$$

The column cannot carry the axial load of 700 kips.

Long Column with Sway

The magnification factor for end moments produced by the loads that cause sway may be calculated in accordance with ACI Sec. 6.6.4.6.2 in one of three ways, either by Eq. (a), by Eq. (b), or by second-order elastic analysis. If $\delta_s > 1.5$, only Eq. (b) or second-order elastic analysis are permitted.

(a) $\delta_s = \dfrac{1}{1 - Q} \geq 1.0$

(b) $\delta_s = \dfrac{1}{1 - \dfrac{\sum P_u}{0.75 \sum P_c}} \geq 1.0$

For Eq. (a) and Eq. (b), the summations extend over all the columns in a story. The sway moments are multiplied by the magnification factor, and the non-sway moments are added, in accordance with ACI Sec. 6.6.4.6.1, to give the final design end moments in the column of

$$M_1 = M_{1ns} + \delta_s M_{1s} \quad \text{[ACI 6.6.4.6.1a]}$$
$$M_2 = M_{2ns} + \delta_s M_{2s} \quad \text{[ACI 6.6.4.6.1b]}$$

Example 2.19

A 24 in diameter, tied column with a 1.5 in cover to the $\frac{3}{8}$ in diameter ties is reinforced with 14 no. 9 grade 60 bars and has a concrete compressive strength of 4000 lbf/in^2. The column is not braced and has an unsupported height of 12 ft, an effective length factor of 1.3, and factored end moments due to sway and non-sway moments of $M_{2s} = 300$ ft-kips and $M_{2ns} = 50$ ft-kips. In the story where the column is located, the sum of the column critical loads is $\Sigma P_c = 29{,}600$ kips, and the sum of the factored column loads is $\Sigma P_u = 2700$ kips. Determine whether the column can carry a factored axial load of 900 kips.

Solution

The slenderness ratio with $r = 6$ in is

$$\frac{kl_u}{r} = \frac{(1.3)(12 \text{ ft})\left(12 \dfrac{\text{in}}{\text{ft}}\right)}{6 \text{ in}} = 31.2$$

In accordance with ACI Sec. 6.2.5, slenderness effects must be considered when

$$\frac{kl_u}{r} > 22$$

$$31.2 > 22 \quad \text{[consider slenderness]}$$

The moment magnification factor for the sway moments is given by ACI Eq. 6.6.4.6.2b as

$$\delta_s = \cfrac{1}{1 - \cfrac{\sum P_u}{0.75 \sum P_c}}$$

$$= \cfrac{1}{1 - \cfrac{2700 \text{ kips}}{(0.75)(29{,}600 \text{ kips})}}$$

$$= 1.14$$

The magnified end moment is given by ACI Eq. 6.6.4.6.1b as

$$M_2 = M_{2ns} + \delta_s M_{2s} = 50 \text{ ft-kips} + (1.14)(300 \text{ ft-kips})$$
$$= 392 \text{ ft-kips}$$

From Ex. 2.17, $\gamma = 0.80$, $\rho_g = 0.031$, and

$$\frac{M_2}{A_g h} = \frac{(392 \text{ ft-kips})\left(12 \frac{\text{in}}{\text{ft}}\right)}{(452 \text{ in}^2)(24 \text{ in})} = 0.43 \text{ kip/in}^2$$

From App. 2.D, with $\gamma = 0.75$, the design axial stress is

$$\frac{P_u}{A_g} = 1.3 \text{ kips/in}^2$$

From App. 2.E, with $\gamma = 0.90$, the design axial stress is

$$\frac{P_u}{A_g} = 1.7 \text{ kips/in}^2$$

By interpolation, for $\gamma = 0.80$, the design axial stress is

$$\frac{P_u}{A_g} = 1.3 \frac{\text{kips}}{\text{in}^2} + \left(0.4 \frac{\text{kip}}{\text{in}^2}\right)\left(\frac{5}{15}\right) = 1.43 \text{ kips/in}^2$$

The maximum factored axial load that the column can carry is

$$P_u = 1.43 A_g = (1.43)(452 \text{ in}^2) = 648 \text{ kips}$$

$$< 900 \text{ kips} \quad \text{[The column is unsatisfactory.]}$$

11. DEVELOPMENT AND SPLICE LENGTH OF REINFORCEMENT

Nomenclature

A_b	area of an individual bar	in^2
A_{tr}	total cross-sectional area of all transverse reinforcement that is within the spacings and that crosses the potential plane of splitting through the reinforcement being developed	in^2
c	spacing or cover dimension	in
c_b	the lesser of the distance from center of bar to nearest concrete surface and one-half of the center-to-center spacing of the bars being developed	in
d_b	nominal diameter of bar	in
f_{ct}	average splitting tensile compressive strength of lightweight aggregate concrete	kips/in^2
h	overall thickness of member	–
K_{tr}	transverse reinforcement index, $40A_{tr}/sn$	in
l_a	additional embedment length at support or at point of inflection	in
l_d	development length, l_{db} applicable modification factors	in
l_{db}	basic development length	in
l_{dh}	development length of standard hook in tension, measured from critical section to outside end of hook (straight embedment length between critical section and start of hook (point of tangency) plus radius of bend and one bar diameter), $l_{hb} \times$ applicable modification factors	in
l_{hb}	basic development length of standard hook in tension	in
l_s	lap splice length	in
n	number of bars or wires being spliced or developed along the plane of splitting	–
s	maximum center-to-center spacing of transverse reinforcement within l_d	in

Symbols

β_b	ratio of area of reinforcement cutoff to total area of tension reinforcement at section	–
λ	lightweight aggregate concrete factor	–
ψ_c	cover factor	–
ψ_e	coating factor	–
ψ_r	confining reinforcement factor	–
ψ_s	reinforcement size factor	–
ψ_t	reinforcement location factor	–

Development Length of Straight Bars in Tension[12]

The development length for tension reinforcement is given by ACI Sec. 25.4.2.3 as

$$\frac{l_d}{d_b} = \frac{0.075 f_y \psi_t \psi_e \psi_s}{\lambda \dfrac{\sqrt{f'_c}\,(c_b + K_{tr})}{d_b}} \quad \text{[ACI 25.4.2.3a]}$$

$$l_d \geq 12 \text{ in}$$

$$\frac{c_b + K_{tr}}{d_b} \leq 2.5$$

As a design simplification, it is permitted to use $K_{tr} = 0$.

From ACI Table 25.4.2.4,

$$\psi_t \psi_e \leq 1.7$$

From ACI Sec. 25.4.1.4,

$$\sqrt{f'_c} \leq 100 \text{ lbf/in}^2$$

Values of the modification factors for the development of straight deformed bars in tension are given in ACI Table 25.4.2.4 and are listed here.

The reinforcement location factor, ψ_t, is equal to

- 1.3 for horizontal reinforcement placed with more than 12 in of concrete below

- 1.0 for all other cases

The reinforcement coating factor, ψ_e, is equal to

- 1.5 for epoxy-coated bars with cover $< 3d_b$ or clear spacing $< 6d_b$

- 1.2 for all other epoxy-coated bars

- 1.0 for uncoated bars

The reinforcement size factor, ψ_s, is equal to

- 1.0 for no. 7 and larger bars

- 0.8 for no. 6 and smaller bars

The lightweight concrete factor, λ, is equal to

- 0.75 for lightweight concrete

- $\dfrac{f_{ct}}{6.7\sqrt{f'_c}} \leq 1.0$ when f_{ct} is specified

- 1.0 for normal weight concrete

The derivation of the transverse reinforcement index is illustrated in Fig. 2.19.

ACI Sec. 25.4.2.2 also provides the following simplified, conservative values.

- Using a minimum clear spacing of $2d_b$, a minimum clear cover to flexural reinforcement of d_b, and in the absence of stirrups,

$$\frac{l_d}{d_b} = \frac{0.04 f_y \psi_t \psi_e}{\lambda \sqrt{f'_c}} \quad \text{[for } d_b \text{ no. 6 or smaller]}$$

$$\frac{l_d}{d_b} = \frac{0.05 f_y \psi_t \psi_e}{\lambda \sqrt{f'_c}} \quad \text{[for } d_b \text{ no. 7 or larger]}$$

- Using a minimum clear spacing of d_b, a minimum clear cover to flexural reinforcement of d_b, and minimum stirrups specified in ACI Table 9.6.3.3,

$$\frac{l_d}{d_b} = \frac{0.04 f_y \psi_t \psi_e}{\lambda \sqrt{f'_c}} \quad \text{[for } d_b \text{ no. 6 or smaller]}$$

$$\frac{l_d}{d_b} = \frac{0.05 f_y \psi_t \psi_e}{\lambda \sqrt{f'_c}} \quad \text{[for } d_b \text{ no. 7 or larger]}$$

- For all other cases,

$$\frac{l_d}{d_b} = \frac{0.06 f_y \psi_t \psi_e}{\lambda \sqrt{f'_c}} \quad \text{[for } d_b \text{ no. 6 or smaller]}$$

$$\frac{l_d}{d_b} = \frac{0.075 f_y \psi_t \psi_e}{\lambda \sqrt{f'_c}} \quad \text{[for } d_b \text{ no. 7 or larger]}$$

Using uncoated, grade 60 reinforcement, where no more than 12 in of normal weight concrete is below the bar gives

$$\psi_e = \psi_t = \lambda = 1.0$$

The resulting values are shown in Table 2.4.

Table 2.4 *Values of l_d/d_b for Grade 60 Bars with $\psi_e = \psi_t = \lambda = 1.0$*

f'_c (lbf/in²)	bar size ≤ no. 6		bar size ≥ no. 7	
	case 1[a]	case 2[b]	case 1[a]	case 2[b]
3000	44	66	55	82
3500	41	61	51	76
4000	38	57	47	71
4500	36	54	45	67
5000	34	51	42	64

[a]For case 1, clear cover ≥ d_b and clear spacing ≥ $2d_b$; or, clear cover ≥ d_b, clear spacing ≥ d_b, and with minimum stirrups.
[b]For case 2, all other conditions apply.

For bundled bars, ACI Sec. 25.6.1.5 specifies that the development length for an individual bar in a bundle is that for an individual bar increased by 20% for a three-bar bundle and 33% for a four-bar bundle. No increase is

Figure 2.19 *Derivation of K_{tr}*

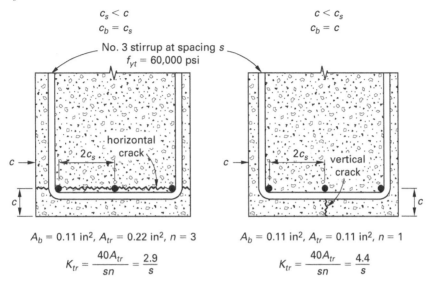

$c_s < c$
$c_b = c_s$

No. 3 stirrup at spacing s
$f_{yt} = 60{,}000$ psi

horizontal
crack

$2c_s$

c

c

$A_b = 0.11$ in^2, $A_{tr} = 0.22$ in^2, $n = 3$

$$K_{tr} = \frac{40A_{tr}}{sn} = \frac{2.9}{s}$$

$c < c_s$
$c_b = c$

$2c_s$

vertical
crack

c

c

$A_b = 0.11$ in^2, $A_{tr} = 0.11$ in^2, $n = 1$

$$K_{tr} = \frac{40A_{tr}}{sn} = \frac{4.4}{s}$$

required for a two-bar bundle. The equivalent diameter d_b of a bundle is specified in ACI Sec. 25.6.1.6 as that of a bar with an area equal to that of the bundle.

ACI Sec. 25.4.10.1 specifies that when excess reinforcement is provided in a member, the development length may be reduced by multiplying by the factor $A_{s(\text{required})}/A_{s(\text{provided})}$. This reduction factor may not be applied for shrinkage and temperature reinforcement, integrity reinforcement, positive moment reinforcement, seismic force resisting systems, headed bars, where bars are required to be continuous, and tension lap splices.

Example 2.20

The simply supported reinforced concrete beam of normal weight concrete shown is reinforced with grade 60 bars and has a concrete compressive strength of 3000 lbf/in^2. The maximum moment in the beam occurs at a point 2 ft from the end of the bars, and 10% more flexural reinforcement is provided than is required. Determine whether the development length available is satisfactory.

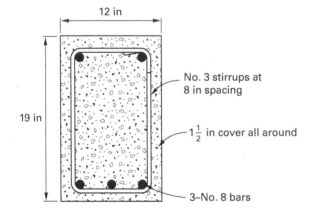

12 in

19 in

No. 3 stirrups at
8 in spacing

$1\frac{1}{2}$ in cover all around

3–No. 8 bars

Solution

From ACI Sec. 25.4.2.4,

$$\psi_t = \psi_e = \psi_s = \lambda = 1.0$$

The excess reinforcement factor is

$$\begin{aligned} E_{xr} &= \frac{A_{s(\text{required})}}{A_{s(\text{provided})}} \\ &= \frac{100 \text{ in}^2}{110 \text{ in}^2} \\ &= 0.91 \end{aligned}$$

The cover dimension to the flexural reinforcement is

$$c = 1.5 \text{ in} + 0.375 \text{ in} + \frac{1.0 \text{ in}}{2} = 2.375 \text{ in}$$

The spacing dimension to the flexural reinforcement is

$$\begin{aligned} c_s &= \frac{12.0 \text{ in} - (2)(1.5 \text{ in}) - (2)(0.375 \text{ in}) - 1.0 \text{ in}}{4} \\ &= 1.81 \text{ in} \quad [\text{horizontal cracking governs}] \\ c_b &= 1.81 \text{ in} \end{aligned}$$

The area of transverse reinforcement crossing the horizontal crack is

$$A_{tr} = (2)(0.11 \text{ in}^2) = 0.22 \text{ in}^2$$

The number of bars being developed along the cracking plane is $n = 3$.

The transverse reinforcement index is

$$K_{tr} = \frac{40A_{tr}}{sn} = \frac{(40)(0.22 \text{ in}^2)}{(8 \text{ in})(3)} = 0.367 \text{ in}$$

$$\frac{c_b + K_{tr}}{d_b} = \frac{1.81 \text{ in} + 0.367 \text{ in}}{1.0 \text{ in}} = 2.18$$

$$< 2.5 \quad [\text{satisfactory}]$$

From ACI Eq. 25.4.2.3a,

$$\frac{l_d}{d_b} = \frac{0.075 f_y \psi_t \psi_e \psi_s E_{xr}}{\lambda \sqrt{f'_c}(c_b + K_{tr})}$$

$$= \frac{(0.075)\left(60{,}000 \ \dfrac{\text{lbf}}{\text{in}^2}\right)(1)(1)(1)(0.91)}{(1)\sqrt{3000 \ \dfrac{\text{lbf}}{\text{in}^2}}\ (2.18)}$$

$$= 34.3$$

The development length required is

$$l_d = 34.3 d_b = (34.3)(1.0 \text{ in})$$

$$= 34.3 \text{ in}$$

$$> 24 \text{ in}$$

The development length available is unsatisfactory for straight bars.

Development Length of Straight Bars in Compression

The basic development length for bars in compression is specified in ACI Sec. 25.4.9.2 as

$$l_{dc} = \frac{0.02 d_b f_y \psi_r}{\lambda \sqrt{f'_c}}$$

$$\geq 0.0003 d_b f_y \begin{bmatrix} \text{governs for normal weight} \\ \text{concrete when } f'_c \geq 4444 \text{ lbf/in}^2 \\ \text{and } \psi_r = 1.0 \end{bmatrix}$$

Table 2.5 shows values of l_{dc} for $\psi_r = 1.0$. The minimum permitted value of l_{dc} is 8 in.

When excess reinforcement is provided as specified in ACI Sec. 25.4.10.1, the development length may be reduced by multiplying by the factor

$$E_{xr} = \frac{A_{s(\text{required})}}{A_{s(\text{provided})}}$$

Table 2.5 *Values of l_{dc} for Grade 60 Bars in Compression for $\psi_r = 1.0$*

bar size no.	concrete strength (lbf/in²)			
	3000	3500	4000	≥ 4444
3	8.2	7.6	7.1	6.8
4	11.0	10.1	9.5	9.0
5	13.7	12.7	11.9	11.3
6	16.4	15.2	14.2	13.5
7	19.2	17.8	16.6	15.8
8	21.9	20.3	19.0	18.0
9	24.7	22.9	21.4	20.3
10	27.8	25.8	24.1	22.9
11	30.9	28.6	26.8	25.4

However, this reduction is not permitted

- at noncontinuous supports

- at locations where anchorage or development for f_y is required

- where bars are required to be continuous

- for headed and mechanically anchored deformed reinforcement

- in seismic-force-resisting systems in structures assigned to seismic design categories D, E, or F

The confining reinforcement factor, ψ_r, is specified in ACI Table 25.4.9.3. ψ_r has a value of 0.75 when any of the following confining reinforcement is provided.

- a spiral

- a circular continuously wound tie with $d_b \geq \frac{1}{4}$ in and pitch = 4 in

- no. 4 bar or D20 wire ties in accordance with ACI Sec. 25.7.2 spaced ≤ 4 in on center

- hoops in accordance with ACI Sec. 25.7.4 spaced ≤ 4 in on center

For other confining configurations, $\psi_r = 1.0$.

The requirements for bundled bars in compression are identical to those applicable to bundled bars in tension.

Example 2.21

A spirally reinforced column of normal weight concrete with a 0.25 in diameter spiral at a pitch of 4 in is reinforced with no. 9 grade 60 bars in bundles of three and has a concrete compressive strength of 4500 lbf/in². The compression reinforcement provided is 15% more than is required. Determine the required development length of an individual bar.

Solution

From ACI Sec. 25.6.1.5, the development length of an individual bar in a three-bar bundle is increased 20%.

From ACI Sec. 25.4.10.1, the excess reinforcement factor is

$$E_{xr} = \frac{100}{115} = 0.87$$

And the confining reinforcement factor is

$$\psi_r = 0.75$$

For a concrete strength in excess of 4444 lbf/in², ACI Sec. 25.4.9.2 and Sec. 25.4.10.1 give the required development length as

$$l_d = 1.2 E_{xr} \psi_r l_{dc} = (1.2)(0.87)(0.75)(0.0003) d_b f_y$$

$$= (1.2)(0.87)(0.75)(0.0003)(1.13 \text{ in}) \left(60{,}000 \ \frac{\text{lbf}}{\text{in}^2} \right)$$

$$= 15.9 \text{ in}$$

Development of Hooked Bars in Tension

The basic development length for hooked bars in tension is specified in ACI Sec. 25.4.3.1 as

$$l_{hb} = \left(\frac{f_y \psi_e \psi_c \psi_r}{50 \lambda \sqrt{f_c'}} \right) d_b$$

Values of the modification factors for the development of hooked bars in tension are given in ACI Table 25.4.3.2 and are listed as follows.

The reinforcement coating factor, ψ_e, is equal to

- 1.2 for epoxy-coated bars
- 1.0 for all other cases

The cover factor, ψ_c, is equal to

- 0.7 for hooked bars no. 11 and smaller with side cover at least 2.5 in and end cover, for a 90° hook, at least 2 in
- 1.0 for all other cases

The confining reinforcement factor, ψ_r, is equal to

- 0.8 for
  - 90° hooks on bars no. 11 and smaller, with ties provided that are perpendicular to the bar being developed over the full development length and at a spacing no greater than $3d_b$

  - 90° hooks on bars no. 11 and smaller, with ties provided that are parallel to the bar being developed along the tail extension of the hook plus bend and at a spacing no greater than $3d_b$

  - 180° hooks on bars no. 11 and smaller, with ties provided that are perpendicular to the bar being developed over the full development length at a spacing no greater than $3d_b$; the first tie must enclose the bent portion of the hook within $2d_b$ from the outside of the bend

- 1.0 for all other cases

Details of the location and spacing of ties is provided in ACI Fig. R25.4.3.2a and Fig. R25.4.3.2b.

The lightweight aggregate concrete factor, λ, is equal to

- 0.75 for lightweight concrete
- 1.0 for normal weight concrete

Under the conditions described earlier in this section, l_{hb} may also be multiplied by the excess reinforcement factor

$$E_{xr} = \frac{A_{s(\text{required})}}{A_{s(\text{provided})}}$$

In accordance with ACI Sec. 25.4.3.1, the minimum allowable value for l_{dh} is the larger of $8d_b$ and 6 in.

Example 2.22

Assume that the reinforced concrete beam for Ex. 2.20 is reinforced with hooked bars. Determine whether the development length provided for the hooked bars is satisfactory.

Solution

Because the bars terminate at the discontinuous end of the beam and the cover provided is less than 2.5 in, the modification factor, ψ_r, is 1.0 and ties must be provided over the full development length at a maximum spacing of

$$s = 3d_b = (3)(1 \text{ in})$$

$$= 3 \text{ in}$$

The excess reinforcement factor derived in Ex. 2.20 is

$$E_{xr} = 0.91$$

All other modification factors are equal to unity.

The required development length is given by ACI Sec. 25.4.3.1 as

$$l_{dh} = E_{xr}l_{hb} = \frac{(0.91)\left(1200 \ \frac{\text{lbf}}{\text{in}^2}\right)d_b}{\lambda\sqrt{f_c'}}$$

$$= \frac{(0.91)\left(1200 \ \frac{\text{lbf}}{\text{in}^2}\right)(1 \ \text{in})}{(1.0)\sqrt{3000 \ \frac{\text{lbf}}{\text{in}^2}}}$$

$$= 20 \ \text{in}$$
$$< 24 \ \text{in}$$

The development length provided is satisfactory.

Curtailment of Reinforcement

ACI Sec. 9.7.3.3 and Sec. 9.7.3.4 specify that reinforcement may extend a distance beyond the theoretical cutoff point not less than the effective depth of the member or twelve times the bar diameter and may extend beyond the point at which it is fully stressed not less than the development length. This is illustrated in Fig. 2.20, where the four no. 9 bars are assumed to be fully stressed at the center of the simply supported beam. These requirements are not necessary at supports of simple spans or at the free ends of cantilevers.

Figure 2.20 *Curtailment of Reinforcement*

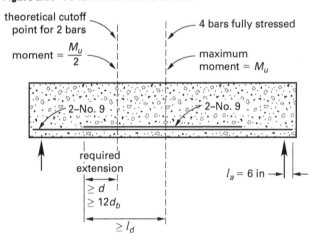

In addition to the previous requirements, ACI Sec. 9.7.3.5 requires that one of the following conditions be satisfied at the physical cutoff point.

- The factored shear force at the cutoff point does not exceed two-thirds of the shear capacity, ϕV_n.

- Additional stirrups with a minimum area of $60b_ws/f_{yt}$ are provided along the terminated bar for a distance of $0.75d$ at a maximum spacing of $d/8\beta_b$.

- For no. 11 bars and smaller, the continuing reinforcement provides twice the flexural capacity required, and the factored shear force at the cutoff point does not exceed three-fourths of the shear capacity, ϕV_n.

Example 2.23

The simply supported reinforced concrete beam, of normal weight concrete, shown in Fig. 2.20 is reinforced with four no. 9 grade 60 bars and has a concrete strength of 3000 lbf/in². The maximum moment in the beam occurs at midspan, and the flexural reinforcement is fully stressed. The bending moment reduces to 50% of its maximum value at a point 3 ft from midspan. The effective depth is 16 in, the beam width is 12 in, and the development length of the no. 9 bars may be taken as $55d_b$. The factored shear force is less than two-thirds the shear capacity of the section everywhere along the span. Determine the distance from midspan at which two of the no. 9 bars may be terminated.

Solution

From ACI Sec. 9.7.3.3 and Sec. 9.7.3.4, the physical cutoff point may be located a minimum distance from midspan given by the largest of

- $12d_b + 36 \ \text{in} = (12)(1.128 \ \text{in}) + 36 \ \text{in} = 49.5 \ \text{in}$

- $d + 36 \ \text{in} = 16 \ \text{in} + 36 \ \text{in} = 52 \ \text{in}$

- $l_d = 55d_b = (55)(1.128 \ \text{in}) = 62 \ \text{in}$ [governs]

Development of Positive Moment Reinforcement

To allow for variations in the applied loads, ACI Sec. 9.7.3.8.1 and Sec. 9.7.3.8.2 require a minimum of one-third of the positive reinforcement in a simply supported beam or one-fourth of the positive reinforcement in a continuous beam to extend not less than 6 in into the support. This is shown in Fig. 2.21.

Figure 2.21 *Positive Moment Reinforcement*

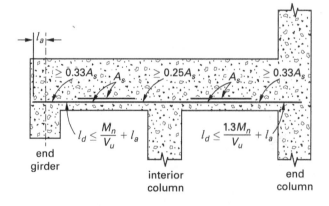

To ensure that allowable bond stresses are not exceeded, ACI Sec. 9.7.3.8.3 requires a bar diameter to be chosen such that its development length, in the case where the end of the reinforcement is not confined by a compressive reaction, is given by

$$l_d \leq \frac{M_n}{V_u} + l_a$$

M_n is the nominal strength, assuming all reinforcement is stressed to the specified yield stress, f_y; V_u is the factored shear force at the section; and l_a is the embedment length beyond the center of the support. It is considered that the reinforcement is unconfined at a point of inflection and when ends of the beam frame into girders.

Alternatively, if the end of the reinforcement is confined by a compressive reaction, ACI Sec. 9.7.3.8.3 requires a bar diameter to be chosen such that its development length is given by

$$l_d \leq 1.3 \frac{M_n}{V_u} + l_a$$

It is considered that a simply supported beam, as shown in Fig. 2.20, provides confinement to the reinforcement. At a point of inflection, ACI Sec. 9.7.3.8.3 limits the embedment length beyond the point of inflection to the greater of d or $12d_b$.

Alternatively, at a simple support, ACI Sec. 9.7.3.8.3 specifies that the reinforcement may terminate in a standard hook or a mechanical anchorage beyond the centerline of the support.

Example 2.24

Assume that the reinforced concrete beam for Ex. 2.23, which is shown in Fig. 2.20, has a factored end reaction of 30 kips and that the beam frames into concrete girders at each end. Determine whether the two no. 9 bars at the support satisfy the requirements for local bond.

Solution

The development length for the no. 9 bars was given in Ex. 2.23 as

$$l_d = 55d_b = 62 \text{ in}$$

The nominal flexural strength at the support is given by ACI Sec. 22.2 as

$$M_n = A_s f_y d \left(1 - \frac{0.59 A_s f_y}{b_w d f'_c} \right)$$

$$= (2 \text{ in}^2)\left(60 \frac{\text{kips}}{\text{in}^2}\right)(16 \text{ in})$$

$$\times \left(1 - \frac{(0.59)(2 \text{ in}^2)\left(60 \dfrac{\text{kips}}{\text{in}^2}\right)}{(12 \text{ in})(16 \text{ in})\left(3 \dfrac{\text{kips}}{\text{in}^2}\right)} \right)$$

$$= 1684 \text{ in-kips}$$

For a beam framing into a concrete girder, the ends of the reinforcement are not restrained by a compressive reaction, and the appropriate factors given by ACI Sec. 9.7.3.8.3 are

$$\frac{M_n}{V_u} + l_a = \frac{1684 \text{ in-kips}}{30 \text{ kips}} + 6 \text{ in} = 62.1 \text{ in}$$

$$> l_d$$

Local bond requirements are satisfied.

Development of Negative Moment Reinforcement

To allow for variation in the applied loads, ACI Sec. 9.7.3.8.4 requires a minimum of one-third of the negative reinforcement at a support to extend past the point of inflection at least the greatest of effective depth of the beam, 12 times the bar diameter, or one-sixteenth of the clear span. This is shown in Fig. 2.22.

Figure 2.22 *Negative Moment Reinforcement*

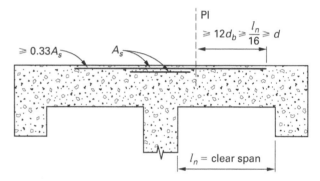

Example 2.25

The reinforced concrete continuous beam shown has an effective depth of 16 in, is reinforced with grade 60 bars, and has a concrete compressive strength of 3000 lbf/in². Determine the minimum length, x, at which the bars indicated may terminate.

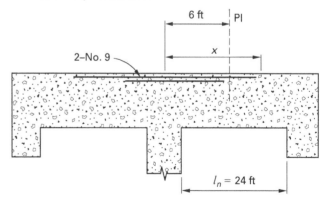

Solution

The cutoff point of the bars may be located a minimum distance beyond the point of inflection given by ACI Sec. 9.7.3.8.4 as the greater of

$$12d_b = (12)(1.128 \text{ in}) = 13.5 \text{ in}$$
$$= 13.5 \text{ in}$$
$$d = 16 \text{ in}$$
$$\frac{l}{16} = \frac{(24 \text{ ft})(12 \text{ in})}{16}$$
$$= 18 \text{ in} \quad [\text{governs}]$$

The minimum allowable length for x is given by

$$x = 6 \text{ ft} + 1.5 \text{ ft} = 7.5 \text{ ft}$$

Splices of Bars in Tension

Lap splices for bars in tension, in accordance with ACI Sec. 25.5.1.1, may not be used either for bars larger than no. 11 or for bundled bars. In flexural members, the transverse spacing between lap splices must not exceed one-fifth of the lap length, or 6 in. Within a bundle, individual bar splices must not overlap, and the lap length for each bar must be increased by 20% for a three-bar bundle and 33% for a four-bar bundle.

The length of a lap splice may not be less than 12 in nor less than the values given by ACI Table 25.5.2.1, which are

$$\text{class A splice length} = 1.0l_d$$
$$\text{class B splice length} = 1.3l_d$$

A class A splice may be used only when the reinforcement area provided is at least twice that required and when, in addition, not more than one-half of the total reinforcement is spliced within the lap length. Otherwise, as shown in Table 2.6, a class B tension lap splice is required.

Table 2.6 *Tension Lap Splices*

$\dfrac{A_s \text{ provided}}{A_s \text{ required}}$	maximum percentage of A_s spliced within lap length	
	50	100
2 or more	class A	class B
less than 2	class B	class B

In determining the relevant development length, all applicable modifiers are used with the exception of that for excess reinforcement. The development length, l_d, is determined by using the values for clear spacing, c_s, indicated in Fig. 2.23 and Fig. 2.24.

Figure 2.23 *Value of c_s for Lap Splices*

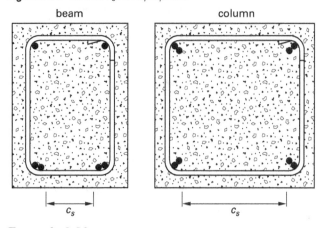

Example 2.26

A reinforced concrete beam is reinforced with two no. 8 grade 60 bars and has a concrete compressive strength of 3000 lbf/in². Both bars are lap spliced at the same location, and the development length of the no. 8 bars may be taken as $50d_b$. Determine the required splice length.

Solution

The development length is given as

$$l_d = 50d_b = (50)(1 \text{ in})$$
$$= 50 \text{ in}$$

Because both bars are spliced at the same location, a class B splice is required and the splice length is

$$l_s = 1.3l_d = (1.3)(50 \text{ in})$$
$$= 65 \text{ in}$$

Splices of Bars in Compression

In accordance with ACI Sec. 25.5.5.1, the length of a lap splice for bars in compression must not be less than 12 in and is given by

$$l_s = 0.0005f_y d_b \quad [f_y \leq 60{,}000 \text{ lbf/in}^2]$$
$$l_s = (0.0009f_y - 24)d_b \quad [f_y > 60{,}000 \text{ lbf/in}^2]$$

An increase in the lap length of 33% is required when the concrete strength is less than 3000 lbf/in². When bars of different sizes are lap spliced, ACI Sec. 25.5.5.4 specifies that the lap length must be the larger of the splice length of the smaller bar or the development length of the larger bar.

In accordance with ACI Sec. 10.7.5.2.1, lap lengths for columns may be reduced by 17% when ties are provided with an effective area of $0.0015hs$ and may be reduced by 25% in a spirally reinforced column. When tensile stress exceeding $0.5f_y$ occurs in a column, a class B

Figure 2.24 *Value of c_s in Slabs and Walls*

non-staggered staggered

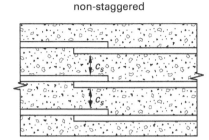 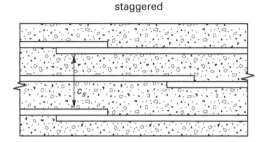

tension lap splice can be used. A class A tension lap splice is adequate provided that the tensile stress does not exceed $0.5f_y$, not more than one-half of the bars are spliced at the same location, and alternate splices are staggered by l_d; otherwise, a class B tension lap splice is required.

Compression lap splices of no. 18, no. 14, and smaller bars are permitted.

Example 2.27

The reinforced concrete column shown is reinforced with grade 60 bars that are fully stressed and has a concrete compressive strength of 4000 lbf/in². The column is subjected to compressive stress only. Determine the required lap splice for the no. 8 and no. 9 bars.

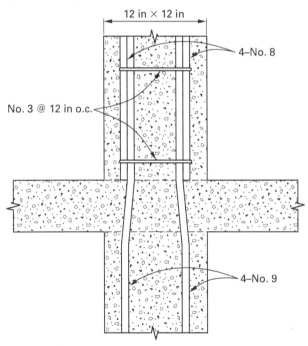

Solution

For the confining reinforcement indicated in the illustration, $\psi_r = 1.0$. The development length of a no. 9 bar is given by ACI Sec. 25.4.9.2 as

$$l_{dc} = \frac{0.02 d_b f_y \psi_r}{\sqrt{f'_c}} = \frac{(0.02)(1.128 \text{ in})\left(60,000 \ \dfrac{\text{lbf}}{\text{in}^2}\right)(1.0)}{\sqrt{4000 \ \dfrac{\text{lbf}}{\text{in}^2}}}$$

$$= 21.4 \text{ in}$$

To qualify for the 17% reduction in lap splice length of the no. 8 bars, ACI Sec. 10.7.5.2.1 requires the two arms of the no. 3 ties to have a minimum effective area of

$$\frac{A_t}{s} = 0.0015h = (0.0015)(12 \text{ in})\left(12 \ \frac{\text{in}}{\text{ft}}\right)$$

$$= 0.22 \text{ in}^2/\text{ft}$$

Two arms of the no. 3 ties provide an area of

$$\frac{A_t}{s} = 0.22 \text{ in}^2/\text{ft}$$

The 17% reduction applies.

The compression lap splice length of the no. 8 bars is given by ACI Sec. 25.5.5.1 and Sec. 12.7.5.2.1 as

$$l_{sc} = 0.0005 f_y d_b (0.83)$$

$$= (0.0005)\left(60,000 \ \frac{\text{lbf}}{\text{in}^2}\right)(1.0 \text{ in})(0.83)$$

$$= 25 \text{ in} \quad \text{[governs]}$$

$$> 21.4 \text{ in}$$

The required lap splice length is 25 in.

12. TWO-WAY SLAB SYSTEMS

Nomenclature

b_o	perimeter of critical section	in
c_1	size of rectangular or equivalent rectangular column, capital, or bracket measured in the direction of the span for which moments are being determined	in
c_2	size of rectangular or equivalent rectangular column, capital, or bracket measured transverse to the direction of the span for which moments are being determined	in
C	cross-sectional constant to define torsional properties $\left(\Sigma \left(1 - 0.63x/y \right) x^3 y \right)/3$; for T- or L-sections, it is permitted to be evaluated by dividing the section into separate rectangular parts and summing the values of C for each part.	–
E_{cb}	modulus of elasticity of beam concrete	lbf/in²
E_{cs}	modulus of elasticity of slab concrete	lbf/in²
I_b	moment of inertia about centroidal axis of gross section of beam	in⁴
I_s	moment of inertia about centroidal axis of gross section of slab $= h^3/12$ times width of slab defined in symbols α and β_t	in⁴
l_n	length of clear span in direction that moments are being determined, measured face-to-face of supports $= l_1 - c_1 \geq 0.65 l_1$	ft or in
l_1	length of span in direction that moments are being determined, measured center-to-center of supports	ft or in
l_2	length of span transverse to l_1, measured center-to-center of supports	ft or in
M_o	total factored static moment	ft-kips
q_u	factored load per unit area	kips/ft²
x	shorter overall dimension of rectangular part of cross section	in
y	longer overall dimension of rectangular part of cross section	in

Symbols

α_f	ratio of flexural stiffness of beam section to flexural stiffness of a slab width bounded laterally by center lines of adjacent panels (if any) on each side of the beam $= E_{cb}I_b/E_{cs}I_s$	–
α_s	constant used to compute V_c	–
α_{f1}	α_f in direction of l_1	–
α_{f2}	α_f in direction of l_2	–
β	ratio of long side to short side of column	–
β_t	ratio of torsional stiffness of edge beam section to flexural stiffness of a slab width equal to span length of beam, center-to-center of supports $= E_{cb}C/2E_{cs}I_s$	–

Design Terminology

The flexural stiffness parameter, α_f, is defined in ACI Sec. 2.2 as the ratio of the flexural stiffness of a beam section to the flexural stiffness of a slab width bounded laterally by the centerlines of adjacent panels on each side of the beam. The flexural stiffness parameter is given in ACI Sec. 8.10.2.7 as

$$\alpha_f = \frac{E_{cb}I_b}{E_{cs}I_s}$$

The moments of inertia of the beam, I_b, and the slab, I_s, are based on the cross sections given in ACI Sec. R8.4.1.8, which are shown in Fig. 2.25. The moduli of elasticity of the concrete in the beam, E_{cb}, and the concrete in the slab, E_{cs}, are obtained from ACI Sec. 19.2.2.1; for normal weight concrete, the modulus of elasticity is

$$E_c = 57,000\sqrt{f'_c}$$

The flexural stiffness parameter, α_f, is used to determine the minimum permitted slab thickness in ACI Sec. 8.3.1.2 and in the distribution of slab moments as indicated in ACI Sec. 8.10.5.

The torsional stiffness parameter, β_t, is defined in ACI Sec. 2.2 as the ratio of the torsional stiffness of an edge beam section to the flexural stiffness of a slab width equal to the span length of the beam, measured center to center from one support to the next. This parameter is used in the distribution of moments in the slab and edge beams as indicated in ACI Sec. 8.10.5. The torsional stiffness parameter given in ACI Sec. 8.10.5.2 is

$$\beta_t = \frac{E_{cb}C}{2E_{cs}I_s}$$

The cross-sectional constant for T and L sections is defined in ACI Sec. 8.10.5.2 as

$$C = \sum \left(1 - 0.63\left(\frac{x}{y} \right) \right)\left(\frac{x^3 y}{3} \right)$$

As shown in Fig. 2.26, the values of x and y are obtained by dividing a cross section into its separate rectangular parts with

$x =$ shorter overall dimension of a rectangular part of cross section

$y =$ longer overall dimension of a rectangular part of cross section

Figure 2.25 *Equivalent Beam and Slab Dimensions for Calculation of* α_f

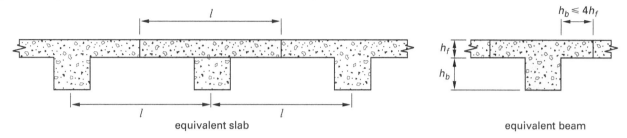

Figure 2.26 *Determination of Cross-Sectional Constant C*

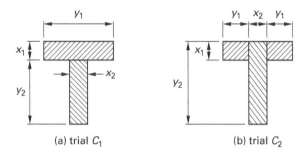

Summing the values of C for each part gives the final value for C. As shown in Fig. 2.26, it is possible to divide a cross section in several ways. The method that produces the maximum value of C is preferred.

Design Techniques[13]

The different types of slab systems are as follows.

- *One-way slabs* are supported by beams on two opposite sides and span in one direction between the beams.

- *Two-way slabs* span in two orthogonal directions and are supported by beams on all four sides. When the ratio of long span to short span exceeds two, the slab acts as a one-way slab spanning in the direction of the long slab.

  Several different methods may be used to design two-way slabs. The *direct design method* divides the slab into a column strip and a middle strip. The *equivalent frame method* divides the slab into rows of columns supporting the tributary slab width. The *yield line method* is based on limit state theory applied to an appropriate collapse mechanism.

- *Flat plates* are supported at the corners by columns without any supporting beams.

- *Flat slabs*, which are similar to flat plates, have increased depth at the columns in order to produce a drop panel.

- *Waffle slabs* are flat slabs constructed with voids formed in the soffit. At the columns, the voids are omitted in order to produce a solid slab similar to a drop panel.

Direct Design Method

ACI Sec. 8.10.2 permits the direct design method to be used provided the following conditions exist.

- A minimum of three continuous spans exist in each direction.

- Panels are rectangular with an aspect ratio not exceeding 2.

- Successive span lengths do not differ by more than one-third of the longer span.

- Columns are not offset by more than 10% of the span.

- Loading consists of uniformly distributed gravity loads with the service live load not exceeding twice the service dead load.

- For beam supported slabs, the ratio of the beam stiffnesses in two perpendicular directions, $\alpha_{f1}l_2^2/\alpha_{f2}l_1^2$, is between 0.2 and 5.0 where the moments of inertia of the equivalent beam and slab are based on the sections shown in Fig. 2.25.

The slab is divided into design strips, as shown in Fig. 2.27, with a column strip extending the lesser of $0.25l_1$ or $0.25l_2$ on each side of a column centerline. A middle strip consists of the remainder of the slab between column strips.

Example 2.28

A 9 in thick flat plate floor has plan dimensions between column centers of 24 ft and 28 ft as shown in Fig. 2.27. Calculate the widths of the column strip and the middle strip in the north-south direction.

Solution

From ACI Sec. 8.4.1.5, the width of the column strip is

$$w_c = 0.5l_{min} = (0.5)(24 \text{ ft}) = 12 \text{ ft}$$

The width of the middle strip in the direction of the longer span is

$$w_m = l_{min} - 0.5l_{min} = 24 \text{ ft} - 12 \text{ ft} = 12 \text{ ft}$$

Figure 2.27 *Details of Design Strips*

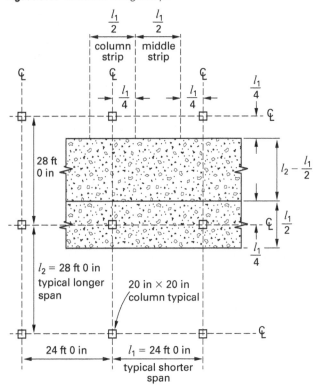

Design for Flexure

The total factored static moment on a panel is calculated by ACI Sec. 8.10.3.2 as

$$M_o = \frac{q_u l_2 l_n^2}{8} \qquad \text{[ACI 8.10.3.2]}$$

The total factored moment is now distributed in accordance with ACI Sec. 8.10.4.1 and Table 8.10.4.2 into positive and negative moments across the full width of the panel, depending on the support conditions, as shown in Fig. 2.28.

These distributed moments are now further subdivided between the column and middle strips as specified in ACI Sec. 8.10.5 through Sec. 8.10.6.3. That portion of the moment not attributed to a column strip is assigned to the corresponding half middle strips. The negative moment at an interior support is distributed to a column strip as shown in Table 2.7.

Table 2.7 *Percentage Distribution of Interior Negative Moment to Column Strip*

l_2/l_1	0.5	1	2
$\alpha_{f1} l_2/l_1 = 0$	75	75	75
$\alpha_{f1} l_2/l_1 \geq 1.0$	90	75	45

The negative moment at an exterior support is distributed to a column strip as shown in Table 2.8.

Table 2.8 *Percentage Distribution of Exterior Negative Moment to Column Strip*

l_2/l_1		0.5	1	2
$\alpha_{f1} l_2/l_1 = 0$	$\beta_t = 0$	100	100	100
	$\beta_t \geq 2.5$	75	75	75
$\alpha_{f1} l_2/l_1 \geq 1.0$	$\beta_t = 0$	100	100	100
	$\beta_t \geq 2.5$	90	75	45

The positive moment at midspan is distributed to a column strip as shown in Table 2.9.

Table 2.9 *Percentage Distribution of Positive Moment to Column Strip*

l_2/l_1	0.5	1	2
$\alpha_{f1} l_2/l_1 = 0$	60	60	60
$\alpha_{f1} l_2/l_1 \geq 1.0$	90	75	45

Beams at panel edges must be assigned the percentage of column strip moment as shown in Table 2.10.

Table 2.10 *Percentage Distribution of Column Strip Moments to Beams Between Supports*

$\alpha_{f1} l_2/l_1$	0	≥ 1.0
% assigned to beam	0	85

Example 2.29

Assume that the flat plate floor for Ex. 2.28, which is shown in Fig. 2.27, supports a total factored distributed load of 200 lbf/ft^2. The column size is 20 in by 20 in. Determine the factored moments in the column strip and middle strip in the north-south direction.

Solution

From ACI Sec. 8.10.3.2.1, the clear span in the north-south direction is

$$l_n = l_2 - c_1 = 28 \text{ ft} - \frac{20 \text{ in}}{12 \frac{\text{in}}{\text{ft}}}$$

$$= 26.33 \text{ ft}$$
$$> 0.65 l_1 \quad \text{[satisfactory]}$$

From ACI Sec. 8.10.3.2, the total factored static moment is

$$M_o = \frac{q_u l_1 l_n^2}{8} = \frac{\left(0.20 \frac{\text{kip}}{\text{ft}^2}\right)(24 \text{ ft})(26.33 \text{ ft})^2}{8}$$

$$= 416 \text{ ft-kips}$$

Figure 2.28 M_o Distribution Factors

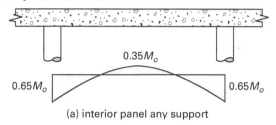

0.35M_o

0.65M_o 0.65M_o

(a) interior panel any support

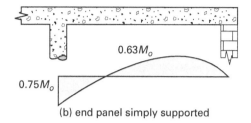

0.63M_o

0.75M_o

(b) end panel simply supported

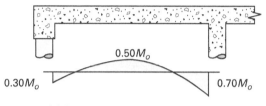

0.50M_o

0.30M_o 0.70M_o

(c) flat plate, end panel with edge beam

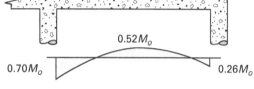

0.52M_o

0.70M_o 0.26M_o

(d) flat plate, end panel without edge beam

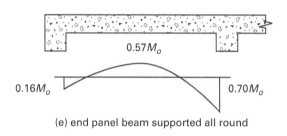

0.57M_o

0.16M_o 0.70M_o

(e) end panel beam supported all round

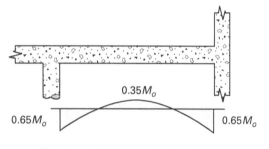

0.35M_o

0.65M_o 0.65M_o

(f) end panel fully restrained end

From ACI Sec. 8.10.4.1 and Fig. 2.28(a), the total positive moment across the panel is

$$M_m = 0.35 M_o = (0.35)(416 \text{ ft-kips})$$
$$= 146 \text{ ft-kips}$$

From ACI Table 8.10.5.5 and Table 2.9, $\alpha_{f1} l_2 / l_1 = 0$ and the column strip positive moment at midspan is

$$M_{cm} = 0.60 M_m = (0.60)(146 \text{ ft-kips})$$
$$= 87 \text{ ft-kips}$$

From ACI Sec. 8.10.6, the middle strip positive moment at midspan is

$$M_{mm} = M_m - M_{cm} = 146 \text{ ft-kips} - 87 \text{ ft-kips}$$
$$= 59 \text{ ft-kips}$$

From ACI Sec. 8.10.4.1 and Fig. 2.28(a), the total negative moment across the panel is

$$M_c = 0.65 M_o = (0.65)(416 \text{ ft-kips})$$
$$= 270 \text{ ft-kips}$$

From ACI Table 8.10.5.1 and Table 2.7, $\alpha_{f1} l_2 / l_1 = 0$ and the column strip negative moment at an interior support is

$$M_{cc} = 0.75 M_c = (0.75)(270 \text{ ft-kips})$$
$$= 203 \text{ ft-kips}$$

From ACI Sec. 8.10.6.1, the middle strip negative moment at an interior support is

$$M_{mc} = M_c - M_{cc}$$
$$= 270 \text{ ft-kips} - 203 \text{ ft-kips}$$
$$= 67 \text{ ft-kips}$$

Design for Shear

The design for shear at column supports must consider both flexural or one-way shear and punching or two-way shear, as shown in Fig. 2.29. The flexural shear capacity of the panel, in a direction parallel to the side l_1, is given by ACI Sec. 22.5.5.1 as

$$\phi V_c = 2\phi d b_w \lambda \sqrt{f_c'}$$

Figure 2.29 *Critical Sections for Shear*

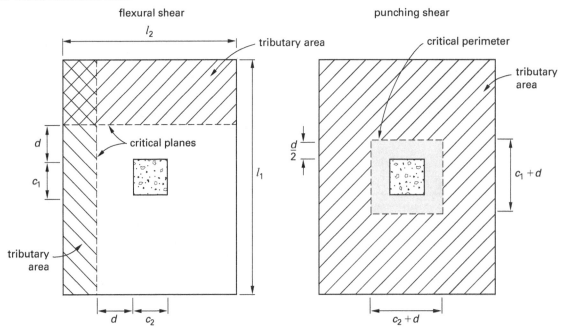

Figure 2.30 *Corner and Edge Columns*

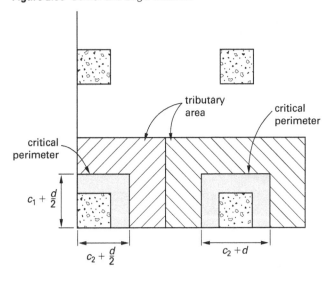

Figure 2.31 *Reduction in Critical Perimeter*

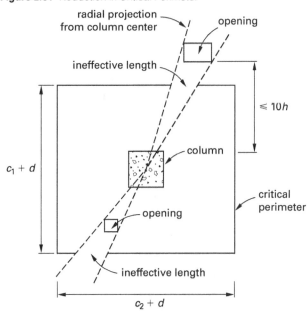

The critical perimeter for punching shear is specified in ACI Sec. 22.6.4.1 as being a distance from the face of the column equal to one-half of the effective depth. The length of the critical perimeter is given by

$$b_o = (2)(c_1 + c_2) + 4d$$

For a corner column located less than $d/2$ from the edge of a panel, the critical perimeter is two sided. For a similarly situated edge column, it is three sided, as shown in Fig. 2.30.

Openings in a panel within 10 times the thickness of the panel from the edge of a column reduce the critical perimeter, as shown in Fig. 2.31.

In accordance with ACI Sec. 8.5.4.2, openings of any size are permitted in the area common to two intersecting middle strips. In the area common to two intersecting column strips, the maximum width of opening is limited to one-eighth of the column strip width. In the area common to one column strip and one middle strip, not more than one-fourth of the reinforcement in either strip should be interrupted by openings. In all cases, the area of reinforcement interrupted by openings must be replaced by an equivalent amount added on the sides of the opening.

The stress corresponding to the punching shear strength of the panel is the smallest of the three values given by ACI Table 22.6.5.2, which are

$$\phi v_c = 4\phi\lambda\sqrt{f_c'} \quad \text{[a]}$$

$$\leq \phi\left(2 + \frac{4}{\beta}\right)\lambda\sqrt{f_c'} \quad \text{[b]}$$

$$\leq \phi\left(2 + \frac{\alpha_s d}{b_o}\right)\lambda\sqrt{f_c'} \quad \text{[c]}$$

$$\alpha_s = 40 \text{ for an interior column}$$
$$= 30 \text{ for an edge column}$$
$$= 20 \text{ for a corner column}$$

Example 2.30

The flat plate floor of normal weight concrete for Ex. 2.28 and Ex. 2.29 has an effective depth of 7.5 in and a concrete compressive strength of 4000 lbf/in². Determine whether the shear capacity of the plate is adequate for an interior column.

Solution

The critical section for flexural shear is located a distance from the center of the panel given by

$$x = \frac{l_1}{2} - \left(d + \frac{c_1}{2}\right)$$

$$= \frac{28 \text{ ft}}{2} - \left(\frac{7.5 \text{ in}}{12 \frac{\text{in}}{\text{ft}}} + \frac{20 \text{ in}}{(2)\left(12 \frac{\text{in}}{\text{ft}}\right)}\right)$$

$$= 12.54 \text{ ft}$$

The factored applied shear at the critical section for flexural shear is

$$V_u = q_u l_2 x = \left(0.2 \frac{\text{kip}}{\text{ft}^2}\right)(24 \text{ ft})(12.54 \text{ ft}) = 60 \text{ kips}$$

The governing flexural shear capacity at the critical section is given by ACI Sec. 22.5.5.1 as

$$\phi V_c = 2\phi d l_2 \lambda \sqrt{f_c'} = \frac{\begin{array}{c}(2)(0.75)(7.5 \text{ in})(24 \text{ ft}) \\ \times \left(12 \frac{\text{in}}{\text{ft}}\right)(1.0)\sqrt{4000 \frac{\text{lbf}}{\text{in}^2}}\end{array}}{1000 \frac{\text{lbf}}{\text{kip}}}$$

$$= 205 \text{ kips}$$
$$> V_u \quad \text{[satisfactory]}$$

The length of one side of the critical perimeter for punching shear is

$$b = c + d = 20 \text{ in} + 7.5 \text{ in} = 27.5 \text{ in}$$

The factored applied shear at the critical perimeter for punching shear is

$$V_u = q_u(l_1 l_2 - b^2)$$

$$= \left(0.2 \frac{\text{kip}}{\text{ft}^2}\right)\left((28 \text{ ft})(24 \text{ ft}) - \frac{(27.5 \text{ in})^2}{\left(12 \frac{\text{in}}{\text{ft}}\right)^2}\right)$$

$$= 133 \text{ kips}$$

The length of the critical perimeter for punching shear is

$$b_o = 4b = (4)(27.5 \text{ in}) = 110 \text{ in}$$

The punching shear capacity of the plate is governed by ACI Table 22.6.5.2(a) as

$$\phi V_c = 4\phi d b_o \lambda \sqrt{f_c'}$$

$$= \frac{(4)(0.75)(7.5 \text{ in})(110 \text{ in})(1.0)\sqrt{4000 \frac{\text{lbf}}{\text{in}^2}}}{1000 \frac{\text{lbf}}{\text{kip}}}$$

$$= 157 \text{ kips}$$
$$> V_u \quad \text{[satisfactory]}$$

Equivalent Frame Method

No limitations are placed on the use of the equivalent frame method. As shown in Fig. 2.32, the structure is divided into a series of equivalent frames by cutting the building along lines midway between columns. The resulting frames are considered separately in the longitudinal and transverse directions of the building.

Figure 2.32 *Equivalent Frame Method*

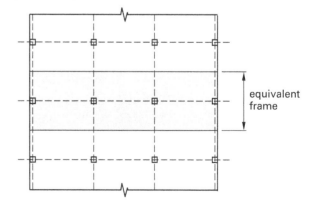

The equivalent frames may be analyzed by any frame analysis technique to determine the positive and negative moments in the slab. These moments are distributed across the slab using the same proportions as are used in the direct design method.

The equivalent frames may be analyzed for both vertical loads and lateral loads.

When the unfactored live load does not exceed three-quarters of the unfactored dead load, it may be assumed that maximum factored moments occur at all sections with the full factored live load on the entire slab system. For other conditions, patch loading must be used with only three-quarters of the full factored live load.

13. ANCHORING TO CONCRETE

Nomenclature

A_{brg}	net bearing area of the head of stud, anchor bolt, or headed deformed bar	in^2
A_{Nc}	projected concrete failure surface of a single anchor or group of anchors	in^2
A_{Nco}	projected concrete failure surface of a single anchor	in^2
$A_{se,N}$	effective cross-sectional area of anchor in tension	in^2
$A_{se,V}$	effective cross-sectional area of anchor in shear	in^2
A_{Vco}	projected concrete failure area of a single anchor for calculation of shear strength in shear, if not limited by corner influences, spacing, or member thickness	in^2
c_{ac}	critical edge distance required to develop the basic concrete breakout strength of a post-installed anchor in uncracked concrete without supplementary reinforcement to control splitting	in
$c_{a,max}$	maximum distance from center of an anchor shaft to the edge of concrete	in
$c_{a,min}$	minimum distance from center of an anchor shaft to the edge of concrete	in
c_{a1}	distance from the center of an anchor shaft to the edge of concrete in one direction. If shear is applied to anchor, c_{a1} is taken in the direction of the applied shear. If tension is applied to the anchor, c_{a1} is the minimum edge distance.	in
c_{a2}	distance from center of an anchor shaft to the edge of concrete in the direction perpendicular to c_{a1}	in
d_a	outside diameter of anchor or shaft diameter of headed stud, headed bolt, or hooked bolt	in
d_a'	value substituted for d_a when an oversized anchor is used	in
e_h	distance from the inner surface of the shaft of a J- or L-bolt to the outer tip of the J- or L-bolt	in
e_N'	distance between resultant tension load on a group of anchors loaded in tension and the centroid of the group of anchors loaded in tension	in
f_{uta}	specified tensile strength of the anchor	lbf/in^2
f_{ya}	specified yield strength of anchor	lbf/in^2
h_a	thickness of member in which an anchor is located, measured parallel to anchor axis	in
h_{ef}	effective embedment depth of anchor	in
I_e	occupancy importance factor	–
k_a	amplification factor for diaphragm flexibility	–
k_{cp}	pryout coefficient	–
l_e	load-bearing length of the anchor for shear	in
n	number of anchors	–
N_b	basic concrete breakout strength in tension of a single anchor in cracked concrete	lbf
N_{cb}	nominal concrete breakout strength in tension of a single anchor	lbf
N_{cbg}	nominal concrete breakout strength in tension of a group of anchors	lbf
N_{cp}	basic concrete pryout strength of a single anchor	lbf
N_n	nominal strength in tension	lbf
N_p	pullout strength in tension of a single anchor in cracked concrete	lbf
N_{pn}	nominal pullout strength in tension of a single anchor	lbf
N_{sa}	nominal strength of a single anchor or group of anchors in tension as governed by the steel strength	lbf
N_{sb}	side-face blowout strength of a single anchor	lbf
N_{sbg}	side-face blowout strength of a group of anchors	lbf
N_{ua}	factored tensile force applied to anchor or group of anchors	lbf
S_{DS}	design response acceleration at a period of 0.2 sec	–
V_{cb}	nominal concrete breakout strength in shear of a single anchor	lbf
V_{cbg}	nominal concrete breakout strength in shear of a group of anchors	lbf
V_{cp}	nominal concrete pryout strength of a single anchor	lbf
V_{cpg}	nominal concrete pryout strength of a group of anchors	lbf
V_n	nominal shear strength	lbf
V_{sa}	nominal strength in shear of a single anchor or group of anchors as governed by the steel strength	lbf
V_{ua}	factored shear force applied to a single anchor or group of anchors	lbf
W_p	weight of wall tributary to the anchor	lbf

Symbols

ϕ	strength reduction factor	–
$\psi_{c,N}$	factor used to modify tensile strength of anchors based on the presence or absence of cracks in concrete	–
$\psi_{c,P}$	factor used to modify pullout strength of anchors based on presence or absence of cracks in concrete	–
$\psi_{c,V}$	factor used to modify shear strength of anchors based on presence or absence of cracks in concrete and presence or absence of supplementary reinforcement	–
$\psi_{cp,N}$	factor used to modify tensile strength of post-installed anchors intended for use in uncracked concrete without supplementary reinforcement	–
$\psi_{ec,N}$	factor used to modify tensile strength of anchors based on eccentricity of applied loads	–
$\psi_{ec,V}$	factor used to modify shear strength of anchors based on eccentricity of applied loads	–
$\psi_{ed,N}$	factor used to modify tensile strength of anchors based on proximity to edges of concrete member	–
$\psi_{ed,V}$	factor used to modify shear strength of anchors based on proximity to edges of concrete member	–
$\psi_{h,V}$	factor used to modify shear strength of anchors located in concrete members with $h_a < 1.5c_{a1}$	–

Code Requirements[14]

ACI Chap. 17 provides design requirements for cast-in and post-installed anchors in concrete that are used to transmit structural loads by means of tension, shear, or a combination of tension and shear. Tensile loading failure modes are shown in Fig. 2.33, and shear loading failure modes are shown in Fig. 2.34.

Seismic Design Requirements

ACI 318 requires that, in calculations for anchorages for seismic loads, anchorage design strengths must be reduced for structures in seismic design categories C, D, E, and F. This requirement is intended to ensure that, for tension loads, anchor strength is governed by yielding of the ductile steel element. IBC Sec. 1905.1.8 modifies the seismic requirements of ACI Sec. 17.2.3.4.2, Sec. 17.2.3.4.3(d), and Sec. 17.2.3.5.2.

In accordance with ACI Sec. 17.2.3.4.2, the anchor design tensile strength is determined by ACI Sec. 17.2.3.4.4 when both

- the seismic component of the total tensile force exceeds 20% of the total tensile force

- one of the qualifying conditions of ACI Sec. 17.2.3.4.3(a) through (d) applies

Figure 2.33 *Tensile Failure Modes*

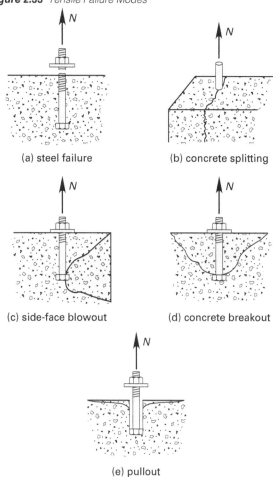

(a) steel failure   (b) concrete splitting

(c) side-face blowout   (d) concrete breakout

(e) pullout

Adapted and reprinted with authorized permission from *Building Code Requirements for Structural Concrete and Commentary* (ACI 318-14), © 2014, by the American Concrete Institute.

Figure 2.34 *Shear Failure Modes*

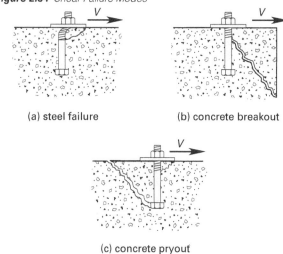

(a) steel failure   (b) concrete breakout

(c) concrete pryout

Adapted with permission from *Building Code Requirements for Structural Concrete and Commentary* (ACI 318-14), copyright © 2014, by the American Concrete Institute.

An exception to these requirements is provided by IBC Sec. 1905.1.8. This exception applies to anchors designed for the wall anchorage force given by ASCE/SEI7 Sec. 12.11.2.1, which is

$$F_p = 0.4 S_{DS} k_a I_e W_p$$

In this case, the design strength for concrete breakout, pullout, and side-face blowout are reduced by 25%, in accordance with ACI Sec. 17.2.3.4.4.

Where anchor reinforcement is provided in accordance with ACI Sec. 17.4.2.9, the design strength of the anchor reinforcement is used instead of the concrete breakout strength.

In light-frame construction, the in-plane shear strength of anchor bolts in wood sill plates or steel track is given by IBC 1905.1.8 as a modification to ACI Sec. 17.2.3.5.2. For these situations, IBC Sec. 1905.1.8 permits the in-plane shear strength to be based solely on the strength of the wood sill plate or the steel track.

Anchors in Tension

The strength reduction factor for tension loads is given by ACI Sec. 17.3.3 as

$$\phi = 0.75 \quad \text{[ductile steel anchor]}$$
$$= 0.65 \quad \text{[brittle steel anchor]}$$

Ductile bolts include ASTM A307 grade A bolts having a minimum specified tensile strength of 60 kips/in^2.

Steel Strength of Anchor

The nominal strength of an anchor in tension is given by ACI Eq. 17.4.1.2 as

$$N_{sa} = A_{se,N} f_{uta}$$

The specified tensile strength of an anchor is given by ACI Sec. 17.4.1.2 as

$$f_{uta} \le 1.9 f_{ya}$$
$$\le 125{,}000 \text{ lbf/in}^2$$

Concrete Breakout Strength

The strength reduction factor for an anchor or anchor group governed by concrete breakout, side-face blowout, pullout, or pryout strength is given by ACI Sec. 17.3.3 as

$$\phi = 0.75 \quad \text{[supplementary reinforcement]}$$
$$= 0.70 \quad \text{[no supplementary reinforcement]}$$

As shown in Fig. 2.35, for a single anchor not near the edges of the concrete element, the failure surface in the concrete is shaped like a pyramid. The pyramid's apex is at the centerline of the anchor at the bearing contact surface of the head. The failure surface radiates outward to the surface at a slope of 1 to 1.5. The maximum projected failure area for a single anchor is given by ACI Eq. 17.4.2.1c as

$$A_{Nco} = 9 h_{ef}^2$$

h_{ef} is the effective embedment depth of an anchor. When the failure surface for a single anchor is limited by adjacent edges, the reduced projected failure area is denoted by A_{Nc}.

Figure 2.35 Concrete Breakout of Anchor in Tension

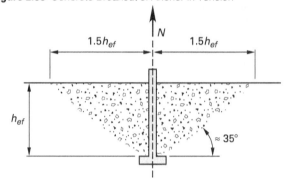

Failure occurs and concrete breakout results when the tensile stress on the failure surface exceeds the tensile strength of the concrete. The nominal concrete breakout strength for a single cast-in anchor in tension is given by ACI Eq. 17.4.2.1a as

$$N_{cb} = \frac{A_{Nc}}{A_{Nco}} \psi_{ed,N} \psi_{c,N} \psi_{cp,N} N_b$$

ACI Sec. 17.4.2.5 gives the modification factor for edge effects, $\psi_{ed,N}$. When $c_{a,\min} \ge 1.5 h_{ef}$,

$$\psi_{ed,N} = 1.0$$

$c_{a,\min}$ is the smallest of the influencing edge distances.

When $c_{a,\min} < 1.5 h_{ef}$,

$$\psi_{ed,N} = 0.7 + 0.3 \left(\frac{c_{a,\min}}{1.5 h_{ef}} \right)$$

The modification factor for cracked concrete from ACI Sec. 17.4.2.6 is

$$\psi_{c,N} = 1.0 \quad \text{[concrete cracked at service load levels]}$$
$$= 1.25 \quad \text{[concrete uncracked at service load levels]}$$

From ACI Sec. 17.4.2.7, the modification factor for cast-in anchors is

$$\psi_{cp,N} = 1.0$$

The basic concrete breakout strength of a single cast-in anchor in tension in cracked concrete as defined in ACI Sec. 17.4.2.2 is

$$N_b = k_c \lambda \sqrt{f_c'} \, h_{ef}^{1.5}$$

The coefficient for concrete breakout strength is

$$k_c = 24 \quad \text{[cast-in anchors]}$$

ACI Eq. 17.4.2.2b applies when $11 \text{ in} < h_{ef} < 25 \text{ in}$, and the basic concrete breakout strength is

$$N_b = 16 \lambda_a \sqrt{f_c'} \, h_{ef}^{5/3}$$

λ_a is the modification factor for lightweight concrete per ACI Sec. 17.2.6 and $\lambda_a = 1.0\lambda$, where λ is determined from ACI Sec. 19.2.4. For normal weight concrete using cast-in anchors, $\lambda_a = 1.0$.

When anchors are spaced closer than three times their embedment depth, the failure surfaces of adjacent anchors intersect. The failure surface for such an anchor group is determined by projecting the failure surface outward from the outer bolts in the group, as shown in Fig. 2.36. From Fig. 2.36, the projected area of this failure surface when distanced from the element edges is

$$A_{Nc} = (a + 3h_{ef})(b + 3h_{ef})$$
$$\leq n A_{Nco}$$

n is the number of anchors in the group, a is the distance between outside anchors in the group, and b is the distance between outside anchors in the group perpendicular to a.

The nominal concrete breakout strength for a cast-in anchor group in tension is given by ACI Eq. 17.4.2.1b as

$$N_{cbg} = \frac{A_{Nc}}{A_{Nco}} \psi_{ec,N} \psi_{ed,N} \psi_{c,N} \psi_{cp,N} N_b$$

Figure 2.36 *Concrete Breakout Surface for an Anchor Group*

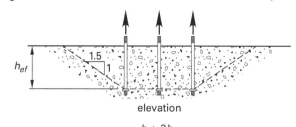

elevation

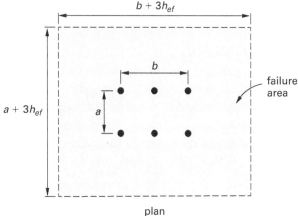

plan

The modification factor for eccentrically loaded anchor groups is given by ACI Eq. 17.4.2.4

$$\psi_{ec,N} = \frac{1}{1 + \dfrac{2e_N'}{3h_{ef}}}$$

$$\leq 1.0 \quad \text{[concentrically loaded groups]}$$

Pullout Strength of Anchor

The nominal concrete pullout strength for a single anchor in tension is given by ACI Eq. 17.4.3.1 as

$$N_{pn} = \psi_{c,P} N_p$$

For a headed bolt or stud, where A_{brg} is the bearing area of the bolt or stud head, the nominal pullout strength in cracked concrete is given by ACI Eq. 17.4.3.4 as

$$N_p = 8 A_{brg} f_c'$$

Bearing areas may be obtained from Cook.[15] For a hooked bolt, the nominal pullout strength in cracked concrete is given by ACI Eq. 17.4.3.5 as

$$N_p = 0.9 e_h d_a f_c'$$

$$3d_a < e_h < 4.5 d_a$$

e_h is the distance from the outer tip of a hooked bolt to the inner surface of the shaft, and d_a is the diameter of the bolt.

The modification factor for cracked concrete is given by ACI Sec. 17.4.3.6 as

$$\psi_{c,P} = 1.0 \quad \text{[concrete cracked at service load levels]}$$

$$= 1.4 \quad \text{[concrete uncracked at service load levels]}$$

Side-Face Blowout Strength of Anchor in Tension

Side-face blowout is caused by spalling of the concrete surface adjacent to the head of an anchor that is close to the face of the concrete. The nominal concrete blowout strength for a single anchor in tension, with $h_{ef} > 2.5c_{a1}$, is given by ACI Eq. 17.4.4.1 as

$$N_{sb} = \left(160c_{a1}\sqrt{A_{brg}}\right)\lambda_a\sqrt{f_c'}$$

c_{a1} is the minimum distance from the center of the anchor shaft to the edge of concrete, and c_{a2} is the distance from the center of the anchor shaft to the edge of the concrete perpendicular to c_{a1}. If c_{a2} for the single headed anchor is less than $3c_{a1}$, the value of N_{sb} is multiplied by the factor $(1 + c_{a2}/c_{a1})/4$, where $1.0 \le c_{a2}/c_{a1} \le 3.0$.

Concrete Splitting

Unless supplementary reinforcement is provided to control splitting and minimum spacing, edge distances for anchors must conform to the requirements of ACI Sec. 17.7. The requirements for cast-in anchors are as follows.

- The minimum center-to-center spacing of anchors is $4d_a$ for untorqued cast-in anchors.

- The minimum center-to-center spacing is $6d_a$ for cast-in anchors that will be torqued.

- The minimum edge distances for cast-in headed anchors that will not be torqued are the same as for normal reinforcement cover requirements as specified in ACI Sec. 20.6.1.

- The minimum edge distance is $6d_a$ for cast-in headed anchors that will be torqued.

Example 2.31

A glulam crosstie between diaphragm chords is supported at one end in a steel beam shoe as shown. The shoe is attached to a tilt-up concrete wall of normal weight concrete with a compressive strength of $f_c' = 4000 \text{ lbf/in}^2$. The four hex ASTM A307 grade A anchor bolts have a $\frac{1}{2}$ in diameter, an effective minimum specified tensile strength of $f_{uta} = 60 \text{ kips/in}^2$, an effective cross-sectional area in tension of $A_{se,N} = 0.142 \text{ in}^2$, an effective cross-sectional area in shear of $A_{se,V} = 0.196 \text{ in}^2$, and an effective bearing area of the bolt head of $A_{brg} = 0.291 \text{ in}^2$. The effective embedment length of the anchor bolts is $5\frac{1}{2}$ in. The anchor bolts are not near

any concrete edges and are not torqued. Supplementary reinforcement is not provided, and the concrete may be considered cracked. The floor diaphragm is flexible, and the wall anchor force is determined from ASCE/SEI7 Sec. 12.11.2.1. The structure is assigned to seismic design category C, and the governing strength load combination gives a tension force on the beam shoe of $N_{ua} = 15$ kips and a shear force of $V_{ua} = 4$ kips. Determine if the anchorage is adequate for the tension force on the shoe if the seismic component exceeds 20% of the total force.

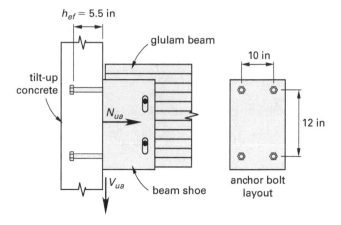

Solution

The effective embedment depth is

$$h_{ef} = 5.5 \text{ in}$$

The edge distance exceeds required cover requirements, and spacing exceeds $4d_a$. In accordance with ACI Sec. 17.7, side-face blowout and splitting are not considered.

The anchor bolts are ductile, so from ACI Sec. 17.3.3, the strength reduction factors are

$$\phi = 0.75 \quad \text{[tension on a ductile anchor bolt]}$$

$$= 0.70 \quad \left[\begin{array}{c} \text{concrete breakout or pullout} \\ \text{without supplemental reinforcement} \end{array}\right]$$

The wall anchor force is determined from ASCE/SEI7 Sec. 12.11.2.1. In accordance with IBC-modified ACI Sec. 17.2.3.4.4, the strength of the anchorage need not be governed by the strength of the steel elements.

Steel Strength of Anchor

The steel strength is based on the effective area of the threaded bolt. For a $\frac{1}{2}$ in diameter threaded anchor, the effective area is $A_{se,N} = 0.142 \text{ in}^2$. The minimum specified tensile strength of the four ASTM A307 grade A anchor bolts is 60 kips/in^2. The design strength of the four $\frac{1}{2}$ in diameter ductile anchors is given by ACI Eq. 17.4.1.2 as

$$4\phi N_{sa} = 4\phi A_{se,N} f_{uta}$$

$$= (4)(0.75)(0.142\text{ in}^2)\left(60\ \frac{\text{kips}}{\text{in}^2}\right)$$

$$= 25.56\text{ kips} > N_{ua} \quad \text{[satisfactory]}$$

Concrete Breakout Strength in Tension

The horizontal spacing of the anchor bolts is

$$s_1 = 10\text{ in} < 3h_{ef} = 16.5\text{ in}$$

The vertical spacing of the anchor bolts is

$$s_2 = 12\text{ in} < 3h_{ef} = 16.5\text{ in}$$

The projected area of the failure surface is

$$A_{Nc} = (s_1 + 3h_{ef})(s_2 + 3h_{ef})$$

$$= (10\text{ in} + 16.5\text{ in})(12\text{ in} + 16.5\text{ in})$$

$$= 755\text{ in}^2$$

$$< 4A_{Nco} \quad \text{[satisfies ACI Sec. 17.4.2.1]}$$

The projection of the failure surface for a single anchor on the concrete outer surface has an area of

$$A_{Nco} = 9h_{ef}^2 = (9)(5.5\text{ in})^2 = 272\text{ in}^2$$

The basic concrete breakout strength in tension of a single anchor in cracked concrete as defined in ACI Sec. 17.4.2.2 and ACI Eq. 17.4.2.2a is

$$N_b = k_c \lambda_a \sqrt{f_c'}\, h_{ef}^{1.5} = \frac{(24)(1.0)\sqrt{4000\ \dfrac{\text{lbf}}{\text{in}^2}}\,(5.5\text{ in})^{1.5}}{1000\ \dfrac{\text{lbf}}{\text{kip}}}$$

$$= 19.58\text{ kips}$$

The modification factor is 1.0 as specified by ACI Sec. 17.2.6.

The nominal concrete breakout strength for the anchor group is given by ACI Eq. 17.4.2.1b as

$$N_{cbg} = \frac{A_{Nc}}{A_{Nco}} \psi_{ec,N} \psi_{ed,N} \psi_{c,N} \psi_{cp,N} N_b$$

$$= \left(\frac{755\text{ in}^2}{272\text{ in}^2}\right)(1.0)(1.0)(1.0)(1.0)(19.58\text{ kips})$$

$$= 54.35\text{ kips}$$

The design concrete breakout strength for the anchor group for seismic loading is given by IBC-modified ACI Sec. 17.2.3.4.2 and 17.2.3.4.3(d) as

$$0.75\phi N_{cbg} = (0.75)(0.7)(54.35\text{ kips})$$

$$= 28.53\text{ kips} > N_{ua} \quad \text{[satisfactory]}$$

Pullout Strength of Anchor in Tension

For a headed bolt, the nominal pullout strength in cracked concrete is

$$N_p = 8A_{brg} f_c'$$

$$= \left(\frac{(8)(0.291\text{ in}^2)}{1000\ \dfrac{\text{lbf}}{\text{kip}}}\right)\left(4000\ \frac{\text{lbf}}{\text{in}^2}\right)$$

$$= 9.31\text{ kips}$$

The modification factor for cracked concrete is 1.0 as specified by ACI Sec. 17.2.6.

The nominal concrete pullout strength for a single anchor in cracked concrete is given by ACI Sec. 17.4.3.1 as as

$$N_{pn} = \psi_{c,P} N_p$$

$$= (1.0)(9.31\text{ kips})$$

$$= 9.31\text{ kips}$$

The design concrete pullout strength for a single anchor in tension is given by IBC-modified ACI Sec. 17.2.3.4.2 and 17.2.3.4.3(d) as

$$0.75\phi N_{pn} = (0.75)(0.7)(9.31\text{ kips})$$

$$= 4.89\text{ kips}$$

The design concrete pullout strength of the four bolts is

$$(4)(0.75)\phi N_{pn} = (4)(4.89\text{ kips})$$

$$= 19.56\text{ kips} > N_{ua} \quad \text{[satisfactory]}$$

Anchors in Shear

Steel Strength of Anchor

The strength reduction factor is given by ACI Sec. 17.3.3 as

$$\phi = 0.65 \quad \text{[ductile steel anchor]}$$

$$= 0.60 \quad \text{[brittle steel anchor]}$$

The nominal strength of a headed stud anchor in shear is given by ACI Eq. 17.5.1.2a as

$$V_{sa} = A_{se,V}f_{uta}$$

The nominal strength of a headed bolt and hooked bolt anchor in shear is given by ACI Eq. 17.5.1.2b as

$$V_{sa} = 0.6A_{se,V}f_{uta}$$

The specified tensile strength of an anchor is given by ACI Sec. 17.5.1.2 as

$$f_{uta} \le 1.9f_{ya}$$
$$\le 125{,}000 \text{ lbf/in}^2$$

f_{ya} is the specific yield strength of the anchor.

Ductile bolts include ASTM A307 grade A bolts with a minimum specified tensile strength of 60 kips/in².

Concrete Breakout Strength in Shear

ACI Sec. 17.3.3 gives the strength reduction factor for a single anchor or an anchor group in shear, governed by concrete breakout or pry-out, as

$$\phi = 0.75 \quad [\text{supplementary reinforcement provided}]$$
$$= 0.70 \quad [\text{supplementary reinforcement not provided}]$$

As shown in Fig. 2.37, for a single anchor not near edges perpendicular to the shear force, the failure surface in the concrete is shaped like a half pyramid. The half pyramid has a side length of $3c_{a1}$ and a depth of $1.5c_{a1}$. The projected area of this failure surface on the concrete outer surface is given by ACI Sec. R17.5.2.1 as

$$A_{Vco} = 4.5c_{a1}^2$$

When the failure surface for a single anchor is limited by corner influences and element thickness, the reduced projected area is denoted by A_{Vc}.

Figure 2.37 *Concrete Breakout Surface in Shear*

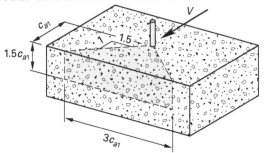

The nominal concrete breakout strength for a single cast-in anchor in shear is given by ACI Eq. 17.5.2.1a as

$$V_{cb} = \frac{A_{Vc}}{A_{Vco}}\psi_{ed,V}\psi_{c,V}\psi_{h,V}V_b$$

The modification factor for edge effects is given by ACI Sec. 17.5.2.6 as

$$\psi_{ed,V} = 1.0 \quad [c_{a2} \ge 1.5c_{a1}]$$
$$= 0.7 + \frac{0.3c_{a2}}{1.5c_{a1}} \quad [c_{a2} < 1.5c_{a1}]$$

The modification factor for cracked concrete is given by ACI Sec. 17.5.2.7 as

$$\psi_{c,V} = 1.0 \quad [\text{no supplementary reinforcement}]$$
$$= 1.4 \quad [\text{with supplementary reinforcement}]$$

The modification factor for anchors located in a member with $h_a < 1.5c_{a1}$ is given by ACI Eq. 17.5.2.8 as

$$\psi_{h,V} = \sqrt{\frac{1.5c_{a1}}{h_a}}$$

ACI Sec. 17.5.2.2 gives the basic concrete breakout strength in shear of a single anchor in cracked concrete as the smaller of

$$V_b = \left(7\left(\frac{l_e}{d_a}\right)^{0.2}\sqrt{d_a}\right)\lambda_a\sqrt{f_c'}\,c_{a1}^{1.5}$$
$$V_b = 9\lambda_a\sqrt{f_c'}\,c_{a1}^{1.5}$$

l_e is the load-bearing length of the anchor and is equal to h_{ef} for anchors with a constant stiffness over the embedded section. In all cases, $l_e \le 8d_a$.

When anchors are spaced closer than three times their edge distance in the direction of the shear, the failure surfaces of adjacent anchors intersect. The failure surface for such an anchor group is determined by projecting the failure surface outward from the outer bolts in the group, as shown in Fig. 2.38. From Fig. 2.38, and where $h_a < c_{a1}$, the projected area of the failure surface shown is

$$A_{Vc} = (3c_{a1} + s_1)h_a$$

Figure 2.38 *Concrete Breakout Surface for an Anchor Group in Shear*

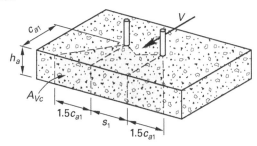

When the projected area is limited by edge distances, the reduced projected area must be calculated; it is also designated by A_{Vc}.

The nominal concrete breakout strength for an anchor group in shear is given by ACI Eq. 17.5.2.1b as

$$V_{cbg} = \frac{A_{Vc}}{A_{Vco}} \psi_{ec,V} \psi_{ed,V} \psi_{c,V} \psi_{h,V} V_b$$

The modification factor for eccentrically loaded anchor groups is given by ACI Eq. 17.5.2.5 as

$$\psi_{ec,V} = \frac{1}{1 + \dfrac{2e_v'}{3c_{a1}}}$$

$$= 1.0 \quad \text{[concentrically loaded groups]}$$

Concrete Pryout Strength of Anchor in Shear

The nominal concrete pryout strength for a single cast-in anchor is given by ACI Eq. 17.5.3.1a as

$$V_{cp} = k_{cp} N_{cp}$$
$$N_{cp} = N_{cb}$$
$$= \frac{A_{Nc}}{A_{Nco}} \psi_{ed,N} \psi_{c,N} \psi_{cp,N} N_b$$

The nominal concrete pryout strength for a group of cast-in anchors is given by ACI Eq. 17.5.3.1b as

$$V_{cpg} = k_{cp} N_{cpg}$$
$$N_{cpg} = N_{cbg}$$
$$= \frac{A_{Nc}}{A_{Nco}} \psi_{ec,N} \psi_{ed,N} \psi_{c,N} \psi_{cp,N} N_b$$

The pryout coefficient is given by ACI Sec. 17.5.3.1 as

$$k_{cp} = 1.0 \quad [h_{ef} < 2.5 \text{ in}]$$
$$= 2.0 \quad [h_{ef} \geq 2.5 \text{ in}]$$

Example 2.32

Using the information provided in Ex. 2.31, determine if the bolts are adequate for the shear force on the shoe.

Solution

The effective embedment depth is

$$h_{ef} = 5.5 \text{ in}$$

The anchor bolts are not near the concrete edges, and concrete breakout in shear is not applicable.

The anchor bolts are ductile. From ACI Sec. 17.3.3, the strength reduction factors are

$$\phi = 0.65 \quad \text{[shear on a ductile anchor bolt]}$$
$$= 0.70 \quad \begin{bmatrix} \text{concrete pryout without} \\ \text{supplemental reinforcement} \end{bmatrix}$$

The wall anchor force is determined from ASCE/SEI7 Sec. 12.11.2.1. In accordance with ACI Sec. 17.2.3.5.3(c), the strength of the anchorage is not governed by the strength of the steel elements.

Strength of Anchor Bolts in Shear

The steel strength in shear is based on the nominal area of the bolt. For a $\frac{1}{2}$ in diameter bolt, the nominal area is $A_{se,V} = 0.196 \text{ in}^2$. The minimum specified tensile strength of the four ASTM A307 grade A headed anchor bolts is 60 kips/in². The design strength of the four $\frac{1}{2}$ in diameter ductile anchor bolts is given by ACI Eq. 17.5.1.2b as

$$4\phi V_{sa} = 4\phi 0.6 A_{se,V} f_{uta}$$
$$= (4)(0.65)(0.6)(0.196 \text{ in}^2)\left(60 \ \frac{\text{kips}}{\text{in}^2}\right)$$
$$= 18.3 \text{ kips} > V_{ua} \quad \text{[satisfactory]}$$

Concrete Pryout Strength in Shear

The design concrete breakout strength for the cast-in anchor group in tension is obtained from Ex. 2.31 as

$$\phi N_{cbg} = \frac{28.53 \text{ kips}}{0.75} = 38.04 \text{ kips}$$

$$k_{cp} = 2.0 \text{ for } h_{ef} > 2.5 \text{ in}$$

The design concrete pryout strength for the cast-in anchor group in shear is given by ACI Eq. 17.5.3.1b as

$$\phi V_{cpg} = k_{cp} \phi N_{cpg} = k_{cp}(\phi N_{cbg})$$
$$= (2.0)(38.04 \text{ kips})$$
$$= 76.08 \text{ kips}$$
$$> V_{ua} \quad \text{[satisfactory]}$$

Interaction of Tensile and Shear Forces

When the factored shear force applied to a single anchor or group of anchors, V_{ua}, is greater than $0.2\phi V_n$, and when the factored tensile force applied to an anchor or a group of anchors, N_{ua}, is greater than $0.2\phi N_n$, the interaction expression of ACI Eq. 17.6.3 applies.

$$\frac{N_{ua}}{\phi N_n} + \frac{V_{ua}}{\phi V_n} \le 1.2$$

ϕN_n is the smallest of one of the following: steel strength of anchor in tension, concrete breakout strength in tension, pullout strength of anchor in tension, or side-face blowout strength. ϕV_n is the smallest one of the following: steel strength of anchor in shear, concrete breakout strength in shear, or the pryout strength.

When $V_{ua} < 0.2\phi V_n$, shear effects are neglected, the full design strength in tension is permitted, and

$$\phi N_n \ge N_{ua}$$

When $N_{ua} < 0.2\phi N_n$, tension effects are neglected, the full design strength in shear is permitted, and

$$\phi V_n \ge V_{ua}$$

Example 2.33

Using the information provided in Ex. 2.31, determine if the bolts are adequate for the combined tension and shear force on the shoe.

Solution

The applied loads are

$$N_{ua} = 15 \text{ kips} \quad \text{[tension]}$$
$$V_{ua} = 4 \text{ kips} \quad \text{[shear]}$$

The governing design strength in tension is the anchor bolt pullout strength, so

$$\phi N_{pn} = \phi N_n = 19.56 \text{ kips}$$

The governing design strength in shear is the anchor bolt shear strength, so

$$\phi V_{sa} = \phi V_n = 18.3 \text{ kips}$$

$$0.2\phi N_n = (0.2)(19.56 \text{ kips})$$
$$= 3.91 \text{ kips} < N_{ua}$$

The full design strength in shear is not permitted.

$$0.2\phi V_n = (0.2)(18.3 \text{ kips})$$
$$= 3.66 \text{ kips} < V_{ua}$$

The full design strength in tension is not permitted.

The interaction expression of ACI Eq. 17.6.3 applies, so

$$\frac{N_{ua}}{\phi N_n} + \frac{V_{ua}}{\phi V_n} \le 1.2$$

$$\frac{15 \text{ kips}}{19.56 \text{ kips}} + \frac{4 \text{ kips}}{18.3 \text{ kips}} = 0.99 < 1.2$$

The anchorage is adequate.

14. DESIGN OF SLENDER WALLS

Nomenclature

a	depth of equivalent rectangular stress block	in
A_s	area of nonprestressed longitudinal tension reinforcement	in²
$A_{se,w}$	effective cross-sectional area of longitudinal reinforcement in tension	in²
b	width of compression face of member, width of member	in
c	distance from extreme compression fiber to neutral axis	in
d	effective depth of section	in
D	effect of service dead load	–
e	eccentricity of applied axial load at top of wall	in
E	effect of horizontal and vertical earthquake-induced forces	–
E_c	modulus of elasticity of concrete	lbf/in²
E_s	modulus of elasticity of reinforcement steel	lbf/in²
f'_c	specified compressive strength of concrete	lbf/in²
f_r	modulus of rupture	lbf/in²
f_s	tensile stress in reinforcement at service loads	lbf/in²
h	overall thickness of wall	in
I_{cr}	moment of inertia of cracked section	in⁴
I_e	importance factor for seismic loads	–
I_g	moment of inertia of gross concrete section	in⁴
l_c	vertical distance between supports	in
l_w	horizontal length of wall	in
L	effect of service live load	–
M_a	maximum moment in wall due to service loads including P-delta effects	in-lbf
M_{cr}	cracking moment	in-lbf
M_e	maximum moment in wall due to eccentricity of service-level roof loads	in-lbf
M_n	nominal flexural strength at section	in-lbf
M_{sa}	service-level moment at the critical section due to seismic lateral force and eccentric roof load, excluding P-delta effects	in-lbf

M_u	maximum moment in wall due to factored loads, including P-delta effects	in-lbf
M_{ua}	maximum moment in wall due to factored loads, excluding P-delta effects	in-lbf
M_{ue}	maximum moment in wall due to eccentricity of strength-level roof loads	in-lbf
M_{uw}	maximum moment in wall due to strength-level lateral seismic force	in-lbf
M_w	maximum moment in wall due to service-level lateral seismic force	in-lbf
n	modular ratio of elasticity, E_s/E_c	–
P_r	unfactored axial load at top of wall	lbf
P_{rD}	unfactored roof dead load	lbf
P_s	unfactored axial load at the location of the critical section	lbf
P_u	factored axial load at the location of the critical section	lbf
P_{ur}	factored axial load at top of wall	lbf
P_{uw}	factored weight of wall above the location of the critical section	lbf
P_w	unfactored weight of wall above the location of the critical section	lbf
q	self-weight of the wall	lbf/ft^2
Q_E	effect of horizontal seismic forces	–
S	effect of snow load	–
S_{DS}	response acceleration for a period of 0.2 sec	–
U	factored loads in such combinations as stipulated in the applicable code	–
w	distributed lateral load	kips/ft
w_u	factored distributed lateral load	kips/ft
y_t	distance from centroidal axis of gross concrete section to tension face	in

Symbols

β_1	compression zone factor, a/c	–
γ_w	specific weight of concrete	lbf/ft^3
δ_1	moment magnification factor	–
Δ_{cr}	out-of-plane deflection at the critical section corresponding to cracking moment M_{cr}	in
Δ_n	out-of-plane deflection at the critical section corresponding to nominal flexural strength M_n	in
Δ_s	out-of-plane deflection at the critical section due to service loads	in
Δ_u	out-of-plane deflection at the critical section due to factored loads	in
ϵ_c	compressive strain in extreme concrete compression fiber	–
ϵ_s	tensile strain in tension reinforcement	–
λ	modification factor related to the unit weight of lightweight concrete	–
ϕ	strength reduction factor	–

Code Requirements

ACI 318-14 provides three alternative procedures for the design of walls subject to lateral forces and axial forces.

- The general procedure of ACI Sec. 11.5.2 may be applied to any wall. This procedure uses the strength design principles of ACI Chap. 22 and the stability requirements of ACI Chap. 6.

- The simplified design procedure is documented in ACI Sec. 11.5.3. This method applies only to solid rectangular cross sections in which the resultant of all factored loads is located within the middle third of the wall thickness.

- The alternative design procedure is documented in ACI Sec. 11.8.

Alternative Design Procedure

The principal requirements for the application of this procedure are given in ACI Sec. 11.8.1.1, ACI Sec. R11.8.1.1, ACI Sec. 11.8.2.1, and ACI Sec. 11.8.3.1 and are as follows.

- The cross section is rectangular and constant over the full height of the wall.

- The wall is restrained against rotation at the top.

- As shown in Fig. 2.39, the wall is analyzed as a simply supported, axially loaded member subject to an out-of-plane uniformly distributed lateral load, with maximum moment and deflection occurring at midheight.

- The reinforcement ratio is limited so as to provide a wall that is tension-controlled for out-of-plane moment. This is defined in ACI Table 21.2.2 as the strain in the extreme tension reinforcement, $\epsilon_s \geq 0.005$, when the concrete reaches its ultimate strain, $\epsilon_c = 0.003$. This corresponds to a neutral axis depth ratio of $c/d \leq 0.375$.

- The design moment strength of the wall is governed by

$$\phi M_n \geq M_c$$

- The factored axial force at the midheight of the wall is governed by

$$P_u \leq 0.06 f_c' A_g$$

- The maximum permissible deflection at the midheight of the wall due to service loads, including second-order effects, is

$$\Delta_s \leq \frac{l_c}{150}$$

Figure 2.39 *Design of Slender Wall*

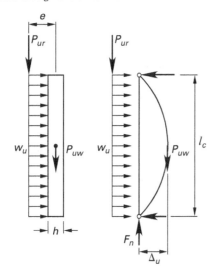

- The calculated factored applied moment at the mid-height of the wall, M_u, must include second-order effects.

Determination of Design Moment Strength of the Wall

Figure 2.40 illustrates the strength design assumptions of ACI Sec. 22.2.2 as applied to a reinforced concrete wall with the reinforcement located in the center of the wall. The effective depth of the wall section is

$$d = h/2$$

An effective area of longitudinal tension reinforcement is used in the design calculations to allow for the axial load on the wall, and this is given by ACI Sec. R11.8.3.1 as

$$A_{se,w} = A_s + \frac{P_u h}{2 f_y d}$$

$$= \frac{P_u + A_s f_y}{f_y} \quad \text{[for one central layer of reinforcement]}$$

Using ACI Sec. 22.2.2.4.1, the depth of the equivalent rectangular stress block in the concrete is

$$a = \frac{A_{se,w} f_y}{0.85 f_c' l_w}$$

For a given concrete strength, the compression zone factor, β_1, is determined from ACI Sec. 22.2.2.4.3.

The depth to the neutral axis is given by ACI Sec. 22.2.2.4.1 as

$$c = \frac{a}{\beta_1}$$

The nominal moment strength is given by[16]

$$M_n = A_{se,w} f_y \left(d - \frac{a}{2} \right)$$

As shown in Fig. 2.40, the strain in the tension reinforcement is given by

$$\epsilon_s = \frac{\epsilon_c (d - c)}{c}$$

To comply with ACI Sec. 11.8.1.1(b), the value of ϵ_s must not be less than 0.005. The wall is tension-controlled, and the strength reduction factor is given by ACI Sec. 21.2.1 as

$$\phi = 0.9$$

The design moment strength of the wall is

$$\phi M_n = \phi A_{se,w} f_y \left(d - \frac{a}{2} \right)$$

Figure 2.40 *Strength Design Assumptions*

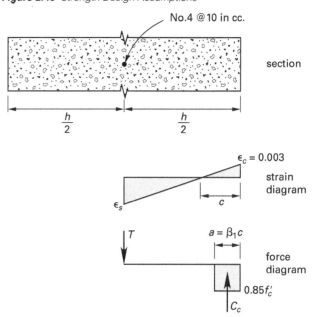

Example 2.34

The 6 in thick wall of a tilt-up concrete warehouse shown has a normal weight concrete cylinder strength, f_c', of 4000 lbf/in² and no. 4 grade 60 reinforcing bars at 10 in centers placed centrally in the wall. The wall has an effective span, l_c, of 20 ft. The unfactored roof live load, P_{rL}, is 100 lbf/ft and the unfactored roof dead load, P_{rD}, is 200 lbf/ft, both acting at an eccentricity, e, of 7 in with respect to the center of the wall section. The structure has a redundancy factor, ρ, of 1.0 and the 5% damped, design spectral response acceleration, S_{DS}, for a period of 0.2 sec is $0.9g$. The structure is assigned to seismic design category D, and the roof diaphragm may be considered flexible. Seismic load combinations govern the design. Determine the design moment strength of the wall and the cracking moment. Check that the factored axial load at the critical section is satisfactory.

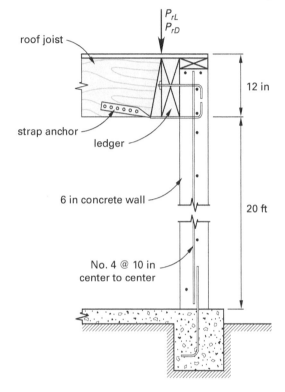

Solution

Determine Design Moment Strength

The wall thickness, h, is 6 in.

For reinforcing bars placed centrally in the wall, the effective depth is

$$\frac{h}{2} = \frac{6 \text{ in}}{2} = 3 \text{ in}$$

Where the effects of dead load and seismic load are additive, the applicable loading case is given by ASCE/SEI7 Sec. 2.3.6.

$$U = 1.2D + 1.0E + 1.0L + 0.2S$$
$$= (1.2 + 0.2S_{DS})D + \rho Q_E + 1.0L + 0.2S$$

Considering a 1 ft width of wall, the factored axial load from the roof, not including roof live load, is

$$P_{ur} = (1.2 + 0.2S_{DS})P_{rD} + \rho Q_E + 1.0L + 0.2S$$
$$= \left(1.2 + (0.2)(0.9)\right)\left(200 \ \frac{\text{lbf}}{\text{ft}}\right)$$
$$+ (1.0)(0) + (1.0)(0) + (0.2)(0)$$
$$= 276 \text{ lbf}$$

The self-weight of the wall is

$$q = \gamma_w h$$
$$= \left(150 \ \frac{\text{lbf}}{\text{ft}^3}\right)\left((6 \text{ in})\left(\frac{1 \text{ ft}}{12 \text{ in}}\right)\right)$$
$$= 75 \text{ lbf/ft}^2$$

The critical section of the wall is midway between supports and, for a 1 ft width, the weight of wall tributary to the critical section is

$$P_w = q\left(\frac{l_c}{2} + 1 \text{ ft}\right)b$$
$$= \left(75 \ \frac{\text{lbf}}{\text{ft}^2}\right)\left(\frac{20 \text{ ft}}{2} + 1 \text{ ft}\right)(1 \text{ ft})$$
$$= 825 \text{ lbf}$$

The factored axial load from the weight of the wall above the critical section is

$$P_{uw} = (1.2 + 0.2S_{DS})P_w + \rho Q_E + 1.0L + 0.2S$$
$$= \left(1.2 + (0.2)(0.9)\right)(825 \text{ lbf})$$
$$+ (1.0)(0) + (1.0)(0) + (0.2)(0)$$
$$= 1139 \text{ lbf}$$

The total factored axial load at the critical section is

$$P_u = P_{ur} + P_{uw} = 276 \text{ lbf} + 1139 \text{ lbf}$$
$$= 1415 \text{ lbf}$$

Considering a 1 ft width of wall, the reinforcement area is

$$A_s = (0.2 \text{ in}^2)\left(\frac{12}{10}\right)$$
$$= 0.24 \text{ in}^2$$

The effective reinforcement area for one central layer of reinforcement is given by

$$A_{se,w} = \frac{P_u + A_s f_y}{f_y}$$

$$= \frac{1.415 \text{ kips} + (0.24 \text{ in}^2)\left(60 \dfrac{\text{kips}}{\text{in}^2}\right)}{60 \dfrac{\text{kips}}{\text{in}^2}}$$

$$= 0.26 \text{ in}^2$$

The depth of the equivalent rectangular stress block is given by ACI Sec. 22.2.2.4.1 as

$$a = \frac{A_{se,w} f_y}{0.85 f'_c l_w}$$

$$= \frac{(0.26 \text{ in}^2)\left(60 \dfrac{\text{kips}}{\text{in}^2}\right)}{(0.85)\left(4 \dfrac{\text{kips}}{\text{in}^2}\right)(12 \text{ in})} \quad \text{[for a 12 in width]}$$

$$= 0.38 \text{ in}$$

The nominal moment strength is

$$M_n = A_{se,w} f_y \left(d - \frac{a}{2}\right)$$

$$= (0.26 \text{ in}^2)\left(60 \dfrac{\text{kips}}{\text{in}^2}\right)\left(3 - \frac{(0.38 \text{ in})}{2}\right)$$

$$= 43.84 \text{ in-kips}$$

For a concrete strength of 4000 lbf/in², the compression zone factor β_1 is given by ACI Sec. 22.2.2.4.3 as $\beta_1 = 0.85$.

The depth to the neutral axis is given by ACI Sec. 22.2.2.4.1 as

$$c = \frac{a}{\beta_1} = \frac{0.38 \text{ in}}{0.85}$$

$$= 0.45 \text{ in}$$

From Fig. 2.40, the strain in the tension reinforcement is

$$\epsilon_s = \frac{\epsilon_c (d - c)}{c}$$

$$= \frac{(0.003)(3 - 0.45)}{0.45}$$

$$= 0.017 > 0.005$$

From ACI Sec. R21.2.2, the section is tension controlled, and ACI Sec. 11.8.1.1(b) is satisfied. In accordance with ACI Sec. 21.2.1, the strength reduction factor is $\phi = 0.9$.

Hence, the design moment strength of the wall is

$$\phi M_n = (0.9)(43.84 \text{ in-kips})$$

$$= 39.46 \text{ in-kips}$$

Determine Cracking Moment

The gross moment of inertia of the concrete section about its centroidal axis is

$$I_g = \frac{bh^3}{12} = \frac{(12 \text{ in})(6 \text{ in})^3}{12}$$

$$= 216 \text{ in}^4$$

The modulus of rupture is given by ACI Eq. 19.2.3.1 as

$$f_r = 7.5 \lambda \sqrt{f'_c}$$

$$= (7.5)(1.0)\sqrt{4000} \quad [\lambda = 1.0 \text{ for normal weight concrete}]$$

$$= 474 \text{ lbf/in}^2$$

The cracking moment is given by ACI Sec. 24.2.3.5 as

$$M_{cr} = \frac{f_r I_g}{y_t} = \frac{\left(474 \dfrac{\text{lbf}}{\text{in}^2}\right)(216 \text{ in}^4)}{(3 \text{ in})\left(1000 \dfrac{\text{lbf}}{\text{kip}}\right)}$$

$$= 34.13 \text{ in-kips} < \phi M_n$$

The required limitation of ACI Sec. 11.8.1(c) is satisfied.

Check the Factored Axial Force at the Critical Section

The factored axial force at the critical section of the wall is

$$P_u = 1415 \text{ lbf}$$

$$0.06 f'_c A_g = (0.06)\left(4000 \dfrac{\text{lbf}}{\text{in}^2}\right)(12 \text{ in})(6 \text{ in})$$

$$= 17{,}280 > P_u$$

The required limitation of ACI Sec. 11.8.1.1(d) satisfied.

Determination of Required Moment Strength of the Wall

The modulus of elasticity of reinforcing steel is given by ACI 20.2.2.2 as

$$E_s = 29{,}000 \text{ kips/in}^2$$

The modulus of elasticity of normal weight concrete is given by ACI 19.2.2.1 as

$$E_c = 57\sqrt{f'_c} \text{ kips/in}^2 \quad [f'_c \text{ in units of lbf/in}^2]$$

The modular ratio of elasticity is

$$n = \frac{E_s}{E_c} \geq 6$$

The moment of inertia of the cracked wall section is given by ACI Eq. 11.8.3.1c as

$$I_{cr} = nA_{se,w}(d - c)^2 + \frac{l_w c^3}{3}$$

The wall displacement at the critical section is obtained from ACI Eq. 11.8.3.1b as

$$\Delta_u = \frac{5M_u l_c^2}{(0.75)48 E_c I_{cr}}$$

The moment at the critical section of the wall due to factored loads, excluding P-delta effects, is

$$M_{ua} = \frac{w_u l_c^2}{8} + \frac{P_{ur} e}{2}$$

The moment at the critical section of the wall due to factored loads, including P-delta effects, is given by ACI Eq. 11.8.3.1a as

$$M_u = M_{ua} + P_u \Delta_u$$

Since the moment at the critical section, M_u, and the wall displacement, Δ_u, are interdependent, an iterative process is required to obtain the value of M_u.

Alternatively, the value of M_u may be obtained directly by using ACI Eq. 11.8.3.1d, which is

$$M_u = \frac{M_{ua}}{1 - \dfrac{5P_u l_c^2}{(0.75)(48 E_c I_{cr})}}$$

Example 2.35

The 6 in thick wall of a tilt-up concrete warehouse, described in Ex. 2.34, has a normal weight concrete cylinder strength, f_c', of 4000 lbf/in² and no. 4 grade 60 reinforcing bars at 10 in centers placed centrally in the wall. The wall has an effective span, l_c, of 20 ft. The unfactored roof live load, P_{rL}, is 100 lbf/ft and the unfactored roof dead load, P_{rD}, is 200 lbf/ft, both acting at an eccentricity, e, of 7 in with respect to the center of the wall section. The structure has a redundancy factor, ρ, of 1.0 and the 5% damped, design spectral response acceleration, S_{DS}, for a period of 0.2 sec is 0.9g. The structure is assigned to seismic design category D, and the roof diaphragm may be considered flexible. Seismic

load combinations govern the design. Determine the required moment strength of the wall, including P-delta effects.

Solution

The importance factor for a warehouse is given by ASCE/SEI7 Table 1.5-2 as $I_e = 1.0$.

The factored seismic lateral force on a 1 ft width of the wall is given by ASCE/SEI7 Sec. 12.11.1 as

$$w_u = 0.40 I_e S_{DS} bq = (0.40)(1.0)(0.9)(1.0 \text{ ft})\left(75 \frac{\text{lbf}}{\text{ft}^2}\right)$$
$$= 27 \text{ lbf/ft}$$

The corresponding factored bending moment at the critical section is

$$M_{uw} = \frac{w_u l_c^2}{8}$$
$$= \frac{\left(27 \dfrac{\text{lbf}}{\text{ft}}\right)(20 \text{ ft})^2}{8}$$
$$= 1350 \text{ ft-lbf}$$

The eccentricity of the roof load about the wall centerline is given as $e = 7$ in.

The factored bending moment at the critical section caused by the eccentricity is

$$M_{ue} = \frac{P_{ur} e}{2}$$
$$= \frac{(276 \text{ lbf})(7 \text{ in})}{(2)\left(12 \dfrac{\text{in}}{\text{ft}}\right)}$$
$$= 81 \text{ ft-lbf}$$

The total factored wall moment at the critical section due to seismic lateral force and the eccentric roof load excluding P-delta effects is

$$M_{ua} = M_{uw} + M_{ue} = 1350 \text{ ft-lbf } + 81 \text{ ft-lbf}$$
$$= 1431 \text{ ft-lbf}$$
$$= 17.17 \text{ in-kips}$$

The modulus of elasticity of normal weight concrete is given by ACI Sec. 19.2.2.1 as

$$E_c = 57\sqrt{f_c'} \quad [f_c' \text{ in units of lbf/in}^2]$$
$$= 57\sqrt{4000}$$
$$= 3605 \text{ kips/in}^2$$

The modular ratio of elasticity is

$$n = \frac{E_s}{E_c} = \frac{29{,}000 \; \dfrac{\text{kips}}{\text{in}^2}}{3605 \; \dfrac{\text{kips}}{\text{in}^2}}$$

$$= 8.0 > 6$$

The moment of inertia of 1 ft width of the cracked wall section is given by ACI Eq. 11.8.3.1c as

$$I_{cr} = nA_{se,w}(d-c)^2 + \frac{l_w c^3}{3}$$

$$= (8.0)(0.26 \text{ in}^2)(3 \text{ in} - 0.45 \text{ in})^2 + \frac{(12 \text{ in})(0.45 \text{ in})^3}{3}$$

$$= 13.89 \text{ in}^4$$

The magnification factor to account for P-delta effects is given by ACI Eq. 11.8.3.1d as

$$B_i = 1 - \frac{5P_u l_c^2}{(0.75)48E_c I_{cr}}$$

$$= 1 - \frac{(5)(1.415 \text{ kips})\left((20 \text{ ft})\left(12 \; \dfrac{\text{in}}{\text{ft}}\right)\right)^2}{(0.75)(48)\left(3605 \; \dfrac{\text{kips}}{\text{in}^2}\right)(13.89 \text{ in}^4)}$$

$$- 1 - 0.23$$

$$= 0.77$$

The factored applied moment including P-delta effects is given by ACI Eq. 11.8.3.1a as

$$M_u = M_{ua} + P_u\Delta_u = \frac{M_{ua}}{B_1}$$

$$= \frac{17.17 \text{ in-kips}}{0.77}$$

$$= 22.30 \text{ in-kips} < \phi M_n \quad \text{[satisfactory]}$$

Determination of Service Load Deflection of the Wall

For calculating service-level lateral deflections due to seismic effects, the appropriate load combination is given by ACI Sec. R11.8.4 as

$$U = D + 0.5L + 0.7E$$

The out-of-plane deflection at the critical section corresponding to the cracking moment, M_{cr}, is given by ACI Eq. 11.8.4.3a as

$$\Delta_{cr} = \frac{5M_{cr} l_c^2}{48E_c I_g}$$

The out-of-plane deflection at the critical section corresponding to the nominal flexural strength, M_n, is given by ACI Eq. 11.8.4.3b as

$$\Delta_n = \frac{5M_n l_c^2}{48E_c I_{cr}}$$

For a service-level moment at the critical section due to seismic lateral force and eccentric roof load, excluding P-delta effects, of M_{sa}, a service-level axial load at the critical section of P_s, and an out-of-plane deflection at the critical section due to service loads of Δ_s, the maximum moment at the critical section due to service loads, including P-delta effects, is given by ACI Eq. 11.8.4.2 as

$$M_a = M_{sa} + P_s\Delta_s$$

Then, for a value of $M_a \leq 2M_{cr}/3$, the deflection at the critical section is given by ACI Table 11.8.4.1(a) as

$$\Delta_s = \frac{M_a \Delta_{cr}}{M_{cr}}$$

For a value of $M_a > 2M_{cr}/3$, the deflection at the critical section is given by ACI Table 11.8.4.1(b) as

$$\Delta_s = \frac{2\Delta_{cr}}{3} + \frac{\left(M_a - \dfrac{2M_{cr}}{3}\right)\left(\Delta_n - \dfrac{2\Delta_{cr}}{3}\right)}{M_n - \dfrac{2M_{cr}}{3}}$$

Since Δ_s and M_a are interdependent, an iterative process is required to obtain the value of Δ_s.

The maximum permissible deflection at the critical section, Δ_s, due to service vertical and lateral loads is given by ACI Sec. 11.8.1.1(e) as

$$\Delta_s = \frac{l_c}{150}$$

Example 2.36

The 6 in thick wall of a tilt-up concrete warehouse, described in Ex. 2.34, has a normal weight concrete cylinder strength, f_c', of 4000 lbf/in² and no. 4 grade 60 reinforcing bars at 10 in centers placed centrally in the wall. The wall has an effective span, l_c, of 20 ft. The unfactored roof live load, P_{rL}, is 100 lbf/ft and the unfactored roof dead load, P_{rD}, is 200 lbf/ft, both acting at an eccentricity, e, of 7 in with respect to the center of the wall section. The structure has a redundancy factor, ρ, of 1.0 and the 5% damped, design spectral response acceleration, S_{DS}, for a period of 0.2 sec is $0.9g$. The structure is assigned to seismic design category D, and the roof diaphragm may be considered flexible. Seismic load combinations govern the design. Determine if the service-level deflection of the wall at the critical section is satisfactory.

Solution

For calculating service-level lateral deflections due to seismic effects, the appropriate load combination is given by ACI Sec. R11.8.4 as

$$U = D + 0.5L + 0.7E$$
$$= \left(1 + (0.2)(0.7S_{DS})\right)D + 0.5L + 0.7\rho Q_E$$

Considering a 1 ft width of wall, the service-level axial load from the roof, not including roof live load, is given by

$$P_r = (1 + 0.14S_{DS})D + 0$$
$$= (1 + (0.14)(0.9))\left(200 \ \frac{\text{lbf}}{\text{ft}}\right) + 0$$
$$= 225 \text{ lbf}$$

The self-weight of the wall is

$$q = \gamma_w h$$
$$= \left(150 \ \frac{\text{lbf}}{\text{ft}^3}\right)(6 \text{ in})\left(\frac{1 \text{ ft}}{12 \text{ in}}\right)$$
$$= 75 \text{ lbf/ft}^2$$

The critical section of the wall is midway between supports and, for a 1 ft width, the weight of wall tributary to the critical section is

$$W = q\left(\frac{l_c}{2} + 1 \text{ ft}\right)b$$
$$= \left(75 \ \frac{\text{lbf}}{\text{ft}^2}\right)\left(\frac{20 \text{ ft}}{2} + 1 \text{ ft}\right)(1 \text{ ft})$$
$$= 825 \text{ lbf}$$

For calculating the service-level lateral deflection, the effective axial load from the weight of the wall above the critical section is

$$P_w = (1 + 0.14S_{DS})W$$
$$= \left(1 + (0.14)(0.9)\right)(825 \text{ lbf})$$
$$= 929 \text{ lbf}$$

The total service-level axial load at the critical section is

$$P_s = P_r + P_w = 225 \text{ lbf} + 929 \text{ lbf}$$
$$= 1154 \text{ lbf}$$

The service-level seismic lateral force on the wall is

$$w = 0.7w_u$$
$$= (0.7)\left(27 \ \frac{\text{lbf}}{\text{ft}}\right)$$
$$= 18.90 \text{ lbf/ft}$$

The moment at the critical section due to service-level lateral seismic force is

$$M_w = \frac{wl_c^2}{8}$$
$$= \frac{\left(18.90 \ \frac{\text{lbf}}{\text{ft}}\right)(20 \text{ ft}^2)}{8}$$
$$= 945 \text{ ft-lbf}$$

The service-level bending moment at the critical section caused by the eccentricity is

$$M_e = \frac{P_r e}{2}$$
$$= \frac{(225 \text{ lbf})(7 \text{ in})\left(\frac{1 \text{ ft}}{12 \text{ in}}\right)}{2}$$
$$= 66 \text{ ft-lbf}$$

The total service-level bending moment at the critical section due to seismic lateral force and the eccentric roof load excluding *P*-delta effects is

$$M_{sa} = M_w + M_e$$
$$= 945 \text{ ft-lbf} + 66 \text{ ft-lbf}$$
$$= 1011 \text{ ft-lbf}$$
$$= 12.13 \text{ in-kips}$$

The maximum permissible deflection at the critical section, Δ_s, due to service vertical and lateral loads is given by ACI Sec. 11.8.1.1(e) as

$$\Delta_s = \frac{l_c}{150}$$
$$= \frac{(20 \text{ ft})\left(12 \ \frac{\text{in}}{\text{ft}}\right)}{150}$$
$$= 1.60 \text{ in}$$

The corresponding moment at this deflection due to service loads including P-delta effects is

$$M_a = M_{sa} + P_s\Delta_s$$
$$= 12.13 \text{ in-kips} + (1.154 \text{ kips})(1.60 \text{ in})$$
$$= 13.98 \text{ in-kips}$$
$$\frac{2M_{cr}}{3} = \frac{(2)(34.13 \text{ in-kips})}{3}$$
$$= 22.75 \text{ in-kips} > M_a \quad [\text{ACI Table 11.8.4.1(a) applies}]$$

The out-of-plane deflection at the critical section corresponding to the cracking moment, M_{cr}, is given by ACI Eq. 11.8.4.3a as

$$\Delta_{cr} = \frac{5M_{cr}l_c^2}{48E_cI_g}$$
$$= \frac{(5)(34.13 \text{ in-kips})\left((20 \text{ ft})\left(12 \frac{\text{in}}{\text{ft}}\right)\right)^2}{(48)\left(3605 \frac{\text{kips}}{\text{in}^2}\right)(216 \text{ in}^4)}$$
$$= 0.26 \text{ in}$$

From ACI Table 11.8.4.1(a), the deflection due to service loads including P-delta effects is

$$\Delta = \frac{M_a\Delta_{cr}}{M_{cr}}$$
$$= \frac{(13.98 \text{ in-kips})(0.26 \text{ in})}{34.13 \text{ in-kips}}$$
$$= 0.11 \text{ in} < \Delta_s$$

Hence, the service-level deflection of the wall at the critical section is less than the maximum permissible allowed, and ACI Sec. 11.8.1.1(e) is satisfied.

15. DESIGN OF SHEAR WALLS

Nomenclature

a	depth of equivalent rectangular stress block	in
A_{ch}	cross-sectional area of a member measured to the outside edges of confinement reinforcement	in^2
A_{cv}	gross area of concrete section bounded by web thickness and length of section in the direction of shear force considered in the case of walls	in^2
A_g	gross area of concrete section	in^2
A_s	area of nonprestressed longitudinal tension reinforcement	in^2
A_{sh}	total area of confinement reinforcement within a spacing s and perpendicular to dimension b_c	in^2
A_{st}	total area of longitudinal reinforcement	in^2
b	width of compression face of member	in
b_c	cross-sectional dimension of member core measured to the outside edges of the transverse reinforcement	in
c	distance from extreme compression fiber to neutral axis	in
C	force in compression reinforcement	kips
C_c	force in compression stress block	kips
C_d	deflection amplification factor	–
d_b	nominal diameter of smallest longitudinal bar	in
D	effect of service dead load	–
e	distance of reinforcing bar from the neutral axis	in
E	effect of horizontal and vertical earthquake-induced forces	–
E_s	modulus of elasticity of reinforcing steel	lbf/in^2
f_c	concrete stress	lbf/in^2
$f_{c,\max}$	maximum concrete fiber stress	lbf/in^2
f_c'	specified compressive strength of concrete	lbf/in^2
f_y	specified yield strength for nonprestressed reinforcement	lbf/in^2
f_{yt}	specified yield strength of confinement reinforcement	lbf/in^2
F	force in reinforcing bar	kips
h	overall thickness, height, or depth of member	in
$h_{\min}$	minimum boundary element dimension	in
h_u	laterally unsupported height at extreme compression fiber of wall	in
h_w	height of entire wall from base to top	in
h_x	maximum center-to-center spacing of longitudinal bars laterally supported by corners of crossties or hoop legs around the perimeter of the boundary element	in
I_e	seismic importance factor	=
l_o	length over which special confinement reinforcement must be provided	in
l_w	length of entire wall in direction of shear force	in
L	effect of service live load	–
M_n	nominal flexural strength at section	in-lbf
M_u	factored moment at section	in-lbf
P_D	unfactored axial dead load at base of wall including self-weight	kips
P_n	nominal axial strength	kips
P_u	factored axial load at base of wall including self-weight	kips
Q	section modulus of gross concrete section	in^3
Q_E	effect of horizontal seismic forces	–
s	center-to-center spacing of reinforcement	in

s_o	confinement reinforcement spacing calculated by ACI Eq. 18.7.5.3	in
S	effect of snow load	–
S_c	section modulus of gross concrete section	in^3
S_{DS}	response acceleration for a period of 0.2 sec	–
U	strength of a member to resist factored loads in such combinations as stipulated in the code	–
V	shear force	kips
V_n	nominal shear strength provided by concrete	kips
V_u	factored shear force	kips

Symbols

α_c	wall shear factor	–
β_1	factor relating depth of equivalent rectangular compressive stress block to depth of neutral axis	–
δ_u	inelastic design displacement of wall	in
δ_{xe}	calculated elastic displacement of wall due to factored forces	in
ϵ_c	assumed maximum compressive strain in concrete	–
ϵ_s	strain in reinforcement	–
ϵ_t	strain in extreme tension steel	–
ϵ_y	strain in reinforcement at yield	–
λ	modification factor for lightweight concrete	–
ρ	redundancy factor	–
ρ_l	ratio of area of distributed longitudinal reinforcement to gross concrete area perpendicular to that reinforcement	–
ρ_t	ratio of area of distributed transverse reinforcement to gross concrete area perpendicular to that reinforcement	–
ϕ	strength reduction factor	–

Structural Walls

ACI Sec. 2.3 defines a structural wall as a wall designed to resist combinations of shear, moment, and axial force in the plane of the wall. A shear wall is classified as a structural wall.

An *ordinary reinforced concrete structural wall* is a shear wall complying with ACI Chap. 11.

A *special structural wall* is defined as a cast-in-place structural wall designed in compliance with ACI Sec. 18.2.3 through 18.2 8 and 18.10; or a precast structural wall in compliance with ACI Sec. 18.2.3 through 18.2 8 and 18.11. Special structural walls are the only shear walls permitted by ASCE/SEI7 in seismic design categories D through F.

Shear Capacity of Special Structural Walls

The nominal shear strength of a shear wall may be determined as specified in ACI Sec. 18.10.4.1 and given in ACI Eq. 18.10.4.1 as

$$V_n = A_{cv}(\alpha_c \lambda \sqrt{f_c'} + \rho_t f_y)$$

For a value of $h_w/l_w \leq 1.5$, $\alpha_c = 3$.

For a value of $h_w/l_w \geq 2.0$, $\alpha_c = 2$.

Linear interpolation may be used in the determination of α_c for values of h_w/l_w between 1.5 and 2.

In accordance with ACI Sec. 18.10.4.4, the maximum allowable nominal shear strength for all vertical wall segments sharing a common lateral force is

$$V_n = 8A_{cv}\sqrt{f_c'} \quad [A_{cv} = \text{gross combined area of all wall segments}]$$

For any individual wall segment, the maximum allowable nominal shear strength is

$$V_n = 10A_{cv}\sqrt{f_c'} \quad [A_{cv} = \text{area of segment considered}]$$

In accordance with ACI Sec. 18.10.2.1, when the design shear force, V_u, exceeds $A_{cv}\lambda\sqrt{f_c'}$, the minimum reinforcement ratios for the horizontal and vertical reinforcement must be

$$\rho_t = 0.0025$$
$$\rho_l = 0.0025$$

Where the design force, V_u, does not exceed $A_{cv}\lambda\sqrt{f_c'}$ the minimum reinforcement ratios for the horizontal and vertical reinforcement must be as specified in ACI Table 11.6.1. For this situation, the minimum required ratios are

$$\rho_l = 0.0012 \quad \begin{bmatrix} \text{for \#5 deformed bars or smaller with} \\ \text{with } f_y \geq 60 \text{ kips/in}^2 \end{bmatrix}$$

$$\rho_l = 0.0015 \quad [\text{for other deformed bars}]$$

$$\rho_t = 0.0020 \quad \begin{bmatrix} \text{for \#5 deformed bars or smaller with} \\ \text{with } f_y \geq 60 \text{ kips/in}^2 \end{bmatrix}$$

$$\rho_t = 0.0025 \quad [\text{for other deformed bars}]$$

ACI Sec. R18.10.4 stipulates that any concentrated longitudinal reinforcement at the wall edges for resisting bending moment is not included in calculating ρ_l.

In accordance with ACI Sec. 18.10.4.3, when $h_w/l_w \leq 2.0$, the ratio $\rho_l \geq \rho_t$.

ACI Sec. 18.10.2.1 requires the spacing of reinforcement to be limited to a maximum of 18 in each way, and reinforcement contributing to the shear strength must be continuous and be distributed across the shear plane.

ACI Sec. 18.10.2.3 requires shear reinforcement at splices to develop the yield strength of the bar in tension.

In accordance with ACI Sec. 11.7.4.1, longitudinal reinforcement must be supported by lateral ties when $A_{st} > 0.01A_g$ or if the reinforcement is required for compressive strength.

For seismic loading, in order to control cracking and inhibit fragmentation of the wall due to cyclical loading in the inelastic range, ACI Sec. 18.10.2.2 specifies that two curtains of reinforcement must be provided when

$$V_u > 2A_{cv}\lambda\sqrt{f_c'} \quad \text{or} \quad \frac{h_w}{l_w} \geq 2.0$$

Walls designed for combined flexure and axial load are designed as columns, in accordance with ACI Sec. 22.4, based on a strain compatibility analysis.

The effective width of flanged sections contributing to the section is specified in ACI Sec. 18.10.5.2 as half the distance between adjacent walls but not more than 25% of the wall height, as shown in Fig. 2.41.

Figure 2.41 *Effective Flange Widths*

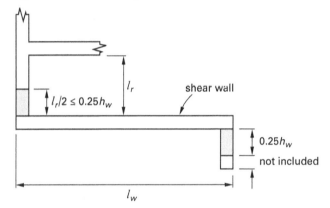

Example 2.37

The 12 in thick shear wall shown is located in the bottom story of a multistory office block and has a normal weight concrete cylinder strength, f_c', of 5000 lbf/in². The wall is braced against sidesway and is bent in single curvature. Longitudinal reinforcement consists of two curtains each of 7 no. 6 grade 60 bars. Transverse reinforcement consists of two curtains each of no. 5 grade 60 bars at 15 in centers. Lateral ties are provided at the wall edges to provide support for the longitudinal compression bars. The unfactored dead load acting at the wall base is 875 kips, and the seismic shear and moment acting at the base of the wall are 170 kips and 2720 ft-kips, respectively. The structure has a redundancy factor, ρ, of 1.0, the 5% damped, design spectral response acceleration, S_{DS}, for a period of 0.2 sec is 0.9g, and the

seismic design category is D. Seismic load combinations govern the design. Determine if the proposed reinforcement details are satisfactory.

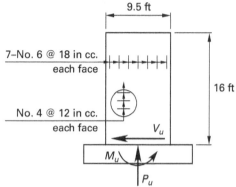

Solution

Stability Check

From ACI Sec. 6.6.4.4.3, the effective length factor, k, for a column braced against sidesway is 1.0.

The radius of gyration, in accordance with ACI Sec. 6.2.5.1, is

$$\begin{aligned}
r &= 0.3l_w \\
&= (0.3)(9.5 \text{ ft})\left(12 \frac{\text{in}}{\text{ft}}\right) \\
&= 34.2 \text{ in}
\end{aligned}$$

The slenderness ratio is

$$\begin{aligned}
\frac{kh_w}{r} &= \frac{(1.0)(16 \text{ ft})\left(12 \dfrac{\text{in}}{\text{ft}}\right)}{34.2 \text{ in}} \\
&= 5.6
\end{aligned}$$

The bending moments at the top and bottom of the wall are

$$\begin{aligned}
M_1 &= 0 \\
M_2 &= 2720 \text{ ft-kips} \\
\frac{M_1}{M_2} &= 0
\end{aligned}$$

In accordance with ACI Sec. 6.2.5(b), the wall, which is bent in single curvature, may be classified as a short column provided that

$$\frac{kh_w}{r} \leq 34 + 12\left(-\frac{M_1}{M_2}\right) \leq 40$$

$$\frac{kh_w}{r} = 34 > 5.6$$

The wall is classified as a short column, and slenderness effects may be ignored.

Applied Loads

Typically, load combination ACI Eq. 5.3.1g governs when the structure is subjected to maximum values of seismic load opposing the effects of dead load and is

$$U = 0.9D + 1.0E$$

When the effects of gravity and seismic loads counteract, E is given by ASCE/SEI7 Eq. 12.4-2, Eq. 12.4-3, and Eq. 12.4-4a as

$$E = \rho Q_E - 0.2 S_{DS} D$$

Load combination ACI Eq. 5.3.1g may now be defined as

$$U = (0.9 - 0.2 S_{DS})D + \rho Q_E$$

The factored wall axial load caused by dead load is

$$\begin{aligned}
P_u &= (0.9 - 0.2 S_{DS})D + \rho Q_E \\
&= \big(0.9 - (0.2)(0.9)\big)(875 \text{ kips}) + (1.0)(0) \\
&= 630 \text{ kips}
\end{aligned}$$

The corresponding factored wall moment is

$$\begin{aligned}
M_u &= (0.9 - 0.2 S_{DS})D + \rho Q_E \\
&= \big(0.9 - (0.2)(0.9)\big)(0) + (1.0)(2720 \text{ ft-kips}) \\
&= 2720 \text{ ft-kips}
\end{aligned}$$

The corresponding factored shear force is

$$\begin{aligned}
V_u &= (0.9 - 0.2 S_{DS})D + \rho Q_E \\
&= \big(0.9 - 0.2(0.9)\big)(0) + (1.0)(170 \text{ kips}) \\
&= 170 \text{ kips}
\end{aligned}$$

Assuming that the wall is tension controlled, for combined moment and axial force, ACI Table 21.2.1 gives the strength reduction factor as $\phi = 0.9$ for both moment and axial force. For shear, ACI Table 21.2.1 gives the strength reduction factor, ϕ, as 0.75. Hence, the required nominal strength values are

$$\begin{aligned}
M_n &= \frac{M_u}{\phi} = \frac{2720 \text{ ft-kips}}{0.9} \\
&= 3022 \text{ in-kips} \\
P_n &= \frac{P_u}{\phi} = \frac{630 \text{ ft-kips}}{0.9} \\
&= 700 \text{ kips} \\
V_n &= \frac{V_u}{\phi} = \frac{170 \text{ ft-kips}}{0.75} \\
&= 227 \text{ kips}
\end{aligned}$$

Wall Parameters

The modification factor for normal weight concrete is given by ACI Table 19.2.4.2 as $\lambda = 1.0$.

The gross area of the concrete section bounded by the web thickness and length of wall in the direction of shear force is

$$\begin{aligned}
A_{cv} &= hl_w = (12 \text{ in})(9.5 \text{ ft})\left(12 \frac{\text{in}}{\text{ft}}\right) \\
&= 1368 \text{ in}^2
\end{aligned}$$

The ratio of the area of distributed longitudinal reinforcement to the gross concrete area perpendicular to that reinforcement is

$$\rho_l = \frac{A_s}{hs} = \frac{(2)(0.44 \text{ in}^2)}{(12 \text{ in})(18 \text{ in})} = 0.00407$$

The ratio of the area of distributed transverse reinforcement to gross concrete area perpendicular to that reinforcement is

$$\rho_t = \frac{A_s}{hs} = \frac{(2)(0.20 \text{ in}^2)}{(12 \text{ in})(12 \text{ in})} = 0.00278$$

Since $h_w/l_w = 1.68 < 2.0$, ACI Sec. 18.10.4.3 requires that the ratio $\rho_l/\rho_t \geq 1.0$.

$$\frac{\rho_l}{\rho_t} = \frac{0.00407}{0.00278} = 1.46 > 1.0 \quad \text{[satisfactory]}$$

Since $V_u = 170$ kips $> A_{cv}\lambda\sqrt{f_c'} = 97$ kips, ACI Sec. 18.10.2.1 requires that both ρ_t and ρ_l equal or exceed the value 0.0025.

$$\begin{aligned}
\rho_t &= 0.00278 > 0.0025 \quad \text{[satisfactory]} \\
\rho_l &= 0.00407 > 0.0025 \quad \text{[satisfactory]}
\end{aligned}$$

Wall Shear Capacity

In accordance with ACI Eq. 18.10.4.1,

$$\begin{aligned}
&\text{for } h_w/l_w \leq 1.5, \quad a_c = 3 \\
&\text{for } h_w/l_w \geq 2.0, \quad a_c = 2
\end{aligned}$$

Hence, for a value of $h_w/l_w = 16 \text{ ft}/9.5 \text{ ft} = 1.68$,

$$\alpha_c = 3 - \frac{(3-2)(1.68 - 1.5)}{2 - 1.5} = 2.64$$

From ACI Eq. 18.10.4.1, the nominal shear strength of the shear wall is

$$V_n = A_{cv}(\alpha_c \lambda \sqrt{f_c'} + \rho_t f_y)$$

$$= \frac{(1368 \text{ in}^2)\begin{pmatrix}(2.64)(1.0)\sqrt{5000} \\ +(0.00278)\left(60{,}000 \ \dfrac{\text{lbf}}{\text{in}^2}\right)\end{pmatrix}}{1000 \ \dfrac{\text{lbf}}{\text{kip}}}$$

$$= 484 \text{ kips} > 227 \text{ kips} \quad \text{[satisfactory]}$$

Wall Flexural Capacity

To determine the flexural strength, the strain compatibility method may be used. As shown in the exhibit, this assumes a linear distribution of strains across the section with the maximum concrete compressive strain specified in ACI 22.2.2.1 as 0.003. The strain in the reinforcement is assumed to be directly proportional to the distance from the neutral axis, and a rectangular stress block is assumed in the concrete. The strain distribution across the section and the forces developed are shown in the exhibit.

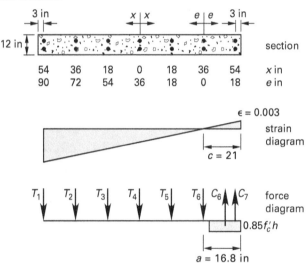

Assuming the depth to the neutral axis is given by $c = 21$ in, the strain produced in a reinforcing bar is

$$\epsilon_x = \frac{\epsilon_c e}{c} = \frac{(0.003)e}{21 \text{ in}}$$
$$= 0.000143e$$

The strain in the extreme tension steel is

$$\epsilon_t = (0.000143 \text{ in}^{-1})(90 \text{ in})$$
$$= 0.013 > 0.005$$

Hence, in accordance with ACI Table 21.2.2, the section is tension-controlled.

The force produced in a reinforcing bar is given by

$$F = \epsilon_s A_s E_s$$
$$= (0.000143)(0.62 \text{ in}^2)\left(29{,}000 \ \frac{\text{kips}}{\text{in}^2}\right)e$$
$$= 2.57e \text{ kips}$$

The strain-producing yield in the reinforcement is

$$\epsilon_y = \frac{f_y}{E_s}$$
$$= \frac{60 \ \dfrac{\text{kips}}{\text{in}^2}}{29{,}000 \ \dfrac{\text{kips}}{\text{in}^2}}$$
$$= 0.00207$$

The maximum force is then produced in the reinforcement and is given by

$$F_{\max} = f_y A_s$$
$$= \left(60 \ \frac{\text{kips}}{\text{in}^2}\right)(0.62 \text{ in}^2)$$
$$= 37.2 \text{ kips}$$

Bars 1 through 5 are stressed to the maximum value of 37.2 kips tension.

Bar 6 has zero stress.

Bar 7 is stressed to the maximum value of 37.2 kips compression, and, as stated in the problem statement, lateral ties are provided to prevent buckling of this bar.

The compression zone factor is obtained from ACI Table 22.2.2.4.3 as $\beta_1 = 0.80$.

The depth of the equivalent rectangular concrete stress block is given by ACI Eq. 22.2.2.4.1 as

$$a = c\beta_1$$
$$= (21 \text{ in})(0.80)$$
$$= 16.8 \text{ in}$$

The force in the concrete stress block is obtained from ACI Sec. 22.2.2.4.1 as

$$C_c = 0.85 f_c'(ah - A_s)$$
$$= (0.85)\left(5 \ \frac{\text{kips}}{\text{in}^2}\right)\left((16.8 \text{ in})(12 \text{ in}) - 0.62 \text{ in}^2\right)$$
$$= 854 \text{ kips}$$

The nominal axial load capacity at this strain condition is

$$P_n = C_c + C_7 - (T_1 + T_2 + T_3 + T_4 + T_5)$$
$$= 854 \text{ kips} + 37.2 \text{ kips} - (5)(37.2 \text{ kips})$$
$$= 705 \text{ kips} \approx 700 \text{ kips}$$

$$\left[\begin{array}{l}\text{the initial estimate of the location}\\\text{of the neutral axis is satisfactory}\end{array}\right]$$

The nominal moment capacity for this neutral axis depth is obtained by summing moments about the mid-depth of the section and is given by

$$M_n = \left(57 \text{ in} - \frac{16.8 \text{ in}}{2}\right) C_c + (54 \text{ in}) T_1 + (36 \text{ in}) T_2$$
$$+ (18 \text{ in}) T_3 - (18 \text{ in}) T_5 - (36 \text{ in}) T_6 + (54 \text{ in}) C_7$$
$$= (48.6 \text{ in})(854 \text{ kips}) + (2)(54 \text{ in})(37.2 \text{ kips})$$
$$= 45{,}522 \text{ in-kips}$$
$$= 3793 \text{ ft-kips} \ > 3022 \text{ ft-kips} \quad \text{[satisfactory]}$$

The wall reinforcement details are satisfactory.

Shear Wall Boundary Elements

In designing shear walls for seismic loading, special boundary elements are provided to prevent spalling of the concrete cover during cyclical deformations in the inelastic range. The ACI code provides two alternative procedures for determining the required reinforcement detailing. These procedures are as follows.

- The procedure of ACI Sec. 18.10.6.2 is applicable to walls with $h_w/l_w \geq 2.0$, and boundary elements are required where the distance from the extreme compression fiber to the neutral axis, at the nominal flexural strength of the wall, is

$$c \geq \frac{l_w}{600\left(\dfrac{1.5\delta_u}{h_w}\right)}$$

$\delta_u =$ inelastic design displacement of the wall
$\quad = C_d \delta_{xe}/I_e$

$C_d = $ deflection amplification factor
$\quad\quad \times$ defined in ASCE/SEI7 Table 12.2-1

$\delta_{xe} = $ theoretical elastic displacement
$\quad\quad$ due to factored forces

$\dfrac{\delta_u}{h_w} \geq 0.005$

The neutral axis depth, c, corresponds to the largest neutral axis depth calculated for the factored axial force and nominal moment strength consistent with the direction of design displacement, δ_u.

The special boundary element is required by ACI Sec. 18.10.6.2(b) to extend vertically, above and below the critical section, a distance of l_w but not less than $M_u/4V_u$.

- The procedure of ACI Sec. 18.10.6.3, in which boundary elements are required, is based on the extreme fiber stress. The stress is calculated for the factored loads acting on the gross concrete section, assuming a linearly elastic response. Boundary elements are required where the maximum extreme concrete fiber stress exceeds $0.2f_c'$. The extreme concrete fiber stress is

$$f_{c,\,max} = \frac{P_u}{A_{cv}} + \frac{M_u}{S_c}$$

The special boundary elements may be discontinued at a level above the base where $f_c < 0.15\,f_c'$.

Design of Special Boundary Element Confinement Reinforcement

As shown in Fig. 2.42, the special boundary element may be discontinued a horizontal distance from the extreme compression fiber given by ACI Sec. 18.10.6.4 (a) as $c - 0.1\,l_w$ but not less than $c/2$. The value of c corresponds to the largest neutral axis depth calculated for the factored axial force and nominal moment strength consistent with the direction of design displacement δ_u.

Figure 2.42 *Special Boundary Element Dimensions for ACI Section 18.10.6.2 Procedure*

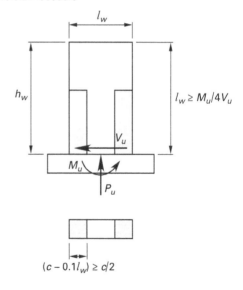

In accordance with ACI Sec. 18.10.6.4(b), over the width of the special boundary element, the width of the flexural compression zone must not be less than

$b = h_u/16$. When $h_w/l_w \geq 2.0$ and $c/l_w \geq 3/8$, ACI Sec. 18.10.6.4(c) requires the width of the flexural compression zone to be not less than 12 in.

The area of rectilinear confinement reinforcement required is given by ACI Table 18.10.6.4(f) as the greater of

$$A_{sh} = \frac{0.09 s b_c f_c'}{f_{yt}}$$

$$A_{sh} = 0.3 s b_c \left(\frac{A_g}{A_{ch}} - 1 \right) \left(\frac{f_c'}{f_{yt}} \right)$$

Confinement reinforcement details are shown in Fig. 2.43 and are provided in ACI Sec. 18.7.5.2(a) through (e), ACI Sec. 18.7.5.3, and ACI Sec. 18.10.6.4(e). The spacing of confinement reinforcement is limited to the minimum value given by

$$s = \frac{h_{\min}}{3} \quad \text{[as modified by ACI Sec. 18.10.6.4(e)]}$$

$$s = 6 d_b$$

$$s_o = 4 + \frac{(14 - h_x)}{3} \geq 4 \text{ in} \leq 6 \text{ in}$$

The maximum horizontal spacing of longitudinal bars on all faces of the boundary element is

$$h_x = 14 \text{ in} \quad \text{[maximum]}$$

$$\leq 2b/3 \quad \text{[as modified by ACI Sec. 18.10.6.4(e)]}$$

As required by ACI Sec. 25.7.2.3(b), no unsupported bar may be more than 6 in clear on each side from a laterally supported longitudinal bar.

The diameter of confinement reinforcement is given in ACI Sec. 25.7.2.2 as the least of

- no. 3 enclosing no. 10 or smaller longitudinal bars
- no. 4 enclosing no. 11 or larger longitudinal bars

Where the special boundary element terminates at a footing, ACI Sec. 18.10.6.4(g) requires the special confinement reinforcement to extend at least 12 in into the footing.

Figure 2.43 *Confinement Reinforcement in Special Boundary Element*

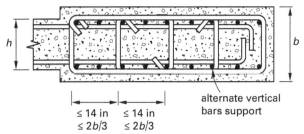

≤ 14 in   ≤ 14 in
≤ 2b/3   ≤ 2b/3

alternate vertical bars support

Example 2.38

The 12 in thick shear wall shown is located in the bottom story of a multistory office block and has a normal weight concrete cylinder strength, f_c', of 5000 lbf/in². The wall is braced against sidesway and is bent in single curvature. Longitudinal reinforcement consists of two curtains each of 7 no. 6 grade 60 bars. Transverse reinforcement consists of two curtains each of no. 5 grade 60 bars at 15 in centers. Lateral ties are provided at the wall edges to provide support for the longitudinal compression bars. The unfactored dead load and live load acting at the wall base are 355 kips and 150 kips, respectively. The seismic moment acting at the base of the wall is 1000 ft-kips. The structure has a redundancy factor, ρ, of 1.0, the 5% damped, design spectral response acceleration, S_{DS}, for a period of 0.2 sec is $0.9g$, and the seismic design category is D. Seismic load combinations govern the design. Using ACI Sec. 18.10.6.3, determine if special boundary elements are required.

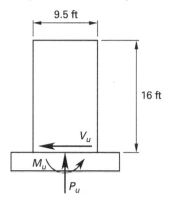

Solution

Applied Loads

To produce maximum compressive stress in the extreme fiber, load combination ACI Eq. 5.3.1e modified by ACI Sec. 5.3.3 governs when the structure is subjected to maximum values of seismic load augmenting the effects of dead load and is

$$U = 1.2D + 1.0E + (0.5 \text{ or } 1.0)L + 0.2S$$

When the effects of gravity and seismic loads are additive, E is given by ASCE/SEI7 Eq. 12.4-1, Eq. 12.4-3, and Eq. 12.4-4a as

$$E = \rho Q_E + 0.2 S_{DS} D$$

Load combination ACI Eq. 5.3.1e may now be defined as

$$U = (1.2 + 0.2 S_{DS})D + \rho Q_E + (0.5 \text{ or } 1.0)L + 0.2S$$

The factored wall axial load caused by dead load and live load in an office building is given by ACI Sec. 5.3.3 as

$$P_u = (1.2 + 0.2S_{DS})D + \rho Q_E + 0.5L + 0.2S$$
$$= \big(1.2 + (0.2)(0.9)\big)(355 \text{ kips})$$
$$\qquad + (1.0)(0) + (0.5)(150 \text{ kips}) + (0.2)(0)$$
$$= 565 \text{ kips}$$

The corresponding factored wall moment is

$$M_u = (1.2 + 0.2S_{DS})D + \rho Q_E + 0.5L + 0.2S$$
$$= \big(1.2 + (0.2)(0.9)\big)(0)$$
$$\qquad + (1.0)(1000 \text{ ft-kips}) + (0.5)(0) + (0.2)(0)$$
$$= 1000 \text{ ft-kips}$$
$$= 12{,}000 \text{ in-kips}$$

The gross area of the concrete section bounded by wall thickness and length of wall in the direction of the shear force considered is

$$A_{cv} = hl_w = (12 \text{ in})(9.5 \text{ ft})\left(12 \frac{\text{in}}{\text{ft}}\right)$$
$$= 1368 \text{ in}^2$$

The section modulus is

$$S_c = \frac{hl_w^2}{6} = \frac{(12 \text{ in})\left((9.5 \text{ ft})\left(12 \frac{\text{in}}{\text{ft}}\right)\right)^2}{6}$$
$$= 25{,}992 \text{ in}^3$$

The extreme concrete fiber stress is

$$f_{c,\max} = \frac{P_u}{A_{cv}} + \frac{M_u}{S_c}$$
$$= \frac{565 \text{ kips}}{1368 \text{ in}^2} + \frac{12{,}000 \text{ in-kips}}{25{,}992 \text{ in}^3}$$
$$= 0.87 \text{ kips/in}^2$$
$$0.2f_c' = (0.2)\left(5 \frac{\text{kips}}{\text{in}^2}\right)$$
$$= 1.0 \text{ kip/in}^2 > f_{c,\max}$$

Hence, in accordance with ACI Sec. 18.10.6.3, special boundary elements are not required.

Nonspecial Boundary Elements

When special boundary elements are not required, some confinement reinforcement may be necessary to prevent buckling of boundary longitudinal bars under cyclic loading.

In accordance with ACI Sec. 18.10.6.5(a), confinement reinforcement must satisfy ACI Sec. 18.7.5.2(a) through (e), and the details are shown in Fig. 2.44.

Figure 2.44 *Nonspecial Boundary Element*

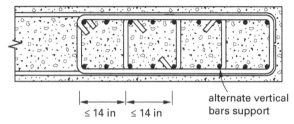

≤ 14 in   ≤ 14 in   alternate vertical bars support

Where the longitudinal reinforcement ratio at the wall boundary is greater than $400/f_y$, ACI Sec. 18.10.6.5 requires confinement reinforcement to extend a horizontal distance from the extreme compression fiber given by $c - 0.1l_w$ but not less than $c/2$. The value of c corresponds to the largest neutral axis depth calculated for the factored axial force and nominal moment strength consistent with the direction of design displacement δ_u.

In accordance with ACI Sec. R18.10.6.5, confinement reinforcement must extend over the entire wall height with a maximum spacing limited by ACI Sec. 18.10.6.5(a) to the lesser of

$$s = 8d_b$$
$$= 8 \text{ in}$$

Over a height equal to the greater of l_w or $M_u/4V_u$, this spacing is reduced to the lesser of 6 in or $6d_b$ where yielding of the longitudinal reinforcement is likely to occur.

The maximum horizontal spacing of longitudinal bars on all faces of the boundary element is given in ACI Sec. 18.7.5.2 as $h_x = 14$ in maximum.

As required by ACI Sec. 25.7.2.3(b), no unsupported bar may be more than 6 in clear on each side from a laterally supported longitudinal bar.

The diameter of confinement reinforcement is given in ACI Sec. 25.7.2.2 as the least of

- no. 3 enclosing no. 10 or smaller longitudinal bars
- no. 4 enclosing no. 11 or larger longitudinal bars

In accordance with ACI Sec. 18.10.6.5(b), where the factored shear force is not less than $A_{cv}\lambda\sqrt{f_c'}$, horizontal reinforcement in the wall web must be anchored at the end of the wall with a standard hook engaging the edge reinforcement. Alternatively, the edge reinforcement must be enclosed in U-stirrups having the same size and spacing as, and spliced to, the horizontal reinforcement. In walls with low in-plane shear, the development of horizontal reinforcement is not necessary.

16. REFERENCES

1. International Code Council. *2018 International Building Code*. Falls Church, VA: International Code Council, 2018.

2. American Concrete Institute. *Building Code Requirements for Structural Concrete and Commentary*. Farmington Hills, MI: American Concrete Institute, 2014.

3. American Concrete Institute. *Commentary on Building Code Requirements for Reinforced Concrete*. Farmington Hills, MI: American Concrete Institute, 1985.

4. Williams, Alan. *Design of Reinforced Concrete Structures*, Fifth ed. Austin, TX: Kaplan, 2012.

5. American Concrete Institute. *Reinforced Concrete Design Manual*. Farmington Hills, MI: American Concrete Institute, 2012.

6. Ghosh, S. K. and August W. Domel. *Design of Concrete Buildings for Earthquake and Wind Forces*. Skokie, IL: Portland Cement Association, 1995.

7. Kamara, Mahmoud and Lawrence Novak. *Notes on ACI 318-11 Building Code Requirements for Structural Concrete: With Design Applications*. Skokie, IL: Portland Cement Association, 2013.

8. Furlong, R. *Reinforced Concrete Shear Strength Analysis with Strut-and-Tie Models*, Engineering Data Report Number 56. Chicago, IL: Concrete Reinforcing Steel Institute, 2005.

9. American Institute of Steel Construction. *Steel Construction Manual*, Fifteenth ed. Chicago, IL: American Institute of Steel Construction, 2017.

10. Structure Point. *spColumn v. 4.60*. (Supports ACI 318-14.)

11. Kamara, Mahmoud and Lawrence Novak. *Slender Column Design Based on ACI 318-08*. Skokie, IL: Portland Cement Association, 2010.

12. Spiker, J. *Development and Splicing of Flexural Reinforcement Based on ACI 318-08*. Skokie, IL: Portland Cement Association, 2008.

13. Fanella, David. and I. M. Alsamsam. "Design of Reinforced Concrete Floor Systems." *Structural Engineer*, 2009.

14. Bartlett, Mark. *ACI 318-08, Appendix D: Anchorage to Concrete*. Pleasanton, CA: Simpson Strong-Tie Anchor Systems, 2013.

15. Cook, Ronald A. *Strength Design of Anchorage to Concrete*. Skokie, IL: Portland Cement Association, 1999.

16. American Concrete Institute and Structural Engineers Association of Southern California. *Report of the Task Committee on Slender Walls*. Los Angeles, CA, 1982.

Reinforced Concrete

Foundations and Retaining Structures

1. STRIP FOOTING

Nomenclature

A_1	loaded area at base of column	ft²
A_2	area of the base of an imaginary truncated pyramid within the footing, its top being the loaded area and with side slopes of 1:2	ft²
b_o	perimeter of critical section for punching shear	in
B	length of strip footing parallel to wall	ft
B	length of short side of a rectangular footing	ft
c	length of side of column	in
D	dead load	kips
e	eccentricity with respect to center of footing	in
e'	eccentricity with respect to edge of footing, $(L/2 - e)$	in
h	depth of footing	in
H	lateral force due to earth pressure	kips
l	distance between column centers	ft
L	length of strip footing perpendicular to wall	ft
L	length of side of a square footing	ft
L	live load	kips
M_u	factored moment at critical section	ft-kips
P	column axial service load	kips
P_{bn}	nominal bearing strength	kips
P_D	column axial service dead load	kips
P_L	column axial service live load	kips
P_u	column axial factored load	kips
q	soil bearing pressure due to service loads	lbf/ft²
q_s	equivalent bearing pressure due to service loads	lbf/ft²
q_u	net factored pressure acting on footing	lbf/ft²
s	spacing of reinforcement	in
V_c	shear capacity of footing	kips
V_u	factored shear force	kips
w_c	specific weight of concrete	lbf/ft³
x	distance from edge of footing or center of column to critical section	ft
x_o	distance from edge of property line to centroid of service loads	ft

Symbols

β	ratio of long side to short side of loaded area	–

Pressure Distribution

To determine soil pressure under a footing, the unfactored self-weight of the footing is added to the unfactored applied load from the wall. The footing dimensions are adjusted to ensure that the soil pressure as calculated from the unfactored loads does not exceed the allowable pressure. To determine the forces acting on a footing, the net pressure is required as determined from the applied wall load only. For a strip footing of length B parallel to the wall, the net pressure acting on the footing, as shown in Fig. 3.1, is given by

$$q = \frac{P}{BL} \quad [e = 0]$$

$$q_{max}, q_{min} = \frac{P\left(1 \pm \dfrac{6e}{L}\right)}{BL} \quad [e \leq L/6]$$

$$q_{max} = \frac{2P}{3Be'} \quad [e > L/6]$$

Design of a footing is based on the net factored applied loads, determined in accordance with ACI Sec. 5.3.

Figure 3.1 *Net Pressure Distribution on a Footing*

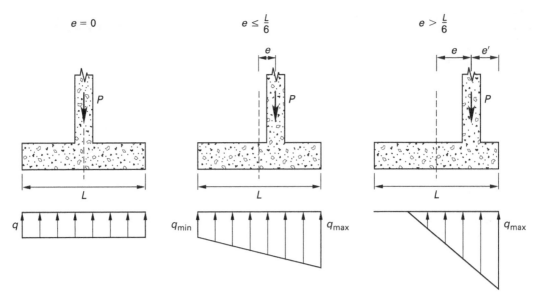

Example 3.1

The 18 in deep strip footing of normal weight concrete shown supports a 12 in concrete wall that is offset 1 ft from the center of the footing. The applied service loads are indicated in the figure, and the allowable soil pressure is 5000 lbf/ft² . Determine the required footing dimensions, the soil pressure under the footing, and the net factored pressure acting on the footing.

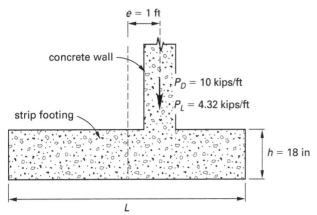

Solution

The total applied service load per foot run is

$$P_1 = P_D + P_L$$
$$= 10{,}000 \ \frac{\text{lbf}}{\text{ft}} + 4320 \ \frac{\text{lbf}}{\text{ft}}$$
$$= 14{,}320 \ \text{lbf/ft}$$

The self-weight of the footing per foot run is

$$P_2 = w_c h L$$
$$= \left(150 \ \frac{\text{lbf}}{\text{ft}^3}\right)(1.5 \ \text{ft}) L$$
$$= 225 L \ \text{lbf/ft}^2$$

Assuming that $e < L/6$, the maximum soil pressure per foot run is given by

$$q_{max} = \frac{P_1\left(1 + \dfrac{6e}{L}\right)}{BL} + \frac{P_2}{BL}$$
$$= \frac{\left(14{,}320 \ \dfrac{\text{lbf}}{\text{ft}}\right)\left(1 + \dfrac{(6)(1 \ \text{ft})}{L}\right)}{(1 \ \text{ft}) L} + \frac{225 L \ \dfrac{\text{lbf}}{\text{ft}^2}}{(1 \ \text{ft}) L}$$
$$= 5000 \ \text{lbf/ft}^2 \ \text{per foot} \quad \text{[as given]}$$

Solving for the footing length gives $L = 6$ ft.

Checking this value, the maximum soil pressure under the footing per foot run is

$$q_{max} = \frac{P_1\left(1 + \dfrac{6e}{L}\right)}{BL} + \frac{P_2}{BL}$$
$$= \frac{\left(14{,}320 \ \dfrac{\text{lbf}}{\text{ft}}\right)\left(1 + \dfrac{(6)(1 \ \text{ft})}{6 \ \text{ft}}\right)}{(1 \ \text{ft})(6 \ \text{ft})} + \frac{\left(225 \ \dfrac{\text{lbf}}{\text{ft}}\right)(6 \ \text{ft})}{(1 \ \text{ft})(6 \ \text{ft})}$$
$$= 5000 \ \text{lbf/ft}^2 \ \text{per foot}$$

The minimum soil pressure under the footing is

$$
\begin{aligned}
q_{min} &= \frac{P_1\left(1 - \dfrac{6e}{L}\right)}{BL} + \frac{P_2}{BL} \\[2mm]
&= \frac{\left(14{,}320\ \dfrac{lbf}{ft}\right)\left(1 - \dfrac{(6)(1\ ft)}{6\ ft}\right)}{(1\ ft)(6\ ft)} + \frac{1350\ \dfrac{lbf}{ft}}{(1\ ft)(6\ ft)} \\[2mm]
&= 225\ lbf/ft^2 \text{ per foot}
\end{aligned}
$$

The net factored load on the footing, in accordance with ACI Table 5.3.1, is

$$
\begin{aligned}
P_u &= 1.2P_D + 1.6P_L \\
&= (1.2)(10\ kips) + (1.6)(4.32\ kips) \\
&= 18.91\ kips
\end{aligned}
$$

The maximum net factored pressure acting on the footing is

$$
\begin{aligned}
q_{u(max)} &= \frac{P_u\left(1 + \dfrac{6e}{L}\right)}{BL} \\[2mm]
&= \frac{(18.91\ kips)\left(1 + \dfrac{(6)(1\ ft)}{6\ ft}\right)}{(1\ ft)(6\ ft)} \\[2mm]
&= 6.30\ kips/ft^2
\end{aligned}
$$

The minimum net factored pressure acting on the footing is

$$
\begin{aligned}
q_{u(min)} &= \frac{P_u\left(1 - \dfrac{6e}{L}\right)}{BL} \\[2mm]
&= \frac{(18.91\ kips)\left(1 - \dfrac{(6)(1\ ft)}{6\ ft}\right)}{(1\ ft)(6\ ft)} \\[2mm]
&= 0
\end{aligned}
$$

Factored Soil Pressure

A reinforced concrete strip footing may be designed for flexure or for flexural (i.e., one-way) shear. A reinforced concrete isolated footing may additionally be designed for punching (i.e., two-way) shear.

The critical section for flexure and shear is located at a different position in the footing, and each must be designed for the applied factored loads. The soil pressure distribution due to factored loads must be determined. Since the self-weight of the footing produces an equal and opposite pressure in the soil, the footing is designed for the net pressure due to the column load only, and the weight of the footing is not included.

Design for Flexural Shear

The critical section for flexural shear is defined in ACI Sec. 8.4.3.2 as being located a distance d from the face of the concrete or masonry wall, as shown in Fig. 3.2. The shear strength of the footing is determined in accordance with ACI Sec. 22.5.

Example 3.2

The strip footing for Ex. 3.1 has an effective depth of 14 in and a concrete strength of 3000 lbf/in². Using *Illustration for Ex. 3.2*, determine whether the shear capacity is adequate.

Solution

The net factored pressure acting on the footing is shown in the illustration, and the pressure at the critical section for shear on the right side of the footing is

$$
\begin{aligned}
q &= \frac{q_u(5.67\ ft)}{6\ ft} \\[2mm]
&= \frac{\left(6.30\ \dfrac{kips}{ft^2}\right)(5.67\ ft)}{6\ ft} \\[2mm]
&= 5.95\ kips/ft^2
\end{aligned}
$$

For a 1 ft strip, the factored shear force at the critical section is

$$
\begin{aligned}
V_u &= \frac{(0.33\ ft)(1\ ft)(q + q_u)}{2} \\[2mm]
&= \frac{(0.33\ ft)(1\ ft)\left(5.95\ \dfrac{kips}{ft^2} + 6.30\ \dfrac{kips}{ft^2}\right)}{2} \\[2mm]
&= 2.02\ kips
\end{aligned}
$$

The pressure at the critical section for shear on the left side of the footing is

$$
\begin{aligned}
q &= \frac{q_u(2.33\ ft)}{6\ ft} \\[2mm]
&= \frac{\left(6.30\ \dfrac{kips}{ft^2}\right)(2.33\ ft)}{6\ ft} \\[2mm]
&= 2.45\ kips/ft^2
\end{aligned}
$$

Figure 3.2 *Critical Sections for Flexure and Flexural Shear*

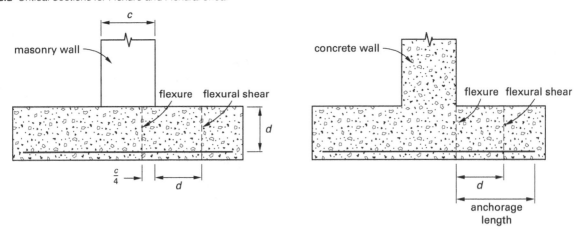

Illustration for Ex. 3.2

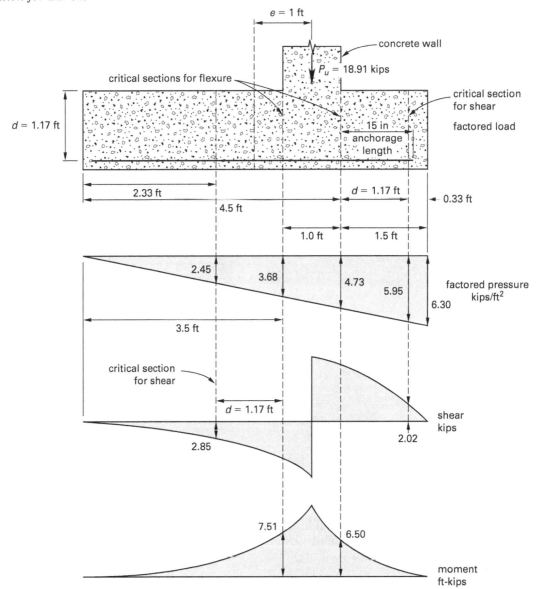

For a 1 ft strip, the factored shear force at the critical section is

$$V_u = \frac{(2.33 \text{ ft})(1 \text{ ft})q}{2}$$

$$= \frac{(2.33 \text{ ft})(1 \text{ ft})\left(2.45 \dfrac{\text{kips}}{\text{ft}^2}\right)}{2}$$

$$= 2.85 \text{ kips} \quad [\text{governs}]$$

The shear capacity of the footing is given by ACI Eq. 22.5.5.1 as

$$\phi V_c = 2\phi bd\lambda\sqrt{f_c'}$$

$$= \frac{(2)(0.75)(12 \text{ in})(14 \text{ in})(1.0)\sqrt{3000 \dfrac{\text{lbf}}{\text{in}^2}}}{1000 \dfrac{\text{lbf}}{\text{kip}}}$$

$$= 13.80 \text{ kips}$$
$$> V_u \quad [\text{satisfactory}]$$

Design for Flexure

The critical section for flexure is defined in ACI Table 13.2.7.1 as being located at the face of a concrete wall and halfway between the center and the face of a masonry wall, as shown in Fig. 3.2. The required reinforcement area is determined in accordance with ACI Sec. 22.2. For a footing, the minimum ratio, $\rho_{\min}$, of reinforcement area to gross concrete area is specified in ACI Table 7.6.1.1, for both main reinforcement and distribution reinforcement, as 0.0018 for grade 60 bars. The maximum spacing of the main reinforcement must not exceed 18 in or three times the footing depth. The diameter of bar provided must be such that the development length does not exceed the available anchorage length. Distribution reinforcement may be spaced at a maximum of 18 in or five times the footing depth.

Example 3.3

Determine the reinforcement required in the strip footing for Ex. 3.1 and Ex. 3.2.

Solution

As shown in *Illustration for Ex. 3.2*, on the right side of the footing, the factored pressure at the critical section for flexure, which is at the right face of the wall, is given by

$$q = \frac{q_u(4.5 \text{ ft})}{6 \text{ ft}} = \frac{\left(6.30 \dfrac{\text{kips}}{\text{ft}^2}\right)(4.5 \text{ ft})}{6 \text{ ft}}$$

$$= 4.73 \text{ kips/ft}^2$$

For a 1 ft strip, the factored moment at the critical section is

$$M_u = \frac{(1.5 \text{ ft})^2(1 \text{ ft})(q + 2q_u)}{6}$$

$$= \frac{(1.5 \text{ ft})^2(1 \text{ ft})\left(4.73 \dfrac{\text{kips}}{\text{ft}^2} + (2)\left(6.30 \dfrac{\text{kips}}{\text{ft}^2}\right)\right)}{6}$$

$$= 6.50 \text{ ft-kips}$$

As shown in *Illustration for Ex. 3.2*, on the left side of the footing the factored pressure at the critical section for flexure, which is at the left face of the wall, is given by

$$q = \frac{q_u(3.5 \text{ ft})}{6 \text{ ft}} = \frac{\left(6.30 \dfrac{\text{kips}}{\text{ft}^2}\right)(3.5 \text{ ft})}{6 \text{ ft}}$$

$$= 3.68 \text{ kips/ft}^2$$

For a 1 ft strip, the factored moment at the critical section is

$$M_u = \frac{(3.5 \text{ ft})^2(1 \text{ ft})q}{6}$$

$$= \frac{(3.5 \text{ ft})^2(1 \text{ ft})\left(3.68 \dfrac{\text{kips}}{\text{ft}^2}\right)}{6}$$

$$= 7.51 \text{ ft-kips} \quad [\text{governs}]$$

Assuming a tension-controlled section, the required reinforcement ratio is given by

$$\rho = \frac{0.85f_c'\left(1 - \sqrt{1 - \dfrac{M_u}{0.383bd^2f_c'}}\right)}{f_y}$$

$$= \frac{(0.85)\left(3 \dfrac{\text{kips}}{\text{in}^2}\right)\times\left(1 - \sqrt{1 - \dfrac{(7.51 \text{ ft-kips})\left(12 \dfrac{\text{in}}{\text{ft}}\right)\times\left(1000 \dfrac{\text{lbf}}{\text{kip}}\right)}{(0.383)(12 \text{ in})(14 \text{ in})^2\times\left(3000 \dfrac{\text{lbf}}{\text{in}^2}\right)}}\right)}{60 \dfrac{\text{kips}}{\text{in}^2}}$$

$$= 0.0007$$

Foundations

The minimum reinforcement area governs and, for a footing, is given by ACI Table 7.6.1.1 as

$$A_s = 0.0018bh$$
$$= (0.0018)(12 \text{ in})(18 \text{ in})$$
$$= 0.39 \text{ in}^2$$

Providing no. 4 bars at 6 in on center for both main and distribution reinforcement gives

$$A_s = 0.40 \text{ in}^2/\text{ft} \quad [\text{satisfactory}]$$
$$\Psi_t = \Psi_e = \lambda$$
$$= 1 \quad \begin{bmatrix} \text{uncoated bottom bars} \\ \text{in normal weight concrete} \end{bmatrix}$$
$$\Psi_s = 0.8 \quad [\text{for no. 4 bars}]$$
$$\frac{c_b + K_{tr}}{d_b} = 2.5 \quad [\text{from ACI Sec. 25.4.2.3}]$$

ACI Eq. 25.4.2.3a for development length reduces to

$$l_d = \frac{(0.075)(0.8d_bf_y)}{2.5\lambda\sqrt{f'_c}}$$
$$= \frac{(0.06d_b)\left(60{,}000 \ \dfrac{\text{lbf}}{\text{in}^2}\right)}{(2.5)(1.0)\sqrt{3000 \ \dfrac{\text{lbf}}{\text{in}^2}}}$$
$$= 26.3d_b$$
$$= 13.2 \text{ in} \quad [\text{for no. 4 bars}]$$
$$< 15 \text{ in anchorage length provided} \quad [\text{satisfactory}]$$

2. ISOLATED COLUMN WITH SQUARE FOOTING

Nomenclature

A_g	gross area of footing cross section	in^2
A_s	area of distribution reinforcement	in^2
b_1	width of the critical perimeter measured in the direction of M_u, $c_1 + d$	in
b_2	width of the critical perimeter measured perpendicular to b_1, $c_2 + d$	in
b_o	length of the critical perimeter, $2(b_1 + b_2)$	in
B	length of the footing measured perpendicular to L	ft
c_1	width of the column measured in the direction of M_u	in
c_2	width of the column measured perpendicular to c_1	in
d	average effective depth	in

d_b	bar diameter	in
J_c	polar moment of inertia of critical perimeter	in^4
J_c/y	$(b_1d(b_1 + 3b_2) + d^3)/3$ [for a footing with interior column as given by PCA Fig. 16-13]	in^3
L	length of the footing measured in the direction of M_u	ft
M_u	moment applied to the column	ft-kips
P_u	factored axial force on column	kips
s	spacing of reinforcement	in
v_u	shear stress at critical perimeter	lbf/in^2
V_u	factored shear force acting on the critical perimeter, $P_u(1 - b_1b_2/BL)$	kips
y	distance from the centroid of the critical perimeter to edge of critical perimeter, $b_1/2$ [for footing with central column]	in

Symbols

α_s	constant used to compute shear in slabs and footings	–
β	ratio of long side to short side of column	–
γ_v	fraction of the applied moment transferred by shear as specified by ACI Sec. 8.4.2.3.2 and Sec. 8.4.4.2.2, $1 - 1/(1 + 0.67\sqrt{b_1/b_2})$	–

Reinforcement Details

The maximum spacing of the principal reinforcement in footings is limited by ACI Sec. 7.7.2.3 to

$$s = 3h \leq 18 \text{ in}$$

The maximum spacing of the distribution reinforcement, which is required to resist shrinkage and temperature stresses, is limited by ACI Sec. 7.7.6.2.1 to

$$s = 5h \leq 18 \text{ in}$$

The minimum spacing of reinforcement must be adequate to allow full consolidation of the concrete around the bars. The minimum clear spacing between parallel bars in a layer is specified by ACI Sec. 25.2.1 as

$$s_{\text{min,clear}} = d_b \geq 1 \text{ in}$$
$$\geq (1.33)(\text{maximum aggregate size})$$

The minimum reinforcement ratio specified for distribution steel in ACI Sec. 7.6.1.1 is

$$\frac{A_s}{A_g} = 0.0018 \quad [\text{grade 60 reinforcement}]$$

Figure 3.3 *Critical Perimeter for Punching Shear*

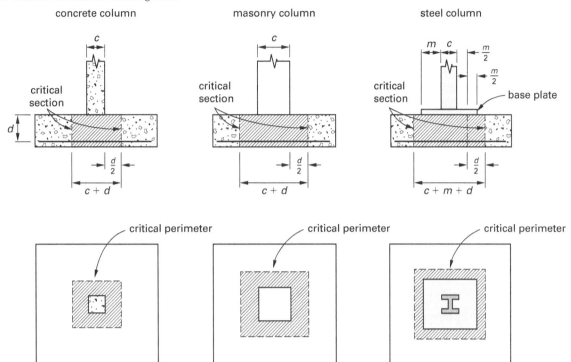

A_s is the area of distribution reinforcement, and A_g is the gross area of footing cross section. When concrete is cast against and exposed to earth (as is the case with the bottom and side of footings), ACI Table 20.6.1.3.1 specifies the minimum concrete cover provided for reinforcement to be 3 in. When concrete is exposed to earth or weather, ACI Table 20.6.1.3.1 specifies the minimum cover to be 1½ in for no. 5 bars and smaller, and 2 in for no. 6 through no. 18 bars.

Design for Punching Shear

The critical perimeter for punching shear is specified in ACI Sec. 22.6.4.1 and illustrated in Fig. 3.3. For a concrete or masonry column, the critical section is a distance from the face of the column equal to one-half the effective depth. For a steel column with a base plate, the critical section is one-half the effective depth from a plane halfway between the face of the column and the edge of the base plate. The punching shear strength of the footing is determined as the smallest of the three values given by ACI Table 22.6.5.2.

$$\phi V_c = 4\phi db_o \lambda \sqrt{f_c'} \quad \text{[a]}$$

$$\phi V_c = \phi db_o \left(2 + \frac{4}{\beta} \right) \lambda \sqrt{f_c'} \quad \text{[b]}$$

$$\phi V_c = \phi \left(\frac{\alpha_s d}{b_o} + 2 \right) \lambda \sqrt{f_c'} \, b_o d \quad \text{[c]}$$

Where,

$$\alpha_s = 40 \quad \text{[interior columns]}$$
$$= 30 \quad \text{[edge columns]}$$
$$= 20 \quad \text{[corner columns]}$$
$$\phi = 0.75$$

For an interior column, Expression (a) governs for a square column, Expression (b) governs for a column when $\beta > 2$, and Expression (c) governs for a column when $b_o > 20d$.

When the column supports only an axial load, shear stress at the critical perimeter is uniformly distributed around the critical perimeter, and is

$$v_u = \frac{V_u}{db_o}$$

The factored shear force acting on the critical perimeter is

$$V_u = P_u \left(1 - \frac{(c_1 + d)(c_2 + d)}{BL} \right)$$

The depth of the footing is typically governed by the punching shear capacity.

When, in addition to the axial load, a bending moment, M_u, is applied to the column, an eccentric shear stress is introduced into the critical section with the maximum value occurring on the face nearest the largest bearing pressure. When both axial load and bending moment occur, the shear stresses from both conditions are combined as specified in ACI Sec. R8.4.4.2 to give a maximum value of

$$v_u = \frac{V_u}{db_o} + \frac{\gamma_v M_u y}{J_c}$$

Example 3.4

A 6 ft square reinforced concrete footing of normal weight concrete with an effective depth of 12 in supports a 12 in square column with a factored axial load of 200 kips. The concrete strength is 3000 lbf/in². Determine whether the punching shear capacity of the footing is satisfactory.

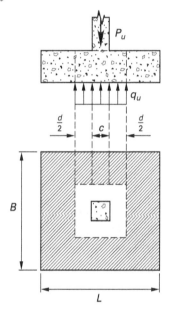

Solution

The net factored pressure on the footing is

$$q_u = \frac{P_u}{LB} = \frac{200 \text{ kips}}{(6 \text{ ft})(6 \text{ ft})}$$

$$= 5.56 \text{ kips/ft}^2$$

The length of the critical perimeter is

$$b_o = 4(c + d)$$

$$= \frac{(4)(12 \text{ in} + 12 \text{ in})}{12 \frac{\text{in}}{\text{ft}}}$$

$$= 8 \text{ ft}$$

Shear at the critical perimeter is

$$V_u = P_u - q_u(c + d)^2$$

$$= 200 \text{ kips} - \left(5.56 \frac{\text{kips}}{\text{ft}^2}\right)\left(\frac{12 \text{ in} + 12 \text{ in}}{12 \frac{\text{in}}{\text{ft}}}\right)^2$$

$$= 177.7 \text{ kips}$$

Shear capacity of the footing is given by ACI Table 22.6.5.2(a) as

$$\phi V_c = 4\phi db_o \lambda \sqrt{f_c'}$$

$$= \frac{(4)(0.75)(12 \text{ in})(96 \text{ in})(1.0)\sqrt{3000 \frac{\text{lbf}}{\text{in}^2}}}{1000 \frac{\text{lbf}}{\text{kip}}}$$

$$= 189.3 \text{ kips}$$

$$> V_u \quad \text{[satisfactory]}$$

Design for Flexural Shear

For concrete and masonry columns, the location of the critical section for flexural shear is identical with that for a strip footing and is shown in Fig. 3.2. As shown in Fig. 3.4, the critical section for a steel column with a base plate is located an effective depth from the plane halfway between the face of the column and edge of the base plate.

Figure 3.4 *Critical Sections for a Footing with Steel Base Plate*

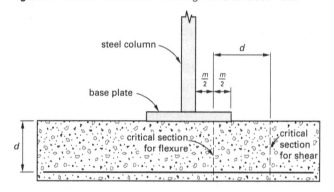

Example 3.5

For the reinforced concrete footing of Ex. 3.4, determine whether the flexural shear capacity is adequate.

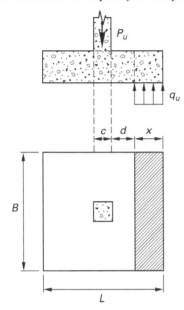

Solution

The distance from the edge of the footing of the critical section for flexural shear is

$$x = \frac{L}{2} - \frac{c}{2} - d$$
$$= \frac{6 \text{ ft}}{2} - \frac{1 \text{ ft}}{2} - 1 \text{ ft}$$
$$= 1.5 \text{ ft}$$

The factored shear force at this section is

$$V_u = q_u B x$$
$$= \left(5.56 \frac{\text{kips}}{\text{ft}^2}\right)(6 \text{ ft})(1.5 \text{ ft})$$
$$= 50.0 \text{ kips}$$

The shear capacity of the footing is given by ACI Eq. 22.5.5.1 as

$$\phi V_c = 2\phi b d \lambda \sqrt{f'_c}$$
$$= \frac{(2)(0.75)(72 \text{ in})(12 \text{ in})(1.0)\sqrt{3000 \frac{\text{lbf}}{\text{in}^2}}}{1000 \frac{\text{lbf}}{\text{kip}}}$$
$$= 71.0 \text{ kips}$$
$$> V_u \quad [\text{satisfactory}]$$

Design for Flexure

For concrete and masonry columns, the location of the critical section for flexure is identical with that for a strip footing and is shown in Fig. 3.2. As shown in Fig. 3.4, the critical section for flexure for a steel column with a base plate is at a plane halfway between the face of the column and the edge of the base plate.

Example 3.6

Determine the reinforcement required in the square footing for Ex. 3.4. The depth of the footing is 15.5 in, and the reinforcement is grade 60.

Solution

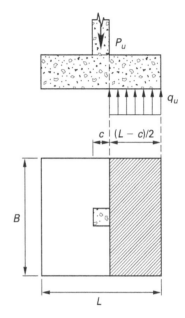

The factored moment at the critical section for flexure, which is at the face of the column, is given by

$$M_u = \frac{q_u B \left(\frac{L}{2} - \frac{c}{2}\right)^2}{2}$$
$$= \frac{\left(5.56 \frac{\text{kips}}{\text{ft}^2}\right)(6 \text{ ft})\left(\frac{6 \text{ ft}}{2} - \frac{1 \text{ ft}}{2}\right)^2}{2}$$
$$= 104.3 \text{ ft-kips}$$

Assuming a tension-controlled section, the required reinforcement ratio is given by

$$\rho = \frac{0.85f'_c\left(1 - \sqrt{1 - \dfrac{M_u}{0.383bd^2f'_c}}\right)}{f_y}$$

$$= \frac{(0.85)\left(3\;\dfrac{\text{kips}}{\text{in}^2}\right)}{60\;\dfrac{\text{kips}}{\text{in}^2}}$$
$$\times\left(1 - \sqrt{1 - \frac{(104.3\text{ ft-kips})\left(12\;\dfrac{\text{in}}{\text{ft}}\right)\times\left(1000\;\dfrac{\text{lbf}}{\text{kip}}\right)}{(0.383)(72\text{ in})(12\text{ in})^2\times\left(3000\;\dfrac{\text{lbf}}{\text{in}^2}\right)}}\right)$$

$$= 0.0023$$

The maximum allowable reinforcement ratio for a tension-controlled section is given by

$$\rho_t = 0.319\beta_1\left(\frac{f'_c}{f_y}\right) = 0.0136$$

$$> \rho \quad \begin{bmatrix}\text{satisfactory, the section}\\ \text{is tension controlled}\end{bmatrix}$$

The required reinforcement area is

$$A_s = \rho bd = (0.0023)(72\text{ in})(12\text{ in}) = 1.99\text{ in}^2$$

Providing 10 no. 4 bars gives a reinforcement area of

$$A_s = 2.0\text{ in}^2 \quad \text{[satisfactory]}$$

The minimum allowable reinforcement area is given by ACI Sec. 7.6.1.1 as

$$A_{s(\min)} = 0.0018bh = (0.0018)(72\text{ in})(15.5\text{ in})$$
$$= 2.0\text{ in}^2 \quad \text{[satisfactory]}$$

From Ex. 3.3, the development length of a no. 4 bar is

$$l_d = 13.2\text{ in}$$

The anchorage length provided (in inches) is

$$l_a = \frac{L}{2} - \frac{c}{2} - \text{end cover} = \frac{72\text{ in}}{2} - \frac{12\text{ in}}{2} - 3\text{ in}$$
$$= 27\text{ in}$$
$$> l_d \quad \text{[satisfactory]}$$

Transfer of Force at Base of Column

Bearing Capacity of the Column Concrete

In accordance with ACI Sec. 16.3.1.1, load transfer between a reinforced concrete column and the footing may be affected both by bearing on concrete and by reinforcement.

The bearing capacity of the column concrete at the interface is given by ACI Table 22.8.3.2.

$$\phi P_{bn} = 0.85\phi f'_c A_1$$
$$= 0.553f'_c A_1 \quad \text{[for } \phi = 0.65\text{]}$$

A_1 is the area of the column.

Bearing Capacity of the Footing Concrete

The bearing capacity of the footing concrete at the interface is given by ACI Table 22.8.3.2. (Refer to Fig. 3.5.)

$$\phi P_{bn} = 0.85\phi f'_c A_1\sqrt{\frac{A_2}{A_1}}$$

$$\leq (0.85\phi f'_c A_1)(2)$$

A_2 is the area of the base of the pyramid, with side slopes of 1:2, formed within the footing by the column base.

In accordance with ACI Sec. 16.3.1.2, when the bearing strength at the base of the column or at the top of the footing is exceeded, reinforcement must be provided to carry the excess load. This reinforcement may be provided by dowels or extended longitudinal bars, and the capacity of this reinforcement is

$$\phi P_s = \phi A_s f_y$$

A minimum area of reinforcement is required across the interface, and this is given by ACI Sec. 16.3.4.1 as

$$A_{s(\min)} = 0.005A_g$$

A_g is the gross area of the supported member.

Figure 3.5 *Bearing on Footing Concrete*

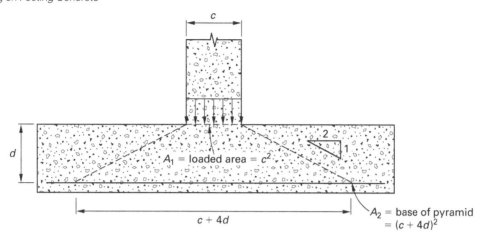

Example 3.7

Assume that the column in Ex. 3.4 has a concrete compressive strength of 3000 lbf/in^2 and carries a factored axial load of 280 kips. Design the dowels required at the interface.

Solution

The bearing capacity of the column concrete is given by ACI Table 22.8.3.2 as

$$
\phi P_{bn} = 0.553 f'_c A_1 = (0.553)\left(3\ \frac{\text{kips}}{\text{in}^2}\right)(144\ \text{in}^2)
$$
$$
= 239\ \text{kips}
$$

Excess column load to be carried by dowels is

$$
\phi P_s = P_u - \phi P_{bn} = 280\ \text{kips} - 239\ \text{kips}
$$
$$
= 41\ \text{kips}
$$

The required area of dowels is given by

$$
A_{s(\text{reqd})} = \frac{\phi P_s}{0.65 f_y} = \frac{41\ \text{kips}}{(0.65)\left(60\ \dfrac{\text{kips}}{\text{in}^2}\right)} = 1.05\ \text{in}^2
$$

Providing four no. 5 dowels gives an area of

$$
A_{s(\text{prov})} = 1.24\ \text{in}^2 \quad [\text{satisfactory}]
$$

The minimum dowel area allowed is given by ACI Sec. 16.3.1.2 as

$$
A_{s(\text{min})} = 0.005 A_1 = (0.005)(144\ \text{in}^2) = 0.72\ \text{in}^2
$$
$$
< A_{s(\text{prov})} \quad [\text{satisfactory}]
$$

Allowing for the excess reinforcement provided, the development length of the dowels in the column and in the footing is given by ACI Sec. 25.4.9.2 as

$$
l_{dc} = \frac{\left(\dfrac{A_{s(\text{reqd})}}{A_{s(\text{prov})}}\right)(0.02 d_b f_y)}{\lambda \sqrt{f'_c}}
$$
$$
= \frac{\left(\dfrac{1.05\ \text{in}^2}{1.24\ \text{in}^2}\right)\left((0.02)(0.63\ \text{in})\left(60{,}000\ \dfrac{\text{lbf}}{\text{in}^2}\right)\right)}{1.0\sqrt{3000\ \dfrac{\text{lbf}}{\text{in}^2}}}
$$
$$
= 12\ \text{in}
$$

This length exceeds the minimum length of 8 in specified in ACI Sec. 25.4.9.1 and is satisfactory. In the footing, the length of the base of the pyramid, with side slopes of 1:2, formed within the footing by the loaded area is

$$
L_p = c + 4d = 12\ \text{in} + (4)(12\ \text{in}) = 60\ \text{in}
$$
$$
< L \quad [\text{satisfactory}]
$$

The area of the base of the pyramid is

$$
A_2 = L_p^2 = (60\ \text{in})^2
$$
$$
= 3600\ \text{in}^2
$$
$$
\sqrt{\frac{A_2}{A_1}} = \sqrt{\frac{3600\ \text{in}^2}{144\ \text{in}^2}}
$$
$$
= 5
$$

Use a maximum value of 2.

Figure 3.6 *Rectangular Footing: Reinforcement Areas*

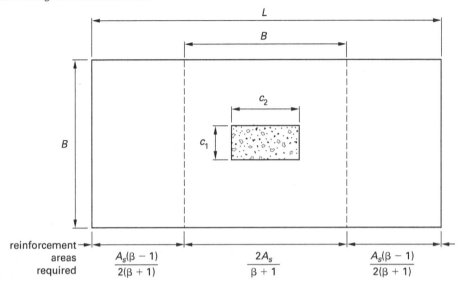

reinforcement areas required

$$\frac{A_s(\beta - 1)}{2(\beta + 1)} \qquad \frac{2A_s}{\beta + 1} \qquad \frac{A_s(\beta - 1)}{2(\beta + 1)}$$

Then, bearing capacity of the footing concrete is given by ACI Table 22.8.3.2 as

$$\phi P_{bn} = 2(0.533 f_c' A_1) = (2)(0.553)\left(3\ \frac{\text{kips}}{\text{in}^2}\right)(144\ \text{in}^2)$$

$$= 478\ \text{kips}$$

$$> P_u \quad [\text{satisfactory}]$$

3. ISOLATED COLUMN WITH RECTANGULAR FOOTING

Nomenclature

A_b	area of reinforcement in central band	in^2
A_s	total required reinforcement area	in^2
B	length of short side of a rectangular footing	ft
c_1	length of short side of a rectangular column	in
c_2	length of long side of a rectangular column	in
L	length of long side of a rectangular footing	ft

Symbols

β	ratio of the long side to the short side of the footing, L/B	–

Design for Flexure

Bending moments are calculated at the critical sections in both the longitudinal and transverse directions. The reinforcement required in the longitudinal direction is distributed uniformly across the width of the footing. Part of the reinforcement required in the transverse direction is concentrated in a central band width equal to the length of the short side of the footing, as shown in Fig. 3.6.

The area of reinforcement required in the central band is given by ACI Sec. 13.3.3.3 as

$$A_b = \frac{2A_s}{\beta + 1}$$

The remainder of the reinforcement required in the transverse direction is

$$A_r = \frac{A_s(\beta - 1)}{\beta + 1}$$

This is distributed uniformly on each side of the center band.

Example 3.8

The reinforced concrete footing shown is 10 ft long and 7 ft wide, has an effective depth of 12 in and an overall depth of 16 in, and has a concrete compressive strength of 5000 lbf/in^2. The footing supports a column with dimensions 12 in by 18 in and is reinforced with grade 60 bars. Determine the transverse reinforcement required when the net factored pressure acting on the footing is 4.8 kips/ft^2.

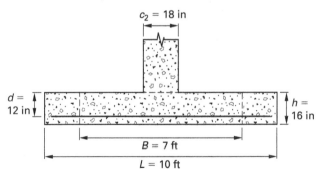

Solution

The factored moment in the transverse direction at the critical section, which is at the face of the column, is

$$M_u = \frac{q_u L \left(\dfrac{B}{2} - \dfrac{c_1}{2} \right)^2}{2}$$

$$= \frac{\left(4.8 \, \dfrac{\text{kips}}{\text{ft}^2} \right)(10 \text{ ft})\left(\dfrac{7 \text{ ft}}{2} - \dfrac{1 \text{ ft}}{2} \right)^2}{2}$$

$$= 216 \text{ ft-kips}$$

Assuming a tension-controlled section, the required reinforcement ratio is given by

$$\rho = \frac{0.85 f'_c \left(1 - \sqrt{1 - \dfrac{M_u}{0.383 b d^2 f'_c}} \right)}{f_y}$$

$$= \frac{(0.85)\left(5 \, \dfrac{\text{kips}}{\text{in}^2} \right)}{60 \, \dfrac{\text{kips}}{\text{in}^2}} \times \left(1 - \sqrt{1 - \dfrac{(216 \text{ ft-kips})\left(12 \, \dfrac{\text{in}}{\text{ft}} \right) \times \left(1000 \, \dfrac{\text{lbf}}{\text{kip}} \right)}{(0.383)(120 \text{ in})(12 \text{ in}^2)^2 \times \left(5000 \, \dfrac{\text{lbf}}{\text{in}^2} \right)}} \right)$$

$$= 0.0028$$

The maximum allowable reinforcement ratio for a tension-controlled section is given by

$$\rho_t = 0.319 \beta_1 \frac{f'_c}{f_y} = 0.0213 > \rho \quad \left[\begin{array}{l} \text{satisfactory, the section} \\ \text{is tension controlled} \end{array} \right]$$

The minimum allowable reinforcement area is given by ACI Sec. 7.6.1.1 as

$$A_{s(\min)} = 0.0018 bh$$

$$= (0.0018)(120 \text{ in})(16 \text{ in})$$

$$= 3.46 \text{ in}^2$$

The required reinforcement area is

$$A_s = \rho b d$$

$$= (0.0028)(120 \text{ in})(12 \text{ in})$$

$$= 4.032 \text{ in}^2$$

$$> A_{s(\min)} \quad [\text{satisfactory}]$$

The reinforcement required in the central 7 ft band width is

$$A_b = \frac{2 A_s}{\beta + 1}$$

$$= \frac{(2)(4.032 \text{ in}^2)}{\dfrac{10 \text{ ft}}{7 \text{ ft}} + 1}$$

$$= 3.32 \text{ in}^2$$

Providing 11 no. 5 bars gives a reinforcement area of

$$A_{b(\text{prov})} = 3.41 \text{ in}^2 \quad [\text{satisfactory}]$$

The remaining reinforcement is

$$A_r = A_s - A_{b(\text{prov})}$$

$$= 4.032 \text{ in}^2 - 3.41 \text{ in}^2$$

$$= 0.62 \text{ in}^2$$

Providing two no. 4 bars on each side of the central band gives a reinforcement area of

$$A_{r(\text{prov})} = (4)(0.20 \text{ in}^2)$$

$$= 0.80 \text{ in}^2 \quad [\text{satisfactory}]$$

By inspection, the anchorage length provided is adequate.

4. COMBINED FOOTING

Nomenclature

A_{1b}	area of reinforcement required in the band width under column no. 1	in^2
P_1	service load for column no. 1	kips
P_2	service load for column no. 2	kips
P_{1u}	factored load for column no. 1	kips
P_{2u}	factored load for column no. 2	kips
q_e	equivalent soil bearing pressure	lbf/ft^2
V_c	shear capacity of concrete footing	kips
V_u	factored shear force	kips

Pressure Distribution

A combined footing is used when a column is located adjacent to a property line. A second column is placed on the combined footing, and the length of the footing is adjusted until its centroid coincides with the centroid of the service loads on the two columns. A uniformly distributed soil pressure is produced under the combined footing, as shown in Fig. 3.7. The footing width is adjusted to ensure that the soil bearing pressure does not exceed the allowable pressure.

Figure 3.7 *Combined Footing with Applied Service Loads*

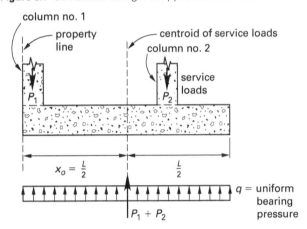

The footing must be designed for punching shear, flexural shear, and flexure. Each of these must be designed for the applied factored loads. The soil pressure distribution due to factored loads must be determined, as shown in Fig. 3.8. The soil pressure will not be uniformly distributed unless the ratios of the factored loads to service loads on both columns are identical. Since the self-weight of the footing produces an equal and opposite pressure in the soil, the footing is designed for the net pressure due to the column load only, and the weight of the footing is not included.

The footing is designed in the longitudinal direction as a beam continuous over two supports. As shown in Fig. 3.8, the maximum negative moment occurs at the section where there is zero shear. The maximum positive moment occurs at the outside face of column no. 2. In the transverse direction, it is assumed that the footing cantilevers about the face of both columns. The reinforcement required is concentrated under each column in a band width equal to the length of the shorter side. The area of reinforcement required in the band width under column no. 1 is given by ACI Sec. 13.3.3.3 as

$$A_{1b} = \frac{2A_s P_{1u}}{(\beta + 1)(P_{1u} + P_{2u})}$$

Figure 3.8 *Combined Footing with Applied Factored Loads*

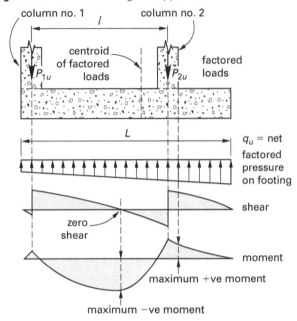

Example 3.9

Determine the plan dimensions required for the combined footing shown in *Illustration for Ex. 3.9* to provide a uniform soil bearing pressure of 4000 lbf/ft² under the service loads indicated.

Solution

Allowing for the self-weight of the footing, the maximum allowable equivalent soil bearing pressure is

$$q_e = q - w_c h$$

$$= \frac{4000 \ \frac{\text{lbf}}{\text{ft}^2} - \frac{\left(150 \ \frac{\text{lbf}}{\text{ft}^3}\right)(27 \ \text{in})}{12 \ \frac{\text{in}}{\text{ft}}}}{1000 \ \frac{\text{lbf}}{\text{kip}}}$$

$$= 3.663 \ \text{kips/ft}^2$$

The centroid of the column service loads is located a distance x_o from the property line, which is obtained by taking moments about the property line and is given by

$$x_o = \frac{0.5P_1 + 15.5P_2}{P_1 + P_2}$$

$$= \frac{(0.5 \ \text{ft})(200 \ \text{kips}) + (15.5 \ \text{ft})(300 \ \text{kips})}{200 \ \text{kips} + 300 \ \text{kips}}$$

$$= 9.5 \ \text{ft}$$

Illustration for Ex. 3.9

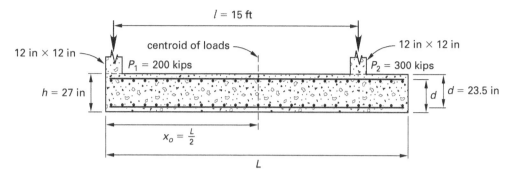

The length of footing required to produce a uniform bearing pressure on the soil is

$$L = 2x_o = (2)(9.5 \text{ ft}) = 19 \text{ ft}$$

The width of footing to produce a uniform pressure on the soil of 4000 lbf/ft² is

$$B = \frac{P_1 + P_2}{q_e L} = \frac{500 \text{ kips}}{\left(3.663 \dfrac{\text{kips}}{\text{ft}^2}\right)(19 \text{ ft})} = 7.2 \text{ ft}$$

Design for Punching Shear

The critical perimeter for punching shear in a combined footing is identical with that in an isolated column footing and is located a distance from the face of the column equal to one-half the effective depth.

For the interior column, the length of the critical perimeter is

$$b_o = 4(c + d)$$

For the end column, the length of the critical perimeter is

$$b_o = (c + d) + 2\left(c + \frac{d}{2}\right)$$

The design punching shear strength of the footing is determined by ACI Table 22.6.5.2.

The net factored pressure on the footing must be determined from the factored applied column loads, as shown in Fig. 3.8. It will not necessarily be uniform unless the ratios of the factored loads to service loads on both columns are identical.

Example 3.10

The combined footing of normal weight concrete for Ex. 3.9 has a concrete strength of 5000 lbf/in² and a factored load on each column that is 1.5 times the service load. Determine whether the punching shear capacity is adequate.

Solution

Because the ratios of the factored loads to service loads on both columns are identical, the net factored pressure on the footing is uniform and has a value of

$$q_u = 1.5q_e = (1.5)\left(3.663 \frac{\text{kips}}{\text{ft}^2}\right)$$
$$= 5.5 \text{ kips/ft}^2$$

For column no. 1, the factored load is

$$P_{1u} = 1.5P_1 = (1.5)(200 \text{ kips}) = 300 \text{ kips}$$

The length of the critical perimeter, as shown in the illustration, is

$$b_o = (c + d) + 2\left(c + \frac{d}{2}\right)$$
$$= (12 \text{ in} + 23.5 \text{ in}) + (2)\left(12 \text{ in} + \frac{23.5 \text{ in}}{2}\right)$$
$$= 83 \text{ in}$$

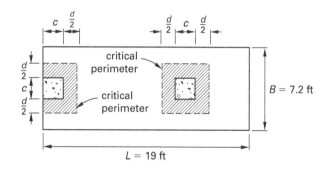

The punching shear force at the critical perimeter is

$$V_u = P_{1u} - q_u(c+d)\left(c+\frac{d}{2}\right)$$

$$= 300 \text{ kips} - \frac{\left(5.5 \dfrac{\text{kips}}{\text{ft}^2}\right)(35.5 \text{ in})(23.75 \text{ in})}{\left(12 \dfrac{\text{in}}{\text{ft}}\right)^2}$$

$$= 268 \text{ kips}$$

The punching shear capacity is given by ACI Table 22.6.5.2(a) as

$$\phi V_c = 4\phi d b_o \lambda \sqrt{f'_c}$$

$$= \frac{(4)(0.75)(23.5 \text{ in})(83 \text{ in})(1.0)\sqrt{5000 \dfrac{\text{lbf}}{\text{in}^2}}}{1000 \dfrac{\text{lbf}}{\text{kip}}}$$

$$= 414 \text{ kips}$$
$$> V_u \quad \text{[satisfactory]}$$

For column no. 2, the factored load is

$$P_{2u} = 1.5 P_2 = (1.5)(300 \text{ kips}) = 450 \text{ kips}$$

The length of the critical perimeter is

$$b_o = 4(c+d)(4)(35.5 \text{ in}) = 142 \text{ in}$$

The punching shear force at the critical perimeter is

$$V_u = P_{2u} - q_u(c+d)^2$$

$$= 450 \text{ kips} - \frac{\left(5.5 \dfrac{\text{kips}}{\text{ft}^2}\right)(35.5 \text{ in})^2}{\left(12 \dfrac{\text{in}}{\text{ft}}\right)^2}$$

$$= 402 \text{ kips}$$

The punching shear capacity is given by ACI Table 22.6.5.2(a) as

$$\phi V_c = 4\phi d b_o \lambda \sqrt{f'_c} = \frac{(414 \text{ kips})(142 \text{ in})}{83 \text{ in}} = 708 \text{ kips}$$
$$> V_u \quad \text{[satisfactory]}$$

Design for Flexural Shear

The critical section for flexural shear in a combined footing is identical with that in an isolated column footing and is located a distance d from the face of the column. The shear force at the critical section is

determined from the shear force diagram, as shown in Fig. 3.8. The depth of the footing is usually governed by flexural shear.

Example 3.11

Determine whether the flexural shear capacity is adequate for the combined footing of Ex. 3.10.

Solution

At the center of column no. 1, the shear force is

$$V_1 = P_{1u} - \frac{q_u B c}{2}$$

$$= 300 \text{ kips} - \frac{\left(5.5 \dfrac{\text{kips}}{\text{ft}^2}\right)(7.2 \text{ ft})(1 \text{ ft})}{2}$$

$$= 280 \text{ kips}$$

At the center of column no. 2, the shear force is

$$V_2 = V_1 - q_u B l$$

$$= 280 \text{ kips} - \left(5.5 \dfrac{\text{kips}}{\text{ft}^2}\right)(7.2 \text{ ft})(15 \text{ ft})$$

$$= -314 \text{ kips}$$

The shear force diagram is shown in *Illustration for Ex. 3.11*, and the critical flexural shear is a distance $(d + c/2)$ from the center of column no. 2.

The critical flexural shear at this section is

$$V_u = V_2 - q_u B\left(d + \frac{c}{2}\right)$$

$$= 314 \text{ kips} - \frac{\left(5.5 \dfrac{\text{kips}}{\text{ft}^2}\right)(7.2 \text{ ft})(29.5 \text{ in})}{12 \dfrac{\text{in}}{\text{ft}}}$$

$$= 217 \text{ kips}$$

The flexural shear capacity of the footing is given by ACI Eq. 22.5.5.1 as

$$\phi V_c = 2\phi b d \lambda \sqrt{f'_c}$$

$$= \frac{(2)(0.75)\left[(7.2 \text{ ft})\left(12 \dfrac{\text{in}}{\text{ft}}\right)\right] \times (23.5 \text{ in})(1.0)\sqrt{5000 \dfrac{\text{lbf}}{\text{in}^2}}}{1000 \dfrac{\text{lbf}}{\text{kip}}}$$

$$= 215 \text{ kips}$$
$$\approx V_u \quad \text{[satisfactory]}$$

Illustration for Ex. 3.11

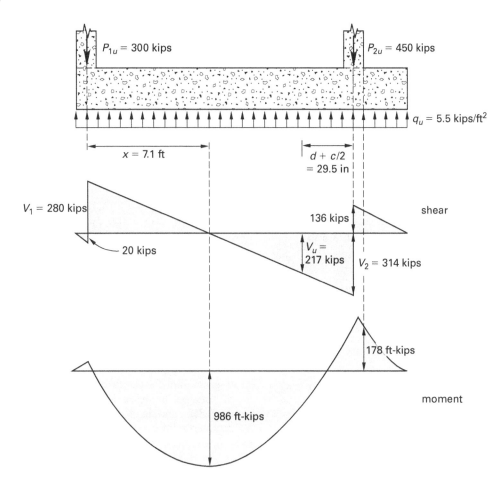

Example 3.12

Determine the required longitudinal and transverse grade 60 reinforcement for the combined footing of Ex. 3.11.

Solution

From Ex. 3.11, the point of zero shear is a distance from the center of column no. 1 given by

$$x = \frac{V_1}{q_u B}$$

$$= \frac{280 \text{ kips}}{\left(5.5 \dfrac{\text{kips}}{\text{ft}^2}\right)(7.2 \text{ ft})}$$

$$= 7.1 \text{ ft}$$

The maximum negative moment at this point is

$$M_u = P_{1u}x - \frac{q_u B\left(x + \dfrac{c}{2}\right)^2}{2}$$

$$= (300 \text{ kips})(7.1 \text{ ft})$$

$$\quad - \frac{\left(5.5 \dfrac{\text{kips}}{\text{ft}^2}\right)(7.2 \text{ ft})\left(7.1 \text{ ft} + \dfrac{1 \text{ ft}}{2}\right)^2}{2}$$

$$= 986 \text{ ft-kips}$$

For the reinforcement in the top of the footing, assuming a tension-controlled section, the reinforcement ratio required is

$$\rho = \frac{0.85f'_c\left(1 - \sqrt{1 - \dfrac{M_u}{0.383bd^2f'_c}}\right)}{f_y}$$

$$= \frac{(0.85)\left(5\ \dfrac{\text{kips}}{\text{in}^2}\right)}{60\ \dfrac{\text{kips}}{\text{in}^2}} \times \left(1 - \sqrt{1 - \dfrac{(986\ \text{ft-kips})\left(12\ \dfrac{\text{in}}{\text{ft}}\right) \times \left(1000\ \dfrac{\text{lbf}}{\text{kip}}\right)}{(0.383)(86.4\ \text{in})(23.5\ \text{in})^2 \times \left(5000\ \dfrac{\text{lbf}}{\text{in}^2}\right)}}\right)$$

$$= 0.0048$$

The maximum allowable reinforcement ratio for a tension-controlled section is given by

$$\rho_t = 0.319\beta_1\left(\frac{f'_c}{f_y}\right) = 0.0213 > \rho \quad \begin{bmatrix}\text{satisfactory, the section}\\ \text{is tension controlled}\end{bmatrix}$$

The required reinforcement area in the top of the footing is

$$A_s = \rho bd = (0.0048)(86.4\ \text{in})(23.5\ \text{in}) = 9.75\ \text{in}^2$$

Providing 10 no. 9 bars gives an area of 10 in² (satisfactory). The minimum permissible reinforcement area is given by ACI Sec. 7.6.1.1 as

$$A_{s(\text{min})} = 0.0018bh = (0.0018)(86.4\ \text{in})(27\ \text{in}) = 4.20\ \text{in}^2$$

$$< A_s \quad \text{[satisfactory]}$$

The maximum positive moment at the outside face of column no. 2 is

$$M_u = \frac{q_u B(L - l - c)^2}{2}$$

$$= \frac{\left(5.5\ \dfrac{\text{kips}}{\text{ft}^2}\right)(7.2\ \text{ft})(19\ \text{ft} - 15\ \text{ft} - 1\ \text{ft})^2}{2}$$

$$= 178\ \text{ft-kips}$$

For the reinforcement in the bottom of the footing, assuming a tension-controlled section, the reinforcement ratio required is

$$\rho = \frac{0.85f'_c\left(1 - \sqrt{1 - \dfrac{M_u}{0.383bd^2f'_c}}\right)}{f_y}$$

$$= \frac{(0.85)\left(5\ \dfrac{\text{kips}}{\text{in}^2}\right)}{60\ \dfrac{\text{kips}}{\text{in}^2}} \times \left(1 - \sqrt{1 - \dfrac{(178\ \text{ft-kips})\left(12\ \dfrac{\text{in}}{\text{ft}}\right) \times \left(1000\ \dfrac{\text{lbf}}{\text{kip}}\right)}{(0.383)(86.4\ \text{in})(23.5\ \text{in})^2 \times \left(5000\ \dfrac{\text{lbf}}{\text{in}^2}\right)}}\right)$$

$$= 0.00083$$

The combined area of reinforcement in the top and bottom of the footing exceeds the minimum required value of $0.0018bh$, and in accordance with ACI Sec. 9.6.1.3, the required reinforcement area in the bottom of the footing is

$$A_s = 1.33\rho bd$$

$$= (1.33)(0.00083)(86.4\ \text{in})(23.5\ \text{in})$$

$$= 2.24\ \text{in}^2$$

Providing 10 no. 5 bars gives an area of 3.10 in² (satisfactory).

The factored moment in the transverse direction at the face of the columns is

$$M_u = \frac{q_u L\left(\dfrac{B}{2} - \dfrac{c}{2}\right)^2}{2}$$

$$= \frac{\left(5.5\ \dfrac{\text{kips}}{\text{ft}^2}\right)(19\ \text{ft})\left(\dfrac{7.2}{2}\ \text{ft} - \dfrac{1}{2}\ \text{ft}\right)^2}{2}$$

$$= 502\ \text{ft-kips}$$

For the transverse reinforcement in the bottom of the footing, assuming a tension-controlled section, the reinforcement ratio required is

$$
\rho = \frac{0.85 f_c'\left(1 - \sqrt{1 - \dfrac{M_u}{0.383 b d^2 f_c'}}\right)}{f_y}
$$

$$
= \frac{(0.85)\left(5 \dfrac{\text{kips}}{\text{in}^2}\right)}{60 \dfrac{\text{kips}}{\text{in}^2}}
$$

$$
\times \left(1 - \sqrt{1 - \dfrac{(502 \text{ ft-kips})\left(12 \dfrac{\text{in}}{\text{ft}}\right) \times \left(1000 \dfrac{\text{lbf}}{\text{kip}}\right)}{(0.383)(228 \text{ in})(23.5 \text{ in})^2 \times \left(5000 \dfrac{\text{lbf}}{\text{in}^2}\right)}}\right)
$$

$$
= 0.00089
$$

The minimum permissible reinforcement area governs, and the reinforcement area required in both the top and bottom of the footing transversely is

$$
\begin{aligned}
A_s &= \frac{0.0018 bh}{2} \\
&= \frac{(0.0018)\left(12 \dfrac{\text{in}}{\text{ft}}\right)(27 \text{ in})}{2} \\
&= 0.29 \text{ in}^2/\text{ft}
\end{aligned}
$$

Providing no. 4 bars at 8 in centers gives an area of 0.30 in^2/ft (satisfactory). Since the minimum distribution reinforcement governs, banding of the reinforcement is not required.

5. STRAP FOOTING

Nomenclature

A_1	base area of pad footing no. 1	in^2
A_2	base area of pad footing no. 2	in^2
B_S	length of short side of strap	ft
B_1	length of short side of pad footing no. 1	ft
B_2	length of short side of pad footing no. 2	ft
h_S	depth of strap	in
h_1	depth of pad footing no. 1	in
h_2	depth of pad footing no. 2	in
l	distance between column centers	ft
l_R	distance between soil reactions	ft
L_S	length of long side of strap	ft
L_1	length of long side of pad footing no. 1	ft
L_2	length of long side of pad footing no. 2	ft
R_1	soil reaction under pad footing no. 1	kips
R_2	soil reaction under pad footing no. 2	kips
w_c	unit weight of concrete	lbf/ft^3
W_S	weight of strap beam	kips
W_1	weight of pad footing no. 1	kips
W_2	weight of pad footing no. 2	kips

Pressure Distribution

The strap footing shown in Fig. 3.9 has the strap beam, which connects the two pad footings, underlaid by a layer of Styrofoam™ so that the soil pressure under the strap may be considered negligible. Because of the stiffness of the strap beam, the strap and pad footings act as a rigid body producing uniform soil pressure under the pad footings. The base areas of the two pad footings may be adjusted to produce equal soil pressure q under both footings.

The total service load acting is

$$
\sum P = P_1 + P_2 + W_1 + W_2 + W_S
$$

$$
q = \frac{\sum P}{A_1 + A_2}
$$

The soil reactions act at the center of the pad footings and are given by

$$
R_1 = q A_1
$$
$$
R_2 = q A_2
$$

Pad footing no. 2 is located symmetrically with respect to column no. 2 so that the lines of action of P_2 and R_2 are coincident.

$$
l_R = l + \frac{c_1}{2} - \frac{B_1}{2}
$$

$$
L_S = l_R - \frac{B_1 + B_2}{2}
$$

Equating vertical forces gives

$$
R_2 = \sum P - R_1 \quad \text{[equilibrium equation no. 1]}
$$

Figure 3.9 *Strap Footing with Applied Service Loads*

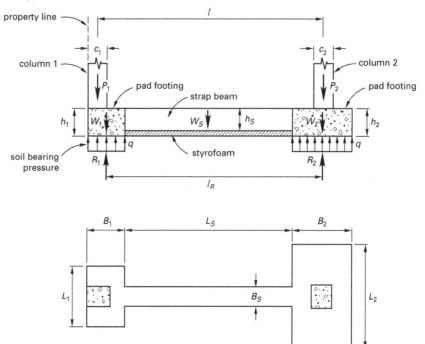

Taking moments about the center of pad footing no. 2 gives

$$R_1 = \frac{P_1 l + W_1 l_R + \dfrac{W_S(L_S + B_2)}{2}}{l_R}$$

[equilibrium equation no. 2]

To determine suitable dimensions that will give a soil bearing pressure equal to the allowable pressure q, suitable values are selected for h_1, h_2, h_S, B_1, B_2, and B_S. l_R and L_S are determined, and

$$W_S = w_c L_S B_S h_S$$

An initial estimate is made of R_1, and

$$A_1 = \frac{R_1}{q}$$

$$W_1 = w_c A_1 h_1$$

An initial estimate is made of R_2, and

$$A_2 = \frac{R_2}{q}$$

$$W_2 = w_c A_2 h_2$$

$$\sum P = P_1 + P_2 + W_1 + W_2 + W_S$$

Substituting in the two equilibrium equations provides revised estimates of R_1 and R_2, and the process is repeated until convergence is reached.

Example 3.13

Determine the plan dimensions required for the strap footing shown to provide a uniform bearing pressure of 3000 lbf/ft² under both pad footings for the service loads indicated.

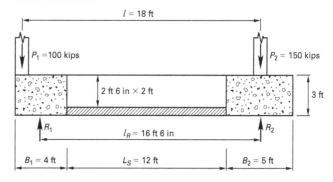

Solution

From the dimensions indicated in the illustration,

$$W_S = w_c L_S B_S h_S$$
$$= \left(0.15 \, \frac{\text{kip}}{\text{ft}^3}\right)(12 \text{ ft})(2 \text{ ft})(2.5 \text{ ft})$$
$$= 9 \text{ kips}$$

Assuming that $R_1 = 134$ kips, then

$$A_1 = \frac{R_1}{q} = \frac{134 \text{ kips}}{3 \ \dfrac{\text{kips}}{\text{ft}^2}} = 44.67 \text{ ft}^2$$

$$
\begin{aligned}
W_1 &= w_c A_1 h_1 \\
&= \left(0.15 \ \frac{\text{kip}}{\text{ft}^3}\right)(44.67 \text{ ft}^2)(3 \text{ ft}) \\
&= 20.1 \text{ kips}
\end{aligned}
$$

Assuming that $R_2 = 171$ kips, then

$$
\begin{aligned}
A_2 &= \frac{R_2}{q} \\
&= \frac{171 \text{ kips}}{3 \ \dfrac{\text{kips}}{\text{ft}^2}} \\
&= 57 \text{ ft}^2
\end{aligned}
$$

$$
\begin{aligned}
W_2 &= w_c A_2 h_2 \\
&= \left(0.15 \ \frac{\text{kip}}{\text{ft}^3}\right)(57 \text{ ft}^2)(3 \text{ ft}) \\
&= 25.7 \text{ kips}
\end{aligned}
$$

$$
\begin{aligned}
\sum P &= P_1 + P_2 + W_1 + W_2 + W_S \\
&= 100 \text{ kips} + 150 \text{ kips} + 20.1 \text{ kips} \\
&\quad + 25.7 \text{ kips} + 9 \text{ kips} \\
&= 304.8 \text{ kips}
\end{aligned}
$$

Equating vertical forces gives

$$
\begin{aligned}
R_2 &= \sum P - R_1 \\
&= 304.8 \text{ kips} - 134 \text{ kips} \\
&= 170.8 \text{ kips} \\
&\approx 171 \text{ kips} \quad \text{[satisfactory]}
\end{aligned}
$$

Taking moments about the center of pad footing no. 2 gives

$$
\begin{aligned}
R_1 &= \frac{P_1 l + W_1 l_R + \dfrac{W_S(L_S + B_2)}{2}}{l_R} \\
&= \frac{\begin{aligned}(100 \text{ kips})(18 \text{ ft}) &+ (20.1 \text{ kips})(16.5 \text{ ft}) \\ &+ \frac{(9 \text{ kips})(12 \text{ ft} + 5 \text{ ft})}{2}\end{aligned}}{16.5 \text{ ft}} \\
&= 133.8 \text{ kips} \\
&\approx 134 \text{ kips} \quad \text{[satisfactory]}
\end{aligned}
$$

The initial estimates were sufficiently accurate, and the required pad footing areas are

$$A_1 = 44.67 \text{ ft}^2$$
$$A_2 = 57 \text{ ft}^2$$

Design of Strap Beam for Shear

The factored forces acting on the footing are shown in Fig. 3.10. The total factored load on the footing is

$$\sum P_u = P_{1u} + P_{2u} + W_{1u} + W_{2u} + W_{Su}$$

Taking moments about the center of pad footing no. 2 gives

$$R_{1u} = \frac{P_{1u} l + W_{1u} l_R + \dfrac{W_{Su}(L_S + B_2)}{2}}{l_R}$$

Equating vertical forces gives

$$R_{2u} = \sum P_u - R_{1u}$$

The shear at the left end of the strap is

$$V_{Su} = R_{1u} - P_{1u} - W_{1u}$$

The shear at the right end of the strap is

$$V'_{Su} = P_{2u} + W_{2u} - R_{2u}$$

Example 3.14

The strap footing of normal weight concrete for Ex. 3.13 has a concrete strength of 3000 lbf/in^2 and a factored load on each column that is 1.5 times the service load. The strap beam has an effective depth of 27.5 in. Determine whether the shear capacity is adequate.

Solution

The factored forces are

$$
\begin{aligned}
P_{1u} &= 1.5 P_1 = 150 \text{ kips} \\
P_{2u} &= 1.5 P_2 = 225 \text{ kips} \\
W_{1u} &= 1.2 W_1 = 24 \text{ kips} \\
W_{2u} &= 1.2 W_2 = 31 \text{ kips} \\
W_{Su} &= 1.2 W_S = 11 \text{ kips} \\
\sum P_u &= 441 \text{ kips}
\end{aligned}
$$

Foundations

Figure 3.10 *Factored Forces on Strap Footing*

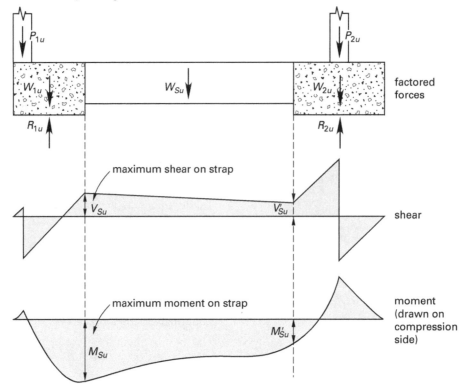

$$R_{1u} = \frac{P_{1u}l + W_{1u}l_R + \dfrac{W_{Su}(L_S + B_2)}{2}}{l_R}$$

$$= \frac{\begin{array}{c}(150 \text{ kips})(18 \text{ ft}) + (24 \text{ kips})(16.5 \text{ ft}) \\[4pt] + \dfrac{(11 \text{ kips})(12 \text{ ft} + 5 \text{ ft})}{2}\end{array}}{16.5 \text{ ft}}$$

$$= 193 \text{ kips}$$

$$R_{2u} = \sum P_u - R_{1u}$$

$$= 248 \text{ kips}$$

The shear at the right end of the strap is

$$V'_{Su} = P_{2u} + W_{2u} - R_{2u}$$
$$= 225 \text{ kips} + 31 \text{ kips} - 248 \text{ kips}$$
$$= 8 \text{ kips}$$

The shear at the left end of the strap is

$$V_{Su} = R_{1u} - P_{1u} - W_{1u}$$
$$= 193 \text{ kips} - 150 \text{ kips} - 24 \text{ kips}$$
$$= 19 \text{ kips} \quad [\text{governs}]$$

The design shear capacity of the strap beam is given by ACI Eq. 22.5.5.1 as

$$\phi V_c = 2\phi bd\lambda\sqrt{f'_c}$$

$$= \frac{(2)(0.75)(24 \text{ in})(27.5 \text{ in})(1.0)\sqrt{3000 \dfrac{\text{lbf}}{\text{in}^2}}}{1000 \dfrac{\text{lbf}}{\text{kip}}}$$

$$= 54 \text{ kips}$$

$$> 2V_{Su} \quad [\text{No shear reinforcement is required.}]$$

Design of Strap Beam for Flexure

From Fig. 3.10, the factored moment at the left end of the strap is

$$M_{Su} = P_{1u}\left(B_1 - \frac{c_1}{2}\right) - \frac{(R_{1u} - W_{1u})B_1}{2}$$

The factored moment at the right end of the strap is

$$M'_{Su} = \frac{(R_{2u} - W_{2u} - P_{2u})B_2}{2}$$

Example 3.15

Determine the required grade 60 flexural reinforcement for the strap beam of Ex. 3.14.

Solution

The factored moment at the right end of the strap is

$$M'_{Su} = \frac{(W_{2u} + P_{2u} - R_{2u})B_2}{2}$$

$$= \frac{(31 \text{ kips} + 225 \text{ kips} - 248 \text{ kips})(5 \text{ ft})}{2}$$

$$= 20 \text{ ft-kips}$$

The factored moment at the left end of the strap is

$$M_{Su} = P_{1u}\left(B_1 - \frac{c_1}{2}\right) - \frac{(R_{1u} - W_{1u})B_1}{2}$$

$$= (150 \text{ kips})\left(4 \text{ ft} - \frac{1 \text{ ft}}{2}\right)$$

$$- \left(\frac{(193 \text{ kips} - 24 \text{ kips})(4 \text{ ft})}{2}\right)$$

$$= 187 \text{ ft-kips} \quad [\text{governs}]$$

Assuming a tension-controlled section, the required reinforcement ratio is

$$\rho = \frac{0.85f'_c\left(1 - \sqrt{1 - \dfrac{M_{Su}}{0.383bd^2f'_c}}\right)}{f_y}$$

$$= \frac{(0.85)\left(3\ \dfrac{\text{kips}}{\text{in}^2}\right)}{60\ \dfrac{\text{kips}}{\text{in}^2}} \times \left(1 - \sqrt{1 - \dfrac{(187 \text{ ft-kips})\left(12\ \dfrac{\text{in}}{\text{ft}}\right) \times \left(1000\ \dfrac{\text{lbf}}{\text{kip}}\right)}{(0.383)(24 \text{ in})(27.5 \text{ in})^2 \times \left(3000\ \dfrac{\text{lbf}}{\text{in}^2}\right)}}\right)$$

$$= 0.0023$$

The controlling minimum reinforcement ratio is given by ACI Sec. 9.6.1.2 and Sec. 9.6.1.3 as the lesser of the following results.

$$\rho_{\min} = \frac{200}{f_y} = \frac{200\ \dfrac{\text{lbf}}{\text{in}^2}}{60{,}000\ \dfrac{\text{lbf}}{\text{in}^2}}$$

$$= 0.0033$$

$$\rho_{\min} = \left(\frac{4}{3}\right)(0.0023)$$

$$= 0.0031 \quad [\text{governs}]$$

The reinforcement required in the top of the strap beam is

$$A_s = bd\rho_{\min}$$

$$= (24 \text{ in})(27.5 \text{ in})(0.0031)$$

$$= 2.05 \text{ in}^2$$

Providing four no. 7 bars gives an area of 2.4 in² (satisfactory).

6. CANTILEVER RETAINING WALL

Nomenclature

F	frictional force at underside of base	kips
$\bar{h}$	equivalent additional height of fill, w/γ_S	ft
h	height from top of stem to bottom of shear key	ft
h_B	depth of base	ft
h_K	height of shear key	ft
h_T	total height of retaining wall, $h_B + L_W$	ft
h_W	stem thickness	ft
H_A	total active earth pressure behind wall	kips
H_L	total pressure behind wall due to live load surcharge	kips
H_P	total passive earth pressure in front of wall	kips
K_A	Rankine coefficient of active earth pressure $(1 - \sin\phi)/(1 + \sin\phi)$	–
K_P	Rankine coefficient of passive earth pressure $(1 + \sin\phi)/(1 - \sin\phi)$	–
L_B	length of base	ft
L_H	length of heel	ft
L_T	length of toe	ft
L_W	height of stem	ft
p_A	active lateral pressure due to a fluid of specific weight γ_S, $K_A\gamma_S$	lbf/ft²
p_L	lateral pressure due to live load surcharge, wK_A	lbf/ft²

Figure 3.11 *Cantilever Retaining Wall with Applied Service Loads*

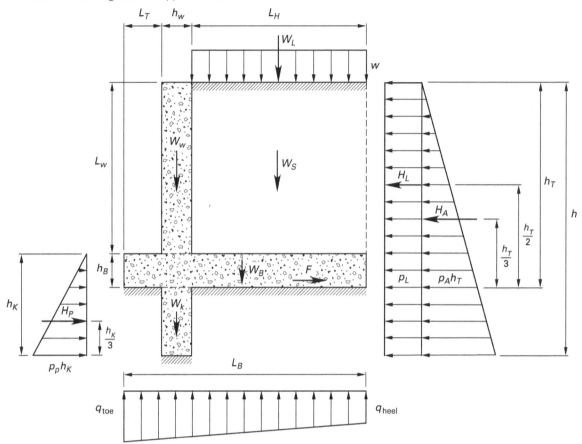

p_P	passive lateral pressure due to a fluid of specific weight γ_S, $K_P\gamma_S$	lbf/ft²
q	earth pressure under the base	lbf/ft²
w	live load surcharge	lbf/ft²
W_B	weight of base	kips
W_K	weight of key	kips
W_L	weight of surcharge	kips
W_S	weight of backfill	kips
W_W	weight of stem	kips
Z	depth below ground surface	ft

Symbols

γ_S	specific weight of backfill	lbf/ft³
μ	coefficient of friction	–
ϕ	angle of internal friction	deg

Pressure Distribution

Figure 3.11 shows the forces acting on a cantilever retaining wall. The total active earth pressure behind the wall is given by Rankine's theory as

$$
\begin{aligned}
H_A &= \frac{p_A h_T^2}{2} = \frac{K_A \gamma_S h_T^2}{2} \\[2mm]
&= \frac{\left(\dfrac{1-\sin\phi}{1+\sin\phi}\right)\gamma_S h_T^2}{2} \\[2mm]
&= \frac{30 h_T^2}{2} \quad \text{[for } \gamma_S = 110 \text{ lbf/ft}^3 \text{ and } \phi = 35°\text{]} \\[2mm]
&= \text{pressure exerted by a fluid} \\
&\quad\ \text{of density 30 lbf/ft}^3
\end{aligned}
$$

The total active earth pressure acts at a height of $h_T/3$ above the base.

The total surcharge pressure behind the wall due to a live load surcharge of w is

$$H_L = p_L h_T$$
$$= w K_A h_T$$
$$= \frac{w p_A h_T}{\gamma_S}$$

The surcharge may be represented by an equivalent height of fill given by

$$\bar{h} = \frac{w}{\gamma_S}$$
$$H_L = p_A \bar{h} h_T$$

The total surcharge pressure acts at a height of $h_T/2$ above the base.

The total passive earth pressure in front of the wall is

$$H_P = \frac{p_P h_K^2}{2}$$
$$= \frac{K_P \gamma_S h_K^2}{2}$$
$$H_P = \frac{\left(\dfrac{1+\sin\phi}{1-\sin\phi}\right)\gamma_S h_K^2}{2}$$
$$= \frac{400 h_K^2}{2} \quad [\text{for } \gamma_S = 110 \text{ lbf/ft}^3 \text{ and } \phi = 35°]$$
$$= \text{pressure exerted by a fluid}$$
$$\text{of density 400 lbf/ft}^3$$

The total passive earth pressure acts at a height of $h_K/3$ above the bottom of the key. The frictional force acting on the underside of the base is given by

$$F = \mu \sum W$$

$\sum W$ is the total weight of the retaining wall plus backfill plus live load surcharge.

A factor of safety of 1.5 against sliding is required, which gives

$$F + H_P \geq 1.5(H_A + H_L)$$

A factor of safety of 1.5 is required for overturning about the toe.

Reinforcement Details

The minimum reinforcement required in the stem wall is specified by ACI Table 11.6.1. For grade 60 bars larger than no. 5, the reinforcement ratios are based on the gross concrete area for vertical and horizontal reinforcement, and are

$$\rho_{\text{vert}} = 0.15\%$$
$$\rho_{\text{hor}} = 0.25\%$$

For grade 60 bars no. 5 and smaller, the corresponding ratios are

$$\rho_{\text{vert}} = 0.12\%$$
$$\rho_{\text{hor}} = 0.20\%$$

A single layer of horizontal reinforcement is permitted in walls not exceeding 10 in thickness. For walls exceeding 10 in thickness, it is customary to provide two layers of horizontal reinforcement, and the limiting reinforcement ratio is based on the total reinforcement ratio for both layers combined. The reinforcement layer for the air face should be between one-half and two-thirds of the total reinforcement. It should be placed between 2 in and one-third the thickness of the wall from the air face. The layer for the earth face, consisting of the balance of the required horizontal reinforcement, should be placed a distance from the earth face not more than one-third the thickness of the wall, but not less than 2 in for bars larger than no. 5, or $1\frac{1}{2}$ in for bars no. 5 or smaller.

The area of vertical reinforcement in the earth face is governed by flexural requirements. Nominal vertical reinforcement is required in the air face to give a total reinforcement ratio of not less than 0.0018% for grade 60 reinforcement.

Vertical and horizontal reinforcement is spaced apart no more than three times the wall thickness and no further apart than 18 in.

Example 3.16

The retaining wall shown in the illustration retains soil with a unit weight of 110 lbf/ft^3 and an equivalent fluid pressure of 30 lbf/ft^2. The live load surcharge behind the wall is equivalent to an additional height of 2 ft of fill. Passive earth pressure may be assumed equivalent to a fluid pressure of 450 lbf/ft^2, and the coefficient of friction at the underside of the base is 0.4. Determine the factors of safety against sliding and overturning and the bearing pressure distribution under the base.

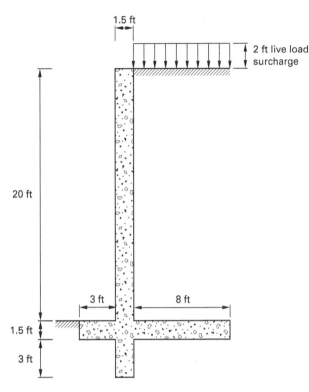

Solution

The lateral pressures from the backfill and the surcharge are

$$H_A = \frac{p_A h_T^2}{2} = \frac{\left(30 \ \frac{\text{lbf}}{\text{ft}^2}\right)(21.5 \ \text{ft})^2}{2}$$
$$= 6934 \ \text{lbf}$$

$$H_L = p_A h_T \overline{h}$$
$$= \left(30 \ \frac{\text{lbf}}{\text{ft}^2}\right)(21.5 \ \text{ft})(2 \ \text{ft})$$
$$= 1290 \ \text{lbf}$$

Taking moments about the toe gives

$$M_o = \frac{H_A h_T}{3} + \frac{H_L h_T}{2}$$
$$= \frac{(6934 \ \text{lbf})(21.5 \ \text{ft})}{3} + \frac{(1290 \ \text{lbf})(21.5 \ \text{ft})}{2}$$
$$= 63{,}561 \ \text{ft-lbf}$$

The gravity loads acting are

$$W_W + W_K = w_c h_W L_W + w_c (h_K - h_B) h_W$$
$$= \left(150 \ \frac{\text{lbf}}{\text{ft}^3}\right)(1.5 \ \text{ft})(20 \ \text{ft})$$
$$+ \left(150 \ \frac{\text{lbf}}{\text{ft}^3}\right)(3 \ \text{ft})(1.5 \ \text{ft})$$
$$= 5175 \ \text{lbf}$$
$$W_B = w_c h_B L_B$$
$$= \left(150 \ \frac{\text{lbf}}{\text{ft}^3}\right)(1.5 \ \text{ft})(12.5 \ \text{ft})$$
$$= 2813 \ \text{lbf}$$

$$W_L + W_S = \gamma_S (h_T - h_B + \overline{h}) L_H$$
$$= \left(110 \ \frac{\text{lbf}}{\text{ft}^3}\right)(20 \ \text{ft} + 2 \ \text{ft})(8 \ \text{ft})$$
$$= 19{,}360 \ \text{lbf}$$
$$\sum W = 5175 \ \text{lbf} + 2813 \ \text{lbf} + 19{,}360 \ \text{lbf}$$
$$= 27{,}348 \ \text{lbf}$$

The distance of the resultant vertical load from the toe is

$$x_o = \frac{\begin{array}{c}(W_W + W_K)\left(L_T + \dfrac{h_W}{2}\right) + \dfrac{W_B L_B}{2} \\[6pt] + (W_L + W_S)\left(L_T + h_W + \dfrac{L_H}{2}\right)\end{array}}{\sum W}$$
$$= \frac{\begin{array}{c}(5175 \ \text{lbf})(3.75 \ \text{ft}) + (2813 \ \text{lbf})(6.25 \ \text{ft}) \\[4pt] + (19{,}360 \ \text{lbf})(8.5 \ \text{ft})\end{array}}{27{,}348 \ \text{lbf}}$$
$$= 7.37 \ \text{ft}$$

The factor of safety against overturning is

$$\frac{x_o \sum W}{M_o} = \frac{(7.37 \ \text{ft})(27{,}348 \ \text{lbf})}{63{,}561 \ \text{ft-lbf}}$$
$$= 3.2$$
$$> 1.5 \quad \text{[satisfactory]}$$

The eccentricity of all applied loads about the toe is

$$e' = \frac{x_o \sum W - M_o}{\sum W}$$
$$= \frac{(7.37 \ \text{ft})(27{,}348 \ \text{lbf}) - 63{,}561 \ \text{lbf}}{27{,}348 \ \text{lbf}}$$
$$= 5.05 \ \text{ft}$$

The eccentricity of all applied loads about the midpoint of the base is

$$e = \frac{L_B}{2} - e'$$
$$= \frac{12.5 \text{ ft}}{2} - 5.05 \text{ ft}$$
$$= 1.20 \text{ ft} \quad \text{[within middle third]}$$

The pressure under the base is given by

$$q = \frac{\sum W \left(1 \pm \frac{6e}{L_B}\right)}{BL_B}$$
$$= \frac{(27{,}348 \text{ lbf})\left(1 \pm \frac{(6)(1.2 \text{ ft})}{12.5 \text{ ft}}\right)}{(1 \text{ ft})(12.5 \text{ ft})}$$
$$q_{\text{toe}} = 3448 \text{ lbf/ft}^2$$
$$q_{\text{heel}} = 928 \text{ lbf/ft}^2$$

The frictional resistance under the base is

$$F = \mu \sum W$$
$$= (0.4)(27{,}348 \text{ lbf})$$
$$= 10{,}939 \text{ lbf}$$

The passive pressure in front of the wall is

$$H_P = \frac{p_P h_K^2}{2}$$
$$= \frac{\left(450 \dfrac{\text{lbf}}{\text{ft}^2}\right)(4.5 \text{ ft})^2}{2}$$
$$= 4556 \text{ lbf}$$

The lateral pressures behind the wall and key from the backfill and surcharge are

$$H_A = \frac{p_A h^2}{2} = \frac{\left(30 \dfrac{\text{lbf}}{\text{ft}^2}\right)(24.5 \text{ ft})^2}{2} = 9004 \text{ lbf}$$
$$H_L = p_A h \bar{h} = \left(30 \dfrac{\text{lbf}}{\text{ft}^2}\right)(2 \text{ ft})(24.5 \text{ ft}) = 1470 \text{ lbf}$$

The factor of safety against sliding is

$$\frac{F + H_P}{H_A + H_L} = \frac{10{,}939 \text{ lbf} + 4556 \text{ lbf}}{9004 \text{ lbf} + 1470 \text{ lbf}} = 1.48$$
$$\approx 1.5 \quad \text{[satisfactory]}$$

Design for Shear and Flexure

To determine the shear and flexure at the critical sections in the wall, the soil pressure under the footing is recalculated by using the factored forces given by ACI Sec. 5.3 as

$$U = 1.2D + 1.6L + 0.5(L_r \text{ or } S \text{ or } R) \quad \text{[ACI 5.3.1b]}$$

Shear is generally not critical. The location of the critical section for flexure in the stem is at the base of the stem; for flexure in the toe, at the front face of the stem; and for flexure in the heel, at the rear face of the stem.

Example 3.17

Determine the reinforcement areas required in the toe, heel, and stem of the retaining wall for Ex. 3.16. The concrete strength is 3000 lbf/in², and grade 60 reinforcement is provided.

Solution

The factored overturning moment about the toe is

$$M_{ou} = 1.6 M_o = (1.6)(63{,}561 \text{ ft-lbf})$$
$$= 101{,}698 \text{ ft-lbf}$$

The factored total vertical load is

$$\sum W_u = 1.2(W_W + W_B + W_S + W_K) + 1.6 W_L$$
$$= (1.2)(5175 \text{ lbf} + 2813 \text{ lbf} + 17{,}600 \text{ lbf})$$
$$\quad + (1.6)(1760 \text{ lbf})$$
$$= 6210 \text{ lbf} + 3376 \text{ lbf} + 21{,}120 \text{ lbf} + 2816 \text{ lbf}$$
$$= 33{,}522 \text{ lbf}$$

The factored restoring moment is

$$M_{Ru} = (6210 \text{ lbf})(3.75 \text{ ft}) + (3376 \text{ lbf})(6.25 \text{ ft})$$
$$\quad + (21{,}120 \text{ lbf})(8.5 \text{ ft}) + (2816 \text{ lbf})(8.5 \text{ ft})$$
$$= 247{,}844 \text{ ft-lbf}$$

The eccentricity of the factored loads about the toe is

$$e_u' = \frac{M_{Ru} - M_{ou}}{\sum W_u}$$
$$= \frac{247{,}844 \text{ ft-lbf} - 101{,}698 \text{ ft-lbf}}{33{,}522 \text{ lbf}}$$
$$= 4.36 \text{ ft}$$

The eccentricity of the factored loads about the midpoint of the base is

$$e_u = \frac{L_B}{2} - e' = \frac{12.5 \text{ ft}}{2} - 4.36 \text{ ft}$$
$$= 1.89 \text{ ft} \quad \text{[within middle third]}$$

The factored pressure under the base is given by

$$q_u = \frac{\sum W_u \left(1 \pm \dfrac{6e_u}{L_B}\right)}{BL_B}$$

$$= \frac{(33{,}522 \text{ lbf})\left(1 \pm \dfrac{(6)(1.89 \text{ ft})}{12.5 \text{ ft}}\right)}{(1 \text{ ft})(12.5 \text{ ft})}$$

$$q_{u(\text{toe})} = 5115 \text{ lbf/ft}^2$$

$$q_{u(\text{heel})} = 249 \text{ lbf/ft}^2$$

Design of Toe

The factored pressure distribution under the base is shown in the illustration, and the maximum factored bending moment in the toe is

$$M_u = \frac{L_T^2 B(q_{uF} + 2q_{u(\text{toe})})}{6} - \frac{1.2 W_B L_T^2}{2L_B}$$

$$= \frac{(3 \text{ ft})^2 (1 \text{ ft})\left(3947 \dfrac{\text{lbf}}{\text{ft}^2} + (2)\left(5115 \dfrac{\text{lbf}}{\text{ft}^2}\right)\right)}{6}$$

$$\quad - \frac{(3376 \text{ lbf})(3 \text{ ft})^2}{(2)(12.5 \text{ ft})}$$

$$= 20{,}050 \text{ ft-lbf}$$

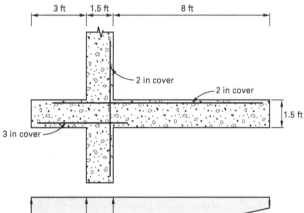

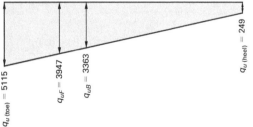

Assuming a tension-controlled section, the required reinforcement ratio is

$$\rho = \frac{0.85 f_c' \left(1 - \sqrt{1 - \dfrac{M_u}{0.383 b d^2 f_c'}}\right)}{f_y}$$

$$= \frac{(0.85)\left(3 \dfrac{\text{kips}}{\text{in}^2}\right)}{60 \dfrac{\text{kips}}{\text{in}^2}}$$

$$\times \left(1 - \sqrt{1 - \frac{(20{,}050 \text{ ft-lbf})\left(12 \dfrac{\text{in}}{\text{ft}}\right)}{(0.383)(12 \text{ in})(14.5 \text{ in})^2 \times \left(3000 \dfrac{\text{lbf}}{\text{ft}^2}\right)}}\right)$$

$$= 0.0018$$

The maximum allowable reinforcement ratio for a tension-controlled section is given by

$$\rho_t = 0.319 \beta_1 \left(\frac{f_c'}{f_y}\right)$$

$$= 0.0136$$

$$> \rho \quad \text{[satisfactory, the section is tension controlled]}$$

$$A_s = \rho bd = (0.0018)(12 \text{ in})(14.5 \text{ in}) = 0.314 \text{ in}^2$$

$$A_{s(\text{min})} = 0.0018 bh = (0.0018)(12 \text{ in})(18 \text{ in})$$

$$= 0.389 \text{ in}^2 \quad \text{[governs]}$$

Design of Heel

The maximum factored bending moment in the heel is

$$M_u = \frac{L_H \left(1.2 W_S + 1.6 W_L + \dfrac{1.2 W_B L_H}{L_B}\right)}{2}$$

$$\quad - \frac{L_H^2 B(q_{uB} + 2q_{u(\text{heel})})}{6}$$

$$= \frac{(8 \text{ ft})\left(\begin{array}{c} 21{,}120 \text{ lbf} + 2816 \text{ lbf} \\ + \dfrac{(3376 \text{ lbf})(8 \text{ ft})}{12.5 \text{ ft}} \end{array}\right)}{2}$$

$$\quad - \frac{(8 \text{ ft})^2 (1 \text{ ft})\left(3363 \dfrac{\text{lbf}}{\text{ft}^2} + (2)\left(249 \dfrac{\text{lbf}}{\text{ft}^2}\right)\right)}{6}$$

$$= 63{,}203 \text{ ft-lbf}$$

Assuming a tension-controlled section, the required reinforcement ratio is

$$\rho = \frac{0.85f'_c\left(1 - \sqrt{1 - \dfrac{M_u}{0.383bd^2f'_c}}\right)}{f_y}$$

$$= \frac{(0.85)\left(3 \dfrac{\text{kips}}{\text{in}^2}\right)}{60 \dfrac{\text{kips}}{\text{in}^2}} \times \left(1 - \sqrt{1 - \dfrac{(63{,}203 \text{ ft-lbf})\left(12 \dfrac{\text{in}}{\text{ft}}\right)}{(0.383)(12 \text{ in})(15.5 \text{ in})^2 \times \left(3000 \dfrac{\text{lbf}}{\text{in}^2}\right)}}\right)$$

$$= 0.0052$$

$$< \rho_t \quad \begin{bmatrix}\text{satisfactory, the section}\\ \text{is tension controlled}\end{bmatrix}$$

$$A_s = \rho bd$$
$$= (0.0052)(12 \text{ in})(15.5 \text{ in})$$
$$= 0.97 \text{ in}^2$$

Design of Stem

The maximum factored bending moment in the stem is

$$M_u = 1.6\left(\frac{p_A L_W^3}{6} + \frac{p_A \bar{h} L_W^2}{2}\right)$$

$$= (1.6)\left(\frac{\left(30 \dfrac{\text{lbf}}{\text{ft}^2}\right)(20 \text{ ft})^3}{6} + \frac{\left(30 \dfrac{\text{lbf}}{\text{ft}^2}\right)(2 \text{ ft})(20 \text{ ft})^2}{2}\right)$$

$$= 83{,}200 \text{ ft-lbf}$$

The required reinforcement ratio is

$$\rho = 0.0070 < \rho_t \quad [\text{satisfactory}]$$
$$A_s = \rho bd$$
$$= (0.0070)(12 \text{ in})(15.5 \text{ in})$$
$$= 1.30 \text{ in}^2$$

7. COUNTERFORT RETAINING WALL

Nomenclature

a	lever-arm of resisting couple in counterfort	ft
b_c	width of counterfort	in
C	compression force in resisting couple	kips
l_c	clear height of counterfort	ft
l_n	clear span between counterforts	ft
q	earth pressure at a depth of $0.6l_c = 0.6p_A l_c$	lbf/ft
s_c	spacing, center to center, of counterforts, $l_n + b_c$	ft
T	tension force in resisting couple	kips

Design of Stem and Base

The stem spans horizontally between counterforts and cantilevers from the base. For ratios of l_n/l_c between 0.5 and 1.0, the stem may be designed for a value of the earth pressure at a depth of $0.6l_c$. The horizontal span moments are given by ACI Table 6.5.2 as $ql_n^2/11$ at counterfort supports and $ql_n^2/16$ between counterforts. The cantilever moment (tension on earth face) at the base is $0.035p_A l_c^3$. The bending moment producing tension on the air face is $0.009p_A l_c^3$. The distribution of bending moment in the stem is shown in Fig. 3.12, and more precise values of moment may be obtained from tabulated coefficients.[1,2]

The base slab is similarly designed for the net factored pressure as a slab spanning longitudinally between counterforts.

Example 3.18

For the counterfort retaining wall shown, determine the design moments in the stem. The fill behind the wall has an equivalent fluid pressure of 40 lbf/ft² per foot.

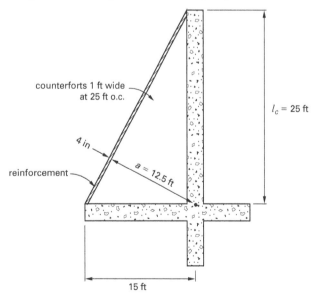

Figure 3.12 *Details of Counterfort Retaining Wall*

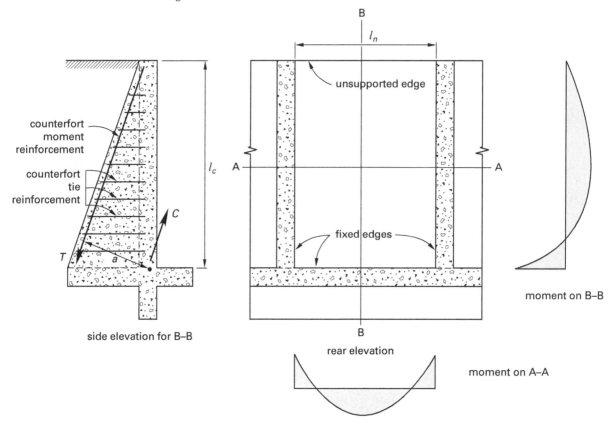

side elevation for B–B

rear elevation

moment on B–B

moment on A–A

Solution

The lateral earth pressure acting on a 1 ft horizontal strip at a depth of 0.6 times the stem height is

$$q = 0.6p_A l_c = (0.6)\left(40 \ \frac{\text{lbf}}{\text{ft}^2}\right)(25 \ \text{ft})$$

$$= 600 \ \text{lbf/ft}$$

At the counterfort supports, the factored design moment is given by ACI Table 6.5.2 as

$$M_u = \frac{1.6ql_n^2}{11} = \frac{(1.6)\left(0.6 \ \dfrac{\text{kip}}{\text{ft}}\right)(24 \ \text{ft})^2}{11}$$

$$= 50.27 \ \text{ft-kips}$$

Between counterforts, the factored design moment is

$$M_u = \frac{1.6ql_n^2}{16} = \frac{(1.6)\left(0.6 \ \dfrac{\text{kip}}{\text{ft}}\right)(24 \ \text{ft})^2}{16}$$

$$= 34.56 \ \text{ft-kips}$$

The factored design cantilever moment at the base is

$$M_u = 1.6(0.035p_A l_c^3)$$

$$= (1.6)(0.035)\left(0.04 \ \frac{\text{kip}}{\text{ft}^2}\right)(25 \ \text{ft})^3$$

$$= 35.00 \ \text{ft-kips}$$

Design of Counterforts

The bending moment produced by the earth pressure at the base of the stem is resisted by the couple produced by the tension in the reinforcement at the rear of the counterfort and the compression in the stem concrete. As shown in Fig. 3.12, the lever-arm of the couple acts at right angles to the reinforcement. The thrust produced by the earth pressure acting on the rear face of the stem is resisted by the horizontal ties in the counterfort.

Example 3.19

For the counterfort retaining wall of Ex. 3.18, determine the reinforcement area required in the rear of the counterfort and the tie area required at the base of the counterfort. Grade 60 reinforcement is provided.

Solution

The factored moment produced by the earth pressure at the base of the stem over one bay is

$$M_u = \frac{1.6 p_A s_c l_c^3}{6}$$

$$= \frac{(1.6)\left(40\ \dfrac{\dfrac{\text{lbf}}{\text{ft}^2}}{\text{ft}}\right)(25\ \text{ft})(25\ \text{ft})^3}{(6)\left(1000\ \dfrac{\text{lbf}}{\text{kip}}\right)}$$

$$= 4167\ \text{ft-kips}$$

The reinforcement area required in the rear of the counterfort to resist this moment is

$$A_s = \frac{M_u}{\phi a f_y}$$

$$= \frac{4167\ \text{ft-kips}}{(0.9)(12.5\ \text{ft})\left(60\ \dfrac{\text{kips}}{\text{in}^2}\right)}$$

$$= 6.17\ \text{in}^2$$

The factored lateral pressure from the backfill on a 1 ft horizontal strip at the base of the stem over one bay is

$$Q_u = 1.6 p_A l_c l_n$$

$$= \frac{(1.6)\left(40\ \dfrac{\text{lbf}}{\text{ft}^2}\right)(25\ \text{ft})(24\ \text{ft})}{1000\ \dfrac{\text{lbf}}{\text{kip}}}$$

$$= 38.40\ \text{kips}$$

The tie reinforcement area required at the base of the counterfort to resist this lateral pressure is

$$A_s = \frac{Q_u}{\phi f_y}$$

$$= \frac{38.40\ \text{kips}}{(0.9)\left(60\ \dfrac{\text{kips}}{\text{in}^2}\right)}$$

$$= 0.72\ \text{in}^2$$

8. GEOTECHNICAL INFORMATION

Geotechnical engineering is the application of rock and soil mechanics in order to determine the appropriate foundation systems for all types of structures.

Before a new building can be designed, or before any change can be made to the distribution of foundation loads for an existing building, a *geotechnical investigation* must usually be conducted. IBC Sec. 1803.1 requires that the geotechnical investigation be conducted by a registered design professional. The requirement to conduct a geotechnical investigation may be waived by the building official where satisfactory data from adjacent areas is available that demonstrates an investigation is not necessary for a particular building site and that local engineering practice may be adopted.

In accordance with IBC Sec. 1803.3, soil classification must be based on observation and tests of material disclosed by borings and test pits. More tests may be needed to evaluate slope stability, soil strength, the adequacy of load-bearing soil, and the effect of moisture variation on soil-bearing capacity, compressibility, liquefaction, and expansiveness. A qualified representative must be on site during all boring and sampling operations.

To determine whether foundations will be subject to hydrostatic pressure, IBC Sec. 1803.5.4 requires that the elevation of the groundwater table be determined.

When deep foundations are proposed, IBC Sec. 1803.5.5 requires that the geotechnical investigation include

- recommended pile types and installed capacities
- recommended pile spacing
- criteria for pile driving
- pile installation procedure
- field inspection procedure
- load test requirements
- designation and suitability of bearing strata
- reductions for group action

Where excavation will remove lateral support from the foundation of existing buildings, IBC Sec. 1803.5.7 requires an investigation to assess potential consequences and address mitigation measures.

9. DEEP FOUNDATIONS

Nomenclature

a	distance from elastic center to resultant of applied loads	ft
a_i	distance from row i to first row of piles in pile cap	ft
d	pile diameter	ft
e	eccentricity of W with respect to centroid of pile group	ft
H	lateral load	kips
I	moment of inertia	ft^4

I	moment of inertia about centroid of pile group	ft^4/ft^2
l	pile length	ft
M	moment acting on pile group	ft-kips
M_p	moment acting on pile	ft-kips
n	number of piles	–
n_i	number of individual piles in row i of pile cap	–
P	axial load	kips
P_e	axial force in pile due to eccentric load or moment	kips
P_v	axial force in pile due to vertical load	kips
q_b	presumptive bearing pressure	lbf/ft^2
q_f	skin friction	lbf/ft^2
Q_b	bearing capacity	kips
Q_f	skin friction capacity	kips
r_i	distance between elastic center and line of action of pile	ft
R	resultant of applied loads	kips
V	vertical force	kips
V_p	shear force acting on pile	kips
W	applied vertical load	kips
x_i	distance from row i to centroid of pile group	ft

Symbols

α	angle of inclination of battered piles	degree

General Principles

A *shallow foundation* is defined by IBC Sec. 202 as an isolated footing, a strip footing, a mat foundation, a slab-on-grade foundation, or any similar foundation element. A shallow foundation is used when the upper soil strata have sufficient strength to support the applied loads.

A *deep foundation* is defined by IBC Sec. 202 as any foundation element that does not satisfy the definition of a shallow foundation. A deep foundation may be provided by driven piles, drilled shafts, drilled piers, helical piles, or caissons, and it may consist of precast concrete, steel HP-shapes, steel pipes, or timber.

Reasons for adopting a deep foundation include

- large design loads
- site constraints such as property lines and adjacent structures
- soil with inadequate bearing capacity at shallow depths
- expansive soil which could cause differential movement in the structure
- soil that is liable to liquefaction during an earthquake

- the possibility of scour undermining a shallow foundation
- the possibility of a future adjacent excavation, which would necessitate underpinning a shallow foundation

Driven Pile Load Carrying Capacity

A *driven pile* is installed using a pile driver as shown in Fig. 3.13.

Figure 3.13 *Typical Pile Driving Rig*

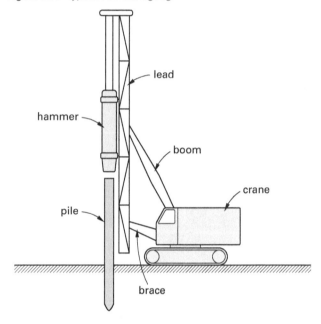

Before selecting driven piles for a project, it is important to make sure they are suitable. Possible disadvantages of installing driven piles include

- the noise of pile driving, which may be unacceptable for urban environments
- vibrations during installation, which could damage adjacent buildings
- ground heave from pile driving, which could affect the foundations of adjacent structures
- limited headroom on the site, which may interfere with the use of pile driving equipment

As shown in Fig. 3.14, the load carrying capacity of a pile is dependent both on the end bearing resistance at the toe of the pile and on the positive *skin friction* (also called *shaft resistance* or *frictional resistance*) acting upward and lengthwise along the perimeter of the pile. In accordance with IBC Sec. 1810.3.3.1.4, end bearing resistance and resistance due to skin friction may not be assumed to act simultaneously unless this is confirmed by a geotechnical investigation.

Figure 3.14 *Pile Resistance Forces*

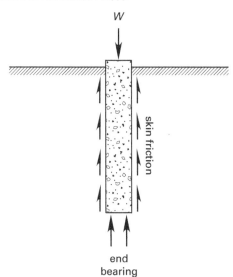

Driven piles are classified as either displacement piles or non-displacement piles. A *displacement pile* increases the soil pressure on its perimeter, and it derives its load carrying capacity mainly from skin friction. Displacement piles include piles made of precast concrete and closed-ended steel pipe. A *non-displacement pile* causes minimal disturbance to the surrounding soil, and it derives its load carrying capacity mainly from end bearing. Non-displacement piles include piles made from open-ended steel pipe and steel HP-shapes.

IBC Sec 1810.3.5.1 requires that precast concrete piles have a minimum lateral dimension of 8 in. To ensure the integrity of piles, IBC Sec. 1810.4.12 requires special inspection of driven piles in conformity with IBC Sec. 1705.9.

The *presumptive load-bearing value* for vertical loads, q_b, that is used in design for a supporting soil near the surface may not exceed the value specified in IBC Table 1806.2 unless data to substantiate the use of a higher value have been submitted and approved. A condensed version of IBC Table 1806.2 is shown in Table 3.1. The assumed frictional resistance, q_f, developed by any uncased cast-in-place deep foundation element is given by IBC Sec. 1810.3.3.1.4 as the minimum of

$$q_f = \frac{q_b}{6}$$

$$q_f = 500 \text{ lbf/ft}^2$$

unless a greater value is determined by a geotechnical investigation.

The load carrying capacity of a pile is

$$P = \frac{q_b \pi d^2}{4} + q_f \pi dl$$

provided that a geotechnical investigation confirms that shaft resistance and end-bearing resistance can act simultaneously.

Table 3.1 *Presumptive Load-Bearing Values*

material	presumptive bearing pressure, q_b (lbf/ft^2)
crystalline bedrock	12,000
sedimentary and foliated rock	4000
gravel, sandy gravel, or both	3000
sand, silty sand, clayey sand, silty gravel, and clayey gravel	2000
clay, sandy clay, silty clay, silt, clayey silt, and sandy silt	1500

Safety Factor

IBC Sec. 1810.3.3.1.7 requires that deep foundation elements must develop ultimate load capacities of at least twice the design service loads in the designated load-bearing layers. Analysis must show that no layer of soil underlying the designated load-bearing layers causes the load-bearing capacity safety factor to be less than two. In accordance with IBC Sec. 1810.2.1, any soil other than fluid soil affords lateral support that is deemed sufficient to prevent deep foundation elements from buckling.

IBC Sec. 1810.3.3.1.5 requires that the uplift capacity of a single deep foundation element must be determined either by an approved method of analysis or by load tests conducted in accordance with ASTM D 3689. The safety factor must be at least three where capacity is determined by analysis and at least two where capacity is determined by load tests. Where uplift is due to wind or seismic loading, a minimum safety factor of two is permitted where capacity is determined by analysis, and one and one-half is permitted where capacity is determined by load tests.

Downdrag

Downdrag, or *negative skin friction*, is the force produced by the downward movement of adjacent soil on a deep foundation element. For a typical case of pile loading, the load applied to the top of the pile causes the pile to move downward with respect to the soil. The resulting positive skin friction acts upward and contributes to the load carrying capacity of the pile. However, if settlement occurs in the soil surrounding the pile, so that the soil moves downward with respect to the pile, the resulting negative skin friction acts downward, producing additional vertical load on the pile. This is shown in Fig. 3.15.

Figure 3.15 *Negative Skin Friction*

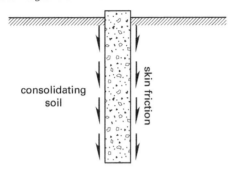

Several factors may contribute to the settlement of surrounding soil.

- Unconsolidated fill on the building site may settle after the piles are driven.

- Compressible material such as clay, silt, or peat may underlie the building site.

- Fill added to the site after the piles are driven may cause the consolidation of surrounding soil.

- The groundwater table may lower.

- Soil previously frozen may thaw.

- Soil liquefaction may occur, such as in an earthquake.

- Reconsolidation of soft soil may occur if the soil has been disturbed during pile driving.

For a pile subject to positive skin friction, settlement of the pile is negligible and poses no problem. When downdrag occurs, however, pile settlement may be considerably greater and can produce excessive differential settlements and severe damage.

The effects of downdrag can be reduced by

- applying a surcharge to the site before driving piles

- constructing drainage to control groundwater

- applying a coating, such as bitumen, to the pile to allow the soil to slide past the pile

- installing a sleeve around the pile to allow the soil to slide past the pile

Expansive Soil

Expansive soil contains clay minerals that can shrink or swell under changing moisture conditions. Clay minerals absorb water molecules, and this causes the soil to expand in wet conditions and shrink in dry conditions. Shrinkage produces large voids in the soil and may lead to subsidence.

Driven piles may be used to counteract the effect of swelling soil. As shown in Fig. 3.16, the piles can pass through the expansive layer and be anchored in the stable zone below where changes in moisture content are negligible.

Figure 3.16 *Skin Friction in Swelling Soil*

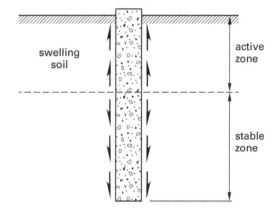

IBC Sec. 1808.6.1 requires that foundations in an active zone of expansive soil must be designed to resist differential volume changes and to prevent structural damage to the supported structure.

- A foundation that penetrates or extends into expansive soil must be designed to prevent uplift of the supported structure.

- A foundation that penetrates expansive soil must be designed to resist the force that may be exerted on the foundation due to changes in soil volume or must be isolated from the expansive soil.

Alternatively, foundation design need not comply with the above provisions when one of the following conditions is satisfied.

- The soil is removed to a depth sufficient to ensure a constant moisture content in the remaining soil.

- The building official approves stabilization of the soil by chemical, dewatering, presaturation, or equivalent techniques.

A drilled pile may be provided with a socketed or belled tip to resist uplift.

Example 3.20

Four symmetrically placed, 12 in diameter driven piles in native soil are shown. The end bearing capacity of the piles is 30,000 lbf/ft². Skin friction is 1000 lbf/ft². The site contains no expansive soil, and downdrag will not occur. It has been determined that frictional resistance and bearing resistance act simultaneously. Group action and pile heave need not be considered. Using a safety factor of two, determine the required length of the piles to support a vertical load of 170 kips.

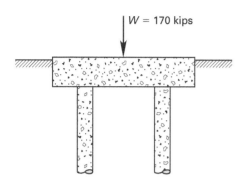

Solution

The bearing capacity of the four piles is

$$Q_b = \frac{4 q_b \pi d^2}{4}$$

$$= \frac{(4)\left(\dfrac{30{,}000 \ \frac{\text{lbf}}{\text{ft}^2}}{1000 \ \frac{\text{lbf}}{\text{kip}}}\right)\pi\left(\dfrac{12 \ \text{in}}{12 \ \frac{\text{in}}{\text{ft}}}\right)^2}{4}$$

$$= 94.2 \ \text{kips}$$

Allowing for a safety factor of two, the skin friction capacity required is

$$Q_f = 2W - Q_b$$

$$= (2)(170 \ \text{kips}) - 94.2 \ \text{kips}$$

$$= 245.8 \ \text{kips}$$

The skin friction capacity provided by four piles of length l is

$$Q_f = 4 q_f \pi d l$$

The required pile length is

$$l = \frac{Q_f}{4 q_f \pi d}$$

$$= \frac{245.8 \ \text{kips}}{4\left(\dfrac{1000 \ \frac{\text{lbf}}{\text{ft}^2}}{1000 \ \frac{\text{lbf}}{\text{kip}}}\right)\pi\left(\dfrac{12 \ \text{in}}{12 \ \frac{\text{in}}{\text{ft}}}\right)}$$

$$= 19.6 \ \text{ft}$$

Pile Groups

For most structures, piles are installed in groups. As required by IBC Sec. 1810.4.1, piles must be installed in a manner that prevents damage to adjacent structures or to piles already in place. In accordance with IBC Sec. 1810.4.6, piles that heave during the driving of adjacent piles must be driven again to develop the required capacity and penetration.

For end-bearing piles on rock, the bearing capacity of the pile group is essentially equal to the bearing capacity of one pile multiplied by the number of piles. This is true regardless of how the piles are spaced.

For closely spaced piles which rely on skin friction, however, the bearing capacity of a pile group may be significantly less than the number of piles multiplied by the bearing capacity of one pile. The reduced bearing capacity of the pile group is due to the group effect. In determining the bearing capacity of a pile group, IBC Sec. 1810.2.5 specifies that the analysis must include group effects on axial behavior whenever the center-to-center spacing of piles is less than three times the least horizontal dimension of a single pile.

IBC Sec. 1810.3.3.1.6 deals with the uplift capacity of pile groups. Where piles have a center-to-center spacing of at least three times the least horizontal dimension of the largest pile, the uplift capacity for the group is the lesser of

- the uplift capacity of an individual pile multiplied by the number of piles in the group

- two-thirds of the effective weight of the group and the soil contained within a block defined by the perimeter of the group and the length of the piles, plus two-thirds of the ultimate shear resistance along the soil block

Figure 3.17 illustrates the demarcation of the soil block and the shear resistance developed on its perimeter.

Analysis of Pile Group with Vertical Piles

ASCE Sec. 12.13.8.4 requires that the forces in individual piles in a pile group be determined in accordance with their relative rigidities and the geometric distribution of the piles within the group. Figure 3.18 shows a typical pile group with an applied load. The applied load, W, includes the weight of the pile cap and has an eccentricity of e with respect to the centroid of the pile group. The number of individual piles in row i of the pile cap is n_i. The distance from row i to the first row of piles is a_i. The distance from row i to the pile group centroid is x_i. The piles are identical.

Figure 3.17 *Uplift on a Pile Group*

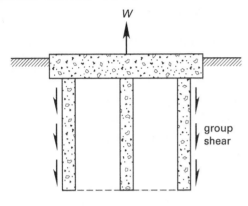

Figure 3.18 *Pile Group with Vertical Piles*

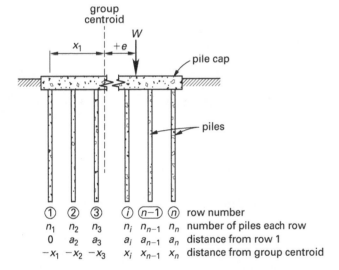

① ② ③ i $n-1$ n row number
n_1 n_2 n_3 n_i n_{n-1} n_n number of piles each row
0 a_2 a_3 a_i a_{n-1} a_n distance from row 1
$-x_1$ $-x_2$ $-x_3$ x_i x_{n-1} x_n distance from group centroid

The distance from row 1 to the group centroid is

$$x_1 = \frac{\sum an}{\sum n}$$

The axial load on each pile in row i is

$$P_i = \frac{W}{\sum n} + \frac{Wex_i}{\sum nx^2}$$

When the location of the applied load coincides with the group centroid, the axial load on each pile is

$$P = \frac{W}{\sum n}$$

Example 3.21

Six identical driven piles, arranged symmetrically in native soil are shown. The applied vertical load, including the weight of the pile cap, is 100 kips, and this load has an eccentricity of 1 ft. Determine the maximum axial force produced in a pile.

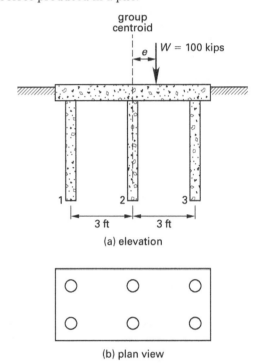

Solution

The pile group is symmetrical about row 2 and has a moment of inertia about row 2 of

$$I = \sum nx^2$$
$$= (4 \text{ piles})(3 \text{ ft})^2$$
$$= 36 \text{ ft}^4/\text{ft}^2$$

The two piles in row 3 are the most heavily loaded in the pile group. Due to vertical load, the axial force in each pile is

$$P_v = \frac{W}{6 \text{ piles}}$$
$$= \frac{100 \text{ kips}}{6 \text{ piles}}$$
$$= 16.67 \text{ kips}$$

Due to the eccentricity of the vertical load, the axial force in each pile in row 3 is

$$P_e = \frac{Wex}{I}$$
$$= \frac{(100 \text{ kips})(1 \text{ ft})(3 \text{ ft})}{36 \, \dfrac{\text{ft}^4}{\text{ft}^2}}$$
$$= 8.33 \text{ kips}$$

The total axial force in each pile in row 3 is

$$P = P_v + P_e$$
$$= 16.67 \text{ kips} + 8.33 \text{ kips}$$
$$= 25 \text{ kips}$$

Analysis of Pile Group in Air, Water, or Fluid Soil

Where piles stand unbraced in air, water, or fluid soil, IBC Sec. 1810.2.1 considers them fixed in position at a point 5 ft into stiff soil or 10 ft into soft soil unless otherwise approved by the building official based on a geotechnical investigation by a registered design professional. This is typically the situation in the case of a pier or wharf where a lateral load, H, is applied to the top of identical piles as shown in Fig. 3.19.

Figure 3.19 Pile Group Supporting a Pier

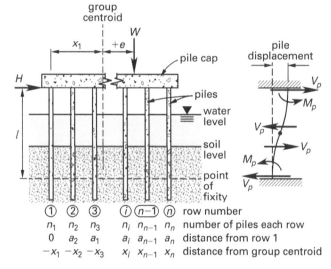

The piles are considered to be fixed-ended at the pile cap and at the point of fixity. The point of contraflexure on each pile occurs at the midpoint between the two fixed supports. The applied vertical load, W, includes the weight of the pile cap and has an eccentricity of e with respect to the centroid of the pile group.

The distance from the pile group centroid to row 1 is

$$x_1 = \frac{\sum an}{\sum n}$$

The applied moment acting about the line through the points of contraflexure is

$$M = We + \frac{Hl}{2}$$

The axial load on each pile in row i is

$$P_i = \frac{W}{\sum n} + \frac{Mx_i}{\sum nx^2}$$

The shear force on each pile is

$$V_p = \frac{H}{\sum n}$$

From the free-body diagram shown in Fig. 3.19, the maximum moment on each pile is

$$M_p = \frac{V_p l}{2}$$

Example 3.22

The illustration shows one line of a group of driven piles supporting a pier. The piles in the pile group are identical and are arranged symmetrically. The applied vertical load of 30 kips includes the weight of the pile cap and acts at the centroid of the line of piles. The lateral load acting on the line of piles is 10 kips. The piles are fixed in stiff soil at a depth of 30 ft below the pile cap soffit. Determine the forces acting in pile 3.

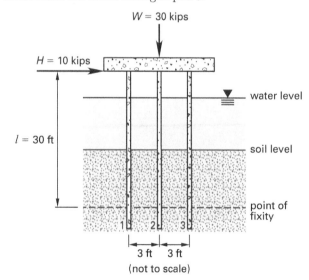

Solution

The pile group is symmetrical about row 2 and has a moment of inertia about row 2 of

$$
\begin{aligned}
I &= \sum x^2 \\
&= (2 \text{ piles})(3 \text{ ft})^2 \\
&= 18 \text{ ft}^4/\text{ft}^2
\end{aligned}
$$

Due to the vertical load, the axial force on pile 3 is

$$
\begin{aligned}
P_v &= \frac{W}{3 \text{ piles}} = \frac{30 \text{ kips}}{3 \text{ piles}} \\
&= 10 \text{ kips}
\end{aligned}
$$

The applied moment acting about the line through the points of contraflexure is

$$
\begin{aligned}
M &= \frac{Hl}{2} \\
&= \frac{(10 \text{ kips})(30 \text{ ft})}{2} \\
&= 150 \text{ ft-kips}
\end{aligned}
$$

Due to the moment, the axial force on pile 3 is

$$
\begin{aligned}
P_e &= \frac{Mx}{I} \\
&= \frac{(150 \text{ ft-kips})(3 \text{ ft})}{18 \dfrac{\text{ft}^4}{\text{ft}^2}} \\
&= 25 \text{ kips}
\end{aligned}
$$

The total axial force on each pile in row 3 is

$$
\begin{aligned}
P &= P_v + P_e \\
&= 10 \text{ kips} + 25 \text{ kips} \\
&= 35 \text{ kips}
\end{aligned}
$$

The shear force on each pile is

$$
\begin{aligned}
V_p &= \frac{H}{\sum n} \\
&= \frac{10 \text{ kips}}{3 \text{ piles}} \\
&= 3.33 \text{ kips}
\end{aligned}
$$

The maximum moment in pile 3 is

$$
\begin{aligned}
M_p &= \frac{V_p l}{2} \\
&= \frac{(3.33 \text{ kips})(30 \text{ ft})}{2} \\
&= 50 \text{ ft-kips}
\end{aligned}
$$

Analysis of Pile Group with Battered Piles

Battered piles, also known as *raked piles*, are useful for resisting large lateral loads. As shown in Fig. 3.20, they may typically be used to support a wharf.

Figure 3.20 *Battered Piles*

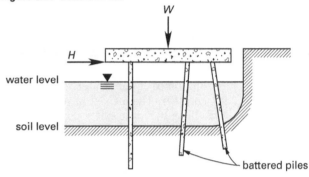

The elastic center method may be used to determine the forces in battered piles. All piles are assumed to be pinned at both ends. The rigid pile cap is assumed to translate horizontally and vertically due to the applied loads and to rotate about the elastic center thus producing axial forces in the piles that balance the external loads. As shown in Fig. 3.21, the elastic center is located at point C, where the lines of action of the forces in the piles intersect.

Figure 3.21 *Location of Elastic Center*

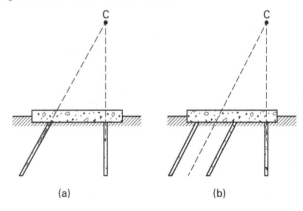

No Rotation of the Pile Cap

When the resultant of the external loads passes through the elastic center, the pile cap does not rotate, and the axial force in each pile is obtained by resolving forces.

Example 3.23

The illustration shows one line of a group of driven piles. All piles in the pile group are identical and have a batter of 1:4. The line of action of the external loads passes through the elastic center. Determine the axial force in the piles.

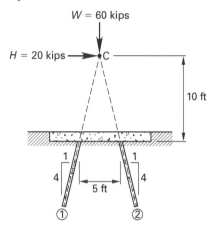

Solution

The angle of inclination of the piles is

$$\alpha = \arctan \frac{1}{4}$$
$$= 14.04°$$

Rotation of the pile cap does not occur. The axial force in pile 1 due to translation of the pile cap is

$$P_1 = \frac{W}{2\cos\alpha} - \frac{H}{2\sin\alpha}$$
$$= \frac{60 \text{ kips}}{2\cos 14.04°} - \frac{20 \text{ kips}}{2\sin 14.04°}$$
$$= -10.30 \text{ kips} \quad [\text{tension}]$$

The axial force in pile 2 due to translation of the pile cap is

$$P_2 = \frac{W}{2\cos\alpha} + \frac{H}{2\sin\alpha}$$
$$= \frac{60 \text{ kips}}{2\cos 14.04°} + \frac{20 \text{ kips}}{2\sin 14.04°}$$
$$= 72.14 \text{ kips} \quad [\text{compression}]$$

Rotation of the Pile Cap

When the line of action of the applied loads does not pass through the elastic center, the pile cap rotates as shown in Fig. 3.22, and the statical design method may be used.[3]

$$M = Ra$$

Figure 3.22 *Rotation of the Pile Cap*

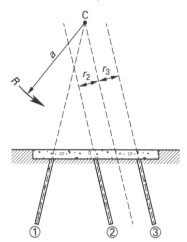

Initially, all piles in the pile group are considered to be vertical. The axial force in a vertical pile or the vertical component of the axial force in a batter pile is determined from the expression

$$Y_i = \frac{W}{\sum n} + \frac{Mx_i}{\sum nx^2}$$

In this equation, the moment acting on the pile group is

$$M = We + Hy$$

W is the vertical load applied to the pile group at an eccentricity e, and H is the horizontal load applied to the pile group at a height of y above the pile group. The axial force in a vertical pile is

$$P_i = Y_i$$

The axial force in a battered pile is

$$P_i = \frac{Y_i}{\cos\alpha}$$

The horizontal component of the axial force in a battered pile is

$$X_i = Y_i \tan\alpha$$

Subtract the sum of the horizontal components of the axial forces in all battered piles, $\sum X_i$, from the applied horizontal load, H, to get the residual horizontal force, H'.

$$H' = H - \sum X_i$$

This residual force is now distributed equally among all the piles in the pile group, and the piles are assumed to resist the lateral force by the soil pressure developed along their lengths.

Alternatively, the batter of the piles may be adjusted so as to exactly resist the applied horizontal load, H.

Example 3.24

The illustration shows one line of a group of driven piles. The piles are identical and have a batter of 1:4. The line of action of the external loads does not pass through the elastic center. The vertical load of 60 kips is located 3 ft from pile number 1, and the horizontal load of 20 kips is located 5 ft above the top of the piles. Determine the axial force in the piles and the residual lateral force, if any.

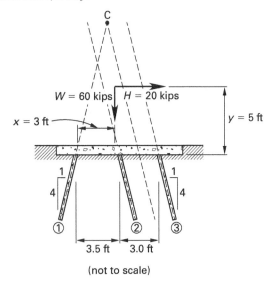

(not to scale)

Solution

The angle of inclination of the piles is

$$\alpha = \arctan \frac{1}{4}$$
$$= 14.04°$$

The distance of the pile group centroid from pile number 1 is

$$x_1 = \frac{\sum an}{\sum n}$$
$$= \frac{3.5 \text{ pile-ft} + 6.5 \text{ pile-ft}}{3 \text{ piles}}$$
$$= 3.33 \text{ ft}$$
$$x_2 = a_2 - x_1 = 3.5 \text{ ft} - 3.33 \text{ ft}$$
$$= 0.17 \text{ ft}$$
$$x_3 = a_3 - x_1 = 6.5 \text{ ft} - 3.33 \text{ ft}$$
$$= 3.17 \text{ ft}$$

The moment of inertia about the centroid of the pile group is

$$\sum nx^2 = x_1^2 + x_2^2 + x_3^2$$
$$= (3.33 \text{ ft})^2 + (0.17 \text{ ft})^2 + (3.17 \text{ ft})^2$$
$$= 21.17 \text{ ft}^4/\text{ft}^2$$

The eccentricity of W with respect to the centroid of the pile group is

$$e = x - x_1 = 3 \text{ ft} - 3.33 \text{ ft} = -0.33 \text{ ft}$$

The moment acting on the pile group is

$$M = We + Hy = (60 \text{ kips})(-0.33 \text{ ft}) + (20 \text{ kips})(5 \text{ ft})$$
$$= 80.20 \text{ ft-kips}$$

The vertical component of the axial force in a battered pile is

$$Y_i = \frac{W}{\sum n} + \frac{Mx_i}{\sum nx^2}$$
$$= \frac{60 \text{ kips}}{3 \text{ piles}} + \frac{(80.20 \text{ ft-kips})x_i}{21.17 \dfrac{\text{ft}^4}{\text{ft}^2}}$$
$$= 20 \text{ kips} + x_i\left(3.79 \frac{\text{kips}}{\text{ft}}\right)$$

For pile 1,

$$P_1 = \frac{Y_1}{\cos \alpha}$$
$$= \frac{20 \text{ kips} + (-3.33 \text{ ft})\left(3.79 \dfrac{\text{kips}}{\text{ft}}\right)}{\cos 14.04°}$$
$$= 7.61 \text{ kips} \quad [\text{compression}]$$
$$X_1 = P_1 \sin \alpha = (7.61 \text{ kips})\sin 14.04°$$
$$= 1.85 \text{ kips} \quad [\text{acting to the right}]$$

For pile 2,

$$P_2 = \frac{Y_2}{\cos \alpha}$$
$$= \frac{20 \text{ kips} + (0.17 \text{ ft})\left(3.79 \dfrac{\text{kips}}{\text{ft}}\right)}{\cos 14.04°}$$
$$= 21.28 \text{ kips} \quad [\text{compression}]$$
$$X_2 = P_2 \sin \alpha = (21.28 \text{ kips})\sin 14.04°$$
$$= 5.16 \text{ kips} \quad [\text{acting to the left}]$$

For pile 3,

$$P_3 = \frac{Y_3}{\cos \alpha}$$

$$= \frac{20 \text{ kips} + (3.17 \text{ ft})\left(3.79 \ \dfrac{\text{kips}}{\text{ft}}\right)}{\cos 14.04°}$$

$$= 33.00 \text{ kips} \quad [\text{compression}]$$

$$X_3 = P_3 \sin \alpha = (33.00 \text{ kips}) \sin 14.04°$$

$$= 8.01 \text{ kips} \quad [\text{acting to the left}]$$

Taking forces acting to the left as positive, the sum of the horizontal components of the forces is

$$\sum X_i = X_1 + X_2 + X_3$$

$$= -1.85 \text{ kips} + 5.16 \text{ kips} + 8.01 \text{ kips}$$

$$= 11.32 \text{ kips}$$

The residual horizontal force is

$$H' = H - \sum X_i = 20 \text{ kips} - 11.32 \text{ kips} = 8.68 \text{ kips}$$

Timber Piles

Driven *timber piles* are an inexpensive, readily available, and renewable alternative to concrete and steel piles. Timber piles provide a convenient foundation for lighter loads, and they are used extensively in harbor works such as piers and wharves.

When fully submerged below the water table, timber piles will last indefinitely. Rotting is induced by oxygen, which is available only above the water table. Timber piles that will be subjected alternately to wetting and drying are pressure treated with an approved preservative and capped with concrete. With this treatment, timber piles will last more than 100 years.

Timber piles are available in diameters from 10 in to 14 in and lengths from 20 ft to 40 ft. They are installed using a standard pile driver. When driving in soft soil, an unprotected point may be used. In hard soil, a steel shoe is added to prevent the crushing of fibers. A steel ring, which fits tightly around the head of the pile, is also added to prevent *crushing*, or *brooming*, from occurring in the head of the pile during driving.

Timber piles are spliced readily as needed. The butting surfaces are cut square at the splice. A 12 in steel tube is driven, for half its length, onto the head of the leader pile, using a drop hammer. The follower pile is then driven into the steel tube, and driving continues.

There are many advantages to using timber piles, as they are

- less expensive than other systems

- readily available and a renewable resource

- light in weight, making them easy to handle and install

- easy to drive using standard equipment

- easy to cut to size, with the cutoff being easily disposable

Design and construction details for timber piling are available.[4] Reference design values are provided in the *National Design Specification for Wood Construction*.[5]

10. ADDITIONAL PILE TYPES

General Principles

In addition to driven piles, several other types of piles are available that may be more suitable in some circumstances. These include

- drilled shafts

- drilled piers

- caissons

- auger cast piles

- Franki piles

- helical piles

- micropiles

Drilled Shafts, Drilled Piers, and Caissons

A *drilled shaft pile* is a cast-in-place pile. The pile may be cased or uncased. As shown in Fig. 3.23(a), an auger is used to drill a hole of the required depth and diameter. If the surrounding soil is liable to collapse into the hole, a steel casing is inserted. When the required depth is attained, the auger is withdrawn. If required, a reinforcement cage is lowered into the hole, and a tremie is then used to fill the hole with grout, which reduces the vertical drop of the grout and prevents segregation. A short, drilled shaft pile is generally referred to as a *drilled pier*.

There are many advantages to using drilled shaft piles.

- The diameter can be large, thus support heavy loads.

- A single pile is often sufficient for supporting a heavy load, eliminating the need for a pile cap.

- Large diameter piles allow for easy visual inspection, making it easy to spot defects before pouring the grout.

- There is very little noise or vibration during construction.

- A bell at the pile tip significantly increases end bearing capacity and uplift resistance.

Figure 3.23 *Drilled Shaft Piles*

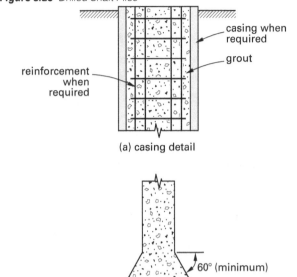

(a) casing detail

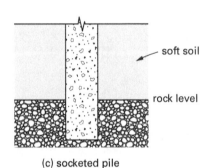

(b) belled pile

(c) socketed pile

In urban environments and in places where noise and vibration during installation are concerns, drilled shaft piles are generally preferred to driven piles.

There are, however, several disadvantages to using drilled shaft piles. Disposal of the drill spoil is an additional cost. Also, drilled shaft piles cannot be used as battered piles because of the risk of caving and the difficulty of placing concrete and reinforcement in a battered hole.

In accordance with IBC Sec. 1810.3.5.2.2, a shaft without a permanent casing must have a diameter of no less than 12 in, and the shaft length must not exceed 30 times the diameter. In accordance with IBC Sec.1810.3.5.2.1, a shaft with a permanent casing must have an outside diameter of no less than 8 in.

As shown in Fig 3.23(b), a *belled pile*, or *under reamed pile*, has an enlarged base in the form of an inverted cone. The base is mechanically formed in stable soil. In accordance with IBC Sec. 1810.3.9.5, the edge thickness of the bell must be no less than the thickness required for the edge of footings. Where the sides of the bell slope at an angle less than 60° from the horizontal, the effects of vertical shear must be considered.

As shown in Fig 3.23(c), a *socketed pile* is a drilled shaft pile with its end drilled into rock. A socketed shaft is generally used to transfer loads through a deep layer of soft soil to underlying bedrock. In accordance with IBC Sec. 1810.3.9.6, a socketed shaft must have a permanent casing extending down to bedrock and an uncased socket drilled into the bedrock, both filled with concrete. The depth of the socket in the bedrock must be sufficient to develop the required load carrying capacity with a minimum safety factor of two, and the minimum socket depth is equal to the outside diameter of the casing. The load carrying capacity is the sum of the end bearing capacity and the bond along the sides of the socket.

An open concrete cylindrical *caisson* is shown in Fig. 3.24. The walls of the caisson are reinforced concrete, and steel shoes at the base provide a cutting edge. The soil within the caisson is excavated either manually or mechanically, which causes the caisson to sink into the ground under its own weight. As the penetration of the caisson increases, the caisson may need more weight to sink further; this weight may be provided by additional concrete or kentledge. When the desired founding level is reached, the caisson is backfilled with concrete.

Figure 3.24 *Concrete Caisson*

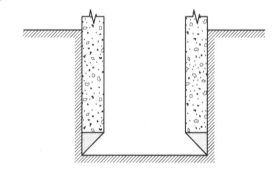

If the foundation is to be constructed underwater, the caisson may be prefabricated with a sealed base. The caisson is then floated to the required location and sunk in place.

Auger Cast Piles

Auger cast piles are also referred to as *continuous flight auger piles* or *auger cast-in-place piles*. The piles are cast in place using a hollow stem continuous flight auger. The auger is drilled into the ground to the design depth, thus filling the flights with soil and providing lateral support to the auger and stability to the hole. After reaching the design depth, the auger is slowly pulled out of the hole, bringing with it the soil confined on its flights. Simultaneously, grout is pumped through the hollow stem to fill the gap left at the toe of the auger. The rate of withdrawal of the auger and the pumping of the grout is carefully controlled to avoid leaving any cavities in the pile. When the auger is fully withdrawn, any required reinforcing cage may be lowered into the wet grout.

The principal advantages that auger cast piles have over drilled shaft piles are that the process is much quicker and that a casing need not be used. The principal disadvantage is that installing auger cast piles requires considerably more torque, which limits their use to softer soil profiles, smaller diameters, and shorter lengths than drilled shafts.

Franki Piles

As shown in Fig. 3.25, a *Franki pile*, or *pressure injected footing*, is a concrete pile that is cast in situ and has an expanded base. The principal advantages to using Franki piles over driven piles are lower noise and vibration levels during construction and the significant increases in both end bearing capacity and uplift resistance provided by their expanded bases. Also, Franki piles can be battered, while drilled shaft piles must be vertical.

Figure 3.25 Franki Piles

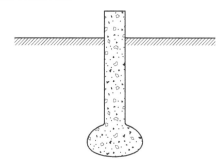

To install a Franki pile, a steel casing is placed in position, and the bottom 3 ft is filled with a dry concrete mix. A drop hammer located inside the casing is then used to consolidate the concrete mix to form a watertight plug and drive the plug into the ground, dragging the steel casing with it. The process continues until the design depth is reached. The steel casing is then restrained, and continued application of the drop hammer expels the dry concrete mix out of the steel casing to form the bulbous base. If required, a reinforcing cage is lowered into the steel casing. More concrete is added and consolidated with the drop hammer working within the reinforcing cage. Concurrently, the steel casing is withdrawn.

Helical Piles

IBC Sec. 202 defines a *helical pile*, or *screw pile*, as a manufactured steel deep foundation element consisting of a central shaft and one or more helical bearing plates. A helical pile is installed by rotating it into the ground. The helix is formed by welding circular flat plates to the central shaft at a specified pitch.

Figure 3.26 shows a typical helical pile. As the shaft penetrates into the soil, additional lengths with or without helical bearing plates are attached by means of couplers. This technique is eminently suited to *underpinning* an existing foundation.

Figure 3.26 Helical Pile

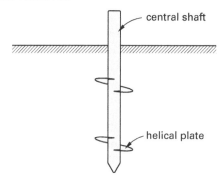

IBC Sec. 1810.3.5.3.5 requires that the dimensions of the central shaft and the number, size, and thickness of the helical bearing plates be sufficient to support the design loads. IBC Eq. 18-4 gives the allowable axial design load as $P_a = 0.5 P_u$, where P_u is the lowest of

- the product of the ultimate bearing capacity of the soil or rock in the bearing stratum and the sum of the areas of the helical bearing plates

- the ultimate capacity as determined from well-documented correlations with installation torque

- the ultimate capacity as determined from load tests

- the ultimate axial capacity of the central shaft

- the ultimate axial capacity of central shaft couplings

- the sum of the ultimate axial capacities of each helical bearing plate affixed to the central shaft

Some advantages to helical piles are that they

- are rapidly installed

- have minimal environmental impact and no drilling spoil to dispose of

- produce no noise or vibration during installation

- may be used in either tension or compression

- may be loaded immediately after installation

- may be battered and used as tie-back anchors

- may be installed in restricted areas

Micropiles

A *micropile*, also called a *minipile*, is defined in IBC Sec. 202 as a deep foundation element that is bored, grouted in place, and limited in diameter. A micropile develops its load carrying capacity by means of a bond zone in soil, bedrock, or a combination of the two. In accordance with IBC Sec. 1810.3.5.2.3, the maximum permitted outside diameter of a micropile is 12 in.

A micropile typically consists of a high-strength steel casing and a threaded bar for reinforcement. The steel casing is installed to the design depth using a technique such as rotary drilling, jetting, impact driving, or jacking. Reinforcement, in the form of an all-threaded steel bar, is inserted into the full depth of the casing, which is injected with grout.

Micropiles are particularly suited to underpinning situations with low headroom and limited access conditions. Because of their small diameters, micropiles are used in locations with congested underground utilities. Micropiles cause minimal disruption to building occupants, allowing operations to continue during construction.

11. CANTILEVER SHEET PILE WALLS

Nomenclature

h_o	exposed wall height	ft
h_p	wall embedment	ft
p_A	active lateral pressure per foot equivalent to a fluid of specific weight γ_S, equal to $K_A\gamma_S$	lbf/ft^2
p_P	passive lateral pressure per foot equivalent to a fluid of specific weight, γ_S, equal to $K_P\gamma_S$	lbf/ft^2
z	depth below ground surface	ft

Symbols

| γ_S | specific weight of soil | lbf/ft^3 |

General Principles

A typical *cantilever steel sheet pile wall* is shown in Fig. 3.27. A *sheet pile wall* is used to retain existing ground on one side of the wall while soil on the other side of the wall is excavated. The wall may be a temporary work to allow the construction of a building foundation or basement, in which case the piling can be removed and reused after the foundation is finished. Alternatively, the wall may form a permanent construction for the protection of a river bank or sea wall.

A steel sheet pile wall consists of individual pile sections with interlocking edges. The interlocking edges of the pile provide a semi-watertight seal and reduce groundwater ingress to the excavation. Prestressed concrete sheet piles may be used in place of steel sheet piles.

Steel sheet piles are installed using a vibratory hammer. In dense soil, a vibratory hammer can be supplemented with an impact hammer. Where vibration is not permitted on site, the piles can be hydraulically pressed into position.

Figure 3.27 *Steel Sheet Pile Wall*

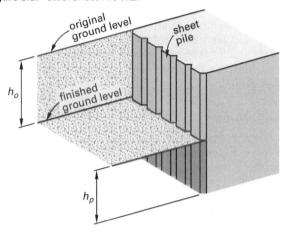

Because of deflection limitations as well as the increase in cost due to exposure, cantilever walls are generally restricted to a maximum height of 15 ft. Penetration depths as much as two times the exposed wall height may be necessary for wall stability. Erosion in front of the wall can affect stability as passive pressure on the embedded portion of the wall gives the wall lateral support.

Design Details

In designing a sheet pile wall, a trial-and-error procedure is generally adopted, and the depth of embedment is initially estimated. The resulting simplified pressure diagram for a homogeneous, non-cohesive soil is shown in Fig. 3.28.

The active pressure tending to overturn the wall is represented by triangle 168. The active pressure is resisted by the passive pressures acting in front of and behind the embedded length of the wall. Point O is the point at which passive pressure is zero and the point about which the wall is assumed to rotate.

An initial trial depth is assumed for the embedment. The following relationships are derived from the Fig. 3.28.[6]

$$23 = p_A h_o$$
$$45 = p_P(h_p - z_1) - p_A(l - z_1)$$
$$67 = p_P l - p_A h_p$$
$$68 = p_A l$$
$$69 = p_P l$$
$$610 = p_P h_p - p_A l$$
$$611 = p_P h_p$$
$$z_1 = \frac{p_P h_p^2 - p_A l^2}{(p_P - p_A)(l + h_p)}$$

Figure 3.28 *Forces Acting on a Cantilever Sheet Pile Wall*

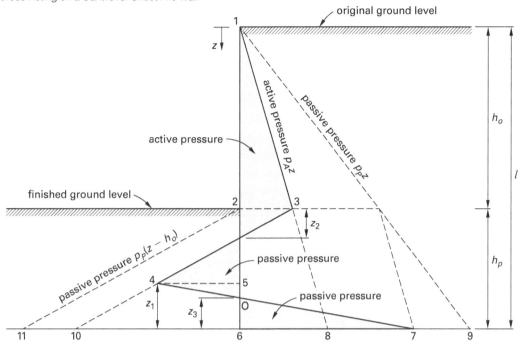

$$z_2 = \frac{p_A h_o}{p_P - p_A}$$

$$z_3 = \frac{z_1(p_P l - p_A h_p)}{(p_P - p_A)(h_o - 2h_p - z_1)}$$

p_A	active lateral pressure per foot equivalent to a fluid of specific weight γ_S, equal to $K_A \gamma_S$	lbf/ft²
p_P	passive lateral pressure per foot equivalent to a fluid of specific weight, γ_S, equal to $K_P \gamma_S$	lbf/ft²

If the embedment has been correctly chosen, the sum of the active and passive forces in the horizontal direction will be zero, and the sum of the moments about any point on the wall will also be zero. Adjust the depth of embedment if this is not the case and repeat until convergence is obtained.

12. ANCHORED SHEET PILE RETAINING WALLS

Nomenclature

F_C	force in compression pile	lbf
F_T	force in tension pile	lbf
h_o	exposed wall height	ft
h_p	wall embedment	ft
h_t	depth of tie-back	ft
h_1	distance between H_A and H_T	ft
h_2	distance between H_P and H_T	ft
H_A	total active pressure on the back of the wall	lbf/ft
H_P	total passive pressure on the front of the wall	lbf/ft
H_T	force in the tie	lbf/ft

Symbols

γ_S	specific weight of soil	lbf/ft³
θ	angle of inclination of anchor pile	deg

General Principles

A typical *anchored sheet pile retaining wall* is shown in Fig. 3.29. The penetration depth needed for stability is less than the depth needed for a non-anchored wall. The tie-back is located at or near the top of the wall and is anchored with a *deadman*, a concrete block that resists the tie-back force by taking advantage of the passive soil resistance that develops in front of the block. As shown in Fig. 3.30, the deadman must be located far enough from the wall to prevent the passive wedge in front of the deadman and the active wedge behind the wall from intersecting.

Figure 3.29 Anchored Sheet Pile Retaining Wall

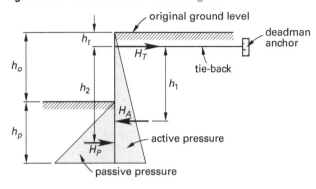

Figure 3.30 Location of Anchor

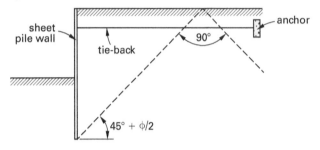

An alternative way of securing the tie-back is to use anchor piles as shown in Fig. 3.31. The forces in the piles are the force in compression, F_C, and the force in the tension pile, F_T. Using the notation in Fig. 3.31,

$$F_C = \frac{H_T}{\sin\theta_1 + \cos\theta_1\tan\theta_2}$$

$$F_T = \frac{H_T}{\sin\theta_2 + \cos\theta_2\tan\theta_1}$$

Figure 3.31 Forces Acting on Anchor Piles

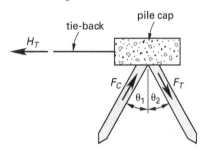

Design Details

In designing anchor piles, the required penetration of the sheet piles, h_p, is determined first, then the forces acting on the sheet piles are obtained.

The total active pressure on the back of the wall is

$$H_A = \frac{p_A(h_o + h_p)^2}{2}$$

The total passive pressure on the front of the wall is

$$H_P = \frac{p_P h_p^2}{2}$$

The distance between H_A and H_T is

$$h_1 = 0.67(h_o + h_p) - h_t$$

The distance between H_P and H_T is

$$h_2 = h_o - h_t + 0.67h_p$$

Taking moments about the tie point, T,

$$
\begin{aligned}
0 &= H_A h_1 - H_P h_2 \\
&= \left(\frac{p_A(h_o + h_p)^2}{2}\right)\left((0.67(h_o + h_p)) - h_t\right) \\
&\quad - \left(\frac{p_P h_p^2}{2}\right)(h_o - h_t + 0.67h_p)
\end{aligned}
$$

Solving this expression for h_p gives the required penetration.

The force in the tie is

$$H_T = H_A - H_P$$

The bending moment and shear force acting on the sheet piles may now be determined by simple statics.

Example 3.25

The anchored sheet pile retaining wall shown has its tie-back located at the top of the wall. The height of the exposed wall is 12 ft. The active earth pressure is equivalent to a fluid pressure of 30 lbf/ft^2, and the passive earth pressure is equivalent to a fluid pressure of 400 lbf/ft^2. Determine the required penetration, h_p, of the sheet piles and the force in the tie-back.

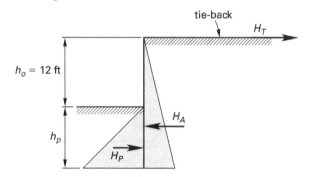

Solution

The total active pressure on the back of the wall is

$$H_A = \frac{p_A(h_o + h_p)^2}{2}$$

$$= \frac{\left(30 \ \frac{\text{lbf}}{\text{ft}^2}\right)(12 \ \text{ft} + h_p)^2}{2}$$

$$= 2160 \ \text{lbf} + \left(360 \ \frac{\text{lbf}}{\text{ft}}\right)h_p + \left(15 \ \frac{\text{lbf}}{\text{ft}^2}\right)h_p^2$$

The total passive pressure on the front of the wall is

$$H_P = \frac{p_P h_p^2}{2}$$

$$= \frac{\left(400 \ \frac{\text{lbf}}{\text{ft}^2}\right)h_p^2}{2}$$

$$= \left(200 \ \frac{\text{lbf}}{\text{ft}^2}\right)h_p^2$$

The distance between H_A and H_T is

$$h_1 = 0.67(h_o + h_p)$$
$$= (0.67)(12 \ \text{ft} + h_p)$$
$$= 8 \ \text{ft} + 0.67h_p$$

The distance between H_P and H_T is

$$h_2 = h_o + 0.67h_p$$
$$= 12 \ \text{ft} + 0.67h_p$$

Taking moments about the tie point gives

$$0 = H_A h_1 - H_P h_2$$
$$= \left(2160 \ \text{lbf} + \left(360 \ \frac{\text{lbf}}{\text{ft}}\right)h_p + \left(15 \ \frac{\text{lbf}}{\text{ft}^2}\right)h_p^2\right)$$
$$\times (8 \ \text{ft} + 0.67h_p) - \left(200 \ \frac{\text{lbf}}{\text{ft}^2}\right)h_p^2$$
$$\times (12 \ \text{ft} + 0.67h_p)$$

Solving for h_p gives the required penetration.

$$h_p = 3.64 \ \text{ft}$$

The force in the tie-back is obtained by adding the horizontal forces.

$$H_A = 2160 \ \text{lbf} + \left(360 \ \frac{\text{lbf}}{\text{ft}}\right)(3.64 \ \text{ft})$$
$$+ \left(15 \ \frac{\text{lbf}}{\text{ft}^2}\right)(3.64 \ \text{ft})^2$$
$$= 3670 \ \text{lbf}$$
$$H_P = \left(200 \ \frac{\text{lbf}}{\text{ft}^2}\right)(3.64 \ \text{ft})^2$$
$$= 2650 \ \text{lbf}$$
$$H_T = H_A - H_P$$
$$= 3670 \ \text{lbf} - 2650 \ \text{lbf}$$
$$= 1020 \ \text{lbf}$$

13. MASS GRAVITY RETAINING WALLS

Nomenclature

b	width at top of wall	ft
B	width at base of wall	ft
e	eccentricity	–
FS	factor of safety	–
h	height of wall	ft
H_A	total active pressure on back of wall	lbf/ft
M_O	overturning moment	ft-lbf
M_R	restoring moment	ft-lbf
R	resultant of H_A and W	lbf
S	base section modulus	–
w_c	specific weight of concrete	lbf/ft³
W	weight of wall	lbf
x_o	distance from wall centroid to heel	ft
z	depth below ground surface	ft

Symbols

μ	coefficient of friction	–

General Principles

A typical mass gravity retaining wall is shown in Fig. 3.32. The stability of the wall and the soil pressure produced are dependent on the size and weight of the wall. If tensile stresses are not permitted in the wall, the resultant of the self-weight of the wall and the lateral soil pressure on the back of the wall must lie within the middle third of the wall section at any point.

In designing the wall, the wall base width is adjusted to provide an acceptable soil bearing pressure and the dimensions of the wall are adjusted to provide factors of safety of 1.5 against overturning and sliding.

Figure 3.32 *Mass Gravity Wall*

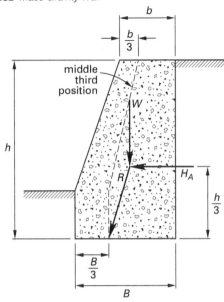

Example 3.26

The illustration shows a mass concrete retaining wall with a height of 12 ft and a width at the top of the wall of 2 ft. Active earth pressure is equivalent to a fluid pressure of 30 lbf/ft^2. The specific weight of the concrete is 144 lbf/ft^3. Allowable soil bearing pressure is 3000 lbf/ft^2. The coefficient of friction is 0.5. Determine the required base width if tensile stress is not allowed in the concrete. Check that the soil bearing pressure and the factors of safety against overturning and sliding are satisfactory. Neglect passive pressure in front of the wall.

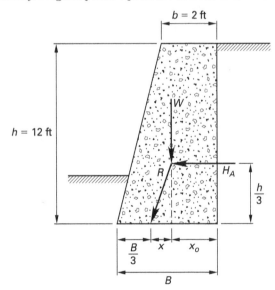

Solution

Base Width

The weight of a 1 ft length of the wall is

$$
\begin{aligned}
W &= \frac{w_c L h (b + B)}{2} \\
&= \frac{\left(144 \ \dfrac{\text{lbf}}{\text{ft}^3}\right)(1 \text{ ft})(12 \text{ ft})(2 \text{ ft} + B)}{2} \\
&= (864 \text{ lbf/ft})(2 \text{ ft} + B)
\end{aligned}
$$

Due to active soil pressure, the force acting on the earth face of the wall 4.0 ft above the base is

$$
H_A = \frac{p_A h^2}{2} = \frac{\left(30 \ \dfrac{\text{lbf}}{\text{ft}^2}\right)(12 \text{ ft})^2}{2} = 2160 \text{ lbf}
$$

From the illustration, the distance from the wall centroid to the heel is

$$
\begin{aligned}
x_o &= \frac{b^2 + bB + B^2}{3(b + B)} \\
&= \frac{(2 \text{ ft})^2 + (2 \text{ ft})B + B^2}{3(2 \text{ ft} + B)} \\
&= \frac{4 \text{ ft}^2 + (2 \text{ ft})B + B^2}{3(2 \text{ ft} + B)}
\end{aligned}
$$

The line of action of the resultant thrust cuts the base at a distance from the wall centroid.

$$
\begin{aligned}
x &= \left(\frac{h}{3}\right)\left(\frac{H_A}{W}\right) \\
&= \left(\frac{12 \text{ ft}}{3}\right)\left(\frac{2160 \text{ lbf}}{(864 \text{ lbf})(2 \text{ ft} + B)}\right) \\
&= \frac{10 \text{ ft}}{2 \text{ ft} + B}
\end{aligned}
$$

For no tensile stress in the base,

$$
x + x_o = \frac{2B}{3}
$$

So,

$$
\begin{aligned}
\frac{2B}{3} &= \frac{10 \text{ ft}}{2 \text{ ft} + B} + \frac{4 \text{ ft}^2 + (2 \text{ ft})B + B^2}{(3)(2 \text{ ft} + B)} \\
B &= 4.92 \text{ ft}
\end{aligned}
$$

The minimum base width required to prevent tensile stress in the concrete is 4.92 ft.

Bearing Pressure

The total weight of a 1 ft length of wall is

$$W = \left(864 \ \frac{\text{lbf}}{\text{ft}}\right)(2 \text{ ft} + 4.92 \text{ ft})$$
$$= 5978 \text{ lbf}$$

Due to active soil pressure, the force acting on the earth face of the wall 4 ft above the base is

$$H_A = 2160 \text{ lbf}$$

The distance from the wall centroid to the heel is

$$x_o = \frac{4 \text{ ft}^2 + (2 \text{ ft})B + B^2}{(3)(2 \text{ ft} + B)}$$
$$= \frac{4 \text{ ft}^2 + (2 \text{ ft})(4.92 \text{ ft}) + (4.92 \text{ ft})^2}{(3)(2 \text{ ft} + 4.92 \text{ ft})}$$
$$= 1.83 \text{ ft}$$

The eccentricity of W about the wall base is

$$e = \frac{B}{2} - x_o = \frac{4.92 \text{ ft}}{2} - 1.83 \text{ ft} = 0.63 \text{ ft}$$

The net moment about the centroid of the wall base is

$$M = \frac{H_A h}{3} - We$$
$$= \frac{(2160 \text{ lbf})(12 \text{ ft})}{3} - (5978 \text{ lbf})(0.63 \text{ ft})$$
$$= 4874 \text{ ft-lbf}$$

The base area is

$$A = (4.92 \text{ ft})(1 \text{ ft}) = 4.92 \text{ ft}^2$$

The base section modulus is

$$S = \frac{(1 \text{ ft})(4.92 \text{ ft})^2}{6} = 4.03 \text{ ft}^3$$

The maximum soil bearing pressure at the toe of the wall is

$$q = \frac{W}{A} + \frac{M}{S}$$
$$= \frac{5978 \text{ lbf}}{4.92 \text{ ft}^2} + \frac{4874 \text{ ft-lbf}}{4.03 \text{ ft}^3}$$
$$= 2424 \text{ lbf/ft}^2$$

This is less than 3000 lbf/ft², so this is satisfactory.

Overturning Factor of Safety

The overturning moment about the toe is

$$M_O = \frac{H_A h}{3} = \frac{(2160 \text{ lbf})(12 \text{ ft})}{3}$$
$$= 8640 \text{ ft-lbf}$$

The restoring moment is

$$M_R = W(B - x_o)$$
$$= (5978 \text{ lbf})(4.92 \text{ ft} - 1.83 \text{ ft})$$
$$= 18{,}472 \text{ ft-lbf}$$

The factor of safety against overturning is

$$FS = \frac{M_R}{M_O} = \frac{18{,}472 \text{ ft-lbf}}{8640 \text{ ft-lbf}} = 2.14$$

This is more than 1.5, so the overturning factor of safety is satisfactory.

Sliding Factor of Safety

The frictional force produced on the underside of the base is

$$F = \mu W = (0.5)(5978 \text{ lbf}) = 2989 \text{ lbf}$$

The factor of safety against sliding is

$$FS = \frac{F}{H_A} = \frac{2989 \text{ lbf}}{2160 \text{ lbf}} = 1.38$$

This is less than 1.5, so the sliding factor of safety is unsatisfactory.

14. BASEMENT WALLS

Nomenclature

h	height of wall	ft
H_O	total at-rest pressure on the back of the wall	lbf/ft
K_A	Rankine coefficient of active earth pressure, $(1 - \sin\phi)/(1 + \sin\phi)$	–
K_O	coefficient for earth pressure at rest, $1 - \sin\phi$	–
K_P	Rankine coefficient of passive earth pressure, $(1 + \sin\phi)/(1 - \sin\phi)$	–
p_o	at-rest lateral pressure per foot equivalent to a fluid of specific weight γ_S, equal to $K_O\gamma_S$	lbf/ft²

Symbols

γ_S	specific weight of soil	lbf/ft³

General Principles

As specified in IBC[7] Sec. 1610.1, *basement walls* in which horizontal movement is restricted at the top must be designed for at-rest pressure. Earth pressure at rest may be as much as twice the value of active earth pressure. In accordance with IBC Sec. 1610.1, basement walls extending no more than 8 ft below grade and supporting flexible floor systems may be designed for active pressure.

A typical basement wall with its top restrained by the ground floor diaphragm is shown in Fig. 3.33.

Figure 3.33 Basement Wall

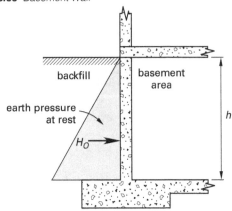

The three types of earth pressure that can act on a retaining wall are active earth pressure, passive earth pressure, and at-rest earth pressure. When the top of a retaining wall is free to move outward, a sliding wedge of soil is produced behind the wall, and active earth pressure develops behind the wall. The Rankine coefficient of active earth pressure is

$$K_A = \frac{1 - \sin\phi}{1 + \sin\phi}$$

When the toe of a retaining wall is below grade and the wall is free to move forward, the soil in front of the toe is compressed and passive earth pressure develops in front of the toe. The Rankine coefficient of passive earth pressure is

$$K_P = \frac{1 + \sin\phi}{1 - \sin\phi}$$

When, prior to placing the backfill, a basement wall is restrained at the top by a floor diaphragm and at the bottom by a basement floor slab, the wall is prevented from displacing laterally, and active earth pressure from a sliding wedge of soil cannot develop. The load on the wall is then the lateral pressure exerted by the soil at rest. The coefficient of at-rest earth pressure is

$$K_O = 1 - \sin\phi$$

The at-rest lateral soil pressure per foot of depth is

$$p_O = K_O \gamma_s$$

Where a specific soil investigation has not been performed, IBC Table 1610.1 gives values for the at-rest lateral soil pressure for a variety of soil types. Typical values are 60 lbf/ft^2 for granular backfill and 100 lbf/ft^2 for inorganic silt and clayey silt.

The wall may be designed as a propped cantilever with load increasing uniformly to the fixed end. The bending moment at the base of the wall is

$$M_b = -\frac{hH_O}{7.5} \quad \text{[tension on the earth face]}$$

The bending moment in the stem of the wall at a depth of $0.45h$ from the top of the wall is

$$M_s = +0.0596hH_O \quad \text{[tension on the air face]}$$

When basement walls extend not more than 8 ft below grade and are laterally supported at the top by flexible diaphragms, IBC Sec. 1610.1 permits the wall to be designed for active pressure.

Example 3.27

The basement wall shown in Fig. 3.33 has a height of 10 ft. The top of the wall is restrained by a rigid floor diaphragm, and the base of the wall may be considered fixed-ended. At-rest earth pressure is equivalent to a fluid pressure of 60 lbf/ft^2. Determine the bending moments produced at the base of the wall and in the stem.

Solution

Consider a 1 ft length of wall.

The total at-rest pressure on the back of the wall is

$$H_O = \frac{p_O h^2}{2}$$
$$= \frac{\left(60 \, \dfrac{\text{lbf}}{\text{ft}^2}\right)(10 \text{ ft})^2}{2}$$
$$= 3000 \text{ lbf}$$

The bending moment produced at the base of the wall is

$$M_b = -\frac{hH_O}{7.5}$$
$$= -\frac{(10 \text{ ft})(3000 \text{ lbf})}{7.5}$$
$$= -4000 \text{ ft-lbf} \quad \text{[tension on earth face]}$$

The bending moment produced in the stem of the wall is

$$M_s = 0.0596hH_O$$
$$= (0.0596)(10 \text{ ft})(3000 \text{ lbf})$$
$$= 1788 \text{ ft-lbf} \quad [\text{tension on air face}]$$

15. SOIL SURCHARGE LOADS

Nomenclature

a	length along which H_I is distributed	ft
h	height of wall	ft
h_e	equivalent height of fill	ft
H_I	lateral thrust due to isolated load	lbf
H_L	lateral line load	lbf/ft
K_A	Rankine coefficient of active earth pressure, $(1 - \sin \phi)/(1+\sin \phi)$	–
l	length of isolated load	ft
p_e	constant lateral pressure	lbf/ft^2
p_v	distributed vertical load	lbf/ft^2
w	uniformly distributed load	lbf/ft^2
W_I	isolated load	lbf
W_L	line load surcharge	lbf/ft
x	distance from top of wall	ft

Symbols

γ_S	specific weight of backfill	lbf/ft^3

General Principles

A *soil surcharge load* is a load that is applied to the top of the retained fill close enough to the retaining wall to cause a lateral pressure to act on the retaining wall. This lateral pressure is additional to the basic earth pressure. Surcharge loads are caused by adjacent buildings, construction machinery, material stockpiles, highways, and parking areas. The four principal types of surcharge loads are uniform load, line load, strip load, and isolated load.

Uniform Load

A uniformly distributed load of w (in pounds per square foot) applied to the retained fill, as shown in Fig. 3.34, is equivalent to an additional height of fill of

$$h_e = w/\gamma_S$$

Figure 3.34 *Lateral Pressure Due to Uniform Surcharge*

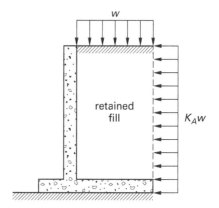

This produces a constant lateral pressure on the back of the wall of

$$p_e = K_A\gamma_S h_e$$
$$= K_A w$$

When the surcharge consists of a highway live load, with the edge of traffic 1 ft or more from the back face of the wall, AASHTO[8] Table 3.11.6.4-2 specifies that the surcharge is equivalent to an additional height of fill of $h_e = 2$ ft. This produces a constant lateral pressure on the back of the wall of

$$p_e = 2K_A\gamma_S$$

Line Load

The construction shown in Fig. 3.35(a) may be used to determine the lateral force produced by a line load surcharge of infinite length.[9] A line load, W_L, parallel to the retaining wall and located a distance x from the top of the wall, produces a lateral line load on the backface of the wall of

$$H_L = K_A W_L$$

This lateral load acts at a depth of $0.84x$ below the top of the wall.

As shown in Fig. 3.35(b), the line load surcharge also produces a distributed vertical load on the heel, or a portion of the heel, of the retaining wall, and this contributes to the overall stability of the wall. The distributed vertical load is

$$p_v = \frac{W_L}{1.15h}$$

Figure 3.35 *Line Load Surcharge*

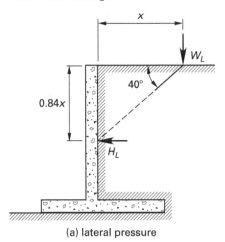

(a) lateral pressure

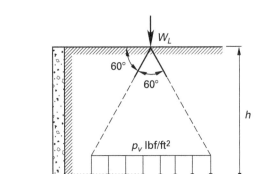

(b) vertical pressure

Strip Load

A strip load may be treated as a series of line loads.

Isolated Load

The effect of an isolated load surcharge may be determined by an adaptation of the line load surcharge method.[10] As shown in Fig. 3.36, the line of action of the lateral thrust is determined as for a line load. The magnitude of the thrust is

$$H_I = K_A W_I$$

For an isolated load with a length l, the thrust is assumed distributed longitudinally along a length of wall given by

$$a = x + l$$

Figure 3.36 *Lateral Pressure Due to an Isolated Load Surcharge*

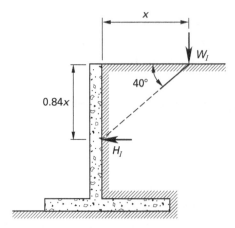

16. MODULUS OF SUBGRADE REACTION

Nomenclature

K_s	modulus of subgrade reaction	lbf/ft^3, psf/ft
p	bearing pressure	lbf/ft^2
s	settlement	ft

General Principles

The *modulus of subgrade reaction*, K_s, represents the ratio of bearing pressure, p, to corresponding settlement, s, of a soil and is expressed mathematically[11] as

$$K_s = \frac{p}{s}$$

In effect, K_s is the pressure per unit of settlement and is determined by means of a plate bearing test. The concept of the modulus of subgrade reaction is applied to the design of raft foundations, settlement of piles, and horizontal deflection of sheet piles.

17. REFERENCES

1. Portland Cement Association. *Rectangular Concrete Tanks*. Skokie, IL: Portland Cement Association, 1998.

2. Reynolds, Charles E., James C. Steedman, and Anthony J. Threlfall. *Reinforced Concrete Designer's Handbook*. Wexham Springs, UK: Cement and Concrete Association, 1981.

3. Hsiao, J. K. Statical Analysis of Pile Groups Containing Batter Piles. University of Melbourne, Australia. Electronic Journal of Structural Engineering, Vol. 12, 2012.

4. American Wood Preservers Institute. *Timber Pile Design and Construction Manual*. Fairfax, VA: American Wood Preservers Institute, 2015.

5. American Wood Council. *National Design Specification Supplement: Design Values for Wood Construction*. Leesburg, VA: American Wood Council, 2018.

6. United States Steel. *Steel Sheet Piling Design Manual*. Pittsburgh, PA: United States Steel, 1984.

7. International Code Council. *International Building Code*. Falls Church, VA: International Code Council, 2018.

8. *AASHTO LRFD Bridge Design Specifications*, Eighth ed., 2017, PE/SE Exam Edition. Washington, DC: American Association of State Highway and Transportation Officials.

9. Terzaghi, K. and Peck, R. B. *Soil Mechanics in Engineering Practice*. New York, NY: J. Wiley & Sons Inc., 1948.

10. C. P. 2. Civil Engineering Code of Practice No. 2. *Earth Retaining Structures*. London, UK: Institution of Civil Engineers, 1951.

11. Tribedi, A. Correlation between Soil Bearing Capacity and Modulus of Subgrade Reaction. *Structure Magazine*, Chicago, IL, December 2013.

Prestressed Concrete Design

1. DESIGN STAGES

Nomenclature

a	depth of equivalent rectangular stress block	in
A_{ct}	area of concrete section between the centroid and extreme tension fiber	in^2
A_g	area of concrete section	in^2
A_{ps}	area of prestressed reinforcement in tension zone	in^2
A_s	area of nonprestressed tension reinforcement	in^2
A_s'	area of compression reinforcement	in^2
b	width of compression face of member	in
c	distance from extreme compression fiber to neutral axis	in
C_u	total compression force in equivalent rectangular stress block	lbf
d	distance from extreme compression fiber to centroid of nonprestressed reinforcement	in
d'	distance from extreme compression fiber to centroid of compression reinforcement	in
d_p	distance from extreme compression fiber to centroid of prestressed reinforcement as defined in Fig. 4.8	in
e	eccentricity of prestressing force	in
E_c	modulus of elasticity of concrete	kips/in^2
E_p	modulus of elasticity of prestressing tendon	kips/in^2
f_{be}	bottom fiber stress at service load after allowance for all prestress losses	lbf/in^2
f_{bi}	bottom fiber stress immediately after prestress transfer and before time-dependent prestress losses	lbf/in^2
f_c'	specified compressive strength of concrete	lbf/in^2

f_{ci}'	compressive strength of concrete at time of prestress transfer	lbf/in^2
f_{ps}	stress in prestressed reinforcement at nominal strength	kips/in^2
f_{pu}	specified tensile strength of prestressing tendons	kips/in^2
f_{py}	specified yield strength of prestressing tendons	kips/in^2
f_r	modulus of rupture of concrete	lbf/in^2
f_s	permissible stress in prestressed reinforcement at the jacking end	kips/in^2
f_{se}	effective stress in prestressed reinforcement after allowance for all prestress losses	kips/in^2
f_{si}	stress in prestressed reinforcement immediately after prestress transfer	kips/in^2
f_{te}	top fiber stress at service loads after allowance for all prestress losses	lbf/in^2
f_{ti}	top fiber stress immediately after prestress transfer and before time-dependent prestress losses	lbf/in^2
f_y	specified yield strength of nonprestressed reinforcement	kips/in^2
h	height of section	in
I_g	moment of inertia of gross concrete section	in^4
l	span length	ft
M_{cr}	cracking moment strength	ft-kips
M_D	bending moment due to superimposed dead load	ft-kips
M_G	bending moment due to self-weight of member	ft-kips
M_L	bending moment due to superimposed live load	ft-kips
M_n	nominal flexural strength	ft-kips
M_S	bending moment due to sustained load	ft-kips
M_T	bending moment due to total load	ft-kips
M_u	factored moment	ft-kips
P_e	prestressing force after all losses	kips
P_i	initial prestressing force	kips
r	distance of tendon from the neutral axis	in
S_b	section modulus of the concrete section referred to the bottom fiber	in^3
S_t	section modulus of the concrete section referred to the top fiber	in^3
w_c	unit weight of concrete	lbf/ft^3
$\bar{y}$	height of centroid of the concrete section	in

Prestressed Concrete

Symbols

β_1	compression zone factor	–
γ_p	factor for type of prestressing tendon	–
ϵ_c	strain at extreme compression fiber at nominal strength, 0.003	–
ϵ_p	prestrain in prestressed reinforcement due to the final prestress	–
ϵ_s	strain produced in prestressed reinforcement by the ultimate loading	–
ϵ_t	strain at extreme tension steel at nominal strength	–
λ	correction factor related to unit weight of concrete as given by ACI Sec. 19.2.4.1	–
ϕ	strength reduction factor	–

General Requirements

There are three major design stages that must be investigated when designing prestressed concrete members.

- *Transfer design stage*

  At the transfer design stage, a prestressing force is transferred from a hydraulic jack to the concrete member. Immediate prestress losses occur due to elastic deformation of the concrete and, in the case of post-tensioned concrete, due to anchor set and friction losses. At this stage, the prestressing force is at its maximum, the concrete compressive strength and the dead load are at their minimums, and the live load is zero. The member must be designed so that stresses produced by the applied forces do not exceed the allowable values. Since the transfer stresses are temporary, the allowable values are higher than they are for the serviceability stage. The applied forces consist of the prestressing force and the self-weight of the member.

- *Serviceability design stage*

  At the serviceability design stage, all time-dependent prestress losses have occurred due to the concrete's creep and shrinkage and the relaxation of the tendon stress. At this stage, the prestressing force is at its minimum, and the concrete compressive strength, dead load, and live load are at their maximums. The member must be designed so that stresses produced by the applied forces do not exceed the allowable values.

- *Strength design stage*

  In the strength design stage, the design strength of the member is calculated. It must not be less than the required strength of the member, which is calculated using the most critical combination of the factored loads. At this stage, a rectangular stress block is assumed with a maximum strain in the concrete of 0.003.

Transfer Design Stage

The permissible stresses at transfer are specified in ACI Table 24.5.3.1 and Table 24.5.3.2. As shown in Fig. 4.1, the initial prestressing force mobilizes the self-weight of the member producing the stresses. The permissible stresses are shown in Fig. 4.2. A minus sign indicates tension.

$$f_{ti} = P_i\left(\frac{1}{A_g} - \frac{e}{S_t}\right) + \frac{M_G}{S_t} = P_i R_t + \frac{M_G}{S_t}$$

$$\geq -6\sqrt{f_{ci}'} \quad \left[\begin{array}{l}\text{at ends of simply supported beams}\\ \text{without auxiliary reinforcement}\end{array}\right]$$

$$\geq -3\sqrt{f_{ci}'} \quad \left[\begin{array}{l}\text{at all other locations without}\\ \text{auxiliary reinforcement}\end{array}\right]$$

P_i = force in prestressing tendon immediately
  after prestress transfer
  $= A_{ps}f_{si}$

$$R_t = \frac{1}{A_g} - \frac{e}{S_t}$$

$$f_{bi} = P_i\left(\frac{1}{A_g} + \frac{e}{S_b}\right) - \frac{M_G}{S_b} = P_i R_b - \frac{M_G}{S_b}$$

$$\leq 0.70 f_{ci}' \quad \text{[at ends of simply supported beams]}$$

$$\leq 0.60 f_{ci}' \quad \text{[at all other locations]}$$

$$R_b = \frac{1}{A_g} + \frac{e}{S_b}$$

In accordance with ACI Table 24.5.3.1 and Table 24.5.3.2, the permissible stress should not exceed the following. At post-tensioning anchorages and couplers, the permissible stress is

$$f_{si} = 0.70 f_{pu}$$

The maximum permissible stress due to the tendon jacking force is

$$f_s = 0.94 f_{py}$$
$$\leq 0.80 f_{pu}$$

The permissible tendon stresses are shown in Fig. 4.3.

Example 4.1

The pretensioned beam of normal weight concrete shown in *Illustration for Ex. 4.1* is simply supported over a span of 20 ft and has a concrete strength at transfer of 4500 lbf/in^2. Determine the magnitude and location of the initial prestressing force required to produce satisfactory stresses at midspan, immediately after transfer, without using auxiliary reinforcement.

Figure 4.1 *Transfer Design Stage*

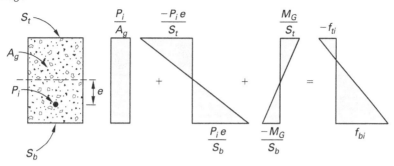

Figure 4.2 *Limiting Permissible Concrete Stress at Transfer*

$-6\sqrt{f'_c}$...at member ends without auxiliary reinforcement
$-3\sqrt{f'_c}$...at other locations without auxiliary reinforcement

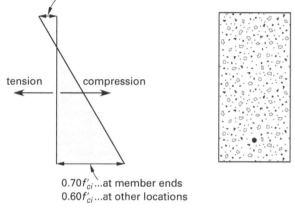

0.70f'_{ci}...at member ends
0.60f'_{ci}...at other locations

Solution

The properties of the concrete section are

$$A_g = 72 \text{ in}^2$$
$$I_g = 863 \text{ in}^4$$
$$\bar{y} = 4.67 \text{ in}$$
$$S_t = 118 \text{ in}^3$$
$$S_b = 185 \text{ in}^3$$

At midspan, the self-weight moment is

$$M_G = \frac{w_c A_g l^2}{8}$$

$$= \frac{\left(150 \dfrac{\text{lbf}}{\text{ft}^3}\right)(72 \text{ in}^2)(20 \text{ ft})^2}{(8)\left(12 \dfrac{\text{in}}{\text{ft}}\right)}$$

$$= 45{,}000 \text{ in-lbf}$$

At midspan, the permissible tensile stress in the top fiber without auxiliary reinforcement is given by ACI Table 24.5.3.1 as

$$f_{ti} = -3\sqrt{f'_{ci}} = -3\sqrt{4500 \dfrac{\text{lbf}}{\text{in}^2}}$$

$$= -201 \text{ lbf/in}^2$$

$$= \frac{P_i}{A_g} - \frac{P_i e}{S_t} + \frac{M_G}{S_t}$$

$$= \frac{P_i}{72 \text{ in}^2} - \frac{P_i e}{118 \text{ in}^3} + \frac{45{,}000 \text{ in-lbf}}{118 \text{ in}^3}$$

$$-582 \frac{\text{lbf}}{\text{in}^2} = \frac{P_i}{72 \text{ in}^2} - \frac{P_i e}{118 \text{ in}^3} \quad [\text{Eq. 1}]$$

At midspan, the permissible compressive stress in the bottom fiber is given by ACI Table 24.5.3.1 as

$$f_{bi} = 0.6f'_{ci} = 2700 \text{ lbf/in}^2$$

$$= \frac{P_i}{A_g} + \frac{P_i e}{S_b} - \frac{M_G}{S_b}$$

$$= \frac{P_i}{72 \text{ in}^2} + \frac{P_i e}{185 \text{ in}^3} - \frac{45{,}000 \text{ in-lbf}}{185 \text{ in}^3}$$

$$2943 \frac{\text{lbf}}{\text{in}^2} = \frac{P_i}{72 \text{ in}^2} + \frac{P_i e}{185 \text{ in}^3} \quad [\text{Eq. 2}]$$

Solving Eq. [1] and Eq. [2] gives

$$P_i = 113{,}056 \text{ lbf}$$
$$e = 2.25 \text{ in}$$

Auxiliary Reinforcement

ACI Sec. 24.5.3.2.1 specifies that when the computed tensile stress exceeds the permissible stress, bonded auxiliary reinforcement must be provided to resist the total tensile force in the concrete. The tensile force is computed by using the properties of the uncracked concrete section, and in accordance with ACI Sec. R24.5.3.2, the permissible stress in the auxiliary reinforcement is $0.6f_y$ or 30 kips/in^2 maximum.

Illustration for Ex. 4.1

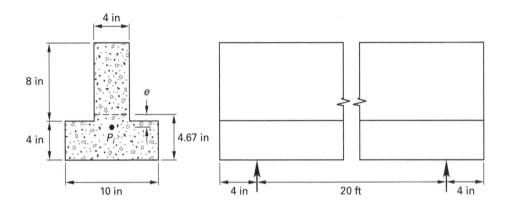

Figure 4.3 *Specified Stress in Prestressing Tendons*

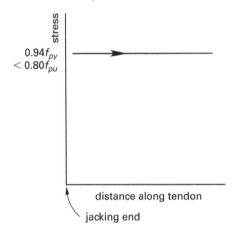

From Fig. 4.4, the depth to the location of zero stress is given by

$$c = \frac{hf_t}{f_t + f_b}$$

Figure 4.4 *Determination of Tensile Force*

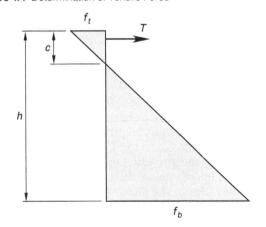

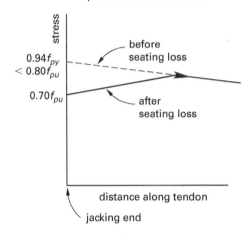

The tensile force in the concrete is

$$T = \frac{cf_t b}{2}$$

The area of auxiliary reinforcement required is given by ACI Sec. R24.5.3.2 as

$$A_s = \frac{T}{0.6f_y} \geq \frac{T}{30 \ \dfrac{\text{kips}}{\text{in}^2}}$$

Example 4.2

The pretensioned beam of Ex. 4.1 is prestressed with tendons providing an initial prestressing force of 110,100 lbf at an eccentricity of 2.37 in at midspan. Determine the area of grade 60 auxiliary reinforcement required.

Solution

$$R_b = \frac{1}{A_g} + \frac{e}{S_b}$$

$$= \frac{1}{72 \text{ in}^2} + \frac{2.37 \text{ in}}{185 \text{ in}^3}$$

$$= 0.0267 \ 1/\text{in}^2$$

$$R_t = \frac{1}{A_g} - \frac{e}{S_t}$$

$$= \frac{1}{72 \text{ in}^2} - \frac{2.37 \text{ in}}{118 \text{ in}^3}$$

$$= -0.0062 \ 1/\text{in}^2$$

The top and bottom fiber stresses are

$$f_t = P_i R_t + \frac{M_G}{S_t}$$

$$= (110{,}100 \text{ lbf})\left(-0.0062 \ \frac{1}{\text{in}^2}\right)$$

$$+ \frac{45{,}000 \text{ in-lbf}}{118 \text{ in}^3}$$

$$= -301 \text{ lbf/in}^2$$

This is less than the minimum permissible value of $-3\sqrt{f_c'} = -201 \text{ lbf/in}^2$ determined in Ex. 4.1, and auxiliary reinforcement is required.

$$f_b = P_i R_b - \frac{M_G}{S_b}$$

$$= (110{,}100 \text{ lbf})\left(0.0267 \ \frac{1}{\text{in}^2}\right) - \frac{45{,}000 \text{ in-lbf}}{185 \text{ in}^3}$$

$$= 2696 \text{ lbf/in}^2$$

This is less than the maximum permissible value of $0.6f_{ci}' = 2700 \text{ lbf/in}^2$ determined in Ex. 4.1 and is satisfactory.

Depth to the neutral axis is obtained from Fig. 4.4 as

$$c = \frac{hf_t}{f_t + f_b}$$

$$= \frac{(12 \text{ in})\left(301 \ \dfrac{\text{lbf}}{\text{in}^2}\right)}{301 \ \dfrac{\text{lbf}}{\text{in}^2} + 2696 \ \dfrac{\text{lbf}}{\text{in}^2}}$$

$$= 1.21 \text{ in}$$

The tensile force in the concrete is

$$T = \frac{cf_t b}{2} = \frac{(1.21 \text{ in})\left(301 \ \dfrac{\text{lbf}}{\text{in}^2}\right)(4 \text{ in})}{2}$$

$$= 728 \text{ lbf}$$

The area of auxiliary reinforcement required is obtained from ACI Sec. R24.5.3.2 as

$$A_s = \frac{T}{30{,}000 \ \dfrac{\text{lbf}}{\text{in}^2}}$$

$$= 0.024 \text{ in}^2$$

Serviceability Design Stage

The limiting stresses under service loads after all prestressing losses have occurred are specified in ACI Table 24.5.4.1 and Table 24.5.2.1. The stress conditions are shown in Fig. 4.5, and the stresses are given by

$$f_{te} = P_e R_t + \frac{M_G + M_D + M_L}{S_t}$$

$$\le 0.45 f_c' \quad \text{[for sustained loads]}$$

$$\le 0.60 f_c' \quad \text{[for total loads]}$$

$$f_{be} = P_e R_b - \frac{M_G + M_D + M_L}{S_b}$$

$$= -7.5\sqrt{f_c'} \quad \text{[for class U member]}$$

$$= -12\sqrt{f_c'} \quad \text{[for class T member]}$$

$$< -12\sqrt{f_c'} \quad \text{[for class C member]}$$

Class U members are assumed to behave as uncracked members. Class C members are assumed to behave as cracked members. Class T members are assumed to be in transition between uncracked and cracked.

In accordance with ACI Sec. 24.5.2.2 and 24.5.2.3, stresses at the serviceability design stage in class U and class T members may be computed using uncracked section properties, and no crack control measures are necessary. Stresses in class C members are computed using cracked section properties, and crack control measures are necessary as specified in ACI Sec. 24.3. Deflections for class U members are based on uncracked section properties, and for class T and class C members are based on the cracked transformed section properties, as specified in ACI Sec. 24.2.3.8 and Sec. 24.2.3.9.

$$P_e = \text{force in prestressing tendon at service}$$
$$\text{loads after allowance for all losses}$$
$$= A_{ps} f_{se}$$

Figure 4.5 *Serviceability Design Stage After all Losses*

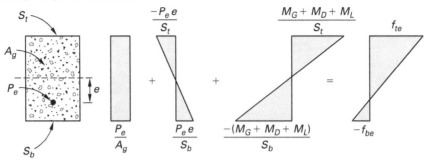

Figure 4.6 *Limiting Permissible Concrete Stress at Service Load*

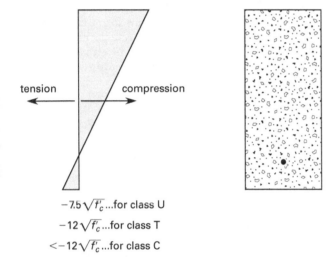

$0.45f'_c$ ...for sustained loads

$0.60f'_c$ ...for total loads

tension ← | → compression

$-7.5\sqrt{f'_c}$ ...for class U

$-12\sqrt{f'_c}$ ...for class T

$<-12\sqrt{f'_c}$ ...for class C

The permissible stresses are shown in Fig. 4.6.

Example 4.3

The class U pretensioned beam of Ex. 4.1 has a long-term loss in prestress of 25% and a 28-day compressive strength of 6000 lbf/in². The initial prestressing force is $P_i = 112{,}850$ lbf with an eccentricity of $e = 2.25$ in. Determine the maximum bending moment the beam can carry if the sustained load is 75% of the total superimposed load.

The relevant parameters are $e = 2.25$ in, $R_b = 0.0261$ in^{-2}, $P_i = 112{,}850$ lbf, $P_i R_b = 2940$ lbf/in², $P_i R_t = -584$ lbf/in², and $R_t = -0.00518$ in^{-2}.

Solution

The permissible tensile stress at midspan, in the bottom fiber, due to the total load is given by ACI Table 24.5.2.1 as

$$f_{be} = -7.5\sqrt{f'_c} = -7.5\sqrt{6000 \ \frac{\text{lbf}}{\text{in}^2}}$$

$$= -581 \ \text{lbf/in}^2 = P_e R_b - \frac{M_G}{S_b} - \frac{M_T}{S_b}$$

$$= (0.75)\left(2940 \ \frac{\text{lbf}}{\text{in}^2}\right) - 243 \ \frac{\text{lbf}}{\text{in}^2} - \frac{M_T}{185 \ \text{in}^3}$$

$$M_T = 470{,}455 \ \text{in-lbf}$$

The permissible compressive stress at midspan, in the top fiber, due to the sustained load is given by ACI Table 24.5.4.1 as

$$f_{te} = 0.45f'_c = (0.45)\left(6000 \ \frac{\text{lbf}}{\text{in}^2}\right)$$

$$= 2700 \ \text{lbf/in}^2$$

$$= P_e R_t + \frac{M_G}{S_t} + \frac{M_S}{S_t}$$

$$= (0.75)\left(-584 \ \frac{\text{lbf}}{\text{in}^2}\right) + 381 \ \frac{\text{lbf}}{\text{in}^2} + \frac{0.75 M_T}{118 \ \text{in}^3}$$

$$M_T = 433{,}770 \ \text{in-lbf}$$

The permissible compressive stress at midspan, in the top fiber, due to the total load is given by ACI Table 24.5.4.1 as

$$f_{te} = 0.60f'_c = (0.60)\left(6000 \ \frac{\text{lbf}}{\text{in}^2}\right)$$

$$= 3600 \ \text{lbf/in}$$

$$= P_e R_t + \frac{M_G}{S_t} + \frac{M_T}{S_t}$$

$$= (0.75)\left(-584 \ \frac{\text{lbf}}{\text{in}^2}\right) + 381 \ \frac{\text{lbf}}{\text{in}^2} + \frac{M_T}{118 \ \text{in}^3}$$

$$M_T = 431{,}530 \ \text{in-lbf} \quad [\text{governs}]$$

Figure 4.7 *Cracking Moment*

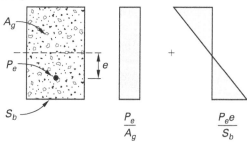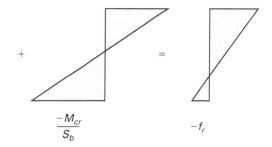

Cracking Moment

The cracking moment is the moment that, when applied to the member after all losses have occurred, will cause cracking in the bottom fiber. From Fig. 4.7, equating the bottom fiber stresses gives a value for the modulus of rupture of

$$f_r = \frac{M_{cr}}{S_b} - P_e R_b = 7.5\lambda\sqrt{f_c'} \qquad \text{[ACI 19.2.3.1]}$$
$$M_{cr} = S_b(P_e R_b + f_r)$$

As specified in ACI Sec. 9.6.2.1, a prestressed beam with bonded tendons must have adequate reinforcement to support an applied factored moment (M_u) at least 1.2 times the cracking moment. This requirement is waived for members with both shear and flexural design strengths at least twice the required strength.

$$\phi M_n \geq 1.2 M_{cr}$$

Example 4.4

For the pretensioned beam of Ex. 4.3, determine the cracking moment strength.

Solution

The modulus of rupture is given by ACI Sec. 19.2.3.1 as

$$f_r = 7.5\lambda\sqrt{f_c'}$$
$$= (7.5)(1.0)\sqrt{6000 \ \frac{\text{lbf}}{\text{in}^2}}$$
$$= 581 \ \text{lbf/in}^2$$

The cracking moment strength is

$$M_{cr} = S_b(P_e R_b + f_r)$$
$$= (185 \ \text{in}^3)\left(\begin{matrix} (0.75)\left(2940 \ \dfrac{\text{lbf}}{\text{in}^2}\right) \\ + \ 581 \ \dfrac{\text{lbf}}{\text{in}^2} \end{matrix} \right)$$
$$= 515{,}410 \ \text{in-lbf}$$

Strength Design Stage

The modulus of rupture is given by ACI Sec. 19.2.3.1 and is assumed to contribute to the ultimate moment of resistance of the section at its yield strength. Equating the longitudinal forces shown in Fig. 4.8 gives

$$0.85 f_c' ab = A_{ps} f_{ps} + A_s f_y - A_s' f_y$$

The nominal flexural strength of the member is

$$M_n = A_{ps} f_{ps}\left(d_p - \frac{a}{2}\right) + A_s f_y\left(d - \frac{a}{2}\right) + A_s' f_y\left(\frac{a}{2} - d'\right)$$

Using the notation in Fig. 4.8,

$$a = \frac{A_{ps} f_{ps} + A_s f_y - A_s' f_y}{0.85 f_c' b}$$

When the section does not contain compression reinforcement

$$M_n = A_{ps} f_{ps}\left(d_p - \frac{a}{2}\right) + A_s f_y\left(d - \frac{a}{2}\right)$$
$$a = \frac{A_{ps} f_{ps} + A_s f_y}{0.85 f_c' b}$$

When the section contains neither compression nor auxiliary reinforcement the depth of the stress block is

$$a = \frac{A_{ps} f_{ps}}{0.85 f_c' b}$$

The nominal flexural strength is

$$M_n = A_{ps} f_{ps}\left(d_p - \frac{a}{2}\right)$$
$$= A_{ps} f_{ps}\left(d_p - \frac{0.59 A_{ps} f_{ps}}{f_c' b}\right)$$

Prestressed Concrete

Figure 4.8 *Strain Distribution and Internal Forces at Flexural Failure*

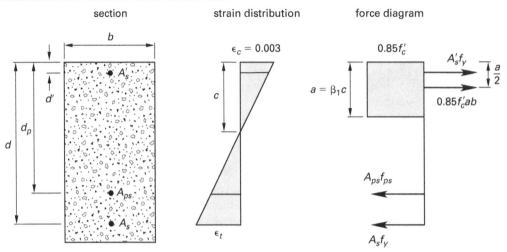

Tension- and Compression-Controlled Sections

In ACI Sec. R21.2.2, a section is defined as tension-controlled if the net tensile strain in the extreme tension steel, ϵ_t, is not less than 0.005 when the concrete reaches its maximum usable compressive strain of 0.003. ACI Table 21.2.2 gives the strength reduction factor as

$$\phi = 0.9$$

The following relationships may be derived from Fig. 4.9 using simple geometry.

$$\epsilon_t = 0.005$$

$$\frac{c}{d_p} = 0.375$$

$$a = 0.375\beta_1 d_p$$

In ACI Sec. R21.2.2, a section is defined as compression-controlled if the net tensile strain in the extreme tension steel is not more than 0.002 when the concrete reaches a strain of 0.003. The strength reduction factor is then given by ACI Table 21.2.2 as

$$\phi = 0.65$$

The following relationships may be derived from Fig. 4.9 using simple geometry.

$$\epsilon_t = 0.002$$

$$\frac{c}{d_p} = 0.600$$

$$a = 0.600\beta_1 d_p$$

Members having a tensile strain between 0.002 and 0.005 are in the transition zone. (See Fig. 4.10.) The strength reduction factor in the transition zone may be interpolated from Fig. 4.10.

Figure 4.9 *Strain Distribution at Nominal Strength*

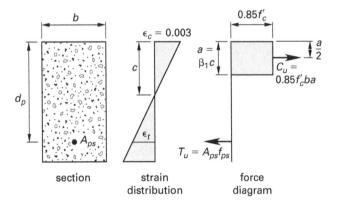

Figure 4.10 *Variation of ϕ with ϵ_t*

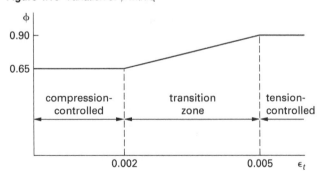

Example 4.5

For the pretensioned beam of Ex. 4.1, which has a 28-day compressive strength of 6000 lbf/in², determine the maximum possible value of the nominal flexural strength for a tension-controlled section.

Solution

The height of the centroid of the section is given in Ex. 4.1 as

$$\bar{y} = 4.67 \text{ in}$$

The eccentricity of the prestressing force is given in Ex. 4.1 as

$$e = 2.25 \text{ in}$$

The height of the section is given in Ex. 4.1 as

$$h = 12 \text{ in}$$

Then, the distance from the extreme compression fiber to the centroid of prestressed reinforcement is given by

$$d_p = h - \bar{y} + e$$
$$= 12 \text{ in} - 4.67 \text{ in} + 2.25 \text{ in}$$
$$= 9.58 \text{ in}$$

The maximum depth of the rectangular stress block for a tension-controlled section is given by ACI Sec. 22.2.2 as

$$a = 0.375\beta_1 d_p$$
$$= (0.375)(0.75)(9.58 \text{ in})$$
$$= 2.69 \text{ in}$$

The maximum nominal flexural strength is

$$M_n = 0.85 f_c' ab\left(d_p - \frac{a}{2}\right)$$
$$= (0.85)\left(6000 \ \frac{\text{lbf}}{\text{in}^2}\right)(2.69 \text{ in})(4 \text{ in})$$
$$\times \left(9.58 \text{ in} - \frac{2.69 \text{ in}}{2}\right)$$
$$= 451{,}903 \text{ in-lbf}$$
$$\phi M_n < 1.2 M_{cr} \quad \text{[unsatisfactory]}$$

Flexural Strength of Members with Bonded Tendons

Approximate values of f_{ps} in terms of the reinforcement index may be determined in accordance with ACI Sec. 20.3.2.3.1, provided that $f_{se} \geq 0.5 f_{pu}$ and that all the prestressing tendons are located in the tensile zone and are bonded.

The reinforcement indices are

ω = reinforcement index of nonprestressed tension reinforcement
$$= \frac{\rho f_y}{f_c'}$$

ω' = reinforcement index of compression reinforcement
$$= \frac{\rho' f_y}{f_c'}$$

The reinforcement ratios are given by ACI Sec. 2.2 as

ρ = ratio of nonprestressed tension reinforcement
$$= \frac{A_s}{bd}$$

ρ' = ratio of compression reinforcement
$$= \frac{A_s'}{bd}$$

ρ_p = ratio of prestressed reinforcement
$$= \frac{A_{ps}}{bd_p}$$

The value of the stress in the prestressed reinforcement at nominal strength is given by a simplified form of ACI Eq. 20.3.2.3.1 as

$$f_{ps} = f_{pu}\left[1 - \left(\frac{\gamma_p}{\beta_1}\right)\left(\rho_p\left(\frac{f_{pu}}{f_c'}\right) + \frac{d(\omega - \omega')}{d_p}\right)\right]$$

The factor for type of prestressing tendon is given by ACI Table 20.3.2.3.1 as

$\gamma_p = 0.55$ for deformed bars with $f_{py}/f_{pu} \geq 0.80$
$\quad = 0.40$ for stress-relieved wire and strands, and plain bars with $f_{py}/f_{pu} \geq 0.85$
$\quad = 0.28$ for low-relaxation wire and strands with $f_{py}/f_{pu} \geq 0.90$

When compression reinforcement is taken into account while calculating f_{ps} by the simplified form of ACI Eq. 20.3.2.3.1,

$$0.17 \leq \left(\rho_p\left(\frac{f_{pu}}{f_c'}\right) + \frac{d(\omega - \omega')}{d_p}\right)$$
$$d' \leq 0.15 d_p$$

When the section contains no auxiliary reinforcement, the value for f_{ps} reduces to

$$f_{ps} = f_{pu}\left(1 - \frac{\gamma_p \rho_p f_{pu}}{\beta_1 f_c'}\right)$$

Example 4.6

The post-tensioned beam with bonded tendons shown is simply supported over a span of 30 ft and has a 28-day concrete strength of 6000 lbf/in². The area of the low-relaxation prestressing tendons provided is 0.765 in² with a specified tensile strength of 270 kips/in², a yield strength of 243 kips/in², and an effective stress of 150 kips/in² after all losses. Determine the nominal flexural strength of the beam.

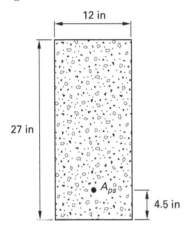

Solution

The relevant properties of the beam are

$$A_g = 324 \text{ in}^2$$
$$S_b = 1458 \text{ in}^3$$
$$e = \frac{h}{2} - 4.5 \text{ in} = \frac{27 \text{ in}}{2} - 4.5 \text{ in}$$
$$= 9 \text{ in}$$
$$R_b = \frac{1}{A_g} + \frac{e}{S_b} = \frac{1}{324 \text{ in}^2} + \frac{9 \text{ in}}{1458 \text{ in}^3}$$
$$= 0.00926 \text{ 1/in}^2$$

The factor for this type of prestressing tendon is given by ACI Table 20.3.2.3.1 as

$$\gamma_p = 0.28 \quad [\text{for } f_{py}/f_{pu} \geq 0.9]$$
$$\rho_p = \frac{A_{ps}}{bd_p} = \frac{0.765 \text{ in}^2}{(12 \text{ in})(22.5 \text{ in})}$$
$$= 0.00283$$

From ACI Table 22.2.2.4.3, the compression zone factor is given by

$$\beta_1 = 0.75$$
$$M_{cr} = S_b(P_e R_b + f_r)$$
$$= (1458 \text{ in}^3)\left(\begin{array}{c}(0.765 \text{ in}^2)\left(150 \dfrac{\text{kips}}{\text{in}^2}\right) \\ \times (0.00926 \text{ in}^{-2}) + 0.581 \dfrac{\text{kips}}{\text{in}^2}\end{array}\right)$$
$$= 2400 \text{ in-kips}$$

From ACI Eq. 20.3.2.3.1 for a member without auxiliary reinforcement,

$$f_{ps} = f_{pu}\left(1 - \frac{\gamma_p \rho_p f_{pu}}{\beta_1 f_c'}\right)$$
$$= \left(270 \dfrac{\text{kips}}{\text{in}^2}\right)\left(1 - \frac{(0.28)(0.00283) \times \left(270 \dfrac{\text{kips}}{\text{in}^2}\right)}{(0.75)\left(6 \dfrac{\text{kips}}{\text{in}^2}\right)}\right)$$
$$= 257 \text{ kips/in}^2$$

The depth of the stress block is given by

$$a = \frac{A_{ps} f_{ps}}{0.85 f_c' b}$$
$$= \frac{(0.765 \text{ in}^2)\left(257 \dfrac{\text{kips}}{\text{in}^2}\right)}{(0.85)\left(6 \dfrac{\text{kips}}{\text{in}^2}\right)(12 \text{ in})}$$
$$= 3.21 \text{ in}$$

The maximum depth of the stress block for a tension-controlled section is given by ACI Sec. R21.2.2 as

$$a_t = 0.375\beta_1 d_p$$
$$= (0.375)(0.75)(22.5 \text{ in})$$
$$= 6.33 \text{ in}$$
$$> a$$

Therefore, the section is tension-controlled and $\phi = 0.9$.

The nominal moment of resistance of the section is given by

$$
\begin{aligned}
M_n &= A_{ps}f_{ps}\left(d_p - \frac{0.59A_{ps}f_{ps}}{bf'_c}\right) \\
&= (0.765 \text{ in}^2)\left(257 \ \frac{\text{kips}}{\text{in}^2}\right) \\
&\quad \times\left(22.5 \text{ in} - \frac{\begin{array}{c}(0.59)(0.765 \text{ in}^2) \\ \times\left(257 \ \dfrac{\text{kips}}{\text{in}^2}\right)\end{array}}{(12 \text{ in})\left(6 \ \dfrac{\text{kips}}{\text{in}^2}\right)}\right) \\
&= 4106 \text{ in-kips} \\
\phi M_n &> 1.2M_{cr} \quad [\text{satisfactory}]
\end{aligned}
$$

Flexural Strength of Members with Unbonded Tendons

When the effective stress after all losses have occurred is not less than half the tensile stress of the tendon, ACI Sec. 20.3.2.4.1 permits the stress in the unbonded tendons at nominal strength to be calculated by one of the following methods.

- For unbonded tendons and a span-to-depth ratio $\leq$ 35, ACI Table 20.3.2.4.1 gives the stress in the tendons at nominal strength as

$$
\begin{aligned}
f_{ps} &= f_{se} + 10{,}000 \ \frac{\text{lbf}}{\text{in}^2} + \frac{f'_c}{100\rho_p} \\
&\leq f_{py} \\
&\leq f_{se} + 60{,}000 \ \frac{\text{lbf}}{\text{in}^2}
\end{aligned}
$$

f_{se}, f_{py}, and f'_c are in lbf/in^2.

- For unbonded tendons and a span-to-depth ratio $>$ 35, ACI Table 20.3.2.4.1 gives

$$
\begin{aligned}
f_{ps} &= f_{se} + 10{,}000 \ \frac{\text{lbf}}{\text{in}^2} + \frac{f'_c}{300\rho_p} \\
&\leq f_{py} \\
&\leq f_{se} + 30{,}000 \ \frac{\text{lbf}}{\text{in}^2}
\end{aligned}
$$

f_{se}, f_{py}, and f'_c are in lbf/in^2.

In accordance with ACI Sec. 9.6.2.3, auxiliary bonded reinforcement is required near the extreme tension fiber in all beams with unbonded tendons. The minimum area required is independent of the grade of steel and is given by

$$
A_s = 0.004A_{ct} \quad [\text{ACI 9.6.2.3}]
$$

A_{ct} is the area of the concrete section between the centroid of the section and the extreme tension fiber, as shown in Fig. 4.11.

Figure 4.11 Bonded Reinforcement Area

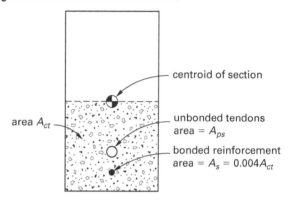

In flat slabs, when the tensile stress due to dead load plus live load is less than $-2\sqrt{f'_c}$, auxiliary reinforcement at a stress of $0.5f_y$ must be provided to resist the total tensile force in the concrete.

Example 4.7

The post-tensioned beam with bonded tendons shown is simply supported over a span of 30 ft and has a 28-day concrete strength of 6000 lbf/in^2. The area of the low-relaxation unbonded tendons provided is 0.765 in^2 with a specified tensile strength of 270 kips/in^2, a yield strength of 243 kips/in^2, and an effective stress of 150 kips/in^2 after all losses. The area of the grade 60 auxiliary reinforcement provided is 0.8 in^2. Determine the nominal flexural strength of the beam.

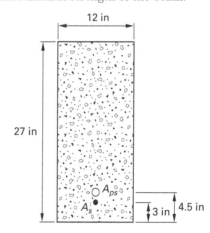

Solution

Because $f_{se}/f_{pu} > 0.5$, the method of ACI Sec. 20.3.2.4.1 may be used. The ratio of prestressed reinforcement is

$$\rho_p = \frac{A_{ps}}{bd_p} = \frac{0.765 \text{ in}^2}{(12 \text{ in})(22.5 \text{ in})}$$

$$= 0.00283$$

$l_n/h < 35$ and from ACI Table 20.3.2.4.1, the stress in the unbonded tendons at nominal strength is

$$f_{ps} = f_{se} + 10 \, \frac{\text{kips}}{\text{in}^2} + \frac{f_c'}{100\rho_p}$$

$$= 150 \, \frac{\text{kips}}{\text{in}^2} + 10 \, \frac{\text{kips}}{\text{in}^2} + \frac{6 \, \dfrac{\text{kips}}{\text{in}^2}}{(100)(0.00283)}$$

$$= 181 \text{ kips/in}^2$$

$$< f_{py} \quad \text{[satisfactory]}$$

$$< f_{se} + 60 \text{ kips/in}^2 \quad \text{[satisfactory]}$$

The minimum area of auxiliary reinforcement required is specified by ACI Sec. 9.6.2.3 as

$$A_s = 0.004 A_{ct} = (0.004)(12 \text{ in})(13.5 \text{ in})$$

$$= 0.648 \text{ in}^2$$

$$< 0.80 \text{ in}^2 \quad \text{[satisfactory]}$$

Assuming full use of the auxiliary reinforcement, the depth of the stress block is

$$a = \frac{A_{ps}f_{ps} + A_s f_y}{0.85 f_c' b}$$

$$= \frac{\begin{array}{c}(0.765 \text{ in}^2)\left(181 \, \dfrac{\text{kips}}{\text{in}^2}\right) \\[2mm] + (0.8 \text{ in}^2)\left(60 \, \dfrac{\text{kips}}{\text{in}^2}\right)\end{array}}{(0.85)\left(6 \, \dfrac{\text{kips}}{\text{in}^2}\right)(12 \text{ in})}$$

$$= 3.04 \text{ in}$$

The maximum depth of the stress block for a tension-controlled section is

$$a_t = 0.375 \beta_1 d_p = (0.375)(0.75)(22.5 \text{ in})$$

$$= 6.33 \text{ in}$$

$$> a \quad \text{[section is tension-controlled]}$$

The nominal flexural strength is

$$M_n = A_{ps}f_{ps}\left(d_p - \frac{a}{2}\right) + A_s f_y\left(d - \frac{a}{2}\right)$$

$$= (0.765 \text{ in}^2)\left(181 \, \frac{\text{kips}}{\text{in}^2}\right)\left(22.5 \text{ in} - \frac{3.04 \text{ in}}{2}\right)$$

$$+ (0.80 \text{ in}^2)\left(60 \, \frac{\text{kips}}{\text{in}^2}\right)\left(24 \text{ in} - \frac{3.04 \text{ in}}{2}\right)$$

$$= 3980 \text{ in-kips}$$

$$\phi M_n > 1.2 M_{cr} \quad \text{[satisfactory]}$$

Flexural Strength of Members Using Strain Compatibility

When the approximate methods of determining flexural strength cannot be used, a member's flexural strength can be determined from the strain compatibility method using Fig. 4.12 and the following steps.

step 1: Per ACI 318, use 0.003 as the concrete's maximum strain, ϵ.

step 2: Make an initial estimate of the depth, c, to the neutral axis.

step 3: Calculate the strain in the tendons and auxiliary reinforcement from

$$\epsilon_s = 0.003 \frac{r}{c}$$

step 4: To the strain in the tendons, add the pre-existing strain due to the final prestress which is given by

$$\epsilon_p = \frac{f_{se}}{E_p}$$

step 5: Determine the stresses in the tendons and auxiliary reinforcement from the appropriate stress-strain curve, and calculate the forces in the tendons and auxiliary reinforcement.

step 6: Compare the total tensile force, T, with the compressive force, C_u, in the concrete, and adjust the location of the neutral axis until

$$\sum T = C_u$$

step 7: Sum the moments of forces about the neutral axis to determine the required flexural strength.

Figure 4.12 *Flexural Strength by Strain Compatibility*

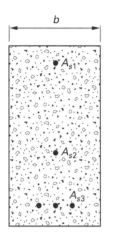
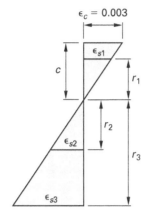
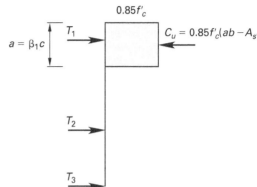

section   strain distribution   force diagram

Example 4.8

The pretensioned beam shown in *Illustration for Ex. 4.8* has a 28-day concrete strength of 6000 lbf/in² and is pretensioned with five ½ in diameter strands. The area of each strand is 0.153 in² with a specified tensile strength of 270 kips/in² and an effective stress of 150 kips/in² after all losses. Using the idealized stress-strain curve shown, determine the nominal flexural strength of the beam.

Solution

From the illustration, the depth to the neutral axis is $c = 4$ in. Then, the depth of the equivalent stress block is

$$a = \beta_1 c = (0.75)(4 \text{ in}) = 3 \text{ in}$$

The total compressive force in the concrete stress block is obtained from Fig. 4.12 as

$$C_u = 0.85 f_c'(ab - A_{s1})$$
$$= (0.85)\left(6 \; \frac{\text{kips}}{\text{in}^2}\right)\left((3 \text{ in})(12 \text{ in}) - 0.153 \text{ in}^2\right)$$
$$= 183 \text{ kips}$$

For an effective final prestress in each tendon of 150 kips/in², the prestrain in each tendon is

$$\epsilon_p = \frac{f_{se}}{E_p} = \frac{150 \; \dfrac{\text{kips}}{\text{in}^2}}{28{,}000 \; \dfrac{\text{kips}}{\text{in}^2}} = 5.36 \times 10^{-3}$$

The total strain in each tendon is given by

$$\epsilon_{\text{tot}} = \epsilon_s + \epsilon_p = 0.003\frac{r}{c} + \epsilon_p$$

The tendons reach their specified tensile strength at a strain of

$$\epsilon_{pu} = 14 \times 10^{-3} \quad \text{[from stress-strain curve]}$$
$$\epsilon_{s1} = (0.003)\left(\frac{-2.5 \text{ in}}{4 \text{ in}}\right) + 5.36 \times 10^{-3}$$
$$= 3.49 \times 10^{-3}$$
$$\epsilon_{s2} = (0.003)\left(\frac{18.5 \text{ in}}{4 \text{ in}}\right) + 5.36 \times 10^{-3}$$
$$= 19.24 \times 10^{-3} \quad \text{[exceeds } \epsilon_{pu}\text{]}$$
$$\epsilon_{s3} = (0.003)\left(\frac{21.5 \text{ in}}{4 \text{ in}}\right) + 5.36 \times 10^{-3}$$
$$= 21.49 \times 10^{-3} \quad \text{[exceeds } \epsilon_{pu}\text{]}$$

The force in each tendon is given by

$$T = A_s f_s$$
$$T_1 = A_{s1}\epsilon_{s1}E_p$$
$$= (0.153 \text{ in}^2)(3.49 \times 10^{-3})\left(28{,}000 \; \frac{\text{kips}}{\text{in}^2}\right)$$
$$= 15 \text{ kips}$$
$$T_2 = (0.153 \text{ in}^2)\left(270 \; \frac{\text{kips}}{\text{in}^2}\right)$$
$$= 41 \text{ kips}$$
$$T_3 = (3)(0.153 \text{ in}^2)\left(270 \; \frac{\text{kips}}{\text{in}^2}\right)$$
$$= 124 \text{ kips}$$
$$\sum T = 15 \text{ kips} + 41 \text{ kips} + 124 \text{ kips}$$
$$= 180 \text{ kips}$$
$$\approx C_u \quad \text{[satisfactory]}$$

Prestressed Concrete

Illustration for Ex. 4.8

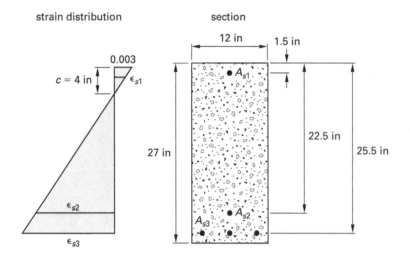

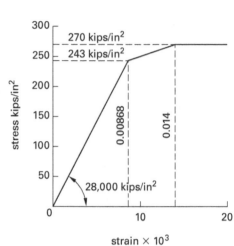

Taking moments about the neutral axis gives

$$M_n = T_{ps1}(-r_1) + T_{s2}r_2 + T_{s3}r_3 + C_u\left(c - \frac{a}{2}\right)$$
$$= (15 \text{ kips})(-2.5 \text{ in}) + (41 \text{ kips})(18.5 \text{ in})$$
$$+ (124 \text{ kips})(21.5 \text{ in})$$
$$+ (183 \text{ kips})\left(4 \text{ in} - \frac{3 \text{ in}}{2}\right)$$
$$= 3845 \text{ in-kips}$$

2. DESIGN FOR SHEAR

Nomenclature

A_{ps}	area of prestressed reinforcement in tension zone	in^2
A_v	area of shear reinforcement within a spacing s	in^2
b_w	web width	in
d	distance from extreme compression fiber to centroid of nonprestressed reinforcement	in
d	distance from extreme compression fiber to centroid of prestressed and nonprestressed reinforcement $\geq 0.8h$, as given in ACI Sec. 22.5.2.1	in
d_p	actual distance from extreme compression fiber to centroid of prestressing tendons as defined in Fig. 4.8 $\geq 0.8h$	in
f_d	tensile stress at bottom fiber of section due to unfactored dead load	kips/in^2
f_{pc}	compressive stress in concrete at the centroid of the section due to final prestressing force	kips/in^2
f_{pe}	compressive stress in concrete at the bottom fiber of the section due to final prestressing force	kips/in^2
f_{pu}	specified strength of prestressing tendons	kips/in^2
f_{se}	effective stress in prestressing reinforcement after allowance for all prestressing losses	kips/in^2
f_{yt}	yield strength of transverse reinforcement	kips/in^2
g	sag of the tendon	in
h	depth of member	in
h	overall thickness of member	in
l	span length	in
M_{cre}	moment causing flexural cracking at section	in-kips
M_{ct}	cracking moment	in-kips
$M_{\max}$	maximum factored moment at section due to externally applied loads	in-kips
M_u	factored moment at section	in-kips
R_u	end reaction	lbf
s	spacing of shear or torsion reinforcement in direction parallel to longitudinal reinforcement	in
S_b	section modulus of the section referred to the bottom fiber	in^3
V_c	nominal shear strength provided by concrete	kips
V_{ci}	nominal shear strength provided by concrete when diagonal cracking results from combined shear and moment	kips
V_{cw}	nominal shear strength provided by concrete when diagonal cracking results from excessive principal tensile stress in the web	kips
V_d	shear force at section due to unfactored dead load	kips

V_i factored shear force at section due to   kips
externally applied loads applied
simultaneously with $M_{\max}$

V_p vertical component of effective prestress   kips
force at section

V_s nominal shear strength provided by   kips
shear reinforcement

V_u factored shear force at section   kips

w_u distributed load   lbf/ft

x distance from support to section   in
considered

Symbols

λ correction factor related to unit weight   –
of concrete

ϕ strength reduction factor, 0.75 for shear   –
and torsion

Critical Section

As shown in Fig. 4.13, the critical section for the calculation of shear in a prestressed beam is located at a distance from the support that is equal to half the overall thickness of the section. The maximum design factored shear force is given by

$$V_u = R_u - w_u\left(\frac{h}{2}\right)$$

Figure 4.13 *Critical Section for Shear*

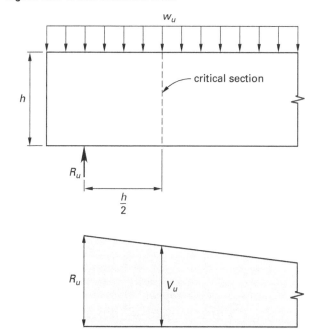

As specified in ACI Sec. 9.4.3.2, sections located less than a distance $h/2$ from the face of the support may be designed for the shear force, V_u. This is permitted

provided that the support reaction produces a compressive stress in the end of the beam, loads are applied at or near the top of the beam, and concentrated loads are not located closer to the support than half the overall depth.

Design for Shear

The nominal shear capacity of shear reinforcement perpendicular to the member is given by ACI Sec. 22.5.10.5.3 as

$$V_s = \frac{A_v f_{yt} d}{s}$$

The nominal shear strength of the shear reinforcement is limited by ACI Sec. 22.5.1.2 to a value of

$$V_s = 8 b_w d \sqrt{f_c'}$$

If additional shear capacity is required, the size of the concrete section must be increased. ACI Table 9.7.6.2.2 limits the spacing of the stirrups to a maximum value of $0.75h$ or 24 in, and when the value of V_s exceeds $4 b_w d \sqrt{f_c'}$, the spacing is reduced to a maximum value of $0.375h$ or 12 in.

In accordance with ACI Sec. 22.5.1.1, the combined shear capacity of the concrete section and the shear reinforcement is

$$\phi V_n = \phi V_c + \phi V_s$$

When the applied factored shear force, V_u, does not exceed $\phi V_c/2$, the concrete section is adequate to carry the shear without any shear reinforcement. Where V_u exceeds $\phi V_c/2$, a minimum area of shear reinforcement must be provided, and this area is specified as the greater of ACI Table 9.6.3.3 Eq. (a) and Eq. (b).

$$A_{v,\min} = 0.75\sqrt{f_c'}\left(\frac{b_w}{f_{yt}}\right)$$

$$A_{v,\min} = 50\left(\frac{b_w}{f_{yt}}\right)$$

For prestressed members with an effective prestressing force not less than 40% of the tensile strength of the flexural reinforcement ($A_{ps}f_{se} \geq 0.4(A_{ps}f_{pu} + A_s f_y)$), the minimum required area of shear reinforcement is specified by ACI Table 9.6.3.3 Eq. (e) as

$$A_{v,\min} = \frac{A_{ps}f_{pu}}{80 f_{yt} d}\sqrt{\frac{d}{b_w}}$$

However, this area need not be taken as more than the greater value from Eq. (a) and Eq. (b).

As stated in ACI Sec. 22.5.2.1, d need not be taken as less than $0.8h$.

When the effective prestressing force is at least 40% of the tensile strength of the flexural reinforcement, the nominal shear strength of the concrete, V_c, is conservatively given by ACI Sec. 22.5.8.2 as the smallest of the following values.

(a) $\quad V_c = \left(0.6\lambda\sqrt{f_c'} + 700\left(\dfrac{V_u d_p}{M_u}\right)\right)b_w d$

(b) $\quad V_c = (0.6\lambda\sqrt{f_c'} + 700)\,b_w d$

(c) $\quad V_c = 5\lambda\sqrt{f_c'}\,b_w d$

However, the value of V_c need not be taken as less than the value calculated by ACI Eq. 22.5.5.1, which is

$$V_c = 2\lambda\sqrt{f_c'}\,b_w d$$

$$\sqrt{f_c'} \le 100 \ \text{lbf/in}^2 \quad \text{[from ACI Sec. 22.5.3.1]}$$

For simply supported beams with uniformly distributed loads, ACI Eq. (a) may be expressed as

$$\frac{V_u d_p}{M_u} = \frac{d_p(l-2x)}{x(l-x)}$$

Example 4.9

The post-tensioned beam shown in *Illustration for Ex. 4.9* has a 28-day concrete strength of 6000 lbf/in² and is tensioned with five ½ in diameter strands. The area of each strand is 0.153 in² with a specified tensile strength of 270 kips/in² and an effective stress of 150 kips/in² after all losses. The cable centroid, as shown, is parabolic in shape, and the value of $V_u d_p/M_u = 1.0$ at section A-A. The effective prestressing force is greater than 40% of the tensile strength of the flexural reinforcement. Determine the nominal shear capacity at section A-A.

Solution

The equation of the parabolic cable profile is

$$y = \frac{gx^2}{a^2} = \frac{(10.5 \ \text{in})x^2}{\left[(15 \ \text{ft})\left(12 \ \dfrac{\text{in}}{\text{ft}}\right)\right]^2}$$

At section A-A, the rise of the cable is given by

$$y_A = \frac{(10.5 \ \text{in})\left[(15 \ \text{ft})\left(12 \ \dfrac{\text{in}}{\text{ft}}\right) - 23 \ \text{in}\right]^2}{\left[(15 \ \text{ft})\left(12 \ \dfrac{\text{in}}{\text{ft}}\right)\right]^2}$$

$$= 8 \ \text{in}$$

The actual depth of the cable is

$$d_p = h - y_A - y_o$$
$$= 27 \ \text{in} - 8 \ \text{in} - 4.5 \ \text{in}$$
$$= 14.5 \ \text{in}$$

The effective depth of the section is

$$d = 0.8h$$
$$= (0.8)(27 \ \text{in})$$
$$= 21.6 \ \text{in} \quad \text{[governs]}$$

Taking $V_u d_p/M_u = 1.0$ as given, the nominal shear capacity of the concrete section is given by ACI Table 22.5.8.2 as

$$V_c = \left(0.6\lambda\sqrt{f_c'} + 700\frac{V_u d_p}{M_u}\right)b_w d$$

$$= \left[(0.6)(1.0)\sqrt{6000 \ \frac{\text{lbf}}{\text{in}^2}} + (700)(1.0)\right]$$

$$\times \left(\frac{(12 \ \text{in})(21.6 \ \text{in})}{1000 \ \dfrac{\text{lbf}}{\text{kip}}}\right)$$

$$= 193 \ \text{kips}$$

$$\le 5\lambda\sqrt{f_c'}\,b_w d$$

$$= \frac{(5)(1.0)\sqrt{6000 \ \dfrac{\text{lbf}}{\text{in}^2}}\,(12 \ \text{in})(21.6 \ \text{in})}{1000 \ \dfrac{\text{lbf}}{\text{kip}}}$$

$$= 100 \ \text{kips} \quad \text{[governs and is not less than ACI Eq. 22.5.5.1]}$$

The nominal shear capacity of the stirrups provided is given by ACI Eq. 22.5.10.5.3 as

$$V_s = \frac{A_v f_{yt} d}{s}$$

$$= \frac{(0.22 \ \text{in}^2)\left(60 \ \dfrac{\text{kips}}{\text{in}^2}\right)(21.6 \ \text{in})}{12 \ \text{in}}$$

$$= 24 \ \text{kips}$$

Illustration for Ex. 4.9

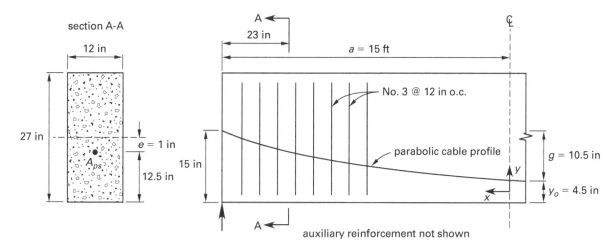

section A-A

The total nominal shear capacity is given by ACI Eq. 22.5.1.1 as

$$V_n = V_c + V_s$$
$$= 100 \text{ kips} + 24 \text{ kips}$$
$$= 124 \text{ kips}$$

Flexure-Shear and Web-Shear Cracking

A more precise value of the shear capacity is obtained by distinguishing between flexure-shear cracking and web-shear cracking. The two types of cracking are shown in Fig. 4.14.

Figure 4.14 *Cracking in Prestressed Concrete Beams*

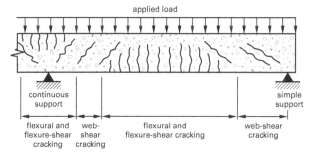

applied load

| continuous support | | | | simple support |

| flexural and flexure-shear cracking | web-shear cracking | flexural and flexure-shear cracking | web-shear cracking |

Adapted and reprinted with authorized permission from *Building Code Requirements for Structural Concrete and Commentary* (ACI 318-14), © 2014, by the American Concrete Institute.

The nominal shear capacity is provided by the lesser value of V_{ci} or V_{cw} given by ACI Sec. 22.5.8.3.1 and Sec. 22.5.8.3.2.

Flexure-Shear Cracking

For flexure-shear cracking, the nominal shear capacity is given by

$$V_{ci} = 0.6 b_w d_p \lambda \sqrt{f'_c} + V_d + \frac{V_i M_{cre}}{M_{\max}} \qquad \text{[ACI 22.5.8.3.1a]}$$

$$\geq 1.7 b_w d \lambda \sqrt{f'_c} \qquad \text{[ACI 22.5.8.3.1b]}$$

V_d is the shear force at the section due to the unfactored dead load, $M_{\max}$ is the maximum factored moment at the section due to the externally applied loads, V_i is the factored shear force at the section associated with $M_{\max}$, and d_p is the distance from the top fiber of the section to the centroid of the prestressing tendons, but not less than $0.8h$. The cracking moment due to the unfactored external applied loads is given by

$$M_{cre} = S_b(6\lambda\sqrt{f'_c} + f_{pe} - f_d) \qquad \text{[ACI 22.5.8.3.1c]}$$

f_d is the tensile stress at the bottom fiber of the section, due to the unfactored dead load, and f_{pe} is the compressive stress in the concrete, due to the final prestressing force, at the bottom fiber of the section.

For a uniformly loaded member, ACI Sec. R22.5.8.3.1 gives the variant of ACI Eq. 22.5.8.3.1.

$$V_{ci} = 0.6 b_w d \lambda \sqrt{f'_c} + \frac{V_u M_{ct}}{M_u}$$

M_u is the total factored moment at the section, V_u is the factored shear force associated with M_u, and the total unfactored moment, including dead load, required to cause cracking is given by

$$M_{ct} = S_b(6\lambda\sqrt{f'_c} + f_{pe})$$

For composite members, in accordance with ACI Sec. R22.5.8.3.1, the applicable equations are ACI Eq. 22.5.8.3.1a and Eq. 22.5.8.3.1c, with the shear force, V_d, and the stress, f_d, determined from the unfactored dead load resisted by the precast unit and the unfactored superimposed dead load resisted by the composite member. Similarly, M_d is the bending moment at the section due to the unfactored dead load acting on the precast unit plus the moment due to the unfactored superimposed dead load acting on the composite member; V_d is the unfactored shear force associated with M_d. Then,

$$V_i = V_u - V_d$$
$$M_{\max} = M_u - M_d$$

Web-Shear Cracking

For web-shear cracking, the nominal shear capacity is given by ACI Sec. 22.5.8.3.2 as

$$V_{cw} = b_w d_p (3.5\lambda\sqrt{f_c'} + 0.3 f_{pc}) + V_p \quad \text{[ACI 22.5.8.3.2]}$$

f_{pc} is the compressive stress in the concrete due to the final prestressing force at the centroid of the section, and V_p is the vertical component of the effective pre-stress force at the section, in kips, and d_p need not be taken less than $0.80h$.

Example 4.10

For the post-tensioned beam of Ex. 4.9, determine the nominal shear capacity of the concrete at section A-A by using ACI Eq. 22.5.8.3.1a and Eq. 22.5.8.3.2. The unfactored bending moment at section A-A due to dead load and live load is 500 in-kips.

Solution

At section A-A, the slope of the cable is given by

$$\frac{dy}{dx} = 2\left(\frac{gx}{a^2}\right)$$
$$= (2)\left(\frac{(10.5\text{ in})\left((15\text{ ft})\left(12\,\frac{\text{in}}{\text{ft}}\right) - 23\text{ in}\right)}{(180\text{ in})^2}\right)$$
$$= 0.102$$

The vertical component of the final effective prestressing force at section A-A is

$$V_p = A_{ps}f_{se}\frac{dy}{dx}$$
$$= (0.765\text{ in}^2)\left(150\,\frac{\text{kips}}{\text{in}^2}\right)(0.102)$$
$$= 11.7\text{ kips}$$

The compressive stress in the concrete, due to the final prestressing force, at the centroid of the section is

$$f_{pc} = \frac{P_e}{A_g}$$
$$= \frac{(0.765\text{ in}^2)\left(150\,\frac{\text{kips}}{\text{in}^2}\right)}{324\text{ in}^2}$$
$$= 0.354\text{ kips/in}^2$$

The nominal web-shear capacity of the concrete at section A-A is given by ACI Eq. 22.5.8.3.2 as

$$V_{cw} = b_w d_p (3.5\lambda\sqrt{f_c'} + 0.3 f_{pc}) + V_p$$
$$= \frac{(12\text{ in})(21.6\text{ in})\left((3.5)(1.0)\sqrt{6000\,\frac{\text{lbf}}{\text{in}^2}} + (0.3)\left(354\,\frac{\text{lbf}}{\text{in}^2}\right)\right) + 11{,}700\text{ lbf}}{1000\,\frac{\text{lbf}}{\text{kip}}}$$
$$= 110\text{ kips}$$

At section A-A, the cable eccentricity is

$$e = \frac{h}{2} - y_A - y_o$$
$$= 13.5\text{ in} - 8\text{ in} - 4.5\text{ in}$$
$$= 1.0\text{ in}$$
$$R_b = \frac{1}{A_g} + \frac{e}{S_b} = \frac{1}{324\text{ in}^2} + \frac{1\text{ in}}{1458\text{ in}^3}$$
$$= 0.00377\text{ 1/in}^2$$

The compressive stress in the bottom fiber, at section A-A, due to the final prestressing force is

$$f_{pe} = P_e R_b = (0.765\text{ in}^2)\left(150\,\frac{\text{kips}}{\text{in}^2}\right)\left(0.00377\,\frac{1}{\text{in}^2}\right)$$
$$= 0.433\text{ kips/in}^2$$

The applied moment required to produce cracking at section A-A is given by modified ACI Eq. 22.5.8.3.1c as

$$M_{ct} = S_b(6\lambda\sqrt{f_c'} + f_{pe})$$

$$= (1458 \text{ in}^3)\left(\frac{(6)(1.0)\sqrt{6000 \dfrac{\text{lbf}}{\text{in}^2}}}{1000 \dfrac{\text{lbf}}{\text{kip}}} + 0.433 \dfrac{\text{kips}}{\text{in}^2}\right)$$

$$= 1309 \text{ in-kips}$$

$$> 500 \text{ in-kips}$$

As this moment exceeds the given unfactored applied moment at section A-A, flexural cracking does not occur at section A-A; ACI Eq. 22.5.8.3.1a is not applicable, and ACI Eq. 22.5.8.3.2 governs. The nominal shear capacity of the concrete section is

$$V_c = V_{cw} = 110 \text{ kips}$$

3. DESIGN FOR TORSION

Nomenclature

A_{cp}	area enclosed by outside perimeter of concrete cross section	in^2
A_l	total area of longitudinal reinforcement to resist torsion	in^2
A_o	gross area enclosed by shear flow $= 0.85A_{oh}$	in^2
A_{oh}	gross area enclosed by centerline of the outermost closed transverse torsional reinforcement	in^2
A_t	area of one leg of a closed stirrup resisting torsion within a spacing s	in^2
d_{bl}	diameter of longitudinal torsional reinforcement	in
f_y	yield strength of longitudinal torsional reinforcement	kips/in^2
p_{cp}	outside perimeter of concrete cross section	in
p_h	perimeter of centerline of outermost closed transverse torsional reinforcement	in
T_n	nominal torsional moment strength	in-kips
T_u	factored torsional moment at section	in-kips

Symbols

τ	shear stress	kips/in^2

General Principles

After torsional cracking occurs, the central core of a prestressed concrete member is largely ineffective in resisting applied torsion, and so it can be neglected. When a member is subjected to torsion, ACI Sec. R22.7 assumes

that it behaves as a thin-walled tube. In order to maintain a consistent approach, a member is also analyzed as a thin-walled tube prior to cracking. As shown in Fig. 4.15, the shear stress in the tube walls produces a uniform shear flow, q, that acts at the midpoint of the walls. The shear flow's magnitude is

$$q = \tau t$$

Figure 4.15 *Thin-Walled Tube Analogy*

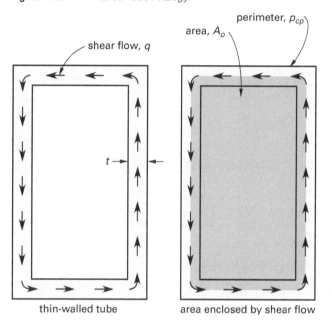

thin-walled tube   area enclosed by shear flow

From ACI Sec. R22.7, the applied torsion is resisted by the moment of the shear flow in the walls about the centroid of the section and is given by

$$T = 2A_o q$$

The gross area enclosed by shear flow, A_o, is the area enclosed by the center line of the walls, A_{cp}, and is given by

$$A_o = \frac{2A_{cp}}{3}$$

The stress in the shear walls is

$$\tau = \frac{T}{2A_o t}$$

For flanged sections, the overhanging flange width used to calculate the values of A_{cp} and p_{cp}, the outside perimeter, is determined from ACI Sec. R9.2.4.4, and shown in Fig. 4.16.

Figure 4.16 *Overhanging Flange Width*

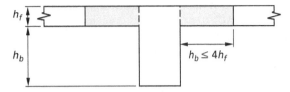

In accordance with ACI Sec. 9.4.4.3, the critical section for the calculation of torsion in a prestressed beam is located at a distance from the support equal to half the overall depth. When a concentrated torsion occurs within this distance, the critical section for design must be at the face of the support.

Cracking is assumed to occur in a member when the principal tensile stress reaches a value of

$$p_t = 4\sqrt{f_c'} = \tau$$

The cracking torsion is given by ACI Table 22.7.5.1 as

$$T_{cr} = 4\lambda\sqrt{f_c'}\left(\frac{A_{cp}^2}{p_{cp}}\right)\sqrt{1 + \frac{f_{pc}}{4\lambda\sqrt{f_c'}}}$$

Design Provisions

The design provisions for torsion in prestressed concrete are similar to those for reinforced concrete. In accordance with ACI Sec. 9.5.4.1 and Table 22.7.4.1(a), torsional effects may be neglected, and closed stirrups and longitudinal torsional reinforcement are not required when $T_u < \phi T_{th}$.

$$T_{th} = \lambda\sqrt{f_c'}\left(\frac{A_{cp}^2}{p_{cp}}\right)\sqrt{1 + \frac{f_{pc}}{4\lambda\sqrt{f_c'}}}$$

When this value is exceeded, reinforcement must be provided to resist the full torsion. When both shear and torsion reinforcements are required, the sum of the individual areas must be provided.

According to ACI Sec. 22.7.6.1, the nominal torsional moment strength is given by the lesser of the following two equations.

$$T_n = \frac{2A_oA_tf_{yt}}{s}\cot\theta \qquad \text{[ACI 22.7.6.1a]}$$

$$T_n = \frac{2A_oA_lf_y}{p_h}\tan\theta \qquad \text{[ACI 22.7.6.1b]}$$

In accordance with ACI Sec. 22.7.6.1.1, for these two equations it is permitted to take

$$A_o = 0.85A_{oh}$$

It is also permitted to assume that the angle between the compression diagonal and the longitudinal axis of the member, θ, is 37.5°.

The required area of one leg of a closed stirrup is derived from ACI Eq. 22.7.6.1a as

$$\frac{A_t}{s} = \frac{T_u}{2\phi A_of_{yt}\cot\theta} = \frac{T_u}{1.7\phi A_{oh}f_{yt}\cot\theta}$$

From ACI Eq. 22.7.6.1a and Eq. 22.7.6.1b, it follows that the corresponding area of longitudinal reinforcement required is

$$A_l = \frac{A_tp_hf_{yt}}{f_ys}\cot^2\theta$$

A_t/s is derived from ACI Eq. 22.7.6.1a.

If torsional reinforcement is required, the minimum permitted area of longitudinal reinforcement, $A_{l,\min}$, is given by ACI Sec. 9.6.4.3 as

$$A_{l,\min} = \frac{5A_{cp}\sqrt{f_c'}}{f_y} - \frac{A_tp_hf_{yt}}{f_ys}$$

In this equation, A_t/s must not be taken as less than $25b_w/f_{yt}$.

The minimum diameter is given by ACI Sec. 9.7.5.2 as

$$d_{bl} = 0.042s \text{ in}$$
$$\geq \text{no. 3 bar}$$

The minimum combined area of transverse closed stirrups for combined shear and torsion is given by ACI Sec. 9.6.4.2,

$$\frac{A_v + 2A_t}{s} = \frac{0.75\sqrt{f_c'}b_w}{f_{yt}}$$
$$\geq \frac{50b_w}{f_{yt}}$$

The maximum spacing of closed stirrups is given by ACI Sec. 9.7.6.3 as

$$s = \frac{p_h}{8} \text{ in}$$
$$\leq 12 \text{ in}$$

When redistribution is possible in an indeterminate structure, the nominal torsional capacity of the member, in accordance with ACI Sec. 22.7.3.2, need not exceed

$$T_n = 4\lambda\sqrt{f_c'}\left(\frac{A_{cp}^2}{p_{cp}}\right)\sqrt{1 + \frac{f_{pc}}{4\lambda\sqrt{f_c'}}}$$

Example 4.11

The post-tensioned beam for Ex. 4.9 is subjected to a factored shear force of 88 kips and a factored torsion of 100 in-kips at section A-A. Determine the combined shear and torsion reinforcement required.

Solution

The area enclosed by the outside perimeter of the beam is

$$A_{cp} = (27 \text{ in})(12 \text{ in}) = 324 \text{ in}^2$$

The length of the outside perimeter of the beam is

$$p_{cp} = (2)(27 \text{ in} + 12 \text{ in}) = 78 \text{ in}$$

The compressive stress at the centroid, due to the final prestressing force, was determined in Ex. 4.10 as

$$f_{pc} = 354 \text{ lbf/in}^2$$

Torsional reinforcement is not required in accordance with ACI Sec. 9.5.4.1 and Table 22.7.4.1(a) when the factored torque is less than

$$T_u = \phi\lambda\sqrt{f_c'}\left(\frac{A_{cp}^2}{p_{cp}}\right)\sqrt{1 + \frac{f_{pc}}{4\lambda\sqrt{f_c'}}}$$

$$= \left((0.75)(1.0)\sqrt{6000 \frac{\text{lbf}}{\text{in}^2}}\left(\frac{(324 \text{ in}^2)^2}{(78 \text{ in})\left(1000 \frac{\text{lbf}}{\text{kip}}\right)}\right)\right)$$

$$\times \sqrt{1 + \frac{354 \frac{\text{lbf}}{\text{in}^2}}{(4)(1.0)\sqrt{6000 \frac{\text{lbf}}{\text{in}^2}}}}$$

$$= 115 \text{ in-kips}$$

$$> 100 \text{ in-kips} \quad \begin{bmatrix} \text{Closed stirrups are} \\ \text{not required.} \end{bmatrix}$$

The shear strength provided by the concrete was determined in Ex. 4.9 as

$$V_c = 100 \text{ kips}$$

From ACI Sec. 22.5.1.1, the required nominal capacity of the shear reinforcement is

$$V_s = \frac{V_u}{\phi} - V_c = \frac{88 \text{ kips}}{0.75} - 100 \text{ kips}$$

$$= 17.33 \text{ kips}$$

For the case of $(A_{ps}f_{se} \geq 0.4(A_{ps}f_{pu} + A_s f_y))$, the minimum permissible area of shear reinforcement is given by ACI Table 9.6.3.3 as

$$\frac{A_{v(\min)}}{s} = \frac{A_{ps}f_{pu}\sqrt{\dfrac{d}{b_w}}}{80 f_{yt} d}$$

$$= \frac{(0.765 \text{ in}^2)\left(270 \dfrac{\text{kips}}{\text{in}^2}\right)\sqrt{\dfrac{21.6 \text{ in}}{12 \text{ in}}}}{(80)\left(60 \dfrac{\text{kips}}{\text{in}^2}\right)(21.6 \text{ in})}$$

$$= 0.0027 \text{ in}^2/\text{in} \quad [\text{governs}]$$

But not more than the greater value from

$$\frac{A_{v(\min)}}{s} = \frac{50 b_w}{f_{yt}} = \frac{(50)(12 \text{ in})}{60,000 \dfrac{\text{lbf}}{\text{in}^2}}$$

$$= 0.010 \text{ in}^2/\text{in}$$

Or,

$$\frac{A_{v(\min)}}{s} = \frac{0.75 b_w \sqrt{f_c'}}{f_{yt}}$$

$$= \frac{(0.75)(12 \text{ in})\sqrt{6000 \dfrac{\text{lbf}}{\text{in}^2}}}{60,000 \dfrac{\text{lbf}}{\text{in}^2}}$$

$$= 0.012 \text{ in}^2/\text{in}$$

From ACI Eq. 22.5.10.5.3, the shear reinforcement required is

$$\frac{A_v}{s} = \frac{V_s}{f_{yt} d} = \frac{17.33 \text{ kips}}{\left(60 \dfrac{\text{kips}}{\text{in}^2}\right)(21.6 \text{ in})}$$

$$= 0.014 \text{ in}^2/\text{in}$$

$$> \frac{A_{v(\min)}}{s} \quad [\text{satisfactory}]$$

Prestressed Concrete

Provide no. 3 stirrups at 15 in spacing, which gives

$$\frac{A_v}{s} = \frac{0.22 \text{ in}^2}{15 \text{ in}} = 0.015 \text{ in}^2/\text{in}$$

$$> 0.014 \quad \text{[satisfactory]}$$

4. PRESTRESS LOSSES

Nomenclature

A_{ps}	area of prestressing tendon	in^2
c	anchor set	in
C	factor for relaxation losses	–
E_{ci}	modulus of elasticity of concrete at time of initial prestress	kips/in^2
E_p	modulus of elasticity of prestressing tendon	kips/in^2
f_{pd}	compressive stress at level of tendon centroid after elastic losses and including sustained dead load	lbf/in^2
f_{pi}	compressive stress at level of tendon centroid after elastic losses	lbf/in^2
f_{pp}	compressive stress at level of tendon centroid before elastic losses	lbf/in^2
g	sag of prestressing tendon	in
H	ambient relative humidity	%
J	factor for relaxation losses	–
K	wobble friction coefficient per foot of prestressing tendon	–
K_{re}	factor for relaxation losses	kips/in^2
K_{sh}	factor for shrinkage losses accounting for elapsed time between completion of casting and transfer of prestressing force	–
l_c	length of prestressing tendon affected by anchor seating loss	ft
l_{px}	length of prestressing tendon from jacking end to any point x measured along the curve	ft
m	loss of force per foot of cable due to friction	kips/ft
n_i	E_p/E_{ci}	–
p_{cp}	outside perimeter of the concrete cross section	in
P_c	prestressing tendon force at a distance of l_c from the jacking end	kips
P_i	prestressing tendon force after elastic losses	kips
P_p	prestressing tendon force before elastic losses	kips
P_{pj}	prestressing tendon force at jacking end	kips
P_{px}	prestressing tendon force at a distance of l_{px} from the jacking end	kips
$P_{\Delta c}$	loss of tendon force due to anchor set	kips
$P_{\Delta cr}$	loss of tendon force due to creep	kips
$P_{\Delta el}$	loss of tendon force due to elastic shortening	kips
$P_{\Delta f}$	friction loss due to curvature and unintentional wobble of the duct profile	kips
$P_{\Delta re}$	loss of tendon force due to relaxation	kips
$P_{\Delta sh}$	loss of tendon force due to shrinkage	kips
R	radius of curvature of tendon profile	ft

Symbols

α	angular change in radians of tendon profile from jacking end to any point x	radians
ϵ_{sh}	basic shrinkage strain	–
μ	curvature friction coefficient	–

Types of Prestress Losses

In a pretensioned concrete member, loss of tendon force is produced by

- elastic shortening of concrete, $P_{\Delta el}$

- long-term creep of concrete, $P_{\Delta cr}$

- long-term shrinkage of concrete, $P_{\Delta sh}$

- long-term relaxation of stress in the prestressed tendons, $P_{\Delta re}$

- anchor seating loss at transfer, $P_{\Delta c}$

The total loss of tendon force in a pretensioned concrete member is

$$P_\Delta = P_{\Delta el} + P_{\Delta cr} + P_{\Delta sh} + P_{\Delta re} + P_{\Delta c}$$

Additional friction losses, $P_{\Delta f}$, are produced in a post-tensioned concrete member, such as losses due to curvature of the duct profile and unintended wobble (i.e., out-of-straightness) of the duct.

The total loss of tendon force in a post-tensioned concrete member is

$$P_\Delta = P_{\Delta el} + P_{\Delta cr} + P_{\Delta sh} + P_{\Delta re} + P_{\Delta c} + P_{\Delta f}$$

To determine the effective stress in the prestressing tendons, f_{se}, all prestress losses must be calculated. ACI Sec. R20.3.2.6.1 does not provide methods to calculate these losses, rather, it refers to the journal article, *Estimating Prestress Losses*.[1]

Friction Losses

Friction losses occur in post-tensioned members due to curvature of the duct profile and wobble of the ducts.

The relationship between tendon force at a distance, l_{px}, from the jack and the tendon force, P_{pj}, at the jack is

$$P_{pj} = P_{px}\exp(Kl_{px} + \mu\alpha)$$

Prestressed Concrete

Illustration for Ex. 4.12

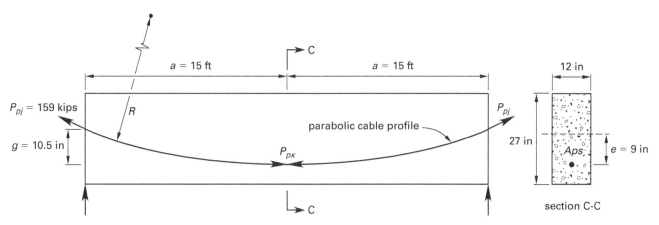

For a value of $Kl_{px} + \mu\alpha$ not greater than 0.3, this expression reduces to

$$P_{pj} = P_{px}(1 + Kl_{px} + \mu\alpha)$$

Values of friction coefficients may be obtained from the manufacturers of the prestressing system.

Example 4.12

The post-tensioned beam shown in *Illustration for Ex. 4.12* has a prestressing cable consisting of five ½ in diameter low-relaxation strands. Each strand has an area of 0.153 in², a yield strength of 243 kips/in², and a tensile strength of 270 kips/in². The cable centroid, as shown, is parabolic in shape and is stressed simultaneously from both ends with a jacking force of 159 kips. The value of the wobble friction coefficient is 0.0015/ft, and the curvature friction coefficient is 0.25. Determine the cable force at midspan of the member before elastic losses.

Solution

The nominal radius of the cable profile is

$$R = \frac{a^2}{2g}$$

$$= \frac{(15\ \text{ft})^2 \left(12\ \dfrac{\text{in}}{\text{ft}}\right)}{(2)(10.5\ \text{in})}$$

$$= 129\ \text{ft}$$

The cable length along the curve, from the jacking end to midspan, is

$$l_{px} = a + \frac{g^2}{3a} = 15\ \text{ft} + \frac{\left(\dfrac{10.5\ \text{in}}{12\ \dfrac{\text{in}}{\text{ft}}}\right)^2}{(3)(15\ \text{ft})}$$

$$= 15.02\ \text{ft}$$

The angular change of the cable profile over this length is

$$\alpha = \frac{l_{px}}{R} = \frac{15.02\ \text{ft}}{129\ \text{ft}}$$

$$= 0.117\ \text{radians}$$

$$Kl_{px} + \mu\alpha = (0.0015)(15.02\ \text{ft})$$
$$+ (0.25)(0.117\ \text{radians})$$
$$= 0.052$$
$$< 0.3$$

The cable force at midspan is given by

$$P_{px} = \frac{P_{pj}}{1 + Kl_{px} + \mu\alpha} = \frac{159\ \text{kips}}{1 + 0.052}$$

$$= 151\ \text{kips}$$

Anchor Seating Loss

Anchor seating loss results from the slip or set that occurs in the anchorage when the prestressing force is transferred to the anchor device. In a pretensioned tendon with an anchor set of c, the loss in prestressing tendon force is constant along the cable, as shown in Fig. 4.17. From Fig. 4.17, the anchor set is obtained as

$$c = \frac{\text{shaded area}}{A_{ps}E_p} = \frac{P_{\Delta c}l}{A_{ps}E_p}$$

Figure 4.17 *Seating Loss in a Pretensioned Tendon*

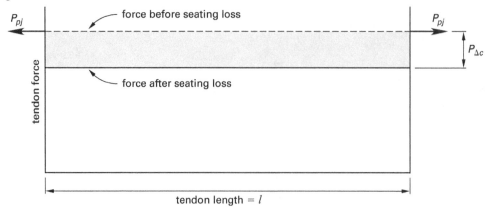

Figure 4.18 *Seating Loss in a Post-Tensioned Tendon*

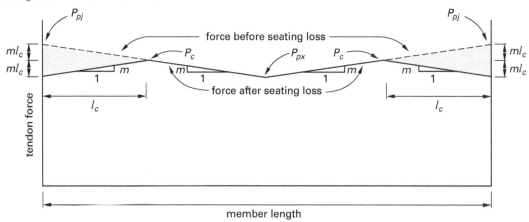

The loss in prestressing tendon force is then found as

$$P_{\Delta c} = \frac{cA_{ps}E_p}{l}$$

In a post-tensioned tendon, friction in the duct resists the inward movement of the tendon[2] and limits the affected zone to the length l_c shown in Fig. 4.18. The anchor set is obtained from Fig. 4.18 as

$$c = \frac{\text{shaded area}}{A_{ps}E_p} = \frac{ml_c^2}{A_{ps}E_p}$$

$$l_c^2 = \frac{cA_{ps}E_p}{m}$$

$$P_c = P_{pj} - ml_c$$

Example 4.13

The post-tensioned beam of Ex. 4.12 has tendon anchorages with a cable set of 0.05 in. Determine the residual tendon force, after anchorage and before elastic and long-term losses occur, at the jacking end and at a distance of l_c from the jacking end.

Solution

From Ex. 4.12, the stress loss per foot due to friction is

$$m = \frac{P_{pj} - P_{px}}{l_{px}} = \frac{159 \text{ kips} - 151 \text{ kips}}{15.02 \text{ ft}}$$

$$= 0.532 \text{ kips/ft}$$

$$l_c^2 = \frac{cA_{ps}E_p}{m}$$

$$= \frac{(0.05 \text{ in})(0.765 \text{ in}^2)\left(28 \times 10^3 \dfrac{\text{kips}}{\text{in}^2}\right)}{\left(0.532 \dfrac{\text{kips}}{\text{ft}}\right)\left(12 \dfrac{\text{in}}{\text{ft}}\right)}$$

$$l_c = 13 \text{ ft}$$

The cable force at a distance of l_c from the jacking end is

$$P_c = P_{pj} - ml_c$$

$$= 159 \text{ kips} - \left(0.532 \frac{\text{kip}}{\text{ft}}\right)(13 \text{ ft})$$

$$= 152 \text{ kips}$$

The cable force at the jacking end after anchoring is

$$P_{pj(\text{anc})} = P_{pj} - 2ml_c = 145 \text{ kips}$$

Elastic Shortening Losses

Losses occur in a prestressed concrete beam at transfer due to the elastic shortening of the concrete at the level of the centroid of the prestressing tendons. The concrete stress at the level of the centroid of the prestressing tendons after elastic shortening is

$$f_{pi} = P_i \left(\frac{1}{A_g} + \frac{e^2}{I_g} \right) - \frac{eM_G}{I_g}$$

Conservatively, the concrete stress at the level of the centroid of the prestressing tendons after elastic shortening is

$$f_{pi} = f_{pp} = P_p \left(\frac{1}{A_g} + \frac{e^2}{I_g} \right) - \frac{eM_G}{I_g}$$

In a pretensioned member with transfer occuring simultaneously in all tendons, the loss of prestressing force is

$$P_{\Delta el} = n_i A_{ps} f_{pi}$$

In a post-tensioned member with only one tendon, no loss from elastic shortening occurs.

In a post-tensioned member with several tendons stressed sequentially, the maximum loss occurs in the first tendon stressed, and no loss occurs in the last tendon stressed. The total loss is then one-half the value for a pretensioned member or

$$P_{\Delta el} = \frac{n_i A_{ps} f_{pi}}{2}$$

Example 4.14

The pretensioned beam shown is simply supported over a span of 30 ft and has a concrete strength at transfer of 4500 lbf/in². Five ½ in diameter low-relaxation strands are provided, each with an area of 0.153 in². The initial force in each tendon after anchor seating loss is 32 kips. Determine the loss of prestressing force due to elastic shortening.

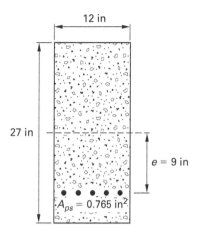

Solution

From ACI Sec. 19.2.2.1, the modulus of elasticity of the concrete at transfer is

$$E_{ci} = 57\sqrt{f'_{ci}} = 57\sqrt{4500 \ \frac{\text{lbf}}{\text{in}^2}} = 3824 \text{ kips/in}^2$$

$$n_i = \frac{E_p}{E_{ci}} = \frac{28 \times 10^3 \ \dfrac{\text{kips}}{\text{in}^2}}{3824 \ \dfrac{\text{kips}}{\text{in}^2}} = 7.32$$

$$\begin{aligned} f_{pp} &= P_p \left(\frac{1}{A_g} + \frac{e^2}{I_g} \right) - \frac{eM_G}{I_g} \\ &= (5)(32 \text{ kips}) \left(\frac{1}{324 \text{ in}^2} + \frac{(9 \text{ in})^2}{19,683 \text{ in}^4} \right) \\ &\quad - \frac{(9 \text{ in})(455 \text{ in-kips})}{19,683 \text{ in}^4} \\ &= 1.152 \ \frac{\text{kips}}{\text{in}^2} - 0.208 \ \frac{\text{kips}}{\text{in}^2} \\ &= 0.944 \text{ kips/in}^2 \end{aligned}$$

The total loss of prestressing force is

$$\begin{aligned} P_{\Delta el} &\approx n_i A_{ps} f_{pp} \\ &= (7.32)(5)(0.153 \text{ in}^2)\left(0.944 \ \frac{\text{kips}}{\text{in}^2} \right) \\ &= 5.3 \text{ kips} \end{aligned}$$

Creep Losses

Creep occurs in a prestressed concrete member as a result of the sustained compressive stress. It causes a prestress loss that is proportional to the concrete's initial stress and inversely proportional to its modulus of elasticity. Creep increases the earlier the stress is applied to the concrete. Half the total creep will take place in the first month after transfer, and three-quarters of the total creep will take place in the first six months after transfer.

The concrete stress at the level of the centroid of the prestressing tendons after elastic shortening is

$$f_{pd} = P_i\left(\frac{1}{A_g} + \frac{e^2}{I_g}\right) - \frac{eM_G}{I_g} - \frac{eM_D}{I_g}$$

For post-tensioned members with transfer at 28 days, the creep loss is given by[1,3]

$$P_{\Delta cr} = 1.6nA_{ps}f_{pd}$$

For pretensioned members with transfer at three days, the creep loss is given by

$$P_{\Delta cr} = 2.0nA_{ps}f_{pd}$$

Example 4.15

For the post-tensioned beam of Ex. 4.12, the cable force at midspan after elastic losses is 151 kips and the 28-day concrete strength is 6000 lbf/in². The superimposed dead load moment is 800 in-kips. Determine the loss of prestressing force due to creep.

Solution

From ACI Sec. 19.2.2.1, the modulus of elasticity of the concrete at 28 days is

$$E_c = 57\sqrt{f_c'} = 57\sqrt{6000\ \frac{\text{lbf}}{\text{in}^2}}$$

$$= 4415\ \text{kips/in}^2$$

$$n = \frac{E_p}{E_c} = \frac{28 \times 10^3\ \dfrac{\text{kips}}{\text{in}^2}}{4415\ \dfrac{\text{kips}}{\text{in}^2}}$$

$$= 6.34$$

$$f_{pd} = P_i\left(\frac{1}{A_g} + \frac{e^2}{I_g}\right) - \frac{eM_G}{I_g} - \frac{eM_D}{I_g}$$

$$= (151\ \text{kips})\left(\frac{1}{324\ \text{in}^2} + \frac{(9\ \text{in})^2}{19{,}683\ \text{in}^4}\right)$$

$$\quad - \frac{(9\ \text{in})(455\ \text{in-kips} + 800\ \text{in-kips})}{19{,}683\ \text{in}^4}$$

$$= 1.087\ \frac{\text{kips}}{\text{in}^2} - 0.574\ \frac{\text{kip}}{\text{in}^2}$$

$$= 0.514\ \text{kip/in}^2$$

The loss of prestressing force due to creep is

$$P_{\Delta cr} = 1.6nA_{ps}f_{pd}$$

$$= (1.6)(6.34)(0.765\ \text{in}^2)\left(0.514\ \frac{\text{kip}}{\text{in}^2}\right)$$

$$= 4.0\ \text{kips}$$

Shrinkage Loss

The shrinkage of a concrete member with time produces a corresponding loss of prestress. The basic shrinkage strain is given by[1,3]

$$\epsilon_{sh} = 8.2 \times 10^{-6}\ \text{in/in}$$

For a pretensioned member, with allowance for the ambient relative humidity, H, and the ratio of the member's volume to surface area, A_g/p_{cp}, the shrinkage loss is

$$P_{\Delta sh} = A_{ps}\epsilon_{sh}E_p\left(1 - \frac{0.06A_g}{p_{cp}}\right)(100 - H)$$

For a post-tensioned member with transfer after some shrinkage has already occurred, the shrinkage loss is

$$P_{\Delta sh} = K_{sh}A_{ps}\epsilon_{sh}E_p\left(1 - \frac{0.06A_g}{p_{cp}}\right)(100 - H)$$

K_{sh} is the factor for shrinkage losses in a post-tensioned member, accounting for elapsed time between completion of casting and transfer of tendon force. These values[4] are given in Table 4.1.

Table 4.1 *Shrinkage Factor Values*

days	K_{sh}
1	0.92
3	0.85
5	0.80
7	0.77
10	0.73
20	0.64
30	0.58
60	0.45

Example 4.16

The post-tensioned beam of Ex. 4.15 is located in an area with an ambient relative humidity of 55% and transfer is effected 7 days after the completion of curing, giving a value[3] of 0.77 for K_{sh}. Determine the loss of prestressing force due to shrinkage.

Solution

The shrinkage loss is given by

$$P_{\Delta sh} = K_{sh} A_{ps} \epsilon_{sh} E_p \left(1 - \frac{0.06 A_g}{p_{cp}}\right)(100 - H)$$

$$= (0.77)(0.765 \text{ in}^2)\left(8.2 \times 10^{-6} \frac{\text{in}}{\text{in}}\right)$$

$$\times \left(28 \times 10^3 \frac{\text{kips}}{\text{in}^2}\right)$$

$$\times \left(1 - \frac{(0.06 \text{ in}^{-1})(324 \text{ in}^2)}{78 \text{ in}}\right)(100 - 55)$$

$$= 4.6 \text{ kips}$$

Relaxation Losses

A prestressing tendon is subjected to relaxation over time. The loss in prestress depends on the tendon properties and the initial force in the tendon and on the losses due to creep, shrinkage, and elastic shortening. The relaxation loss is given by[1,3]

$$P_{\Delta re} = \left(A_{ps} K_{re} - J(P_{\Delta cr} + P_{\Delta sh} + P_{\Delta el})\right)C$$

Values of the relaxation parameters may be obtained from Zia[1] and Kelley.[4]

Example 4.17

For the post-tensioned beam in Ex. 4.15, the values of the relevant parameters are $K_{re} = 5$ kips/in^2, $J = 0.04$, and $C = 0.90$.[3] $P_{\Delta cr} = 4.0$ kips, $P_{\Delta sh} = 4.6$ kips, and $P_{\Delta el} = 2.7$ kips. Determine the loss of prestressing force due to relaxation.

Solution

The loss due to relaxation is

$$P_{\Delta re} = \left(A_{ps} K_{re} - J(P_{\Delta cr} + P_{\Delta sh} + P_{\Delta el})\right)C$$

$$= \left(\begin{array}{l}(0.765 \text{ in}^2)\left(5 \frac{\text{kips}}{\text{in}^2}\right) \\ -(0.04)(4.0 \text{ kips} + 4.6 \text{ kips} + 2.7 \text{ kips})\end{array}\right)$$

$$\times (0.90)$$

$$= 3.0 \text{ kips}$$

5. COMPOSITE CONSTRUCTION

Nomenclature

A_c	area of precast surface or area of contact surface for horizontal shear	in^2
A_v	area of ties within a distance s	in^2
A_{vf}	area of friction reinforcement	in^2
b	actual flange width	in
b_f	effective flange width	in
$b_{f(\text{tran})}$	transformed flange width	in
b_v	width of girder at contact surface	in
b_w	width of girder web	in
E_f	modulus of elasticity of flange concrete	kips/in^2
E_w	modulus of elasticity of precast girder concrete	kips/in^2
f_{cb}	stress in the bottom fiber of composite section	kips/in^2
$f_{ci(\text{flan})}$	stress in the flange at interface of composite section	kips/in^2
$f_{ci(\text{web})}$	stress in the girder at interface of composite section	kips/in^2
f_{ct}	stress in the top fiber of composite section	kips/in^2
h	depth of composite section	in
h_f	depth of flange of composite section	in
h_w	depth of precast girder	in
I_{cc}	moment of inertia of composite section	in^4
l	span length	ft
l_n	clear span measured face to face of supports	ft
M_F	moment due to flange concrete	ft-kips
M_G	moment due to precast girder self-weight	ft-kips
M_{prop}	moment due to removal of props	ft-kips
M_{Sht}	moment due to formwork	ft-kips
M_W	moment due to superimposed dead plus live load	ft-kips
n	modular ratio E_w/E_f	–
P_Δ	total loss of prestress	kips
s	spacing of ties	in
s_w	clear distance between adjacent webs	ft
S_{cb}	section modulus at bottom of composite section	in^3
S_{ci}	section modulus at interface of composite section	in^3
S_{ct}	section modulus at top of composite section	in^3
V_{nh}	nominal horizontal shear strength	lbf
V_u	factored shear force at section	lbf

Symbols

ρ_v	ratio of tie reinforcement area to area of contact surface $A_v/b_v s$	–

Figure 4.19 *Effective Flange Width*

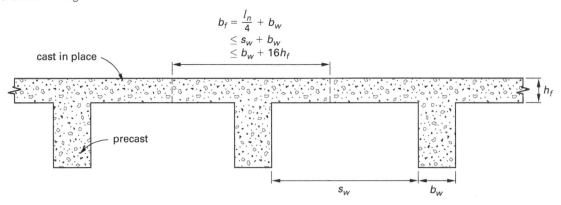

Section Properties

The effective width of the flange of a composite section, as shown in Fig. 4.19, is limited by ACI Table 6.3.2.1 to the least of

- $\dfrac{l_n}{4} + b_w$

- $b_w + 16h_f$

- $s_w + b_w$

When the 28-day compressive strengths of the precast section and the flange are different, the transformed section properties are obtained, as shown in Fig. 4.20, by dividing by the modular ratio $n = E_w/E_f$. The stresses calculated in the flange by using the transformed section properties are converted to actual stresses by dividing by n.

Figure 4.20 *Transformed Flange Width*

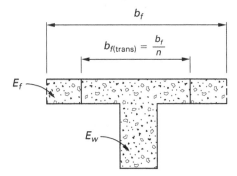

Note: $n = E_w/E_f$

Example 4.18

A composite beam of normal weight concrete with a clear span of 25 ft is shown in *Illustration for Ex. 4.18*. The beam is an interior beam in the floor of a commercial building with a clear distance between webs of $s_w = 26$ in. The precast, pretensioned girder has a 28-day concrete strength of 6000 lbf/in², and the flange has a 28-day concrete strength of 3000 lbf/in². Determine the transformed section properties of the composite section.

Solution

From ACI Sec. 19.2.2.1,

$$E_w = 57\sqrt{f_c'} = 57\sqrt{6000 \ \frac{\text{lbf}}{\text{in}^2}}$$
$$= 4415 \ \text{kips/in}^2$$

$$E_f = 57\sqrt{f_c'} = 57\sqrt{3000 \ \frac{\text{lbf}}{\text{in}^2}}$$
$$= 3122 \ \text{kips/in}^2$$

$$n = \frac{E_w}{E_f} = \frac{4415 \ \dfrac{\text{kips}}{\text{in}^2}}{3122 \ \dfrac{\text{kips}}{\text{in}^2}} = 1.41$$

The effective flange width is limited to the least of

- $b_f = \dfrac{l_n}{4} + b_w = \dfrac{(25 \ \text{ft})\left(12 \ \dfrac{\text{in}}{\text{ft}}\right)}{4} + 4 \ \text{in}$
  $= 79 \ \text{in}$

- $b_f = b_w + 16h_f = 4 \ \text{in} + (16)(2 \ \text{in})$
  $= 36 \ \text{in}$

- $b_f = s_w + b_w = 26 \ \text{in} + 4 \ \text{in} = 30 \ \text{in}$ [governs]

The transformed flange width is

$$b_{f(\text{tran})} = \frac{b_f}{n}$$
$$= \frac{30 \ \text{in}}{1.41}$$
$$= 21 \ \text{in}$$

Illustration for Ex. 4.18

actual section

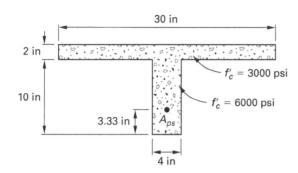

transformed section

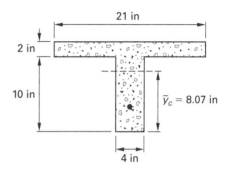

The relevant properties of the precast girder are

$$A_g = 40 \text{ in}^2$$

$$I_g = 333 \text{ in}^4$$

$$S_b = S_t = 66.67 \text{ in}^3$$

The properties of the transformed section are obtained as shown in the following table.

part	A (in²)	y (in)	I (in⁴)	Ay (in³)	Ay^2 (in⁴)
girder	40	5	333	200	1000
flange	42	11	14	462	5082
total	82	–	347	662	6082

$$\bar{y}_c = \frac{\sum Ay}{\sum A} = \frac{662 \text{ in}^3}{82 \text{ in}^2} = 8.07 \text{ in}$$

$$I_{cc} = \sum I + \sum Ay^2 + \bar{y}_c^2 \sum A - 2\bar{y}_c \sum Ay$$
$$= 347 \text{ in}^4 + 6082 \text{ in}^4 + 5344 \text{ in}^4$$
$$\quad -(16.14 \text{ in})(662 \text{ in}^3)$$
$$= 1088 \text{ in}^4$$

$$S_{ct} = \frac{I_{cc}}{h - \bar{y}_c} = \frac{1088 \text{ in}^4}{3.93 \text{ in}}$$
$$= 277 \text{ in}^3$$

$$S_{ci} = \frac{I_{cc}}{h_w - \bar{y}_c} = \frac{1088 \text{ in}^4}{1.93 \text{ in}}$$
$$= 564 \text{ in}^3$$

$$S_{cb} = \frac{I_{cc}}{\bar{y}_c} = \frac{1088 \text{ in}^4}{8.07 \text{ in}}$$
$$= 135 \text{ in}^3$$

Horizontal Shear Requirements

To ensure composite action, full transfer of horizontal shear at the interface is necessary, and ACI Sec. 16.4.3.1 specifies that the factored shear force at a section must not exceed

$$V_u = \phi V_{nh}$$

ACI Table 16.4.4.2 specifies that when the interface is intentionally roughened, the nominal horizontal shear strength is given by

$$V_{nh} = 80 b_v d$$

ACI Table 16.4.4.2 specifies that when the interface is smooth, with minimum ties provided across the interface (as required by ACI Sec. 16.4.6.1), to give $A_v/s = 50 b_w/f_y$,

$$V_{nh} = 80 b_v d$$

ACI Table 16.4.4.2 specifies that when the interface is roughened to $\frac{1}{4}$ in amplitude with minimum ties provided across the interface to give $A_v/s = 50 b_w/f_y$, but not less than $0.75 \sqrt{f_c'}\, b_w s/f_y$.

$$V_{nh} = (260 + 0.6\rho_v f_y)\lambda b_v d$$
$$\leq 500 b_v d$$

The correction factor related to the unit weight of concrete is given by ACI Table 19.2.4.2 as

$$\lambda = 1.0 \quad \text{[for normal weight concrete]}$$
$$= 0.75 \quad \text{[for all lightweight concrete]}$$

ACI Sec. 16.4.4.1 requires that when the factored shear force exceeds $\phi(500 b_v d)$, the design must be based on the shear-friction method given in ACI Sec. 22.9. Where the shear-friction reinforcement is perpendicular to the interface, which is not roughened, the nominal horizontal shear strength given ACI Table 22.9.4.4 is

$$V_{nh} = A_{vf}f_y\mu$$
$$\le 0.2f_c'A_c$$
$$\le 800A_c$$

Where the interface is intentionally roughened to an amplitude of $\frac{1}{4}$ in, the shear strength must additionally not exceed $(480 + 0.08f_c')A_c$.

ACI Table 22.9.4.2 gives the coefficient of friction as

$$\mu = 1.0\lambda \quad \begin{bmatrix} \text{interface roughened to} \\ \text{an amplitude of } 1/4 \text{ in} \end{bmatrix}$$
$$= 0.6\lambda \quad [\text{interface not roughened}]$$

In accordance with ACI Sec. 16.4.7.2, the tie spacing must not exceed four times the least dimension of the supported element, nor exceed 24 in.

Example 4.19

The composite section of Ex. 4.18 has a factored shear force of $V_u = 14$ kips at the critical section. Determine the area of grade 60 ties at a spacing of 12 in required at the interface that is intentionally roughened to an amplitude of $\frac{1}{4}$ in.

Solution

From ACI Sec. 16.4.4.3 and Sec. 16.4.4.2,

$$d = 0.8h = (0.8)(12 \text{ in}) = 9.6 \text{ in}$$
$$500\phi b_v d = \left(500 \frac{\text{lbf}}{\text{in}^2}\right)(0.75)(4 \text{ in})(9.6 \text{ in})$$
$$= 14{,}400 \text{ lbf}$$
$$> V_u \text{ [ACI Table 16.4.4.2(a) applies.]}$$
$$V_u = 14{,}000 \text{ lbf}$$
$$= \phi(260 + 0.6\rho_v f_y)\lambda b_v d$$
$$= (0.75)\left(260 \frac{\text{lbf}}{\text{in}^2} + 0.6\rho_v\left(60{,}000 \frac{\text{lbf}}{\text{in}^2}\right)\right)$$
$$\times(1.0)(4 \text{ in})(9.6 \text{ in})$$
$$\rho_v = 0.0063$$

The required area of vertical ties is

$$A_v = \rho_v b_v s = (0.0063)(4 \text{ in})(12 \text{ in}) = 0.30 \text{ in}^2$$

Nonpropped Construction

In nonpropped construction, the precast section supports its own self-weight, the formwork required to support the cast-in-place flange, and the weight of the flange. It may be conservatively assumed that all prestress losses occur before the flange is cast. As shown in Fig. 4.21, the composite section is subjected to the forces produced by removal of the formwork and by the superimposed applied load.

Figure 4.21 *Nonpropped Construction*

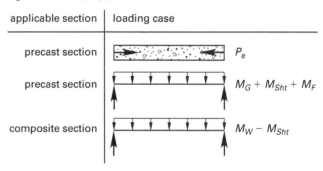

In accordance with ACI Sec. 22.3.3.3, the design of the composite section for the strength limit state is identical for both shored and unshored construction.

Example 4.20

The precast, pretensioned girder of the composite section of Ex. 4.18 is prestressed with an initial prestressing force of 65 kips. The total loss of prestress is 20% and may be assumed to occur before the flange is cast. The weight of the formwork to support the flange is 25 lbf/ft, the superimposed applied load is 250 lbf/ft, and the precast section is not propped. Determine the stresses at midspan in the composite section.

Solution

$$M_G = \frac{wl^2}{8} = \frac{\left(150 \frac{\text{lbf}}{\text{ft}^3}\right)(40 \text{ in}^2)(25 \text{ ft})^2}{(8)\left(12 \frac{\text{in}}{\text{ft}}\right)}$$
$$= 39{,}060 \text{ in-lbf}$$

$$M_{Sht} = \frac{wl^2}{8} = \frac{\left(25 \frac{\text{lbf}}{\text{ft}}\right)(25 \text{ ft})^2\left(12 \frac{\text{in}}{\text{ft}}\right)}{8}$$
$$= 23{,}440 \text{ in-lbf}$$

$$M_F = \frac{wl^2}{8} = \frac{\left(150 \frac{\text{lbf}}{\text{ft}^3}\right)(60 \text{ in}^2)(25 \text{ ft})^2}{(8)\left(12 \frac{\text{in}}{\text{ft}}\right)}$$
$$= 58{,}600 \text{ in-lbf}$$

$$M_W = \frac{wl^2}{8} = \frac{\left(250 \frac{\text{lbf}}{\text{ft}}\right)(25 \text{ ft})^2\left(12 \frac{\text{in}}{\text{ft}}\right)}{8}$$
$$= 234{,}400 \text{ in-lbf}$$
$$P_e = 0.8P_i = (0.8)(65 \text{ kips})$$
$$= 52 \text{ kips}$$

The prestressing force is applied at a height of $h_w/3$, and the stresses in the precast section after casting the flange are

$$f_t = \frac{M_G + M_{Sht} + M_F}{S_t}$$

$$= \frac{121{,}100 \text{ in-lbf}}{66.67 \text{ in}^3}$$

$$= 1816 \text{ lbf/in}^2$$

$$f_b = \frac{2P_e}{A_g} - \frac{M_G + M_{Sht} + M_F}{S_b}$$

$$= \frac{(2)(52{,}000 \text{ lbf})}{40 \text{ in}^2} - 1816 \frac{\text{lbf}}{\text{in}^2}$$

$$= 784 \text{ lbf/in}^2$$

The stresses in the composite section due to all loads are

$$f_{ct} = \frac{M_W - M_{Sht}}{nS_{ct}} = \frac{210{,}960 \text{ in-lbf}}{(1.41)(277 \text{ in}^3)}$$

$$= 540 \text{ lbf/in}^2$$

$$f_{cb} = f_b - \frac{M_W - M_{Sht}}{S_{cb}} = 784 - \frac{210{,}960 \text{ in-lbf}}{135 \text{ in}^3}$$

$$= -779 \text{ lbf/in}^2$$

$$f_{ci \text{ (flan)}} = \frac{M_W - M_{Sht}}{nS_{ci}} = \frac{210{,}960 \text{ in-lbf}}{(1.41)(564 \text{ in}^3)}$$

$$= 265 \text{ lbf/in}^2$$

$$f_{ci \text{ (web)}} = f_t + \frac{M_W - M_{Sht}}{S_{ci}} = 1816 + \frac{210{,}960 \text{ in-lbf}}{564 \text{ in}^3}$$

$$= 2190 \text{ lbf/in}^2$$

Propped Construction

In propped construction, the weight of the formwork and the flange act on the propped precast girder, producing moments in the girder and reactions in the props. As shown in Fig. 4.22, removal of the props is equivalent to applying forces, equal and opposite to the reactions in the props, to the composite section. The superimposed load is carried by the composite section.

When four or more props are used, the precast section may be considered continuously supported, and no stresses are produced in the precast girder by the weight of the formwork and the flange. Similarly, no stresses are produced by the removal of the formwork. On the removal of the props, the weight of the flange is carried by the composite section, as shown in Fig. 4.23.

Figure 4.22 *Propped Construction*

applicable section	loading case
precast section	P_e
precast section	M_G
propped precast section	$M_{Sht} + M_F$
composite section	M_{prop}
composite section	$M_W - M_{Sht}$

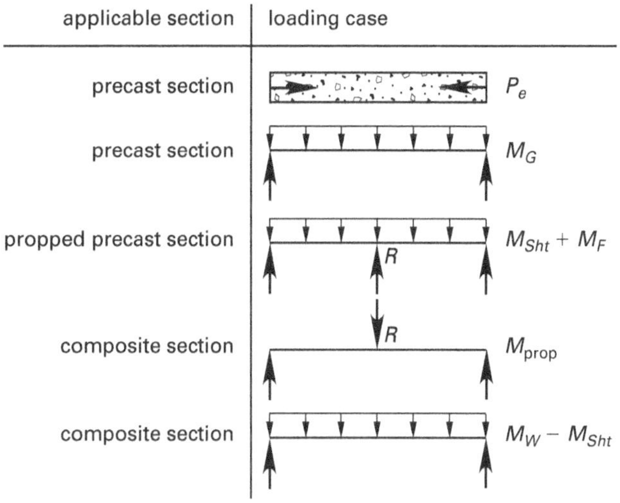

Figure 4.23 *Continuously Supported Section*

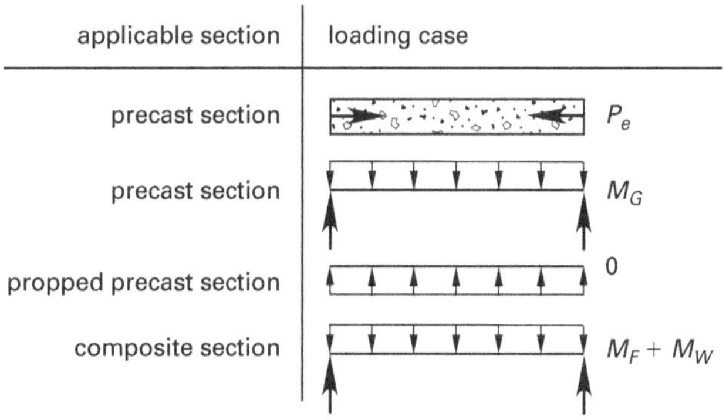

applicable section	loading case
precast section	P_e
precast section	M_G
propped precast section	0
composite section	$M_F + M_W$

Example 4.21

Before placing the formwork to support the flange and casting the flange, the precast, pretensioned girder of the composite section of Ex. 4.20 is propped at midspan. Determine the stresses at midspan in the composite section.

Solution

The prestressing force is applied at a height of $h/3$, and the stresses in the precast girder before propping are

$$f_t = \frac{M_G}{S_t} = \frac{39{,}060 \text{ in-lbf}}{66.67 \text{ in}^3}$$

$$= 586 \text{ lbf/in}^2$$

$$f_b = \frac{2P_e}{A_c} - \frac{M_G}{S_b}$$

$$= \frac{(2)(52{,}000 \text{ lbf})}{40 \text{ in}^2} - 586 \frac{\text{lbf}}{\text{in}^2}$$

$$= 2014 \text{ lbf/in}^2$$

The central prop creates a continuous beam with two spans of 12.5 ft each. The reaction on the prop due to the formwork and the flange concrete is

$$R = 1.25wl$$
$$= (1.25)\left(25\ \frac{\text{lbf}}{\text{ft}} + 62.5\ \frac{\text{lbf}}{\text{ft}}\right)(12.5\ \text{ft})$$
$$= 1367\ \text{lbf}$$

The moment in the precast girder at midspan, due to the formwork and flange concrete, is

$$M_{Sht} + M_F = \frac{wl^2}{8}$$
$$= \frac{\left(87.5\ \frac{\text{lbf}}{\text{ft}}\right)(12.5\ \text{ft})^2\left(12\ \frac{\text{in}}{\text{ft}}\right)}{8}$$
$$= 20{,}510\ \text{in-lbf}$$

The stresses in the precast girder after casting the flange are

$$f_t' = 586\ \frac{\text{lbf}}{\text{in}^2} - \frac{20{,}510\ \text{in-lbf}}{66.67\ \text{in}^3}$$
$$= 278\ \text{lbf/in}^2$$
$$f_b' = 2014\ \frac{\text{lbf}}{\text{in}^2} + \frac{20{,}510\ \text{in-lbf}}{66.67\ \text{in}^3}$$
$$= 2321\ \text{lbf/in}^2$$

Removing the prop produces a moment at midspan of

$$M_{\text{prop}} = \frac{Rl}{4}$$
$$= \frac{(1367\ \text{lbf})(25\ \text{ft})\left(12\ \frac{\text{in}}{\text{ft}}\right)}{4}$$
$$= 102{,}530\ \text{in-lbf}$$

The stresses in the composite section due to all loads are

$$f_{ct} = \frac{M_{\text{prop}} + M_W - M_{Sht}}{nS_{ct}}$$
$$= \frac{313{,}485\ \text{in-lbf}}{(1.41)(277\ \text{in}^3)}$$
$$= 802\ \text{lbf/in}^2$$

$$f_{cb} = f_b' - \frac{M_{\text{prop}} + M_W - M_{Sht}}{S_{cb}}$$
$$= 2321\ \frac{\text{lbf}}{\text{in}^2} - \frac{313{,}485\ \text{in-lbf}}{135\ \text{in}^3}$$
$$= -1\ \text{lbf/in}^2$$
$$f_{ci\,(\text{flan})} = \frac{M_{\text{prop}} + M_W - M_{Sht}}{nS_{ci}}$$
$$= \frac{313{,}485\ \text{in-lbf}}{(1.41)(564\ \text{in}^3)}$$
$$= 394\ \text{lbf/in}^2$$
$$f_{ci\,(\text{web})} = f_t' + \frac{M_{\text{prop}} + M_W - M_{Sht}}{S_{ci}}$$
$$= 278\ \frac{\text{lbf}}{\text{in}^2} + \frac{313{,}485\ \text{in-lbf}}{564\ \text{in}^3}$$
$$= 834\ \text{lbf/in}^2$$

6. LOAD BALANCING PROCEDURE

Nomenclature

f_c	concrete stress	lbf/in^2
g	sag of prestressing tendon	in
M_B	balancing load moment due to w_B	in-lbf
M_O	out-of-balance moment due to w_O, $M_W - M_B$	in-lbf
M_W	applied load moment due to w_W	in-lbf
P	prestressing force	kips
w_B	balancing load produced by prestressing tendon	kips/ft
w_O	out-of-balance load, $w_W - w_B$	kips/ft
w_W	superimposed applied load	kips/ft

Design Technique

The prestressing tendon of the beam shown in Fig. 4.24 has a parabolic profile and produces a uniform upward pressure of

$$w_B = \frac{8Pg}{l^2}$$

If the total downward load on the beam is equal to w_B, the net load is zero and a uniform compressive stress of $f_c = P/A_g$ is produced in the beam. If the downward load is not fully balanced by the upward force, the out-of-balance moment is

$$M_O = M_W - M_B$$

Figure 4.24 *Load Balancing Method*

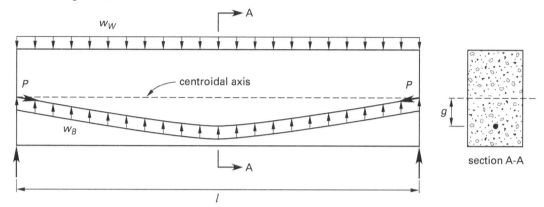

section A-A

The stress in the concrete is then given by

$$f_c = \frac{P}{A_g} \pm \frac{M_O}{S}$$

Balancing loads produced by alternative tendon profiles are available[5,6,7] and are shown in Fig. 4.25. This technique also facilitates the calculation of deflections.

Example 4.22

The post-tensioned beam shown in *Illustration for Ex. 4.22* supports a uniformly distributed load, including the weight of the beam, of 0.75 kip/ft. The tendon has a parabolic profile.

(a) Determine the prestressing force required in the tendon to balance the applied load exactly, and determine the resulting stress in the beam.

(b) Determine the stresses in the beam at midspan when an additional distributed load of 0.75 kip/ft is applied to the beam.

Solution

(a) The sag of the tendon is

$$g = 13.5 \text{ in} - 4.5 \text{ in} = 9 \text{ in}$$

The prestressing force required to balance the applied load exactly is

$$P = \frac{w_W l^2}{8g} = \frac{\left(0.75 \dfrac{\text{kip}}{\text{ft}}\right)(30 \text{ ft})^2 \left(12 \dfrac{\text{in}}{\text{ft}}\right)}{(8)(9 \text{ in})}$$

$$= 112.5 \text{ kips}$$

$$f_c = \frac{P}{A_g} = \frac{112{,}500 \text{ lbf}}{324 \text{ in}^2}$$

$$= 347 \text{ lbf/in}^2$$

(b) The out-of-balance moment produced at midspan by an additional load of 0.75 kip/ft is

$$M_O = \frac{w_O l^2}{8}$$

$$= \frac{\left(0.75 \dfrac{\text{kip}}{\text{ft}}\right)(30 \text{ ft})^2 \left(12 \dfrac{\text{in}}{\text{ft}}\right)}{8}$$

$$= 1013 \text{ in-kips} \quad (1{,}013{,}000 \text{ in-lbf})$$

The resultant stresses at midspan are

$$f_{be} = f_c - \frac{M_O}{S}$$

$$= 347 \frac{\text{lbf}}{\text{in}^2} - \frac{1{,}013{,}000 \text{ in-lbf}}{1458 \text{ in}^3}$$

$$= 347 \frac{\text{lbf}}{\text{in}^2} - 694 \frac{\text{lbf}}{\text{in}^2}$$

$$= -347 \text{ lbf/in}^2$$

$$f_{te} = f_c + \frac{M_O}{S} = 347 \frac{\text{lbf}}{\text{in}^2} + 694 \frac{\text{lbf}}{\text{in}^2}$$

$$= 1041 \text{ lbf/in}^2$$

7. STATICALLY INDETERMINATE STRUCTURES

Nomenclature

e'	resultant cable eccentricity	in
m	moment produced by unit value of the redundant	in-kips
M_R	resultant moment due to prestressing force and secondary effects, $Pe + M_S$	in-kips
M_S	moment produced by secondary effects	in-kips
R	reaction, support restraint	kips

Figure 4.25 *Alternative Tendon Profiles*

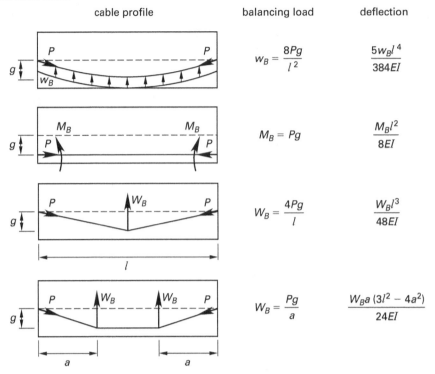

Illustration for Ex. 4.22

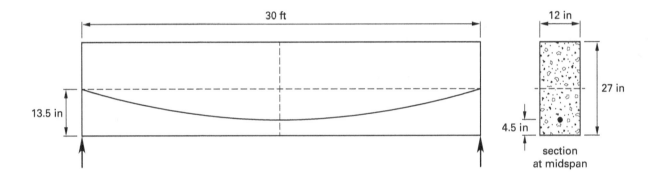

Design Principles

Prestressing an indeterminate structure may result in secondary moments, due to the support restraints, that produce the resultant moment

$$M_R = Pe + M_S$$

In the two-span beam shown in Fig. 4.26, the support restraint R_2 is taken as the redundant and a release introduced at 2 to produce the cut-back structure. Applying the prestressing force to the cut-back structure produces the primary moment $M_P = Pe$. Applying the unit value of R_2 to the cut-back structure produces the moment diagram, m, and the secondary moment is $M_S = R_2 m$. From the compatibility of displacements,

$$\frac{Pe}{EI} \int m \, dx = -\frac{R_2}{EI} \int m^2 \, dx$$

$$m = \frac{x}{2}$$

$$\frac{Pel^2}{2} = -\frac{R_2 l^3}{6}$$

$$R_2 = -\frac{3Pe}{l}$$

Figure 4.26 *Continuous Beam*

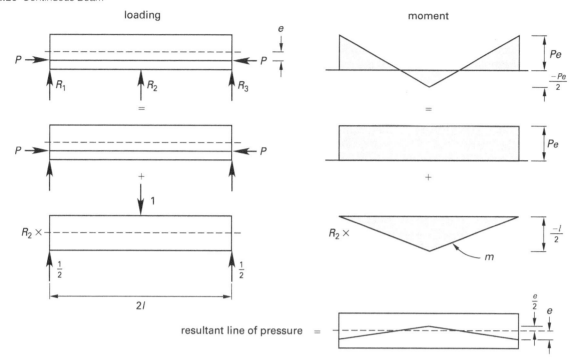

The secondary moment is given by

$$M_S = R_2 m = -\frac{3Pex}{2l}$$

The primary moment is given by

$$M_P = Pe$$

The resultant moment is given by

$$M_R = M_P + M_S = Pe - \frac{3Pex}{2l}$$

The resultant line of pressure, as shown in Fig. 4.26, is given by

$$e' = \frac{M_R}{P} = e - \frac{3ex}{2l}$$

At midspan,

$$M_{S(\text{midspan})} = -\frac{3Pe}{2}$$

$$M_{R(\text{midspan})} = Pe - \frac{3Pe}{2} = -\frac{Pe}{2}$$

$$e'_{(\text{midspan})} = e - \frac{3e}{2} = -\frac{e}{2}$$

A tendon with an initial eccentricity of e' produces no secondary effects in the member and is termed the concordant cable. Similarly, as shown in Fig. 4.27, the bending moment diagram for the external loads on a continuous beam is also a concordant profile because no support restraints will be produced. In addition, a concordant profile may be modified by means of a linear transformation by varying the location of the tendon at interior supports, as shown in Fig. 4.27, without changing the resultant moment.

Figure 4.27 *Concordant Tendon Profile*

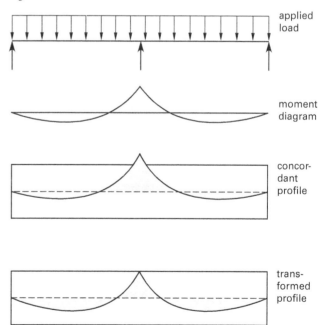

Example 4.23

The post-tensioned two-span beam shown in *Illustration for Ex. 4.23* supports a uniformly distributed load, including the weight of the beam, of 0.75 kip/ft. The tendon profile is parabolic in each span and is located in span 12 as indicated.

(a) Determine the prestressing force required in the tendon and the required sag of the tendon in span 23 to balance the applied load exactly. Then, determine the resulting stress in the beam.

(b) Determine the stresses in the beam at the central support when an additional distributed load of 0.75 kip/ft is applied to the beam. Then, determine the location of the resultant line of pressure at the central support.

Solution

(a) The sag of the tendon in span 12 is given by

$$g_{12} = e_4 + \frac{e_2}{2} = 3.65 \text{ in} + \frac{10.7 \text{ in}}{2}$$
$$= 9 \text{ in}$$

The prestressing force required to balance exactly the applied load in span 12 is

$$P = \frac{w_w l_{12}^2}{8g_{12}}$$
$$= \frac{\left(0.75 \dfrac{\text{kip}}{\text{ft}}\right)(30 \text{ ft})^2\left(12 \dfrac{\text{in}}{\text{ft}}\right)}{(8)(9 \text{ in})}$$
$$= 112.5 \text{ kips}$$

The required sag of the tendon in span 23 is given by

$$g_{23} = \frac{w_w l_{23}^2}{8P}$$
$$= \frac{\left(0.75 \dfrac{\text{kip}}{\text{ft}}\right)(40 \text{ ft})^2\left(12 \dfrac{\text{in}}{\text{ft}}\right)}{(8)(112.5 \text{ kips})}$$
$$= 16 \text{ in}$$
$$e_5 = g_{23} - \frac{e_2}{2} = 16 \text{ in} - \frac{10.7 \text{ in}}{2}$$
$$= 10.65 \text{ in}$$

The uniform compressive stress throughout the beam is

$$f_c = \frac{P}{A_g} = \frac{112{,}500 \text{ lbf}}{324 \text{ in}^2}$$
$$= 347 \text{ lbf/in}^2$$

(b) Allowing for the hinges at supports 1 and 3, the fixed-end moments produced by the additional load of 0.75 kip/ft are

$$M_{F21} = \frac{w_w l_{12}^2}{8} \quad \text{[clockwise]}$$
$$= Pg_{12}$$
$$= (112.5 \text{ kips})(9 \text{ in})$$
$$= 1013 \text{ in-kips}$$
$$M_{F23} = -Pg_{23} \quad \text{[counterclockwise]}$$
$$= -(112.5 \text{ kips})(16 \text{ in})$$
$$= -1800 \text{ in-kips}$$

The fixed-end moments are distributed as shown in the table, allowing for the hinges at the supports to eliminate carryover to ends 1 and 3.

joint	2	
member	21	23
relative $\dfrac{EI}{l}$	$\dfrac{3}{30}$	$\dfrac{3}{40}$
distribution factors	$\dfrac{4}{7}$	$\dfrac{3}{7}$
FEM	1013	−1800
distribution	449	+337
final moments	1463	−1463

The final moment at support 2 due to the distributed load is

$$M_{O2} = 1463 \text{ in-kips} \quad (1{,}463{,}000 \text{ in-lbf})$$

The resultant stresses at the central support are

$$f_{te} = f_c - \frac{M_{O2}}{S}$$
$$= 347 \frac{\text{lbf}}{\text{in}^2} - \frac{1{,}463{,}000 \text{ in-lbf}}{1458 \text{ in}^3}$$
$$= -656 \text{ lbf/in}^2$$

$$f_{be} = f_c + \frac{M_{O2}}{S}$$
$$= 347 \frac{\text{lbf}}{\text{in}^2} + \frac{1{,}463{,}000 \text{ in-lbf}}{1458 \text{ in}^3}$$
$$= 1350 \text{ lbf/in}^2$$

The location of the resultant line of pressure at the central support is

$$e_2' = -\frac{M_{O2}}{P} = -\frac{1463 \dfrac{\text{in}}{\text{kips}}}{112.5 \text{ kips}} = -13 \text{ in}$$

Illustration for Ex. 4.23

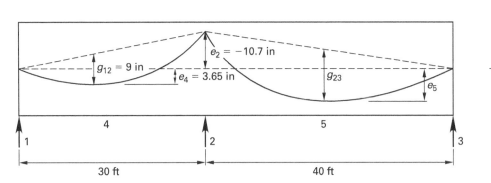
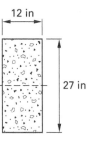

8. REFERENCES

1. Zia, Paul et al. "Estimating Prestress Losses." *Concrete International: Design and Construction* 1, no. 6 (1979): 32–38.

2. The Concrete Society. *Post-Tensioned Flat Slab Design Handbook.* Concrete Society Technical Report no. 25. London: The Concrete Society, 1984.

3. Kamara, Mahmoud E., Lawrence C. Novak, and Basile G. Rabbat. *Notes on ACI 318–08: Building Code Requirements for Structural Concrete with Design Applications*, Tenth ed. Skokie, IL: Portland Cement Association, 2008.

4. Kelley, Gail S. "Prestress Losses in Post-Tensioned Structures." *PTI Technical Note* 10, no. 9 (2000): 1–6.

5. Lin, T. Y. "Load Balancing Method for Design and Analysis of Prestressed Concrete Structures." *ACI Journal Proceedings* 60, no. 6 (1963): 719–742.

6. Prestressed Concrete Institute. *PCI Design Handbook: Precast and Prestressed Concrete.* Chicago, IL: Prestressed Concrete Institute, 2010.

7. Freyermuth, Clifford. L. and Robert A. Shoolbred. *Post-Tensioned Prestressed Concrete.* Skokie, IL: Portland Cement Association, 1967.

Prestressed Concrete

5

Structural Steel Design

1. INTRODUCTION

The *Specification for Structural Steel Buildings*[1] (AISC 360) permits the design of steel buildings by either the load and resistance factor design (LRFD) method or the allowable strength design (ASD) method.

In the LRFD method, factored loads using LRFD load combinations are applied to a member to determine the required ultimate strength. This required ultimate strength must not exceed the design strength, which is calculated as the member nominal strength multiplied by a resistance factor.

The *allowable stress design* method was traditionally used to design steel structures. However, this method has been superseded by the *allowable strength design* method. In the allowable stress design method, nominal loads are applied to a member, and the calculated stress in the member cannot exceed the member's specified allowable stress. In general, the allowable stress is determined as the yield stress of the member divided by a safety factor.

In the *allowable strength design* (ASD) method, factored loads, using ASD load combinations, are applied to a member to determine the required strength. This required strength must not exceed the allowable strength, which is calculated as the member nominal strength divided by a safety factor. The LRFD *resistance factor* is designated by the symbol ϕ, and the ASD *safety factor* is designated by the symbol Ω. The load factors in the two methods are calibrated to provide similar results at a live-to-dead load ratio of 3. The relationship between the resistance factor and safety factor is

$$\Omega = \frac{1.5}{\phi}$$

2. LOAD COMBINATIONS

Nomenclature

D	dead loads	kips or kips/ft
E	earthquake load	kips or kips/ft
H	load due to lateral pressure	kips/ft^2
L	live loads due to occupancy	kips or kips/ft
L_r	roof live load	kips or kips/ft
Q	load effect produced by service load	kips
R	load due to rainwater or ice	kips or kips/ft
R_n	nominal strength	–
S	snow load	kips or kips/ft
W	wind load	kips or kips/ft

Symbols

γ	load factor	–
ϕ	resistance factor	–
Ω	safety factor	–

LRFD Required Strength

The required ultimate strength of a member consists of the most critical combination of factored loads applied to the member. Factored loads consist of nominal, or service, loads multiplied by the appropriate load factors.

Seismic and wind loads specified in the IBC are at the strength design level, in contrast to other loads that are at the service level. In the ASD load combinations, the load factor for seismic loads is 0.7 and the load factor for wind loads is 0.6, which reduces them to service level values. In the LRFD combinations, the load factor for both seismic and wind loads is 1.0.

In accordance with AISC 360[1] Sec. B2, load combinations must be as stipulated by the applicable building code. The required strength, $\Sigma \gamma Q$, is defined by seven

Steel

combinations in IBC[2] Sec. 1605.2. The combinations, with uncommon loads (self-straining loads and fluid pressure) omitted, are as follows.

$$\sum \gamma Q = 1.4D \qquad \text{[IBC 16-1]}$$

$$\sum \gamma Q = 1.2D + 1.6(L + H) \qquad \text{[IBC 16-2]}$$
$$+ 0.5(L_r \text{ or } S \text{ or } R)$$

$$\sum \gamma Q = 1.2D + 1.6(L_r \text{ or } S \text{ or } R) \quad \text{[IBC 16-3]}$$
$$+ 1.6H + (f_1 L \text{ or } 0.5W)$$

$$\sum \gamma Q = 1.2D + 1.0W + f_1 L + 1.6H \quad \text{[IBC 16-4]}$$
$$+ 0.5(L_r \text{ or } S \text{ or } R)$$

$$\sum \gamma Q = 1.2D + 1.0E + f_1 L \qquad \text{[IBC 16-5]}$$
$$+ f_2 S + 1.6H$$

$$\sum \gamma Q = 0.9D + 1.0W + 1.6H \qquad \text{[IBC 16-6]}$$

$$\sum \gamma Q = 0.9D + 1.0E + 1.6H \qquad \text{[IBC 16-7]}$$

For IBC Eq. 16-3, Eq. 16-4, and Eq. 16-5, use $f_1 = 1.0$ for garages, places of public assembly, and areas where $L > 100$ lbf/ft^2. Use $f_1 = 0.5$ for all other live loads. For IBC Eq. 16-5, use $f_2 = 0.7$ for roof configurations that do not shed snow, and use 0.2 for other roof configurations.

ASD Required Strength

The required strength of a member consists of the most critical combination of factored loads applied to the member. Factored loads consist of nominal, or service, loads multiplied by the appropriate load factors. In accordance with AISC 360 Sec. B2, load combinations must be as stipulated by the applicable building code. The required strength, $\sum \gamma Q$, is defined by nine combinations in IBC Sec. 1605.3.1. The combinations, with uncommon load conditions omitted, are as follows.

$$\sum \gamma Q = D \qquad \text{[IBC 16-8]}$$

$$\sum \gamma Q = D + L + H \qquad \text{[IBC 16-9]}$$

$$\sum \gamma Q = D + (L_r \text{ or } S \text{ or } R) + H \qquad \text{[IBC 16-10]}$$

$$\sum \gamma Q = D + 0.75L + H \qquad \text{[IBC 16-11]}$$
$$+ 0.75(L_r \text{ or } S \text{ or } R)$$

$$\sum \gamma Q = D + H + (0.6W \text{ or } 0.7E) \qquad \text{[IBC 16-12]}$$

$$\sum \gamma Q = D + H + 0.75(0.6W) + 0.75L \quad \text{[IBC 16-13]}$$
$$+ 0.75(L_r \text{ or } S \text{ or } R)$$

$$\sum \gamma Q = D + H + 0.75(0.7E) \qquad \text{[IBC 16-14]}$$
$$+ 0.75L + 0.75S$$

$$\sum \gamma Q = 0.6D + 0.6W + H \qquad \text{[IBC 16-15]}$$

$$\sum \gamma Q = 0.6D + 0.7E + H \qquad \text{[IBC 16-16]}$$

Example 5.1

A typical frame of a six-story office building is shown. The loading on the frame is as follows.

roof dead load, including cladding and columns, w_{Dr}	$= 1.2$ kips/ft
roof live load, w_{Lr}	$= 0.4$ kip/ft
floor dead load, including cladding and columns, w_D	$= 1.6$ kips/ft
floor live load, w_L	$= 1.25$ kips/ft
horizontal wind pressure, p_h	$= 1.0$ kip/ft
vertical wind pressure, p_v	$= 0.5$ kip/ft

Determine the maximum and minimum required loads on the columns.

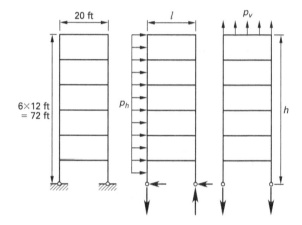

Solution

The axial load on one column due to the dead load is

$$D = \frac{l(w_{Dr} + 5w_D)}{2}$$
$$= \frac{(20 \text{ ft})\left(1.2 \, \dfrac{\text{kips}}{\text{ft}} + (5 \text{ stories})\left(1.6 \, \dfrac{\text{kips}}{\text{ft}}\right)\right)}{2}$$
$$= 92 \text{ kips}$$

The axial load on one column due to the roof live load is

$$L_r = \frac{l w_{Lr}}{2}$$
$$= \frac{(20 \text{ ft})\left(0.4 \, \dfrac{\text{kip}}{\text{ft}}\right)}{2}$$
$$= 4 \text{ kips}$$

The axial load on one column due to the floor live load is

$$L = \frac{5lw_L}{2}$$

$$= \frac{(5)(20 \text{ ft})\left(1.25 \frac{\text{kips}}{\text{ft}}\right)}{2}$$

$$= 62.5 \text{ kips}$$

The axial load on one column due to the horizontal wind pressure is

$$W_h = \pm\frac{p_h h^2}{2l}$$

$$= \pm\frac{\left(1.0 \frac{\text{kip}}{\text{ft}}\right)(72 \text{ ft})^2}{(2)(20 \text{ ft})}$$

$$= \pm 130 \text{ kips}$$

The axial load on one column due to the vertical wind pressure is

$$W_v = \frac{-p_v l}{2} = \frac{-\left(0.5 \frac{\text{kip}}{\text{ft}}\right)(20 \text{ ft})}{2}$$

$$= -5 \text{ kips}$$

LRFD Method

The terms H, S, and R are not applicable. From IBC Eq. 16-2, the maximum design load on a column is

$$\sum \gamma Q = 1.2D + 1.6L + 0.5L_r$$

$$= (1.2)(92 \text{ kips}) + (1.6)(62.5 \text{ kips})$$

$$\quad + (0.5)(4 \text{ kips})$$

$$= 212 \text{ kips} \quad [\text{compression}]$$

Alternatively, from IBC Eq. 16-4,

$$\sum \gamma Q = 1.2D + 1.0W + 0.5L + 0.5L_r$$

$$= (1.2)(92 \text{ kips}) + (1.0)(130 \text{ kips} - 5 \text{ kips})$$

$$\quad + (0.5)(62.5 \text{ kips}) + (0.5)(4 \text{ kips})$$

$$= 269 \text{ kips} \quad [\text{governs}]$$

From IBC Eq. 16-6, the minimum design load on a column is

$$\sum \gamma Q = 0.9D + 1.0W_h + 1.0W_v$$

$$= (0.9)(92 \text{ kips}) + (1.0)(-130 \text{ kips})$$

$$\quad + (1.0)(-5 \text{ kips})$$

$$= -52 \text{ kips} \quad [\text{tension}]$$

ASD Method

The terms H, S, and R are not applicable. From IBC Eq. 16-9, the maximum design load on a column is

$$\sum \gamma Q = D + L$$

$$= 92 \text{ kips} + 62.5 \text{ kips}$$

$$= 155 \text{ kips} \quad [\text{compression}]$$

Alternatively, from IBC Eq. 16-13,

$$\sum \gamma Q = D + 0.75(0.6W) + 0.75L + 0.75L_r$$

$$= 92 \text{ kips} + (0.75)(0.6)(130 \text{ kips} - 5 \text{ kips})$$

$$\quad + (0.75)(62.5 \text{ kips}) + (0.75)(4 \text{ kips})$$

$$= 198 \text{ kips} \quad [\text{governs}]$$

From IBC Eq. 16-15, the minimum design load on a column is

$$\sum \gamma Q = 0.6D + 0.6W_h + 0.6W_v$$

$$= (0.6)(92 \text{ kips}) + (0.6)(-130 \text{ kips})$$

$$\quad + (0.6)(-5 \text{ kips})$$

$$= -26 \text{ kips} \quad [\text{tension}]$$

LRFD Design Strength

The design strength of a member consists of the nominal, or theoretical ultimate, strength of the member, R_n, multiplied by the appropriate resistance factor, ϕ. The resistance factor is defined in AISC 360 as

$\phi_b = 0.90$ [for flexure]

$\phi_v = 1.0$ [for shear in webs of rolled I-shaped members]

$\phi_v = 0.90$ [for shear in all other flexural conditions]

$\phi_c = 0.90$ [for compression]

$\phi_t = 0.90$ [for tensile yielding]

$\phi_t = 0.75$ [for tensile fracture]

$\phi = 0.75$ [for shear rupture of bolts]

To ensure structural safety, AISC 360 Sec. B3 specifies that the design strength is

$$\phi R_n \geq \sum \gamma Q$$

The *AISC Manual*[3] load tables incorporate the appropriate values of ϕ and provide a direct value of the design strength.

Steel

ASD Allowable Strength

The allowable strength of a member consists of the nominal strength of the member, R_n, divided by the appropriate safety factor, Ω. The safety factor is defined in AISC 360 as

$$\Omega_b = 1.67 \quad \text{[for flexure]}$$

$$\Omega_v = 1.5 \quad \text{[for shear in webs of rolled I-shaped members]}$$

$$\Omega_v = 1.67 \quad \text{[for shear in all other conditions]}$$

$$\Omega_c = 1.67 \quad \text{[for compression]}$$

$$\Omega_t = 1.67 \quad \text{[for tensile yielding]}$$

$$\Omega_t = 2.00 \quad \text{[for tensile fracture]}$$

$$\Omega = 2.00 \quad \text{[for shear rupture of bolts]}$$

To ensure structural safety, AISC 360 Sec. B3 specifies that the allowable stress is

$$\frac{R_n}{\Omega} \geq \sum \gamma Q$$

The *AISC Manual*[3] load tables incorporate the appropriate values of Ω and provide a direct value of the allowable strength.

Example 5.2

A pin-ended column of grade A50 steel 14 ft long is subjected to a factored axial load of $\sum \gamma Q = 450$ kips (LRFD) or 300 kips (ASD). Determine the lightest adequate W10 shape.

Solution

LRFD Method

From *AISC Manual* Table 4-1a, for an effective height of 14 ft, a W10 × 49 column provides the design axial strength.

$$\phi R_n = \phi_c P_n = 471 \text{ kips}$$
$$> \sum \gamma Q \quad \text{[satisfactory]}$$

ASD Method

From *AISC Manual* Table 4-1a, for an effective height of 14 ft, a W10 × 49 column provides the allowable axial strength.

$$\frac{R_n}{\Omega} = \frac{P_n}{\Omega_c} = 313 \text{ kips}$$
$$> \sum \gamma Q \quad \text{[satisfactory]}$$

3. DESIGN FOR FLEXURE

Nomenclature

b_f	flange width	in
BF	tabulated factor used to calculate the design flexural strength for unbraced lengths between L_p and L_r	–
C_b	lateral-torsional buckling modification factor	–
F_y	specified minimum yield stress	kips/in^2
I_y	moment of inertia about the y-axis	in^4
L_b	length between braces	ft or in
L_m	limiting laterally unbraced length for full plastic bending capacity ($C_b > 1.0$)	ft or in
L_p	limiting laterally unbraced length for full plastic bending capacity ($C_b = 1.0$)	ft or in
L_r	limiting laterally unbraced length for inelastic lateral-torsional buckling	ft or in
M_a	required flexural strength	ft-kips
M_A	absolute value of moment at quarter point of the unbraced beam segment	ft-kips
M_B	absolute value of moment at centerline of the unbraced beam segment	ft-kips
M_C	absolute value of moment at three-quarter point of the unbraced beam segment	ft-kips
$M_{\max}$	absolute value of maximum moment in the unbraced beam segment	ft-kips
M_n	nominal flexural strength	ft-kips
M_p	plastic bending moment	ft-kips
M_r	$0.7F_yS_x$	ft-kips
M_r	required bending moment	ft-kips
M_u	required flexural strength	ft-kips
M_y	yield moment	ft-kips
S	elastic section modulus	in^3
t_f	flange thickness	in
Z	plastic section modulus	in^3

Symbols

λ_{pf}	limiting slenderness parameter for compact element	–
λ_{rf}	limiting slenderness parameter for noncompact element	–

Plastic Moment of Resistance

When a compact, laterally braced steel beam is loaded to the stage when the extreme fibers reach yield, as shown in Fig. 5.1(a), the applied moment, ignoring residual stresses, is given by

$$M_y = SF_y$$

Figure 5.1 *Stress Distribution in W Shape*

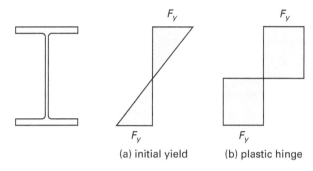

(a) initial yield   (b) plastic hinge

Taking into account the residual stress in the beam, the applied moment at first yielding is given by

$$M_r = 0.7 F_y S$$

Continued loading eventually results in the stress distribution shown in Fig. 5.1(b). A plastic hinge is formed, and the beam cannot sustain any further increase in loading. The nominal strength of the member is given by

$$M_n = M_p = ZF_y$$

The shape factor is defined as

$$\frac{M_p}{M_y} = \frac{Z}{S} \approx 1.12 \quad \text{[for a W shape]}$$

A shape that is compact ensures that full plasticity will be achieved prior to flange or web local buckling. Compactness criteria for flexure are given in AISC 360 Table B4.1b. Most W shapes are compact, and tabulated values of $\phi_b M_p$, in Part 3 of the *AISC Manual*, allow for any reduction due to noncompactness. Adequate lateral bracing of a member ensures that full plasticity will be achieved prior to lateral-torsional buckling occurring.

Example 5.3

Determine the plastic section modulus and the shape factor for the steel section shown in *Illustration for Ex. 5.3.* Assume that the section is compact and adequately braced.

Solution

The properties of the elastic section are obtained as given in the following table.

part	A (in^2)	y (in)	I (in^4)	Ay (in^3)	Ay^2 (in^4)
flange	45.0	18.5	34	832	15,401
web	25.5	8.5	614	217	1842
total	70.5	–	648	1049	17,243

$$\bar{y} = \frac{\sum Ay}{\sum A} = \frac{1049 \text{ in}^3}{70.5 \text{ in}^2}$$
$$= 14.88 \text{ in}$$

$$I = \sum I + \sum Ay^2 - \bar{y}^2 \sum A$$
$$= 648 \text{ in}^4 + 17,243 \text{ in}^4 - 15,610 \text{ in}^4$$
$$= 2281 \text{ in}^4$$

$$S_b = \frac{I}{\bar{y}} = \frac{2281 \text{ in}^4}{14.88 \text{ in}}$$
$$= 153 \text{ in}^3$$

The location of the plastic neutral axis is obtained by equating areas above and below the axis. The depth of the plastic neutral axis is given by

$$yb_f = (h_f - y)b_f + h_w b_w$$

$$y = \frac{\sum A}{2b_f} = \frac{70.5 \text{ in}^2}{(2)(15 \text{ in})}$$
$$= 2.35 \text{ in}$$

The plastic section modulus is obtained by taking moments of areas about the plastic neutral axis.

$$Z = \frac{y^2 b_f}{2} + \frac{(h_f - y)^2 b_f}{2} + A_w \left(\frac{h_w}{2} + h_f - y \right)$$
$$= \frac{(2.35 \text{ in})^2 (15 \text{ in})}{2} + \frac{(0.65 \text{ in})^2 (15 \text{ in})}{2}$$
$$\quad + (25.5 \text{ in}^2)(9.15 \text{ in})$$
$$= 278 \text{ in}^3$$

The shape factor is

$$\frac{Z}{S} = \frac{278 \text{ in}^3}{153 \text{ in}^3} = 1.82$$

Nominal Flexural Strength

Nominal flexural strength is influenced by several factors.

- *Plastic moment strength:* $M_n \leq M_p$

- *Flange local buckling:* a slender flange is prone to local buckling and most rolled I-shapes are compact for a yield stress of $F_y \leq 50$ kips/in^2; the exceptions are W21 × 48, W14 × 99, W14 × 90, W12 × 65, W10 × 12, W8 × 31, and W8 × 10

Illustration for Ex. 5.3

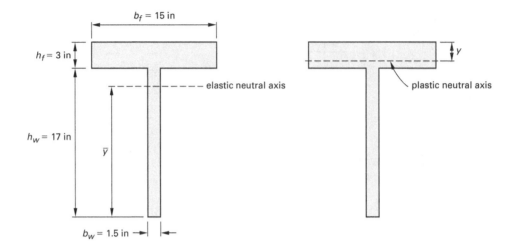

- *Web local buckling:* a slender web is prone to local buckling and all rolled I-shapes are compact for a yield stress of $F_y \leq 70$ kips/in^2; more common with thin web plate girders

- *Lateral-torsional buckling:* lateral bracing applied to the compression flange restrains the flange and prevents local buckling; governs when L_b exceeds L_p and $C_b = 1.0$

- *Lateral-torsional buckling modification factor, C_b:* a beam with a uniform bending moment has $C_b = 1.0$; for other moment gradients, $C_b > 1.0$ and lateral-buckling strength is obtained by multiplying basic strength by C_b, with a maximum permitted value of M_p for the lateral-buckling strength

Tabulated values of $\phi_b M_p$ and M_p/Ω_b in *AISC Manual* Table 3-2 are based on a value of $C_b = 1.0$ and allow for reduction in M_n due to slenderness effects.

Compact, Noncompact, and Slender Sections

Steel beams are classified as compact, noncompact, and slender, in accordance with the slenderness criteria in AISC 360 Table B4.1b. The flexural capacity of an adequately braced beam depends on the slenderness ratio of the compression flange and the web. When the slenderness ratios are sufficiently small, the beam can attain its full plastic moment, and the cross section is classified as compact. When the slenderness ratios are larger, the compression flange or the web may buckle locally before a full plastic moment is attained, and the cross section is classified as noncompact. When the slenderness ratios are sufficiently large, local buckling will occur before the yield stress of the material is reached, and the cross section is classified as slender. The flexural response of the three classifications is shown in Fig. 5.2.

Figure 5.2 *Variation of M_n with λ*

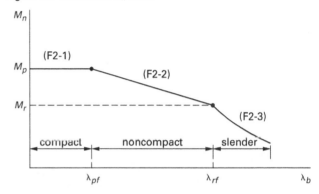

The limiting slenderness parameters for three flange buckling modes are as follows. (See AISC 360 Table B4.1b, Case 10.)

- *Compact flange:* a beam flange with a slenderness parameter of $\lambda = b_f/2t_f < \lambda_{pf}$. The limiting slenderness parameter for a compact flange is

$$\lambda_{pf} = 0.38\sqrt{\frac{E}{F_y}}$$

An adequately braced beam will develop its plastic moment of resistance before the onset of lateral-torsional and local buckling. The nominal flexural strength of the compact section is given by AISC 360 Eq. F2-1 as

$$M_n = M_p = F_y Z_x$$

- *Noncompact flange:* a beam flange with a slenderness parameter of $\lambda_{pf} \leq \lambda \leq \lambda_{rf}$. The limiting slenderness parameter for a noncompact flange is

$$\lambda_{rf} = 1.0\sqrt{\frac{E}{F_y}}$$

A beam with an unbraced length, L_r, will develop its yield moment, M_r, coincident with yielding of the beam's extreme fibers and before the onset of lateral-torsional and local buckling. Allowing for residual stresses in the beam, the yield moment is

$$M_r = 0.7F_yS_x$$

The limits of M_n are given by

$$M_r \leq M_n < M_p$$

The nominal flexural strength of the noncompact section is given by AISC 360 Eq. F3-1 as

$$M_n = M_p - (M_p - 0.7F_yS_x)\left(\frac{\lambda - \lambda_{pf}}{\lambda_{rf} - \lambda_{pf}}\right)$$

- *Slender flange:* a beam flange with the slenderness parameter defined by $\lambda > \lambda_{rf}$. A slender section is one that cannot develop the yield stress prior to flange local buckling. The nominal flexural strength of a section with slender flanges, with adequate lateral bracing, is given by AISC 360 Eq. F3-2 as

$$M_n = \frac{0.9Ek_cS_x}{\lambda^2}$$

The value of k_c is limited to

$$k_c = \frac{4}{\sqrt{\dfrac{h}{t_w}}}$$
$$\geq 0.35$$
$$\leq 0.76$$

Example 5.4

Determine the design flexural strength and allowable flexural strength of a W14 × 90 with a yield stress of 50 kips/in^2 and bent about its major axis with $C_b = 1.0$. $L_b < L_p$, so lateral-torsional buckling will not govern.

Solution

The limiting slenderness parameter for the flange of a rolled I-shape is given by AISC 360 Table B4.1b as

$$\lambda_{pf} = 0.38\sqrt{\frac{E}{F_y}}$$
$$= 0.38\sqrt{\frac{29{,}000\ \dfrac{\text{kips}}{\text{in}^2}}{50\ \dfrac{\text{kips}}{\text{in}^2}}}$$
$$= 9.15$$

From *AISC Manual* Table 1-1, a W14 × 90 has a value of

$$\lambda = \frac{b_f}{2t_f} = 10.2$$
$$> 9.15 \quad \text{[flange is not compact]}$$

A W14 × 90 with a yield stress of 50 kips/in^2 is not a compact section. From *AISC Manual* Table 1-1, the elastic section modulus about the x-axis is

$$S_x = 143\ \text{in}^3$$

The yield moment considering residual stresses is

$$M_r = 0.7F_yS_x$$
$$= \frac{(0.7)\left(50\ \dfrac{\text{kips}}{\text{in}^2}\right)(143\ \text{in}^3)}{12\ \dfrac{\text{in}}{\text{ft}}}$$
$$= 417\ \text{ft-kips}$$

The plastic moment of resistance is given by

$$M_p = Z_xF_y$$
$$= \frac{(157\ \text{in}^3)\left(50\ \dfrac{\text{kips}}{\text{in}^2}\right)}{12\ \dfrac{\text{in}}{\text{ft}}}$$
$$= 654\ \text{ft-kips}$$

From AISC 360 Table B4.1b, the limiting slenderness parameter for a noncompact flange is

$$\lambda_r = 1.0\sqrt{\frac{E}{F_y}} = 1.0\sqrt{\frac{29{,}000\ \dfrac{\text{kips}}{\text{in}^2}}{50\ \dfrac{\text{kips}}{\text{in}^2}}}$$
$$= 24.08$$

Steel

From AISC 360 Eq. F3-1, the nominal flexural strength is given by

$$M_n = M_p - (M_p - 0.7F_yS_x)\left(\frac{\lambda - \lambda_{pf}}{\lambda_{rf} - \lambda_{pf}}\right)$$

$$= 654 \text{ ft-kips} - (654 \text{ ft-kips} - 417 \text{ ft-kips})$$

$$\times \left(\frac{10.2 - 9.15}{24.08 - 9.15}\right)$$

$$= 637.3 \text{ ft-kips}$$

LRFD Method

The design flexural strength is

$$\phi_b M_n = (0.9)(637.3 \text{ ft-kips})$$
$$= 574 \text{ ft-kips}$$

ASD Method

The allowable flexural strength is

$$\frac{M_n}{\Omega_b} = \frac{637.3 \text{ ft-kips}}{1.67}$$
$$= 382 \text{ ft-kips}$$

These values are given in *AISC Manual* Tables 3-2 and 3-6 as $\phi_b M_{px}$ and M_{px}/Ω_b.

Lateral-Torsional Buckling with C_b = 1.0

Lateral-torsional buckling of an I-shape occurs along the length of the beam between lateral supports. The compression flange tends to buckle similarly to a long column while the tension flange provides restraint. This results in lateral displacement of the compression flange and a torsional twisting of the cross section. An I-shape bent about its minor axis does not buckle. An HSS section bent about any axis does not buckle.

The maximum nominal moment capacity of a compact rolled I-shape is $M_n = M_p$. As shown in Fig. 5.3, when the unbraced length, L_b, between points of lateral support on a compact beam with $C_b = 1.0$ increases beyond the length, L_p, the nominal flexural strength of the beam decreases. The beam passes through the three phases: *plastic*, *inelastic*, and *elastic*. L_p is the limiting laterally unbraced length for full plastic flexural capacity with $C_b = 1.0$. L_r is the limiting laterally unbraced length for inelastic lateral-torsional buckling. M_r is the limiting moment reached when $L_b = L_r$ and $M_r = 0.7F_yS_x$.

Values of L_p, L_r, $\phi_b M_{px}$, M_{px}/Ω_b, $\phi_b M_{rx}$, and M_{rx}/Ω_b for beams with a yield stress of 50 kips/in² and with $C_b = 1.0$ are given in *AISC Manual* Table 3-2.

Figure 5.3 *Variation of M_n with L_b for $C_b = 1.0$*

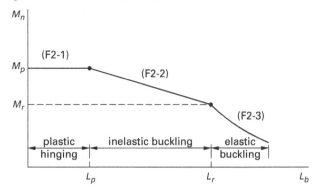

Plastic Phase: $L_b \le L_p$

A compact section subjected to uniform bending moment will develop its full plastic moment capacity, provided that the laterally unsupported segment length is

$$L_b \le L_p$$
$$M_n = M_p$$

The value of L_p is defined in AISC 360 Sec. F2.2 and is tabulated in *AISC Manual* Part 3. The values of $\phi_b M_{px}$ and M_{px}/Ω_b are also tabulated in Part 3.

Example 5.5

A simply supported beam of grade 50 steel is laterally braced at 4 ft intervals. If the beam is subjected to a uniform factored bending moment of 270 ft-kips (LRFD) or 180 ft-kips (ASD), with $C_b = 1.0$, determine (a) the lightest adequate W shape, and (b) the W shape with the minimum allowable depth.

Solution

LRFD Method

(a) From *AISC Manual* Table 3-2, the lightest satisfactory section is a W16 × 40, which has

$$\phi M_{px} = 274 \text{ ft-kips}$$
$$> 270 \text{ ft-kips} \quad [\text{satisfactory}]$$

$$L_p = 5.55 \text{ ft}$$
$$> 4 \text{ ft} \quad [\text{satisfactory}]$$

(b) From *AISC Manual* Table 3-2, the W shape with the minimum depth is a W10 × 60, which has

$$\phi_b M_{px} = 280 \text{ ft-kips}$$
$$> 270 \text{ ft-kips} \quad [\text{satisfactory}]$$
$$L_p = 9.08 \text{ ft}$$
$$> 4 \text{ ft} \quad [\text{satisfactory}]$$

ASD Method

(a) From *AISC Manual* Table 3-2, the lightest satisfactory section is a W16 × 40, which has

$$\frac{M_{px}}{\Omega_b} = 182 \text{ ft-kips}$$

$$> 180 \text{ ft-kips} \quad [\text{satisfactory}]$$

$$L_p = 5.55 \text{ ft}$$

$$> 4 \text{ ft} \quad [\text{satisfactory}]$$

(b) From *AISC Manual* Table 3-2, the W shape with the minimum depth is a W10 × 60, which has

$$\frac{M_{px}}{\Omega_b} = 186 \text{ ft-kips}$$

$$> 180 \text{ ft-kips} \quad [\text{satisfactory}]$$

$$L_p = 9.08 \text{ ft}$$

$$> 4 \text{ ft} \quad [\text{satisfactory}]$$

Inelastic Phase: $L_p < L_b \leq L_r$

When the laterally unsupported segment length equals L_r, the nominal flexural strength is given by

$$M_n = M_r = 0.7F_y S_x$$

The value of L_r is defined in AISC 360 Eq. F2-6 and is tabulated in *AISC Manual* Part 3. The values of $\phi_b M_{rx}$ and M_{rx}/Ω_b are also tabulated in *AISC Manual* Part 3.

The nominal flexural strength for an unbraced length between L_p and L_r is obtained by linear interpolation between M_p and M_r and is given by AISC 360 Eq. F2-2 as

$$M_n = C_b \left(M_p - \frac{(M_p - M_r)(L_b - L_p)}{L_r - L_p} \right)$$

For LRFD,

$$\phi_b M_n = C_b \big(\phi_b M_{px} - (\phi_b BF)(L_b - L_p) \big) \leq \phi_b M_{px}$$

$$BF = \frac{(M_p - M_r)}{L_r - L_p}$$

For ASD,

$$\frac{M_n}{\Omega_b} = C_b \left(\frac{M_{px}}{\Omega_b} - (BF/\Omega_b)(L_b - L_p) \right) \leq \frac{M_{px}}{\Omega_b}$$

$$BF = \frac{M_p - M_r}{(L_r - L_p)}$$

Values of ϕBF and BF/Ω_b are tabulated in *AISC Manual* Part 3 for values of $C_b = 1.0$. The variation of nominal flexural strength with unbraced length is shown in Fig. 5.3.

Example 5.6

A simply supported W16 × 40 beam of grade 50 steel is laterally braced at 6 ft intervals and is subjected to a uniform bending moment with $C_b = 1.0$. Determine the available flexural strength of the beam.

Solution

LRFD Method

From *AISC Manual* Table 3-2, a W16 × 40 has

$$\phi_b M_{px} = 274 \text{ ft-kips}$$

$$\phi_b M_{rx} = 170 \text{ ft-kips}$$

$$L_p = 5.55 \text{ ft}$$

$$< 6 \text{ ft}$$

$$L_r = 15.9 \text{ ft}$$

$$> 6 \text{ ft}$$

$$\phi_b BF = 10.0 \text{ kips}$$

$$\phi_b M_n = C_b \big(\phi_b M_{px} - (\phi_b BF)(L_b - L_p) \big)$$

$$= (1.0) \big(274 \text{ ft-kips} - (10.0 \text{ kips})(6 \text{ ft} - 5.55 \text{ ft}) \big)$$

$$= 269 \text{ ft-kips} \quad [\text{design flexural strength}]$$

This value is given in *AISC Manual* Table 6-2 as $\phi_b M_{nx}$.

ASD Method

From *AISC Manual* Table 3-2, a W16 × 40 has

$$\frac{M_{px}}{\Omega_b} = 182 \text{ ft-kips}$$

$$\frac{M_{rx}}{\Omega_b} = 113 \text{ ft-kips}$$

$$L_p = 5.55 \text{ ft}$$

$$< 6 \text{ ft}$$

$$L_r = 15.9 \text{ ft}$$

$$> 6 \text{ ft}$$

$$BF/\Omega_b = 6.67 \text{ kips}$$

$$\frac{M_n}{\Omega_b} = C_b \left(\frac{M_{px}}{\Omega_b} - \left(\frac{BF}{\Omega_b} \right)(L_b - L_p) \right)$$

$$= (1.0) \big(182 \text{ ft-kips} - (6.67 \text{ kips})(6 \text{ ft} - 5.55 \text{ ft}) \big)$$

$$= 179 \text{ ft-kips} \quad [\text{allowable flexural strength}]$$

This value is given in *AISC Manual* Table 6-2 as M_{nx}/Ω_b.

Elastic Phase: $L_b > L_r$

When the laterally unsupported segment length exceeds L_r, the nominal flexural strength is governed by elastic lateral-torsional buckling. The nominal flexural strength is equal to the critical elastic moment M_n and is defined in AISC 360 Eq. F2-3 as

$$M_n = F_{cr}S_x \leq M_p$$

Values of ϕM_n and M_n/Ω for values of $C_b = 1.0$ are graphed in *AISC Manual* Table 3-10.

Example 5.7

A simply supported beam of grade 50 steel has an unbraced length of 31 ft. Determine the lightest adequate W shape if the beam is subjected to a uniform factored bending moment of 190 ft-kips (LRFD) or 127 ft-kips (ASD) with $C_b = 1.0$.

Solution

LRFD Method

From *AISC Manual* Table 3-10, a W12 × 58 braced at 31 ft intervals has

$$\phi_b M_n = 196 \text{ ft-kips} > 190 \text{ ft-kips} \quad \text{[satisfactory]}$$

ASD Method

From *AISC Manual* Table 3-10, a W12 × 58 braced at 31 ft intervals has

$$\frac{M_n}{\Omega_b} = 131 \text{ ft-kips} > 127 \text{ ft-kips} \quad \text{[satisfactory]}$$

Lateral-Torsional Buckling Modification Factor

The lateral-torsional buckling modification factor, C_b, accounts for the influence of moment gradient on lateral-torsional buckling. The value of C_b is defined in AISC 360 Sec. F1 as

$$C_b = \frac{12.5 M_{\max}}{2.5 M_{\max} + 3 M_A + 4 M_B + 3 M_C} \quad \text{[AISC F1-1]}$$

A beam segment bent in single curvature and subjected to a uniform bending moment has a C_b value of 1.0. Other moment gradients increase the C_b value and increase the resistance of the beam to lateral-torsional buckling. For any loading condition, the C_b value may conservatively be taken as 1.0; this is the value adopted in *AISC Manual* tables. When the C_b value exceeds 1.0, tabulated values of the design moment may be multiplied by C_b, with a maximum permitted value of the design moment of $\phi_b M_p$. For a beam with a compression flange continuously braced along its entire length, such as a beam supporting a composite deck slab, $C_b = 1.0$.

The terms used in determining the C_b value are illustrated in Fig. 5.4, and typical values are shown in Fig. 5.5.

Figure 5.4 Determination of C_b

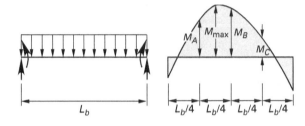

Figure 5.5 Typical Values of C_b

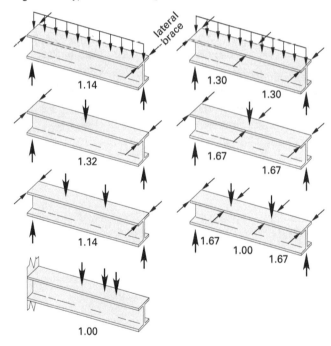

Example 5.8

The strength design loads acting on a simply supported beam with a cantilever overhang, including the beam self-weight, are shown. The beam is laterally braced at the supports. The free end of the cantilever is unbraced and warping is prevented at the support. Determine the relevant C_b values.

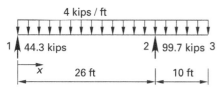

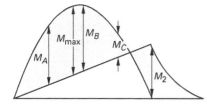

Solution

The bending moment at support 2 is

$$M_2 = \frac{wL^2}{2} = \frac{\left(4 \, \frac{\text{kips}}{\text{ft}}\right)(10 \text{ ft})^2}{2}$$
$$= 200 \text{ ft-kips}$$

The free bending moment in span 12 is

$$M_{\text{span}} = \frac{wL^2}{8} = \frac{\left(4 \, \frac{\text{kips}}{\text{ft}}\right)(26 \text{ ft})^2}{8}$$
$$= 338 \text{ ft-kips}$$

For the cantilever overhang with free end unbraced, AISC 360 Sec. F1 specifies that

$$C_b = 1.0$$

For span 12, the relevant terms are

$$M_{12} = V_1 x - \frac{wx^2}{2}$$
$$= (44.3 \text{ kips})x - \frac{\left(4 \, \frac{\text{kips}}{\text{ft}}\right)x^2}{2}$$
$$\frac{dM_{12}}{dx} = 44.3 \text{ kips} - \left(4 \, \frac{\text{kips}}{\text{ft}}\right)x$$

M_{12} is a maximum at $x = 11.1$ ft.

$$M_{\text{max}} = (44.3 \text{ kips})(11.1 \text{ ft}) - \frac{\left(4 \, \frac{\text{kips}}{\text{ft}}\right)(11.1 \text{ ft})^2}{2}$$
$$= 245 \text{ ft-kips}$$
$$M_A = 0.75 M_{\text{span}} - 0.25 M_2$$
$$= (0.75)(338 \text{ ft-kips}) - (0.25)(200 \text{ ft-kips})$$
$$= 204 \text{ ft-kips}$$

$$M_B = M_{\text{span}} - 0.5 M_2$$
$$= 338 \text{ ft-kips} - (0.5)(200 \text{ ft-kips})$$
$$= 238 \text{ ft-kips}$$
$$M_C = 0.75 M_{\text{span}} - 0.75 M_2$$
$$= (0.75)(338 \text{ ft-kips}) - (0.75)(200 \text{ ft-kips})$$
$$= 104 \text{ ft-kips}$$

The lateral-torsional buckling modification factor is given by AISC 360 Eq. F1-1 as

$$C_b = \frac{12.5 M_{\text{max}}}{2.5 M_{\text{max}} + 3 M_A + 4 M_B + 3 M_C}$$
$$= \frac{(12.5)(245 \text{ ft-kips})}{\begin{array}{c}(2.5)(245 \text{ ft-kips}) + (3)(204 \text{ ft-kips}) \\ + (4)(238 \text{ ft-kips}) + (3)(104 \text{ ft-kips})\end{array}}$$
$$= 1.23$$

Lateral-Torsional Buckling with $C_b > 1.0$

Figure 5.6 shows the effect of $C_b > 1.0$ on the relationship between L_b and M_n. The nominal moment in the elastic and inelastic regions is obtained by multiplying the tabulated available strength values in *AISC Manual* tables by C_b. The maximum permitted value of the nominal flexural capacity is limited to M_p.

Figure 5.6 *Variation of M_n with L_b for $C_b > 1.0$*

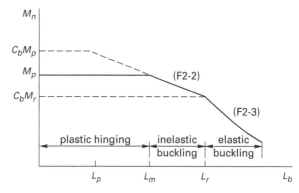

Plastic Phase: $L_b \leq L_m$

For a beam with a value of C_b greater than 1.0, the unbraced length for a full plastic moment of resistance is extended beyond L_p to L_m, which is given by

$$L_m = L_p + \frac{(C_b M_p - M_p)(L_r - L_p)}{C_b(M_p - M_r)}$$

$$= L_p + \frac{\phi_b M_{px}(C_b - 1.0)}{C_b(\phi_b BF)} \quad \text{[LRFD]}$$

$$= L_p + \frac{M_{px}(C_b - 1.0)}{C_b \Omega_b \left(\dfrac{BF}{\Omega_b}\right)} \quad \text{[ASD]}$$

Inelastic Phase: $L_m < L_b \leq L_r$

When L_b exceeds L_m and does not exceed L_r, collapse of the beam occurs prior to the development of the full plastic moment. When $L_b = L_r$, the nominal flexural strength of the beam is

$$M_n = C_b M_r = 0.7 C_b F_y S_x$$

For a value of L_b between L_m and L_r, the nominal flexural strength is obtained by linear interpolation between M_r and M_p, and the design flexural strength is given by

$$\phi_b M_n = C_b \big(\phi_b M_{px} - (\phi_b BF)(L_b - L_p)\big)$$

$$\frac{M_n}{\Omega_b} = C_b \left(\frac{M_{px}}{\Omega_b} - \left(\frac{BF}{\Omega_b}\right)(L_b - L_p)\right)$$

Elastic Phase: $L_b > L_r$

Collapse of the beam occurs by elastic lateral-torsional buckling and when $M_n = F_{cr} S_x$.

Values of ϕM_n and M_n/Ω are graphed in *AISC Manual* Table 3-10.

Example 5.9

Determine the lightest adequate W shape for the beam of Ex. 5.8 using grade 50 steel. The ASD load is 2.67 kips/ft, and the LRFD load is 4 kips/ft.

Solution

LRFD Method

For section 23, from *AISC Manual* Table 3-2, a W18 × 40 has

$$\phi_b M_{px} = 294 \text{ ft-kips}$$
$$\phi_b M_{rx} = 180 \text{ ft-kips}$$

$$L_p = 4.49 \text{ ft}$$
$$< 10 \text{ ft}$$
$$L_r = 13.1 \text{ ft}$$
$$> 10 \text{ ft}$$
$$\phi_b BF = 13.2 \text{ kips}$$

$$\phi_b M_n = C_b \big(\phi_b M_{px} - (\phi_b BF)(L_b - L_p)\big)$$
$$= (1.0)\big(294 \text{ ft-kips} - (13.2 \text{ kips})(10 \text{ ft} - 4.49 \text{ ft})\big)$$
$$= 221 \text{ ft-kips}$$
$$> 200 \text{ ft-kips} \quad \text{[satisfactory]}$$

For section 12, the equivalent design flexural strength required is

$$\phi_b M_n = \frac{M_{\max}}{C_b} = \frac{245 \text{ ft-kips}}{1.23}$$
$$= 199 \text{ ft-kips}$$

From *AISC Manual* Table 3-10, a W12 × 58 with an unbraced length of 26 ft has

$$\phi_b M_n = 227 \text{ ft-kips}$$
$$> 199 \text{ ft-kips} \quad \text{[satisfactory]}$$

Therefore, the W12 × 58 governs the design.

ASD Method

$$M_2 = 133 \text{ ft-kips}$$
$$M_{\max} = 163 \text{ ft-kips}$$

For section 23, from *AISC Manual* Table 3-2, a W18 × 40 has

$$\frac{M_{px}}{\Omega_b} = 196 \text{ ft-kips}$$

$$\frac{M_{rx}}{\Omega_b} = 119 \text{ ft-kips}$$

$$L_p = 4.49 \text{ ft}$$
$$< 10 \text{ ft}$$
$$L_r = 13.1 \text{ ft}$$
$$> 10 \text{ ft}$$

$$\frac{BF}{\Omega_b} = 8.94 \text{ kips}$$

$$\frac{M_n}{\Omega_b} = C_b \left(\frac{M_{px}}{\Omega_b} - \left(\frac{BF}{\Omega_b}\right)(L_b - L_p)\right)$$
$$= (1.0)\big(196 \text{ ft-kips} - (8.94 \text{ kips})(10 \text{ ft} - 4.49 \text{ ft})\big)$$
$$= 147 \text{ ft-kips}$$
$$> 133 \text{ ft-kips} \quad \text{[satisfactory]}$$

For section 12, the equivalent design flexural strength required is

$$\frac{M_n}{\Omega_b} = \frac{M_{max}}{C_b} = \frac{163 \text{ ft-kips}}{1.23}$$
$$= 133 \text{ ft-kips}$$

From *AISC Manual* Table 3-10, a W12 × 58 with an unbraced length of 26 ft has

$$\frac{M_n}{\Omega_b} = 151 \text{ ft-kips}$$
$$> 133 \text{ ft-kips} \quad [\text{satisfactory}]$$

Therefore, the W12 × 58 governs the design.

Moment Redistribution in Continuous Beams

AISC 360 Sec. B3.3 allows for the additional capacity that occurs in continuous beams after plastic hinges have formed because of redistribution of the bending moment. In an indeterminate structure, the formation of a single plastic hinge in the structure does not cause collapse of the structure. The structure can continue to support increasing load, while the moment at the hinge remains constant at a value of M_p and the moments at other locations in the structure continue to increase.

The method is applicable to beams with fixed ends and beams continuous over supports but not to rigid or pin-jointed frames. It is not applicable to simply supported beams or to cantilevers, as redistribution cannot occur in these members. Similarly, the method may be applied only to beams with compact sections, as noncompact sections have inadequate plastic hinge rotation capacity to permit redistribution of moments. The method may be applied to a beam-column, provided that the axial force in the member does not exceed $0.15\phi_c F_y A_g$ for LRFD or $0.15 F_y A_g / \Omega_c$ for ASD. Redistribution applies only to moments computed from an elastic analysis. An inelastic analysis automatically accounts for redistribution; for this reason, additional redistribution of moments is not applicable. The method is not applicable to members with F_y exceeding 65 kips/in².

The method is applied in accordance with AISC 360 Sec. B3.3, which states that negative moments at supports that are produced by gravity loads computed by an elastic analysis may be reduced by 10%, provided span moments are increased by 10% of the average adjacent support moments.

Example 5.10

The factored loading, including the beam self-weight, acting on a three-span continuous beam is shown. Continuous lateral support is provided to the beam. Determine the lightest adequate W shape using grade 50 steel.

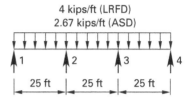

Solution

LRFD Method

From *AISC Manual* Table 3-23,

$$M_2 = \text{moment at interior support} = -0.10wL^2$$
$$= (-0.10)\left(4 \frac{\text{kips}}{\text{ft}}\right)(25 \text{ ft})^2$$
$$= -250 \text{ ft-kips}$$

$$M_{12} = 0.08wL^2 = (0.08)\left(4 \frac{\text{kips}}{\text{ft}}\right)(25 \text{ ft})^2$$
$$= 200 \text{ ft-kips}$$

Allowing for redistribution in accordance with AISC 360 Sec. B3.3, the required flexural strengths are

$$M_{u12} = 200 \text{ ft-kips} + \frac{(0.1)(0 + 250 \text{ ft-kips})}{2}$$
$$= 212.5 \text{ ft-kips}$$
$$M_{u2} = (0.9)(-250 \text{ ft-kips})$$
$$= -225 \text{ ft-kips} \quad [\text{governs}]$$

From *AISC Manual* Table 3-2, a W18 × 35 has

$$\phi_b M_p = 249 \text{ ft-kips}$$
$$> 225 \text{ ft-kips} \quad [\text{satisfactory}]$$

ASD Method

From *AISC Manual* Table 3-23,

$$M_2 = \text{moment at interior support}$$
$$= -0.10wL^2$$
$$= (-0.10)\left(2.67 \frac{\text{kips}}{\text{ft}}\right)(25 \text{ ft})^2$$
$$= -167 \text{ ft-kips}$$

$$M_{12} = 0.08wL^2$$

$$= (0.08)\left(2.67 \frac{\text{kips}}{\text{ft}}\right)(25 \text{ ft})^2$$

$$= 134 \text{ ft-kips}$$

Allowing for redistribution in accordance with AISC 360 Sec. B3.3, the required flexural strengths are

$$M_{a12} = 134 \text{ ft-kips} + \frac{(0.1)(0 \text{ ft-kips} + 167 \text{ ft-kips})}{2}$$

$$= 142.4 \text{ ft-kips}$$

$$M_{a2} = (0.9)(-167 \text{ ft-kips})$$

$$= -150 \text{ ft-kips} \quad [\text{governs}]$$

From *AISC Manual* Table 3-2, a W18 × 35 has

$$\frac{M_p}{\Omega_b} = 166 \text{ ft-kips}$$

$$> 150 \text{ ft-kips} \quad [\text{satisfactory}]$$

Biaxial Bending

A beam subjected to bending moment about both the x- and y-axes may be designed in accordance with AISC 360 Sec. H1.1 by using the following interaction expressions.

$$\frac{M_{rx}}{\phi_b M_{nx}} + \frac{M_{ry}}{\phi_b M_{ny}} \leq 1.00 \quad [\text{LRFD}]$$

$$\frac{M_{rx}\Omega_b}{M_{nx}} + \frac{M_{ry}\Omega_b}{M_{ny}} \leq 1.00 \quad [\text{ASD}]$$

Example 5.11

A simply supported W16 × 36 beam with a span of 15 ft is subjected to a uniformly distributed factored vertical load of 4 kips/ft and a horizontal lateral factored concentrated load of 3 kips applied at midspan for LRFD combinations. The corresponding ASD values are 2.67 kips/ft vertical and 2 kips horizontal. Determine whether the beam is adequate if the beam of grade 50 steel is laterally braced at the supports.

Solution

LRFD Method

The maximum bending moments due to the factored loads are

$$M_{rx} = \frac{wL^2}{8}$$

$$= \frac{\left(4 \frac{\text{kips}}{\text{ft}}\right)(15 \text{ ft})^2}{8}$$

$$= 112.5 \text{ ft-kips}$$

$$M_{ry} = \frac{WL}{4}$$

$$= \frac{(3 \text{ kips})(15 \text{ ft})}{4}$$

$$= 11.3 \text{ ft-kips}$$

For bending about the x-axis with an unbraced length of 15 ft, from *AISC Manual* Table 3-10, a W16 × 36 has

$$\phi_b M_{nx} = C_b(150 \text{ ft-kips})$$

$$= (1.14)(150 \text{ ft-kips})$$

$$= 171 \text{ ft-kips}$$

For bending about the y-axis, from *AISC Manual* Table 3-4, a W16 × 36 has

$$\phi_b M_{ny} = 40.5 \text{ ft-kips}$$

The left side of the interaction equation is

$$\frac{M_{rx}}{\phi_b M_{nx}} + \frac{M_{ry}}{\phi_b M_{ny}} = \frac{112.5 \text{ ft-kips}}{171 \text{ ft-kips}} + \frac{11.3 \text{ ft-kips}}{40.5 \text{ ft-kips}}$$

$$= 0.94$$

$$< 1.0 \quad [\text{satisfactory}]$$

ASD Method

The maximum bending moments due to the factored loads are

$$M_{rx} = \frac{wL^2}{8} = \frac{\left(2.67 \frac{\text{kips}}{\text{ft}}\right)(15 \text{ ft})^2}{8}$$

$$= 75 \text{ ft-kips}$$

$$M_{ry} = \frac{WL}{4} = \frac{(2 \text{ kips})(15 \text{ ft})}{4}$$

$$= 7.5 \text{ ft-kips}$$

For bending about the x-axis with an unbraced length of 15 ft, from *AISC Manual* Table 3-10, a W16 × 36 has

$$\frac{M_{nx}}{\Omega_b} = C_b(100 \text{ ft-kips})$$
$$= (1.14)(100 \text{ ft-kips})$$
$$= 114 \text{ ft-kips}$$

For bending about the y-axis, from *AISC Manual* Table 3-4, a W16 × 36 has

$$\frac{M_{ny}}{\Omega_b} = 26.9 \text{ ft-kips}$$

The left side of the interaction equation is

$$\frac{M_{rx}\Omega_b}{M_{nx}} + \frac{M_{ry}\Omega_b}{M_{ny}} = \frac{75 \text{ ft-kips}}{114 \text{ ft-kips}} + \frac{7.5 \text{ ft-kips}}{26.9 \text{ ft-kips}}$$
$$= 0.94$$
$$< 1.0 \quad \text{[satisfactory]}$$

4. DESIGN FOR SHEAR

Nomenclature

A_e	effective net area	in²
A_{gv}	gross area subject to shear	in²
A_{nt}	net area subject to tension	in²
A_{nv}	net area subject to shear	in²
A_w	web area	in²
C_{v1}	web shear coefficient	–
d	overall depth of member	in
d_b	nominal bolt diameter	in
d_h	diameter of bolt hole	in
F_u	specified minimum tensile strength	kips/in²
F_{yw}	specified minimum yield stress of the web material	kips/in²
h	for rolled shapes, the distance between flanges less the corner radius; for built-up sections, the clear distance between flanges	in
k	distance from outer face of flange to web toe of fillet	in
l_b	length of bearing	in
P_n	rupture strength in tension	kips
R	nominal reaction	kips
s	bolt spacing	in
t_f	flange thickness	in
t_w	web thickness	in
U_{bs}	reduction coefficient	–
V_a	required shear strength	kips
V_{nx}	nominal shear strength	kips
V_u	required shear strength	kips

Shear in Beam Webs

The shear in rolled W shape beams is resisted by the area of the web that is defined as

$$A_w = dt_w$$

It is assumed that the shear stress is uniformly distributed over this area, and for a slenderness ratio $h/t_w \leq 2.24\sqrt{E/F_y}$, the nominal shear strength is governed by yielding of the web. This is the case for most W shapes, and the nominal shear strength is given by AISC 360 Sec. G2.1 as

$$V_n = 0.6F_y A_w C_{v1} \qquad \text{[AISC G2-1]}$$
$$C_{v1} = 1.0$$

The design shear strength is

$$\phi_v V_{nx} = 1.0\, V_{nx}$$

The allowable shear strength is

$$\frac{V_{nx}}{\Omega_v} = \frac{V_{nx}}{1.5}$$

Example 5.12

Check the adequacy in shear of the W12 × 53 grade 50 beam in Ex. 5.8 at support 2.

Solution

LRFD Method

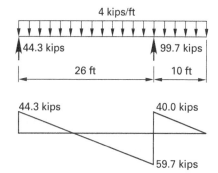

The shear force diagram is shown in the illustration, and the required shear strength is

$$V_u = 59.7 \text{ kips}$$

The design shear strength is obtained from *AISC Manual* Table 3-2 as

$$\phi_v V_{nx} = 125 \text{ kips}$$
$$> V_u \quad \text{[satisfactory]}$$

ASD Method

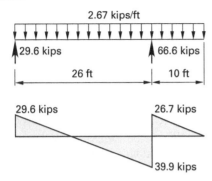

The shear force diagram is shown in the illustration, and the required shear strength is

$$V_a = 39.9 \text{ kips}$$

The allowable shear strength is obtained from *AISC Manual* Table 3-2 as

$$\frac{V_{nx}}{\Omega_v} = 83.5 \text{ kips}$$

$$> V_a \quad \text{[satisfactory]}$$

Block Shear

As shown in Fig. 5.7, failure may occur by block shear in a coped beam. Block shear is a combination of shear along the vertical plane and tension along the horizontal plane. Block shear strength is the sum of the strengths of the shear area and tension area. Tension failure occurs by rupture in the net tension area. Shear failure occurs either by rupture in the net shear area or by shear yielding in the gross shear area, with the minimum value governing.

Figure 5.7 *Block Shear in a Coped Beam*

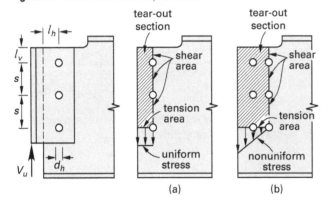

The nominal strengths are given in AISC 360 Sec. J4 and AISC 360 Sec. D2 as follows.

- Rupture strength in tension

$$P_n = F_u A_e \quad \text{[AISC J4-2]}$$
$$= U_{bs} F_u A_{nt}$$

- Yield strength in shear

$$V_n = 0.6 F_y A_{gv} \quad \text{[AISC J4-3]}$$

- Rupture strength in shear

$$V_n = 0.6 F_u A_{nv} \quad \text{[AISC J4-4]}$$

The nominal resistance to block shear is given by AISC 360 Eq. J4-5 as

$$R_n = 0.6 F_u A_{nv} + U_{bs} F_u A_{nt}$$
$$\leq 0.6 F_y A_{gv} + U_{bs} F_u A_{nt}$$

The reduction coefficient is

$$U_{bs} = 1.0 \text{ for uniform tension stress}$$
$$= 0.5 \text{ for nonuniform tension stress}$$

The resistance factor is

$$\phi = 0.75$$

The safety factor is

$$\Omega = 2.00$$

Example 5.13

Determine the resistance to block shear of the coped W16 × 40 grade A36 beam shown in Fig. 5.7(a). The relevant dimensions are $l_h = l_v = 1.5$ in and $s = 3$ in. The bolt diameter is ³⁄₄ in.

Solution

The effective hole diameter for a ³⁄₄ in diameter bolt is defined in AISC 360 Sec. B4.3b and AISC 360 Table J3.3 as

$$d_h = d_b + \frac{1}{8} \text{ in} = 0.75 \text{ in} + 0.125 \text{ in}$$
$$= 0.875 \text{ in}$$
$$t_w = 0.305 \text{ in}$$
$$A_{nv} = t_w (l_v + 2s - 2.5 d_h)$$
$$= t_w \big(1.5 \text{ in} + (2)(3.0 \text{ in}) - (2.5)(0.875 \text{ in})\big)$$
$$= 5.31 t_w \text{ in}^2$$
$$A_{nt} = t_w (l_h - 0.5 d_h)$$
$$= t_w \big(1.5 \text{ in} - (0.5)(0.875 \text{ in})\big)$$
$$= 1.06 t_w \text{ in}^2$$

The tensile stress is uniform, and the reduction coefficient is

$$U_{bs} = 1.0$$

$$U_{bs}F_u A_{nt} = \left(58 \ \frac{\text{kips}}{\text{in}^2}\right)(1.06t_w \ \text{in}^2)$$
$$= 61.48t_w \ \text{kips}$$

$$0.6F_u A_{nv} = (0.6)\left(58 \ \frac{\text{kips}}{\text{in}^2}\right)(5.31t_w \ \text{in}^2)$$
$$= 184.79t_w \ \text{kips}$$

$$A_{gv} = t_w(l_v + 2s)$$
$$= t_w\big(1.5 \ \text{in} + (2)(3 \ \text{in})\big)$$
$$= 7.5t_w \ \text{in}^2$$

$$0.6F_y A_{gv} = (0.6)\left(36 \ \frac{\text{kips}}{\text{in}^2}\right)(7.5t_w \ \text{in}^2)$$
$$= 162t_w \ \text{kips} \quad [\text{governs}]$$
$$< 0.6F_u A_{nv}$$

LRFD Method

Shear yielding governs and the design strength for block shear is given by AISC 360 Eq. J4-5 as

$$\phi R_n = \phi(0.6F_y A_{gv} + U_{bs}F_u A_{nt})$$
$$= (0.75)(0.305 \ \text{in})\left(162 \ \frac{\text{kips}}{\text{in}} + 61.48 \ \frac{\text{kips}}{\text{in}}\right)$$
$$= 51.12 \ \text{kips}$$

ASD Method

Shear yielding governs, and the allowable strength for block shear is given by AISC 360 Eq. J4-5 as

$$\frac{R_n}{\Omega} = \frac{0.6F_y A_{gv} + U_{bs}F_u A_{nt}}{\Omega}$$
$$= \frac{(0.305 \ \text{in})\left(162 \ \frac{\text{kips}}{\text{in}} + 61.48 \ \frac{\text{kips}}{\text{in}}\right)}{2}$$
$$= 34.08 \ \text{kips}$$

Web Local Yielding

As shown in Fig. 5.8, a bearing plate may be used to distribute concentrated loads applied to the flange to prevent web local yielding. The load is assumed to be dispersed, at a gradient of 2.5 to 1.0, to the web toe of fillet. For loads applied at a distance of not more than d from the end of the beam, the nominal strength is given by AISC 360 Sec. J10.2 as

$$R_n = F_{yw}t_w(2.5k + l_b) \quad [\text{AISC J10-3}]$$

Figure 5.8 Web Local Yielding

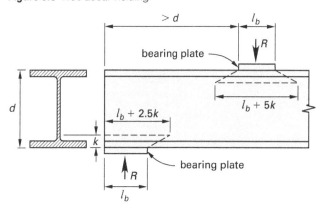

For loads applied at a distance of more than d from the end of the beam, the nominal strength is given as

$$R_n = F_{yw}t_w(5k + l_b) \quad [\text{AISC J10-2}]$$

The design strength is given by ϕR_n with $\phi = 1.0$. *AISC Manual* Table 9-4 tabulates values of

$$\phi R_1 = \phi(2.5kF_y t_w)$$
$$\phi R_2 = \phi(F_y t_w)$$

The allowable strength is given by R_n/Ω with $\Omega = 1.5$. *AISC Manual* Table 9-4 tabulates values of

$$\frac{R_1}{\Omega} = \frac{2.5kF_y t_w}{\Omega}$$
$$\frac{R_2}{\Omega} = \frac{F_y t_w}{\Omega}$$

Example 5.14

Determine the resistance to local web yielding of a W40 × 331 grade 50 beam with a 4.0 in long bearing plate at the end of the beam.

Solution

LRFD Method

AISC 360 Eq. J10-3 is applicable. Using values from *AISC Manual* Table 9-4, the design strength is

$$\phi R_n = \phi F_y t_w(2.5k + l_b) = \phi(R_1 + l_b R_2)$$
$$= 505 \ \text{kips} + (4.0 \ \text{in})\left(61.0 \ \frac{\text{kips}}{\text{in}}\right)$$
$$= 749 \ \text{kips}$$

Steel

ASD Method

AISC 360 Eq. J10-3 is applicable. Using values from *AISC Manual* Table 9-4, the allowable strength is

$$\frac{R_n}{\Omega} = \frac{F_y t_w(2.5k + l_b)}{\Omega} = \frac{R_1 + l_b R_2}{\Omega}$$

$$= 337 \text{ kips} + (4.0 \text{ in})\left(40.7 \frac{\text{kips}}{\text{in}}\right)$$

$$= 500 \text{ kips}$$

Web Crippling

For a concentrated load applied at a distance of not less than $d/2$ from the end of the beam, the nominal strength against web crippling is given by AISC 360 Sec. J10.3 as

$$R_n = 0.80 t_w^2\left[1 + 3\left(\frac{l_b}{d}\right)\left(\frac{t_w}{t_f}\right)^{1.5}\right]\sqrt{\frac{EF_{yw}t_f}{t_w}}$$

[AISC J10-4]

For loads applied at a distance of less than $d/2$ from the end of the beam and for $l_b/d \le 0.2$, the nominal strength is given by

$$R_n = 0.40 t_w^2\left[1 + 3\left(\frac{l_b}{d}\right)\left(\frac{t_w}{t_f}\right)^{1.5}\right]\sqrt{\frac{EF_{yw}t_f}{t_w}}$$

[AISC J10-5a]

For loads applied at a distance of less than $d/2$ from the end of the beam and for $l_b/d > 0.2$, the nominal strength is given by

$$R_n = 0.40 t_w^2\left[1 + \left(4\left(\frac{l_b}{d}\right) - 0.2\right)\left(\frac{t_w}{t_f}\right)^{1.5}\right]\sqrt{\frac{EF_{yw}t_f}{t_w}}$$

[AISC J10-5b]

The design strength is given by $\phi_r R_n$ with $\phi_r = 0.75$. Using values of $\phi_r R_3$, $\phi_r R_4$, $\phi_r R_5$, and $\phi_r R_6$ tabulated in *AISC Manual* Table 9-4 reduces AISC 360 Eq. J10-5a to

$$\phi_r R_n = \phi_r R_3 + l_b(\phi_r R_4)$$

AISC 360 Eq. J10-5b becomes

$$\phi_r R_n = \phi_r R_5 + l_b(\phi_r R_6)$$

The allowable strength is given by R_n/Ω_r with

$\Omega_r = 2.00$. Using values of R_3/Ω_r, R_4/Ω_r, R_5/Ω_r, and R_6/Ω_r tabulated in *AISC Manual* Table 9-4 reduces AISC 360 Eq. J10-5a to

$$\frac{R_n}{\Omega_r} = \frac{R_3}{\Omega_r} + l_b\left(\frac{R_4}{\Omega_r}\right)$$

AISC 360 Eq. J10-5b becomes

$$\frac{R_n}{\Omega_r} = \frac{R_5}{\Omega_r} + l_b\left(\frac{R_6}{\Omega_r}\right)$$

Example 5.15

Determine the design web crippling strength of a W40 × 331 grade 50 beam with a 3.25 in long bearing plate at the end of the beam.

Solution

LRFD Method

For $l_b/d \le 0.2$, using values from *AISC Manual* Table 9-4, the applicable expression is

$$\phi_r R_n = \phi_r R_3 + l_b(\phi_r R_4)$$

$$= 710 \text{ kips} + (3.25 \text{ in})\left(22.6 \frac{\text{kips}}{\text{in}}\right)$$

$$= 783 \text{ kips}$$

ASD Method

For $l_b/d \le 0.2$, using values from *AISC Manual* Table 9-4, the applicable expression is

$$\frac{R_n}{\Omega_r} = \frac{R_3}{\Omega_r} + l_b\left(\frac{R_4}{\Omega_r}\right)$$

$$= 474 \text{ kips} + (3.25 \text{ in})\left(15.1 \frac{\text{kips}}{\text{in}}\right)$$

$$= 523 \text{ kips}$$

5. DESIGN OF COMPRESSION MEMBERS

Nomenclature

A	area of member	in²
A_g	gross area of member	in²
b_f	flange width of rolled beam	in
b_x	$8/9M_{cx} \times 10^3$	(ft-kips)⁻¹
b_y	$8/9M_{cy} \times 10^3$	(ft-kips)⁻¹
B	base plate width	in
B_1	moment magnification factor applied to the primary moments to account for the curvature of the members, with lateral translation inhibited, as defined in AISC 360 Eq. A-8-3	–

B_2 — moment magnification factor applied to the primary moments to account for the translation of the members, with lateral translation permitted as defined in AISC 360 Eq. A-8-6 — –

C_b — bending coefficient — –

C_m — reduction factor given by AISC 360 Eq. A-8-4 — –

d — depth of rolled beam — in

E — modulus of elasticity — kips/in^2

EA^* — reduced value of EA — kips

EI^* — reduced value of EI — kips-in^2

F_{cr} — critical stress — kips/in^2

F_e — elastic critical buckling stress — kips/in^2

F_y — yield stress — kips/in^2

G — ratio of total column stiffness framing into a joint to total girder stiffness framing into the joint — –

H — horizontal force — kips

H — story shear produced by the lateral forces used to compute Δ_H — kips

H_p — lateral load required to produce the design story drift — kips

I — moment of inertia — in^4

K — effective length factor — –

K_1 — effective length factor in the plane of bending — –

K_2 — effective length factor in the plane of bending for an unbraced frame — –

L — actual unbraced length in the plane of bending — ft

L_c — effective length, KL — ft

L_{c1} — effective length for non-sway column, K_1L — ft

L — story height — ft

m — cantilever dimension for base plate along the length of the plate — in

M_a — augmented moment — ft-kips

M_c — available flexural strength — ft-kips

M_{lt} — calculated first-order factored moment in a member, due to lateral translation of the frame only — ft-kips

M_{nt} — calculated first-order factored moment assuming no lateral translation of the frame — ft-kips

M_{nx} — nominal flexural strength about strong axis in the absence of axial load — ft-kips

M_{ny} — nominal flexural strength about weak axis in the absence of axial load — ft-kips

M_r — required second-order factored flexural strength for LRFD or ASD load combinations — ft-kips

M_{rx} — required second-order factored bending moment about strong axis — ft-kips

M_{ry} — required second-order factored bending moment about weak axis — ft-kips

M_1 — smaller moment at end of unbraced length of member, calculated from a first-order analysis — ft-kips

M_2 — larger moment at end of unbraced length of member, calculated from a first-order analysis — ft-kips

n — cantilever dimension for base plate along the width of the plate — in

N — base plate length — in

N_i — notional lateral load applied at level i — kips

p — $1/P_c \times 10^3$ — kips^{-1}

P_a — augmented axial force — kips

P_c — available axial compression strength — kips

$P_{e,story}$ — sum, for all columns in a story of a moment frame, of the Euler buckling strength — kips

P_e — Euler buckling strength — kips

P_{e1} — Euler buckling strength of the member in the plane of bending as defined in AISC 360 Eq. A-8-5 — kips

P_{lt} — calculated first-order factored axial force in a member, due to lateral translation of the frame only — kips

P_{mf} — total vertical load in columns in the story — kips

P_n — nominal axial strength — kips

P_{nt} — first-order factored axial force assuming no lateral translation of the frame — kips

P_r — required second-order axial strength for LRFD or ASD load combinations — kips

P_{story} — total factored vertical load supported by the story, including gravity columns loads — kips

P_y — axial yield strength — kips

r — radius of gyration — in

R — restraint — kips

R_M — system coefficient to account for the influence of $P-\delta$ on $P-\Delta$ — –

t_{req} — required base plate thickness — in

Y_i — gravity load applied at level i independent of loads from above — kips

Symbols

α — force level adjustment factor — –

Δ_a — interstory drift due to applied loads — in

Δ_H — first-order interstory drift due to lateral forces — in

Δ_{oh} — translational deflection of the story under consideration — in

Δ_p — permissible interstory drift — in

Δ_{1st} — first-order drift — in

Δ_{2nd} — second-order drift — in

λ — F_y/F_e — –

$\sum(I_c/L_c)$ — the sum of the I/L values for all columns meeting at a joint — in^3

$\sum (I_g/L_g)$	the sum of the I/L values for all girders meeting at a joint	in^3
τ_b	stiffness reduction coefficient	–
ϕ_b	resistance factor for flexure	–
ϕ_c	resistance factor for compression	–
Ω_b	safety factor for flexure	–
Ω_c	safety factor for compression	–

Effective Length

In the design of compression members, the effective length factor K is used to account for the influence of restraint conditions at each end of a column. The K factor is used to equate the nominal strength of a compression member of length L to that of an equivalent pin-ended member of length KL. The effective length of a member is denoted by $KL = L_c$. The nominal strength of a compression member is dependent on the slenderness ratio, L_c/r, which is limited to a maximum recommended value of 200.

AISC 360 Comm. Table C-A-7.1 specifies effective length factors for well-defined conditions of restraint; these are illustrated in Fig. 5.9. These values may only be used in simple cases when the tabulated end conditions are closely approximated in practice.

For compression members in a braced frame, an effective length factor of 1.0 is used. For load-bearing web stiffeners on a girder, AISC 360 Sec. J10.8 specifies an effective length factor of 0.75. For the braces in an X-braced frame that are attached at midspan, AISC 360 Comm. App. 7.2 specifies a value for the effective length factor of 0.5. For a moment frame, when the sidesway amplification factor $B_2 \geq 1.1$, an effective length factor of $K = 1.0$ may be used. For compression members forming part of a frame with rigid joints, AISC 360 Comm. App. 7.2 presents alignment charts for determining the effective length for the two conditions of sidesway prevented and sidesway permitted; these charts are illustrated in Fig. 5.10.

To use the alignment charts, the stiffness ratio at the two ends of the column under consideration must be determined. This ratio is defined as

$$G = \frac{\sum \left(\dfrac{I_c}{L_c} \right)}{\sum \left(\dfrac{I_g}{L_g} \right)}$$

For a braced frame with rigid joints, the girders are bent in single curvature and the alignment charts are based on a stiffness value of $2EI/L$. If one end of a girder is pinned, its stiffness is $3EI/L$; if one end is fixed, its stiffness is $4EI/L$. Therefore, for these two cases, the I_g/L_g

Figure 5.9 *Effective Length Factors*

illus.	end conditions	theoretical	design
		K	
(a)	both ends pinned	1	1.00
(b)	both ends built in	0.5	0.65
(c)	one end pinned, one end built in	0.7	0.8
(d)	one end built in, one end free	2	2.10
(e)	one end built in, one end fixed against rotation but free to translate	1	1.20
(f)	one end pinned, one end fixed against rotation but free to translate	2	2.0

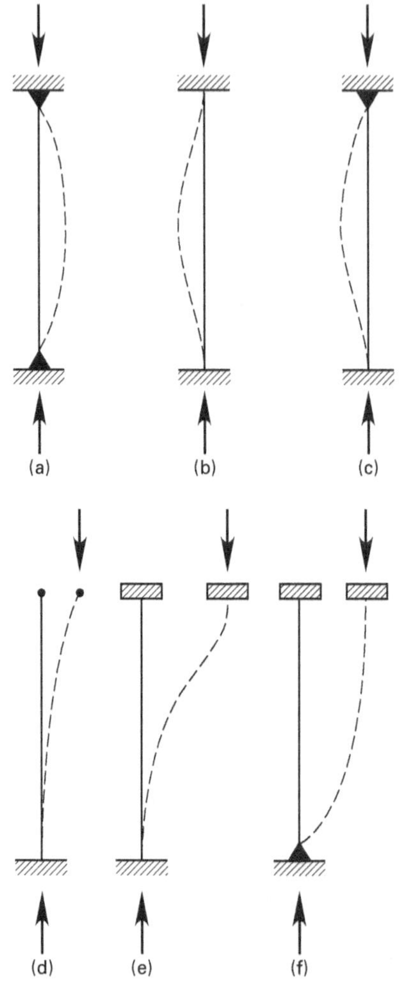

values are multiplied by 1.5 and 2.0, respectively. For a sway frame with rigid joints, the girders are bent in double curvature and the alignment charts are based on a stiffness value of $6EI/L$. If one end of a girder is pinned, its stiffness is $3EI/L$ and the I_g/L_g values are multiplied by 0.5. For a column with a pinned base, AISC 360 Comm. App. 7.2 specifies a stiffness ratio of $G = 10$. For a column with a fixed base, AISC 360 Comm. App. 7.2 specifies a stiffness ratio of $G = 1$.

Figure 5.10 *Alignment Charts for Effective Length Factors*

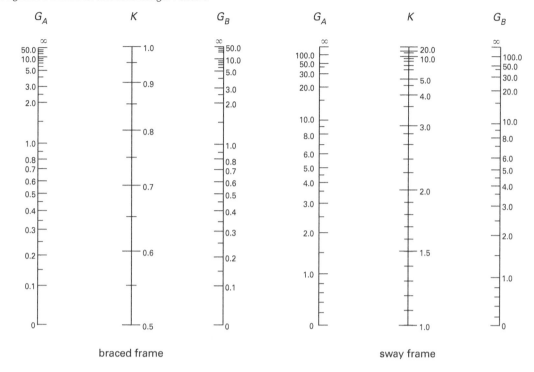

braced frame   sway frame

Adapted from American Institute of Steel Construction, *Specifications for Structural Steel Buildings*, Commentary Fig. C-A-7.1 and Fig. C-A-7.2, 2012.

Example 5.16

The sway frame shown consists of members with identical I/L values. Determine the effective length factors of columns 12 and 34.

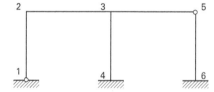

Solution

For the pinned connection at joint 1, AISC 360 Comm. App. 7.2 specifies a stiffness ratio of $G_1 = 10$.

At joint 2,

$$G_2 = \frac{\sum\left(\dfrac{I_c}{L_c}\right)}{\sum\left(\dfrac{I_g}{L_g}\right)} = \frac{1.0 \text{ in}^3}{1.0 \text{ in}^3} = 1$$

From the alignment chart for sway frames, the effective length factor is

$$K_{12} = 1.9$$

Allowing for the pinned end at joint 5, the sum of the adjusted relative stiffness values for the two girders connected to joint 3 is

$$\sum\left(\frac{I_g}{L_g}\right) = 1.0 \text{ in}^3 + 0.5 \text{ in}^3 = 1.5 \text{ in}^3$$

The stiffness ratio at joint 3 is given by

$$G_3 = \frac{\sum\left(\dfrac{I_c}{L_c}\right)}{\sum\left(\dfrac{I_g}{L_g}\right)} = \frac{1.0 \text{ in}^3}{1.5 \text{ in}^3}$$
$$= 0.67$$

For the fixed connection at joint 4, AISC 360 Comm. App. 7.2 specifies a stiffness ratio of $G_4 = 1.0$. From the alignment chart for a sway frame, the effective length factor for column 34 is

$$K_{34} = 1.27$$

Axially Loaded Members

The design strength in compression is given by

$$\phi_c P_n = 0.90 A_g F_{cr}$$

The allowable strength in compression is given by

$$\frac{P_n}{\Omega_c} = \frac{A_g F_{cr}}{1.67}$$

Short Column

For a short column with $L_c/r \leq 4.71(E/F_y)^{0.5}$ and $F_e \geq 0.44 F_y$, inelastic buckling governs, and AISC 360 Eq. E3-2 defines the critical stress as

$$F_{cr} = (0.658^\lambda) F_y$$

The parameter λ is defined by

$$\lambda = \frac{F_y}{F_e}$$

The elastic critical buckling stress is given by AISC 360 Eq. E3-4 as

$$F_e = \frac{\pi^2 E}{\left(\dfrac{L_c}{r}\right)^2}$$

Long Column

For a long column with $L_c/r > 4.71(E/F_y)^{0.5}$ and $F_e < 0.44 F_y$, elastic buckling governs, and the critical stress is given by AISC 360 Eq. E3-3 as

$$F_{cr} = 0.877 F_e$$

Once the governing slenderness ratio of a column is established, the design stress, $\phi_c F_{cr}$, and the allowable stress, F_{cr}/Ω_c, may be obtained directly from *AISC Manual* Table 4-14 for steel members with a yield stress of 35, 36, 46, 50, 65, or 70 kips/in^2.

Values of the design axial strength and the allowable axial strength are tabulated in *AISC Manual* Tables 4-1a ($F_y = 50$ kips/in^2), 4-1b ($F_y = 65$ kips/in^2), and 4-1c ($F_y = 70$ kips/in^2) for rolled sections W14 and smaller with respect to r_y for varying effective lengths. These tabulated values may be used directly when $(L_c/r)_y$ exceeds $(L_c/r)_x$.

Example 5.17

Determine the lightest W12 grade 50 column that will support a factored load of 850 kips (LRFD) or 570 kips (ASD). The column is 12 ft high, is pinned at each end, and has no intermediate bracing about either axis.

Solution

LRFD Method

From *AISC Manual* Table 4-1a, a W12 × 79 column with an effective length of 12 ft has a design axial strength of

$$\phi_c P_n = 887 \text{ kips} \quad \text{[satisfactory]}$$

ASD Method

From *AISC Manual* Table 4-1a, a W12 × 79 column with an effective length of 12 ft has an allowable axial strength of

$$\frac{P_n}{\Omega_c} = 590 \text{ kips} \quad \text{[satisfactory]}$$

Buckling About Major Axis

When the minor axis of a W-shape is braced at closer intervals than the major axis, the slenderness ratio about both axes must be investigated to determine which governs. The larger of the two values will control the design. When the slenderness ratio about the minor axis, $(L_c/r)_y$, governs, the available axial strength values tabulated in *AISC Manual* Tables 4-1a ($F_y = 50$ kips/in^2), 4-1b ($F_y = 65$ kips/in^2), and 4-1c ($F_y = 70$ kips/in^2) may be utilized directly. When the slenderness ratio about the major axis, $(L_c/r)_x$, governs, the effective length about the major axis is divided by r_x/r_y to give an equivalent effective length about the minor axis, which has the same load carrying capacity as the actual effective length about the major axis. *AISC Manual* Tables 4-1a ($F_y = 50$ kips/in^2), 4-1b ($F_y = 65$ kips/in^2), and 4-1c ($F_y = 70$ kips/in^2) may then be used to obtain the available design strength of the member in compression by the equivalent length $(L_c)_x/(r_x/r_y)$.

Equivalent Effective Length

When the effective lengths of a column about the x- and y-axes are different, the strength of the column must be investigated with respect to both axes. Dividing the effective length about the x-axis by the ratio r_x/r_y provides an equivalent effective length about the y-axis.

Example 5.18

Determine the available axial strength of a W12 × 106 grade 50 column that is 12 ft high, pinned at each end, and braced at midheight about the y-axis.

Solution

The effective length about the y-axis is

$$(L_c)_y = 6 \text{ ft}$$

The effective length about the x-axis is

$$(L_c)_x = 12 \text{ ft}$$

From *AISC Manual* Table 4-1a, a W12 × 106 column has a value of

$$\frac{r_x}{r_y} = 1.76$$

The equivalent effective length about the major axis with respect to the y-axis is

$$(L_c)_{y,\text{equiv}} = \frac{(L_c)_x}{\dfrac{r_x}{r_y}} = \frac{12 \text{ ft}}{1.76}$$
$$= 6.8 \text{ ft} \quad [\text{governs}]$$
$$> (L_c)_y$$

LRFD Method

From *AISC Manual* Table 4-1a, a W12 × 106 column with an effective length, $(L_c)_{y,\text{equiv}}$, of 6.8 ft has a design axial strength of

$$\phi_c P_n = 1334 \text{ kips}$$

ASD Method

From *AISC Manual* Table 4-1a, a W12 × 106 column with an effective length, $(L_c)_{y,\text{equiv}}$, of 6.8 ft has an allowable axial strength of

$$\frac{P_n}{\Omega_c} = 888 \text{ kips}$$

Built-Up Sections

For built-up sections and laced compression members, *AISC Manual* Table 4-14 tabulates $\phi_c F_{cr}$ and F_{cr}/Ω_c against L_c/r for steel with yield stresses of 35 kips/in², 36 kips/in², 46 kips/in², 50 kips/in², 65 kips/in², and 70 kips/in², respectively.

Example 5.19

A laced column consisting of four 5 × 5 × ½ angles of grade A36 steel is shown. The column may be considered a single integral member and is 20 ft high with pinned ends. Determine the maximum design axial load.

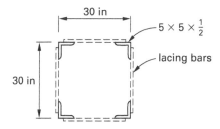

Solution

The relevant properties of a 5 × 5 × ½ angle are

$$A = 4.79 \text{ in}^2$$
$$I = 11.3 \text{ in}^4$$
$$y = 1.42 \text{ in}$$

The relevant properties of the laced column are

$$\sum A = 4A$$
$$= (4)(4.79 \text{ in}^2)$$
$$= 19.16 \text{ in}^2$$
$$\sum I = 4I + \sum A\left(\frac{d}{2} - y\right)^2$$
$$= (4)(11.3 \text{ in}^4) + (19.16 \text{ in}^2)(15 \text{ in} - 1.42 \text{ in})^2$$
$$= 3579 \text{ in}^4$$

The radius of gyration of the laced column is

$$r = \sqrt{\frac{\sum I}{\sum A}} = \sqrt{\frac{3579 \text{ in}^4}{19.16 \text{ in}^2}}$$
$$= 13.67 \text{ in}$$

The slenderness ratio of the laced column is

$$\frac{L_c}{r} = \frac{(1.0)(20 \text{ ft})\left(12 \dfrac{\text{in}}{\text{ft}}\right)}{13.67 \text{ in}}$$
$$= 17.56$$
$$< 200 \quad [\text{satisfactory}]$$

From *AISC Manual* Table 4-14, the design stress is

$$\phi_c F_{cr} = 31.9 \text{ kips/in}^2$$

The allowable stress is

$$\frac{F_{cr}}{\Omega_c} = 21.2 \text{ kips/in}^2$$

The design axial strength is

$$\phi_c P_n = \phi_c F_{cr} \sum A = \left(31.9 \frac{\text{kips}}{\text{in}^2}\right)(19.16 \text{ in}^2)$$
$$= 611 \text{ kips}$$

The allowable axial strength is

$$\frac{P_n}{\Omega_c} = \frac{F_{cr}\sum A}{\Omega_c} = \left(21.2 \frac{\text{kips}}{\text{in}^2}\right)(19.16 \text{ in}^2)$$
$$= 406 \text{ kips}$$

Composite Columns

Concrete filled hollow structural sections and concrete encased rolled steel sections reinforced with longitudinal and lateral reinforcing bars are designed by using AISC 360 Sec. I2. Values of the design axial strength for typical sizes of HSS columns are tabulated in *AISC Design Examples*[4] Part IV.

Example 5.20

Determine the least weight square composite column using an HSS section, with a yield strength of 50 kips/in², filled with 5000 lbf/in² concrete that can support a factored load of 730 kips (LRFD) or 485 kips (ASD). The column is 15 ft high and is pinned at each end.

Solution

LRFD Method

From *AISC Design Examples* Table IV-2B, an HSS 12 × 12 × ⁵⁄₁₆ in with an effective length of 15 ft has a design axial strength of

$$\phi_c P_n = 805 \text{ kips} \quad [\text{satisfactory}]$$

ASD Method

From *AISC Design Examples* Table IV-2B, an HSS 12 × 12 × ⁵⁄₁₆ in with an effective length of 15 ft has an allowable axial strength of

$$\frac{P_n}{\Omega_c} = 537 \text{ kips} \quad [\text{satisfactory}]$$

Second-Order Effects

In accordance with AISC 360 Sec. C2, the design of compression members must take into account secondary effects determined from an elastic analysis. The secondary moments and axial forces caused by the *P*-delta effects must be added to the primary moments and axial forces in a member, which were obtained by a first-order analysis. The *P*-delta effects are the result of the two separate effects *P*-δ and *P*-Δ, as shown in Fig. 5.11. The final forces in a frame (including secondary effects) may be obtained as the summation of the two analyses, sway and non-sway.

Figure 5.11 *P-delta Effects*

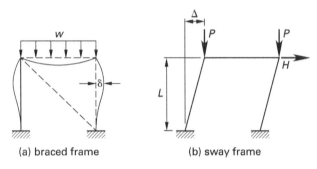

(a) braced frame   (b) sway frame

The *P*-δ effect produces an amplified moment due to the eccentricity of the axial force with respect to the displaced centerline of the member. This is termed the *member effect*. The moment magnification factor which, when applied to the primary moments, accounts for the *P*-δ effect, is termed B_1.

The *P*-Δ effect produces an amplified moment due to the drift in a sway frame. This is termed the *sidesway effect*. The moment magnification factor which, when applied to the primary moments, accounts for the *P*-Δ effect, is termed B_2.

When sidesway is inhibited,

$$B_2 = 0$$

For a member not subject to axial compression,

$$B_1 = 1.0$$

Approximate Second-Order Analysis

As shown in Fig. 5.12, two first-order analyses are required in order to determine M_{nt} and M_{lt}. In the first analysis, imaginary horizontal restraints are introduced at each floor level to prevent lateral translation. The factored loads are then applied, the primary moments, M_{nt}, are calculated, and the magnitudes of the imaginary restraints, R, are determined. In the second analysis, the reverse of the imaginary restraints for the frame is analyzed in order to determine the primary moments, M_{lt}, and axial forces, P_{lt}.

Figure 5.12 *Determination of Secondary Effects*

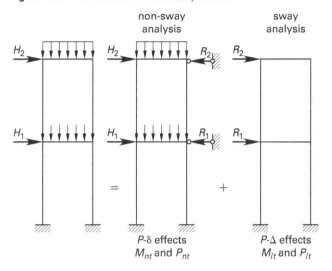

The required final second-order forces are then given by

$$M_r = B_1 M_{nt} + B_2 M_{lt} \quad \text{[AISC A-8-1]}$$

$$P_r = P_{nt} + B_2 P_{lt} \quad \text{[AISC A-8-2]}$$

The determination of the B_1 and B_2 multipliers is detailed in AISC 360 App. 8, and as follows.

Multiplier B_1 for P-δ Effects

The moment magnification factor to account for the P-δ effect, assuming no lateral translation of the frame, is defined in AISC 360 Eq. A-8-3 as

$$B_1 = \frac{C_m}{1 - \dfrac{\alpha P_r}{P_{e1}}} \geq 1$$

$\alpha = 1$ for LRFD load combinations, and $\alpha = 1.6$ for ASD load combinations.

The Euler buckling strength of the member in the plane of bending is defined in AISC 360 Eq. A-8-5 as

$$P_{e1} = \frac{\pi^2 EI^*}{(L_{c1})^2}$$

EI^* = flexural rigidity used in the analysis

$\quad = 0.8\tau_b EI \quad$ [for direct analysis method]

$\quad = EI \quad$ [for effective length and first-order methods]

To assist in the determination of the Euler buckling strength of a member, *AISC Manual* Tables 4-1a, 4-1b, and 4-1c tabulate values of $P_e(L_c)^2/10^4$ for W-shapes with a yield stress of 50 kips/in², 65 kips/in², and 70 kips/in².

The effective-length factor in the plane of bending, for a member in a frame with lateral translation inhibited, is given by

$$K_1 = 1.0$$

$$L_{c1} = K_1 L = L$$

The reduction factor is given by AISC 360 Eq. A-8-4 as

$$C_m = 0.6 - 0.4\left(\frac{M_1}{M_2}\right) \begin{bmatrix} \text{for a member not subjected} \\ \text{to transverse loading} \\ \text{between supports} \end{bmatrix}$$

$$= 1.0 \begin{bmatrix} \text{for a member transversely loaded} \\ \text{between supports} \end{bmatrix}$$

$$= 1.0 \begin{bmatrix} \text{for a member bent in single curvature} \\ \text{under uniform bending moment} \end{bmatrix}$$

M_1/M_2 is positive for a member bent in reverse curvature and negative for a member bent in single curvature.

As specified in AISC 360 App. 8.2.1, in applying AISC 360 Eq. A-8-3, it is permitted to use the first-order estimate of

$$P_r = P_{nt} + P_{lt}$$

In accordance with AISC 360 Sec. C2.1(2), P-δ effects may be neglected when all of the following three conditions apply.

- The structure supports gravity loads primarily through nominally vertical columns, walls, or frames.

- The ratio $\Delta_{2nd}/\Delta_{1st}$, calculated using LRFD load combinations or 1.6 times ASD load combinations with stiffness adjusted as specified in AISC 360 Sec. C2.3, is

$$\frac{\Delta_{2nd}}{\Delta_{1st}} \leq 1.7$$

- No more than one-third of the total gravity load on the structure is supported by columns that are part of moment-resisting frames.

AISC 360 Eq. A-8-1 is then

$$M_r = M_{nt} + B_2 M_{lt}$$

Multiplier B_2 for P-Δ Effects

The moment magnification factor to account for the P-Δ effect, with lateral translation of the frame allowed, is defined in AISC 360 Eq. A-8-6 as

$$B_2 = \frac{1}{1 - \dfrac{\alpha P_{\text{story}}}{P_{e,\text{story}}}}$$

$$= \frac{\Delta_{2\text{nd}}}{\Delta_{1\text{st}}}$$

$\alpha = 1.00$ for LRFD combinations and $\alpha = 1.60$ for ASD combinations.

P_{story} is the total vertical load supported by the story using LRFD or ASD load combinations, as applicable, including loads in columns that are not part of the lateral force resisting system. $P_{e,\text{story}}$ is the elastic critical buckling strength for the story in the direction of translation being considered. $P_{e,\text{story}}$ is determined by either sidesway buckling analysis or by using

$$P_{e,\text{story}} = R_M \frac{HL}{\Delta_H} \quad \text{[AISC A-8-7]}$$

$$R_M = 1 - 0.15\left(\frac{P_{mf}}{P_{\text{story}}}\right) \quad \text{[AISC A-8-8]}$$

$$= 1.00 \quad \text{[braced frame]}$$

$$= 0.85 \quad \begin{bmatrix} \text{moment frames} \\ \text{and combined systems} \end{bmatrix}$$

P_{mf} is the total vertical load in columns in the story that are part of moment frames, in the direction of translation being considered, if any. $P_{mf} = 0$ for braced frames. H is the story shear in the direction of translation being considered that is produced by the lateral forces used to compute Δ_H.

When calculating AISC 360 Eq. A-8-6, use inches for the height of the story, L. Δ_H is computed using the stiffness required to be used in the analysis (stiffness is reduced as provided in AISC 360 Sec. C2.3 when the direct analysis method is used). When Δ_H varies over the plan area of the structure, it is the average drift weighted in proportion to the vertical load or the maximum drift. When a limit is placed on the drift index, Δ_H/L, the amplification factor, B_2, is determined by using this limit in AISC 360 Eq. A-8-7.

Alternatively, P-delta effects may be directly determined in a rigorous second-order frame analysis, and the members may then be designed directly for the calculated axial force and bending moment. The principle of superposition is not valid in a second-order analysis, and separate analyses are necessary for each combination of factored loads.

Analysis Methods[5,6,7]

AISC 360 Sec. C2 requires the design of compression members to take into account secondary effects. The factors that must be considered include

- flexural, shear, and axial deformations of members
- P-Δ second-order effects caused by structure displacements
- P-δ second-order effects caused by member deformations
- geometric imperfections caused by initial out-of-plumbness of the columns
- reduction in member stiffness due to inelasticity and residual stresses
- uncertainty in stiffness and strength

The first three of these issues are covered in the analysis of the structure.

Geometric imperfections are caused by the permitted tolerances in the plumbness of columns. The out-of-plumbness effects are duplicated by applying notional fictitious lateral loads to the structure that produce an equivalent effect. As specified by the American Institute of Steel Construction's *Code of Standard Practice for Steel Buildings and Bridges* (AISC 303) Sec. 7.13.1.1, the maximum tolerance on out-of-plumbness of a column is $1/500$ of the height of the column. This produces a moment in a column of height L of

$$M = P_\Delta = \frac{PL}{500}$$

The same effect may be produced by applying a notional load of $P/500$ at the top of the column. This produces an identical moment at the bottom of the column of

$$M = \frac{PL}{500} = 0.002PL$$

As shown in Fig. 5.13, to account for initial imperfections in the members, notional lateral loads are applied at each story, in accordance with AISC 360 Sec. C2.2b, and are given by

$$N_i = 0.002\alpha Y_i$$

$$\alpha = 1.0 \quad \text{[LRFD]}$$

$$= 1.6 \quad \text{[ASD]}$$

N_i is the notional lateral load applied at level i and Y_i is the gravity load applied at level i.

Figure 5.13 *Notional Loads*

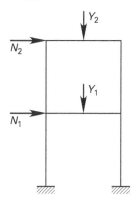

The notional loads are applied solely in gravity-only load combinations when

$$B_2 = \frac{\Delta_{2\text{nd}}}{\Delta_{1\text{st}}} \le 1.7$$

Residual stresses and *plastic yielding* cause increased deformations in the structure. These effects are compensated for by reducing the stiffness of members prior to analysis and adopting a value of $K = 1.0$ in design of the columns. Alternatively, the nominal stiffness of members is used in the analysis, and traditional K values are used in design of the columns.

In accordance with AISC 360 Sec. C2.3 the flexural and axial stiffness of members that contribute to the lateral stability of the structure are reduced to give

$$EI^* = 0.8\tau_b EI$$
$$EA^* = 0.8EA$$

The stiffness reduction parameter is

$$\tau_b = 1.0 \quad [\text{for } \alpha P_r \le 0.5 P_y]$$
$$= 4\left(\frac{\alpha P_r}{P_y}\right)\left(1 - \left(\frac{\alpha P_r}{P_y}\right)\right) \quad [\alpha P_r > 0.5 P_y]$$

The force level adjustment factor is

$$\alpha = 1.0 \quad [\text{LRFD}]$$
$$= 1.6 \quad [\text{ASD}]$$

P_r is the required second-order axial strength. P_y is the member yield strength and is equal to $F_y A$.

The stiffness reduction is applied to all members in the structure, not only to those members contributing to the stability of the structure. This prevents unintended distortion of the structure and redistribution of forces.

Uncertainties in stiffness and strength of members are accounted for by the resistance factors and safety factors adopted.

Four methods are presented in AISC 360 and the *AISC Manual* for determining secondary effects.

- effective length method, detailed in AISC 360 App. 7.2

- direct analysis method, detailed in AISC 360 Sec. C2

- first-order analysis method, detailed in AISC 360 App. 7.3

- simplified method, detailed in *AISC Manual* Part 2

The effective length method and the direct analysis method both require a second-order frame analysis of the structure. This may be accomplished using a rigorous, second-order computer analysis. Alternatively, the approximate B_1-B_2 procedure specified in AISC 360 App. 8.2 may be used.

Effective Length Method

This method is restricted by AISC 360 App. 7.2 to structures with a sidesway amplification factor of

$$B_2 = \frac{\Delta_{2\text{nd}}}{\Delta_{1\text{st}}} \le 1.5$$

Drift is determined for LRFD load combinations or is 1.6 times the ASD load combinations using nominal stiffness values.

In accordance with AISC 360 App. 7.2, the design forces may be determined either by a rigorous second-order computer analysis or by amplifying the results of a first-order analysis.

In applying the effective length method, the nominal stiffness of all members is used in the analysis with no reduction for inelasticity and residual stress. These effects are accounted for in the design of the columns by using an appropriate value for the effective length factor, K. The available strength of columns is determined using an effective length factor as defined in AISC 360 Comm. Table C-A-7.1, or is calculated in accordance with AISC 360 Comm. App. 7.2. The empirical column curve then accounts for inelasticity and residual stress. A value of 1.0 may be used for the effective length factor of members in a braced frame. For columns that do not contribute to the lateral resistance of the structure, a value of $K = 1.0$ may be used for the effective length factor. As specified in AISC 360 App. 7.2.3(b), the value of $K = 1.0$ may also be used for all columns when the structure is sufficiently stiff that

$$B_2 = \frac{\Delta_{2\text{nd}}}{\Delta_{1\text{st}}} \le 1.1$$

Drift is determined for LRFD load combinations or 1.6 times the ASD load combinations.

To account for initial imperfections in the members, minimum lateral loads are applied at each story in accordance with AISC 360 Sec. C2.2b. These loads are given by

$$N_i = 0.002 \alpha Y_i$$

The notional loads are applied solely in gravity-only load combinations in the effective length method.

Example 5.21

The first-order member forces produced in the outer column of a moment frame by the governing factored load combination are shown in *Illustration for Ex. 5.21*. Determine the second-order member forces in the column using the effective length method. The column consists of a W12 × 79 section with a yield stress of 50 kips/in². The bay length is 25 ft and the beams consist of W21 × 62 sections with a yield stress of 50 kips/in². No intermediate bracing is provided to the column about either axis.

Solution

LRFD Method

The loading condition is not a gravity-only load combination, and the notional lateral load is not applicable.

Determine B_1.

The results of the first-order, non-sway analysis are shown in *Illustration for Ex. 5.21*, and the reduction factor, with M_1/M_2 positive, is given by AISC 360 Eq. A-8-4 as

$$C_m = 0.6 - 0.4\left(\frac{M_1}{M_2}\right) = 0.6 - (0.4)\left(\frac{60 \text{ ft-kips}}{100 \text{ ft-kips}}\right)$$
$$= 0.36$$

The effective length factor in the plane of bending for a non-sway frame is given by

$$K_1 = 1.0$$

$$L_{c1} = K_1 L = L$$

From *AISC Manual* Table 4-1a, the Euler buckling strength of a W12 × 79 column in the plane of bending is given by

$$P_{e1} = \frac{\pi^2 EI^*}{(L_{c1})^2} = \frac{(18{,}900 \text{ in-kips}^2)(10^4)}{(144 \text{ in})^2}$$
$$= 9115 \text{ kips}$$

EI^* is the flexural rigidity used in the analysis. For the effective length method, $EI^* = EI$.

As specified in AISC 360 App. 8.2.1, use the value

$$P_r = P_{nt} + P_{lt}$$
$$= 300 \text{ kips} + 100 \text{ kips}$$
$$= 400 \text{ kips}$$

The moment magnification factor to account for the $P\text{-}\delta$ effect, assuming no lateral translation of the frame, is defined in AISC 360 Eq. A-8-3 as

$$B_1 = \frac{C_m}{1 - \dfrac{\alpha P_r}{P_{e1}}}$$

$$= \frac{0.36}{1 - \dfrac{(1.0)(400 \text{ kips})}{9115 \text{ kips}}}$$

$$= 0.38$$

$$\text{use } B_1 = 1.0 \quad [\text{minimum}]$$

Determine B_2.

For a moment frame, $R_M = 0.85$.

The results of the first-order sway analysis are shown in the *Illustration for Ex. 5.21*, and the Euler buckling strength for the story is given by AISC 360 Eq. A-8-7 as

$$P_{e,\text{story}} = \left(\frac{R_M L}{\Delta_H}\right) H = \left(\frac{(0.85)(144 \text{ in})}{0.50 \text{ in}}\right)(100 \text{ kips})$$
$$= 24{,}480 \text{ kips}$$

The moment magnification factor to account for the sidesway effect, with lateral translation of the frame allowed, is given by AISC 360 Eq. A-8-6 as

$$\frac{\Delta_{2\text{nd}}}{\Delta_{1\text{st}}} = B_2 = \frac{1}{1 - \dfrac{\alpha P_{\text{story}}}{P_{e,\text{story}}}}$$

$$= \frac{1}{1 - \dfrac{(1.0)(3000 \text{ kips})}{24{,}480 \text{ kips}}}$$

$$= 1.14 < 1.5$$

The effective length method is applicable.

Illustration for Ex. 5.21

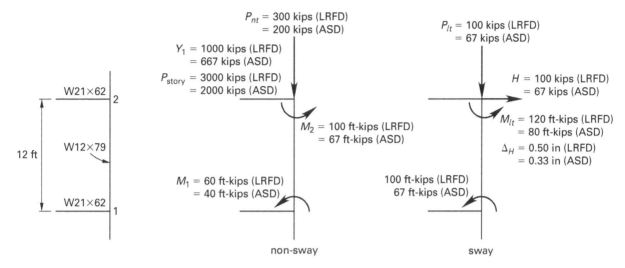

non-sway   sway

Calculate the second-order member forces.

The second-order member forces for the frame are given by AISC 360 Eq. A-8-1 and Eq. A-8-2 as

$$M_r = B_1 M_{nt} + B_2 M_{lt}$$
$$= (1.0)(100 \text{ ft-kips}) + (1.14)(120 \text{ ft-kips})$$
$$= 237 \text{ ft-kips}$$
$$P_r = P_{nt} + B_2 P_{lt}$$
$$= 300 \text{ kips} + (1.14)(100 \text{ kips})$$
$$= 414 \text{ kips}$$

ASD Method

The notional lateral load is not applicable.

Determine B_1.

The results of the first-order, non-sway analysis are shown in the *Illustration for Ex. 5.21*, and the reduction factor is given by AISC 360 Eq. A-8-4 as

$$C_m = 0.6 - 0.4 \left(\frac{M_1}{M_2} \right)$$
$$= 0.6 - (0.4) \left(\frac{40 \text{ ft-kips}}{67 \text{ ft-kips}} \right)$$
$$= 0.36$$

The effective length factor in the plane of bending for a non-sway frame is given by

$$K_1 = 1.0$$

$$L_{c1} = K_1 L = L$$

From *AISC Manual* Table 4-1, the Euler buckling strength of a W12 × 79 column in the plane of bending is given by

$$P_{e1} = \frac{\pi^2 EI}{(K_1 L)^2}$$
$$= \frac{(18{,}900 \text{ in-kips}^2)(10^4)}{(144 \text{ in})^2}$$
$$= 9115 \text{ kips}$$

As specified in AISC 360 App. 8.2.1, use the value

$$P_r = P_{nt} + P_{lt}$$
$$= 200 \text{ kips} + 67 \text{ kips}$$
$$= 267 \text{ kips}$$

The moment magnification factor to account for the member effect, assuming no lateral translation of the frame, is defined in AISC 360 Eq. A-8-3 as

$$B_1 = \frac{C_m}{1 - \dfrac{\alpha P_r}{P_{e1}}}$$
$$= \frac{0.36}{1 - \dfrac{(1.6)(267 \text{ kips})}{9115 \text{ kips}}}$$
$$= 0.38$$
$$\text{use } B_2 = 1.0 \quad \text{[minimum]}$$

Determine B_2.

For a moment frame, $R_M = 0.85$.

The results of the first-order sway analysis are shown in the *Illustration for Ex. 5.21*, and the Euler buckling strength for the story is given by AISC 360 Eq. A-8-7 as

$$P_{e,\,\text{story}} = \left(\frac{R_M L}{\Delta_H} \right) H = \left(\frac{(0.85)(144 \text{ in})}{0.33 \text{ in}} \right)(67 \text{ kips})$$
$$= 24{,}850 \text{ kips}$$

The moment magnification factor to account for the frame effect, with lateral translation of the frame allowed, is given by AISC 360 Eq. A-8-6 as

$$\frac{\Delta_{2\text{nd}}}{\Delta_{1\text{st}}} = B_2 = \frac{1}{1 - \dfrac{\alpha P_{\text{story}}}{P_{e,\,\text{story}}}}$$
$$= \frac{1}{1 - \dfrac{(1.6)(2000 \text{ kips})}{24{,}850 \text{ kips}}}$$
$$= 1.15$$

Calculate the second-order member forces.

The second-order member forces for the frame are given by AISC 360 Eq. A-8-1 and Eq. A-8-2 as

$$M_r = B_1 M_{nt} + B_2 M_{lt}$$
$$= (1.0)(67 \text{ ft-kips}) + (1.15)(80 \text{ ft-kips})$$
$$= 159 \text{ ft-kips}$$
$$P_r = P_{nt} + B_2 P_{lt}$$
$$= 200 \text{ kips} + (1.15)(67 \text{ kips})$$
$$= 277 \text{ kips}$$

Direct Analysis Method

The direct analysis method is applicable to all types of structures and must be used when

$$B_2 = \frac{\Delta_{2\text{nd}}}{\Delta_{1\text{st}}} > 1.5$$

The design forces may be determined either by a rigorous second-order computer analysis, or by amplifying the results of a first-order analysis.

In applying the method, the factored loads are applied to the structure using reduced flexural and axial stiffness of members that contribute to the lateral stability of the structure. The reduced stiffnesses account for elastic instability and inelastic softening effects, and are given by AISC 360 Sec. C2.3 as

$$EI^* = 0.8\tau_b EI$$
$$EA^* = 0.8EA$$

The stiffness reduction coefficient is given by

$$\tau_b = 1.0 \quad [\alpha P_r \le 0.5 P_y]$$
$$= 4 \left(\frac{\alpha P_r}{P_y} \right) \left(1 - \frac{\alpha P_r}{P_y} \right) \quad [\alpha P_r > 0.5 P_y]$$

Alternatively, when $\alpha P_r > 0.5 P_y$, the stiffness reduction factor, τ_b, may be taken as 1.0, provided that the actual lateral loads are increased by a notional lateral load of

$$N_i = 0.001 \alpha Y_i$$

To account for out-of-plumbness in the columns, minimum lateral loads are applied at each story, in accordance with AISC 360 Sec. C2.2b. These loads are given by

$$N_i = 0.002 \alpha Y_i$$

As specified in AISC 360 Sec. C2.2b(d), the notional loads are additive to the applied lateral loads when the sidesway amplification ratio is

$$\frac{\Delta_{2\text{nd}}}{\Delta_{1\text{st}}} > 1.7 \text{ using reduced stiffness}$$

Drift is calculated using the reduced elastic stiffness for LRFD load combinations, or 1.6 times the ASD load combinations.

Otherwise, it is permissible to apply the notional loads solely in gravity-only load combinations.

In designing the members of the frame for the calculated second-order forces, the appropriate effective length factor is specified in AISC 360 Sec. C3, as $K = 1.0$ for all members.

Example 5.22

The first-order member forces produced in the outer column of a moment frame by the governing factored load combination are shown. Reduced stiffness values of $EI^* = 0.8EI$ were used in the analyses. Determine the second-order member forces in the column using the direct analysis method. The column consists of a W12 × 79 section with a yield stress of 50 kips/in^2. The bay length is 25 ft and the beams consist of W21 × 62 sections with a yield stress of 50 kips/in^2. No intermediate bracing is provided to the column about either axis.

Illustration for Ex. 5.22

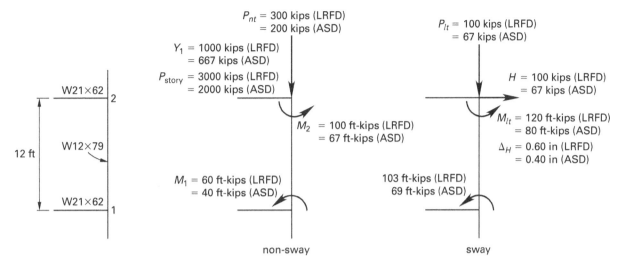

Solution

LRFD Method

The load combination is not a gravity-only combination; using the reduced stiffness and assuming $B_2 < 1.7$, notional lateral loads are not applicable.

Determine B_1.

In accordance with AISC 360 Sec. C2.3, for the direct analysis method, the flexural and axial stiffness of members in the structure are reduced to

$$EI^* = 0.8\tau_b EI$$
$$EA^* = 0.8EA$$

Assuming $\alpha P_r \le 0.5 P_y$, the stiffness reduction parameter is $\tau_b = 1.0$, and the flexural stiffness becomes

$$EI^* = 0.8EI$$

From Ex. 5.21, the Euler buckling strength is

$$P_{e1} = \frac{\pi^2 EI^*}{(L_{c1})^2}$$
$$= \frac{(0.8)(18{,}900 \text{ in-kips}^2)(10^4)}{(144 \text{ in})^2}$$
$$= 7292 \text{ kips}$$
$$C_m = 0.36$$

As specified in AISC 360 App. 8.2.1, use the value

$$P_r = P_{nt} + P_{lt}$$
$$= 300 \text{ kips} + 100 \text{ kips}$$
$$= 400 \text{ kips}$$

$$B_1 = \frac{C_m}{1 - \dfrac{\alpha P_r}{P_{e1}}}$$
$$= \frac{0.36}{1 - \dfrac{(1.0)(400 \text{ kips})}{7292 \text{ kips}}}$$
$$= 0.38$$

use $B_1 = 1.0$ [minimum]

Determine B_2.

For a moment frame, $R_M = 0.85$.

The results of the first-order sway analysis are shown in the *Illustration for Ex. 5.22*, and the Euler buckling strength for the story is given by AISC 360 Eq. A-8-7 as

$$P_{e,\text{story}} = \left(\frac{R_M L}{\Delta_H}\right) H$$
$$= \left(\frac{(0.85)(12 \text{ ft})\left(12 \dfrac{\text{in}}{\text{ft}}\right)}{0.60 \text{ in}}\right)(100 \text{ kips})$$
$$= 20{,}400 \text{ kips}$$

The moment magnification factor to account for the sidesway effect, with lateral translation of the frame allowed, is given by AISC 360 Eq. A-8-6 as

$$\frac{\Delta_{2nd}}{\Delta_{1st}} = B_2 = \cfrac{1}{1 - \cfrac{\alpha P_{story}}{P_{e,story}}}$$

$$= \cfrac{1}{1 - \cfrac{(1.0)(3000 \text{ kips})}{20{,}400 \text{ kips}}}$$

$$= 1.17$$

$$< 1.7 \quad \text{[notional lateral loads are not applicable]}$$

Calculate the second-order member forces.

The second-order member forces for the frame are given by AISC 360 Eq. A-8-1 and AISC 360 Eq. A-8-2 as

$$M_r = B_1 M_{nt} + B_2 M_{lt}$$
$$= (1.0)(100 \text{ ft-kips}) + (1.17)(120 \text{ ft-kips})$$
$$= 240 \text{ ft-kips}$$
$$P_r = P_{nt} + B_2 P_{lt}$$
$$= 300 \text{ kips} + (1.17)(100 \text{ kips})$$
$$= 417 \text{ kips}$$

ASD Method

The load combination is not a gravity-only combination and, using the reduced stiffness and assuming $B_2 < 1.7$, notional lateral loads are not applicable.

Determine B_1.

From Ex. 5.21, the Euler buckling strength is

$$P_{e1} = \frac{\pi^2 EI^*}{(L_{c1})^2} = \frac{(0.8)(18{,}900 \text{ in-kips}^2)(10^4)}{(144 \text{ in})^2}$$
$$= 7292 \text{ kips}$$

As specified in AISC 360 App. 8.2.1, use the value

$$P_r = P_{nt} + P_{lt}$$
$$= 200 \text{ kips} + 67 \text{ kips}$$
$$= 267 \text{ kips}$$
$$B_1 = \cfrac{C_m}{1 - \cfrac{\alpha P_r}{P_{e1}}}$$
$$= \cfrac{0.36}{1 - \cfrac{(1.6)(267 \text{ kips})}{7292 \text{ kips}}}$$
$$= 0.38$$
$$\text{use } B_1 = 1.0 \quad \text{[minimum]}$$

Determine B_2.

For a moment frame, $R_M = 0.85$.

The results of the first-order sway analysis are shown in the *Illustration for Ex. 5.22*, and the sum of the Euler buckling strength, for all columns in the story, is given by AISC 360 Eq. A-8-7 as

$$P_{e,story} = \left(\frac{R_M L}{\Delta_H}\right) H$$

$$= \left(\cfrac{(0.85)(12 \text{ ft})\left(12 \, \cfrac{\text{in}}{\text{ft}}\right)}{0.40 \text{ in}}\right)(67 \text{ kips})$$

$$= 20{,}502 \text{ kips}$$

The moment magnification factor to account for the sidesway effect, with lateral translation of the frame allowed, is given by AISC 360 Eq. A-8-6 as

$$\frac{\Delta_{2nd}}{\Delta_{1st}} = B_2 = \cfrac{1}{1 - \cfrac{\alpha P_{story}}{P_{e,story}}}$$

$$= \cfrac{1}{1 - \cfrac{(1.6)(2000 \text{ kips})}{20{,}502 \text{ kips}}}$$

$$= 1.19$$

$$< 1.7 \quad \text{[notional lateral loads are not applicable]}$$

Calculate the second-order member forces.

The second-order member forces for the frame are given by AISC 360 Eq. A-8-1 and AISC 360 Eq. A-8-2 as

$$M_r = B_1 M_{nt} + B_2 M_{lt}$$
$$= (1.0)(67 \text{ ft-kips}) + (1.19)(80 \text{ ft-kips})$$
$$= 162 \text{ ft-kips}$$
$$P_r = P_{nt} + B_2 P_{lt} = 200 \text{ kips} + (1.19)(67 \text{ kips})$$
$$= 280 \text{ kips}$$

First-Order Elastic Analysis

The first-order elastic analysis method is specified in AISC 360 App. 7.3. This method is restricted by AISC 360 App. 7.3.1(b) to structures with a sidesway amplification factor of

$$B_2 = \frac{\Delta_{2nd}}{\Delta_{1st}} \leq 1.5$$

Drift is determined for LRFD load combinations, or 1.6 times the ASD load combinations.

The design forces are determined by a first-order analysis with notional loads applied to the structure using the nominal (unreduced) member stiffnesses.

In addition, a limit is placed on the required axial compressive strength such that

$$\alpha P_r \leq 0.5 P_{ns}$$
$$P_{ns} = \text{member compressive strength}$$
$$= A_g F_y \quad \text{[nonslender section]}$$
$$= A_e F_y \quad \text{[slender section]}$$

A_e is defined in AISC Sec. E7.

To account for initial imperfections in the members, notional lateral loads are applied at each story and are given by

$$N_i = \frac{2.1\Delta_{1st}\alpha Y_i}{L}$$
$$\geq 0.0042 Y_i$$

The notional loads are additive to all load combinations and eliminate the need for a second-order analysis.

The non-sway amplification of column moments is considered by applying the B_1 amplifier to the total member moments.

In designing the members of the frame for the calculated second-order forces, the appropriate effective length factor is specified in AISC 360 App. 7.3.3 as $K = 1.0$ for all members.

Example 5.23

The factored loads acting on the outer column of a moment frame are shown. Determine the required strength of the column using the first-order elastic analysis method. The column consists of a W12 × 79 section with a yield stress of 50 kips/in^2.

The bay length is 25 ft and the beams consist of W21 × 62 sections with a yield stress of 50 kips/in^2. No intermediate bracing is provided to the column about either axis.

The interstory drift is $\Delta_{1st} = 0.50$ in (LRFD) or $\Delta_{1st} = 0.33$ in (ASD).

Solution

LRFD Method

For a compact section, the column compressive strength is

$$P_{ns} = A_g F_y$$
$$= (23.2 \text{ in}^2)\left(50 \ \frac{\text{kips}}{\text{in}^2}\right)$$
$$= 1160 \text{ kips}$$

Assuming that the required axial load is

$$P_r = 417 \text{ kips}$$
$$\frac{\alpha P_r}{P_{ns}} = \frac{(1.0)(417 \text{ kips})}{1160 \text{ kips}}$$
$$= 0.36$$
$$< 0.5$$

In accordance with AISC 360 App. 7.3.1(c), a first-order elastic analysis is permissible.

Determine the augmented loads.

The first-order interstory drift due to the design loads is

$$\Delta_{1st} = 0.5 \text{ in}$$

The sum of the gravity loads, applied at the story, independent of loads from the upper stories is

$$Y_1 = 1000 \text{ kips}$$

The notional lateral load on the story is given by AISC 360 App. 7.3.2(a) as

$$N_i = 0.0042 Y_i$$
$$= (0.0042)(1000 \text{ kips})$$
$$= 4.2 \text{ kips}$$

However, this value may not be less than

$$N_i = \frac{2.1\Delta_{1st}\alpha Y_i}{L}$$
$$= \frac{(2.1)(0.5 \text{ in})(1.0)(1000 \text{ kips})}{144 \text{ in}}$$
$$= 7.3 \text{ kips} \quad \text{[governs]}$$

This is additive to the applied lateral loads, and the resulting augmented forces on the column are indicated in the *Illustration for Ex. 5.23*.

Determine B_1.

The moment magnification factor to account for the P-δ effect, assuming no lateral translation of the frame, is determined in Ex. 5.21 as

$$B_1 = 1.0$$

Illustration for Ex. 5.23

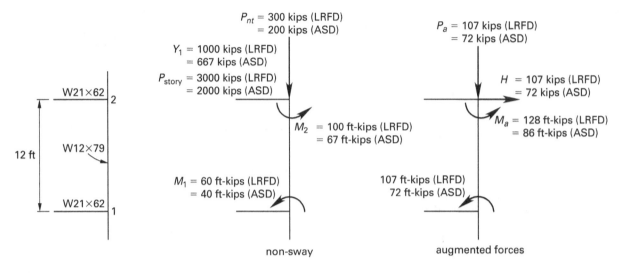

non-sway   augmented forces

Calculate the required forces.

The required forces are given by

$$M_r = B_1(M_{nt} + M_a)$$
$$= (1.0)(100 \text{ ft-kips} + 128 \text{ ft-kips})$$
$$= 228 \text{ ft-kips}$$
$$P_r = P_{nt} + P_a = 300 \text{ kips} + 107 \text{ kips}$$
$$= 407 \text{ kips}$$

ASD Method

For a compact section, the column compressive strength is

$$P_{ns} = A_g F_y = (23.2 \text{ in}^2)\left(50 \ \frac{\text{kips}}{\text{in}^2}\right)$$
$$= 1160 \text{ kips}$$

Assuming that the required axial load is

$$P_r = 278 \text{ kips}$$
$$\frac{\alpha P_r}{P_{ns}} = \frac{(1.6)(278 \text{ kips})}{1160 \text{ kips}}$$
$$= 0.38$$
$$< 0.5$$

In accordance with AISC 360 App. 7.3.1(c), a first-order elastic analysis is permissible.

Determine the augmented loads.

The first-order interstory drift due to the lateral loads is

$$\Delta_{1st} = 0.33 \text{ in}$$

The sum of the gravity loads, applied at the story, independent of loads from the upper stories is

$$Y_1 = 667 \text{ kips}$$

The notional lateral load on the story is given by AISC 360 App. 7.3.2(a) as

$$N_i = 0.0042\,Y_i = (0.0042)(667 \text{ kips})$$
$$= 2.8 \text{ kips}$$

However, this value may not be less than

$$N_i = \frac{2.1\Delta_{1st}\alpha Y_i}{L}$$
$$= \frac{(2.1)(0.33 \text{ in})(1.6)(667 \text{ kips})}{144 \text{ in}}$$
$$= 5.1 \text{ kips} \quad [\text{governs}]$$

This is additive to the applied lateral loads, and the augmented loads and the resulting forces on the column are indicated in the *Illustration Ex. 5.23*.

The moment magnification factor to account for the member effect, assuming no lateral translation of the frame, is determined in Ex. 5.21 as

$$B_1 = 1.0$$

Calculate the required forces.

The required forces are given by

$$M_r = B_1(M_{nt} + M_a)$$
$$= (1.0)(67 \text{ ft-kips} + 86 \text{ ft-kips})$$
$$= 153 \text{ ft-kips}$$
$$P_r = P_{nt} + P_a$$
$$= 200 \text{ kips} + 72 \text{ kips}$$
$$= 272 \text{ kips}$$

Simplified Method

The simplified method is specified in *AISC Manual* Part 2 and requires a first-order analysis only. The method is restricted to structures with a sidesway amplification factor of

$$\frac{\Delta_{2nd}}{\Delta_{1st}} \le 1.5$$

In addition, the ratio of the sway and non-sway amplification factors is restricted to

$$\frac{B_1}{B_2} \le 1.0$$

For members not subjected to transverse loading, it is unlikely that B_1 will be greater than B_2.

In applying the method, the factored loads are applied to the structure using the nominal stiffness of the members, and a first-order analysis is performed. To account for initial imperfections in the members, minimum lateral loads are applied at each story in accordance with AISC 360 Sec. C2.2b(a). These loads are given by

$$N_i = 0.002\alpha Y_i$$

This is only applied in gravity load cases.

Required strengths are determined by multiplying the forces obtained from the first-order analysis by tabulated values of B_2. These tabulated values are a function of the required story drift limit and the ratio of the total story gravity load to the lateral load that produces the drift limit.

From the first-order elastic analysis, the lateral load required to produce the required story drift is determined, and the ratio of the total story gravity load to the lateral load that produces this drift limit is calculated. Using α times this value and the value of the required story drift, the appropriate amplification factor, B_2, may be obtained from the table in *AISC Manual* Part 2 or from *AISC Basic Design Values Cards.*[8] (See Table 5.1.)

Table 5.1 Amplification Factor B_2 for Use with the Simplified Method

design story drift limit	load ratio (times 1.6 for ASD, 1.0 for LRFD)										
	0	5	10	20	30	40	50	60	80	100	120
$H/100$	1	1.1	1.1	1.3	1.5/1.4	When ratio exceeds 1.5, simplified method requires a stiffer structure.					
$H/200$	1	1	1.1	1.1	1.2	1.3	1.4/1.3	1.5/1.4			
$H/300$	1	1	1	1.1	1.1	1.2	1.2	1.3	1.5/1.4		
$H/400$	1	1	1	1.1	1.1	1.1	1.2	1.2	1.3	1.4/1.3	1.5
$H/500$	1	1	1	1	1.1	1.1	1.1	1.2	1.2	1.3	1.4

Note: Where two values are provided, the value in bold is the value associated with $R_m = 0.85$.

In designing the members of the frame for the calculated forces, the appropriate effective length factor is specified as $K = 1.0$ for all members, provided that the amplification factor does not exceed 1.1.

For cases where the amplification factor exceeds 1.1, the effective length factors for the members are determined by analysis.

Example 5.24

The factored loads acting on the outer column of a moment frame are shown. Determine the required strength of the column using the simplified analysis method. The column consists of a W12 × 79 section with a yield stress of 50 kips/in². The bay length is 25 ft and the beams consist of W21 × 62 sections with a yield stress of 50 kips/in². No intermediate bracing is provided to the column about either axis. The required story drift is limited to the product of the story height and 1/240 for LRFD (1/360 for ASD).

Solution

LRFD Method

This is not a gravity-only loading case and the notional lateral load is not applicable.

The factored loads produce the first-order elastic forces, which are derived in Ex. 5.21 and are shown in the *Illustration for Ex. 5.24.* The interstory drift produced is

$$\Delta_a = 0.50 \text{ in}$$

For a design story drift of 1/240, the permissible deflection is

$$\Delta_p = \frac{144 \text{ in}}{240}$$
$$= 0.60 \text{ in}$$

Illustration for Ex. 5.24

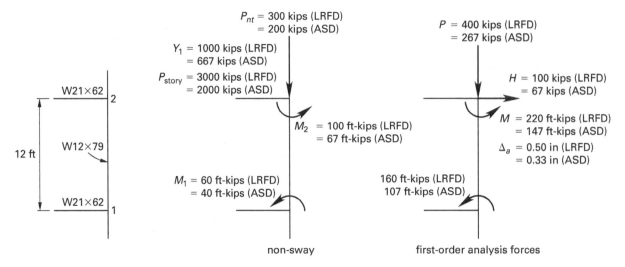

The lateral load required to produce the required story drift is

$$H_p = \frac{(100 \text{ kips})\Delta_p}{\Delta_a} = \frac{(100 \text{ kips})(0.60 \text{ in})}{0.50 \text{ in}}$$
$$= 120 \text{ kips}$$

The ratio of α times the total story gravity load to the lateral load that produces the drift limit is

$$\frac{\alpha P_{story}}{H_p} = \frac{(1.0)(3000 \text{ kips})}{120 \text{ kips}}$$
$$= 25$$

Interpolate from Table 5.1. For a drift limit of 1/240 and a load ratio of 25, the amplification factor is

$$B_2 = 1.13$$

The required forces are given by

$$M_r = B_2 M = (1.13)(220 \text{ ft-kips})$$
$$= 249 \text{ ft-kips}$$
$$P_r = B_2 P = (1.13)(400 \text{ kips})$$
$$= 452 \text{ kips}$$

ASD Method

This is not a gravity-only loading case and the notional lateral load is not applicable.

The factored loads produce the first-order elastic forces, which are derived in Ex. 5.21 and are shown in the *Illustration for Ex. 5.24*. The interstory drift produced is

$$\Delta_a = 0.33 \text{ in}$$

For a design story drift of 1/360, the permissible deflection is

$$\Delta_p = \frac{144 \text{ in}}{360}$$
$$= 0.40 \text{ in}$$

The lateral load required to produce the required story drift is

$$H_p = \frac{(67 \text{ kips})\Delta_p}{\Delta_a}$$
$$= \frac{(67 \text{ kips})(0.40 \text{ in})}{0.33 \text{ in}}$$
$$= 81 \text{ kips}$$

The ratio of α times the total story gravity load to the lateral load that produces the drift limit is

$$\frac{\alpha P_{story}}{H_p} = \frac{(1.6)(2000 \text{ kips})}{81 \text{ kips}}$$
$$= 39.5$$

Interpolate from Table 5.1. For a drift limit of 1/360 and a load ratio of 39.5, the amplification factor is

$$B_2 = 1.14$$

The required forces are given by

$$M_r = B_2 M = (1.14)(147 \text{ ft-kips})$$
$$= 168 \text{ ft-kips}$$
$$P_r = B_2 P = (1.14)(267 \text{ kips})$$
$$= 304 \text{ kips}$$

Combined Compression and Flexure[9]

The adequacy of a member to sustain combined compression and flexure is determined by means of the interaction equations given in AISC 360 Sec. H1.1 as follows.

For $P_r/P_c \geq 0.2$,

$$\frac{P_r}{P_c} + \left(\frac{8}{9}\right)\left(\frac{M_{rx}}{M_{cx}} + \frac{M_{ry}}{M_{cy}}\right) \leq 1.0 \quad \text{[AISC H1-1a]}$$

For $P_r/P_c < 0.2$,

$$\frac{P_r}{2P_c} + \left(\frac{M_{rx}}{M_{cx}} + \frac{M_{ry}}{M_{cy}}\right) \leq 1.0 \quad \text{[AISC H1-1b]}$$

Values are tabulated in *AISC Manual* Table 6-2 for W shapes with a yield stress of 50 kips/in^2 and assuming a bending coefficient of $C_b = 1.0$ for the functions shown in Table 5.2.

Table 5.2 *Available Strength for Members Subject to Combined Forces*

LRFD	ASD
$\phi_c P_n = P_c$	$P_n/\Omega_c = P_c$
$\phi_b M_{nx} = M_{cx}$	$M_{nx}/\Omega_b = M_{cx}$
$\phi_b M_{ny} = M_{cy}$	$M_{ny}/\Omega_b = M_{cy}$

Example 5.25

Determine the adequacy of the W12 × 79 column in the frame analyzed by the effective length method in Ex. 5.21. The beams consist of W21 × 62 sections with a length of 25 ft. All members have a yield stress of 50 kips/in^2, and no intermediate bracing is provided to the column about either axis.

Solution

The frame was analyzed for second-order effects in Ex. 5.21 using the effective length method. The required forces in the column are

$$P_r = 414 \text{ kips (LRFD), 277 kips (ASD)}$$
$$M_{rx} = 237 \text{ ft-kips (LRFD), 159 ft-kips (ASD)}$$
$$M_{ry} = 0 \text{ ft-kips}$$

Calculate the effective column length.

At both joint 1 and joint 2, the stiffness ratio is

$$\begin{aligned} G_2 &= \frac{\sum \dfrac{I_c}{L_c}}{\sum \dfrac{I_g}{L_g}} \\[2mm] &= \frac{(2)\left(\dfrac{662 \text{ in}^4}{12 \text{ ft}}\right)}{\dfrac{1330 \text{ in}^4}{25 \text{ ft}}} \\[2mm] &= 2.07 \end{aligned}$$

From the alignment chart for sway frames (see Fig. 5.10), the effective length factor about the x-axis of column 12 is

$$K_x = 1.65$$

The effective length of the column about the x-axis is

$$(L_c)_x = (1.65)(12 \text{ ft}) = 19.8 \text{ ft}$$

The effective length of the column about the y-axis is

$$(L_c)_y = (1.0)(12 \text{ ft}) = 12 \text{ ft}$$

From *AISC Manual* Table 6-2, a W12 × 79 column has a value of

$$\frac{r_x}{r_y} = 1.75$$

The equivalent effective length about the major axis with respect to the y-axis is

$$\begin{aligned} (L_c)_{y,\text{equiv}} &= \frac{(L_c)_x}{\dfrac{r_x}{r_y}} = \frac{19.8 \text{ ft}}{1.75} \\[2mm] &= 11.31 \text{ ft} \quad \text{[does not govern]} \\ &< K_y L_y \end{aligned}$$

The effective length about the minor axis governs.

$$(L_c)_y = 12 \text{ ft}$$

Apply the interaction equation.

LRFD Method

From *AISC Manual* Table 6-2, a W12 × 79 column with an effective length of $(L_c)_y = 12$ ft has a design axial strength of

$$\phi_c P_n = 887 \text{ kips}$$
$$\frac{P_r}{\phi_c P_n} = \frac{414 \text{ kips}}{887 \text{ kips}}$$
$$= 0.467$$
$$> 0.2$$

AISC 360 Eq. H1-1a applies.

From *AISC Manual* Table 6-2, a W12 × 79 member with an unbraced length of 12 ft has an available strength of

$$M_{cx} = 439 \text{ ft-kips}$$

The appropriate interaction equation is AISC 360 H1-1a, which is

$$\frac{P_r}{P_c} + \left(\frac{8}{9}\right)\left(\frac{M_{rx}}{M_{cx}} + \frac{M_{ry}}{M_{cy}}\right) \leq 1.0$$

Substituting in the left side of this equation gives

$$\frac{414 \text{ kips}}{887 \text{ kips}} + \left(\frac{8}{9}\right)\left(\frac{237 \text{ ft-kips}}{439 \text{ ft-kips}} + 0\right) = 0.95$$
$$< 1.0 \quad \text{[satisfactory]}$$

ASD Method

From *AISC Manual* Table 6-2, a W12 × 79 column with an effective length of $(L_c)_y = 12$ ft has a design axial strength of

$$\frac{P_n}{\Omega_c} = 590 \text{ kips}$$
$$\frac{P_r}{\dfrac{P_n}{\Omega_c}} = \frac{277 \text{ kips}}{590 \text{ kips}}$$
$$= 0.469$$
$$> 0.2$$

AISC 360 Eq. H1-1a applies.

From *AISC Manual* Table 6-2, a W12 × 79 member with an unbraced length of 12 ft has an available strength of

$$M_{cx} = 292 \text{ ft-kips}$$

The appropriate interaction equation is AISC 360 H1-1a, which is

$$\frac{P_r}{P_c} + \left(\frac{8}{9}\right)\left(\frac{M_{rx}}{M_{cx}} + \frac{M_{ry}}{M_{cy}}\right) \leq 1.0$$

Substituting in the left side of this equation gives

$$\frac{277 \text{ kips}}{590 \text{ kips}} + \left(\frac{8}{9}\right)\left(\frac{159 \text{ ft-kips}}{292 \text{ ft-kips}} + 0\right) = 0.95$$
$$< 1.0 \quad \text{[satisfactory]}$$

Column Base Plates

The design of column base plates is covered in *AISC Manual* Part 14 and AISC 360 Sec. J8. As shown in Fig. 5.14, the base plate is assumed to cantilever about axes a distance m or n from the edge of the plate. The required base plate thickness is given by the largest of the three values obtained from

$$t_{\text{req}} = m\sqrt{\frac{2P_u}{0.9F_y BN}} \quad \text{[LRFD]}$$

$$t_{\text{req}} = n\sqrt{\frac{2P_u}{0.9F_y BN}} \quad \text{[LRFD]}$$

$$t_{\text{req}} = \lambda n'\sqrt{\frac{2P_u}{0.9F_y BN}} \quad \text{[LRFD]}$$

$$t_{\text{req}} = m\sqrt{\frac{3.33P_a}{F_y BN}} \quad \text{[ASD]}$$

$$t_{\text{req}} = n\sqrt{\frac{3.33P_a}{F_y BN}} \quad \text{[ASD]}$$

$$t_{\text{req}} = \lambda n'\sqrt{\frac{3.33P_a}{F_y BN}} \quad \text{[ASD]}$$

In all the preceding equations,

$$n' = \frac{\sqrt{db_f}}{4}$$

$$\lambda = \frac{2\sqrt{X}}{1 + \sqrt{1 - X}} \leq 1$$

For LRFD,

$$X = \left(\frac{4db_f}{(d + b_f)^2}\right)\left(\frac{P_u}{\phi_c P_p}\right)$$

Figure 5.14 *Column Base Plate*

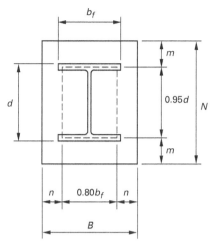

For ASD,

$$X = \left(\frac{4 d b_f}{(d + b_f)^2} \right) \left| \frac{P_a}{\dfrac{P_p}{\Omega_c}} \right|$$

λ may conservatively be taken as 1.0.

Example 5.26

A 19 in $\times$ 19 in grade A36 base plate is proposed for a W12 $\times$ 106 column that supports a factored load of 850 kips (LRFD) or 570 kips (ASD). Determine the minimum required base plate thickness.

Solution

Use a value of $\lambda = 1.0$.

From *AISC Manual* Table 1-1, the relevant dimensions are as follows.

$$d = 12.9 \text{ in}$$
$$b_f = 12.2 \text{ in}$$
$$\lambda n' = \frac{\lambda \sqrt{d b_f}}{4} = \frac{1.0 \sqrt{(12.9 \text{ in})(12.2 \text{ in})}}{4}$$
$$= 3.14 \text{ in}$$
$$m = \frac{N - 0.95 d}{2} = \frac{19 \text{ in} - (0.95)(12.9 \text{ in})}{2}$$
$$= 3.37 \text{ in}$$
$$n = \frac{B - 0.8 b_f}{2} = \frac{19 \text{ in} - (0.8)(12.2 \text{ in})}{2}$$
$$= 4.62 \text{ in} \quad \text{[governs]}$$

LRFD Method

$$t_{\text{req}} = n \sqrt{\frac{2 P_u}{0.9 F_y B N}}$$
$$= 4.62 \text{ in} \sqrt{\frac{(2)(850 \text{ kips})}{(0.9)\left(36 \dfrac{\text{kips}}{\text{in}^2}\right)(19 \text{ in})(19 \text{ in})}}$$
$$= 1.76 \text{ in}$$

ASD Method

$$t_{\text{req}} = n \sqrt{\frac{3.33 P_a}{F_y B N}}$$
$$= 4.62 \text{ in} \sqrt{\frac{(3.33)(570 \text{ kips})}{\left(36 \dfrac{\text{kips}}{\text{in}^2}\right)(19 \text{ in})(19 \text{ in})}}$$
$$= 1.77 \text{ in}$$

6. PLASTIC DESIGN

Nomenclature

A	area of section	in^2
D	degree of indeterminacy of a structure	–
E	modulus of elasticity	kips/in^2
F_y	yield stress	kips/in^2
h	clear distance between flanges less the corner radius at each flange	in
H	horizontal force	kips
l	length of span	ft
L	unbraced length	ft
L_{pd}	limiting laterally unbraced length for plastic analysis	ft or in
m_i	number of independent collapse mechanisms in a structure	–
M_s	bending moment produced by factored loads acting on the cut-back structure	ft-kips
M_y	yield moment	ft-kips
M_1	smaller moment at end of unbraced length of beam	ft-kips
M_2	larger moment at end of unbraced length of beam	ft-kips
M_1/M_2	positive when moments cause reverse curvature and negative for single curvature	–
p	number of possible hinge locations in a structure	–
P	axial force	kips
P_u	required axial strength in compression	kips
P_y	axial yield strength	kips
r	radius of gyration	in

Steel

t_w	web thickness	in
V	shear force	kips
w_u	factored uniformly distributed load	kips/ft
W	total load	lbf

Symbols

ϕ_b	resistance factor for flexure	–
ϕ_c	resistance factor for compression	–

Design Considerations

The plastic method of structural analysis is used to determine the maximum loads a structure can support prior to collapse. The plastic method has several advantages over the ASD and LRFD design techniques, because it

- produces a more economical structure

- provides a simple and direct design technique

- accurately models the structure at ultimate loads

- realistically predicts the ultimate strength

The plastic method is applicable to structures constructed with a ductile material possessing ideal elastic-plastic characteristics. As shown in Fig. 5.15, such a material initially exhibits a linear relationship between stress and strain until the yield point is reached.

Figure 5.15 *Elastic-Plastic Material*

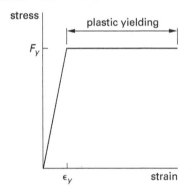

After this point, the stress remains constant at the yield stress, F_y, while the strain continues to increase indefinitely as plastic yielding of the material occurs. The plastic hinge, formed at the location where yielding occurs, has a plastic moment of resistance, M_p, and rotation continues at the hinge without any increase in the resisting moment. As shown in Fig. 5.16, increasing the applied bending moment on a steel beam eventually

causes the extreme fibers to reach the yield stress. The resisting moment developed in the section is the yield moment and is given by

$$M_y = SF_y$$

Figure 5.16 *Plastic Moment of Resistance*

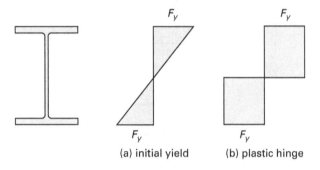

(a) initial yield (b) plastic hinge

As the moment on the section continues to increase, the yielding at the extreme fibers progresses towards the equal area axis, until finally the whole of the section has yielded. The resisting moment developed in the section is the plastic moment and is given by

$$M_p = ZF_y$$

Hinge Formation

A plastic hinge is formed in a structure as the bending moment at a specific location reaches the plastic moment of resistance. In Fig. 5.17, a fixed-ended beam supports a uniformly distributed service total load W. The bending moments produced in the beam are shown in Fig. 5.17(a) with moments at the ends of the beam twice that of the center's. As the load W is progressively increased to W', plastic hinges are formed simultaneously at both ends of the beam, and the bending moments in the beam are shown in Fig. 5.17(b). The system is now equivalent to a simply supported beam with an applied load W' and moment M_p at both ends, as shown in Fig. 5.17(c). Progressively increasing the applied load causes the two plastic hinges to rotate while the moments at both ends remain constant. Finally, as the applied load is increased to the value λW, a third plastic hinge forms in the center of the span, giving the distribution of bending moment shown in Fig. 5.17(d). The system is now an unstable mechanism, shown in Fig. 5.17(e), and collapse occurs under λW. Immediately prior to collapse, the system is statically determinate and the ultimate load may be calculated.

Figure 5.17 *Formation of Plastic Hinges*

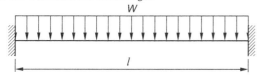

(a)

(b)

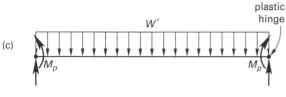
(c)

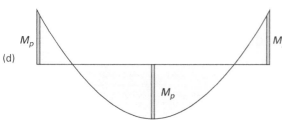
(d)

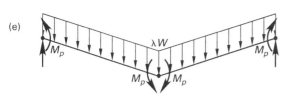
(e)

Taking moments about the center of the span for the left half of the beam gives

$$2M_p = \left(\frac{\lambda W}{2}\right)\left(\frac{l}{2}\right) - \left(\frac{\lambda W}{2}\right)\left(\frac{l}{4}\right)$$

$$\lambda W = \frac{16M_p}{l}$$

The ratio of the collapse load to the service load, where λ is the load factor, is

$$\frac{\lambda W}{W} = \lambda$$

Design Procedure

In accordance with AISC 360 Comm. App. 1.3, plastic design is permitted using LRFD principles only.

Statical Design Method

The statical design procedure is a simple and convenient method for applying inelastic analysis to continuous beams. Figure 5.18 illustrates the procedure that follows for a three-span continuous beam.

step 1: Cut back the continuous beam to three simply supported, statically determinate beams as shown in Fig. 5.18(a), and apply the factored load, λw, to each beam.

step 2: Draw the free bending moment diagram for each beam as shown in Fig. 5.18(b). The maximum moment in each beam is

$$M_s = \frac{\lambda w l^2}{8}$$

step 3: Superimpose the fixing moment line as shown in Fig. 5.18(c). Adjust this line to make the moments at supports 2 and 3 and in spans 12 and 34 equal to M_p.

step 4: The collapse mechanism is shown in Fig. 5.18(d). Collapse occurs simultaneously in the two end spans, with the plastic hinges $0.414l$ from the end supports, and at the interior supports.

This is a partial collapse mechanism, as collapse does not occur in the center span. From the geometry of the figure, $M_p = 0.686M_s$.

To produce a complete collapse mechanism, use a non-uniform beam section as shown in Fig. 5.18(e). The end spans have a plastic moment of resistance of $M_{p2} = 0.766M_s$ that is greater than the plastic moment of resistance $M_{p1} = 0.5M_s$ of the center span. Fig. 5.18(e) shows the superimposed fixing moment line, and the complete collapse mechanism is shown in Fig. 5.18(f).

Example 5.27

The factored loading, including the beam self-weight, acting on a three-span continuous beam of uniform section is shown. Assuming that adequate lateral support is provided to the beam, determine the lightest adequate W12 shape using grade 50 steel.

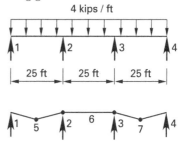

Figure 5.18 *Statical Design Method*

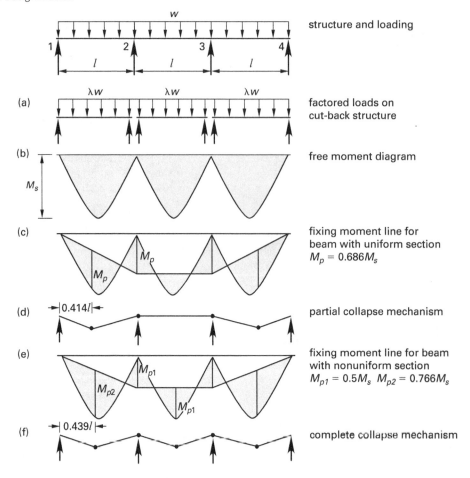

structure and loading

(a) factored loads on cut-back structure

(b) free moment diagram

(c) fixing moment line for beam with uniform section $M_p = 0.686 M_s$

(d) partial collapse mechanism

(e) fixing moment line for beam with nonuniform section $M_{p1} = 0.5 M_s$ $M_{p2} = 0.766 M_s$

(f) complete collapse mechanism

Solution

The free moment in each span is

$$M_s = \frac{w_u l^2}{8} = \frac{\left(4 \ \frac{\text{kips}}{\text{ft}}\right)(25 \ \text{ft})^2}{8}$$

$$= 313 \ \text{ft-kips}$$

Partial collapse occurs as shown in Fig. 5.18(c), with hinges forming in the end spans, and the required plastic moment of resistance is

$$M_p = 0.686 M_s$$

$$= 214 \ \text{ft-kips}$$

The required plastic section modulus is

$$Z = \frac{M_p}{\phi F_y} = \frac{(214 \ \text{ft-kips})\left(12 \ \frac{\text{in}}{\text{ft}}\right)}{(0.9)\left(50 \ \frac{\text{kips}}{\text{in}^2}\right)}$$

$$= 57 \ \text{in}^3$$

From *AISC Manual* Table 3-6, a W12 × 40 has a plastic section modulus of

$$Z = 57 \ \text{in}^3 \quad \text{[satisfactory]}$$

Steel

Beam Design Requirements

To ensure adequate ductility at plastic hinges, AISC 360 App. 1.3.2a specifies that the yield stress of members subject to plastic hinging must not exceed 65 kips/in². In addition, at plastic hinge locations, members must be doubly symmetric and compact with additional slenderness requirements given by AISC 360 Eq. A-1-1 and Eq. A-1-2.

When $P_u/\phi_c P_y \leq 0.125$,

$$\lambda_{pd} = 3.76\sqrt{\frac{E}{F_y}}\left(1 - \frac{2.75 P_u}{\phi_c P_y}\right) \quad \text{[AISC A-1-1]}$$

When $P_u/\phi_c P_y > 0.125$,

$$\lambda_{pd} = 1.12\sqrt{\frac{E}{F_y}}\left(2.33 - \frac{P_u}{\phi_c P_y}\right) \geq 1.49\sqrt{\frac{E}{F_y}}$$
$$\text{[AISC A-1-2]}$$

For member segments that contain plastic hinges, the unbraced length of the segment is restricted to L_{pd}, as given by AISC 360 Eq. A-1-5.

$$L_{pd} = \left(0.12 - 0.076\frac{M_1'}{M_2}\right)\left(\frac{E}{F_y}r_y\right)$$

M_2 is the larger moment at the end of the unbraced length and is taken as positive in all cases. When the magnitude of the bending moment at any location within the unbraced length exceeds M_2, Case 1 applies, and

$$\frac{M_1'}{M_2} = +1 \quad \text{[AISC A-1-6a]}$$

Otherwise, when $M_{\text{mid}} \leq (M_1 + M_2)/2$, Case 2 applies, and

$$M_1' = M_1 \quad \text{[AISC A-1-6b]}$$

When $M_{\text{mid}} > (M_1 + M_2)/2$, Case 3 applies, and

$$M_1' = 2M_{\text{mid}} - M_2 < M_2 \quad \text{[AISC A-1-6c]}$$

M_1 is the smaller moment at the end of the unbraced length, M_{mid} is the moment at the middle of the unbraced length, and M_1' is the effective moment at the end of the unbraced length opposite from M_2. The moments M_1 and M_{mid} are individually taken as positive when they cause compression in the same flange as the moment M_2, and negative otherwise.

In the case of the last hinge to form, rotation does not occur, and the bracing requirements of AISC 360 Sec. F2.2 are applicable. AISC 360 Sec. F2.2 also applies to segments remote from a plastic hinge.

In accordance with AISC 360 App. 1.33, continuous beams not subject to axial compression may be analyzed using a first-order inelastic procedure, and P-δ effects may be ignored. Traditionally, first-order plastic analysis has also been applied to low-rise frames with small axial loads.

Example 5.28

The three-span continuous beam of Ex. 5.27 is laterally braced at the midpoints of the central span and end spans, at supports, and at the locations of plastic hinges. Determine whether this bracing is adequate.

Solution

From *AISC Manual* Table 1-1 and *AISC Manual* Table 3-6, the relevant properties of a W12 × 40 are

$$r_y = 1.94 \text{ in}$$
$$L_p = 6.85 \text{ ft}$$
$$L_r = 21.1 \text{ ft}$$
$$\phi_b M_p = M_2 = 214 \text{ ft-kips}$$

The relevant unbraced lengths in the end spans are

$$L_{15} = 0.414 L_{12} = (0.414)(25 \text{ ft})$$
$$= 10.35 \text{ ft}$$
$$L_{25} = 0.586 L_{12} = (0.586)(25 \text{ ft})$$
$$= 14.65 \text{ ft}$$

For Segment 15

The hinges at 5 and 7 are the last to form, rotation does not occur at these hinges, and the bracing requirements of AISC 360 Sec. F2.2 are applicable to segments 15 and 47. The limiting laterally unbraced length for full plastic bending capacity is

$$L_p = 6.85 \text{ ft}$$
$$< L_{15} = 10.35 \text{ ft}$$

Bracing is inadequate in segments 15 and 47 and an additional brace is required 5 ft from each end support.

For Segment 25

Apply AISC 360 App. 1.3.2c(a) to segment 25.

The larger moment at the end of segment 25 is

$$M_2 = M_{52}$$
$$= +M_p$$

The smaller moment at the end of segment 25 is

$$M_1 = M_{25}$$
$$= -M_p \quad \text{[compression in opposite flange]}$$
$$\frac{M_1 + M_2}{2} = \frac{-M_p + M_p}{2}$$
$$= 0$$

For $x = L_{25}/2 = 7.33$ ft and $L = L_{12} = 25$ ft, the free moment at the midpoint of segment 25 is

$$M_{ms} = \frac{wx(L - x)}{2}$$
$$= -\left(\frac{\left(4\,\frac{\text{kips}}{\text{ft}} \right)(7.33\text{ ft})(25\text{ ft} - 7.33\text{ ft})}{2} \right)$$
$$= -259 \text{ ft-kips}$$

The fixing moment at the midpoint of segment 25 is

$$M_{mf} = \frac{M_p(L_{12} - x)}{L}$$
$$= \frac{(214\text{ ft-kips})(25\text{ ft} - 7.33\text{ ft})}{25\text{ ft}}$$
$$= 151 \text{ ft-kips}$$

The moment at the middle of segment 25 is

$$M_{\text{mid}} = M_{ms} + M_{mf}$$
$$= -259\text{ ft-kips} + 151\text{ ft-kips}$$
$$= -108\text{ ft-kips}$$

Since $M_{\text{mid}} < (M_1 + M_2)/2$, Case 2 applies.

$$M_1' = M_1$$
$$= -M_p$$
$$\frac{M_1'}{M_2} = \frac{-M_p}{M_p}$$
$$= -1$$

The required unbraced length is given by AISC 360 Eq. A-1-5 as

$$L_{pd} = \left(0.12 - 0.076\frac{M_1'}{M_2} \right)\frac{E}{F_y} r_y$$
$$= \left(0.12 - (0.076)(-1) \right) \left(\frac{29{,}000\,\frac{\text{kips}}{\text{in}^2}}{50\,\frac{\text{kips}}{\text{in}^2}} \right) \left(\frac{1.94\text{ in}}{12\,\frac{\text{in}}{\text{ft}}} \right)$$
$$= 18.4 \text{ ft}$$
$$> L_{25} \quad \text{[satisfactory]}$$

Segment 25 is adequately braced.

The maximum moment in the central span is

$$M_{62} = -M_s + M_p$$
$$= -313\text{ ft-kips} + 214\text{ ft-kips}$$
$$= -99 \text{ ft-kips}$$

The unbraced length in the central span is

$$L_{26} = 12.5 \text{ ft}$$

For Segment 26

Apply AISC 360 App. 1.3.2c(a) to segment 26. The larger moment at the end of segment 26 is

$$M_2 = M_{26} = +M_p$$
$$= +214 \text{ ft-kips}$$

The smaller moment at the end of segment 26 is

$$M_1 = M_{62}$$
$$= -99 \text{ ft-kips}$$
$$\frac{M_1 + M_2}{2} = \frac{(-99\text{ ft-kips} + 214\text{ ft-kips})}{2}$$
$$= +58 \text{ ft-kips}$$

$$M_{\text{mid}} = M_2 - 0.75M_s$$
$$= +214\text{ ft-kips} - (0.75)(313\text{ ft-kips})$$
$$= -20.75 \text{ ft-kips}$$

Since $M_{\text{mid}} < (M_1 + M_2)/2$, Case 2 applies.

$$M_1' = M_1$$
$$= -99 \text{ ft-kips}$$
$$\frac{M_1'}{M_2} = \frac{-99\text{ ft-kips}}{214\text{ ft-kips}}$$
$$= -0.46$$

The required unbraced length is given by

$$L_{pd} = \left(0.12 - 0.076\frac{M_1'}{M_2} \right)\frac{E}{F_y} r_y$$
$$= \left(0.12 - (0.076)(-0.46) \right) \left(\frac{29{,}000\,\frac{\text{kips}}{\text{in}^2}}{50\,\frac{\text{kips}}{\text{in}^2}} \right) \left(\frac{1.94\text{ in}}{12\,\frac{\text{in}}{\text{ft}}} \right)$$
$$= 14.5 \text{ ft}$$
$$> L_{26} \quad \text{[satisfactory]}$$

Segment 26 is adequately braced.

Mechanism Design Method

The locations that plastic hinges may form in a structure include the

- ends of a member
- point of application of a concentrated load
- point of zero shear in a member
- weaker of two members meeting at a joint
- end of each member when three or more members meet at a joint

An independent collapse mechanism corresponds to a condition of unstable equilibrium in a structure. The number of possible independent mechanisms is

$$m_i = p - D$$

p is the number of possible hinge locations in the structure and D is the degree of indeterminacy in the structure.

For the rigid frame shown in Fig. 5.19, the number of possible independent mechanisms is

$$m_i = p - D$$
$$= 5 - 3$$
$$= 2$$

Figure 5.19 *Mechanism Design Method*

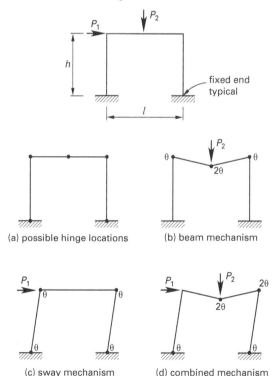

(a) possible hinge locations

(b) beam mechanism

(c) sway mechanism

(d) combined mechanism

These independent mechanisms are the beam mechanism shown in Fig. 5.19(b) and the sway mechanism shown in Fig. 5.19(c). In addition, these may be combined to form the combined mechanism shown in Fig. 5.19(d).

Applying a virtual displacement to each of these mechanisms in turn and equating internal and external work yields three equations, from each of which a value of M_p may be obtained. The largest value of M_p governs.

For the beam mechanism,

$$4M_p\theta = P_2\left(\frac{l}{2}\right)\theta$$
$$M_p = \frac{P_2 l}{8}$$

For the sway mechanism,

$$4M_p\theta = P_1 h\theta$$
$$M_p = \frac{P_1 h}{4}$$

For the combined mechanism,

$$6M_p\theta = P_2\left(\frac{l}{2}\right)\theta + P_1 h\theta$$
$$M_p = \frac{P_1 h + \dfrac{P_2 l}{2}}{6}$$

For the situation where $P_2 = 2P_1$ and $h = l$, the combined mechanism controls and

$$M_p = \frac{P_1 l}{3}$$

Example 5.29

The rigid frame shown is fabricated from members of a uniform section in grade 50 steel. For the factored loading indicated, ignoring the member self-weight and assuming adequate lateral support, determine the lightest adequate W shape.

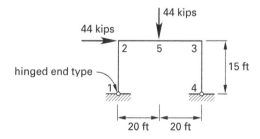

Solution

The three possible collapse mechanisms are shown.

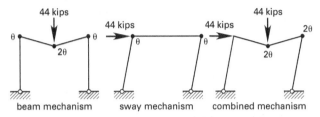

beam mechanism   sway mechanism   combined mechanism

The beam mechanism gives

$$4M_p = (44 \text{ kips})(20 \text{ ft})$$
$$M_p = 220 \text{ ft-kips}$$

The sway mechanism gives

$$2M_p = (44 \text{ kips})(15 \text{ ft})$$
$$M_p = 330 \text{ ft-kips}$$

The combined mechanism gives

$$4M_p = 880 \text{ ft-kips} + 660 \text{ ft-kips}$$
$$M_p = 385 \text{ ft-kips} \quad \text{[governs]}$$

From *AISC Manual* Table 3-6, a W21 × 50 has

$$\phi_b M_p = 413 \text{ ft-kips}$$
$$> 385 \text{ ft-kips} \quad \text{[satisfactory]}$$

A W21 × 48 is noncompact and may not be used.

Static Equilibrium Check

Mechanism methods lead to upper bounds on the collapse load. To confirm that the correct mechanism has been selected, it is necessary to check that the assumed plastic moment is not anywhere exceeded by constructing a moment diagram obtained by static equilibrium methods.

Example 5.30

Draw the bending moment diagram for the assumed collapse mechanism of Ex. 5.29.

Solution

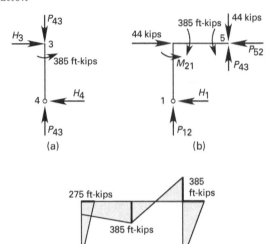

(a)   (b)

(c)

For member 34, as shown at (a) in the illustration,

$$H_4 = \frac{385 \text{ ft-kips}}{15 \text{ ft}} = 25.67 \text{ kips}$$
$$= H_3$$
$$= P_{52}$$

For member 125, as shown at (b) in the figure,

$$H_1 = 44 \text{ kips} - P_{52}$$
$$= 44 \text{ kips} - 25.67 \text{ kips}$$
$$= 18.33 \text{ kips}$$
$$P_{12} = \frac{M_p - H_1 h}{l_{25}}$$
$$= \frac{385 \text{ ft-kips} - (18.33 \text{ kips})(15 \text{ ft})}{20 \text{ ft}}$$
$$= 5.50 \text{ kips}$$
$$M_{21} = H_1 h$$
$$= (18.33 \text{ kips})(15 \text{ ft})$$
$$= 275 \text{ ft-kips}$$
$$P_{43} = 44 \text{ kips} - P_{12}$$
$$= 44 \text{ kips} - 5.50 \text{ kips}$$
$$= 38.50 \text{ kips}$$

The bending moment diagram is shown at (c) in the illustration. Because $M_p = 385$ ft-kips is not exceeded at any point in the frame, the combined mechanism is the correct failure mode.

Column Design Requirements

Flanges and webs of members subjected to combined flexure and compression must be compact with width-thickness ratios not exceeding the values defined in

AISC 360 Table B4.1b. In addition, the webs of W sections must also comply with AISC 360 Eq. A-1-1 and Eq. A-1-2, which are as follows.

For $P_u/\phi_b P_y \leq 0.125$,

$$\frac{h}{t_w} \leq 3.76 \sqrt{\frac{E}{F_y}} \left(1 - \frac{2.75 P_u}{\phi_b P_y} \right)$$

For $P_u/\phi_b P_y > 0.125$,

$$\frac{h}{t_w} \leq 1.12 \sqrt{\frac{E}{F_y}} \left(2.33 - \frac{P_u}{\phi_b P_y} \right)$$

$$\geq 1.49 \sqrt{\frac{E}{F_y}}$$

The member yield strength is

$$P_y = A F_y$$

The resistance factor for flexure is

$$\phi_b = 0.90$$

The maximum permitted slenderness ratio of a column is specified in AISC 360 App. 1.3.2c(b) as

$$\frac{L}{r} = 4.71 \sqrt{\frac{E}{F_y}}$$

$$= 113 \quad [\text{for } F_y = 50 \text{ kips/in}^2]$$

In accordance with AISC 360 App. 1.3.2d, the design strength in compression in a column with plastic hinges may not exceed $0.75 A_g F_y$. As for beams, the maximum unbraced length is controlled by AISC 360 Eq. A-1-5, and for combined axial force and flexure, the interaction expressions of AISC 360 Eq. H1-1a and Eq. H1-1b govern. In practice, it has been the custom that second-order effects may be neglected for low-rise frames with small axial loads.

Example 5.31

Determine whether column 34 of the rigid frame in Ex. 5.29 is satisfactory. The column consists of a grade 50 W21 × 50 section and is laterally braced about its weak axis at 3.75 ft centers and at joint 3. Neglect secondary effects.

Solution

For the pinned connection at joint 4, AISC 360 Comm. App. 7.2 specifies a stiffness ratio of $G_4 = 10$.

At joint 3,

$$G_3 = \frac{\sum \dfrac{I_c}{L_c}}{\sum \dfrac{I_g}{L_g}} = \frac{\dfrac{I}{15}}{\dfrac{I}{40}}$$

$$= 2.7$$

From the alignment chart for sway frames, the effective length factor is

$$K_{34} = 2.2$$

From *AISC Manual* Table 1-1, a W21 × 50 has

$$A_g = 14.7 \text{ in}^2$$
$$r_y = 1.30 \text{ in}$$
$$r_x = 8.18 \text{ in}$$

From AISC 360 App. 1.3.2c(b),

$$\frac{L}{r} < 4.71 \sqrt{\frac{E}{F_y}} = 113$$

The slenderness ratio about the x-axis is

$$\frac{K_{34} L_x}{r_x} = \frac{(2.2)(15 \text{ ft})\left(12 \dfrac{\text{in}}{\text{ft}} \right)}{8.18 \text{ in}}$$

$$= 48.4 \quad [\text{governs for axial load}]$$

$$\frac{L_x}{r_x} = \frac{180 \text{ in}}{8.18 \text{ in}}$$

$$= 22.0$$

$$< 113 \quad [\text{satisfies AISC 360 App. 1.3.2c(b)}]$$

The slenderness ratio about the y-axis is

$$\frac{K_{34} L_y}{r_y} = \frac{(1.0)(3.75 \text{ ft})\left(12 \dfrac{\text{in}}{\text{ft}} \right)}{1.30 \text{ in}}$$

$$= 34.6$$

$$\frac{L_y}{r_y} = \frac{45 \text{ in}}{1.30 \text{ in}}$$

$$= 34.6$$

$$< 113 \quad [\text{satisfies AISC 360 App. 1.3.2c(b)}]$$

In accordance with AISC 360 App. 1.3.2d, the maximum axial load in the column is restricted to

$$P_{max} = 0.75 A_g F_y$$
$$= (0.75)(14.7 \text{ in}^2)\left(50 \ \frac{\text{kips}}{\text{in}^2}\right)$$
$$= 551 \text{ kips}$$
$$> P_{43} \quad \text{[satisfactory]}$$

From *AISC Manual* Table 3-6, for a W21 × 50,

$$L_p = 4.59 \text{ ft}$$
$$> 3.75 \text{ ft} \quad \text{[full plastic bending capacity available]}$$
$$\phi_b M_{nx} = \phi_b M_p$$
$$= 413 \text{ ft-kips}$$

From *AISC Manual* Table 4-14, for a $K_{34}L_x/r_x$ value of 48.4, the design stress for axial load is

$$\phi_c F_{cr} = 37.9 \text{ kips/in}^2$$

The design axial strength is

$$\phi_c P_n = \phi_c F_{cr} A_g = \left(37.9 \ \frac{\text{kips}}{\text{in}^2}\right)(14.7 \text{ in}^2)$$
$$= 557 \text{ kips}$$

$$\frac{P_{34}}{\phi_c P_n} = \frac{38.50 \text{ kips}}{557 \text{ kips}}$$
$$= 0.07$$
$$< 0.20 \quad \text{[AISC Eq. H1-1b governs]}$$

Since secondary effects may be neglected, AISC 360 Eq. H1-1b reduces to

$$\frac{P_{34}}{2\phi_c P_n} + \frac{M_p}{\phi_b M_{nx}} \le 1.0$$
$$= \frac{38.50 \text{ kips}}{(2)(557 \text{ kips})} + \frac{385 \text{ ft-kips}}{413 \text{ ft-kips}}$$
$$= 0.97$$
$$< 1.0 \quad \text{[satisfactory]}$$

7. DESIGN OF TENSION MEMBERS

Nomenclature

A_e	effective net area	in^2
A_g	gross area	in^2
A_n	net area	in^2
A_R	area of section required for fatigue loading	in^2
C_f	fatigue constant	–
d	nominal bolt diameter	in
d_h	effective hole diameter	in
f_{max}	maximum tensile stress in member at service load	kips/in^2
f_{min}	minimum stress in member at service load (compression negative)	kips/in^2
f_{SR}	actual stress range	kips/in^2
F_{SR}	allowable stress range for fatigue loading	kips/in^2
F_{TH}	threshold stress range	kips/in^2
F_u	specified minimum tensile strength	kips/in^2
g	transverse center-to-center spacing between fasteners (gage)	in
l	length of connection	in
l	length of weld	in
n_{SR}	number of stress range fluctuations	–
P_n	nominal axial strength	kips
P_u	required axial strength	kips
s	bolt spacing in direction of load	in
t	plate thickness	in
T_{max}	maximum tensile force in member at service load	kips
T_{min}	minimum force in member (compression negative) at service load	kips
U	shear lag factor used in calculating effective net area	–
w	plate width, distance between welds	in
$\overline{x}$	connection eccentricity	in

Plates in Tension

The available strength for plates in tension is either the limit state of yielding or the limit state of rupture, whichever is lower. The *limit state of yielding* occurs when the gross cross-section becomes excessively elongated, which causes the whole structure to be unstable. The *limit state of rupture* occurs when the net cross-section fails at bolted joints, which causes sudden and catastrophic failure.

For yielding of the gross section, AISC 360 Sec. D2 gives the design strength as

$$P_c = \phi_t P_n = 0.9 F_y A_g \quad \text{[LRFD]}$$

The allowable strength is

$$P_c = \frac{P_n}{\Omega_t} = \frac{F_y A_g}{1.67} \quad \text{[ASD]}$$

As shown in Fig. 5.20, the gross area is given by

$$A_g = wt$$

Figure 5.20 *Effective Net Area of Bolted Connection*

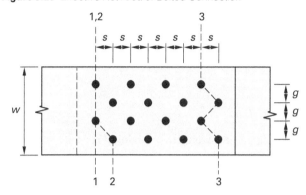

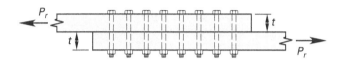

For tensile rupture at the connection, AISC 360 Sec. D2 gives the design strength as

$$P_c = \phi_t P_n = 0.75 F_u A_e \quad \text{[LRFD]}$$

The allowable strength is

$$P_c = \frac{P_n}{\Omega_t} = \frac{F_u A_e}{2.00} \quad \text{[ASD]}$$

Effective Net Area

The shear lag factor, given in AISC 360 Table D3.1, accounts for the effects of eccentricity and shear lag on tension members connected through only part of their cross-sectional elements. The effective net area, A_e, resisting the tensile force is given by AISC 360 Eq. D3-1 as

$$A_e = U A_n$$

A_n is the net area of the member and U is the shear lag factor.

Bolt Hole Diameter

The nominal diameter of a standard hole is detailed in AISC 360 Table J3.3 as $\frac{1}{16}$ in larger than the bolt diameter, for a bolt diameter less than 1 in. As the hole is formed, some deterioration occurs in the surrounding material, and AISC 360 Sec. B4.3b specifies that the effective hole diameter must be taken as $\frac{1}{16}$ in larger than the nominal hole diameter. The effective hole diameter when d is the diameter of the fastener is

$$d_h = d + \frac{1}{16} \text{ in} + \frac{1}{16} \text{ in} = d + \frac{1}{8} \text{ in}$$

For bolts 1 in and greater in diameter, the nominal diameter of a standard hole is detailed in AISC 360 Table J3.3 as $\frac{1}{8}$ in larger than the bolt diameter. The effective hole diameter is then

$$d_h = d + \frac{1}{8} \text{ in} + \frac{1}{16} \text{ in} = d + \frac{3}{16} \text{ in}$$

Plates with Bolted Connections

For a flat plate with bolted connection, the total net cross-section is assumed to transfer the load without shear lag, and AISC 360 Table D3.1, Case 1 indicates that $U = 1.0$. For a flat plate with bolted connection,

$$A_e = U A_n = A_n$$

For the straight perpendicular fracture 1-1 of the bolted plates shown in Fig. 5.20, the effective net area is

$$A_e = t(w - 2 d_h)$$

The effective hole diameter for a bolt diameter of less than 1 in is

$$d_h = d + \frac{1}{8} \text{ in}$$

For a staggered fracture, the effective net width is obtained by deducting from the gross plate width the sum of the bolt holes in the failure path and adding, for each gage space traversed by a diagonal portion of the failure path, the quantity $s^2/(4g)$. g is the transverse center-to-center spacing between fasteners (gage) and s is the bolt spacing in direction of load (pitch).

For the staggered fracture 2-2 shown in Fig. 5.20, the effective net area of the plate is

$$A_e = t\left(w - 3 d_h + \frac{s^2}{4g}\right)$$

For the staggered fracture 3-3 shown in Fig. 5.20, the effective net area of the plate is

$$A_e = t\left(w - 4 d_h + \frac{3s^2}{4g}\right)$$

Plates with Welded Connections

For a flat plate with a welded connection, the effective net area is given by AISC 360 Eq. D3-1 as

$$A_e = A_n U = A_g U$$

For a flat plate with a transverse fillet welded connection as shown in Fig. 5.21(a), the total cross-section is assumed to transfer the load without shear lag, and AISC 360 Table D3.1, Case 3 states that $U = 1.0$. For a flat plate with a transverse fillet welded connection,

$$A_e = A_n$$

For the longitudinal fillet welded connection shown in Fig. 5.21(b), shear lag occurs at the ends of the plate and the shear lag factor is defined in AISC 360 Table D3.1, Case 4 as

$$U = \left(\frac{3l^2}{3l^2 + w^2}\right)\left(1 - \frac{\bar{x}}{l}\right)$$

$$l = \left(\frac{l_1 + l_2}{2}\right) \geq 4 \times (\text{weld size})$$

Figure 5.21 *Welded Connections for Plates*

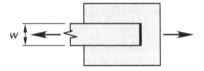

(a) transverse weld

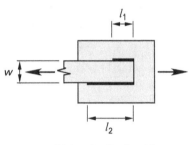

(b) longitudinal welds

Example 5.32

As shown, two plates each $\frac{1}{2}$ in thick $\times$ 9 in wide are connected by three rows of bolts. The distance between rows is 3 in, the distance between bolts in a row is 3 in, and the center row of bolts is staggered. Determine the available axial strength of the plates in direct tension. The relevant properties of the plates are $F_y = 36$ kips/in^2, $F_u = 58$ kips/in^2, and effective hole diameter $d_h = 1.0$ in.

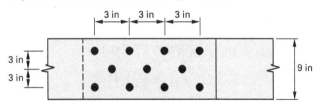

Solution

The gross area of each plate is given by

$$A_g = wt = (9 \text{ in})(0.5 \text{ in})$$
$$= 4.5 \text{ in}^2$$

LRFD Method

The design axial strength for yielding is

$$\phi_t P_n = 0.9 F_y A_g$$
$$= (0.9)\left(36 \frac{\text{kips}}{\text{in}^2}\right)(4.5 \text{ in}^2)$$
$$= 146 \text{ kips}$$

For a straight perpendicular fracture, the effective net area of the plate is given by

$$A_e = t(w - 2d_h)$$
$$= (0.5)\big(9 \text{ in} - (2)(1.0 \text{ in})\big)$$
$$= 3.5 \text{ in}^2$$

The design axial strength for tensile rupture is

$$\phi_t P_n = 0.75 F_u A_e$$
$$= (0.75)\left(58 \frac{\text{kips}}{\text{in}^2}\right)(3.5 \text{ in}^2)$$
$$= 152 \text{ kips}$$

For a staggered fracture, the effective net area of the plate is given by

$$A_e = t\left(w - 3d_h + \frac{2s^2}{4g}\right)$$
$$= (0.5 \text{ in})\left(9 \text{ in} - (3)(1.0 \text{ in}) + \frac{(2)(1.5 \text{ in})^2}{(4)(3 \text{ in})}\right)$$
$$= 3.19 \text{ in}^2$$

The corresponding design axial strength is

$$\phi_t P_n = 0.75 F_u A_e = (0.75)\left(58 \frac{\text{kips}}{\text{in}^2}\right)(3.19 \text{ in}^2)$$
$$= 139 \text{ kips} \quad [\text{governs}]$$

ASD Method

The allowable axial strength for yielding is

$$\frac{P_n}{\Omega_t} = \frac{F_y A_g}{1.67} = \frac{\left(36 \ \dfrac{\text{kips}}{\text{in}^2}\right)(4.5 \ \text{in}^2)}{1.67}$$

$$= 97 \ \text{kips}$$

For a straight perpendicular fracture, the effective net area of the plate is given by

$$A_e = t(w - 2d_h) = (0.5)\big(9 \ \text{in} - (2)(1.0 \ \text{in})\big)$$

$$= 3.5 \ \text{in}^2$$

The allowable axial strength for tensile rupture is

$$\frac{P_n}{\Omega_t} = \frac{F_u A_e}{2.00} = \frac{\left(58 \ \dfrac{\text{kips}}{\text{in}^2}\right)(3.5 \ \text{in}^2)}{2.00}$$

$$= 102 \ \text{kips}$$

For a staggered fracture, the effective net area of the plate is given by

$$A_e = t\left(w - 3d_h + \frac{2s^2}{4g}\right)$$

$$= (0.5 \ \text{in})\left(9 \ \text{in} - (3)(1.0 \ \text{in}) + \frac{(2)(1.5 \ \text{in})^2}{(4)(3 \ \text{in})}\right)$$

$$= 3.19 \ \text{in}^2$$

The corresponding allowable axial strength is

$$\frac{P_n}{\Omega_t} = \frac{F_u A_e}{2.00} = \frac{\left(58 \ \dfrac{\text{kips}}{\text{in}^2}\right)(3.19 \ \text{in}^2)}{2.00}$$

$$= 93 \ \text{kips} \quad [\text{governs}]$$

Rolled Sections in Tension

When the tensile load is transmitted directly to all the cross-sectional elements of a member by fasteners or welds, the shear lag factor is defined by AISC 360 Table D3.1, Case 1 (except as in Cases 4, 5, and 6) as

$$U = 1.0$$

When rolled structural shapes are connected through only part of their cross-sectional elements, the value of the shear lag factor is defined in AISC 360 Table D3.1, Case 2 as

$$U = 1 - \frac{\bar{x}}{l}$$

The length of the connection l is defined in AISC 360 Comm. Sec. D3.3 as the distance, parallel to the line of force, between the first and last fasteners in a line. The connection eccentricity $\bar{x}$ is defined as the distance from the connection plane to the centroid of the member resisting the connection force. In lieu of applying this expression for U, AISC 360 Table D3.1, Cases 7 and 8 permit the adoption of the following values for the shear lag factor.

$U = 0.90$ for T, W, M, HP, and S shapes with

$b_f \geq 2d/3$, connected by the flange,

with not fewer than three bolts
in line in the direction of stress

$U = 0.85$ for T, W, M, HP, and S shapes with

$b_f < 2d/3$ connected by the flange,

with not fewer than three bolts
in line in the direction of stress

For W, M, HP, and S shapes, d is defined as the depth of section. For T shapes, d is defined as the depth of section from which the T shape is cut.

$U = 0.70$ for T, W, M, HP, and S shapes
connected by the web, with
not less than four bolts in line
in the direction of stress

$U = 0.80$ for single or double angles with not
less than four bolts in line in the direction
of stress

$U = 0.60$ for single or double angles with three
bolts in line in the direction of stress

For a welded connection, when the axial force is transmitted only by transverse welds, as shown in Fig. 5.22(a), the effective net area is given by AISC 360 Table D3.1, Case 3 as

$$A_e = \text{area of directly connected elements}$$

Figure 5.22 *Welded Connections for Rolled Sections*

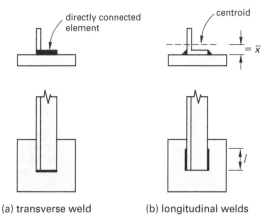

(a) transverse weld   (b) longitudinal welds

For a welded connection, when the axial force is transmitted through only part of the member's cross-sectional elements by longitudinal welds in combination with transverse welds, as shown in Fig. 5.22(b), the effective net area is given by AISC 360 Sec. D3 as

$$A_e = A_g U \quad \text{[AISC D3-1]}$$

The value of the shear lag factor is defined in AISC 360 Table D3.1, Case 2 as

$$U = 1 - \frac{\bar{x}}{l}$$

The length of the connection l is shown in Fig. 5.22(b) and is defined in AISC 360 Comm. Sec. D3.3 as the length of the weld, parallel to the line of force.

For welded connections, when rolled structural shapes are connected through only part of their cross-sectional elements by longitudinal welds, as shown in Fig. 5.21(b), the shear lag factor is defined in AISC 360 Table D3.1, Case 4 as

$$U = \left(\frac{3l^2}{3l^2 + w^2}\right)\left(1 - \frac{\bar{x}}{l}\right)$$

$$l = \left(\frac{l_1 + l_2}{2}\right) \geq 4 \times \text{weld size}$$

Example 5.33

Assuming that the welds are adequate, determine the design axial strength of the grade A50 W12 × 65 member connected as shown.

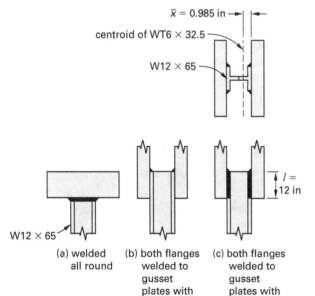

(a) welded all round

(b) both flanges welded to gusset plates with transverse welds

(c) both flanges welded to gusset plates with longitudinal welds and transverse welds

Solution

LRFD Method

The relevant properties of the W12 × 65 are obtained from *AISC Manual* Table 1-1 and are

$$A_g = 19.1 \text{ in}^2$$
$$b_f = 12 \text{ in}$$
$$t_f = 0.605 \text{ in}$$

(a) The W12 × 65 is welded all around to the supporting member.

$$A_e = \text{area of directly connected elements}$$
$$= A_g$$

The available axial strength for rupture is

$$\phi_t P_n = 0.75 F_u A_e$$
$$= (0.75)\left(65 \frac{\text{kips}}{\text{in}^2}\right)(19.1 \text{ in}^2)$$
$$= 931 \text{ kips}$$

The available axial strength for yielding is

$$\phi_t P_n = 0.9 F_y A_g$$
$$= (0.9)\left(50 \frac{\text{kips}}{\text{in}^2}\right)(19.1 \text{ in}^2)$$
$$= 860 \text{ kips} \quad \text{[governs]}$$

(b) Both flanges are welded by transverse welds to gusset plates.

$$A_e = \text{area of directly connected elements}$$
$$= 2 b_f t_f$$
$$= (2)(12 \text{ in})(0.605 \text{ in})$$
$$= 14.5 \text{ in}^2$$

The available axial strength for rupture is

$$\phi_t P_n = 0.75 F_u A_e$$
$$= (0.75)\left(65 \frac{\text{kips}}{\text{in}^2}\right)(14.5 \text{ in}^2)$$
$$= 707 \text{ kips}$$

The available axial strength for yielding is

$$\phi_t P_n = 0.9 F_y A_e$$
$$= (0.9)\left(50 \ \frac{\text{kips}}{\text{in}^2}\right)(14.5 \ \text{in}^2)$$
$$= 653 \ \text{kips} \quad \text{[governs]}$$

(c) Both flanges are welded by longitudinal welds and transverse welds to gusset plates.

In accordance with AISC 360 Comm. Sec. D3.3, the W section is treated as two WT sections as shown in the figure. The centroidal height of a WT6 × 32.5 cut from a W12 × 65 is obtained from *AISC Manual* Table 1-8 and is

$$\overline{x} = 0.985 \ \text{in}$$

The value of the shear lag factor is defined in AISC 360 Table D3.1, Case 2 as

$$U = 1 - \frac{\overline{x}}{l} = 1 - \frac{0.985 \ \text{in}}{12 \ \text{in}}$$
$$= 0.92$$

The effective net area is

$$A_e = A_g U = (19.1 \ \text{in}^2)(0.92)$$
$$= 17.6 \ \text{in}^2$$

The available axial strength for rupture is

$$\phi_t P_n = 0.75 F_u A_e$$
$$= (0.75)\left(65 \ \frac{\text{kips}}{\text{in}^2}\right)(17.6 \ \text{in}^2)$$
$$= 858 \ \text{kips}$$

The available axial strength for yielding is

$$\phi_t P_n = 0.9 F_y A_e$$
$$= (0.9)\left(50 \ \frac{\text{kips}}{\text{in}^2}\right)(17.6 \ \text{in}^2)$$
$$= 792 \ \text{kips} \quad \text{[governs]}$$

ASD Method

The relevant properties of the W12 × 65 are obtained from *AISC Manual* Table 1-1 and are

$$A_g = 19.1 \ \text{in}^2$$
$$b_f = 12 \ \text{in}$$
$$t_f = 0.605 \ \text{in}$$

(a) The W12 × 65 is welded all around to the supporting member.

$$A_e = \text{area of directly connected elements}$$
$$= A_g$$

The available axial strength for rupture is

$$\frac{P_n}{\Omega_t} = \frac{F_u A_e}{2.00} = \frac{\left(65 \ \frac{\text{kips}}{\text{in}^2}\right)(19.1 \ \text{in}^2)}{2.00}$$
$$= 620 \ \text{kips}$$

The available axial strength for yielding is

$$\frac{P_n}{\Omega_t} = \frac{F_y A_g}{1.67} = \frac{\left(50 \ \frac{\text{kips}}{\text{in}^2}\right)(19.1 \ \text{in}^2)}{1.67}$$
$$= 572 \ \text{kips} \quad \text{[governs]}$$

(b) Both flanges are welded by transverse welds to gusset plates.

$$A_e = \text{area of directly connected elements}$$
$$= 2 b_f t_f$$
$$= (2)(12 \ \text{in})(0.605 \ \text{in})$$
$$= 14.5 \ \text{in}^2$$

The available axial strength for rupture is

$$\frac{P_n}{\Omega_t} = \frac{F_u A_e}{2.00}$$
$$= \frac{\left(65 \ \frac{\text{kips}}{\text{in}^2}\right)(14.5 \ \text{in}^2)}{2.00}$$
$$= 471 \ \text{kips}$$

The available axial strength for yielding is

$$\frac{P_n}{\Omega_t} = 435 \ \text{kips} \quad \text{[governs]}$$

(c) Both flanges are welded by longitudinal welds and transverse welds to gusset plates.

In accordance with AISC 360 Comm. Sec. D3.3, the W section is treated as two WT sections as shown in the figure. The centroidal height of a WT6 × 32.5 cut from a W12 × 65 is obtained from *AISC Manual* Table 1-8 and is

$$\overline{x} = 0.985 \ \text{in}$$

The value of the shear lag factor is defined in AISC 360 Table D3.1, Case 2 as

$$U = 1 - \frac{\overline{x}}{l} = 1 - \frac{0.985 \text{ in}}{12 \text{ in}}$$
$$= 0.92$$

The effective net area is

$$A_e = A_g U = (0.92)(19.1 \text{ in}^2)$$
$$= 17.6 \text{ in}^2$$

The available axial strength for rupture is

$$\frac{P_n}{\Omega_t} = \frac{F_u A_e}{2.00} = \frac{\left(65 \dfrac{\text{kips}}{\text{in}^2}\right)(17.6 \text{ in}^2)}{2.00}$$
$$= 572 \text{ kips}$$

The available axial strength for yielding is

$$\frac{P_n}{\Omega_t} = 528 \text{ kips} \quad \text{[governs]}$$

Design for Fatigue

Fatigue failure is caused by fluctuations of tensile stress that cause crack propagation in the parent metal. Fatigue must be considered for tensile stresses, stress reversals, and shear when the number of loading cycles exceeds 20,000 and is based on the stress level at nominal loads. Fatigue effects are analyzed using nominal loads. The maximum permitted stress due to nominal loads is $0.66F_y$.

Fatigue failure is influenced by the number of applied load cycles, the magnitude of the stress range produced by nominal live loads, and stress concentrations produced by the fabrication details.

Fatigue will not occur when the

- number of applied load cycles is less than 20,000
- live load stress range is less than the threshold stress range
- fluctuations in stress do not involve tensile stress
- members are HSS sections in buildings subjected to wind loads

The design procedure is given in AISC 360 App. 3 and consists of establishing the applicable loading condition from AISC 360 Table A-3.1.

The applicable values of the fatigue constant, the threshold stress range, and the stress category are obtained from the table.

The stress range is defined as the magnitude of the change in stress due to the application or removal of the unfactored live load. Fatigue must be considered if the stress range in the member exceeds the threshold stress range. The actual stress range, at service level values, is given by

$$f_{SR} = f_{\max} - f_{\min}$$
$$< F_{SR}$$

Eleven stress categories are defined in AISC 360 Table A-3.1. For stress categories A, B, B′, C, D, E, and E′, the design stress range in the member must not exceed the value given by AISC 360 Eq. A-3-1 as

$$F_{SR} = 1000 \left(\frac{C_f}{n_{SR}}\right)^{0.333}$$
$$\geq F_{TH}$$

For stress category F, the design stress range in the member must not exceed the value given by AISC 360 Eq. A-3-2.

$$F_{SR} = 100 \left(\frac{1.5}{n_{SR}}\right)^{0.167}$$
$$\geq 8 \frac{\text{kips}}{\text{in}^2}$$

Example 5.34

A tie member in a steel truss consists of a pair of grade A36 5 in × 5 in × ⅜ in angles fillet welded to a gusset plate. The force in the member, due to dead load only, is 90 kips tension. The additional force in the member, due to live load only, varies from a compression of 7 kips to a tension of 50 kips. During the design life of the structure, the live load may be applied 600,000 times. Determine whether fatigue effects are a concern.

Solution

From AISC 360 Table A-3.1, the loading condition of Sec. 4.1 is applicable and the relevant factors are

$$E = \text{stress category}$$
$$F_{SR} = \text{allowable stress range}$$
$$= 1000 \left(\frac{C_f}{n_{SR}}\right)^{0.333} = (1000) \left(\frac{1.1}{6 \times 10^5}\right)^{0.333}$$
$$= 12.29 \text{ kips/in}^2$$

The area of the tie is

$$A_s = 7.30 \text{ in}^2$$

The actual stress range is

$$f_{SR} = \frac{T_{max} - T_{min}}{A_s} = \frac{50 \text{ kips} - (-7 \text{ kips})}{7.30 \text{ in}^2}$$

$$= 7.8 \text{ kips/in}^2 \quad [\text{exceeds } F_{TH}]$$

$$< F_{SR}$$

This is within the allowable stress range, so fatigue effects are not a concern.

8. DESIGN OF BOLTED CONNECTIONS

Nomenclature

A_b	nominal unthreaded body area of bolt	in^2
C	coefficient for eccentrically loaded bolt and weld groups	–
d	nominal bolt diameter	in
d_m	moment arm between resultant tensile and compressive forces due to an eccentric force	in
d_n	nominal hole diameter	–
D_u	a multiplier that reflects the ratio of the mean installed bolt tension to the specified minimum bolt pretension	–
f_{rv}	required shear stress	kips/in^2
f_v	computed shear stress	kips/in^2
F_{nt}	nominal tensile stress of bolt	kips/in^2
F_{nt}'	nominal tensile stress of a bolt subjected to combined shear and tension	kips/in^2
F_{nv}	nominal shear stress of bolt	kips/in^2
F_u	specified minimum tensile strength	kips/in^2
h_{sc}	modification factor for type of hole	–
k_s	slip-critical combined tension and shear coefficient	–
l_c	clear distance, in the direction of force, between the edge of the hole and the edge of the adjacent hole or edge of the material	in
L_e	edge distance between the bolt center and the edge of the connected part	in
n	number of bolts in a connection	–
n'	number of bolts above the neutral axis (in tension)	–
n_b	number of bolts carrying strength level tension T_u	–
n_s	number of slip planes	–
P_r	load on connection	kips
R_n	nominal strength	kips
s	center-to-center pitch of two consecutive bolts	in
t	thickness of connected part	in
T_a	applied tensile force (ASD)	kips
T_b	minimum pre-tension force	kips
T_u	applied tensile force (LRFD)	kips

Symbols

μ	mean slip coefficient for the applicable surface	–
ϕ	resistance factor	–

Types of Bolts

There are two categories of bolts: common bolts and high-strength bolts. High-strength bolts are additionally grouped by strength levels into three categories: group A bolts (A325, F1852, and A354 grade BC), group B bolts (A490, F2280, and A354 grade BD), and group C bolts (F3043 and F3111).

Common bolts of grade A307 with a nominal tensile strength of 45 kips/in^2 are used in snug-tight (bearing-type) connections only.

High-strength bolts in group A with a nominal tensile strength of 90 kips/in^2, or group B with a nominal tensile strength of 113 kips/in^2, may be used in snug-tight, pretensioned, and slip-critical connections.

Where loosening or fatigue due to vibration or load fluctuations are not design considerations, snug-tight installation is permitted for

- group A bolts used in tension or in combined tension and shear applications
- group B bolts used in shear only applications

Group C bolts, with a nominal tensile strength of 150 kips/in^2, are subdivided into two categories, grade 1 and grade 2. Grade 1 bolts are restricted to use in snug-tight conditions. Grade 2 bolts may be used in snug-tight, pretensioned, and slip-critical conditions. Group C bolts may not be used in situations that would subject the bolts to hydrogen embrittlement, corrosive conditions, high humidity, or submerging in soil or water.

AISC 360 Table J3.2 lists the nominal tensile strength of bolts. AISC 360 Table J3.1 lists the required minimum pretension of high-strength bolts.

Bolts are installed in the following three types of connections.

1. *Bearing-type* or *snug-tight connections* require the bolts to be tightened sufficiently to bring the plies into firm contact. Levels of installed tension are not specified. Transfer of the load from one connected part to another depends on the bearing of the bolts against the side of the holes. This type may be used when pretensioned or slip-critical connections are not required.

2. *Pretensioned connections* require the bolts to be pretensioned to a minimum value given in AISC 360 Table J3.1, and the faying surfaces may be uncoated, coated, or galvanized regardless of the slip coefficient. In accordance with RCSC[10]

Sec. 4.2, transfer of the load from one connected part to another depends on the bearing of the bolts against the side of the holes. Pretensioned connections are required when bearing-type connections are used in

- column splices in buildings with high ratios of height to width

- bracing members in tall buildings

- structures carrying cranes of over 5 ton capacity

- supports of machinery causing impact or stress reversal

- joints with group A bolts subject to tensile fatigue

- joints with group B bolts subject to tension or combined tension and shear

- end connections of built-up compression members

3. *Slip-critical connections* require the bolts to be pretensioned to a minimum value given in AISC 360 Table J3.1, and the faying surfaces must be prepared to produce a specific value of the slip coefficient. In accordance with RCSC Sec. 4.3, transfer of the load from one connected part to another depends on the friction induced between the parts. Slip-critical connections are required where

- fatigue load occurs with reversal of the load direction

- bolts are used in oversize holes or slotted holes parallel to the direction of load

- slip at the faying surfaces will affect the performance of the structure

Bearing-Type Bolts in Shear

The minimum permissible distance and the preferred distance between the centers of holes is given by AISC 360 Sec. J3.3 as

$$s_{\min} = 2.67d$$
$$s_{\mathrm{pref}} = 3.0d$$

The minimum clear distance between bolt holes is

$$s_{\mathrm{clear}} = d$$

The nominal shear strength is based on the nominal unthreaded cross-sectional area of the bolt, A_b, and the nominal shear strength, F_{nv}. Nominal shear strength of

fasteners and threaded parts is given in AISC 360 Table J3.2, and for high-strength bolts a reduced nominal strength is applicable when threads are not excluded from the shear planes. No reduction is made for A307 bolts. For connections longer than 38 in, the nominal strength is reduced to 83.3% of the tabulated values. The bolt's available shear capacity is obtained from AISC 360 Eq. J3-1.

$$\phi R_n = \phi F_{nv} A_b$$
$$= 0.75 F_{nv} A_b \quad [\text{LRFD}]$$
$$\frac{R_n}{\Omega} = \frac{F_{nv} A_b}{\Omega}$$
$$= \frac{F_{nv} A_b}{2.00} \quad [\text{ASD}]$$

Values of ϕR_n and R_n/Ω are given in *AISC Manual* Table 7-1.

Bearing-Type Bolts in Tension and Combined Shear and Tension

The available strength in tension is given by AISC 360 Sec. J3.6 as

$$\phi R_n = \phi F_{nt} A_b$$
$$= 0.75 F_{nt} A_b \quad [\text{LRFD}]$$
$$\frac{R_n}{\Omega} = \frac{F_{nt} A_b}{\Omega}$$
$$= \frac{F_{nt} A_b}{2.00} \quad [\text{ASD}]$$

Values of the nominal tensile stress F_{nt} are given in AISC 360 Table J3.2 for all types of bolts. Values of ϕR_n and R_n/Ω are given in *AISC Manual* Table 7-2.

When a bearing-type bolt is subjected to combined shear and tension, the available strength in shear is unaffected, and the available strength in tension is reduced in accordance with AISC 360 Sec. J3.7.

The value of the reduced available tensile capacity is

$$\phi R_n = \phi F'_{nt} A_b$$
$$= 0.75 F'_{nt} A_b \quad [\text{LRFD}]$$
$$\frac{R_n}{\Omega} = \frac{F'_{nt} A_b}{\Omega}$$
$$= \frac{F'_{nt} A_b}{2.00} \quad [\text{ASD}]$$

F'_{nt}, the modified nominal tensile stress, is calculated using AISC 360 Eq. J3-3a for the LRFD method or AISC 360 Eq. J3-3b for the ASD method.

$$F'_{nt} = 1.3F_{nt} - \frac{F_{nt}}{\phi F_{nv}}f_{rv} \quad \text{[LRFD]}$$

$$= 1.3F_{nt} - \frac{\Omega F_{nt}}{F_{nv}}f_{rv} \quad \text{[ASD]}$$

The required shear stress, f_{rv}, is determined using appropriate load combinations, and must be equal to or less than the available shear stress. Values of the nominal shear stress F_{nv} are given in AISC 360 Table J3.2 for all types of bolts.

When either $f_{rv} \le 30\%$ of the available shear stress or $f_t \le 30\%$ of the available tensile stress, the effects of combined stress do not need to be considered.

Example 5.35

The connection analyzed in Ex. 5.32 consists of 11 grade A307 $^7/_8$ in diameter bolts. Determine the design shear strength of the bolts in the connection.

Solution

From *AISC Manual* Table 7-1, the available strength of the 11 bolts in single shear is

LRFD Method

$$\phi R_n = \phi F_{nv}A_b n$$
$$= \left(12.2 \, \frac{\text{kips}}{\text{bolt}}\right)(11 \text{ bolts})$$
$$= 134 \text{ kips}$$

ASD Method

$$\frac{R_n}{\Omega} = \frac{F_{nv}A_b n}{\Omega}$$
$$= \left(8.11 \, \frac{\text{kips}}{\text{bolt}}\right)(11 \text{ bolts})$$
$$= 89 \text{ kips}$$

Slip-Critical Bolts in Shear

Slip-critical bolts are high strength group A, group B, or group C grade 2 bolt pretensioned to the value specified in AISC 360 Table J3.1.

The pretension produces a clamping force between the parts, and transfers the shear load from one connected part to another by friction. At the strength limit state, the connection may slip sufficiently to place the bolts in bearing and AISC 360 Sec. J3.8 requires slip-critical connections to also comply with the requirements of snug-tight connections.

The *frictional resistance* developed in a slip-critical connection depends on the condition of the faying surfaces. The values of the slip coefficient, μ, for two types of surface conditions (class A and class B) are given in AISC 360 Sec. J3.8.

- Class A surface conditions consist of unpainted clean mill scale surfaces or blast-cleaned surfaces or hot-dipped galvanized and roughened surfaces with class A coatings. The slip coefficient is

$$\mu = 0.30$$

- Class B surface conditions consist of unpainted blast-cleaned surfaces or blast-cleaned surfaces with class B coatings. The slip coefficient is

$$\mu = 0.50$$

The *nominal slip resistance* is identical for the cases of threads included or excluded from the shear plane, and is given by AISC 360 Eq. J3-4 as

$$R_n = \mu D_u h_f T_b n_s$$

The *bolt tension multiplier* reflects the ratio of the mean installed bolt tension to the specified minimum bolt pretension and is given by

$$D_u = 1.13$$

The *modification factor for fillers*, h_f, is

- 1.00 where bolts are added to distribute loads in the filler

- 1.00 for one filler between connected parts

- 0.85 for two or more fillers between connected parts

The resistance factor and safety factor adopted depend on the type of hole used in the connection. Connections allowing a large amount of slip may cause unacceptable slip and will require a higher safety factor than a connection with little slip. The resistance factors and safety factors that modify the available slip resistance are specified in AISC 360 Sec. J3.8 as follows.

- For standard size and short-slotted holes perpendicular to the direction of the load,

$$\phi = 1.00 \quad \text{[LRFD]}$$
$$\Omega = 1.50 \quad \text{[ASD]}$$

- For oversized and short-slotted holes parallel to the direction of the load,

$$\phi = 0.85 \quad \text{[LRFD]}$$
$$\Omega = 1.76 \quad \text{[ASD]}$$

• For long-slotted holes,

$$\phi = 0.70 \quad \text{[LRFD]}$$
$$\Omega = 2.14 \quad \text{[ASD]}$$

Values for the available slip-critical shear resistance for a class A faying surface for group A, group B, and group C grade 2 bolts are given in *AISC Manual* Table 7-3. These values are multiplied by 1.67 for a class B faying surface. The values assume no more than one filler is provided or bolts have been added to distribute the load in the fillers.

Slip-Critical Bolts in Tension and Combined Shear and Tension

The available tensile strength of slip-critical bolts is independent of the pretension in the bolt.

The nominal tensile strength is given by AISC 360 Eq. J3-1 as

$$R_n = F_{nt}A_b$$

The available tensile strength is

$$\phi R_n = 0.75 F_{nt} A_b \quad \text{[LRFD]}$$
$$\frac{R_n}{\Omega} = \frac{F_{nt} A_b}{2.00} \quad \text{[ASD]}$$

Values of the nominal tensile stress, F_{nt}, are given in AISC 360 Table J3.2 for all types of bolts. Values of ϕR_n and R_n/Ω are given in *AISC Manual* Table 7-2.

When a slip-critical bolt is subjected to combined shear and tension, the available strength in tension is unaffected. However, in accordance with AISC 360 Sec. J3.9, the available resistance to shear is reduced by being multiplied by the factor

$$k_{sc} = 1 - \frac{T_u}{D_u T_b n_b} \quad \text{[LRFD]} \qquad \text{[AISC J3-5a]}$$

$$k_{sc} = 1 - \frac{1.5\,T_a}{D_u T_b n_b} \quad \text{[ASD]} \qquad \text{[AISC J3-5b]}$$

Example 5.36

The connection analyzed in Ex. 5.32 consists of 11 grade A490 $^7/_8$ in diameter slip-critical bolts. Determine the available resistance to shear of the bolts in the connection. The bolts are in standard holes with a class A faying surface.

Solution

For group B bolts in standard holes and a class A faying surface, *AISC Manual* Table 7-3 gives the available single shear strength of the 11 bolts in shear.

LRFD Method

$$\phi R_n = (16.6 \text{ kips})(11 \text{ bolts})$$
$$= 183 \text{ kips}$$

ASD Method

$$\frac{R_n}{\Omega} = (11.1 \text{ kips})(11 \text{ bolts})$$
$$= 122 \text{ kips}$$

Bolts in Bearing

The bearing strength of connected parts is specified in AISC 360 Sec. J3.10 and is dependent on the diameter, spacing, and edge distance of a bolt; the material of the connected parts; and the acceptable deformation at the bolt hole. For all equations in AISC 360 Sec. J3.10, use

$$\phi = 0.75$$
$$\Omega = 2.00$$

High-strength bolts in slip-critical connections must also be checked for bearing strength since the connection at the strength limit state may slip sufficiently to place the bolts in bearing.

When deformation at the hole is a design consideration, the nominal bearing strength of connected parts is given by AISC 360 Eq. J3-6a and Eq. J3-6c as

$$R_n = 1.2 l_c t F_u \quad \text{[when tear-out strength governs]}$$
$$\leq 2.4 dt F_u \quad \text{[when bearing strength governs]}$$

The clear distance between adjacent holes is

$$l_c = s - d_n$$

The clear distance between the edge of a hole and the edge of a connected part is

$$l_c = L_e - 0.5 d_n$$

L_e is the edge distance between the bolt center and the edge of the connected part.

The nominal hole diameter given in AISC 360 Table J3.3 is

$$d_n = d + \frac{1}{16} \text{ in}$$

When deformation at the hole is not a design consideration, the nominal bearing capacity of each bolt is given by AISC 360 Eq. J3-6d and Eq. J3-6b as

$$R_n = 1.5 l_c t F_u \quad \text{[when tear-out strength governs]}$$
$$\leq 3.0 dt F_u \quad \text{[when bearing strength governs]}$$

For long-slotted holes with the slot perpendicular to the direction of force, the nominal bearing capacity of each bolt is given by AISC 360 Eq. J3-6f and Eq. J3-6e as

$$R_n = 1.0 l_c t F_u \quad \text{[when tear-out strength governs]}$$
$$\leq 2.0 dt F_u \quad \text{[when bearing strength governs]}$$

The available bearing capacity at bolt holes that considers deformation of the connected parts is given in *AISC Manual* Table 7-4 and is based on bolt center-to-center spacing. The available bearing capacity that considers deformation of the connected parts is given in *AISC Manual* Table 7-5 based on bolt edge distances, measured from the center of bolt to the edge of the connected part. Tables are not provided for the bearing capacity when hole deformation is not considered.

Example 5.37

The connection analyzed in Ex. 5.32 consists of 11 grade A307 $^7/_8$ in diameter bolts in standard holes. Determine the available bearing strength of the bolts in the A36 plates ($F_u = 58$ kips/in^2) if the edge distance is $L_e = 2.5$ in and $s = 3$ in. Hole deformation is considered.

Solution

From *AISC Manual* Table 7-5, the minimum edge distance for full bearing strength is

$$L_e = 2.25 \text{ in}$$
$$< 2.5 \text{ in provided}$$

The edge distance does not govern.

AISC Manual Table 7-4 gives the available strength of the 11 bolts in bearing.

LRFD Method

$$\phi R_n = \left(91.4 \ \frac{\frac{\text{kips}}{\text{in}}}{\text{bolt}}\right)\left(\frac{1}{2} \text{ in}\right)(11 \text{ bolts})$$
$$= 503 \text{ kips}$$

ASD Method

$$\frac{R_n}{\Omega} = \left(60.9 \ \frac{\frac{\text{kips}}{\text{in}}}{\text{bolt}}\right)\left(\frac{1}{2} \text{ in}\right)(11 \text{ bolts})$$
$$= 335 \text{ kips}$$

Bolt Group Eccentrically Loaded in Plane of Faying Surface

Eccentrically loaded bolt groups of the type shown in Fig. 5.23 may be conservatively designed by means of the elastic unit area method. The moment of inertia of the bolt group about the x-axis is

$$I_x = \sum y^2$$

Figure 5.23 *Eccentrically Loaded Bolt Group*

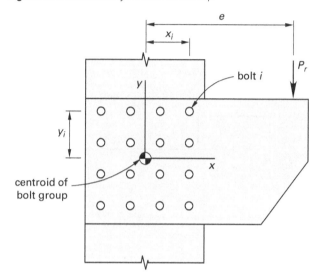

The moment of inertia of the bolt group about the y-axis is

$$I_y = \sum x^2$$

The polar moment of inertia of the bolt group about the centroid is

$$I_o = I_x + I_y$$

The vertical force on bolt i due to the applied load, P_r, is

$$V_p = \frac{P_r}{n}$$

The vertical force on bolt i due to the eccentricity, e, is

$$V_e = \frac{P_r e x_i}{I_o}$$

The horizontal force on bolt i due to the eccentricity, e, is

$$H_e = \frac{P_r e y_i}{I_o}$$

The resultant force on bolt i is

$$R = \sqrt{(V_p + V_e)^2 + H_e^2}$$

The instantaneous center of rotation method of analyzing eccentrically loaded bolt groups affords a more realistic estimate of a bolt group's capacity. *AISC Manual* Table 7-6 through Table 7-13 provide a means of designing common bolt group patterns by this method.

Example 5.38

Determine the diameter of the A325 bearing-type bolts required in the bolted bracket shown. Use the elastic unit area method and compare with the instantaneous center of rotation method.

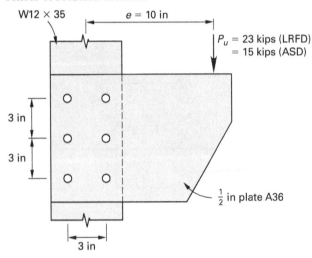

Solution

The geometric properties of the bolt group are obtained by applying the unit area method.

The moment of inertia about the x-axis is

$$I_x = \sum y^2 = (4)(3 \text{ in})^2$$
$$= 36 \text{ in}^4/\text{in}^2$$

The moment of inertia about the y-axis is

$$I_y = \sum x^2 = (6)(1.5 \text{ in})^2$$
$$= 13.5 \text{ in}^4/\text{in}^2$$

The polar moment of inertia about the centroid is

$$I_o = I_x + I_y = 49.5 \text{ in}^4/\text{in}^2$$

The top right bolt is the most heavily loaded, and the coexistent forces on this bolt are as follows.

LRFD Method

Elastic Method

- vertical force due to applied load

$$V_p = \frac{P_r}{n}$$
$$= \frac{23 \text{ kips}}{6}$$
$$= 3.83 \text{ kips}$$

- vertical force due to eccentricity

$$V_e = \frac{P_r e x_i}{I_o}$$
$$= \frac{(23 \text{ kips})(10 \text{ in})(1.5 \text{ in})}{49.5 \dfrac{\text{in}^4}{\text{in}^2}}$$
$$= 6.97 \text{ kips}$$

- horizontal force due to eccentricity

$$H_e = \frac{P_r e y_i}{I_o}$$
$$= \frac{(23 \text{ kips})(10 \text{ in})(3 \text{ in})}{49.5 \dfrac{\text{in}^4}{\text{in}^2}}$$
$$= 13.94 \text{ kips}$$

- resultant force

$$R = \sqrt{(V_p + V_e)^2 + H_e^2}$$
$$= \sqrt{(3.83 \text{ kips} + 6.97 \text{ kips})^2 + (13.94 \text{ kips})^2}$$
$$= 17.6 \text{ kips}$$

Shear controls, and from *AISC Manual* Table 7-1 the design shear strength of a ³⁄₄ in diameter A325N bolt in a standard hole in single shear is

$$\phi R_n = 17.9 \text{ kips}$$
$$> 17.6 \text{ kips} \quad \text{[satisfactory]}$$

Instantaneous Center of Rotation Method

From *AISC Manual* Table 7-7, the coefficient C is given as 1.46, and the required design strength of an individual bolt, based on the instantaneous center of rotation method, is

$$\phi R_n = \frac{P_r}{C}$$
$$= \frac{23 \text{ kips}}{1.46}$$
$$= 15.8 \text{ kips}$$

Shear controls, and from *AISC Manual* Table 7-1 the design shear strength of a $\frac{3}{4}$ in diameter A325N bolt in a standard hole in single shear is

$$\phi R_n = 17.9 \text{ kips}$$
$$> 15.8 \text{ kips} \quad [\text{satisfactory}]$$

ASD Method

Elastic Method

- vertical force due to applied load

$$V_p = \frac{P_r}{n}$$
$$= \frac{15 \text{ kips}}{6}$$
$$= 2.50 \text{ kips}$$

- vertical force due to eccentricity

$$V_e = \frac{P_r e x_i}{I_o} = \frac{(15 \text{ kips})(10 \text{ in})(1.5 \text{ in})}{49.5 \frac{\text{in}^4}{\text{in}^2}}$$
$$= 4.55 \text{ kips}$$

- horizontal force due to eccentricity

$$H_e = \frac{P_r e y_i}{I_o}$$
$$= \frac{(15 \text{ kips})(10 \text{ in})(3 \text{ in})}{49.5 \frac{\text{in}^4}{\text{in}^2}}$$
$$= 9.09 \text{ kips}$$

- resultant force

$$R = \sqrt{(V_p + V_e)^2 + H_e^2}$$
$$= \sqrt{(2.50 \text{ kips} + 4.55 \text{ kips})^2 + (9.09 \text{ kips})^2}$$
$$= 11.50 \text{ kips}$$

Shear controls, and from *AISC Manual* Table 7-1 the allowable shear strength of a $\frac{3}{4}$ in diameter A325N bolt in a standard hole in single shear is

$$\frac{R_n}{\Omega} = 11.9 \text{ kips}$$
$$> 11.50 \text{ kips} \quad [\text{satisfactory}]$$

Instantaneous Center of Rotation Method

From *AISC Manual* Table 7-7, the coefficient C is given as 1.46, and the required allowable strength of an individual bolt, based on the instantaneous center of rotation method, is

$$\frac{R_n}{\Omega} = \frac{P_r}{C}$$
$$= \frac{15 \text{ kips}}{1.46}$$
$$= 10.3 \text{ kips}$$

Shear controls, and from *AISC Manual* Table 7-1 the design shear strength of a $\frac{3}{4}$ in diameter A325N bolt in a standard hole in single shear is

$$\frac{R_n}{\Omega} = 11.9 \text{ kips}$$
$$> 10.3 \text{ kips} \quad [\text{satisfactory}]$$

Bolt Group Eccentrically Loaded Normal to the Faying Surface (LRFD)

Eccentrically loaded bolt groups of the type shown in Fig. 5.24 may be conservatively designed by assuming that the neutral axis is located at the centroid of the bolt group and that a plastic stress distribution is produced in the bolts. The tensile force in each bolt above the neutral axis due to the eccentricity is given by

$$T_u = \frac{P_u e}{n' d_m}$$

The shear force in each bolt due to the applied load is given by

$$V_p = \frac{P_u}{n}$$

Example 5.39

Determine whether the $\frac{7}{8}$ in diameter A325N bearing-type bolts in the bolted bracket shown are adequate. Prying action may be neglected.

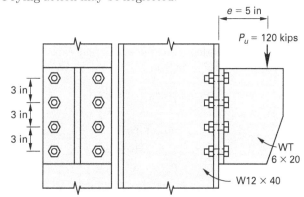

Steel

Figure 5.24 *Bolt Group Eccentrically Loaded Normal to Faying Surface (LRFD)*

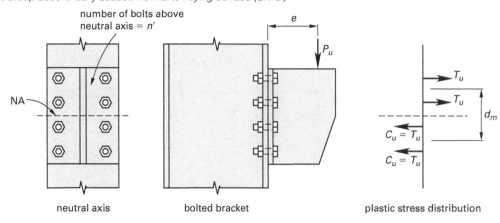

Solution

The tensile force in each bolt above the neutral axis due to the eccentricity is given by

$$T_u = \frac{P_u e}{n' d_m} = \frac{(120 \text{ kips})(5 \text{ in})}{(4)(6 \text{ in})}$$
$$= 25 \text{ kips}$$

The shear force in each bolt due to the applied load is given by

$$V_p = \frac{P_u}{n}$$
$$= \frac{120 \text{ kips}}{8}$$
$$= 15 \text{ kips}$$

The required shear stress on each bolt is

$$f_{rv} = \frac{V_p}{A_b}$$
$$= \frac{15 \text{ kips}}{0.601 \text{ in}^2}$$
$$= 25 \text{ kips/in}^2$$

The design shear stress for grade A325 bolts, with threads not excluded from the shear plane, is obtained from AISC 360 Table J3.2 as

$$\phi F_{nv} = (0.75)\left(54 \ \frac{\text{kips}}{\text{in}^2}\right)$$
$$= 40.5 \text{ kips/in}^2$$
$$> f_{rv} \quad \text{[satisfactory]}$$
$$f_{rv} > 0.3\phi F_{nv}$$

The design tensile stress for grade A325 bolts is obtained from AISC 360 Table J3.2 as

$$\phi F_{nt} = (0.75)\left(90 \ \frac{\text{kips}}{\text{in}^2}\right)$$
$$= 67.5 \text{ kips/in}^2$$

The strength design tensile stress in each $^7/_8$ in diameter bolt is

$$f_t = \frac{T_u}{A_b}$$
$$= \frac{25 \text{ kips}}{0.601 \text{ in}^2}$$
$$= 41.60 \text{ kips/in}^2$$
$$< \phi F_{nt} \quad \text{[satisfactory]}$$
$$> 0.3\phi F_{nt}$$

Therefore, it is necessary to investigate the effects of the combined shear and tensile stress. The nominal tensile stress, F_{nt}', of a bolt subjected to combined shear and tension is given by AISC 360 Eq. J3-3a as

$$F_{nt}' = 1.3 F_{nt} - \frac{f_{rv} F_{nt}}{\phi F_{nv}}$$

$$= (1.3)\left(90 \ \frac{\text{kips}}{\text{in}^2}\right) - \left(\frac{90 \ \dfrac{\text{kips}}{\text{in}^2}}{40.5 \ \dfrac{\text{kips}}{\text{in}^2}}\right)\left(25 \ \frac{\text{kips}}{\text{in}^2}\right)$$

$$= 61.44 \text{ kips/in}^2$$

The design tensile stress, $\phi F'_{nt}$, of a bolt subjected to combined shear and tension is given by AISC 360 Eq. J3-2 as

$$\phi F'_{nt} = (0.75)\left(61.44 \ \frac{\text{kips}}{\text{in}^2}\right)$$
$$= 46.08 \ \text{kips/in}^2$$
$$> f_t \quad [\text{satisfactory}]$$

Bolt Group Eccentrically Loaded Normal to the Faying Surface (ASD)

Eccentrically loaded bolt groups of the type shown in Fig. 5.25 may be conservatively designed by assuming that the neutral axis is located at the centroid of the bolt group. The tensile force in a bolt a distance y_i from the neutral axis is

$$T_i = \frac{P_a e y_i}{I_x}$$
$$I_x = \sum y^2$$

The shear force in each bolt due to the applied load is

$$V_p = \frac{P_a}{n}$$

Example 5.40

Determine whether the $\frac{7}{8}$ in diameter A325N bearing-type bolts in the bolted bracket shown are adequate. Prying action may be neglected.

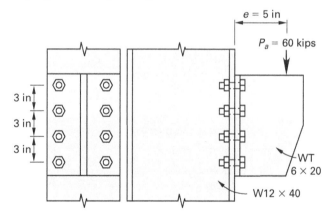

Solution

$$I_x = \sum y^2 = (4)(4.5 \ \text{in})^2 + (4)(1.5 \ \text{in})^2$$
$$= 90 \ \text{in}^4/\text{in}^2$$

The tensile force in each of the top bolts due to the eccentricity is given by

$$T_i = \frac{P_a e y_i}{I_x} = \frac{(60 \ \text{kips})(5 \ \text{in})(4.5 \ \text{in})}{90 \ \dfrac{\text{in}^4}{\text{in}^2}}$$
$$= 15 \ \text{kips}$$

The shear force in each bolt due to the applied load is given by

$$V_p = \frac{P_a}{n} = \frac{60 \ \text{kips}}{8}$$
$$= 7.5 \ \text{kips}$$

The required shear stress on each bolt is

$$f_{rv} = \frac{V_p}{A_b} = \frac{7.5 \ \text{kips}}{0.601 \ \text{in}^2}$$
$$= 12.5 \ \text{kips/in}^2$$

The allowable shear stress for grade A325 bolts, with threads not excluded from the shear plane, is obtained from AISC 360 Table J3.2 as

$$\frac{F_{nv}}{\Omega} = \frac{54 \ \dfrac{\text{kips}}{\text{in}^2}}{2.00}$$
$$= 27 \ \text{kips/in}^2$$
$$> f_{rv} \quad [\text{satisfactory}]$$
$$f_{rv} > \frac{0.3 F_{nv}}{\Omega}$$

The allowable tensile stress for grade A325 bolts is obtained from AISC 360 Table J3.2 as

$$\frac{F_{nt}}{\Omega} = \frac{90 \ \dfrac{\text{kips}}{\text{in}^2}}{2.00}$$
$$= 45 \ \text{kips/in}^2$$

The tensile stress in each of the top $\frac{7}{8}$ in diameter bolts is

$$f_t = \frac{T_i}{A_b} = \frac{15 \ \text{kips}}{0.601 \ \text{in}^2}$$
$$= 25.0 \ \text{kips/in}^2$$
$$< \frac{F_{nt}}{\Omega} \quad [\text{satisfactory}]$$
$$> \frac{0.3 F_{nt}}{\Omega}$$

Figure 5.25 *Bolt Group Eccentrically Loaded Normal to the Faying Surface (ASD)*

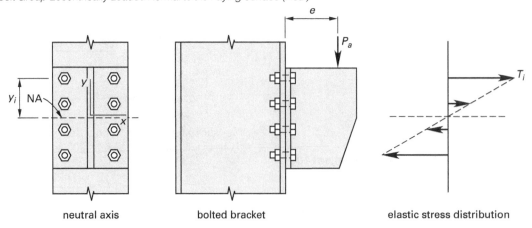

| neutral axis | bolted bracket | elastic stress distribution |

It is necessary to investigate the effects of the combined shear and tensile stress. The nominal tensile stress, F'_{nt}, of a bolt subjected to combined shear and tension is given by AISC 360 Eq. J3-3b as

$$F'_{nt} = 1.3F_{nt} - \frac{\Omega F_{nt}}{F_{nv}} f_{rv}$$

$$= (1.3)\left(90 \ \frac{\text{kips}}{\text{in}^2}\right)$$

$$- \left(\frac{(2.00)\left(90 \ \frac{\text{kips}}{\text{in}^2}\right)}{54 \ \frac{\text{kips}}{\text{in}^2}}\right)\left(12.5 \ \frac{\text{kips}}{\text{in}^2}\right)$$

$$= 75.33 \text{ kips/in}^2$$

The allowable tensile stress, $\phi F'_{nt}$, of a bolt subjected to combined shear and tension is given by AISC 360 Eq. J3-2 as

$$\frac{F'_{nt}}{\Omega} = \frac{75.33 \ \frac{\text{kips}}{\text{in}^2}}{2.00}$$

$$= 37.67 \text{ kips/in}^2$$

$$> f_t \quad \text{[satisfactory]}$$

9. DESIGN OF WELDED CONNECTIONS

Nomenclature

a	coefficient for eccentrically loaded weld group	–
A_{BM}	effective area for base metal	in^2
A_{we}	effective area of the weld	in^2
C	coefficient for eccentrically loaded weld group	–
D	number of sixteenths-of-an-inch in the weld size	–
F_{EXX}	classification of weld metal	kips/in^2
F_{nBM}	nominal stress for base metal	kips/in^2
F_{nw}	nominal strength of weld electrode	kips/in^2
k	coefficient for eccentrically loaded weld group	–
l	characteristic length of weld group used in tabulated values of instantaneous center method	in
$\bar{l}$	total length of weld	in
q	allowable strength	kips/in
q_u	design strength	kips/in
R_n	nominal strength	–
R_{wl}	total nominal strength of longitudinally loaded fillet welds, as determined in accordance with AISC 360 Table J2.5	–
R_{wt}	total nominal strength of transversely loaded fillet welds, as determined in accordance with AISC 360 Table J2.5 without the amplification of the weld shear strength given by AISC 360 Eq. J2-5	–
t_e	effective throat thickness	in
t_e	effective plate thickness	in
w	fillet weld size	in
w	width of plate	in

Symbols

| θ | angle of inclination of loading measured from the weld longitudinal axis | degree |
| ϕ | resistance factor | – |

Weld Design Strength

The strength of a welded connection depends on both the strength of the base metal and the strength of the weld metal. Welded connections must be made using a "matching" weld metal of sufficient strength, given the type of base metal used. These weld strengths are listed in the user note in AISC 360 Sec. J2.6. For typical structural steel applications that use A36 and grade 50, the matching weld metal is E70XX, which has a tensile strength of 70 ksi. Lower strength welded metal can be used in certain situations as specified in AISC 360 Sec. J2.6. Weld nominal stress values, effective areas, resistance factors, and safety factors are tabulated in AISC 360 Table J2.5.

The nominal strength of the base metal is the product of the nominal stress and the effective area of the base metal, and is given by AISC 360 Eq. J2-2 as

$$R_n = F_{nBM}A_{BM}$$

The nominal strength of the weld metal is the product of the nominal stress and the effective area of the weld metal, and is given by AISC 360 Eq. J2-3 as

$$R_n = F_{nw}A_{we}$$

Complete-Penetration Groove Weld

In accordance with AISC 360 Table J2.5, the nominal strength of a complete-penetration groove weld is governed by the base metal and computation of the strength of the weld is not required. The effective thickness of the joint is the thickness of the thinner part joined, shown as t_e in Fig. 5.26.

Figure 5.26 *Complete-Penetration Groove Weld*

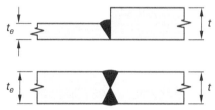

Example 5.41

The two $\frac{1}{2}$ in thick by 2 in wide plates shown are connected with a complete-penetration groove weld as indicated. The plates are grade A36 steel and the electrodes are E70XX. Determine the available tensile capacity of the connection.

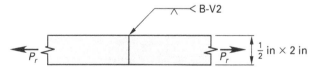

Solution

The effective area of the joint is

$$
\begin{aligned}
A_{BM} &= wt_e \\
&= (2 \text{ in})\left(\frac{1}{2} \text{ in}\right) \\
&= 1.0 \text{ in}^2
\end{aligned}
$$

LRFD Method

The design axial strength for tensile rupture is

$$
\begin{aligned}
\phi R_n &= \phi F_u A_{BM} \\
&= (0.75)\left(58 \ \frac{\text{kips}}{\text{in}^2}\right)(1.0 \text{ in}^2) \\
&= 43.5 \text{ kips}
\end{aligned}
$$

The design axial strength for yielding is

$$
\begin{aligned}
\phi R_n &= \phi F_y A_{BM} \\
&= (0.90)\left(36 \ \frac{\text{kips}}{\text{in}^2}\right)(1.0 \text{ in}^2) \\
&= 32.4 \text{ kips} \quad \text{[governs]}
\end{aligned}
$$

ASD Method

The allowable axial strength for tensile rupture is

$$
\begin{aligned}
\frac{R_n}{\Omega} &= \frac{F_u A_{BM}}{\Omega} \\
&= \frac{\left(58 \ \dfrac{\text{kips}}{\text{in}^2}\right)(1.0 \text{ in}^2)}{2.00} \\
&= 29 \text{ kips}
\end{aligned}
$$

The allowable axial strength for yielding is

$$
\begin{aligned}
\frac{R_n}{\Omega} &= \frac{F_y A_{BM}}{\Omega} \\
&= \frac{\left(36 \ \dfrac{\text{kips}}{\text{in}^2}\right)(1.0 \text{ in}^2)}{1.67} \\
&= 21.6 \text{ kips} \quad \text{[governs]}
\end{aligned}
$$

Steel

Partial-Penetration Groove Weld

The nominal strength of a partial-penetration groove weld is governed by the effective throat thickness, given in AISC 360 Table J2.1 and shown as t_e in Fig. 5.27.

Figure 5.27 *Partial-Penetration Groove Weld*

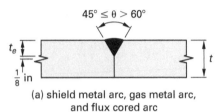

(a) shield metal arc, gas metal arc, and flux cored arc

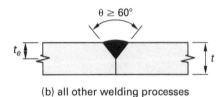

(b) all other welding processes and positions

Example 5.42

The two $4 \times 1\frac{1}{2}$ in plates shown are connected with a double V-groove weld with a $\frac{1}{2}$ in penetration into each face, as indicated. The plates are grade A36 steel and the electrodes are E70XX. Determine the available tensile capacity of the connection.

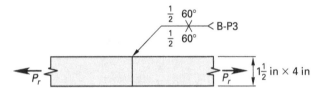

Solution

For an included angle of 60°, the effective throat thickness equals the depth of the weld, which is $\frac{1}{2}$ in. The total effective area of the weld is

$$A_{we} = wt_e$$
$$= (4 \text{ in})\left(\frac{1}{2} \text{ in}\right)(2 \text{ sides})$$
$$= 4.0 \text{ in}^2$$

The effective area of the base metal is

$$A_{BM} = wt_e$$
$$= (4 \text{ in})\left(1\frac{1}{2} \text{ in}\right)$$
$$= 6.0 \text{ in}^2$$

LRFD Method

From AISC 360 Table J2.5, the design axial strength for rupture in the base metal is

$$\phi R_n = \phi F_u A_{BM}$$
$$= (0.75)\left(58 \ \frac{\text{kips}}{\text{in}^2}\right)(6.0 \text{ in}^2)$$
$$= 261.0 \text{ kips}$$

The design tensile capacity of the weld, normal to the effective area, is given by AISC 360 Table J2.5 as

$$\phi R_n = \phi(0.60 F_{EXX} A_{we})$$
$$= (0.8)\left[(0.6)\left(70 \ \frac{\text{kips}}{\text{in}^2}\right)(4.0 \text{ in}^2)\right]$$
$$= 134.4 \text{ kips} \quad \text{[governs]}$$

ASD Method

From AISC 360 Table J2.5, the allowable axial strength for rupture in the base metal is

$$\frac{R_n}{\Omega} = \frac{F_u A_{BM}}{\Omega} = \frac{\left(58 \ \frac{\text{kips}}{\text{in}^2}\right)(6.0 \text{ in}^2)}{2.0}$$
$$= 174.0 \text{ kips}$$

The allowable tensile capacity of the weld, normal to the effective area, is given by AISC 360 Table J2.5 as

$$\frac{R_n}{\Omega} = \frac{0.60 F_{EXX} A_{we}}{\Omega} = \frac{(0.60)\left(70 \ \frac{\text{kips}}{\text{in}^2}\right)(4.0 \text{ in}^2)}{1.88}$$
$$= 89.4 \text{ kips} \quad \text{[governs]}$$

Fillet Weld

The leg length, w, of a fillet weld is used to designate the nominal size of the weld. For fillet welds made by the shielded metal arc process, the effective throat thickness, t_e, is given in AISC 360 Sec. J2.2a and shown in Fig. 5.28. The *effective throat thickness* is the shortest distance between the root and the weld face, and is given by

$$t_e = 0.707 w$$

Figure 5.28 *Fillet Weld*

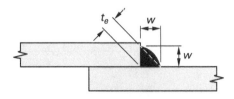

The minimum permitted size of fillet welds is specified in AISC 360 Sec. J2.2b and shown in Table 5.3.

Table 5.3 *Minimum Size of Fillet Welds*

thickness of thinner part joined (in)	minimum size of fillet weld (in)
$t \leq \frac{1}{4}$	$w \geq \frac{1}{8}$
$\frac{1}{4} < t \leq \frac{1}{2}$	$w \geq \frac{3}{16}$
$\frac{1}{2} < t \leq \frac{3}{4}$	$w \geq \frac{1}{4}$
$t > \frac{3}{4}$	$w \geq \frac{5}{16}$

The maximum permitted size of fillet welds along the edges of connected parts is specified in AISC 360 Sec. J2.2b and shown in Table 5.4.

Table 5.4 *Maximum Size of Fillet Welds*

thickness of part (in)	maximum size·of fillet weld (in)
$t < \frac{1}{4}$	$w \leq t$
$t \geq \frac{1}{4}$	$w \leq t - \frac{1}{16}$

The minimum permissible length of a fillet weld is four times the nominal weld size.

Available Strength of a $\frac{1}{16}$ in Fillet Weld

To simplify calculations, it is appropriate to determine the available strength of a $\frac{1}{16}$ in fillet weld per inch run of E70XX grade electrodes.

LRFD Method

The design strength of a $\frac{1}{16}$ in fillet weld per inch run of E70XX grade electrodes is

$$q_u = \phi F_{nw} A_{we}$$
$$= (0.75)(0.60)\left(70 \ \frac{\text{kips}}{\text{in}^2}\right)(0.707)\left(\frac{1}{16} \ \text{in}\right)(1 \ \text{in})$$
$$= 1.39 \ \text{kips/in per } \frac{1}{16} \text{ in}$$

ASD Method

The allowable strength of a $\frac{1}{16}$ in fillet weld per inch run of E70XX grade electrodes is

$$q = \frac{F_{nw} A_{we}}{\Omega}$$
$$= \frac{(0.60)\left(70 \ \frac{\text{kips}}{\text{in}^2}\right)(0.707)\left(\frac{1}{16} \ \text{in}\right)(1 \ \text{in})}{2.00}$$
$$= 0.928 \ \text{kips/in per } \frac{1}{16} \text{ in}$$

Example 5.43

The $\frac{1}{2}$ in plate shown is connected to a $\frac{5}{8}$ in gusset plate with $\frac{1}{4}$ in E70XX fillet welds as indicated. Both plates are A36 material. Determine the available strength of the welds.

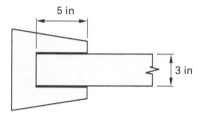

Solution

The total length of weld provided on the $\frac{1}{2}$ in plate is

$$l = (2)(5 \ \text{in}) = 10 \ \text{in}$$

From Table 5.3, the minimum permitted size of fillet weld for the $\frac{1}{2}$ in thick gusset plate is

$$w_{\min} = \frac{1}{4} \ \text{in}$$
$$w_{\text{provided}} = \frac{1}{4} \ \text{in} \quad [\text{satisfactory}]$$

From Table 5.4, the maximum permitted size of fillet weld along the edge of the $\frac{1}{2}$ in thick plate is

$$w_{\max} = \frac{1}{2} \ \text{in} - \frac{1}{16} \ \text{in}$$
$$= \frac{7}{16} \ \text{in}$$
$$> w_{\text{provided}} \quad [\text{satisfactory}]$$

LRFD Method

The design shear capacity of the total length of weld is

$$\phi R_n = l D q_u$$
$$= (10 \ \text{in})(4 \ \text{sixteenths})\left(1.39 \ \frac{\text{kips}}{\text{in}} \ \text{per } \frac{1}{16} \ \text{in}\right)$$
$$= 55.6 \ \text{kips}$$

ASD Method

The allowable shear capacity of the total length of weld is

$$\frac{R_n}{\Omega} = l D q$$
$$= (10 \ \text{in})(4 \ \text{sixteenths})\left(0.928 \ \frac{\text{kips}}{\text{in}} \ \text{per } \frac{1}{16} \ \text{in}\right)$$
$$= 37.1 \ \text{kips}$$

Fillet Weld Size Governed by Base Metal Thickness

In accordance with AISC 360 Sec. J2.4, the design of a welded connection is governed by the capacity of the weakest shear plane. This is either through the weld or through the base material. Providing a weld size with a strength in excess of the base material strength will not increase the strength of the connection as the base material strength governs. The design shear strength of the weld per linear inch is

$$Q_{uw} = q_u D = \left(1.39 \, \frac{\text{kips}}{\text{in}}\right) D$$

The design shear rupture strength per linear inch of grade 50 base material of thickness, t, is derived from AISC 360 Eq. J4-4 as

$$
\begin{aligned}
Q_{uBM} &= \phi(0.60 F_u A_{nv}) \\
&= (0.75)\left[(0.60)\left(65 \, \frac{\text{kips}}{\text{in}^2}\right)(1.0 \text{ in})\right] t \\
&= (29.25 \text{ kips/in}) t
\end{aligned}
$$

The largest effective weld size is given by

$$Q_{uw} = Q_{uBM}$$

$$\left(1.39 \, \frac{\text{kips}}{\text{in}}\right) D = \left(29.25 \, \frac{\text{kips}}{\text{in}}\right) t$$

$$D = (21.04 \text{ sixteenths}) t$$

The corresponding values of D for A36 base material and for welds on both sides of the base material are shown in Table 5.5.

Table 5.5 *Effective Weld Size*

base metal	welds on one side of base	welds on both sides of base
A36	$(18.75 \text{ sixteenths}) t$	$(9.38 \text{ sixteenths}) t$
grade 50	$(21.04 \text{ sixteenths}) t$	$(10.52 \text{ sixteenths}) t$

Example 5.44

For the welded connection analyzed in Ex. 5.43, determine the maximum effective weld size applicable.

Solution
For A36 base metal with welds on one side only of the $^5\!/_8$ in gusset plate, the maximum effective weld size is obtained from Table 5.5 as

$$
\begin{aligned}
w &= (18.75 \text{ sixteenths}) t \\
&= \left(\frac{18.75 \text{ sixteenths}}{16 \text{ sixteenths}}\right)(0.625 \text{ in}) \\
&= 0.73 \text{ in} \\
&> {}^1\!/_4 \text{ in} \quad [\text{satisfactory}]
\end{aligned}
$$

Strength of Fillet Weld Groups

AISC 360 Sec. J2.4 describes three methods for determining the strength of fillet weld groups, depending on the configuration of the weld elements.

- For a linear weld group with a uniform leg size loaded through the center of gravity, use AISC 360 Sec. J2.4 (b)(1).

- For weld elements within a weld group that are analyzed using the instantaneous center of rotation method, use AISC 360 Sec. J2.4(b)(2).

- For fillet weld groups with concentric loading that are made of elements with a uniform leg size and oriented either longitudinally or transversely to the direction of the applied load use AISC 360 Sec. J2.4(b)(2).

The strength of linear weld groups, in which all the elements are in line or are parallel, may be analyzed by the method specified in AISC 360 Sec. J2.4(b)(1). This method accounts for the angle of inclination of the applied loading to the longitudinal axis of the weld. For an angle of inclination θ, the nominal strength in shear is given by AISC 360 Eq. J2-4 as

$$R_n = F_{nw} A_{we}$$

The nominal strength of the weld metal is given by AISC 360 Eq. J2-5 as

$$F_{nw} = 0.60 F_{\text{EXX}}(1.0 + 0.50 \sin^{1.5}\theta)$$

For the usual situation with $\theta = 0°$, $F_{nw} = 0.60 F_{\text{EXX}}$.

The resistance factor is given by

$$\phi = 0.75$$

The safety factor is given by

$$\Omega = 2.00$$

For concentrically loaded weld groups, with elements oriented both longitudinally or transversely to the direction of the applied load, the strength is determined as specified in AISC 360 Sec. J2.4(b)(2). The combined nominal strength of the weld group is given by the greater of

$$R_n = R_{wl} + R_{wt} \qquad [\text{AISC J2-6a}]$$
$$R_n = 0.85 R_{wl} + 1.5 R_{wt} \qquad [\text{AISC J2-6b}]$$

Example 5.45

The two grade A36 plates shown are connected by E70XX fillet welds as indicated. Determine the size of weld required to develop the full available axial strength of the $\frac{5}{8}$ in plate.

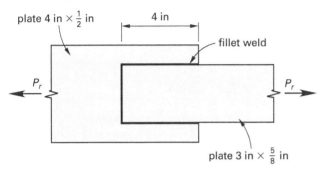

plate 4 in × $\frac{1}{2}$ in

4 in

fillet weld

P_r

P_r

plate 3 in × $\frac{5}{8}$ in

Solution

The total length of the longitudinally loaded weld is

$$l_{wl} = (2)(4 \text{ in})$$
$$= 8 \text{ in}$$

The total length of the transversely loaded weld is

$$l_{wt} = 3 \text{ in}$$

LRFD Method

The design axial strength of the $\frac{5}{8}$ in plate is

$$P_u = \phi_t P_n$$
$$= 0.9 F_y A_g$$
$$= (0.9)\left(36 \ \frac{\text{kips}}{\text{in}^2}\right)(3 \text{ in})(0.625 \text{ in})$$
$$= 60.75 \text{ kips}$$

The design shear capacity of a $\frac{1}{4}$ in fillet weld is

$$Q_w = q_u D$$
$$= \left(1.39 \ \frac{\text{kips}}{\text{in}} \text{ per } \frac{1}{16} \text{ in}\right)(4 \text{ sixteenths})$$
$$= 5.56 \text{ kips/in}$$

Applying AISC 360 Eq. J2-10a, the design strength of the connection is

$$\phi R_n = \phi(R_{wl} + R_{wt})$$
$$= l_{wl} Q_w + l_{wt} Q_w$$
$$= (8 \text{ in})\left(5.56 \ \frac{\text{kips}}{\text{in}}\right) + (3 \text{ in})\left(5.56 \ \frac{\text{kips}}{\text{in}}\right)$$
$$= 61.16 \text{ kips}$$

Applying AISC 360 Eq. J2-6b, the design strength of the connection is

$$\phi R_n = \phi(0.85 R_{wl}) + \phi(1.5 R_{wt})$$
$$= 0.85 l_{wl} Q_w + 1.5 l_{wt} Q_w$$
$$= (0.85)(8 \text{ in})\left(5.56 \ \frac{\text{kips}}{\text{in}}\right) + (1.5)(3 \text{ in})\left(5.56 \ \frac{\text{kips}}{\text{in}}\right)$$
$$= 62.83 \text{ kips} \quad \text{[governs]}$$
$$> 61.16 \text{ kips}$$
$$> P_u$$

ASD Method

The allowable axial strength of the $\frac{5}{8}$ in plate is

$$P_a = \frac{P_n}{\Omega} = F_y A_g$$
$$= \frac{\left(36 \ \frac{\text{kips}}{\text{in}^2}\right)(3 \text{ in})(0.625 \text{ in})}{1.67}$$
$$= 40.42 \text{ kips}$$

The allowable shear capacity of a $\frac{1}{4}$ in fillet weld is

$$Q_w = qD$$
$$= \left(0.928 \ \frac{\text{kips}}{\text{in}} \text{ per } \frac{1}{16} \text{ in}\right)(4 \text{ sixteenths})$$
$$= 3.71 \text{ kips/in}$$

Applying AISC 360 Eq. J2-6a, the allowable strength of the connection is

$$\frac{R_n}{\Omega} = \frac{R_{wl} + R_{wt}}{\Omega} = l_{wl} Q_w + l_{wt} Q_w$$
$$= (8 \text{ in})\left(3.71 \ \frac{\text{kips}}{\text{in}}\right) + (3 \text{ in})\left(3.71 \ \frac{\text{kips}}{\text{in}}\right)$$
$$= 40.81 \text{ kips}$$

Applying AISC 360 Eq. J2-6b, the allowable strength of the connection is

$$\frac{R_n}{\Omega} = \frac{0.85 R_{wl}}{\Omega} + \frac{1.5 R_{wt}}{\Omega}$$
$$= 0.85 l_{wl} Q_w + 1.5 l_{wt} Q_w$$
$$= (0.85)(8 \text{ in})\left(3.71 \ \frac{\text{kips}}{\text{in}}\right) + (1.5)(3 \text{ in})\left(3.71 \ \frac{\text{kips}}{\text{in}}\right)$$
$$= 41.92 \text{ kips} \quad \text{[governs]}$$
$$> 40.81 \text{ kips}$$
$$> P_a$$

Steel

From AISC 360 Table J2.4, the minimum size of fillet weld required for the ½ in plate is

$$w_{\min} = \frac{3}{16} \text{ in}$$
$$< \tfrac{1}{4} \text{ in} \quad [\text{satisfactory}]$$

From AISC 360 Sec. J2.2b, the maximum size of fillet weld permitted at the edge of the ⅝ in plate is

$$w_{\max} = \frac{5}{8} \text{ in} - \frac{1}{16} \text{ in}$$
$$= \tfrac{9}{16} \text{ in}$$
$$> \tfrac{1}{4} \text{ in} \quad [\text{satisfactory}]$$

A ¼ in fillet weld is adequate.

Weld Group Eccentrically Loaded in Plane of Faying Surface

Eccentrically loaded weld groups of the type shown in Fig. 5.29 may be conservatively designed by means of the elastic vector analysis technique assuming unit size of weld. The polar moment of inertia of the weld group about the centroid is

$$I_o = I_x + I_y$$

Figure 5.29 *Eccentrically Loaded Weld Group*

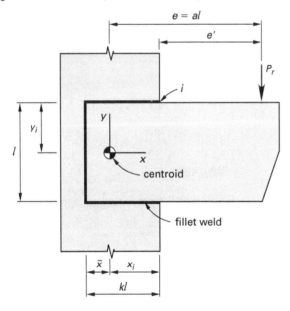

For a total length of weld $\bar{l}$, the vertical force per linear inch of weld due to the applied load, P_r, is

$$V_p = \frac{P_r}{\bar{l}}$$

The vertical force at point i due to the eccentricity, e, is

$$V_e = \frac{P_r e x_i}{I_o}$$

The horizontal force at point i due to the eccentricity, e, is

$$H_e = \frac{P_r e y_i}{I_o}$$

The resultant force at point i is

$$R = \sqrt{(V_p + V_e)^2 + H_e^2}$$

The instantaneous center of rotation method of analyzing eccentrically loaded weld groups affords a more realistic estimate of a weld group's capacity. *AISC Manual* Table 8-4 through Table 8-11a provide a means of designing common weld group patterns by this method.

Example 5.46

Determine the size of E70XX fillet weld required in the welded bracket shown. Use the elastic unit area method and compare with the instantaneous center of rotation method.

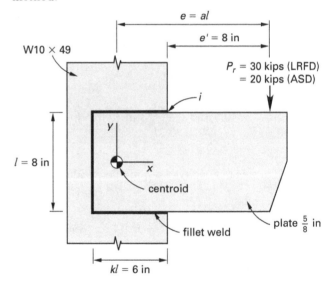

Solution

Assuming unit size of weld, the properties of the weld group are obtained by applying the elastic vector technique. The total length of the weld is

$$\bar{l} = l + 2kl = 8 \text{ in} + (2)(6 \text{ in})$$
$$= 20 \text{ in}$$
$$k = \frac{kl}{l} = \frac{6}{8}$$
$$= 0.75$$

For a value of $k = 0.75$, the centroid location is given by *AISC Manual* Table 8-8 as

$$\bar{x} = xl = (0.225)(8 \text{ in})$$
$$= 1.8 \text{ in}$$

The moment of inertia about the x-axis is

$$I_x = \frac{l^3}{12} + 2(kl)\left(\frac{l}{2}\right)^2$$
$$= \frac{(8 \text{ in})^3}{12} + (2)(6 \text{ in})(4 \text{ in})^2$$
$$= 235 \text{ in}^4/\text{in}$$

The moment of inertia about the y-axis is

$$I_y = \frac{2(kl)^3}{12} + 2(kl)\left(\frac{kl}{2} - \bar{x}\right)^2 + l\,\bar{x}^2$$
$$= \frac{(2)(6 \text{ in})^3}{12} + (2)(6 \text{ in})(1.2 \text{ in})^2 + (8 \text{ in})(1.8 \text{ in})^2$$
$$= 79 \text{ in}^4/\text{in}$$

The polar moment of inertia is

$$I_o = I_x + I_y$$
$$= 235 \frac{\text{in}^4}{\text{in}} + 79 \frac{\text{in}^4}{\text{in}}$$
$$= 314 \text{ in}^4/\text{in}$$

The eccentricity of the applied load about the centroid of the weld profile is

$$e = e' + kl - \bar{x} = 8 \text{ in} + 6 \text{ in} - 1.8 \text{ in}$$
$$= 12.2 \text{ in}$$
$$a = \frac{e}{l} = \frac{12.2}{8}$$
$$= 1.53$$

The top right corner of the weld profile is the most highly stressed, and the coexistent forces acting at this point in the x-direction and y-direction are as follows.

LRFD Method

Elastic Method

- vertical force due to applied load

$$V_p = \frac{P_r}{\bar{l}}$$
$$= \frac{30 \text{ kips}}{20 \text{ in}}$$
$$= 1.5 \text{ kips/in}$$

- vertical force due to eccentricity

$$V_e = \frac{P_r e x_i}{I_o}$$
$$= \frac{(30 \text{ kips})(12.2 \text{ in})(4.2 \text{ in})}{314 \frac{\text{in}^4}{\text{in}}}$$
$$= 4.9 \text{ kips/in}$$

- horizontal force due to eccentricity

$$H_e = \frac{P_r e y_i}{I_o}$$
$$= \frac{(30 \text{ kips})(12.2 \text{ in})(4 \text{ in})}{314 \frac{\text{in}^4}{\text{in}}}$$
$$= 4.7 \text{ kips/in}$$

- resultant force

$$R = \sqrt{(V_p + V_e)^2 + H_e^2}$$
$$= \sqrt{\left(1.5 \frac{\text{kips}}{\text{in}} + 4.9 \frac{\text{kips}}{\text{in}}\right)^2 + \left(4.7 \frac{\text{kips}}{\text{in}}\right)^2}$$
$$= 7.9 \text{ kips/in}$$

The required fillet weld size per $\frac{1}{16}$ in is

$$D = \frac{R}{q_u} = \frac{7.9 \text{ kips}}{1.39 \frac{\text{kips}}{\text{in}} \text{ per } \frac{1}{16} \text{ in}}$$
$$= 5.7 \text{ sixteenths}$$

ASD Method

Elastic Method

- vertical force due to applied load

$$V_p = \frac{P_r}{\bar{l}} = \frac{20 \text{ kips}}{20 \text{ in}}$$
$$= 1.0 \text{ kips/in}$$

- vertical force due to eccentricity

$$V_e = \frac{P_r e x_i}{I_o}$$
$$= \frac{(20 \text{ kips})(12.2 \text{ in})(4.2 \text{ in})}{314 \frac{\text{in}^4}{\text{in}}}$$
$$= 3.3 \text{ kips/in}$$

- horizontal force due to eccentricity

$$H_e = \frac{P_r e y_i}{I_o}$$

$$= \frac{(20 \text{ kips})(12.2 \text{ in})(4 \text{ in})}{314 \frac{\text{in}^4}{\text{in}}}$$

$$= 3.1 \text{ kips/in}$$

- resultant force

$$R = \sqrt{(V_p + V_e)^2 + H_e^2}$$

$$= \sqrt{\left(1.0 \frac{\text{kips}}{\text{in}} + 3.3 \frac{\text{kips}}{\text{in}}\right)^2 + \left(3.1 \frac{\text{kips}}{\text{in}}\right)^2}$$

$$= 5.3 \text{ kips/in}$$

The required fillet weld size per $1/16$ in is

$$D = \frac{R}{q} = \frac{5.3 \text{ kips}}{0.928 \frac{\text{kips}}{\text{in}} \text{ per } \frac{1}{16} \text{ in}}$$

$$= 5.7 \text{ sixteenths}$$

Use a weld size of

$$w = \tfrac{3}{8} \text{ in}$$

The flange thickness of the W10 × 49 is

$$t_f = 0.560 \text{ in}$$

From AISC 360 Table J2.4, the minimum size of fillet weld is

$$w_{\min} = \tfrac{1}{4} \text{ in}$$
$$< \tfrac{3}{8} \text{ in} \quad [\text{satisfactory}]$$

From AISC 360 Sec. J2.2b, the maximum size of fillet weld for the $5/8$ in plate is

$$w_{\max} = \frac{5}{8} \text{ in} - \frac{1}{16} \text{ in}$$
$$> \tfrac{3}{8} \text{ in} \quad [\text{satisfactory}]$$

Instantaneous Center of Rotation Method

From *AISC Manual* Table 8-3, $C_1 = 1.0$. From *AISC Manual* Table 8-8, for values of $a = 1.53$ and $k = 0.75$, the coefficient C is given as 1.59, and the required fillet weld size per $1/16$ in, based on the instantaneous center of rotation method, is as follows.

LRFD Method

$$D = \frac{P_r}{\phi C l C_1} = \frac{30 \text{ kips}}{(0.75)\left(1.59 \frac{\text{kips}}{\text{in}} \text{ per } \frac{1}{16} \text{ in}\right)(8 \text{ in})(1.0)}$$

$$= 3.1 \text{ sixteenths}$$

ASD Method

$$D = \frac{\Omega P_r}{C l C_1} = \frac{(2.00)(20 \text{ kips})}{\left(1.59 \frac{\text{kips}}{\text{in}} \text{ per } \frac{1}{16} \text{ in}\right)(8 \text{ in})(1.0)}$$

$$= 3.1 \text{ sixteenths}$$

Use a weld size of

$$w = \tfrac{1}{4} \text{ in}$$

Weld Group Eccentrically Loaded Normal to Faying Surface

Eccentrically loaded weld groups of the type shown in Fig. 5.30 may be conservatively designed by means of the elastic vector analysis technique assuming unit size of weld. For a total length of weld $\bar{l}$, the vertical force per linear inch of weld due to the applied load, P_r, is

$$V_p = \frac{P_r}{\bar{l}} = \frac{P_r}{2l}$$

Moment of inertia about the x-axis is

$$I_x = \frac{2l^3}{12} = \frac{l^3}{6}$$

The horizontal force at point i due to the eccentricity, e, is

$$H_e = \frac{P_r e y_i}{I_x} = \frac{3P_r e}{l^2}$$

The resultant force at point i is

$$R = \sqrt{V_p^2 + H_e^2}$$

The instantaneous center of rotation method of analyzing eccentrically loaded weld groups may also be used to determine a weld group's capacity. *AISC Manual* Table 8-4, with $k=0$, provides a means of designing weld groups by this method.

Figure 5.30 *Weld Group Eccentrically Loaded Normal to Faying Surface*

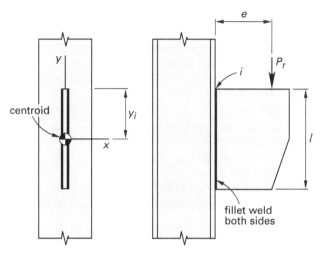

Example 5.47

Determine the size of E70XX fillet weld required in the welded gusset plate shown. Use the elastic unit area method and compare with the instantaneous center of rotation method.

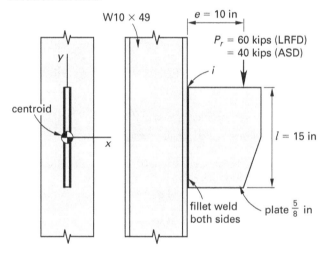

Solution

Assuming unit size of weld, the properties of the weld group are obtained by applying the elastic vector technique. The total length of the weld is

$$\bar{l} = 2l = 30 \text{ in}$$

Moment of inertia about the x-axis is

$$I_x = \frac{2l^3}{12} = \frac{(2)(15 \text{ in})^3}{12} = 563 \text{ in}^3$$

LRFD Method

Elastic Method

The vertical force per linear inch of weld due to the applied load, P_r, is

$$V_p = \frac{P_r}{2l} = \frac{60 \text{ kips}}{(2)(15 \text{ in})} = 2.0 \text{ kips/in}$$

The horizontal force at point i due to the eccentricity, e, is

$$\begin{aligned} H_e &= \frac{P_r e y_i}{I_x} \\ &= \frac{(60 \text{ kips})(10 \text{ in})(7.5 \text{ in})}{563 \text{ in}^3} \\ &= 8.0 \text{ kips/in} \end{aligned}$$

The resultant force at point i is

$$\begin{aligned} R &= \sqrt{V_p^2 + H_e^2} = \sqrt{\left(2.0 \ \frac{\text{kips}}{\text{in}}\right)^2 + \left(8.0 \ \frac{\text{kips}}{\text{in}}\right)^2} \\ &= 8.2 \text{ kips/in} \end{aligned}$$

The required fillet weld size per $\frac{1}{16}$ in is

$$\begin{aligned} D &= \frac{R}{q_u} = \frac{8.2 \ \dfrac{\text{kips}}{\text{in}}}{1.39 \ \dfrac{\text{kips}}{\text{in}} \text{ per } \dfrac{1}{16} \text{ in}} \\ &= 5.9 \text{ sixteenths} \end{aligned}$$

Use a weld size of

$$w = \tfrac{3}{8} \text{ in}$$

Instantaneous Center of Rotation Method

From *AISC Manual* Table 8-3, $C_1 = 1.0$. From *AISC Manual* Table 8-4, for values of $a = 0.67$ and $k = 0$, the coefficient C is given as 1.83, and the required fillet weld size per $\frac{1}{16}$ in, based on the instantaneous center of rotation method, is

$$\begin{aligned} D &= \frac{P_r}{\phi C l C_1} = \frac{60 \text{ kips}}{(0.75)\left(1.83 \ \dfrac{\text{kips}}{\text{in}} \text{ per } \dfrac{1}{16} \text{ in}\right)(15 \text{ in})(1.0)} \\ &= 2.9 \text{ sixteenths} \end{aligned}$$

Use $w_{\min} = \frac{1}{4}$ in.

ASD Method

Elastic Method

The vertical force per linear inch of weld due to the applied load, P_r, is

$$V_p = \frac{P_r}{2l} = 1.3 \text{ kips/in}$$

The horizontal force at point i due to the eccentricity, e, is

$$H_e = \frac{P_r e y_i}{I_x} = \frac{(40 \text{ kips})(10 \text{ in})(7.5 \text{ in})}{563 \text{ in}^3}$$
$$= 5.3 \text{ kips/in}$$

The resultant force at point i is

$$R = \sqrt{V_p^2 + H_e^2} = \sqrt{\left(1.3 \, \frac{\text{kips}}{\text{in}}\right)^2 + \left(5.3 \, \frac{\text{kips}}{\text{in}}\right)^2}$$
$$= 5.5 \text{ kips/in}$$

The required fillet weld size per $\frac{1}{16}$ in is

$$D = \frac{R}{q} = \frac{5.5 \, \dfrac{\text{kips}}{\text{in}}}{0.928 \, \dfrac{\text{kips}}{\text{in}} \text{ per } \dfrac{1}{16} \text{ in}}$$
$$= 5.9 \text{ sixteenths}$$

Use a weld size of

$$w = \tfrac{3}{8} \text{ in}$$

Instantaneous Center of Rotation Method

$$a = \frac{al}{l} = \frac{e}{l} = \frac{10}{15}$$
$$= 0.67$$

From *AISC Manual* Table 8-4, for values of $a = 0.67$ and $k = 0$, the coefficient C is given as 1.83, and the required fillet weld size per $\frac{1}{16}$ in, based on the instantaneous center of rotation method, is

$$D = \frac{\Omega P_r}{C l C_1} = \frac{(2.00)(40 \text{ kips})}{\left(1.83 \, \dfrac{\text{kips}}{\text{in}} \text{ per } \dfrac{1}{16} \text{ in}\right)(15 \text{ in})(1.0)}$$
$$= 2.9 \text{ sixteenths}$$

Use $w_{\min} = \tfrac{1}{4}$ in.

10. PLATE GIRDERS

Nomenclature

a	clear distance between transverse stiffeners	in
a_w	ratio of two times the web area in compression to the compression flange area	–
A_{fc}	compression flange area	in^2
A_{ft}	tension flange area	in^2
A_{pb}	bearing area of stiffener after allowing for corner snip	in^2
A_{sc}	cross-sectional area of a stud shear connector	in^2
A_{st}	area of transverse stiffener	in^2
A_T	area of compression flange plus $\frac{1}{6}$ web	in^2
A_w	web area	in^2
b_f	flange width	in
b_{st}	width of transverse stiffener	in
C_b	bending coefficient	–
C_{v1}	web shear strength coefficient	–
C_{v2}	web shear buckling coefficient	–
d	overall depth	in
D_s	factor dependent on the type of transverse stiffener used	–
E_c	modulus of elasticity of concrete	kips/in^2
f_c'	specified compressive strength of concrete	kips/in^2
F_{cr}	critical column axial compression stress	kips/in^2
F_{cr}	critical plate girder compression flange stress	kips/in^2
F_r	compressive residual stress in the flange	kips/in^2
F_u	minimum specified tensile strength of stud shear connector	kips/in^2
F_{uv}	required design shear strength of the stiffener-to-web weld	kips/in
F_{yf}	specified minimum yield stress of flange material	kips/in^2
F_{yst}	specified minimum yield stress of the stiffener material	kips/in^2
F_{yw}	specified minimum yield stress of the web material	kips/in^2
h	clear distance between flanges of a welded plate girder	in
I_{oy}	moment of inertia of flange plus $\frac{1}{6}$ web referred to the y-axis	in^4
I_{st}	moment of inertia of transverse stiffener	in^4
I_x	moment of inertia referred to the x-axis	in^4
j	factor used to define moment of inertia of transverse stiffener	–
k_v	web plate shear buckling coefficient	–
K	effective length factor	–
l	largest unbraced length along either flange at the point of load	in
l	laterally unbraced length of column	in
L_b	unbraced length of compression flange	in or ft

L_p	maximum unbraced length for the limit state of yielding	ft
L_r	maximum unbraced length for the limit state of inelastic lateral-torsional buckling	ft
M_n	nominal flexural capacity	ft-kips
P_u	factored end reaction	kips
q_u	design strength of a $\frac{1}{16}$ in fillet weld per inch run of E70XX grade electrodes per $\frac{1}{16}$ in	kips/in per $\frac{1}{16}$ in
r_{st}	radius of gyration of bearing stiffener	in
r_t	radius of gyration of compression flange plus $\frac{1}{6}$ web referred to the y-axis	in
R_g	stud group coefficient	–
R_p	stud position coefficient	–
R_{pg}	plate girder flexural coefficient	–
R_u	nominal bearing strength	kips/in^2
S_x	elastic section modulus referred to the x-axis	in^3
t_f	flange thickness	in
t_{st}	stiffener thickness	in
t_w	web thickness	in
V_n	nominal shear capacity	kips
w	fillet weld size	in
w	unit weight of concrete	lbf/ft^3

Symbols

λ	slenderness parameter	–
λ_p	limiting slenderness parameter for compact element	–
λ_r	limiting slenderness parameter for noncompact element	–
ϕ_b	resistance factor for flexure	–
ϕ_v	resistance factor for shear	–

Girder Proportions

The typical components of a welded, doubly symmetric, nonhybrid plate web girder are shown in Fig. 5.31.

Figure 5.31 *Plate Web Girder*

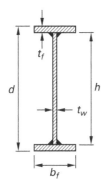

Girder proportions are given in AISC 360 Sec. F5 and Sec. G2 and are summarized as follows.

- Overall girder depth is usually in the range $L/12 < d < L/10$ where L is the span length.

- Assuming the flanges provide all the moment of resistance, flange area is approximately given by

$$A_f \approx \frac{M_u}{\phi_b h F_y} \quad \text{[LRFD]}$$

$$A_f \approx \frac{\Omega M_a}{h F_y} \quad \text{[ASD]}$$

- The flange width is usually in the range $h/5 < b_f < h/3$.

- To ensure a compact compression flange the slenderness parameter is limited by AISC 360 Table B4.1b, Case 11 to

$$\lambda = \frac{b_f}{2t_f} = 0.38\sqrt{\frac{E}{F_y}}$$

- For an unstiffened web, $h/t_w \leq 260$ and $ht_w/A_f \leq 10$ in accordance with AISC 360 Sec. F13.2.

- In accordance with AISC 360 Sec. G2.3a, intermediate stiffeners are not required, provided

$$\frac{h}{t_w} \leq 2.46\sqrt{\frac{E}{F_y}}$$

- When intermediate stiffeners are provided and the panel aspect ratio is $a/h > 1.5$, where a is the clear distance between stiffeners, the limiting web depth-to-thickness ratio is

$$\left(\frac{h}{t_w}\right)_{\max} = \frac{0.40E}{F_y} \quad \text{[AISC F13-4]}$$

- When intermediate stiffeners are provided and $a/h \leq 1.5$, the limiting web depth-to-thickness ratio is

$$\left(\frac{h}{t_w}\right)_{\max} = 12.0\sqrt{\frac{E}{F_y}} \quad \text{[AISC F13-3]}$$

- To utilize tension field action in the web, the panel aspect ratio is limited by AISC 360 Comm. Sec. G2.2 to a maximum value given by

$$\frac{a}{h} = 3.0$$

Example 5.48

For the welded plate girder of grade A36 steel shown, determine if the web thickness is adequate. The panel aspect ratio is $a/h = 2$ and tension field action is not utilized.

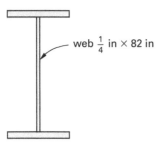

web $\frac{1}{4}$ in × 82 in

Solution

From the illustration,

$$\frac{h}{t_w} = \frac{82 \text{ in}}{\frac{1}{4} \text{ in}} = 328$$

For an aspect ratio of $a/h = 2$, AISC 360 Eq. F13-4 applies and the maximum permitted web slenderness ratio is

$$\left(\frac{h}{t_w}\right)_{max} = \frac{0.40F_y}{F_y} = \frac{(0.40)\left(29{,}000 \ \dfrac{\text{kips}}{\text{in}^2}\right)}{36 \ \dfrac{\text{kips}}{\text{in}^2}}$$

$$= 322$$
$$< 328$$

The web thickness is not adequate.

Design for Flexure

Doubly symmetric, nonhybrid beams with slender webs and values of $h/t_w > 5.7\sqrt{E/F_y}$, are classified as slender by AISC 360 Table B4.1b, Case 15, and the nominal flexural capacity is given by AISC 360 Sec. F5 as

$$M_n = S_x R_{pg} F_{cr} \qquad \text{[AISC F5-2]}$$

$$R_{pg} = 1 - \frac{a_w\left(\dfrac{h}{t_w} - 5.7\sqrt{\dfrac{E}{F_y}}\right)}{1200 + 300a_w} \qquad \text{[AISC F5-6]}$$

$$\leq 1.0$$
$$a_w \leq 10$$

$a_w = ht_w/A_{fc}$ for a doubly symmetrical built up section.

For the limit state of lateral-torsional buckling, the relevant parameters are

$$L_p = 1.1r_t\sqrt{\frac{E}{F_y}} \qquad \text{[AISC F4-7]}$$

$$L_r = \pi r_t\sqrt{\frac{E}{0.7F_y}} \qquad \text{[AISC F5-5]}$$

Compression Flange Yielding Governs

For an unbraced length of $L_b \leq L_p$, the critical stress is obtained from AISC 360 Eq. F5-1 as

$$F_{cr} = F_y$$

Inelastic Buckling Governs

When $L_p < L_b \leq L_r$, the critical stress is given by

$$F_{cr} = C_b F_y\left(1 - \frac{0.3(L_b - L_p)}{L_r - L_p}\right) \qquad \text{[AISC F5-3]}$$

$$\leq F_y$$

Elastic Buckling Governs

When $L_b > L_r$, the critical stress is given by

$$F_{cr} = \frac{C_b\pi^2 E}{\left(\dfrac{L_b}{r_t}\right)^2} \leq F_y \qquad \text{[AISC F5-4]}$$

For the limit state of flange local buckling, the relevant parameters are given by AISC 360 Sec. F5.3 as

$$\lambda = \frac{b_f}{2t_f}$$

$$\lambda_p = 0.38\sqrt{\frac{E}{F_y}} \qquad \text{[AISC Table B4.1b, Case 11]}$$

$$= 11 \quad \text{[for } F_y = 36 \text{ kips/in}^2\text{]}$$

$$\lambda_r = 0.95\sqrt{\frac{k_c E}{0.7F_y}} \qquad \begin{bmatrix}\text{AISC Table B4.1b,}\\ \text{Case 11}\end{bmatrix}$$

$$k_c = \frac{4}{\sqrt{\dfrac{h}{t_w}}}$$

$$\geq 0.35$$
$$\leq 0.76$$

Compact Flange

Flange local buckling does not occur in a compact flange with $\lambda \le \lambda_p$, and the critical compression flange stress is given by AISC 360 Sec. F5.3(a) as

$$F_{cr} = F_y$$

Noncompact Flange

Inelastic local buckling of the flange occurs in a noncompact flange with $\lambda_p < \lambda \le \lambda_r$, and the critical compression flange stress is given by AISC 360 Sec. F5.3(b) as

$$F_{cr} = F_y\left(1 - \frac{0.3(\lambda - \lambda_p)}{\lambda_r - \lambda_p}\right)$$

Slender Flange

Elastic local buckling of the flange occurs in a slender flange with $\lambda > \lambda_r$, and the critical compression flange stress is given by AISC 360 Sec. F5.3(c) as

$$F_{cr} = \frac{0.9Ek_c}{\lambda^2}$$

The available flexural capacity is given by AISC 360 Sec. F1 as

$$\phi_b M_n = 0.90 M_n \quad \text{[LRFD]}$$
$$\frac{M_n}{\Omega_b} = \frac{M_n}{1.67} \quad \text{[ASD]}$$

Example 5.49

Determine the available flexural capacity for the welded plate girder of grade A36 steel shown. Lateral support to the compression flange is provided at $L_b = 10$ ft centers, and $C_b = 1.0$.

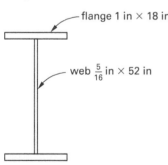

flange 1 in × 18 in

web $\frac{5}{16}$ in × 52 in

Solution

From the illustration,

$$\frac{h}{t_w} = \frac{52 \text{ in}}{\frac{5}{16} \text{ in}} = 166$$

$$< 260 \quad \begin{bmatrix} \text{stiffeners not mandatory} \\ \text{by AISC 360 Sec. F13.2} \end{bmatrix}$$

$$> 5.70\sqrt{\frac{E}{F_y}}$$

The web is classified as slender by AISC 360 Table B4.1b, Case 15, and AISC 360 Sec. F5 applies.

$$ht_w = (52 \text{ in})\left(\frac{5}{16} \text{ in}\right) = 16.25 \text{ in}^2$$

$$A_{fc} = b_f t_f = (18 \text{ in})(1.0 \text{ in}) = 18 \text{ in}^2$$

$$a_w = \frac{ht_w}{A_{fc}} = \frac{16.25 \text{ in}^2}{18 \text{ in}^2}$$
$$= 0.903$$

The moment of inertia of the flange plus $\frac{1}{6}$ web about the y-axis is

$$I_{oy} = \frac{t_f b_f^3}{12} = \frac{(1.0 \text{ in})(18 \text{ in})^3}{12}$$
$$= 486 \text{ in}^4$$

$$A_T = A_{fc} + \frac{ht_w}{6} = 18 \text{ in}^2 + \frac{16.25 \text{ in}^2}{6}$$
$$= 20.71 \text{ in}^2$$

$$r_t = \sqrt{\frac{I_{oy}}{A_T}} = \sqrt{\frac{486 \text{ in}^4}{20.71 \text{ in}^2}}$$
$$= 4.84 \text{ in}$$

The section modulus referred to the x-axis is

$$S_x = \frac{(2)\left(\begin{array}{c}\left(\dfrac{1}{12}\right)(18 \text{ in})(54 \text{ in})^3 \\ -\left(\dfrac{1}{12}\right)\left(18 \text{ in} - \dfrac{5}{16} \text{ in}\right)(52 \text{ in})^3\end{array}\right)}{54 \text{ in}}$$
$$= 1072 \text{ in}^3$$

For the limit state of lateral-torsional buckling,

$$L_p = 1.1 r_t \sqrt{\frac{E}{F_y}}$$

$$= (1.1)(4.84 \text{ in}) \sqrt{\frac{29{,}000 \dfrac{\text{kips}}{\text{in}^2}}{36 \dfrac{\text{kips}}{\text{in}^2}}}$$

$$= 151 \text{ in}$$
$$> 120 \text{ in} \quad [L_b = 120 \text{ in}]$$

Compression flange yielding governs.

$$F_{cr} = F_y$$

For the limit state of flange local buckling,

$$\lambda = \frac{b_f}{2t_f} \quad \text{[AISC Sec. F5.3]}$$

$$= \frac{18 \text{ in}}{(2)(1 \text{ in})}$$

$$= 9$$
$$< \lambda_p = 11$$

Flange local buckling does not occur. Therefore,

$$R_{pg} = 1 - \frac{a_w \left(\dfrac{h}{t_w} - 5.7 \sqrt{\dfrac{E}{F_y}} \right)}{1200 + 300 a_w}$$

$$= 1 - \frac{(0.903) \left(166 - 5.7 \sqrt{\dfrac{29{,}000 \dfrac{\text{kips}}{\text{in}^2}}{36 \dfrac{\text{kips}}{\text{in}^2}}} \right)}{1200 + (300)(0.903)}$$

$$= 0.997$$

LRFD Method

The design flexural capacity is given by AISC 360 Sec. F5.1 as

$$\phi_b M_n = \phi_b S_x R_{pg} F_{cr} = (0.9)(1072 \text{ in}^3)(0.997)\left(36 \frac{\text{kips}}{\text{in}^2} \right)$$

$$= 34{,}629 \text{ in-kips}$$

ASD Method

The allowable flexural capacity is given by AISC 360 Sec. F5.1 as

$$\frac{M_n}{\Omega} = \frac{S_x R_{pg} F_{cr}}{\Omega} = \frac{(1072 \text{ in}^3)(0.997)\left(36 \dfrac{\text{kips}}{\text{in}^2} \right)}{1.67}$$

$$= 23{,}040 \text{ in-kips}$$

Design for Shear

Providing intermediate transverse stiffeners in a girder increases the elastic buckling strength of the web. When the enhanced elastic critical load is reached, tension field action is induced. Additional load can be carried by the girder through diagonal tension in the web and a corresponding compression in the stiffeners. The top and bottom flanges of the girder act as chords and produce an equivalent Pratt truss as shown in Fig. 5.32.

Figure 5.32 *Tension Field Action*

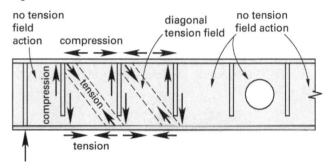

Design using tension field action is not permitted in end panels as a web panel is not available to anchor the tension field. Similarly, design using tension field action is not permitted in a panel with a large hole or in the adjacent panels. Large panel aspect ratios inhibit tension field action and design using tension field action is not permitted when $a/h > 3.0$ or when $a/h > (260 t_w/h)^2$. The design of plate girders for shear may be accomplished, using AISC 360 tables, by the four following methods.

- *Design using an unstiffened web: AISC Manual* Table 3-16a and Table 3-17a provide values of available shear stress for a range of values of h/t_w and a/h for steel with a yield stress of 36 kips/in² and 50 kips/in². By entering these tables with a value of 3.0 for a/h the available shear stress in an unstiffened web may be obtained.

- *Design using a stiffened web without utilizing tension field action: AISC Manual* Table 3-16a and Table 3-17a provide values of available shear stress for a range of values of h/t_w and a/h for steel with a yield stress of 36 kips/in² and 50 kips/in². By entering these tables with the appropriate values of h/t_w and a/h the available shear stress in a stiffened web may be obtained.

- *Design using a stiffened web with tension field action utilized when $2A_w/(A_{fc} + A_{ft}) \leq 2.5$, $h/b_{fc} \leq 6.0$, and $h/b_{ft} \leq 6.0$:AISC Manual Table 3-16b and Table 3-17b provide values of available shear stress for a range of values of h/t_w and a/h for steel with a yield stress of 36 kips/in^2 and 50 kips/in^2. By entering these tables with the appropriate values of h/t_w and a/h the available shear stress may be obtained.*

- *Design using a stiffened web with tension field action utilized when $2A_w/(A_{fc} + A_{ft}) > 2.5$ or $h/b_{fc} > 6.0$ or $h/b_{ft} > 6.0$: AISC Manual Table 3-16c and Table 3-17c provide values of available shear stress for a range of values of h/t_w and a/h for steel with a yield stress of 36 kips/in^2 and 50 kips/in^2. By entering these tables with the appropriate values of h/t_w and a/h, the available shear stress may be obtained.*

Design for Shear Without Utilizing Tension Field Action

For values of $h/t_w \leq 1.10\sqrt{k_v E/F_y}$, the nominal shear capacity, based on shear yielding of the stiffened or unstiffened web, is given by AISC 360 Sec. G2.1 as

$$V_n = 0.6F_y A_w C_{v1} \quad \text{[AISC G2-1]}$$

$$C_{v1} = 1.0 \quad \text{[AISC G2-3]}$$

For values of $1.10 \ h/t_w > 1.10\sqrt{k_v E/F_y}$, the nominal shear capacity, based on inelastic buckling of the web, is given by AISC 360 Sec. G2.1 as

$$V_n = 0.6F_y A_w C_{v1}$$

$$C_{v1} = \dfrac{1.10\sqrt{\dfrac{k_v E}{F_y}}}{\dfrac{h}{t_w}} \quad \text{[AISC G2-4]}$$

The web plate shear buckling coefficient is given by AISC 360 Sec. G2.1(b)(2).

$$k_v = 5 + \dfrac{5}{\left(\dfrac{a}{h}\right)^2} \quad \text{[for stiffened webs]}$$

$$= 5.34 \text{ when } a/h > 3 \quad \text{[for stiffened webs]}$$

$$= 5.34 \text{ for unstiffened webs}$$

Example 5.50

Determine the available shear capacity for the welded plate girder of Ex. 5.49 by using AISC 360 Sec. G2.1. Check the solution by using *AISC Manual* Table 3-16a.

Solution

From Ex. 5.49, for an unstiffened web,

$$A_w = dt_w = (54 \text{ in})(0.313 \text{ in})$$
$$= 16.90 \text{ in}^2$$
$$k_v = 5.34$$
$$\frac{h}{t_w} = 166$$

$$1.10\sqrt{\frac{Ek_v}{F_y}} = (1.10)\sqrt{\frac{\left(29{,}000 \ \dfrac{\text{kips}}{\text{in}^2}\right)(5.34)}{36 \ \dfrac{\text{kips}}{\text{in}^2}}}$$
$$= 72.15 < h/t_w$$

From AISC 360 Eq. G2-4, the web shear strength coefficient is

$$C_{v1} = \frac{1.10\sqrt{\dfrac{Ek_v}{F_y}}}{\left(\dfrac{h}{t_w}\right)} = \frac{72.15}{166} = 0.43$$

$$V_n = 0.6F_y A_w C_{v1} = (0.6)\left(36 \ \frac{\text{kips}}{\text{in}^2}\right)(16.9 \text{ in}^2)(0.43)$$
$$= 157 \text{ kips}$$

LRFD Method

The design shear capacity is then

$$\phi_v V_n = 0.9 V_n = (0.9)(157 \text{ kips})$$
$$= 141 \text{ kips}$$

From *AISC Manual* Table 3-16a, for a value of $a/h > 3.0$ and a value of $h/t_w = 166$,

$$\phi_v V_n = 8.75 A_w = \left(8.75 \ \frac{\text{kips}}{\text{in}^2}\right)(16.90 \text{ in}^2)$$
$$= 148 \text{ kips}$$

ASD Method

The allowable shear capacity is then

$$\frac{V_n}{\Omega} = \frac{V_n}{1.67} = \frac{157 \text{ kips}}{1.67}$$
$$= 94 \text{ kips}$$

Steel

From *AISC Manual* Table 3-16a, for a value of $a/h > 3.0$ and a value of $h/t_w = 166$,

$$\frac{V_n}{\Omega} = 5.9 A_w = \left(5.9 \ \frac{\text{kips}}{\text{in}^2}\right)(16.90 \ \text{in}^2)$$
$$= 100 \ \text{kips}$$

Design for Shear with Tension Field Action Utilized

When tension field action is used, the nominal shear strength is determined in accordance with AISC 360 Sec. G2.2.

For a girder with the property

$$\frac{h}{t_w} \leq 1.10 \sqrt{\frac{k_v E}{F_y}}$$

- shear yielding of the web governs, and the nominal shear strength is given by AISC 360 Eq. G2-6 as

$$V_n = 0.60 F_y A_w$$

- the web plate shear buckling coefficient for a stiffened web is given by AISC 360 Sec. G2.1(b)(2) as

$$k_v = 5 + \frac{5}{\left(\dfrac{a}{h}\right)^2}$$

For a girder with the property

$$\frac{h}{t_w} > 1.10 \sqrt{\frac{k_v E}{F_y}}$$

- when $2A_w/(A_{fc} + A_{ft}) \leq 2.5$, $h/b_{fc} \leq 6.0$, and $h/b_{ft} \leq 6.0$, AISC 360 Eq. G2-7 applies, and the nominal shear strength is

$$V_n = 0.6 F_y A_w \left(C_{v2} + \frac{1 - C_{v2}}{1.15 \sqrt{1 + \left(\dfrac{a}{h}\right)^2}} \right)$$

Otherwise, AISC 360 Eq. G2-8 applies, and the nominal shear strength is

$$V_n = 0.6 F_y A_w \left(C_{v2} + \frac{1 - C_{v2}}{1.15 \left(\dfrac{a}{h} + \sqrt{1 + \left(\dfrac{a}{h}\right)^2} \right)} \right)$$

The web shear buckling coefficient, C_{v2}, is obtained from the following equations.

For inelastic buckling of the web with

$$1.10 \sqrt{\frac{k_v E}{F_y}} < \frac{h}{t_w} \leq 1.37 \sqrt{\frac{k_v E}{F_y}}$$

the web shear buckling coefficient is given by AISC 360 Eq. G2-10 as

$$C_{v2} = \frac{1.10 \sqrt{\dfrac{k_v E}{F_y}}}{\left(\dfrac{h}{t_w}\right)}$$

For elastic buckling of the web with

$$\frac{h}{t_w} > 1.37 \sqrt{\frac{k_v E}{F_{yw}}}$$

the web shear buckling coefficient is given by AISC 360 Eq. G2-11 as

$$C_{v2} = \frac{1.51 k_v E}{\left(\dfrac{h}{t_w}\right)^2 F_y}$$

Example 5.51

The welded plate web girder of Ex. 5.49 has intermediate stiffeners provided at 100 in centers. Determine the design shear capacity using *AISC Manual* tables.

Solution

From previous examples,

$$\frac{h}{t_w} = 166$$

$$h = 52 \text{ in}$$

$$A_w = 16.90 \text{ in}^2$$

$$A_{fc} = A_{ft} = 18 \text{ in}^2$$

$$b_{fc} = b_{ft} = 18 \text{ in}$$

$$2\left(\frac{A_w}{(A_{fc} + A_{ft})}\right) = (2)\left(\frac{16.90 \text{ in}^2}{18 \text{ in}^2 + 18 \text{ in}^2}\right)$$

$$= 0.94$$

$$< 2.5$$

$$\frac{h}{b_{ft}} = \frac{h}{b_{fc}} = \frac{52 \text{ in}}{18 \text{ in}}$$

$$= 2.89$$

$$< 6.0$$

Hence, *AISC Manual* Table 3-16b is applicable.

For intermediate stiffeners provided at 100 in centers,

$$\frac{a}{h} = \frac{100 \text{ in}}{52 \text{ in}}$$

$$= 1.92$$

From *AISC Manual* Table 3-16b, for a value of $a/h = 1.92$ and a value of $h/t_w = 166$, the available shear strength is

LRFD Method

$$V_c = \phi_v V_n = 11.3 A_w = \left(11.3 \ \frac{\text{kips}}{\text{in}^2}\right)(16.90 \text{ in}^2)$$

$$= 191 \text{ kips}$$

ASD Method

$$V_c = \frac{V_n}{\Omega} = 7.6 A_w = \left(7.6 \ \frac{\text{kips}}{\text{in}^2}\right)(16.90 \text{ in}^2)$$

$$= 128 \text{ kips}$$

Design of Intermediate Stiffeners

Tension Field Action Excluded

The required moment of inertia of a single stiffener about the face in contact with the web plate or of a pair of stiffener plates about the web centerline is given by AISC 360 Sec. G2.15 as

$$I_{st2} = b_p t_w^3 j$$

$$j = \frac{2.5}{\left(\dfrac{a}{h}\right)^2} - 2$$

$$\geq 0.5$$

b_p is the smaller of the dimensions a and h.

The maximum allowable width-to-thickness ratio of a stiffener plate is given by AISC 360 Table B4.1a, Case 1 as

$$\frac{b_{st}}{t_{st}} = 0.56\sqrt{\frac{E}{F_{yst}}}$$

$$= 15.89 \quad [\text{for } F_{yst} = 36 \text{ kips/in}^2]$$

As specified in AISC 360 Sec. G2.3, the weld used to attach the stiffener to the web must terminate between four times and six times the web thickness from the near toe of the web-to-flange weld.

Tension Field Action Included

The maximum width-to-thickness ratio of a stiffener plate subject to tension field action is given by AISC 360 Eq. G2-12 as

$$\left(\frac{b}{t}\right)_{st} \leq 0.56\sqrt{\frac{E}{F_{yst}}}$$

The minimum required moment of inertia of a single stiffener about the face in contact with the web plate or of a pair of stiffener plates about the web centerline is specified by AISC 360 Eq. G2-13 as

$$I_{st} \geq I_{st2} + (I_{st1} - I_{st2})\left(\frac{V_r - V_{c2}}{V_{c1} - V_{c2}}\right)$$

V_{c1} is the available shear strength calculated with V_n as defined in AISC 360 Sec. G2.1 (tension field action not considered) or AISC Sec. G2.2 (tension field action considered) as applicable.

V_{c2} is the available shear strength calculated with $V_n = 0.6 F_y A_w C_{v2}$.

V_r is the required shear strength in the panel considered.

I_{st2} is the moment of inertia required for the development of the web shear buckling resistance as defined in AISC 360 Eq. G2-15.

$$I_{st2} = b_p t_w^3 j$$

$$j = \frac{2.5}{\left(\dfrac{a}{h}\right)^2} - 2$$

$$\geq 0.5$$

b_p is the smaller of the dimensions a and h.

I_{st1} is the moment of inertia required for the development of the web shear buckling resistance plus the web tension field resistance as given in AISC 360 Eq. G2-14.

$$I_{st1} = \frac{h^4 \rho_{st}^{1.3}}{40}\left(\frac{F_{yw}}{E}\right)^{1.5}$$

ρ_{st} is the larger of F_{yw}/F_{yst} and 1.0. F_{yw} is the yield stress of the web material.

Example 5.52

Design the intermediate stiffeners, using a pair of stiffener plates, for the plate web girder of Ex. 5.49, which excludes tension field action. Stiffeners are provided at 100 in centers. The required shear strength is

$$V_r = 120 \text{ kips} \quad [\text{LRFD}]$$
$$V_r = 80 \text{ kips} \quad [\text{ASD}]$$

Solution

For a value of $a/h = 1.92$, the moment of inertia factor is given by AISC 360 Eq. G2-15 as

$$j = \frac{2.5}{\left(\dfrac{a}{h}\right)^2} - 2 = \frac{2.5}{(1.92)^2} - 2$$
$$= -1.324 \quad [\text{0.5 minimum}]$$
$$b_p = h = 52 \text{ in}$$

The required moment of inertia of a pair of stiffener plates about the web centerline is given by AISC 360 Sec. G2.15 as

$$I_{st2} = b_p t_w^3 j = (52 \text{ in})\left(\frac{5}{16} \text{ in}\right)^3 (0.5)$$
$$= 0.80 \text{ in}^4$$

For a pair of 4 in × ⅜ in stiffener plates, the width-to-thickness ratio is

$$\frac{b_{st}}{t_{st}} = \frac{4 \text{ in}}{\dfrac{3}{8} \text{ in}} = 10.7$$
$$< 0.56\sqrt{\frac{E}{F_y}} \quad [\text{satisfactory}]$$

In accordance with AISC 360 Sec. G2.3, the stiffener may terminate 2 in above the bottom flange.

The moment of inertia provided by the pair of plates is

$$I_{st2} = \frac{t_{st}(2b_{st} + t_w)^3}{12} = \frac{\left(\dfrac{3}{8} \text{ in}\right)\left((2)(4 \text{ in}) + \dfrac{5}{16} \text{ in}\right)^3}{12}$$
$$= 17.95 \text{ in}^4$$
$$> 0.80 \text{ in}^4 \quad [\text{satisfactory}]$$

Design of Bearing Stiffeners

Bearing stiffeners are required on a plate girder when an applied load exceeds the web's yielding, crippling, or sidesway buckling capacity. The following provisions on the design of bearing stiffeners are stipulated by AISC 360 Sec. J10.8.

- Stiffeners are placed in pairs on opposite sides of the web at the location of the load. (See Fig. 5.33.)

- The column section is composed of the two stiffener plates plus a strip of web having a width of $25t_w$ at interior stiffeners and $12t_w$ at end stiffeners. (See Fig. 5.34.)

Figure 5.33 *Bearing Stiffeners*

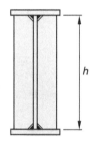

Figure 5.34 *Stiffener Cross-Section*

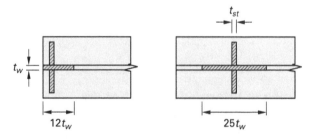

- In accordance with AISC 360 Sec. J10.8, the stiffener is designed as an axially loaded cruciform column.

- The effective length of the column is given by AISC 360 Sec. J10.8 as $L_c = 0.75h$ and the slenderness ratio is

$$\frac{L_c}{r} = 0.75\frac{h}{r}$$

When $L_c/r \leq 25$, the nominal axial strength is given by AISC 360 Eq. J4-6 as

$$P_n = F_y A_g$$

The available axial strength is

$$\phi P_n = 0.90 P_n \quad \text{[LRFD]}$$
$$\frac{P_n}{\Omega} = \frac{P_n}{1.67} \quad \text{[ASD]}$$

When $L_c/r > 25$, the provisions of AISC 360 Chap. E apply.

- Bearing stiffeners must extend the full height of the web and provide close bearing on, or be welded to, the loaded flange.

- The capacity of the fillet weld between the stiffener plate and the web must be sufficient to transmit the applied compressive force.

- In accordance with AISC 360 Table B4.1a, Case 3, the limiting width-thickness ratio of each plate is given by

$$\frac{b_{st}}{t_{st}} = 0.45 \sqrt{\frac{E}{F_y}}$$

- The nominal bearing strength on the area of the stiffener plate, A_{pb}, in contact with the flange is given by AISC 360 Sec. J7 as

$$R_n = 1.8 F_y A_{pb}$$

- The available bearing strength is

$$\phi R_n = 0.75 R_n \quad \text{[LRFD]}$$
$$\frac{R_n}{\Omega} = \frac{R_n}{2.00} \quad \text{[ASD]}$$

Example 5.53

The welded plate web girder of Ex. 5.49 is provided with bearing stiffeners at each end consisting of a pair of $\frac{1}{2}$ in $\times$ 8 in plates of grade A36 steel. Determine the maximum reaction that may be applied to the girder.

Solution

Allowing for a 1 in corner snip to clear the weld, the nominal bearing area of the stiffener plates is

$$A_{pb} = 2 t_{st}(b_{st} - 1) = (2)\left(\frac{1}{2} \text{ in}\right)(8 \text{ in} - 1 \text{ in})$$
$$= 7 \text{ in}^2$$

The available bearing strength is given by AISC 360 Sec. J7 as follows.

For LRFD,

$$\phi R_n = \phi(1.8 F_y A_{pb}) = (0.75)(1.8)\left(36 \ \frac{\text{kips}}{\text{in}^2}\right)(7 \text{ in}^2)$$
$$= 340 \text{ kips}$$

For ASD,

$$\frac{R_n}{\Omega} = \frac{1.8 F_y A_{pb}}{\Omega} = \frac{(1.8)\left(36 \ \dfrac{\text{kips}}{\text{in}^2}\right)(7 \text{ in}^2)}{2.00}$$
$$= 227 \text{ kips}$$

The moment of inertia provided by the pair of plates is

$$I_{st2} = \frac{t_{st}(2 b_{st} + t_w)^3}{12}$$
$$= \frac{\left(\dfrac{1}{2} \text{ in}\right)\left((2)(8 \text{ in}) + \dfrac{5}{16} \text{ in}\right)^3}{12}$$
$$= 181 \text{ in}^4$$

The effective area of the bearing stiffener is

$$A_{st} = 2 t_{st} b_{st} + 12 t_w^2$$
$$= (2)\left(\frac{1}{2} \text{ in}\right)(8 \text{ in}) + (12)\left(\frac{5}{16} \text{ in}\right)^2$$
$$= 9.17 \text{ in}^2$$

The radius of gyration of the bearing stiffener is

$$r_{st} = \sqrt{\frac{I_{st}}{A_{st}}} = \sqrt{\frac{181 \text{ in}^4}{9.17 \text{ in}^2}}$$
$$= 4.44 \text{ in}$$

The slenderness ratio of the bearing stiffener is

$$\frac{L_c}{r_{st}} = \frac{(0.75)(52 \text{ in})}{4.44 \text{ in}} = 8.78 < 25$$

From AISC 360 Sec. J4.4, the nominal axial strength is

$$P_n = F_y A_g$$

Steel

For LRFD, the maximum end reaction is

$$P_u = \phi_c F_y A_{st}$$
$$= (0.9)\left(36 \ \frac{\text{kips}}{\text{in}^2}\right)(9.18 \ \text{in}^2)$$
$$= 297 \ \text{kips} \quad \text{[governs]}$$
$$< 340 \ \text{kips}$$

For ASD, the maximum end reaction is

$$P_a = \frac{F_y A_{st}}{\Omega_c}$$
$$= \frac{\left(36 \ \frac{\text{kips}}{\text{in}^2}\right)(9.18 \ \text{in}^2)}{1.67}$$
$$= 198 \ \text{kips} \quad \text{[governs]}$$
$$< 227 \ \text{kips}$$

11. COMPOSITE BEAMS

Nomenclature

a	depth of compression block	in
a	distance between connectors	in
A_c	area of concrete slab within the effective width	in^2
A_s	cross-sectional area of structural steel	in^2
A_{sa}	cross-sectional area of a stud shear connector	in^2
b	effective concrete flange width	in
C_{con}	compressive force in slab at ultimate load	kips
d	depth of steel beam	in
d	diameter of stud shear connector	in
E_c	modulus of elasticity of concrete	kips/in^2
$e_{\text{mid-ht}}$	distance from the edge of the stud shank to the steel deck web, measured at mid-height of the deck rib, in the direction of maximum moment for a simply supported beam	–
f_c'	specified compressive strength of the concrete	kips/in^2
F_u	minimum specified tensile strength of stud shear connector	kips/in^2
F_y	specified minimum yield stress of the structural steel section	kips/in^2
h_r	nominal steel deck rib height	in
H_s	length of shear connector, not to exceed $(h_r + 3 \ \text{in})$ in computations	in
I_{LB}	lower bound moment of inertia	in^4
L	span length	ft
M_n	nominal flexural strength of member	in-kips or ft-kips

n	number of shear connectors between point of maximum positive moment and point of zero moment	–
N_r	number of studs in one rib at a beam intersection, not to exceed 3 in calculations	–
Q_n	nominal shear strength of single shear connector	kips
R_g	stud group coefficient	–
R_p	stud position coefficient	–
s	beam spacing	ft or in
t_c	actual slab thickness	in
T_{stl}	tensile force in steel at ultimate load	kips
V'	total factored horizontal shear between point of maximum moment and point of zero moment	kips
w	unit weight of concrete	lbf/ft^3
w_r	average width of concrete rib	in
y	moment arm between centroids of tensile force and compressive force	in
Y_{con}	distance from top of steel beam to top of concrete	in
Y_1	distance from top of steel beam to plastic neutral axis	in
Y_2	distance from top of steel beam to concrete flange force	in

Symbols

ρ	reduction factor for studs in ribbed steel deck	–
$\sum Q_n$	summation of Q_n between point of maximum moment and point of zero moment on either side	kips

Section Properties

The composite beam shown in Fig. 5.35 consists of a concrete slab supported by a formed metal deck, with the slab acting compositely with a steel beam. In accordance with AISC 360 Sec. I3.1, the effective width of the concrete slab on either side of the beam centerline must not exceed

- one-eighth of the beam span

- one-half of the beam spacing

- the distance to the edge of the slab

For the composite beam shown in Fig. 5.35, at the ultimate load the depth of the concrete stress block is less than the depth of the slab.[12] For this situation, the plastic neutral axis is located at the top of the steel beam and

$$Y_1 = 0$$
$$Y_2 = Y_{\text{con}} - \frac{a}{2}$$

Figure 5.35 *Fully Composite Beam Section Properties*

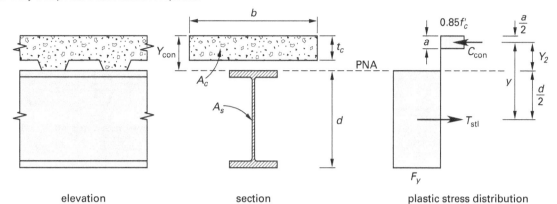

elevation	section	plastic stress distribution

When sufficient shear connectors are provided to ensure full composite action, the depth of the stress block is given by

$$a = \frac{F_y A_s}{0.85 f_c' b}$$

When insufficient shear connectors are provided to ensure full composite action, the depth of the stress block is given by

$$a = \frac{\sum Q_n}{0.85 f_c' b}$$

Using this value of the depth of the stress block to define an equivalent slab depth provides a lower bound on the actual moment of inertia based on elastic principles. *AISC Manual* Table 3-20 provides values of I_{LB} for a range of values of Y_1 and Y_2. This moment of inertia is used to determine the deflection of the composite member.

Example 5.54

A simply supported composite beam consists of a 3 in concrete slab cast on a 3 in formed steel deck over a W21 × 50 grade 50 steel beam. The beams are spaced at 8 ft centers and span 30 ft; the slab consists of 4000 lbf/in² normal weight concrete. Determine the lower bound moment of inertia if full composite action is provided.

Solution

The effective width of the concrete slab is the lesser of

$$s = (8 \text{ ft})\left(12 \frac{\text{in}}{\text{ft}}\right) = 96 \text{ in}$$

$$\frac{L}{4} = \frac{(30 \text{ ft})\left(12 \frac{\text{in}}{\text{ft}}\right)}{4}$$
$$= 90 \text{ in} \quad [\text{governs}]$$

For full composite action, the depth of the stress block is

$$a = \frac{F_y A_s}{0.85 f_c' b} = \frac{\left(50 \frac{\text{kips}}{\text{in}^2}\right)(14.7 \text{ in}^2)}{(0.85)\left(4 \frac{\text{kips}}{\text{in}^2}\right)(90 \text{ in})}$$

$$= 2.40 \text{ in} \quad [\text{within the slab}]$$

The distance from the top of the steel beam to the line of action of the concrete slab force is

$$Y_2 = Y_{\text{con}} - \frac{a}{2}$$
$$= 3 \text{ in} + 3 \text{ in} - \frac{2.40 \text{ in}}{2}$$
$$= 4.80 \text{ in}$$
$$Y_1 = 0$$

From *AISC Manual* Table 3-20,

$$I_{LB} = 2686 \text{ in}^4$$

Shear Connection

Shear connectors are provided to transfer the horizontal shear force across the interface. The nominal shear strengths Q_n of different types of shear connectors are given in *AISC Manual* Table 3-21. The required number of connectors may be uniformly distributed between the point of maximum moment and the support on either side, with the total horizontal shear being determined by the lesser value given by AISC 360 Eq. I3-1a and Eq. I3-1b as

$$V' = 0.85 f_c' A_c \quad [\text{AISC I3-1a}]$$
$$V' = F_y A_s \quad [\text{AISC I3-1b}]$$

Figure 5.36 *Placement of Shear Connectors*

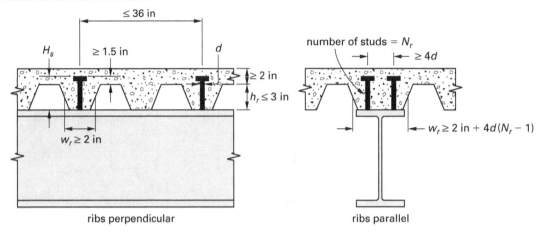

To provide complete shear connection and full composite action, the required number of connectors on either side of the point of maximum moment is given by

$$n = \frac{V'}{Q_n}$$

If a smaller number of connectors is provided, only partial composite action can be achieved, and the nominal flexural strength of the composite member is reduced. The number of shear connectors placed between a concentrated load and the nearest support must be sufficient to develop the moment required at the load point.

The nominal strength of one stud shear connector embedded in a solid slab is given by AISC 360 Eq. I8-1 as

$$Q_n = 0.5 A_{sa}\sqrt{f'_c E_c}$$
$$\leq R_g R_p A_{sa} F_u$$

The minimum tensile strength of a type B shear stud connector made from ASTM A108 material is

$$F_u = 65 \text{ kips/in}^2$$

The stud group coefficient for flat soffit, solid slabs with the stud welded directly to the girder flange is

$$R_g = 1.0$$

The stud position coefficient for flat soffit, solid slabs with the stud welded directly to the girder flange is

$$R_p = 0.75$$

The modulus of elasticity of concrete is given by

$$E_c = w_c^{1.5}\sqrt{f'_c}$$

The unit weight of normal weight concrete is given by

$$w_c = 145 \text{ lbf/ft}^3$$
$$Q_n = 0.5 A_{sa}(w_c f'_c)^{0.75}$$

When the concrete is cast on a formed metal deck, the limitations imposed on the spacing and placement of shear connectors are given in AISC 360 Sec. I3.2c and are summarized in Fig. 5.36.

When the concrete slab is cast on a formed metal deck, the values of R_g and R_p are modified as detailed in AISC 360 Sec. I8.2a.

Deck Ribs Parallel to Steel Beam

When the deck ribs are parallel to the steel beam as shown in Fig. 5.37, the stud group coefficient for any number of studs welded in a row through the steel deck is given by

$$R_g = 1.0 \quad [\text{when } w_r \geq 1.5h_r]$$
$$R_g = 0.85 \quad [\text{when } w_r < 1.5h_r]$$

When the deck ribs are parallel to the steel beam as shown in Fig. 5.37, the stud position coefficient for studs welded through the steel deck is given by

$$R_p = 0.75$$

Figure 5.37 *Deck Ribs Parallel to Steel Beam, R_g and R_p Values*

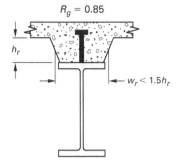

$R_g = 0.85$

$w_r < 1.5h_r$

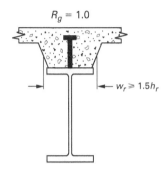

$R_g = 1.0$

$w_r \geq 1.5h_r$

$R_p = 0.75$

Figure 5.38 *Deck Ribs Perpendicular to Steel Beams, R_g Values*

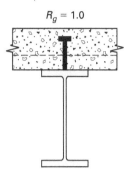

$R_g = 1.0$

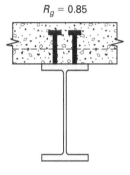

$R_g = 0.85$

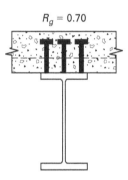

$R_g = 0.70$

Deck Ribs Perpendicular to Steel Beam

When the deck ribs are perpendicular to the steel beam as shown in Fig. 5.38, the stud group coefficient for studs welded through the steel deck is given by

$R_g = 1.0$ [for one stud welded in a steel deck rib]

$R_g = 0.85$ [for two studs welded in a steel deck rib]

$R_g = 0.70$ $\begin{bmatrix} \text{for three or more studs} \\ \text{welded in a steel deck rib} \end{bmatrix}$

When the deck ribs are perpendicular to the steel beam as shown in Fig. 5.39, the stud position coefficient for studs welded through the steel deck is given by

$R_p = 0.75$ $\begin{bmatrix} \text{for studs welded in a steel deck rib} \\ \text{with } e_{\text{mid-ht}} \geq 2 \text{ in} \end{bmatrix}$

$R_p = 0.60$ $\begin{bmatrix} \text{for studs welded in a steel deck rib} \\ \text{with } e_{\text{mid-ht}} < 2 \text{ in} \end{bmatrix}$

Figure 5.39 *Deck Ribs Perpendicular to Steel Beams, R_p Values*

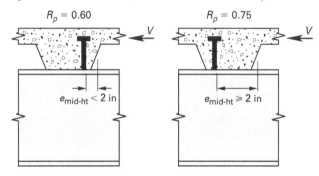

$R_p = 0.60$ $\qquad$ $R_p = 0.75$

$e_{\text{mid-ht}} < 2$ in $\qquad$ $e_{\text{mid-ht}} \geq 2$ in

The nominal strength of different stud diameters in 3 kips/in² and 4 kips/in² normal weight and lightweight concrete is given in *AISC Manual* Table 3-21.

Example 5.55

Determine the number of $\frac{3}{4}$ in diameter stud shear connectors required in the composite beam of Ex. 5.54 to provide full composite action. The ribs of the formed steel deck are perpendicular to the steel beams with $h_r = 3$ in, $w_r = 3\frac{1}{2}$ in, $H_s = 5$ in, and $N_r = 2$. The beam is loaded with a uniformly distributed load.

Solution

The total horizontal shear is given by

$$V' = F_y A_s = \left(50\ \frac{\text{kips}}{\text{in}^2}\right)(14.7\ \text{in}^2)$$
$$= 735\ \text{kips}$$

Two $\frac{3}{4}$ in diameter studs are located in each rib in the weak position.

The nominal shear strength of each $\frac{3}{4}$ in diameter stud is obtained from *AISC Manual* Table 3-21, as

$$Q_n = 14.6\ \text{kips}$$

The required number of studs in the beam is

$$2n = \frac{2V'}{Q_n} = \frac{(2)(735\ \text{kips})}{14.6\ \text{kips}}$$
$$= 100\ \text{studs}$$

Design for Flexure

AISC Manual Table 3-19 provides values of ϕM_n for a range of values of Y_1, Y_2, and $\sum Q_n$. The value of $\sum Q_n$ is given by AISC 360 Sec. I3.2d as the least of

- $0.85 f_c' A_c$

- $F_y A_s$

- $n Q_n$

Because of redistribution of stresses at the ultimate load, the composite section is designed to support the total factored loads, due to all dead and live loads, for both shored and unshored construction. In addition, for unshored construction, the steel beam alone must be adequate to support all loads applied before the concrete has attained 75% of its required strength.

Example 5.56

Determine the available flexural strength of the composite beam of Ex. 5.55.

Solution

Because sufficient shear connectors are provided to ensure full composite action,

$$\sum Q_n = F_y A_s = 735\ \text{kips}$$

From Ex. 5.54,

$$Y_2 = 4.80\ \text{in}$$
$$Y_1 = 0$$

From *AISC Manual* Table 3-19, the available strength is

$$\phi M_n = 839\ \text{ft-kips} \quad [\text{LRFD}]$$

$$\frac{M_n}{\Omega} = 558\ \text{ft-kips} \quad [\text{ASD}]$$

12. REFERENCES

1. American Institute of Steel Construction. *Specification for Structural Steel Buildings*. Chicago, IL: American Institute of Steel Construction, 2016.

2. International Code Council. *International Building Code*. Falls Church, VA: International Code Council, 2018.

3. American Institute of Steel Construction. *Steel Construction Manual*, Fifteenth ed. Chicago, IL: American Institute of Steel Construction, 2017.

4. American Institute of Steel Construction. *Design Examples*: Version 15.0. Chicago, IL: American Institute of Steel Construction, 2017.

5. White, D. W. and L. G. Griffis. "Stability Design of Steel Buildings: Highlights of a New AISC Design Guide." *Proceedings, North American Steel Construction Conference*. New Orleans, LA: American Institute of Steel Construction, 2007.

6. Carter, Charles J. and Louis F. Geschwindner. "A Comparison of Frame Stability Analysis Methods in ANSI/AISC 360-05." *Engineering Journal* (third quarter 2008).

7. Schwinger, Clifford. "Stability Analysis and Design." *Modern Steel Construction*, vol. 4 (2013).

8. American Institute of Steel Construction. *AISC Basic Design Values Cards*. Chicago, IL: American Institute of Steel Construction, 2017.

9. Aminmansour, Abbas. "Design of Structural Steel Members Subject to Combined Loading." *Structure Magazine*, vol. 2 (2007).

10. Research Council for Structural Connections (RCSC). *Specification for Structural Joints Using High-Strength Bolts*. AISC, Chicago, IL, 2014.

11. American Institute of Steel Construction. *Code of Standard Practice for Steel Buildings and Bridges*. Chicago, IL: American Institute of Steel Construction, 2017.

12. Vogel, Ron. *LRFD-Composite Beam Design with Metal Deck*. Walnut Creek, CA: Steel Committee of California, 1991.

Steel

6 Wood Design

H	load due to lateral pressure	kips/ft^2
L	live load	kips or kips/ft
L_r	roof live load	kips or kips/ft
Q	load effect due to service load	kips
R	rain load	kips or kips/ft
S	snow load	kips or kips/ft
W	wind load	kips or kips/ft

Symbols

ϕ	resistance factor	–
γ	load factor	–

1. ASD AND LRFD METHODS

In accordance with the *National Design Specification* (NDS) *for Wood Construction*[1] Sec. 1.4, wood structures must be designed using either the provisions for load and resistance factor design (LRFD) or the provisions for allowable stress design (ASD).

The ASD method is the traditional method of designing wood structures. Working loads, also referred to as nominal or service loads, are the actual dead loads plus the live loads applied to a member. The ASD method, based on elastic theory, is used to calculate the stresses produced by the factored working loads. Factored working loads, using ASD load combinations, are applied to the member; the stresses produced in the member must not exceed the specified allowable stress. The allowable stress is determined by applying adjustment factors to the reference design values tabulated in the *NDS Supplement*[2]. The application of load duration factors to the reference design values allow for the effects of transient loads, such as wind and seismic loads.

In the LRFD method, working loads are factored using LRFD load combinations before being applied to the member to determine the required strength. This required strength must not exceed the design strength, which is calculated as the member nominal strength multiplied by a resistance factor, ϕ.

2. LOAD COMBINATIONS

Nomenclature

D	dead load	kips or kips/ft
E	earthquake load	kips or kips/ft

ASD Required Strength

The required strength of a member consists of the most critical combination of factored loads applied to the member. In accordance with NDS Sec. 1.4.4, load combinations must be as specified in the applicable building code. The required strength, $\Sigma\gamma Q$, is defined by nine combinations given in IBC[3] Sec. 1605.3.1. The seismic and wind loads specified in the IBC are at the strength design level in contrast to other loads, which are at the service level. In ASD load combinations, the load factor for seismic loads is 0.7, and the load factor for wind loads is 0.6 to reduce the factored loads to service level values. The combinations, with uncommon load conditions (self-straining loads and fluid pressure) omitted, are as follows.

$$\Sigma\gamma Q = D \qquad \text{[IBC 16-8]}$$
$$\Sigma\gamma Q = D + H + L \qquad \text{[IBC 16-9]}$$
$$\Sigma\gamma Q = D + H + (L_r \text{ or } S \text{ or } R) \qquad \text{[IBC 16-10]}$$
$$\Sigma\gamma Q = D + H + 0.75L + 0.75(L_r \text{ or } S \text{ or } R) \quad \text{[IBC 16-11]}$$
$$\Sigma\gamma Q = D + H + (0.6W \text{ or } 0.7E) \qquad \text{[IBC 16-12]}$$
$$\Sigma\gamma Q = D + H + 0.75(0.6W) + 0.75L \qquad \text{[IBC 16-13]}$$
$$+ 0.75(L_r \text{ or } S \text{ or } R)$$
$$\Sigma\gamma Q = D + H + 0.75(0.7E) + 0.75L \qquad \text{[IBC 16-14]}$$
$$+ 0.75S$$
$$\Sigma\gamma Q = 0.6D + 0.6W + H \qquad \text{[IBC 16-15]}$$
$$\Sigma\gamma Q = 0.6D + 0.7E + H \qquad \text{[IBC 16-16]}$$

Wood

LRFD Required Strength

The required ultimate strength of a member consists of the most critical combination of factored loads applied to the member. The required strength, $\sum \gamma Q$, is defined by seven combinations given in IBC Sec. 1605.2. The combinations, with uncommon load conditions (self-straining loads and fluid pressure) omitted, are as follows.

$$\sum \gamma Q = 1.4D \qquad \text{[IBC 16-1]}$$

$$\sum \gamma Q = 1.2D + 1.6(L+H) \qquad \text{[IBC 16-2]}$$
$$+ 0.5(L_r \text{ or } S \text{ or } R)$$

$$\sum \gamma Q = 1.2D + 1.6(L_r \text{ or } S \text{ or } R) \qquad \text{[IBC 16-3]}$$
$$+ 1.6H + (f_1L \text{ or } 0.5W)$$

$$\sum \gamma Q = 1.2D + 1.0W + f_1L \qquad \text{[IBC 16-4]}$$
$$+ 1.6H + 0.5(L_r \text{ or } S \text{ or } R)$$

$$\sum \gamma Q = 1.2D + 1.0E + f_1L + 1.6H + f_2S \quad \text{[IBC 16-5]}$$

$$\sum \gamma Q = 0.9D + 1.0W + 1.6H \qquad \text{[IBC 16-6]}$$

$$\sum \gamma Q = 0.9D + 1.0E + 1.6H \qquad \text{[IBC 16-7]}$$

For IBC Eq. 16-3, 16-4, and 16-5, use $f_1 = 1.0$ for garages, places of public assembly, and areas where $L > 100 \text{ lbf/ft}^2$. Use $f_1 = 0.5$ for all other live loads. For IBC Eq. 16-5, use $f_2 = 0.7$ for roof configurations that do not shed snow, and use 0.2 for other roof configurations. Where the effect of H resists the primary variable load effect, a load factor of 0.9 must be included with H where H is permanent. H must be set to zero for all other conditions.

3. DEFINITIONS AND TERMINOLOGY

A description of the wood products available, and of the terminology used, is as follows.

A *composite panel* is a wood structural panel consisting of wood veneer and reconstituted wood material bonded with waterproof adhesive.

Cross-laminated timber consists of layers of lumber stacked at right angles and glued together on their wide faces.

Decking consists of solid sawn lumber or glued laminated members with 2–4 in nominal thickness, and 4 in or more wide. For 2 in thicknesses, it is usually single tongue and groove, and for 3–4 in thicknesses, it may be double tongue and groove.

Dimension lumber consists of solid sawn lumber members with 2–4 in nominal thickness, and 2 in or more wide.

Dressed size refers to the dimensions of a lumber member after it has been surfaced with a planing machine. It is usually $\frac{1}{2}$–$\frac{3}{4}$ in less than nominal size.

Grade indicates the classification wood products are given with respect to strength in accordance with specific grading rules.

A *joist* is a lumber member with 2–4 in nominal thickness, and 5 in or wider. A joist is typically loaded on the narrow face and used as framing in floors and roofs.

Laminated veneer lumber consists of wood veneer sheet elements with the wood fiber oriented along the length of the member.

Lumber is cut to size in the sawmill and surfaced in a planing machine, and it is not further processed.

Mechanically graded lumber is dimension lumber that has been individually evaluated in a testing machine. Load is applied to the piece of lumber, the deflection is measured, and the modulus of elasticity is calculated. The strength characteristics of the lumber are directly related to the modulus of elasticity and can be determined. A visual check is also made on the lumber to detect visible flaws.

Nominal size is the term used to specify the undressed size of a lumber member. The finished size of a member after dressing is normally $\frac{1}{2}$–$\frac{3}{4}$ in smaller than the original size. Thus, a 2 in nominal × 4 in nominal member has actual dimensions of $1\frac{1}{2}$ in × $3\frac{1}{2}$ in.

Oriented strand board is a wood structural panel consisting of thin rectangular wood strands arranged in cross-aligned layers.

Parallel strand lumber consists of a composite of wood strands oriented along the length of the member and formed into a large mat and pressed together.

Plywood is a wood structural panel consisting of wood veneer plies arranged in cross-aligned layers

Structural glued laminated timber, or *glulams*, are built up from wood laminations bonded together with adhesives. The grain of all laminations is parallel to the length of the beam, and the laminations are typically $1\frac{1}{2}$ in thick.

Timbers are lumber members of nominal 5 in × 5 in or larger.

Visually stress-graded lumber are lumber members that have been graded visually to detect flaws and defects, and to assess the inherent strength of the member.

Wood structural panels are manufactured from veneers or wood strands bonded together with adhesives. Examples are plywood, oriented strand board, and composite panels.

Wood

4. REFERENCE DESIGN VALUES

Nomenclature

A_{eff}	effective area of a cross-laminated timber member	in^2/ft width
b	width of rectangular bending member	in
c	column parameter	–
C_b	bearing area factor	–
C_c	curvature factor for structural glued laminated member	–
C_D	load duration factor	–
C_F	size factor for sawn lumber	–
C_{fu}	flat use factor	–
C_H	shear stress adjustment factor	–
C_i	incising factor	–
C_I	stress interaction factor	–
C_L	beam stability factor	–
C_M	wet service factor	–
C_P	column stability factor	–
C_r	repetitive member factor for dimension lumber	–
C_t	temperature factor	–
C_T	buckling stiffness factor	–
C_V	volume factor for structural glued laminated timber	–
C_{vr}	shear reduction factor	–
d	depth (width) of bending member	in
E, E'	reference and adjusted modulus of elasticity	lbf/in^2
E_{min}, E'_{min}	reference and adjusted modulus of elasticity for beam stability and column stability calculations	lbf/in^2
F	ratio of F_{bE} to F_b^*	–
F'	ratio of F_{cE} to F_c^*	–
F_b^*	tabulated bending design value multiplied by all applicable adjustment factors except C_{fu}, C_V, and C_L	lbf/in^2
F_b, F_b'	reference and adjusted bending design value	lbf/in^2
F_{bE}	critical buckling design value for bending member	lbf/in^2
F_c^*	tabulated compressive design value multiplied by all applicable adjustment factors except C_P	lbf/in^2
F_c, F_c'	reference and adjusted compression design value parallel to grain	lbf/in^2
$F_{c\perp}, F_{c\perp}'$	reference and adjusted compression design value perpendicular to grain	lbf/in^2
F_{cE}	critical buckling design value	lbf/in^2
F_s, F_s'	reference and adjusted shear in the plane design value for wood structural panels and cross-laminated timber	lbf/in^2
F_t, F_t'	reference and adjusted tension design value	lbf/in^2
F_v, F_v'	reference and adjusted shear design value	lbf/in^2
I_{eff}	effective moment of inertia of a cross-laminated timber member	in^4/ft width
K_{cr}	time dependent deformation (creep) factor	–
K_F	format conversion factor	–
K_s	shear deformation adjustment factor	–
l_b	length of bearing parallel to the grain of the wood	in
l_e	effective length of compression member	ft or in
l_e	effective span length of bending member	ft or in
l_u	laterally unsupported length of beam	ft or in
L	span length of bending member	ft or in
R	radius of curvature of inside face of lamination	in
R_B	slenderness ratio of bending member	–
t	thickness of lamination	in
x	species parameter for volume factor	–

Symbols

λ	time effect factor	–
ϕ	resistance factor	–

NDS Supplement

The *NDS Supplement* provides reference design values for various wood species and for several types of wood products.

NDS Supplement

table	reference design values for wood
Table 4A	visually graded dimension lumber, all species except southern pine
Table 4B	visually graded southern pine dimension lumber
Table 4C	mechanically graded dimension lumber
Table 4D	visually graded timbers
Table 4E	visually graded decking
Table 4F	non-North American visually graded dimension lumber
Table 5A	structural glued laminated softwood timber, stressed primarily in bending
Table 5B	structural glued laminated softwood timber, stressed primarily in axial tension or compression
Table 5C	structural glued laminated hardwood timber, stressed primarily in bending
Table 5D	structural glued laminated hardwood timber, stressed primarily in axial tension or compression

Wood

The allowable, or adjusted, design values for a wood member depend on the application of the member and on the service conditions under which it is utilized. The reference design values are applicable to specified moisture conditions of service and normal load duration as specified in NDS Sec. 2.2. To determine the relevant adjusted values for other conditions of service, the reference design values are multiplied by adjustment factors specified in NDS Sec. 2.3. The applicability of each adjustment factor to the reference design values for sawn lumber, glued laminated timber, or cross-laminated timber is given in NDS Table 4.3.1, Table 5.3.1, and Table 10.3.1.

5. ADJUSTMENT OF REFERENCE DESIGN VALUES

Adjusted design values for wood members and connections must be appropriate for the conditions under which the wood is used. These values take into account the differences in wood strength properties with different moisture content, load durations, and treatment types. Reference design values are for normal load durations under the moisture service conditions specified in NDS Sec. 2.2. The applicability of adjustment factors to the reference design values for sawn lumber and glued laminated timber is summarized in Table 6.1.

Three additional factors are used solely in the LRFD method. These are the

- format conversion factor, K_F
- resistance factor, ϕ
- time effect factor, λ

Adjusted design values are obtained by multiplying reference design values by the applicable adjustment factors in accordance with the equations given.

Sawn Lumber

Adjusted design values for sawn lumber are as follows.

- bending

$$F'_b = F_b C_F C_r C_i C_D C_M C_t C_{fu} C_L \quad \text{[ASD]}$$
$$F'_b = F_b C_F C_r C_i C_M C_t C_{fu} C_L K_F \lambda \phi_b \quad \text{[LRFD]}$$

- tension

$$F'_t = F_t C_F C_i C_D C_M C_t \quad \text{[ASD]}$$
$$F'_t = F_t C_F C_i C_M C_t K_F \lambda \phi_t \quad \text{[LRFD]}$$

- shear

$$F'_v = F_v C_i C_D C_M C_t \quad \text{[ASD]}$$
$$F'_v = F_v C_i C_M C_t K_F \lambda \phi_v \quad \text{[LRFD]}$$

- compression parallel to the grain

$$F'_c = F_c C_F C_i C_D C_M C_t C_P \quad \text{[ASD]}$$
$$F'_c = F_c C_F C_i C_M C_t C_P K_F \lambda \phi_c \quad \text{[LRFD]}$$

- compression perpendicular to the grain

$$F'_{c\perp} = F_{c\perp} C_b C_i C_M C_t \quad \text{[ASD]}$$
$$F'_{c\perp} = F_{c\perp} C_b C_i C_M C_t K_F \phi_c \quad \text{[LRFD]}$$

The adjusted modulus of elasticity for sawn lumber is

$$E' = E C_i C_M C_t \quad \text{[ASD]}$$
$$E' = E C_i C_M C_t \quad \text{[LRFD]}$$

The adjusted modulus of elasticity for beam and column stability calculations for sawn timber is

$$E'_{\min} = E_{\min} C_i C_M C_t C_T \quad \text{[ASD]}$$
$$E'_{\min} = E_{\min} C_i C_M C_t C_T K_F \phi_s \quad \text{[LRFD]}$$

Glued Laminated Timber

Adjusted design values for glued laminated timber are as follows.

- bending

$$F'_b = F_b C_c C_I C_D C_M C_t C_{fu}(C_L \text{ or } C_V) \quad \text{[ASD]}$$
$$F'_b = F_b C_c C_I C_M C_t C_{fu}(C_L \text{ or } C_V) K_F \lambda \phi_b \quad \text{[LRFD]}$$

- tension

$$F'_t = F_t C_D C_M C_t \quad \text{[ASD]}$$
$$F'_t = F_t C_M C_t K_F \lambda \phi_t \quad \text{[LRFD]}$$

- shear

$$F'_v = F_v C_{vr} C_D C_M C_t \quad \text{[ASD]}$$
$$F'_v = F_v C_{vr} C_M C_t K_F \lambda \phi_v \quad \text{[LRFD]}$$

- compression parallel to the grain

$$F'_c = F_c C_D C_M C_t C_P \quad \text{[ASD]}$$
$$F'_c = F_c C_M C_t C_P K_F \lambda \phi_c \quad \text{[LRFD]}$$

Table 6.1 *Applicability of Adjustment Factors*

adjustment factor		sawn lumber	glued laminated	F_b	F_t	F_v	$F_{c\perp}$	F_c	E	E_{min}
C_F	size	yes	no	$\checkmark$	$\checkmark$	–	–	$\checkmark$	–	–
C_r	repetitive member	yes	no	$\checkmark$	–	–	–	–	–	–
C_i	incising	yes	no	$\checkmark$	$\checkmark$	$\checkmark$	$\checkmark$	$\checkmark$	$\checkmark$	$\checkmark$
C_V	volume*	no	yes	$\checkmark$	–	–	–	–	–	–
C_c	curvature	no	yes	$\checkmark$	–	–	–	–	–	–
C_D	load duration	yes	yes	$\checkmark$	$\checkmark$	$\checkmark$	–	$\checkmark$	–	–
C_M	wet service	yes	yes	$\checkmark$	$\checkmark$	$\checkmark$	$\checkmark$	$\checkmark$	$\checkmark$	$\checkmark$
C_b	bearing area	yes	yes	–	–	–	$\checkmark$	–	–	–
C_t	temperature	yes	yes	$\checkmark$	$\checkmark$	$\checkmark$	$\checkmark$	$\checkmark$	$\checkmark$	$\checkmark$
C_{fu}	flat use	yes	yes	$\checkmark$	–	–	–	–	–	–
C_T	buckling	yes	no	–	–	–	–	–	–	$\checkmark$
C_L	beam stability*	yes	yes	$\checkmark$	–	–	–	–	–	–
C_P	column stability	yes	yes	–	–	–	–	$\checkmark$	–	–
C_I	stress interaction	no	yes	$\checkmark$	–	–	–	–	–	–
C_{vr}	shear reduction	no	yes	–	–	$\checkmark$	–	–	–	–

*When applied to glued laminated members, only the lesser value of C_L or C_V is applicable. C_D is applicable to the ASD method only.

- compression perpendicular to the grain

$$F'_{c\perp} = F_{c\perp} C_b C_M C_t \quad \text{[ASD]}$$

$$F'_{c\perp} = F_{c\perp} C_b C_M C_t K_F \phi_c \quad \text{[LRFD]}$$

The adjusted modulus of elasticity for glued laminated timber is

$$E' = E C_M C_t \quad \text{[ASD]}$$

$$E' = E C_M C_t \quad \text{[LRFD]}$$

The adjusted modulus of elasticity for beam and column stability calculations for glued laminated timber is

$$E'_{min} = E_{min} C_M C_t \quad \text{[ASD]}$$

$$E'_{min} = E_{min} C_M C_t K_F \phi_s \quad \text{[LRFD]}$$

6. ADJUSTMENT FACTORS

Adjustment Factors Applicable to Sawn Lumber and Glued Laminated Members

Format Conversion Factor, K_F (LRFD Method)

The format conversion factor converts ASD reference design values to LRFD reference resistances. Values of K_F are specified in NDS Table 2.3.5 and are given in Table 6.2.

Table 6.2 *Format Conversion Factor, K_F*

property	F_b	F_t	F_v	$F_{c\perp}$	F_c	E	E_{min}
format conversion factor, K_F*	2.54	2.70	2.88	1.67	2.40	1.00	1.76

*For connections (all design) values, $K_F = 3.32$.

Resistance Factor, ϕ (LRFD Method)

The resistance factor converts nominal values to design values. Values of ϕ are specified in NDS Table 2.3.6 and are given in Table 6.3.

Table 6.3 *Resistance Factor, ϕ*

property	F_b	F_t	F_v	$F_{c\perp}$	F_c	E	E_{min}
resistance factor, ϕ*	0.85	0.80	0.75	0.90	0.90	1.0	0.85

*For connections (all design) values, $\phi = 0.65$.

Time-Effect Factor, λ (LRFD Method)

The time-effect factor provides higher allowable stresses for short-term transient loads. Values of λ are specified in NDS Table N3.3 for the load combinations indicated and are given in Table 6.4.

The time-effect factor is applicable to bending, tension, shear, and compression parallel to the grain values.

Wood

Table 6.4 Time-Effect Factor, λ

load combination	λ
$1.4D$	0.6
$1.2D + 1.6(L + H) + 0.5$ $(L_r \text{ or } S \text{ or } R)$	0.7 when L is from storage
	0.8 when L is from occupancy
	1.25 when L is from impact
$1.2D + 1.6(L_r \text{ or } S \text{ or } R) + 1.6H +$ $(f_1L \text{ or } 0.5W)$	0.8
$1.2D + 1.0W + f_1L + 1.6H + 0.5$ $(L_r \text{ or } S \text{ or } R)$	1.0
$1.2D + 1.0E + f_1L + 0.2S + 1.6H$	1.0
$0.9D + 1.0W + 1.6H$	1.0
$0.9D + 1.0E + 1.6H$	1.0

Note:
Where the effect of H resists the primary variable load effect, a load factor of 0.9 must be included with H where H is permanent. H must be set to zero for all other conditions.
Replace f_1L with $1.0L$ for garages, places of public assembly, and areas where $L > 100$ lbf/ft^2. Use $f_1L = 0.5L$ for all other live loads.
Replace $0.2S$ with $0.7S$ for roof configurations that do not shed snow.

Load Duration Factor, C_D (ASD Method)

The load duration factor, C_D, is applicable to all reference design values with the exception of compression perpendicular to the grain and modulus of elasticity. Values of the load duration factor are given in Table 6.5.

Table 6.5 Load Duration Factors

design load	C_D
dead load	0.90
occupancy live load	1.00
snow load	1.15
construction load	1.25
wind or earthquake load	1.60
impact load	2.00

Load duration factors > 1.6 may not be used in the design of connections or members pressure treated with preservatives or fire retardant chemicals. In a combination of loads, the load duration factor for the shortest duration load applies for that combination.

Wet Service Factor, C_M

When the moisture content of sawn lumber exceeds 19%, the adjustment factors given in NDS Supp. Table 4A and Table 4B are applicable to visually graded dimension lumber; in NDS Supp. Table 4C to machine-graded dimension lumber; in NDS Supp. Table 4D to visually graded timbers; in NDS Supp. Table 4E to decking; and in NDS Supp. Table 4F to non-North

American visually graded dimension lumber. When the moisture content of glued laminated members exceeds 16%, the adjustment factors given in NDS Supp. Table 5A through Table 5D are applicable.

The applicability of the wet service factor to the reference design values for sawn lumber and glued laminated timber is summarized in Table 6.6.

Table 6.6 Wet Service Factor, C_M

classification	F_b	F_t	F_v	$F_{c\perp}$	F_c	E	E_{min}
sawn lumber $<$ $5 \times$ member	0.85[a]	1.00	0.97	0.67	0.8[b]	0.90	0.90
sawn lumber $\geq$ $5 \times$ member	1.00	1.00	1.00	0.67	0.91	1.00	1.00
decking	0.85[a]	–	–	0.67	–	0.90	0.90
glued laminated member	0.80	0.80	0.875	0.53	0.73	0.833	0.833

[a]When $F_bC_F \leq 1150$ lbf/in^2, $C_M = 1.00$.
[b]When $F_cC_F \leq 750$ lbf/in^2, $C_M = 1.00$.

For sawn lumber, C_M is applicable when the moisture content exceeds 19%. For glued laminated members, C_M is applicable when the moisture content exceeds 16%.

Example 6.1

A 2×10 visually graded, select structural, southern pine, sawn lumber member's moisture content exceeds 19%. The governing load combination is the sum of the dead load, the live load, and the wind load. Determine the allowable shear capacity (ASD) and the factored shear capacity (LRFD) of the member.

Solution

The reference design value for shear stress from NDS Supp. Table 4B is

$$F_v = 175 \text{ lbf/in}^2$$

The applicable wet service factor for shear stress from Table 6.6 is

$$C_M = 0.97$$

The temperature factor and the incising factor are

$$C_t = 1.0$$
$$C_i = 1.0$$

ASD Method

The applicable load duration factor for a load combination including the wind load from Table 6.5 is

$$C_D = 1.60$$

The adjusted allowable design value in shear is

$$F_v' = F_v C_i C_D C_M C_t$$
$$= \left(175 \ \frac{\text{lbf}}{\text{in}^2}\right)(1.0)(1.60)(0.97)(1.0)$$
$$= 271.60 \ \text{lbf/in}^2$$

By rearranging NDS Eq. 3.4-2, the allowable shear capacity of the member is

$$V = \frac{2bdF_v'}{3} = \frac{(2)(1.5 \ \text{in})(9.25 \ \text{in})\left(271.60 \ \dfrac{\text{lbf}}{\text{in}^2}\right)}{3}$$
$$= 2512 \ \text{lbf}$$

LRFD Method

From Table 6.4, the applicable time effect factor for IBC Eq. 16-4 is

$$\lambda = 1.0$$

From Table 6.3, the applicable resistance factor for shear is

$$\phi_v = 0.75$$

From Table 6.2, the applicable format conversion factor for shear is

$$K_F = 2.88$$

The adjusted factored design value in shear is

$$F_v' = F_v C_i C_M C_t K_F \lambda \phi_v$$
$$= \left(175 \ \frac{\text{lbf}}{\text{in}^2}\right)(1.0)(0.97)(1.0)(2.88)(1.0)(0.75)$$
$$= 366.66 \ \text{lbf/in}^2$$

By rearranging NDS Eq. 3.4-2, the factored shear capacity of the member is

$$V = \frac{2bdF_v'}{3} = \frac{(2)(1.5 \ \text{in})(9.25 \ \text{in})\left(366.66 \ \dfrac{\text{lbf}}{\text{in}^2}\right)}{3}$$
$$= 3392 \ \text{lbf}$$

Bearing Area Factor, C_b

For bearings less than 6 in long and not less than 3 in from the end of a member, the reference design values for compression perpendicular to the grain are modified by the adjustment factor C_b. This is specified in NDS Sec. 3.10.4 as

$$C_b = \frac{l_b + 0.375}{l_b} \qquad \text{[NDS 3.10-2]}$$

For round bearing areas, the bearing length, l_b, equals the diameter.

Beam Stability Factor, C_L

The beam stability factor is applicable to the reference bending design value for sawn lumber and glued laminated members. For glued laminated members, C_L is not applied simultaneously with the volume factor C_V, and the lesser of these two factors is applicable. The beam stability factor is given by NDS Sec. 3.3.3 as

$$C_L = \frac{1.0 + F}{1.9} - \sqrt{\left(\frac{1.0 + F}{1.9}\right)^2 - \frac{F}{0.95}} \qquad \text{[NDS 3.3-6]}$$

The variables are defined as

$$F = \frac{F_{bE}}{F_b^*}$$

$F_b^* =$ reference bending design value multiplied by all applicable adjustment factors except C_{fu}, C_V (when $C_V \leq 1$), and C_L

$\qquad = F_b C_D C_M C_t C_i C_F C_r C_c \quad$ [ASD]

$\qquad = F_b C_M C_i C_t C_F C_r C_c K_F \lambda \phi_b \quad$ [LRFD]

$$\left[\begin{array}{l} C_F \text{ applies to visually graded sawn} \\ \text{lumber only}, C_r \text{ applies to dimension} \\ \text{lumber only}, C_c \text{ applies to curved glued} \\ \text{laminated members only, and } C_i \\ \text{applies to sawn lumber only.} \end{array} \right]$$

$F_{bE} =$ critical buckling design value

$$= \frac{1.20 E'_{\min}}{R_B^2}$$

Wood

E'_{min} = adjusted modulus of elasticity for stability calculations

$\quad = E_{min}C_MC_tC_iC_T$ [ASD]

$\quad = E_{min}C_MC_TC_tC_iK_F\phi_s$ [LRFD]

$\quad\quad$ [C_i and C_T apply to sawn lumber only.]

R_B = slenderness ratio

$\quad = \sqrt{\dfrac{l_e d}{b^2}} \leq 50$ [NDS 3.3-5]

As specified in NDS Sec. 3.3.3, $C_L = 1$ when the depth of the beam does not exceed its breadth, or when continuous lateral restraint is provided to the compression edge of a beam with the ends restrained against rotation.

The effective span length, l_e, is determined in accordance with NDS Table 3.3.3. The value of l_e depends on the loading configuration and the distance between lateral restraints, l_u. Typical values for l_e are given in Fig. 6.1.

In accordance with NDS Sec. 4.4.1, $C_L = 1.0$ when, based on nominal dimensions, any of the following conditions are met,

- $d/b \leq 2$

- $2 < d/b \leq 4$ and full depth bracing is provided at the ends of the member

- $4 < d/b \leq 5$ and the compression edge is continuously restrained

- $5 < d/b \leq 6$ and the compression edge is continuously restrained with full depth bracing provided at a maximum of 8 ft centers

- $6 < d/b \leq 7$ and both edges are continuously restrained

Column Stability Factor, C_P

The column stability factor is applicable to the reference compression design values parallel to the grain, and it is specified by NDS Sec. 3.7.1 as

$$C_P = \frac{1.0 + F'}{2c} - \sqrt{\left(\frac{1.0 + F'}{2c}\right)^2 - \frac{F'}{c}}$$

[NDS 3.7-1]

The variables are defined as

$F' = \dfrac{F_{cE}}{F_c^*}$

F_c^* = reference compression design value multiplied by all applicable adjustment factors except C_P

$\quad = F_c C_D C_M C_t C_i C_F$ [ASD]

$\quad = F_c C_M C_t C_i C_F K_F \lambda \phi_c$ [LRFD]

$\quad\quad \left[\begin{array}{c} C_F \text{ applies to visually graded sawn} \\ \text{lumber only, and } C_i \text{ applies to sawn} \\ \text{lumber only} \end{array}\right]$

F_c = reference compression design value parallel to grain

F_{cE} = critical buckling design value

$\quad = \dfrac{0.822 E'_{min}}{\left(\dfrac{l_e}{d}\right)^2}$

E'_{min} = adjusted modulus of elasticity for stability calculations

$\quad = E_{min}C_MC_TC_iC_t$ [ASD]

$\quad = E_{min}C_MC_TC_tC_iK_F\phi_s$ [LRFD]

$\quad\quad$ [C_i and C_T apply to sawn lumber only.]

E_{min} = reference modulus of elasticity

c = column parameter

$\quad = 0.8$ [for sawn lumber]

$\quad = 0.9$ [for glued laminated timber, structural composite lumber, and cross-laminated timber]

$\quad = 0.85$ [for round timber poles and piles]

Temperature Factor, C_t

The temperature factor is applicable to all reference design values for members exposed to sustained temperatures up to 150°F and is specified by NDS Sec. 2.3.3.

Values for the temperature factor are given in Table 6.7.

Example 6.2

A glued laminated beam of combination 24F-V10 western species with 1.5 in thick laminations has a width of 6¾ in and a depth of 30 in. The beam has a moisture content exceeding 16% and is subjected to sustained temperatures between 100°F and 125°F. The governing load combination is dead load plus live load plus wind load. Determine the allowable design value for the modulus of elasticity, $E_{x\,app}$, about the x-x axis.

Figure 6.1 *Typical Values of Effective Length, l_e*

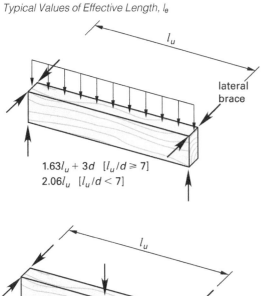

$1.63l_u + 3d \quad [l_u/d \geq 7]$
$2.06l_u \quad [l_u/d < 7]$

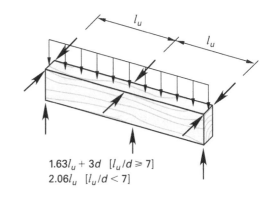

$1.63l_u + 3d \quad [l_u/d \geq 7]$
$2.06l_u \quad [l_u/d < 7]$

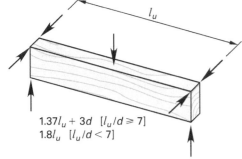

$1.37l_u + 3d \quad [l_u/d \geq 7]$
$1.8l_u \quad [l_u/d < 7]$

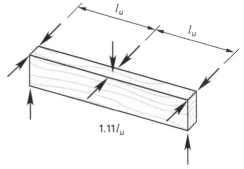

$1.11l_u$

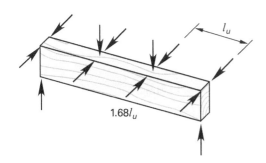

$1.68l_u$

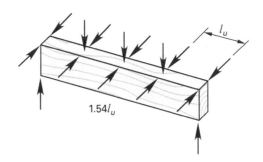

$1.54l_u$

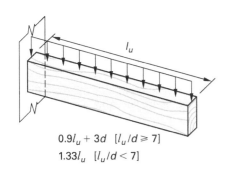

$0.9l_u + 3d \quad [l_u/d \geq 7]$
$1.33l_u \quad [l_u/d < 7]$

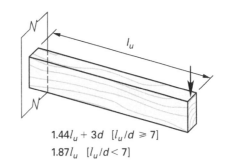

$1.44l_u + 3d \quad [l_u/d \geq 7]$
$1.87l_u \quad [l_u/d < 7]$

Wood

Table 6.7 *Temperature Factor, C_t*

reference design values	in-service moisture conditions	temperature factor, C_t		
		$T \leq 100°F$	$100°F < T \leq 125°F$	$125°F < T \leq 150°F$
F_t, E, $E_{\min}$	wet or dry	1.0	0.9	0.9
F_b, F_v, $F_{c\perp}$, F_c	dry	1.0	0.8	0.7
	wet	1.0	0.7	0.5

Solution

The reference design value for the modulus of elasticity about the x-x axis is tabulated in NDS Supp. Table 5A as

$$E_{x\,\text{app}} = 1.8 \times 10^6 \text{ lbf/in}^2$$

The applicable adjustment factors for the modulus of elasticity are C_M, for wet service conditions, and C_t, for elevated temperature in wet conditions.

$$C_M = 0.833 \quad \text{[NDS Supp. Table 5A]}$$
$$C_t = 0.9 \quad \text{[NDS Table 2.3.3]}$$

The adjusted modulus of elasticity is

$$
\begin{aligned}
E'_{x\,\text{app}} &= E_x C_M C_t \\
&= \left(1.8 \times 10^6 \frac{\text{lbf}}{\text{in}^2}\right)(0.833)(0.9) \\
&= 1.35 \times 10^6 \text{ lbf/in}^2
\end{aligned}
$$

Flat Use Factor, C_{fu}

When wood members are loaded flatwise, flat use adjustment factors are applied to the bending stress. The adjustment factors given in NDS Supp. Table 4A and Table 4B are applicable to visually graded dimension lumber, in NDS Supp. Table 4C to machine-graded dimension lumber, in NDS Supp. Table 4F to non-North American visually graded dimension lumber, and in NDS Supp. Table 5A through Table 5D to glued laminated members.

When members classified as beams and stringers are loaded on the wide face, the flat use adjustment factors given in NDS Supp. Table 4D are applied to the bending stress and the modulus of elasticity (E or $E_{\min}$).

The reference design values for visually graded decking given in NDS Supp. Table 4E already incorporate the applicable flat use factor.

Adjustment Factors Applicable to Sawn Lumber

Size Factor, C_F

The size factor is applicable to visually graded dimension lumber, visually graded timbers, and visually graded decking. It is not applied to mechanically graded dimension lumber. The reference design values for bending, tension, and compression are multiplied by the size factor, C_F, to give the appropriate design values. For visually graded dimension lumber, 2 in to 4 in thick, values of the size factor are given in NDS Supp. Table 4A, Table 4B, and Table 4F. For visually graded decking, values of the size factor are given in NDS Supp. Table 4E. For visually graded timbers exceeding 12 in depth and 5 in thickness, the size factor is specified in NDS Sec. 4.3.6 as

$$C_F = \left(\frac{12}{d}\right)^{1/9} \quad \text{[NDS 4.3-1]}$$
$$\leq 1.0$$

Repetitive Member Factor, C_r

The repetitive member factor is applicable to visually graded dimension lumber and mechanically graded dimension lumber. It is not applied to visually graded timbers. The design values for visually graded decking in NDS Supp. Table 4E already incorporate the applicable repetitive member factor. The reference design value for bending is multiplied by the repetitive member factor, C_r, when three or more sawn lumber elements, not more than 4 in thick and spaced not more than 24 in apart, are joined by a transverse load distributing element. The value of the repetitive member factor is given in NDS Supp. Table 4A to Table 4C, and Table 4F, and in NDS Sec. 4.3.9, as

$$C_r = 1.15$$

Incising Factor, C_i

Values of the incising factor, C_i, for a prescribed incising pattern are provided in NDS Sec. 4.3.8. These values are applicable to all reference design values for all sawn lumber. The prescribed incising pattern consists of incisions made parallel to the grain at a maximum depth of 0.4 in, a maximum length of $\frac{3}{8}$ in, and at a density of 1100/ft^2.

Wood

Values of the incising factor are given in Table 6.8.

Table 6.8 Incising Factor, C_i

design value	C_i
E, $E_{\min}$	0.95
F_b, F_t, F_c, F_v	0.80
$F_{c\perp}$	1.00

As an alternative, incising factors may be obtained from the company providing the incising.

Buckling Stiffness Factor, C_T

The buckling stiffness factor applies to 2×4 and smaller compression chords in trusses with $\frac{3}{8}$ in or thicker wood structural panel sheathing nailed to the narrow face of the chord. Multiply $E_{\min}$ by the buckling stiffness factor to account for the increased chord stiffness. The effective chord length is represented by l_e.

When $l_e < 96$ in, NDS Eq. 4.4-1 defines the buckling stiffness factor as

$$C_T = 1 + \frac{K_M l_e}{K_T E}$$

The value of K_M is 2300 for wood seasoned to 19% moisture content or less at the time of plywood attachment. The value of K_M is 1200 for unseasoned or partially seasoned wood at the time of plywood attachment. Values of K_T are

- $1 - 1.645(\text{COV}_E)$
- 0.59 for visually graded lumber
- 0.75 for machine evaluated lumber
- 0.82 for products with $\text{COV}_E \leq 0.11$

When $l_e > 96$ in, C_T is calculated based on $l_e = 96$ in.

COV_E is the coefficient of variation for the modulus of elasticity.

Example 6.3

Selected Douglas fir-larch, visually graded 3×6 decking is incised with the prescribed incising pattern. Normal service live load, temperature, and dry service conditions are applicable. Determine the adjusted allowable bending stress and the adjusted factored bending stress.

Solution

The reference design value for bending stress tabulated in NDS Supp. Table 4E, which includes the repetitive member and flat use adjustment factors, is

$$F_b = 2000 \text{ lbf/in}^2$$

From Table 6.8, the applicable incising factor for bending is

$$C_i = 0.80$$

From NDS Supp. Table 4E, the applicable size factor for a 3 in thick member is

$$C_F = 1.04$$

From Table 6.7, the applicable temperature factor for normal temperature conditions is

$$C_t = 1.0$$

From NDS Supp. Table 4E, the applicable wet service factor for decking where moisture content does not exceed 19% is

$$C_M = 1.0$$

In accordance with NDS Sec. 3.3.3, the beam stability factor is

$$C_L = 1.0$$

ASD Method

For normal occupancy live load, the load duration factor given by Table 6.5 is

$$C_D = 1.00$$

The adjusted allowable bending stress is

$$\begin{aligned} F_b' &= F_b C_F C_r C_i C_D C_M C_t C_{fu} C_L \\ &= \left(2000 \frac{\text{lbf}}{\text{in}^2}\right)(1.04)(1.0)(0.8)(1.00) \\ &\quad \times (1.0)(1.0)(1.0)(1.0) \\ &= 1664 \text{ lbf/in}^2 \end{aligned}$$

LRFD Method

From Table 6.4, the applicable time effect factor for dead load plus occupancy live load is

$$\lambda = 0.8$$

From Table 6.3, the applicable resistance factor for bending is

$$\phi_b = 0.85$$

From Table 6.2, the applicable format conversion factor for bending is

$$K_F = 2.54$$

The adjusted factored bending stress is

$$F_b' = F_b C_F C_r C_i C_M C_t C_{fu} C_L K_F \lambda \phi_b$$

$$= \left(2000 \ \frac{\text{lbf}}{\text{in}^2}\right)(1.04)(1.0)(0.8)(1.0)$$

$$\times (1.0)(1.0)(1.0)(2.54)(0.8)(0.85)$$

$$= 2874 \ \text{lbf/in}^2$$

Adjustment Factors Applicable to Glued Laminated Members Only

Volume Factor, C_V

The volume factor is applicable to the reference design value for bending and is not applied simultaneously with the beam stability factor, C_L; the lesser of these two factors is applicable. The volume factor is dimensionless and defined in NDS Sec. 5.3.6 as

$$C_V = \left(\frac{1291.5}{b_{\text{in}} d_{\text{in}} L_{\text{ft}}}\right)^{1/x} \qquad \text{[NDS 5.3-1]}$$

The variables are defined as

L = length of beam between points of zero moment, in feet

b = beam width, in inches

d = beam depth, in inches

x = 20 [for southern pine]

  = 10 [for all other species]

Pressure-Preservative Treatment

Reference design values apply to glued laminated members treated by an approved process and preservative.[4] Load duration factors greater than 1.6 may not apply to members treated with a water-based preservative.

Curvature Factor, C_c

To account for residual stresses in curved, glued laminated members, the curvature factor is specified in NDS Sec. 5.3.8 as

$$C_c = 1 - 2000\left(\frac{t}{R}\right)^2 \qquad \text{[NDS 5.3-3]}$$

The variables are defined as

t = thickness of lamination, in inches

R = radius of curvature of inside face of lamination in inches

$\dfrac{t}{R} \le \dfrac{1}{100}$ [for hardwoods and southern pine]

$\quad \le \dfrac{1}{125}$ [for other softwoods]

The curvature factor does not apply to design values in the straight portion of a member, regardless of curvature elsewhere.

Stress Interaction Factor, C_I

NDS Sec. 5.3.9 provides values for the stress interaction factor for tapered bending members. The stress interaction factor, C_I, is applied to the reference bending design value, F_b. For members tapered on the compression face, the stress interaction factor does not apply simultaneously with the volume factor, and the lesser factor applies. For members tapered on the tension face, the stress interaction factor does not apply simultaneously with the beam stability factor, and the lesser factor applies.

Shear Reduction Factor, C_{vr}

The shear reduction factor, $C_{vr} = 0.72$, is applied to the reference shear design value, F_v, where members are nonprismatic, subject to impact or repetitive cyclic loading, at notches, or at connections.

Example 6.4

A curved, glued laminated beam of stress class 24F-V10 western species with 1.5 in thick laminations has a radius of curvature of 30 ft, a width of 6 ¾ in, and a depth of 30 in. The beam has continuous lateral support, a moisture content exceeding 16%, and is subjected to sustained temperatures between 100°F and 125°F. The governing loading combination is dead plus live load. The span is 40 ft. The beam is simply supported, and all loading is uniformly distributed. Determine the adjusted allowable bending stress and the adjusted factored bending stress.

Solution

The reference design value for bending is tabulated in NDS Supp. Table 5A and is

$$F_b = 2400 \ \text{lbf/in}^2$$

The applicable adjustment factors for bending stress are as follows.

C_M = wet service factor

  = 0.8 [Table 6.6]

C_t = temperature factor for wet conditions

  = 0.7 [Table 6.7]

C_V = volume factor

$$= \left(\frac{1291.5}{b_{\text{in}} d_{\text{in}} L_{\text{ft}}}\right)^{1/x} \qquad \text{[NDS 5.3-1]}$$

$$= \left(\frac{1291.5 \ \text{in}^2\text{-ft}}{(6.75 \ \text{in})(30 \ \text{in})(40 \ \text{ft})}\right)^{1/10}$$

  = 0.832

C_L = stability factor

  = 1.0 [continuous lateral support]

  > volume factor

The volume factor governs.

$$C_c = \text{curvature factor}$$

$$= 1 - 2000\left(\frac{t}{R}\right)^2 \qquad \text{[NDS Sec. 5.3.3]}$$

$$= 1 - (2000)\left(\frac{1.5 \text{ in}}{(30 \text{ ft})\left(12 \; \frac{\text{in}}{\text{ft}}\right)}\right)^2$$

$$= 0.965$$

$$\frac{t}{R} = \frac{1.5}{360}$$

$$< \frac{1}{125} \qquad \text{[satisfactory]}$$

The beam is not tapered and the stress interaction factor is

$$C_I = 1.0$$

The flat use factor is not applicable and

$$C_{fu} = 1.0$$

ASD Method

For normal occupancy live load, the load duration factor given by Table 6.5 is

$$C_D = 1.00$$

The adjusted allowable bending stress is

$$F_b' = F_b C_c C_I C_D C_M C_t C_{fu}(C_L \text{ or } C_V)$$

$$= \left(2400 \; \frac{\text{lbf}}{\text{in}^2}\right)(0.965)(1.0)(1.00)(0.8)(0.7)(1.0)(0.832)$$

$$= 1079 \text{ lbf/in}^2$$

LRFD Method

From Table 6.4, the applicable time effect factor for dead load plus occupancy live load is

$$\lambda = 0.8$$

From Table 6.3, the applicable resistance factor for bending is

$$\phi_b = 0.85$$

From Table 6.2, the applicable format conversion factor for bending is

$$K_F = 2.54$$

The adjusted factored bending stress is

$$F_b' = F_b C_c C_I C_M C_t C_{fu}(C_L \text{ or } C_V)K_F\lambda\phi_b$$

$$= \left(2400 \; \frac{\text{lbf}}{\text{in}^2}\right)(0.965)(1.0)(0.8)(0.7)$$

$$\times (1.0)(0.832)(2.54)(0.8)(0.85)$$

$$= 1864 \text{ lbf/in}^2$$

Adjustment Factors Applicable to Cross-Laminated Timber

Reference Design Values

Reference design values for cross-laminated timber are provided by the timber manufacturer or a code evaluation report. The applicability of adjustment factors to the reference design values for cross-laminated timber is summarized in Table 6.9. Cross-laminated timber cannot be used with a moisture content exceeding 16% unless specifically permitted by the manufacturer.

Table 6.9 *Applicability of Adjustment Factors*

adjustment factor	F_b	F_t	F_y	$F_{c\perp}$	F_c	F_S	EI	$(EI)_{\min}$
load duration, C_D	√	√	√	-	√	-	-	-
wet service, C_M	√	√	√	√	√	√	√	√
bearing area, C_b	-	-	-	√	-	-	-	-
beam stability, C_L	√	-	-	-	-	-	-	-
column stability, C_P	-	-	-	-	√	-	-	-
temperature, C_t	√	√	√	√	√	√	√	√
format conversion, K_F	2.54	2.70	2.88	1.67	2.40	2.00	-	1.76
resistance, ϕ	0.85	0.80	0.75	0.90	0.90	0.75	-	0.85
time effect, λ	λ	λ	λ	-	λ	-	-	-

Load Duration Factor, C_D (ASD Method)

The values for the load duration factor, C_D, given in Table 6.5 are applicable to all reference design values except EI, $(EI)_{\min}$, F_S, and $F_{c\perp}$.

Wet Service Factor, C_M

The wet service factor, C_M, is equal to 1.0 in dry service conditions, which NDS Sec. 10.1.5 defines as having a moisture content up to 16%. When this moisture content is exceeded, the value of C_M is taken in accordance with information provided by the cross-laminated timber manufacturer.

Bearing Area Factor, C_b

Reference design values of compression perpendicular to the grain, $F_{c\perp}$, may be multiplied by the bearing area factor, C_b, as specified in NDS Sec. 3.10.4.

Beam Stability Factor, C_L

The beam stability factor, C_L, is applicable to the reference bending design value of cross-laminated timber. The beam stability factor is specified in NDS Sec. 3.3.3.

Column Stability Factor, C_P

The column stability factor, C_P, is applicable to the reference design value of compression parallel to the grain, F_c. The column stability factor is specified in NDS Sec. 3.7.1.

Temperature Factor, C_t

The temperature factor is applicable to reference design values of cross-laminated timber elements exposed to sustained temperatures up to 150°F. The temperature factor is specified by NDS Sec. 2.3.3 and is shown in Table 6.7.

Pressure-Preservative Treatment

Reference design values apply to cross-laminated timber treated by an approved process and preservative.[4] Load duration factors greater than 1.6 may not apply to members treated with water-based preservative.

Format Conversion Factor, K_F (LRFD Method)

The format conversion factor, K_F, converts ASD reference design values to LRFD values. Values of the format conversion factor are given in Table 6.9.

Resistance Factor, ϕ (LRFD Method)

The resistance factor, ϕ, converts nominal values to design values. Values of the resistance factor are given in Table 6.3.

Time Effect Factor, λ (LRFD Method)

Values of the time effect factor, λ, are provided in Table 6.4 for the load combinations indicated.

7. DESIGN FOR FLEXURE

General Requirements Applicable to Sawn Lumber, Glued Laminated Members, and Cross-Laminated Timber

For all flexural members, in accordance with NDS Sec. 3.2.1, the beam span is taken as the clear span plus one-half the required bearing length at each end.

When the depth of a beam does not exceed its breadth, or when continuous lateral restraint is provided to the compression edge of a beam with the ends restrained against rotation, the beam stability factor, C_L, is 1.0. For other situations, the value of C_L is calculated in accordance with NDS Sec. 3.3.3, and the effective span length, l_e, is determined in accordance with NDS Table 3.3.3. The value of l_e depends on the loading configuration and the distance between lateral restraints, l_u. Typical values for l_e are given in Fig. 6.1.

Requirements Applicable to Sawn Lumber

For sawn lumber, both the stability factor, C_L, and the size factor, C_F, must be applied concurrently.

Example 6.5

A select structural Douglas fir-larch 4×12 beam is simply supported over a span of 20 ft. The governing load combination is a uniformly distributed dead plus live load, and the beam is laterally braced at midspan and at the ends of the beam.

The beam is incised with the prescribed pattern specified in NDS Sec. 4.3.8. Determine the adjusted allowable bending stress and the adjusted factored bending stress.

Solution

The reference design values for bending and modulus of elasticity for beam stability calculations are tabulated in NDS Supp. Table 4A and are

$$F_b = 1500 \text{ lbf/in}^2$$
$$E_{\min} = 0.69 \times 10^6 \text{ lbf/in}^2$$
$$C_M = 1.0, \ C_t = 1.0, \ C_r = 1.0, \ C_{fu} = 1.0, \ C_T = 1.0$$

The applicable incising factor for *modulus of elasticity* is obtained from NDS Sec. 4.3.8 and is

$$C_i = 0.95$$

The applicable incising factor for flexure is obtained from NDS Sec. 4.3.8 and is

$$C_i = 0.80$$

The applicable size factor for flexure is obtained from NDS Supp. Table 4A and is

$$C_F = 1.1$$

The distance between lateral restraints is

$$l_u = \frac{20 \text{ ft}}{2}$$
$$= 10 \text{ ft}$$

$$\frac{l_u}{d} = \frac{(10 \text{ ft})\left(12 \frac{\text{in}}{\text{ft}}\right)}{11.25 \text{ in}}$$
$$= 10.7$$
$$> 7$$

For a uniformly distributed load and a l_u/d ratio > 7, the effective length is obtained from Fig. 6.1 as

$$l_e = 1.63 l_u + 3d$$
$$= (1.63)(10 \text{ ft})\left(12 \frac{\text{in}}{\text{ft}}\right) + (3)(11.25 \text{ in})$$
$$= 229.4 \text{ in}$$

The slenderness ratio is given by NDS Sec. 3.3.3 as

$$R_B = \sqrt{\frac{l_e d}{b^2}}$$
$$= \sqrt{\frac{(229.4 \text{ in})(11.25 \text{ in})}{(3.5 \text{ in})^2}}$$
$$= 14.52$$
$$< 50 \quad \begin{bmatrix} \text{satisfies criteria} \\ \text{of NDS Sec. 3.3.3} \end{bmatrix}$$

ASD Method

The adjusted modulus of elasticity for stability calculations with $C_t = C_T = C_M = 1.0$ is

$$E'_{\min} = E_{\min} C_M C_t C_i C_T$$
$$= \left(0.69 \times 10^6 \frac{\text{lbf}}{\text{in}^2}\right)(1.0)(1.0)(0.95)(1.0)$$
$$= 0.656 \times 10^6 \text{ lbf/in}^2$$

The critical buckling design value is

$$F_{bE} = \frac{1.20 E'_{\min}}{R_B^2}$$
$$= \frac{(1.20)\left(0.656 \times 10^6 \frac{\text{lbf}}{\text{in}^2}\right)}{(14.52)^2}$$
$$= 3734 \text{ lbf/in}^2$$

The load duration factor for occupancy live load is

$$C_D = 1.00$$

The reference flexural design value multiplied by all applicable adjustment factors except C_L and C_{fu} is

$$F_b^* = F_b C_D C_M C_t C_F C_i C_r$$
$$= \left(1500 \frac{\text{lbf}}{\text{in}^2}\right)(1.00)(1.0)(1.0)(1.1)(0.8)(1.0)$$
$$= 1320 \text{ lbf/in}^2$$

$$F = \frac{F_{bE}}{F_b^*}$$
$$= \frac{3734 \frac{\text{lbf}}{\text{in}^2}}{1320 \frac{\text{lbf}}{\text{in}^2}}$$
$$= 2.83$$

The beam stability factor is given by NDS Sec. 3.3.3 as

$$C_L = \frac{1.0 + F}{1.9} - \sqrt{\left(\frac{1.0 + F}{1.9}\right)^2 - \frac{F}{0.95}}$$
$$= 0.974$$

The adjusted allowable flexural stress is

$$F'_b = F_b C_M C_T C_L C_F C_{fu} C_i C_r$$
$$= \left(1500 \frac{\text{lbf}}{\text{in}^2}\right)(1.0)(1.0)(0.974)(1.1)(1.0)(0.80)(1.0)$$
$$= 1286 \text{ lbf/in}^2$$

LRFD Method

From Table 6.4, the time effect factor for dead load plus occupancy live load is

$$\lambda = 0.8$$

The adjusted modulus of elasticity for LRFD stability calculations with $C_t = C_T = 1.0$ is

$$
\begin{aligned}
E'_{min} &= E_{min} C_M C_t C_i C_T K_F \phi \\
&= \left(0.69 \times 10^6 \ \frac{\text{lbf}}{\text{in}^2}\right)(1.0)(1.0)(0.95)(1.0)(1.76)(0.85) \\
&= 0.981 \times 10^6 \ \text{lbf/in}^2
\end{aligned}
$$

The critical buckling LRFD design value is

$$
\begin{aligned}
F_{bE} &= \frac{1.20 E'_{min}}{R_B^2} = \frac{(1.20)\left(0.981 \times 10^6 \ \dfrac{\text{lbf}}{\text{in}^2}\right)}{(14.52)^2} \\
&= 5584 \ \text{lbf/in}^2
\end{aligned}
$$

The reference flexural design value multiplied by all applicable LRFD adjustment factors except C_L and C_{fu} is

$$
\begin{aligned}
F_b^* &= F_b C_M C_t C_F C_i C_r K_F \phi_b \lambda \\
&= \left(1500 \ \frac{\text{lbf}}{\text{in}^2}\right)(1.0)(1.0)(1.1)(0.80)(1.0) \\
&\quad \times (2.54)(0.85)(0.8) \\
&= 2280 \ \text{lbf/in}^2
\end{aligned}
$$

$$
\begin{aligned}
F &= \frac{F_{bE}}{F_b^*} = \frac{5584 \ \dfrac{\text{lbf}}{\text{in}^2}}{2280 \ \dfrac{\text{lbf}}{\text{in}^2}} \\
&= 2.45
\end{aligned}
$$

The LRFD beam stability factor given by NDS Sec. 3.3.3 is

$$
\begin{aligned}
C_L &= \frac{1.0 + F}{1.9} - \sqrt{\left(\frac{1.0 + F}{1.9}\right)^2 - \frac{F}{0.95}} \\
&= \frac{1.0 + 2.45}{1.9} - \sqrt{\left(\frac{1.0 + 2.45}{1.9}\right)^2 - \frac{2.45}{0.95}} \\
&= 0.968
\end{aligned}
$$

The adjusted factored bending stress is

$$
\begin{aligned}
F_b' &= F_b C_M C_t C_L C_F C_{fu} C_i C_r K_F \phi_b \lambda \\
&= \left(1500 \ \frac{\text{lbf}}{\text{in}^2}\right)(1.0)(1.0)(0.968)(1.1)(1.0)(0.8) \\
&\quad \times (1.0)(2.54)(0.85)(0.80) \\
&= 2207 \ \text{lbf/in}^2
\end{aligned}
$$

Requirements Applicable to Glued Laminated Members Only

For glued laminated members, both the stability factor, C_L, and the volume factor, C_V, must be determined. Only the lesser of these two factors is applicable in determining the allowable design value in bending.

Example 6.6

A glued laminated $6\frac{3}{4} \times 30$ beam of stress class 24F-1.7E western species is simply supported over a span of 40 ft. The governing load combination is a uniformly distributed dead plus live load, and the beam is laterally braced at midspan. Determine the adjusted allowable bending stress and the adjusted factored bending stress.

Solution

$$
\begin{aligned}
C_M &= 1.0, \ \ C_t = 1.0, \\
C_c &= 1.0, \ \ C_{fu} = 1.0, \ \ C_I = 1.0
\end{aligned}
$$

From NDS Supp. Table 5A,

$$
\begin{aligned}
F_b &= 2400 \ \text{lbf/in}^2 \\
E_{y(min)} &= 0.69 \times 10^6 \ \text{lbf/in}^2
\end{aligned}
$$

From Ex. 6.4, the volume factor is

$$
\begin{aligned}
C_V &= 0.832 \\
&< 1.0
\end{aligned}
$$

The distance between lateral restraints is

$$
\begin{aligned}
l_u &= \frac{40 \ \text{ft}}{2} \\
&= 20 \ \text{ft} \\
\frac{l_u}{d} &= \frac{(20 \ \text{ft})\left(12 \ \dfrac{\text{in}}{\text{ft}}\right)}{30 \ \text{in}} \\
&= 8.0 \\
&> 7
\end{aligned}
$$

For a uniformly distributed load and an l_u/d ratio > 7, the effective length is obtained from Fig. 6.1 as

$$
\begin{aligned}
l_e &= 1.63 l_u + 3d \\
&= (1.63)(20 \ \text{ft})\left(12 \ \frac{\text{in}}{\text{ft}}\right) + (3)(30 \ \text{in}) \\
&= 481 \ \text{in}
\end{aligned}
$$

The slenderness ratio is given by NDS Eq. 3.3-5.

$$R_B = \sqrt{\frac{l_e d}{b^2}}$$
$$= \sqrt{\frac{(481 \text{ in})(30 \text{ in})}{(6.75 \text{ in})^2}}$$
$$= 17.80$$
$$< 50 \begin{bmatrix} \text{satisfies criteria} \\ \text{of NDS Sec. 3.3.3} \end{bmatrix}$$

ASD Method

The adjusted modulus of elasticity for stability calculations with $C_M = C_t = 1.0$ is

$$E'_{\min} = E_{y(\min)} C_M C_t$$
$$= \left(0.69 \times 10^6 \frac{\text{lbf}}{\text{in}^2}\right)(1.0)(1.0)$$
$$= 0.69 \times 10^6 \text{ lbf/in}^2$$

The critical buckling design value is

$$F_{bE} = \frac{1.20 E'_{\min}}{R_B^2}$$
$$= \frac{(1.20)\left(0.69 \times 10^6 \frac{\text{lbf}}{\text{in}^2}\right)}{(17.80)^2}$$
$$= 2613 \text{ lbf/in}^2$$

The load duration factor for occupancy live load is

$$C_D = 1.00$$

The reference flexural design value multiplied by all applicable adjustment factors except C_L, C_V (when $C_V \leq 1$), and C_{fu} is

$$F_b^* = F_b C_M C_t C_c C_I$$
$$= \left(2400 \frac{\text{lbf}}{\text{in}^2}\right)(1.0)(1.0)(1.0)(1.0)(1.0)$$
$$= 2400 \text{ lbf/in}^2$$

$$F = \frac{F_{bE}}{F_b^*}$$
$$= \frac{2613 \frac{\text{lbf}}{\text{in}^2}}{2400 \frac{\text{lbf}}{\text{in}^2}}$$
$$= 1.09$$

The beam stability factor is given by NDS Sec. 3.3.3 as

$$C_L = \frac{1.0 + F}{1.9} - \sqrt{\left(\frac{1.0 + F}{1.9}\right)^2 - \frac{F}{0.95}}$$
$$= \frac{1.0 + 1.09}{1.9} - \sqrt{\left(\frac{1.0 + 1.09}{1.9}\right)^2 - \frac{1.09}{0.95}}$$
$$= 0.85$$
$$> \text{volume factor derived in Ex. 6.4}$$

The volume factor governs. The adjusted allowable flexural stress is

$$F_b' = F_b C_D C_M C_t C_{fu} C_c C_I C_V$$
$$= \left(2400 \frac{\text{lbf}}{\text{in}^2}\right)(1.00)(1.0)(1.0)(1.0)(1.0)(1.0)$$
$$\times (0.832)$$
$$= 1997 \text{ lbf/in}^2$$

LRFD Method

From Table 6.4, the time effect factor for dead load plus occupancy live load is

$$\lambda = 0.8$$

The adjusted modulus of elasticity for LRFD stability calculations with $C_t = C_M = 1.0$ is

$$E'_{\min} = E_{\min} C_M C_t K_F \phi$$
$$= \left(0.69 \times 10^6 \frac{\text{lbf}}{\text{in}^2}\right)(1.0)(1.0)(1.76)(0.85)$$
$$= 1.032 \times 10^6 \text{ lbf/in}^2$$

The critical buckling LRFD design value is

$$F_{bE} = \frac{1.20 E'_{\min}}{R_B^2} = \frac{(1.2)\left(1.032 \times 10^6 \frac{\text{lbf}}{\text{in}^2}\right)}{(17.80)^2}$$
$$= 3909 \text{ lbf/in}^2$$

The reference flexural design value multiplied by all applicable LRFD adjustment factors except C_V (when $C_V \leq 1$), C_L, and C_{fu} is

$$F_b^* = F_b C_M C_t C_c C_I K_F \phi_b \lambda$$
$$= \left(2400 \frac{\text{lbf}}{\text{in}^2}\right)(1.0)(1.0)(1.0)(1.0)(2.54)$$
$$\times (0.85)(0.8)$$
$$= 4145 \text{ lbf/in}^2$$

$$F = \frac{F_{bE}}{F_b^*} = \frac{3909 \; \frac{\text{lbf}}{\text{in}^2}}{4145 \; \frac{\text{lbf}}{\text{in}^2}}$$

$$= 0.943$$

The LRFD beam stability factor given by NDS Sec. 3.3.3 is

$$C_L = \frac{1.0 + F}{1.9} - \sqrt{\left(\frac{1.0 + F}{1.9}\right)^2 - \frac{F}{0.95}}$$

$$= \frac{1.0 + 0.943}{1.9} - \sqrt{\left(\frac{1.0 + 0.943}{1.9}\right)^2 - \frac{0.943}{0.95}}$$

$$= 0.792$$

$$< 0.832 \quad \text{[volume factor derived from Ex. 6.4]}$$

The stability factor governs, and the adjusted factored bending stress is

$$F_b' = F_b C_M C_t C_L C_{fu} C_c C_I K_F \phi_b \lambda$$

$$= \left(2400 \; \frac{\text{lbf}}{\text{in}^2}\right)(1.0)(1.0)(0.792)(1.0)(1.0)(1.0)$$

$$\times (2.54)(0.85)(0.8)$$

$$= 3283 \; \text{lbf/in}^2$$

Deflection Calculations for Cross-Laminated Timber

To allow for the creep that occurs under long-term loading, the total deflection of a member is given by NDS Eq. 3.5-1 as

$$\Delta_T = K_{cr}\Delta_{LT} + \Delta_{ST}$$

In this equation, K_{cr} is the time-dependent creep factor and is equal to 2.0 for cross-laminated timber used in dry service conditions. Δ_{LT} is the immediate deflection due to the long-term loads, and Δ_{ST} is the deflection due to the short-term loads.

To account for the effects of shear deformation in a cross-laminated member, an apparent value of the bending stiffness is used, which is given by NDS Eq. 10.4-1 as

$$(EI)_{\text{app}} = \frac{EI_{\text{eff}}}{1 + \dfrac{K_s EI_{\text{eff}}}{GA_{\text{eff}} L^2}}$$

In this equation,

$E =$ reference modulus of elasticity (psi)

$G =$ modulus of rigidity

$\quad = E/16$ [NDS Sec. C3.5.1]

$EI_{\text{eff}} =$ effective bending stiffness of the cross-laminated timber section (lbf-in^2 per foot of panel width)

$K_s =$ shear deformation adjustment factor provided in Table 6.10

$GA_{\text{eff}} =$ effective shear stiffness of the cross-laminated timber section (lbf per foot of panel width)

$L =$ span of the cross-laminated timber (in)

Table 6.10 *Values for Shear Deformation Adjustment Factor, K_s*

loading	end condition	K_s
uniformly distributed	pinned	11.5
	fixed	57.6
line load at midspan	pinned	14.4
	fixed	57.6
line load at quarter points	pinned	10.5
constant moment pinned	pinned	0
uniformly distributed	cantilevered	4.8
line load at free end	cantilevered	3.6
column buckling pinned	pinned	11.8
	fixed	47.4

8. DESIGN FOR SHEAR

Nomenclature

d	depth of unnotched bending member	in
d_e	depth of member, less the distance from the unloaded edge of the member to the nearest edge of the nearest split ring or shear plate connector	in
d_e	depth of member, less the distance from the unloaded edge of the member to the center of the nearest bolt or lag screw	in
d_n	depth of member remaining at a notch	in
e	distance a notch extends past the inner edge of a support	in
f_v	actual shear stress parallel to grain	lbf/in^2
l_n	length of notch	in
V	shear force	lbf
V_r, V_r'	reference and adjusted design shear	lbf
x	distance from beam support face to load	in

General Requirements

The shear stress in a rectangular beam is defined in NDS Sec. 3.4.2 as

$$f_v = \frac{3V}{2bd} \qquad \text{[NDS 3.4-2]}$$

In determining the shear force on the member, uniformly distributed loads applied to the top of the beam within a distance from either support equal to the depth of the beam are ignored, as shown in Fig. 6.2.

Figure 6.2 *Shear Determination in a Beam*

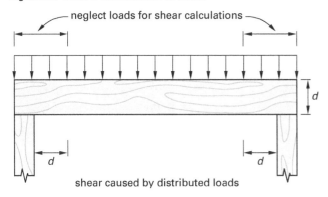

shear caused by distributed loads

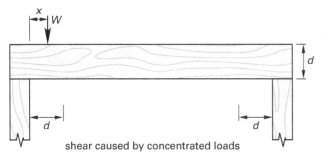

shear caused by concentrated loads

The procedure for a beam with concentrated loads is also illustrated in Fig. 6.2. Concentrated loads within a distance from either support equal to the depth of the beam are multiplied by x/d, where x is the distance from the support to the load, to give an equivalent shear force.

Example 6.7

A glued laminated beam of combination 24F-V10 western species with a width of $6\frac{3}{4}$ in and a depth of 30 in is loaded with the service loads shown. The loading consists of a distributed dead load of 2.0 kips/ft, which includes the self-weight of the beam, plus a concentrated

occupancy live load of 15 kips. Normal dry use conditions of service are applicable. Determine whether the beam is adequate in shear.

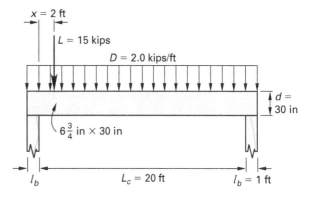

Solution

From NDS Supp. Table 5A, the reference design value for shear stress is

$$F_v = 215 \ \text{lbf/in}^2$$

From the illustration, the clear span is

$$L_c = 20 \ \text{ft}$$

Also from the illustration, the bearing length is

$$l_b = 1.0 \ \text{ft}$$

The design span is defined in NDS Sec. 3.2.1 as

$$L = L_c + 2\left(\frac{l_b}{2}\right)$$
$$= 20 \ \text{ft} + (2)\left(\frac{1.0 \ \text{ft}}{2}\right)$$
$$= 21 \ \text{ft}$$

ASD Method

The adjustment factors for shear stress are

$$C_D = C_M = C_t = C_{vr} = 1.0$$

$$F_v' = F_v = 215 \ \text{lbf/in}^2$$

Applying IBC Eq. 16-9, the factored distributed dead load is

$$w = 1.0D = (1.0)\left(2.0 \ \frac{\text{kips}}{\text{ft}}\right)$$
$$= 2 \ \text{kips/ft}$$

In calculating shear force due to distributed load, the distributed load within a distance, d, from either support is ignored. In accordance with NDS Sec. 3.4.3.1(a), the shear caused by the distributed load at the location of the concentrated load is given by

$$V_D = 0.5w(L_c - 2d)$$

$$= (0.5)\left(2\,\frac{\text{kips}}{\text{ft}}\right)\left(20\text{ ft} - (2)\left(\frac{30\text{ in}}{12\,\frac{\text{in}}{\text{ft}}}\right)\right)$$

$$= 15\text{ kips}$$

Applying IBC Eq. 16-9, the factored concentrated live load is

$$W = 1.0L = (1.0)(15\text{ kips})$$
$$= 15\text{ kips}$$

The concentrated load is less than a distance, d, from the face of the left support. In accordance with NDS Sec. 3.4.3.1, this is equivalent to a load of

$$W' = \frac{Wx}{d} = \frac{(15\text{ kips})(2\text{ ft})\left(12\,\frac{\text{in}}{\text{ft}}\right)}{30\text{ in}}$$
$$= 12\text{ kips}$$

The shear at the left support caused by the equivalent load is

$$V_C = \frac{W'(L - x - 0.5l_b)}{L}$$

$$= \frac{(12\text{ kips})\left(21\text{ ft} - 2\text{ ft} - (0.5)(1.0\text{ ft})\right)}{21\text{ ft}}$$

$$= 10.57\text{ kips}$$

The total shear at the left support is

$$V = V_D + V_C$$
$$= 15\text{ kips} + 10.57\text{ kips}$$
$$= 25.57\text{ kips}$$

The shear stress parallel to the grain given by NDS Sec. 3.4.2 is

$$F_v = \frac{3V}{2bd} = \frac{(3)(25.57\text{ kips})\left(1000\,\frac{\text{lbf}}{\text{kip}}\right)}{(2)(6.75\text{ in})(30\text{ in})}$$

$$= 189\text{ lbf/in}^2$$

$$< F_v'\quad\text{[satisfactory]}$$

The beam is adequate.

LRFD Method

The adjustment factors for shear stress are

$$C_D = C_M = C_t = C_{vr} = 1.0$$

From Table 6.2, the format conversion factor is

$$K_F = 2.88$$

From Table 6.4, the time effect factor for occupancy live load is

$$\lambda = 0.8$$

From Table 6.3, the resistance factor for shear is

$$\phi_v = 0.75$$

The adjusted factored shear stress is

$$F_v' = F_v C_M C_t C_{vr} K_F \phi_v \lambda$$

$$= \left(215\,\frac{\text{lbf}}{\text{in}^2}\right)(1.0)(1.0)(1.0)(2.88)(0.75)(0.8)$$

$$= 372\text{ lbf/in}^2$$

Applying IBC Eq. 16-2, the factored distributed dead load is

$$w = 1.2D = (1.2)\left(2.0\,\frac{\text{kips}}{\text{ft}}\right) = 2.4\text{ kips/ft}$$

The shear at the left support caused by the distributed load is given by NDS Sec. 3.4.3.1(a) as

$$V_D = 0.5w(L_c - 2d)$$

$$= (0.5)\left(2.4\,\frac{\text{kips}}{\text{ft}^2}\right)\left(20\text{ ft} - (2)\left(\frac{30\text{ in}}{12\,\frac{\text{in}}{\text{ft}}}\right)\right)$$

$$= 18\text{ kips}$$

Applying IBC Eq. 16-2, the factored concentrated live load is

$$W = 1.6L = (1.6)(15\text{ kips}) = 24\text{ kips}$$

The concentrated load is less than a distance, d, from the face of the left support. In accordance with NDS Sec. 3.4.3.1, this is equivalent to a load of

$$W' = \frac{Wx}{d}$$

$$= \frac{(24 \text{ kips})(2 \text{ ft})\left(12 \dfrac{\text{in}}{\text{ft}}\right)}{30 \text{ in}}$$

$$= 19.2 \text{ kips}$$

The shear at the left support caused by the equivalent live load is

$$V_C = \frac{W'(L - x - 0.5l_b)}{L}$$

$$= \frac{(19.2 \text{ kips})\big(21 \text{ ft} - 2 \text{ ft} - (0.5)(1.0 \text{ ft})\big)}{21 \text{ ft}}$$

$$= 16.91 \text{ kips}$$

The total shear at the left support is

$$V = V_D + V_C$$

$$= 18 \text{ kips} + 16.91 \text{ kips}$$

$$= 34.91 \text{ kips}$$

The shear stress parallel to the grain given by NDS Sec. 3.4.2 is

$$F_v = \frac{3V}{2bd}$$

$$= \frac{(3)(34.91 \text{ kips})\left(1000 \dfrac{\text{lbf}}{\text{kip}}\right)}{(2)(6.75 \text{ in})(30 \text{ in})}$$

$$= 259 \text{ lbf/in}^2$$

$$< F_v' \quad [\text{satisfactory}]$$

The beam is adequate.

Notched Beams

Notches in a beam reduce the shear capacity, and NDS Sec. 4.4.3 and Sec. 5.4.5 impose restrictions on their size and location, as shown in Fig. 6.3. The adjusted design shear at a notch on the tension side of a beam is given by NDS Sec. 4.4.3 and Sec. 5.4.5 as

$$V_r' = \left(\tfrac{2}{3}F_v'bd_n\right)\left(\frac{d_n}{d}\right)^2 \qquad [\text{NDS 3.4-3}]$$

When $e \le d_n$, the adjusted design shear at a notch on the compression side of a beam is given by NDS Sec. 3.4.3.2 as

$$V_r' = \left(\tfrac{2}{3}F_v'b\right)\left(d - \left(\frac{d - d_n}{d_n}\right)e\right) \qquad [\text{NDS 3.4-5}]$$

When $e > d_n$,

$$V_r' = \tfrac{2}{3}F_v'bd_n$$

Shear at Connections

For a connection less than five times the depth of the member from its end, as shown in Fig. 6.4, the adjusted design shear is given by NDS Sec. 3.4.3.3 as

$$V_r' = \left(\tfrac{2}{3}F_v'bd_e\right)\left(\frac{d_e}{d}\right)^2 \qquad [\text{NDS 3.4-6}]$$

When the connection is at least five times the depth of the member from its end, the adjusted design shear is given by

$$V_r' = \tfrac{2}{3}F_v'bd_e \qquad [\text{NDS 3.4-7}]$$

To facilitate the selection of glued laminated beam sections, tables are available[5] that provide shear and bending capacities of sections. For lumber joists, tables are available[6] that assist in the selection of a joist size for various span and live load combinations.

Example 6.8

A glued laminated $6\frac{3}{4} \times 30$ beam of stress class 24F-1.7E western species is notched and loaded as shown in *Illustration for Ex. 6.8*. The beam has a moisture content exceeding 16% and is subjected to sustained temperatures between 100°F and 125°F. The governing load combination is dead plus occupancy live load. Determine the maximum allowable shear force at each support and at the hanger connection.

Solution

The reference design value for shear stress, tabulated in NDS Supp. Table 5A, is

$$F_v = 210 \text{ lbf/in}^2$$

The applicable shear reduction factor for the design of members at a notch is given by NDS Sec.5.3.10 as

$$C_{vr} = 0.72$$

Figure 6.3 *Notched Beams*

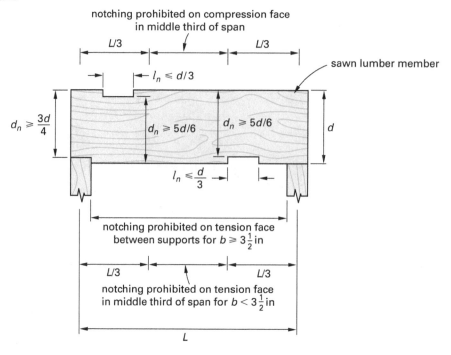

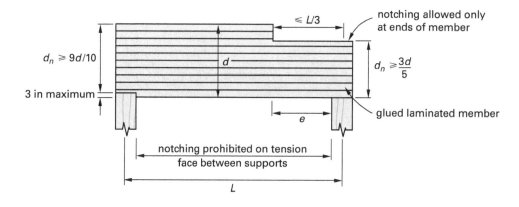

Figure 6.4 *Bolted Connections*

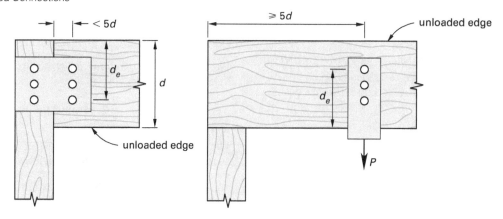

Illustration for Ex. 6.8

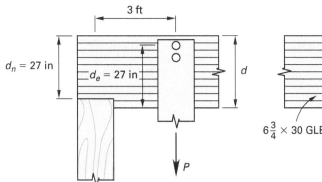
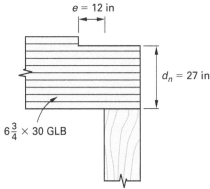

From Table 6.7, the applicable temperature factor for wet conditions between 100°F and 125°F is

$$C_t = 0.7$$

From Table 6.6, the applicable wet service factor for moisture content exceeding 16% is

$$C_M = 0.875$$

ASD Method

For normal occupancy live load, the load duration factor given in Table 6.5 is

$$C_D = 1.00$$

The adjusted allowable shear stress is

$$
\begin{aligned}
F'_v &= F_v C_D C_M C_t C_{vr} \\
&= \left(210\ \frac{\text{lbf}}{\text{in}^2}\right)(1.00)(0.875)(0.7)(0.72) \\
&= 93\ \text{lbf/in}^2
\end{aligned}
$$

At the left support and from NDS Eq. 3.4-6, the allowable shear force is

$$
\begin{aligned}
V'_r &= \left(\tfrac{2}{3} F'_v b d_n\right)\left(\frac{d_n}{d}\right)^2 \\
&= \left(\frac{(2)\left(93\ \dfrac{\text{lbf}}{\text{in}^2}\right)(6.75\ \text{in})(27\ \text{in})}{3}\right)\left(\frac{27\ \text{in}}{30\ \text{in}}\right)^2 \\
&= 9153\ \text{lbf}
\end{aligned}
$$

At the right support, $e < d_n$, and from NDS Eq. 3.4-5, the allowable shear force is

$$
\begin{aligned}
V'_r &= \tfrac{2}{3} F'_v b\left(d - \left(\frac{d - d_n}{d_n}\right)e\right) \\
&= \left(\frac{(2)\left(93\ \dfrac{\text{lbf}}{\text{in}^2}\right)(6.75\ \text{in})}{3}\right) \\
&\quad \times\left(30\ \text{in} - \left(\frac{30\ \text{in} - 27\ \text{in}}{27\ \text{in}}\right)(12\ \text{in})\right) \\
&= 11{,}997\ \text{lbf}
\end{aligned}
$$

The hanger connection is less than $5d$ from the end of the beam, and from NDS Eq. 3.4-6, the adjusted factored force is

$$
\begin{aligned}
V'_r &= \left(\tfrac{2}{3} F'_v b d_e\right)\left(\frac{d_e}{d}\right)^2 \\
&= \left(\frac{(2)\left(93\ \dfrac{\text{lbf}}{\text{in}^2}\right)(6.75\ \text{in})(27\ \text{in})}{3}\right)\left(\frac{27\ \text{in}}{30\ \text{in}}\right)^2 \\
&= 9153\ \text{lbf}
\end{aligned}
$$

LRFD Method

From Table 6.4, the applicable time effect factor for dead load plus occupancy live load is

$$\lambda = 0.8$$

From Table 6.3, the applicable resistance factor for shear is

$$\phi = 0.75$$

From Table 6.2, the applicable format conversion factor for shear is

$$K_F = 2.88$$

The adjusted factored shear stress is

$$
\begin{aligned}
F_v' &= F_v C_M C_t C_{vr} K_F \lambda \phi_v \\
&= \left(210 \ \frac{\text{lbf}}{\text{in}^2}\right)(0.875)(0.7)(0.72) \\
&\quad \times (2.88)(0.8)(0.75) \\
&= 160 \ \text{lbf/in}^2
\end{aligned}
$$

At the left support and from NDS Eq. 3.4-3, the allowable shear force is

$$
\begin{aligned}
V_r' &= \left(\frac{2}{3} F_v' b d_n\right)\left(\frac{d_n}{d}\right)^2 \\
&= \left(\frac{(2)\left(160 \ \dfrac{\text{lbf}}{\text{in}^2}\right)(6.75 \ \text{in})(27 \ \text{in})}{3}\right)\left(\frac{27 \ \text{in}}{30 \ \text{in}}\right)^2 \\
&= 15{,}747 \ \text{lbf}
\end{aligned}
$$

At the right support, $e < d_n$, and from NDS Eq. 3.4-5, the adjusted factored force is

$$
\begin{aligned}
V_r' &= \frac{2}{3} F_v' b\left[d - \left(\frac{d - d_n}{d_n}\right)e\right] \\
&= \left(\frac{(2)\left(160 \ \dfrac{\text{lbf}}{\text{in}^2}\right)(6.75 \ \text{in})}{3}\right) \\
&\quad \times \left(30 \ \text{in} - \left(\frac{30 \ \text{in} - 27 \ \text{in}}{27 \ \text{in}}\right)(12 \ \text{in})\right) \\
&= 20{,}640 \ \text{lbf}
\end{aligned}
$$

The hanger connection is less than $5d$ from the end of the beam, and from NDS Eq. 3.4-6, the adjusted factored force is

$$
\begin{aligned}
V_r' &= \left(\frac{2}{3} F_v' b d_e\right)\left(\frac{d_e}{d}\right)^2 \\
&= \left(\frac{(2)\left(160 \ \dfrac{\text{lbf}}{\text{in}^2}\right)(6.75 \ \text{in})(27 \ \text{in})}{3}\right)\left(\frac{27 \ \text{in}}{30 \ \text{in}}\right)^2 \\
&= 15{,}747 \ \text{lbf}
\end{aligned}
$$

9. DESIGN FOR COMPRESSION

Nomenclature

A	area of cross section	in^2
C_{m1}	moment magnification factor for biaxial bending and axial compression, $1.0 - f_c/F_{cE1}$	–
C_{m2}	moment magnification factor forbiaxial bending and axial compression, $1.0 - f_c/F_{cE2} - (f_{b1}/F_{bE})^2$	–
C_{m3}	moment magnification factor for axial compression and flexure with load applied to narrow face, $1.0 - f_c/F_{cE1}$	–
C_{m4}	moment magnification factor for axial compression and flexure with load applied to wide face, $1.0 - f_c/F_{cE2}$	–
C_{m5}	moment magnification factor for biaxial bending, $1.0 - (f_{b1}/F_{bE})^2$	–
d_1	dimension of wide face	in
d_2	dimension of narrow face	in
f_{b1}	actual edgewise bending stress for load applied to the narrow face	lbf/in^2
f_{b2}	actual flatwise bending stress for load applied to the wide face	lbf/in^2
f_c	actual compression stress parallel to grain	lbf/in^2
F_{bE}	critical buckling design value for bending member, $1.20E_{\min}'/R_B^2$	lbf/in^2
F_{b1}'	allowable bending design value for load applied to the narrow face, including adjustment for slenderness ratio	lbf/in^2
F_{b2}'	allowable bending design value for load applied to the wide face, including adjustment for slenderness ratio	lbf/in^2
F_c'	allowable compression design value, including adjustment for largest slenderness ratio	lbf/in^2
F_{cE1}	critical buckling design value in plane of bending for load applied to the narrow face, $0.822E_{\min}'/(l_{e1}/d_1)^2$	lbf/in^2
F_{cE2}	critical buckling design value in plane of bending for load applied to the wide face, $0.822E_{\min}'/(l_{e2}/d_2)^2$	lbf/in^2
K_e	buckling length coefficient for compression members	–

l_1	distance between points of lateral support restraining buckling about the strong axis of compression member	ft or in
l_2	distance between points of lateral support restraining buckling about the weak axis of compression member	ft or in
l_{e1}	effective length between supports restraining buckling in plane of bending from load applied to narrow face of compression member, $K_e l_1$	ft or in
l_{e2}	effective length between supports restraining buckling in plane of bending from load applied to wide face of compression member, $K_e l_2$	ft or in
l_{e1}/d_1	slenderness ratio about the strong axis of compression member	–
l_{e2}/d_2	slenderness ratio about the weak axis of compression member	–
P	total concentrated load or total axial load	lbf or kips

Axial Load Only

The effective length of a column is defined in NDS Sec. 3.7.1.2 as $l_e = K_e l$ where l is the distance between lateral supports. The slenderness ratio is defined in NDS Sec. 3.7.1.3 as $l_e/d \le 50$. When the distance between lateral supports about the x-x axis and the y-y axis is different, as shown in Fig. 6.5, two values of the slenderness ratio are obtained. These values are l_{e1}/d_1 and l_{e2}/d_2. The larger of these values governs.

Figure 6.5 *Axially Loaded Column*

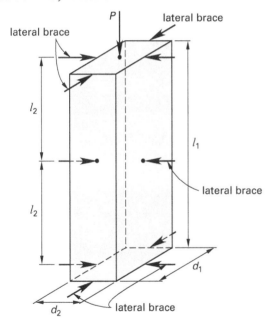

Values of the buckling length coefficient, K_e, for various end restraint conditions are given in NDS Table G1 and are summarized in Fig. 6.6.

Figure 6.6 *Buckling Length Coefficients*

illus.	end conditions	K_e theoretical	design
(a)	both ends pinned	1	1.00
(b)	both ends built in	0.5	0.65
(c)	one end pinned, one end built in	0.7	0.8
(d)	one end built in, one end free	2	2.10
(e)	one end built in, one end fixed against rotation but free to translate	1	1.20
(f)	one end pinned, one end fixed against rotation but free to translate	2	2.40

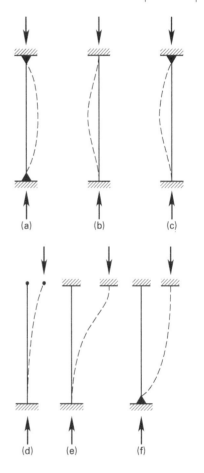

The maximum slenderness ratio of a column determines the critical buckling design value and the adjusted compressive stress, F_c'. The allowable axial column load is

$$P = AF_c'.$$

Example 6.9

The select structural 2×6 Douglas fir-larch top chord of a truss is loaded with the service level loads shown in the illustration. The governing load combination consists of dead plus occupancy live load, and the moisture content exceeds 19%. The chord is laterally braced at midlength about the weak axis, and the self-weight of the chord and bracing members may be neglected. Determine whether the member is adequate.

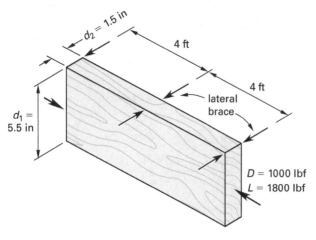

Solution

The reference design values for compression and modulus of elasticity are tabulated in NDS Supp. Table 4A and are

$$F_c = 1700 \text{ lbf/in}^2$$
$$E_{\min} = 0.69 \times 10^6 \text{ lbf/in}^2$$
$$C_T = 1.0, \ C_t = 1.0, \ C_i = 1.0$$

The applicable adjustment factors for compression and modulus of elasticity are as follows.

C_M = wet service factor from Table 6.6

   = 0.8   [compression member for $F_c C_F > 750$ lbf/in^2]

   = 0.9   [modulus of elasticity]

C_F = size factor from NDS Supp. Table 4A

   = 1.1   [2 in × 6 in compression member]

From the illustration, the slenderness ratio about the strong axis is

$$\frac{K_e l_1}{d_1} = \frac{(1.0)(8 \text{ ft})\left(12 \ \frac{\text{in}}{\text{ft}}\right)}{5.5 \text{ in}}$$
$$= 17.46$$

From the illustration, the slenderness ratio about the weak axis is

$$\frac{K_e l_2}{d_2} = \frac{(1.0)(4 \text{ ft})\left(12 \ \frac{\text{in}}{\text{ft}}\right)}{1.5 \text{ in}}$$
$$= 32.00 \quad \text{[governs]}$$

ASD Method

The adjusted modulus of elasticity for stability calculations is

$$E'_{\min} = E_{\min} C_M = \left(0.69 \times 10^6 \ \frac{\text{lbf}}{\text{in}^2}\right)(0.9)$$
$$= 0.62 \times 10^6 \text{ lbf/in}^2$$

The load duration factor for dead load plus occupancy live load from Table 6.5 is

$$C_D = 1.00$$

The reference compression design value multiplied by all applicable adjustment factors except C_P is given by

$$F_c^* = F_c C_D C_M C_F$$
$$= \left(1700 \ \frac{\text{lbf}}{\text{in}^2}\right)(1.00)(0.8)(1.1)$$
$$= 1496 \text{ lbf/in}^2$$

The critical buckling design value is

$$F_{cE2} = \frac{0.822 E'_{\min}}{\left(\dfrac{l_{e2}}{d_2}\right)^2} = \frac{(0.822)\left(0.62 \times 10^6 \ \dfrac{\text{lbf}}{\text{in}^2}\right)}{\left(\dfrac{(4 \text{ ft})\left(12 \ \dfrac{\text{in}}{\text{ft}}\right)}{1.5 \text{ in}}\right)^2}$$
$$= 498 \text{ lbf/in}^2$$

The ratio of F_{cE2} to F_c^* is

$$F' = \frac{F_{cE2}}{F_c^*} = \frac{498 \ \dfrac{\text{lbf}}{\text{in}^2}}{1496 \ \dfrac{\text{lbf}}{\text{in}^2}}$$
$$= 0.333$$

The column parameter is obtained from NDS Sec. 3.7.1.5 as

$$c = 0.8 \quad \text{[for sawn lumber]}$$

The column stability factor is specified by NDS Sec. 3.7.1 as

$$C_P = \frac{1.0 + F'}{2c} - \sqrt{\left(\frac{1.0 + F'}{2c}\right)^2 - \frac{F'}{c}}$$
$$= \frac{1.0 + 0.333}{(2)(0.8)} - \sqrt{\left(\frac{1.0 + 0.333}{(2)(0.8)}\right)^2 - \frac{0.333}{0.8}}$$
$$= 0.31$$

The allowable compression design value parallel to grain is

$$
\begin{aligned}
F_c' &= F_c C_M C_F C_P \\
&= \left(1700 \ \frac{\text{lbf}}{\text{in}^2}\right)(0.8)(1.1)(0.31) \\
&= 464 \ \text{lbf/in}^2
\end{aligned}
$$

Applying IBC Eq. 16-9, the factored load is

$$
\begin{aligned}
P &= D + L \\
&= 1000 \ \text{lbf} + 1800 \ \text{lbf} \\
&= 2800 \ \text{lbf}
\end{aligned}
$$

The actual compression stress on the chord is given by

$$
\begin{aligned}
f_c &= \frac{P}{A} = \frac{2800 \ \text{lbf}}{(1.5 \ \text{in})(5.5 \ \text{in})} \\
&= 339 \ \text{lbf/in}^2 \\
&< F_c'
\end{aligned}
$$

The chord is adequate.

LRFD Method

The adjusted factored modulus of elasticity for stability calculations is

$$
\begin{aligned}
E_{\min}' &= E_{\min} C_M C_t K_F \phi_s \\
&= \left(0.69 \times 10^6 \ \frac{\text{lbf}}{\text{in}^2}\right) \\
&\quad \times (0.9)(1.0)(1.76)(0.85) \\
&= 0.93 \times 10^6 \ \text{lbf/in}^2
\end{aligned}
$$

The time effect factor for dead load plus occupancy live load is obtained from Table 6.4 as $\lambda = 0.8$.

The reference compression design value multiplied by all applicable adjustment factors except C_P is given by

$$
\begin{aligned}
F_c^* &= F_c C_M C_F C_t C_i K_F \lambda \phi_c \\
&= \left(1700 \ \frac{\text{lbf}}{\text{in}^2}\right)(0.8)(1.1)(1.0)(1.0)(2.40)(0.8)(0.90) \\
&= 2585 \ \text{lbf/in}^2
\end{aligned}
$$

The critical buckling design value is

$$
F_{cE2} = \frac{0.822 E_{\min}'}{\left(\dfrac{l_{e2}}{d_2}\right)^2} = \frac{(0.822)\left(0.93 \times 10^6 \ \dfrac{\text{lbf}}{\text{in}^2}\right)}{\left(\dfrac{(4 \ \text{ft})\left(12 \ \dfrac{\text{in}}{\text{ft}}\right)}{1.5 \ \text{in}}\right)^2}
$$

$$
= 747 \ \text{lbf/in}^2
$$

The ratio of F_{cE2} to F_c^* is

$$
\begin{aligned}
F' &= \frac{F_{cE2}}{F_c^*} = \frac{747 \ \dfrac{\text{lbf}}{\text{in}^2}}{2585 \ \dfrac{\text{lbf}}{\text{in}^2}} \\
&= 0.289
\end{aligned}
$$

The column parameter is obtained from NDS Sec. 3.7.1.5 as

$$
c = 0.8 \quad \text{[for sawn lumber]}
$$

The column stability factor is specified by NDS Sec. 3.7.1 as

$$
\begin{aligned}
C_P &= \frac{1.0 + F'}{2c} - \sqrt{\left(\frac{1.0 + F'}{2c}\right)^2 - \frac{F'}{c}} \\
&= \frac{1.0 + 0.289}{(2)(0.8)} - \sqrt{\left(\frac{1.0 + 0.289}{(2)(0.8)}\right)^2 - \frac{0.289}{0.8}} \\
&= 0.27
\end{aligned}
$$

The adjusted factored compression design value parallel to grain is

$$
\begin{aligned}
F_c' &= F_c C_M C_F C_P C_i C_t K_F \lambda \phi_c \quad \text{[λ is not applicable]} \\
&= \left(1700 \ \frac{\text{lbf}}{\text{in}^2}\right)(0.8)(1.1)(0.27)(1.0)(1.0)(2.40) \\
&\quad \times (0.8)(0.90) \\
&= 698 \ \text{lbf/in}^2
\end{aligned}
$$

Applying IBC Eq. 16-2, the factored load is

$$
\begin{aligned}
P &= 1.2D + 1.6L \\
&= (1.2)(1000 \ \text{lbf}) + (1.6)(1800 \ \text{lbf}) \\
&= 4080 \ \text{lbf}
\end{aligned}
$$

The actual compression stress on the chord is given by

$$f_c = \frac{P}{A} = \frac{4080 \text{ lbf}}{(1.5 \text{ in})(5.5 \text{ in})}$$
$$= 495 \text{ lbf/in}^2$$
$$< F_c'$$

The chord is adequate.

Combined Axial Compression and Flexure

Members subjected to combined compression and flexural stresses due to axial and transverse loading must satisfy the interaction equations given in NDS Sec. 3.9.2 as

$$\left(\frac{f_c}{F_c'}\right)^2 + \frac{f_{b1}}{F_{b1}'C_{m1}} + \frac{f_{b2}}{F_{b2}'C_{m2}} \leq 1.00 \qquad \text{[NDS 3.9-3]}$$

For bending load applied to the narrow face of the member and concentric axial compression load, the interaction equation reduces to

$$\left(\frac{f_c}{F_c'}\right)^2 + \frac{f_{b1}}{F_{b1}'C_{m3}} \leq 1.00$$

For bending load applied to the wide face of the member and concentric axial compression load, the equation reduces to

$$\left(\frac{f_c}{F_c'}\right)^2 + \frac{f_{b2}}{F_{b2}'C_{m4}} \leq 1.00$$

For bending loads applied to the narrow and wide faces of the member and no concentric axial load, the equation reduces to

$$\frac{f_{b1}}{F_{b1}'} + \frac{f_{b2}}{F_{b2}'C_{m5}} \leq 1.00$$

In addition, members must satisfy the interaction expression given by NDS Eq. 3.9-4, which is

$$\frac{f_c}{F_{cE2}} + \left(\frac{f_{b1}}{F_{bE}}\right)^2$$

Example 6.10

The select structural 2×6 Douglas fir-larch top chord of a truss is loaded with the service level loads shown in the illustration. The governing load combination consists of dead plus live load, and the moisture content exceeds 19%. The chord is laterally braced at midlength about the weak axis, and the self-weight of the chord and bracing members may be neglected. Determine whether the member is adequate.

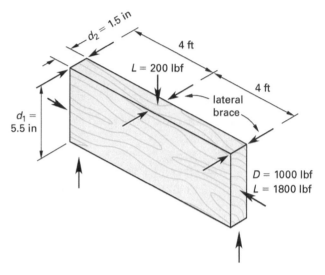

Solution

The reference design values for bending and modulus of elasticity are tabulated in NDS Supp. Table 4A, and they are

$$F_b = 1500 \text{ lbf/in}^2$$
$$E_{\min} = 0.69 \times 10^6 \text{ lbf/in}^2$$
$$C_{fu} = 1.0, \quad C_t = 1.0, \quad C_i = 1.0, \quad C_r = 1.0$$

The distance between lateral restraints is

$$l_u = \frac{8 \text{ ft}}{2}$$
$$= 4 \text{ ft}$$

From Fig. 6.1, for a concentrated load at midspan and with lateral restraint at midspan, the effective length for flexure is

$$l_e = 1.11 l_u$$
$$= (1.11)(4 \text{ ft})\left(12 \frac{\text{in}}{\text{ft}}\right)$$
$$= 53.28 \text{ in}$$

The slenderness ratio for flexure is given by NDS Sec. 3.3.3 as

$$R_B = \sqrt{\frac{l_e d_1}{d_2^2}} = \sqrt{\frac{(53.28 \text{ in})(5.5 \text{ in})}{(1.5 \text{ in})^2}}$$
$$= 11.41$$
$$< 50 \qquad \begin{bmatrix} \text{satisfies criteria of} \\ \text{NDS Sec. 3.3.3} \end{bmatrix}$$

From Table 6.6, for $F_b C_F > 1150$ lbf/in^2, the applicable wet service factor for flexure is

$$C_M = 0.85$$

From NDS Supp. Table 4A, the applicable size factor for flexure is

$$C_F = 1.3$$

ASD Method

From Table 6.5, the load duration factor for dead load plus occupancy live load is

$$C_D = 1.00$$

From Ex. 6.9, the adjusted modulus of elasticity for stability calculations is

$$E'_{\min} = E_{\min} C_M = \left(0.69 \times 10^6 \ \frac{\text{lbf}}{\text{in}^2}\right)(0.9)$$
$$= 0.62 \times 10^6 \ \text{lbf/in}^2$$

The critical buckling design value for flexure is

$$F_{bE} = \frac{1.20 E'_{\min}}{R_B^2}$$
$$= \frac{(1.20)\left(0.62 \times 10^6 \ \frac{\text{lbf}}{\text{in}^2}\right)}{(11.41)^2}$$
$$= 5715 \ \text{lbf/in}^2$$

The reference flexural design value multiplied by all applicable adjustment factors except C_L is

$$F_b^* = F_b C_M C_F$$
$$= \left(1500 \ \frac{\text{lbf}}{\text{in}^2}\right)(0.85)(1.3)$$
$$= 1657 \ \text{lbf/in}^2$$
$$F = \frac{F_{bE}}{F_b^*}$$
$$= \frac{5715 \ \dfrac{\text{lbf}}{\text{in}^2}}{1657 \ \dfrac{\text{lbf}}{\text{in}^2}}$$
$$= 3.45$$

The beam stability factor is given by NDS Sec. 3.3.3 as

$$C_L = \frac{1.0 + F}{1.9} - \sqrt{\left(\frac{1.0 + F}{1.9}\right)^2 - \frac{F}{0.95}}$$
$$= \frac{1.0 + 3.45}{1.9} - \sqrt{\left(\frac{1.0 + 3.45}{1.9}\right)^2 - \frac{3.45}{0.95}}$$
$$= 0.98$$

The allowable flexural design value for load applied to the narrow face is

$$F'_{b1} = F_b C_M C_L C_F$$
$$= \left(1500 \ \frac{\text{lbf}}{\text{in}^2}\right)(0.85)(0.98)(1.3)$$
$$= 1624 \ \text{lbf/in}^2$$

Applying IBC Eq. 16-9, the factored vertical load is

$$W = 1.0L$$
$$= (1.0)(200 \ \text{lbf})$$
$$= 200 \ \text{lbf}$$

The actual edgewise bending stress is

$$f_{b1} = \frac{WL}{4S}$$
$$= \frac{(200 \ \text{lbf})(8 \ \text{ft})\left(12 \ \dfrac{\text{in}}{\text{ft}}\right)}{(4)(7.56 \ \text{in}^3)}$$
$$= 635 \ \text{lbf/in}^2$$

From Ex. 6.9,

$$\frac{K_e l_1}{d_1} = 17.46$$
$$F'_c = 464 \ \text{lbf/in}^2 \quad [\text{LRFD}]$$
$$f_c = 339 \ \text{lbf/in}^2 \quad [\text{LRFD}]$$

Wood

The critical buckling design value, in the plane of bending, for load applied to the narrow face is

$$F_{cE1} = \frac{0.822 E'_{min}}{\left(\dfrac{l_{e1}}{d_1}\right)^2}$$

$$= \frac{(0.822)\left(0.62 \times 10^6 \ \dfrac{lbf}{in^2}\right)}{\left(\dfrac{(8 \ ft)\left(12 \ \dfrac{in}{ft}\right)}{5.5 \ in}\right)^2}$$

$$= 1673 \ lbf/in^2$$

The moment magnification factor for axial compression and flexure with load applied to the narrow face is

$$C_{m3} = 1.0 - \frac{f_c}{F_{cE1}}$$

$$= 1.0 - \frac{339 \ \dfrac{lbf}{in^2}}{1673 \ \dfrac{lbf}{in^2}}$$

$$= 0.797$$

The interaction equation for bending load applied to the narrow face of the member and concentric axial compression load is given in NDS Sec. 3.9.2 as

$$\left(\frac{f_c}{F'_c}\right)^2 + \frac{f_{b1}}{F'_{b1} C_{m3}} \leq 1.0$$

The left side of the expression is

$$\left(\frac{339 \ \dfrac{lbf}{in^2}}{464 \ \dfrac{lbf}{in^2}}\right)^2 + \frac{635 \ \dfrac{lbf}{in^2}}{\left(1624 \ \dfrac{lbf}{in^2}\right)(0.797)} = 0.534 + 0.491$$

$$= 1.025$$
$$\approx 1.0 \quad \text{[satisfactory]}$$

In addition, the post must satisfy the interaction expression given by NDS Eq. 3.9-4, which is

$$\frac{f_c}{F_{cE2}} + \left(\frac{f_{b1}}{F_{bE}}\right)^2 < 1.0$$

From Ex. 6.9,

$$f_c = 339 \ lbf/in^2$$
$$F_{cE2} = 498 \ lbf/in^2$$

The left-hand side of NDS Eq. 3.9-4 is

$$\frac{339 \ \dfrac{lbf}{in^2}}{498 \ \dfrac{lbf}{in^2}} + \left(\frac{635 \ \dfrac{lbf}{in^2}}{5715 \ \dfrac{lbf}{in^2}}\right)^2 = 0.69$$

$$< 1.0 \quad \text{[satisfactory]}$$

The chord is adequate.

LRFD Method

From Table 6.4, the time effect factor for dead load plus occupancy live load is

$$\lambda = 0.8$$

From Ex. 6.9, the adjusted modulus of elasticity for stability calculations is

$$E'_{min} = E_{min} C_M C_t K_F \phi_s$$

$$= \left(0.69 \times 10^6 \ \frac{lbf}{in^2}\right)$$

$$\times (0.9)(1.0)(1.76)(0.85)$$

$$= 0.93 \times 10^6 \ lbf/in^2$$

The critical buckling design value for flexure is

$$F_{bE} = \frac{1.20 E'_{min}}{R_B^2}$$

$$= \frac{(1.20)\left(0.93 \times 10^6 \ \dfrac{lbf}{in^2}\right)}{(11.41)^2}$$

$$= 8572 \ lbf/in^2$$

Wood

The reference flexural design value multiplied by all applicable adjustment factors except C_L is

$$F_b^* = F_b C_M C_F C_t C_{fu} C_i C_r K_F \lambda \phi_b$$

$$= \left(1500 \ \frac{\text{lbf}}{\text{in}^2}\right)(0.85)(1.3)(1.0)(1.0)(1.0)(1.0)$$

$$\times (2.54)(0.8)(0.85)$$

$$= 2863 \ \text{lbf/in}^2$$

$$F = \frac{F_{bE}}{F_b^*}$$

$$= \frac{8572 \ \dfrac{\text{lbf}}{\text{in}^2}}{2863 \ \dfrac{\text{lbf}}{\text{in}^2}}$$

$$= 3.0$$

The beam stability factor is given by NDS Sec. 3.3.3 as

$$C_L = \frac{1.0 + F}{1.9} - \sqrt{\left(\frac{1.0 + F}{1.9}\right)^2 - \frac{F}{0.95}}$$

$$= \frac{1.0 + 3.0}{1.9} - \sqrt{\left(\frac{1.0 + 3.0}{1.9}\right)^2 - \frac{3.0}{0.95}}$$

$$= 0.98$$

The adjusted factored flexural design value for load applied to the narrow face is

$$F_{b1}' = F_b C_M C_L C_F C_t C_{fu} C_i C_r K_F \lambda \phi_b$$

$$= \left(1500 \ \frac{\text{lbf}}{\text{in}^2}\right)(0.85)(0.98)(1.3)(1.0)(1.0)(1.0)(1.0)$$

$$\times (2.54)(0.8)(0.85)$$

$$= 2806 \ \text{lbf/in}^2$$

Applying IBC Eq. 16-2, the factored vertical load is

$$W = 1.6L = (1.6)(200 \ \text{lbf})$$

$$= 320 \ \text{lbf}$$

The actual edgewise bending stress is

$$f_{b1} = \frac{WL}{4S}$$

$$= \frac{(320 \ \text{lbf})(8 \ \text{ft})\left(12 \ \dfrac{\text{in}}{\text{ft}}\right)}{(4)(7.56 \ \text{in}^3)}$$

$$= 1016 \ \text{lbf/in}^2$$

From Ex. 6.9,

$$\frac{K_e l_1}{d_1} = 17.46$$

$$F_c' = 698 \ \text{lbf/in}^2 \quad [\text{LRFD}]$$

$$f_c = 495 \ \text{lbf/in}^2 \quad [\text{LRFD}]$$

The critical buckling design value, in the plane of bending, for load applied to the narrow face is

$$F_{cE1} = \frac{0.822 E_{\min}'}{\left(\dfrac{l_{e1}}{d_1}\right)^2}$$

$$= \frac{(0.822)\left(0.93 \times 10^6 \ \dfrac{\text{lbf}}{\text{in}^2}\right)}{\left(\dfrac{(8 \ \text{ft})\left(12 \ \dfrac{\text{in}}{\text{ft}}\right)}{5.5 \ \text{in}}\right)^2}$$

$$= 2509 \ \text{lbf/in}^2$$

The moment magnification factor for axial compression and flexure with load applied to the narrow face is

$$C_{m3} = 1.0 - \frac{f_c}{F_{cE1}}$$

$$= 1.0 - \frac{495 \ \dfrac{\text{lbf}}{\text{in}^2}}{2509 \ \dfrac{\text{lbf}}{\text{in}^2}}$$

$$= 0.803$$

The interaction equation for bending load applied to the narrow face of the member and concentric axial compression load is given in NDS Sec. 3.9.2 as

$$\left(\frac{f_c}{F_c'}\right)^2 + \frac{f_{b1}}{F_{b1}' C_{m3}} \leq 1.0$$

The left side of the expression is

$$\left(\frac{495 \ \dfrac{\text{lbf}}{\text{in}^2}}{698 \ \dfrac{\text{lbf}}{\text{in}^2}}\right)^2 + \frac{1016 \ \dfrac{\text{lbf}}{\text{in}^2}}{\left(2806 \ \dfrac{\text{lbf}}{\text{in}^2}\right)(0.803)} = 0.503 + 0.451$$

$$= 0.95$$

$$< 1.0 \quad [\text{satisfactory}]$$

Wood

In addition, the post must satisfy the interaction expression given by NDS Eq. 3.9-4, which is

$$\frac{f_c}{F_{cE2}} + \left(\frac{f_{b1}}{F_{bE}}\right)^2 < 1.0$$

From Ex. 6.9,

$$f_c = 495 \text{ lbf/in}^2 \quad [\text{LRFD}]$$

$$F_{cE2} = 747 \text{ lbf/in}^2 \quad [\text{LRFD}]$$

The left-hand side of NDS Eq. 3.9-4 is

$$\frac{495 \dfrac{\text{lbf}}{\text{in}^2}}{747 \dfrac{\text{lbf}}{\text{in}^2}} + \left(\frac{1016 \dfrac{\text{lbf}}{\text{in}^2}}{8572 \dfrac{\text{lbf}}{\text{in}^2}}\right)^2 = 0.68$$

$$< 1.0 \quad [\text{satisfactory}]$$

The chord is adequate.

10. DESIGN FOR TENSION

Nomenclature

f_t	actual tension stress parallel to grain	lbf/in^2
F_b^*	reference bending design value multiplied by all applicable adjustment factors except C_L	lbf/in^2
F_b^{**}	reference bending design value multiplied by all applicable adjustment factors except C_V	lbf/in^2
F_t, F_t'	reference and adjusted tension design value parallel to grain	lbf/in^2
T	tensile force on member	lbf

Combined Axial Tension and Flexure

Members subjected to combined tension and flexural stresses due to axial and transverse loading must satisfy the two expressions given in NDS Sec. 3.9.1 as

$$\frac{f_t}{F_t'} + \frac{f_b}{F_b^*} \leq 1.0 \qquad [\text{NDS 3.9-1}]$$

$$\frac{f_b - f_t}{F_b^{**}} \leq 1.0 \qquad [\text{NDS 3.9-2}]$$

Example 6.11

The select structural 2×6 Douglas fir-larch bottom chord of a truss is loaded with the service level loads shown in the illustration. The governing load combination consists of dead plus occupancy live load, and the moisture content exceeds 19%. The chord is laterally braced at midlength about the weak axis, and the self-weight of the chord and bracing members may be neglected. Determine whether the member is adequate.

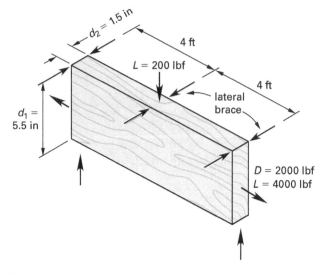

Solution

The reference design value for tension is tabulated in NDS Supp. Table 4A and is

$$F_t = 1000 \text{ lbf/in}^2$$

$$C_t = 1.0, \quad C_i = 1.0$$

The applicable adjustment factors for tension are as follows.

$$C_M = 1.00 \quad [\text{wet service factor from Table 6.6}]$$

$$C_F = 1.3 \quad [\text{size factor from NDS Supp. Table 4A}]$$

The applicable adjustment factors for flexure are as follows.

$$C_M = 0.85 \quad [\text{wet service factor from Table 6.6}]$$

$$C_F = 1.3 \quad [\text{size factor from NDS Supp. Table 4A}]$$

ASD Method

From Table 6.5, the load duration factor for dead load plus occupancy live load is

$$C_D = 1.00$$

The adjusted allowable tension design value parallel to grain is

$$F_t' = F_t C_M C_F C_D$$
$$= \left(1000 \ \frac{\text{lbf}}{\text{in}^2}\right)(1.00)(1.3)(1.00)$$
$$= 1300 \ \text{lbf/in}^2$$

Applying IBC Eq. 16-9, the factored tensile load is

$$T = 1.0D + 1.0L$$
$$= (1.0)(2000 \ \text{lbf}) + (1.0)(4000 \ \text{lbf})$$
$$= 6000 \ \text{lbf}$$

The actual tension stress on the chord is given by

$$f_t = \frac{T}{A}$$
$$= \frac{6000 \ \text{lbf}}{8.25 \ \text{in}^2}$$
$$= 727 \ \text{lbf/in}^2$$
$$< F_t' \quad \text{[satisfactory]}$$

From Ex. 6.10, the beam stability factor is

$$C_L = 0.98$$

The actual edgewise bending stress is

$$f_{b1} = 635 \ \text{lbf/in}^2$$

The reference bending design value multiplied by all applicable adjustment factors except C_L is

$$F_b^* = F_b C_M C_F C_D$$
$$= \left(1500 \ \frac{\text{lbf}}{\text{in}^2}\right)(0.85)(1.3)(1.0)$$
$$= 1658 \ \text{lbf/in}^2$$

The reference bending design value multiplied by all applicable adjustment factors is

$$F_b^{**} = F_b C_M C_L C_F C_D$$
$$= \left(1500 \ \frac{\text{lbf}}{\text{in}^2}\right)(0.85)(0.98)(1.3)(1.0)$$
$$= 1624 \ \text{lbf/in}^2$$

Substituting in the two expressions given in NDS Sec. 3.9.1 gives

$$\frac{f_t}{F_t'} + \frac{f_{b1}}{F_b^*} = \frac{727 \ \dfrac{\text{lbf}}{\text{in}^2}}{1300 \ \dfrac{\text{lbf}}{\text{in}^2}} + \frac{635 \ \dfrac{\text{lbf}}{\text{in}^2}}{1658 \ \dfrac{\text{lbf}}{\text{in}^2}}$$
$$= 0.942$$
$$< 1.0 \quad \text{[satisfactory]}$$

$$\frac{f_{b1} - f_t}{F_b^{**}} = \frac{635 \ \dfrac{\text{lbf}}{\text{in}^2} - 727 \ \dfrac{\text{lbf}}{\text{in}^2}}{1624 \ \dfrac{\text{lbf}}{\text{in}^2}}$$
$$< 1.0 \quad \text{[satisfactory]}$$

The chord is adequate.

LRFD Method

From Table 6.4, the time effect factor for dead load and occupancy live load is

$$\lambda = 0.8$$

The adjusted factored tension design value parallel to grain is

$$F_t' = F_t C_M C_F C_t C_i K_F \lambda \phi_t$$
$$= \left(1000 \ \frac{\text{lbf}}{\text{in}^2}\right)(1.0)(1.3)(1.0)(1.0)(2.70)(0.8)(0.80)$$
$$= 2246 \ \text{lbf/in}^2$$

Applying IBC Eq. 16-2, the factored tension load is

$$T = 1.2D + 1.6L$$
$$= (1.2)(2000 \ \text{lbf}) + (1.6)(4000 \ \text{lbf})$$
$$= 8800 \ \text{lbf}$$

The actual tension stress on the chord is given by

$$f_t = \frac{T}{A}$$
$$= \frac{8800 \ \text{lbf}}{(1.5 \ \text{in})(5.5 \ \text{in})}$$
$$= 1067 \ \text{lbf/in}^2$$
$$< F_t' \quad \text{[satisfactory]}$$

From Ex. 6.10, the beam stability factor is

$$C_L = 0.98$$

Wood

The actual edgewise bending stress is

$$f_{b1} = 1016 \text{ lbf/in}^2$$

The reference bending design value multiplied by all applicable adjustment factors except C_L is

$$F_b^* = F_b C_M C_F C_t C_{fu} C_i C_r K_F \lambda \phi_b$$

$$= \left(1500 \frac{\text{lbf}}{\text{in}^2}\right)(0.85)(1.3)(1.0)(1.0)(1.0)(1.0)$$

$$\times (2.54)(0.8)(0.85)$$

$$= 2863 \text{ lbf/in}^2$$

The reference bending design value multiplied by all applicable adjustment factors except C_V is

$$F_b^{**} = F_b C_M C_L C_F C_t C_{fu} C_i C_r K_F \lambda \phi_b$$

$$= \left(1500 \frac{\text{lbf}}{\text{in}^2}\right)(0.85)(0.98)(1.3)(1.0)(1.0)(1.0)(1.0)$$

$$\times (2.54)(0.8)(0.85)$$

$$= 2806 \text{ lbf/in}^2$$

Substituting in the two expressions given in NDS Sec. 3.9.1 gives

$$\frac{f_t}{F_t'} + \frac{f_{b1}}{F_b^*} = \frac{1067 \dfrac{\text{lbf}}{\text{in}^2}}{2246 \dfrac{\text{lbf}}{\text{in}^2}} + \frac{1016 \dfrac{\text{lbf}}{\text{in}^2}}{2863 \dfrac{\text{lbf}}{\text{in}^2}}$$

$$= 0.83$$

$$< 1.0 \quad \text{[satisfactory]}$$

$$\frac{f_{b1} - f_t}{F_b^{**}} = \frac{1016 \dfrac{\text{lbf}}{\text{in}^2} - 1067 \dfrac{\text{lbf}}{\text{in}^2}}{2806 \dfrac{\text{lbf}}{\text{in}^2}}$$

$$< 1.0 \quad \text{[satisfactory]}$$

The chord is adequate.

11. DESIGN OF CONNECTIONS

Nomenclature

a	center-to-center spacing between adjacent rows of fasteners	in
a_e	minimum edge distance with load parallel to grain	in
a_p	minimum end distance with load parallel to grain	in
a_q	minimum end distance with load perpendicular to grain	in

A	area of cross section	in^2
A_m	gross cross-sectional area of main wood member(s)	in^2
A_n	net area of member	in^2
A_s	sum of gross cross-sectional areas of side member(s)	in^2
C_d	penetration depth factor for connections	–
C_{di}	diaphragm factor for nailed connections	–
C_{eg}	end grain factor for connections	–
C_g	group action factor for connections	–
C_{st}	metal side plate factor for 4 in shear plate connections	–
C_{tn}	toe-nail factor for nailed connections	–
C_Δ	geometry factor for connections	–
d	pennyweight of nail or spike	–
d_e	effective depth of member at a connection	in
D	diameter	in
D_H	fastener head diameter	in
e_p	minimum edge distance unloaded edge	in
e_q	minimum edge distance loaded edge	in
E	length of tapered tip	in
g	gage of screw	–
G	specific gravity	–
l_m	length of bolt in wood main member	in
l_s	total length of bolt in wood side member(s)	in
L	length of nail	in
L	length of screw	in
n	number of fasteners in a row	–
n	number of shear plates	–
N, N'	reference and adjusted lateral design values at an angle of α to the grain for a single split ring connector unit or shear plate connector unit	lbf
p	depth of fastener penetration into wood member, not including tapered tip	in
p_t	length of thread penetration for withdrawal calculations	in
P, P'	reference and adjusted lateral design values parallel to grain for a single split ring connector unit or shear plate connector unit	lbf
Q, Q'	reference and adjusted lateral design values perpendicular to grain for a single split ring connector unit or shear plate connector unit	lbf
R_W	withdrawal design resistance	lbf
s	center-to-center spacing between adjacent fasteners in a row	in
S	unthreaded shank length	in
t_m	thickness of main member	in
t_{ns}	net side member thickness	in
t_s	thickness of side member	in
T	thread length	in
TL	length of threaded shank	in
W, W'	reference and adjusted withdrawal design values for fastener	lbf/in
Z_α'	allowable design values for lag screw with load applied at an angle α to the wood surface	lbf
Z, Z'	reference and adjusted lateral design values for a single fastener connection	lbf

$Z_{\parallel}$ reference lateral design value for a single   lbf
bolt or lag screw connection with all wood
members loaded parallel to grain

$Z_{m\perp}$ reference lateral design value for a single   lbf
bolt or lag screw wood-to-wood connection
with main member loaded perpendicular to
grain and side member loaded parallel to
grain

$Z_{s\perp}$ reference lateral design value for a single   lbf
bolt or lag screw wood-to-wood connection
with main member loaded parallel to grain
and side member loaded perpendicular to
grain

$Z_{\perp}$ reference lateral design value for a single   lbf
bolt or lag screw, wood-to-wood, wood-to-
metal, or wood-to-concrete connection with
all wood member(s) loaded perpendicular to
grain

Symbols

α angle between wood surface and direction of   degree
applied load

γ load/slip modulus for a connection   lbf/in

θ angle between direction of load and   degree
direction of grain (longitudinal axis of
member)

Adjustment of Design Values

The reference design values for fasteners are given in NDS Part 11 through Part 14. These design values are applicable to single fastener connections and normal conditions of use as defined in NDS Sec. 2.2. For other conditions of use, these values are multiplied by adjustment factors, specified in NDS Sec. 11.3, to determine the relevant design values. A summary of the adjustment factors follows, and the applicability of each to the nominal design values is shown in Table 6.11.

Load Duration Factor, C_D

With the exception of the impact load duration factor, values of the load duration factor given in Table 6.5 are applicable to connections. Load duration factors do not apply when the capacity of the connection is governed by metal strength or the strength of concrete or masonry.

Wet Service Factor, C_M

Wet service adjustment factors are given in NDS Table 11.3.3.

Temperature Factor, C_t

The temperature factor is applicable to all connectors and is specified by NDS Table 11.3.4.

Group Action Factor, C_g

The group action factors for various connection geometries and fastener types are given in NDS Table 11.3.6A through Table 11.3.6D. This factor is dependent on the ratio of the area of the side members in a connection to the area of the main member, A_s/A_m. A_m and A_s are calculated by using gross areas without deduction for holes. When adjacent rows of fasteners are staggered, as shown in Fig. 6.7, the adjacent rows are considered a single row.

Figure 6.7 *Staggered Fasteners*

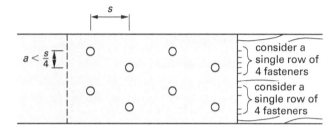

Geometry Factor, C_Δ

The geometry factor applies to bolts, lag screws, split rings, and shear plates. The factor is applied, in accordance with NDS Sec. 12.5.1 and Sec. 13.3.2, when end or edge distances or spacing are less than the specified minimum.

The geometry factor and group action factor are not applied to nails or screws. NDS Comm. Table C12.1.5.7 provides recommended spacing requirements for screws. NDS Comm. Table C12.1.6.6 provides recommended spacing requirements for nails.

Penetration Depth Factor, C_d

The penetration depth factor applies to lag screws, split rings, shear plates, screws, and nails. The factor is applied in accordance with NDS Table 13.2.3 and footnotes to NDS Table 12J through Table 12T when the penetration is less than the minimum specified.

End Grain Factor, C_{eg}

The end grain factor applies to lag screws, screws, and nails. The factor is applied in accordance with NDS Sec. 12.5.2 when the fastener is inserted in the end grain of a member.

C_{eg} is 0.75 for lag screws loaded in withdrawal. C_{eg} is 0.67 for laterally loaded dowel-type fasteners. Spacing and edge distance requirements for laterally loaded fasteners in the narrow edge of cross-laminated timber are given in NDS Table 12.5.1G and NDS Fig. 12I. For lag screws installed in cross-laminated timber end grain or side grain and loaded laterally, $C_{eg} = 0.67$.

Wood screws and nails must not be loaded in withdrawal from the end grain of wood or cross-laminated timber.

Wood

Table 6.11 *Adjustment Factors for Connections**

adjustment factor	bolts	lag screws		split rings and shear plates		screws		nails	
design value	Z	W	Z	P	Q	W	Z	W	Z
C_D load duration factor	✓	✓	✓	✓	✓	✓	✓	✓	✓
C_M wet service factor	✓	✓	✓	✓	✓	✓	✓	✓	✓
C_t temperature factor	✓	✓	✓	✓	✓	✓	✓	✓	✓
C_g group action factor	✓	–	✓	✓	✓	–	–	–	–
C_Δ geometry factor	✓	–	✓	✓	✓	–	–	–	–
C_d penetration depth factor	–	–	✓	✓	✓	–	✓	–	✓
C_{eg} end grain factor	–	✓	✓	–	–	–	✓	–	✓
C_{st} metal side plate factor	–	–	–	✓	–	–	–	–	–
C_{di} diaphragm factor	–	–	–	–	–	–	–	–	✓
C_{tn} toe-nail factor	–	–	–	–	–	–	–	✓	✓

*Z = lateral design value; W = withdrawal design value; P = parallel to grain design value; Q = perpendicular to grain design value

Metal Side Plate Factor, C_{st}

The metal side plate factor is applicable to split rings and shear plates. The factor is applied in accordance with NDS Sec. 13.2.4 when metal side plates are used instead of wood side members.

The effect of metal side plates on the lateral design values of bolts, lag screws, wood screws, and nails are incorporated into the appropriate tables of reference design values.

Diaphragm Factor, C_{di}

The diaphragm factor applies to nails and spikes. The factor is applied in accordance with NDS Sec. 12.5.3 when the fasteners are used in diaphragm construction and $C_{di} = 1.1$.

Toe-Nail Factor, C_{tn}

The toe-nail factor applies to nails and spikes. The factor is applied in accordance with NDS Sec. 12.5.4 when toe-nailed connections are used and $C_{tn} = 0.83$ for lateral design values.

LRFD Factors

Three additional factors are applied in the LRFD method. The format conversion factor given by NDS Table 11.3.1 is $K_F = 3.32$. The resistance factor given by NDS Table 11.3.1 is $\phi = 0.65$. The time effect factor, λ, is given in Table 6.4.

Example 6.12

A bolted connection in tension consists of a single row of eight $\frac{3}{4}$ in diameter bolts in two select structural 2×6 Douglas fir-larch members in single shear. The governing load combination consists of dead plus occupancy live load, and the moisture content exceeds 19%. The bolt spacing and end distance are 4 in. Determine the capacity of the connection.

Solution

The $\frac{3}{4}$ in diameter bolt reference design value for single shear is tabulated in NDS Table 12A as

$$Z_{\parallel} = 720 \text{ lbf}$$
$$C_t = 1.0$$
$$A_s = A_m = 8.25 \text{ in}^2$$

The specified minimum end distance for the full bolt design value is specified in NDS Table 12.5.1A as

$$\begin{aligned} a_p &= 7D \\ &= (7)(0.75 \text{ in}) \\ &= 5.25 \text{ in} \end{aligned}$$

The applicable adjustment factors for the bolts are as follows.

$$C_M = 0.7 \quad \text{[wet service factor from NDS Table 11.3.3]}$$
$$C_g = 0.71 \quad \text{[group action factor from NDS Table 11.3.6A]}$$
$$\begin{aligned} C_\Delta &= \frac{\text{actual end distance}}{\text{specified minimum end distance}} \\ &= \frac{4 \text{ in}}{5.25 \text{ in}} \\ &= 0.76 \quad \text{[geometry factor from NDS Sec. 12.5.1]} \end{aligned}$$

ASD Method

From Table 6.5,

$$C_D = 1.00$$

From Ex. 6.11, the allowable tension capacity of the members is

$$
\begin{aligned}
T &= F_t' A_n \\
&= F_t'\left(A - \left(D + \frac{1}{16}\right)b\right) \\
&= \left(1300\ \frac{\text{lbf}}{\text{in}^2}\right)\left(8.25\ \text{in}^2 - \left(\frac{3}{4}\ \text{in} + \frac{1}{16}\ \text{in}\right)(1.5\ \text{in})\right) \\
&= 9140\ \text{lbf}
\end{aligned}
$$

The allowable capacity for eight bolts is

$$
\begin{aligned}
T &= nZ_{\parallel}C_M C_g C_\Delta C_D C_t \\
&= (8)(720\ \text{lbf})(0.7)(0.71)(0.76)(1.00)(1.0) \\
&= 2176\ \text{lbf}\quad[\text{governs}]
\end{aligned}
$$

LRFD Method

From Table 6.4, for dead load plus occupancy live load, λ is 0.8. From Ex. 6.11, the strength capacity of the members in tension is

$$
\begin{aligned}
T &= F_t' A_n \\
&= F_t'\left(A - \left(D + \frac{1}{16}\right)b\right) \\
&= \left(2246\ \frac{\text{lbf}}{\text{in}^2}\right)\left(8.25\ \text{in}^2 - \left(\frac{3}{4}\ \text{in} + \frac{1}{16}\ \text{in}\right)(1.5\ \text{in})\right) \\
&= 15{,}791\ \text{lbf}
\end{aligned}
$$

The strength lateral capacity for eight bolts is

$$
\begin{aligned}
T &= nZ_{\parallel}C_M C_g C_\Delta C_t K_F \lambda \phi \\
&= (8)(720\ \text{lbf})(0.7)(0.71)(0.76)(1.0)(3.32)(0.8)(0.65) \\
&= 3756\ \text{lbf}\quad[\text{governs}]
\end{aligned}
$$

Bolted Connections

Installation Requirements

In accordance with NDS Sec. 12.1.3, bolt holes must be $\frac{1}{32}$ in to $\frac{1}{16}$ in larger than the bolt diameter, and a metal washer or plate is required between the wood and the nut and bolt head. To ensure that the full design values of bolts are attained, spacing and edge and end distances are specified in NDS Sec. 12.5 and are illustrated in Fig. 6.8.

Single Shear Connection

Reference design values for single shear connections are specified in NDS Sec. 12.3 and are tabulated in NDS Table 12A for two sawn lumber members of identical species, in NDS Table 12.B for a sawn lumber member with a steel side plate, in NDS Table 12.C for a glued laminated member with sawn lumber side member, in NDS Table 12D for a glued laminated member with a steel side plate, and in NDS Table 12.E for connections to concrete.

Example 6.13

A 3×8 select structural Douglas fir-larch ledger attached to a concrete wall with $\frac{3}{4}$ in hook bolts at 4 ft centers is shown. What is the maximum dead load plus occupancy live load that the ledger can support?

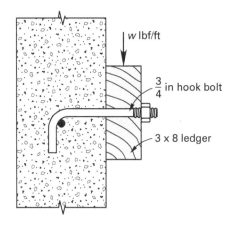

Solution

The applicable adjustment factors are

$$C_M = C_g = C_t = 1.0$$

From NDS Table 12E, the reference lateral design value of a $\frac{3}{4}$ in diameter bolt in a $2\frac{1}{2}$ in thick member loaded perpendicular to the grain and attached to a concrete wall is

$$Z_\perp = 800\ \text{lbf}$$

From NDS Table 12.5.1C, the minimum edge distance for the full bolt design value, for loading perpendicular to grain, is

$$e_{q,\text{full}} = 4D = (4)(0.75\ \text{in}) = 3.0\ \text{in}$$

The actual edge distance is

$$
\begin{aligned}
e_q &= \frac{7.5\ \text{in}}{2} \\
&= 3.75\ \text{in} > e_{q,\text{full}}\quad[\text{satisfactory}]
\end{aligned}
$$

The geometry factor given by NDS Sec. 12.5.1 is

$$C_\Delta = 1.0$$

Figure 6.8 *Bolt Spacing Requirements for Full Design Values*

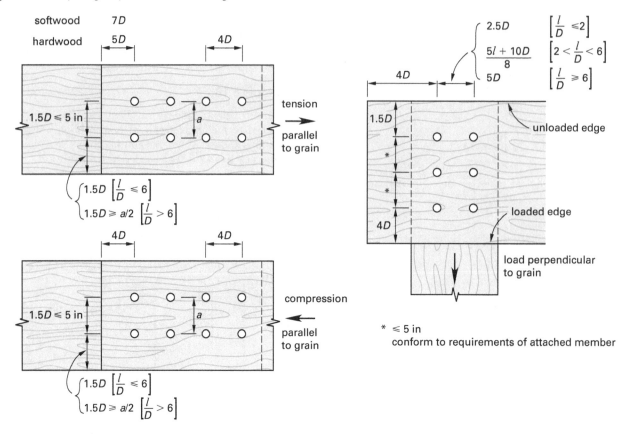

l = lesser of length of bolt in main member or total length of bolt in side member(s)
D = diameter of bolt

ASD Method

From Table 6.5, for dead load plus occupancy live load, the load duration factor is

$$C_D = 1.00$$

The adjusted allowable lateral design value is

$$\begin{aligned}
Z'_\perp &= Z_\perp C_M C_g C_\Delta C_t C_D \\
&= (800 \text{ lbf})(1.0)(1.0)(1.0)(1.0)(1.00) \\
&= 800 \text{ lbf}
\end{aligned}$$

The maximum allowable load that the ledger can support is

$$\begin{aligned}
w &= \frac{Z'_\perp}{4} = \frac{800 \text{ lbf}}{4 \text{ ft}} \\
&= 200 \text{ lbf/ft}
\end{aligned}$$

LRFD Method

From Table 6.4, for dead load plus occupancy live load, the time effect factor is

$$\lambda = 0.8$$

From NDS Table 11.3.1, the format conversion factor is

$$K_F = 3.32$$

The resistance factor given by NDS Table 11.3.1 is

$$\phi = 0.65$$

The adjusted factored lateral design value is

$$\begin{aligned}
Z'_\perp &= Z_\perp C_M C_g C_\Delta C_t K_F \lambda \phi \\
&= (800 \text{ lbf})(1.0)(1.0)(1.0)(1.0)(3.32)(0.8)(0.65) \\
&= 1381 \text{ lbf}
\end{aligned}$$

The maximum strength level load that the ledger can support is

$$w = \frac{Z'_\perp}{4} = \frac{1381 \text{ lbf}}{4 \text{ ft}} = 345 \text{ lbf/ft}$$

Double Shear Connection

Reference design values for double shear connections are tabulated in NDS Table 12F for three sawn lumber members of identical species, in NDS Table 12G for a sawn lumber member with steel side plates, in NDS Table 12H for a glued laminated member with sawn lumber side members, and in NDS Table 12I for a glued laminated member with steel side plates.

Example 6.14

Determine the minimum values for the dimensions A, B, C, and D, shown in the illustration, that will allow the full design values to be applied to the $\frac{3}{4}$ in diameter bolts. Determine the maximum tensile force, T, due to wind load that can be resisted by the connection. The $5\frac{1}{8} \times 12$ glued laminated member is of Douglas fir-larch species.

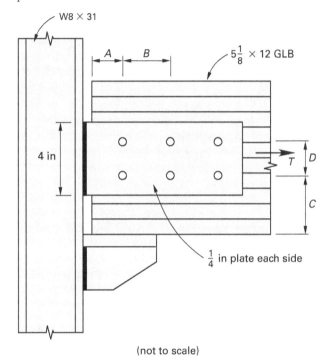

(not to scale)

Solution

The $\frac{3}{4}$ in diameter bolt reference design value for double shear is tabulated in NDS Table 12I as

$$Z_\| = 3340 \text{ lbf}$$
$$C_M = 1.0, \;\; C_t = 1.0$$
$$A_s = 2 \text{ in}^2, \;\; A_m = 61.5 \text{ in}^2$$
$$\frac{A_m}{A_s} = 30.75$$

The applicable adjustment factors for the bolts are as follows.

C_g = group action factor from NDS Table 11.3.6C

$\qquad = 0.99$

C_Δ = geometry factor from NDS Sec. 12.5.1

$\qquad = 1.0 \begin{bmatrix} \text{All dimensions conform} \\ \text{to the specified minimums.} \end{bmatrix}$

The specified minimum end distance, A, for the full bolt design value is specified in NDS Table 12.5.1A as

$$a_p = 7D$$
$$= (7)(0.75 \text{ in})$$
$$= 5.25 \text{ in}$$

The specified minimum spacing, B, for the full bolt design value is specified in NDS Table 12.5.1B as

$$s = 4D$$
$$= (4)(0.75 \text{ in})$$
$$= 3.00 \text{ in}$$

The specified minimum spacing between rows, D, for the full bolt design value is specified in NDS Table 12.5.1D as

$$a = 1.5D$$
$$= (1.5)(0.75 \text{ in})$$
$$= 1.125 \text{ in}$$

The ratio of the length of the bolt in the main member to the bolt diameter is

$$\frac{l_m}{D} = \frac{5.125 \text{ in}}{0.75 \text{ in}}$$
$$= 6.83$$
$$> 6$$

The specified minimum edge distance, C, for the full bolt design value is specified in NDS Table 12.5.5C as the greater of

- $a_e = \dfrac{a}{2}$

$\qquad = \dfrac{1.125 \text{ in}}{2}$

$\qquad = 0.563 \text{ in}$

- $a_e = 1.5D$

$\qquad = (1.5)(0.75 \text{ in})$

$\qquad = 1.125 \text{ in} \quad [\text{governs}]$

ASD Method

C_D = load duration factor for wind load

$\qquad$ from Table 6.5

$\qquad = 1.60$

The allowable lateral capacity for six bolts is

$$T = nZ_{\|}C_D C_g C_\Delta C_M C_t$$
$$= (6)(3340 \text{ lbf})(1.60)(0.99)(1.0)(1.0)(1.0)$$
$$= 31{,}743 \text{ lbf}$$

LRFD Method

From Table 6.4, the time effect factor for wind load is

$$\lambda = 1.0$$

From NDS Table 11.3.1, the format conversion factor is

$$K_F = 3.32$$

From NDS Table 11.3.1, the resistance factor is

$$\phi = 0.65$$

The strength lateral capacity for six bolts is

$$T = nZ_{\|}C_M C_t C_g C_\Delta K_F \lambda \phi$$
$$= (6)(3340 \text{ lbf})(1.0)(1.0)(0.99)(1.0)(3.32)(1.0)(0.65)$$
$$= 42{,}814 \text{ lbf}$$

Lag Screw Connections

Installation Requirements

In accordance with NDS Sec. 12.1.4, a clearance hole matching the diameter of the shank must be bored for the full length of the unthreaded shank in the member. A lead hole at least equal in length to the threaded portion of the screw must be provided. For wood with a specific gravity greater than 0.6, the lead hole diameter must equal 65–85% of the shank diameter. For wood with a specific gravity of 0.5 or less, the lead hole diameter must equal 40–70% of the shank diameter. For wood with an intermediate specific gravity, the lead hole diameter must equal 60–75% of the shank diameter.

Lag screws are inserted into the lead hole using a wrench. Lead holes and clearance holes are not required when lag screws with a diameter $\frac{3}{8}$ in or less are loaded primarily in withdrawal in wood that has a specific gravity of 0.5 or less provided that edge distance, end distance, and spacing are sufficient to prevent splitting.

Lateral Design Values in Side Grain

Minimum edge distances, end distances, spacing, and geometry factors are identical with those for bolts with a diameter equal to the shank diameter of the lag screw. As specified in NDS Table 12J, for full design values to

be applicable, the depth of lag screw penetration, not including the length of the tapered tip, must not be less than

$$p = 8D$$

The minimum allowable penetration is $4D$. When the penetration is between $4D$ and $8D$, the reference design value is multiplied by the penetration factor, which is defined in NDS Table 12J as

$$C_d = \frac{p}{8D}$$
$$\leq 1.0$$

Reference design values, for $p - 8D$, for single shear connections are specified in NDS Sec. 12.3 and are tabulated in NDS Table 12J for connections with a wood side member and in NDS Table 12K for connections with a steel side plate.

For lateral loads, the design value is governed by the total length of penetration of the lag screw into the main member, less the length of the tapered tip, and is given by

$$p = S + T - E - t_s$$

The notation is defined in NDS App. Table L2 and is

S = unthreaded shank length
$T - E$ = thread length less the length of the tapered tip
t_s = thickness of side member

Withdrawal Design Values in Side Grain Without Lateral Load

Minimum edge distance, end distance, and spacing are specified in NDS Table 12.5.1E and are

$$\text{edge distance} = 1.5D$$
$$\text{end distance} = 4D$$
$$\text{spacing} = 4D$$

For withdrawal loads, the design value is governed by the total length of thread penetration of the lag screw into the main member, less the length of the tapered tip, and is given by

$$p_t = T - E$$

Reference withdrawal design values in pounds per inch of thread penetration (not including the length of the tapered tip) are tabulated in NDS Table 12.2A.

Where lag screws are loaded in withdrawal from the narrow edge of cross-laminated timber, the reference withdrawal value, W, is multiplied by the end grain factor, $C_{eg} = 0.75$, regardless of grain orientation.

Combined Lateral and Withdrawal Design Values

When the load applied to a lag screw is at an angle, α, to the wood surface, the lag screw is subjected to combined lateral and withdrawal loading. The design value is determined by the Hankinson formula given by NDS Sec. 12.4.1 as

$$Z'_\alpha = \frac{W'p_t Z'}{W'p_t \cos^2\alpha + Z'\sin^2\alpha}$$

Example 6.15

A 3 in long, $\frac{3}{8}$ in diameter lag screw inserted into a Douglas fir-larch joist with a 10-gage steel side plate is subjected to a force inclined at an angle of 30° to the wood surface. Determine the maximum force, due to occupancy live load, that may be applied.

Solution

From NDS Table 12K, the reference lateral design value for load applied parallel to the grain is

$$Z_\| = 220 \text{ lbf}$$
$$C_M = 1.0, \;\; C_t = 1.0, \;\; C_g = 1.0, \;\; C_\Delta = 1.0$$

From NDS App. Table L2, the penetration into the main member of the screw shank, plus the threaded length and less the length of the tapered tip is

$$p = S + (T - E) - t_s$$
$$= 1.0 \text{ in} + 1.781 \text{ in} - 0.134 \text{ in}$$
$$= 2.647 \text{ in}$$

From NDS Table 12K, footnote 3, the penetration factor is obtained as

$$C_d = \frac{p}{8D} = \frac{2.647 \text{ in}}{(8)(0.375 \text{ in})}$$
$$= 0.882$$

From NDS Table 12K, the reference lateral design value is $Z_\| = 220$ lbf.

From NDS Table 12.3.3A, the specific gravity of the Douglas fir-larch joist is

$$G = 0.50$$

From NDS App. Table L2, the penetration into the main member of the threaded length, less the length of the tapered tip, is

$$p_t = T - E$$
$$= 1.781 \text{ in}$$

ASD Method

For occupancy live load, $C_D = 1.0$.

The adjusted allowable lateral design value is

$$Z'_\| = Z_\| C_d C_D C_M C_t C_g C_\Delta$$
$$= (220 \text{ lbf})(0.882)(1.0)(1.0)(1.0)(1.0)(1.0)$$
$$= 194 \text{ lbf}$$

From NDS Table 12.2A, the reference withdrawal design value is

$$W = 305 \text{ lbf/in}$$

The adjusted allowable withdrawal design value is

$$W' = W C_D C_M C_t$$
$$= \left(305 \; \frac{\text{lbf}}{\text{in}}\right)(1.0)(1.0)(1.0)$$
$$= 305 \text{ lbf/in}$$

The maximum allowable force that may be applied is determined by NDS Sec. 12.4.1 as

$$Z'_\alpha = \frac{(W'p_t)Z'_\|}{(W'p_t)\cos^2\alpha + Z'_\| \sin^2\alpha}$$
$$= \frac{\left(305 \; \dfrac{\text{lbf}}{\text{in}}\right)(1.781 \text{ in})(194 \text{ lbf})}{\left(305 \; \dfrac{\text{lbf}}{\text{in}}\right)(1.781 \text{ in})(\cos^2 30°) + (194 \text{ lbf})(\sin^2 30°)}$$
$$= 231 \text{ lbf}$$

LRFD Method

From Table 6.4, the time effect factor for occupancy live load is

$$\lambda = 0.8$$

The format conversion factor from NDS Table 11.3.1 is

$$K_F = 3.32$$

The resistance factor given by NDS Table 11.3.1 is

$$\phi = 0.65$$

The adjusted factored lateral design value is

$$Z'_\| = Z_\| C_M C_t C_g C_\Delta C_d K_F \lambda \phi$$
$$= (220 \text{ lbf})(1.0)(1.0)(1.0)(1.0)(0.882)$$
$$\times (3.32)(0.8)(0.65)$$
$$= 335 \text{ lbf}$$

Wood

The adjusted factored withdrawal design value is

$$W' = W C_M C_t K_F \lambda \phi$$
$$= (305 \text{ lbf})(1.0)(1.0)(3.32)(0.8)(0.65)$$
$$= 527 \text{ lbf}$$

From NDS Sec. 12.4.1, the maximum strength level force that may be applied is

$$Z'_\alpha = \frac{(W'p_t)Z'_\parallel}{(W'p_t)\cos^2\alpha + Z'_\parallel \sin^2\alpha}$$

$$= \frac{\left(527 \dfrac{\text{lbf}}{\text{in}}\right)(1.781 \text{ in})(335 \text{ lbf})}{\left(527 \dfrac{\text{lbf}}{\text{in}}\right)(1.781 \text{ in})\cos^2 30° + (335 \text{ lbf})\sin^2 30°}$$

$$= 399 \text{ lbf}$$

Split Ring and Shear Plate Connections

Installation Requirements

Edge and end distances, spacing, and geometry factors, C_Δ, for various sizes of split ring and shear plate connectors are specified in NDS Table 13.3.2.2 and Table 13.3.3.1-1 through Table 13.3.3.1-4. When lag screws are used instead of bolts, nominal design values should, where appropriate, be multiplied by the penetration depth factors specified in NDS Table 13.2.3 for various sizes of connectors and wood species. NDS Table 13.2.4 provides metal side plate factors, C_{st}, for 4 in shear plate connectors, loaded parallel to the grain, when metal side plates are substituted for wood side members. Group action factors, C_g, for 4 in split ring or shear plate connectors with wood side members are tabulated in NDS Table 11.3.6B. Group action factors, C_g, for 4 in shear plate connectors with steel side plates are tabulated in NDS Table 11.3.6D. Dimensions for split ring and shear plate connectors are provided in NDS App. K.

In accordance with NDS Sec. 13.1.1, provisions for the design of split rings and shear plates are not applicable to cross-laminated timber.

Lateral Design Values

Reference design values for split ring connectors are provided in NDS Table 13.2A and for shear plate connectors in NDS Table 13.2B. When a load acts in the plane of the wood surface at an angle θ to the grain, the allowable design value is given by NDS Sec. 13.2.5 as

$$N' = \frac{P'Q'}{P'\sin^2\theta + Q'\cos^2\theta} \qquad \text{[NDS 13.2-1]}$$

Example 6.16

The Douglas fir-larch select structural members shown in the illustration are connected with $2\frac{5}{8}$ in shear plate connectors. The governing load combination consists of dead plus occupancy live loads. The connector spacing and end distances are as shown. Determine the capacity of the connection.

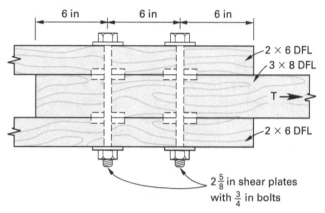

Solution

The specific gravity of Douglas fir-larch is obtained from NDS Table 12.3.3A as $G = 0.5$. From NDS Table 13A, the species group is B. The reference $2\frac{5}{8}$ in shear plate design value for the $2\frac{1}{2}$ in thick main member of group B species with a connector on two faces is tabulated in NDS Table 13.2B as

$$P_{\text{main}} = 2860 \text{ lbf}$$

The reference $2\frac{5}{8}$ in shear plate design value for a $1\frac{1}{2}$ in thick side member of group B species with a connector on one face is tabulated in NDS Table 13.2B as

$$P_{\text{side}} = 2670 \text{ lbf} \quad \text{[governs]}$$
$$C_M = 1.0, \quad C_t = 1.0$$
$$A_s = (2)(8.25 \text{ in}^2) = 16.5 \text{ in}^2$$
$$A_m = 18.13 \text{ in}^2$$
$$\frac{A_s}{A_m} = 0.91$$

The specified minimum spacing for the full shear plate design value is given in NDS Table 13.3 as

$$s = 6.75 \text{ in}$$

The applicable adjustment factors for the bolts are as follows.

$$C_g = \text{group action factor from NDS}$$
$$\text{Table 11.3.6B}$$
$$= 0.98$$

C_Δ = geometry factor by linear interpolation from NDS Sec. 13.3.2.1 for a spacing of 6 in

$$= 0.5 + \frac{(0.5)(6 \text{ in} - 3.5 \text{ in})}{6.75 \text{ in} - 3.5 \text{ in}}$$

$$= 0.885$$

ASD Method

From Table 6.5, the load duration factor is

$$C_D = 1.0$$

The allowable design value for four shear plates is

$$\begin{aligned}
T &= n P_{\text{side}} C_g C_\Delta C_D C_M C_t \\
&= (4)(2670 \text{ lbf})(0.98)(0.885)(1.00)(1.0)(1.0) \\
&= 9263 \text{ lbf}
\end{aligned}$$

LRFD Method

From Table 6.4, the time effect factor for occupancy live load is

$$\lambda = 0.8$$

The format conversion factor from NDS Table 11.3.1 is

$$K_F = 3.32$$

The resistance factor given by NDS Table 11.3.1 is

$$\phi = 0.65$$

The strength level design value for four shear plates is

$$\begin{aligned}
T &= n P_{\text{side}} C_g C_\Delta C_M C_t K_F \lambda \phi \\
&= (4)(2670 \text{ lbf})(0.98)(0.885)(1.0)(1.0)(3.32) \\
&\quad \times (0.8)(0.65) \\
&= 15{,}991 \text{ lbf}
\end{aligned}$$

Wood Screw Connections

Installation

In accordance with NDS Sec. 12.1.5.3, wood screws loaded laterally in wood with a specific gravity in excess of 0.6 must have a clearance hole approximately equal in diameter to the diameter of the shank. The clearance hole must be bored in the member for the full length of the unthreaded shank. The lead hole receiving the threaded portion of the screw must have a diameter approximately equal to the wood screw root diameter. Wood with a specific gravity not exceeding 0.6 must have a clearance hole approximately equal in diameter to $^7/_8$ the diameter of the shank. The clearance hole must be bored in the member for the full length of the unthreaded shank. The lead hole receiving the threaded portion of the screw must have a diameter approximately equal to $^7/_8$ the diameter to the wood screw root diameter.

Lateral Design Values in Side Grain

Recommended edge distances, end distances, and spacing are tabulated in NDS Comm. Table C12.1.5.7 for wood and steel side plates with and without pre-bored holes. Wood screws are not subject to the group action factor, C_g, or the geometry factor.

As specified in NDS Table 12L, for full design values to be applicable, the depth of penetration must not be less than

$$p = 10D$$

The minimum allowable penetration is $6D$. When the penetration is between $6D$ and $10D$, the reference design value is multiplied by the penetration factor, which is defined in NDS Table 12L as

$$\begin{aligned}
C_d &= \frac{p}{10D} \\
&\leq 1.0
\end{aligned}$$

Reference design values, for $p - 8D$, for single shear connections are specified in NDS Sec. 12.3 and tabulated in NDS Table 12L for connections with a wood side member and in NDS Table 12M for connections with a steel side plate.

Withdrawal Design Values in Side Grain

Withdrawal design values in pounds per inch of thread penetration are tabulated in NDS Table 12.2B. The length of thread is specified in App. L as two-thirds the total screw length or four times the screw diameter, whichever is greater.

Wood screws must not be loaded in withdrawal from the end grain of wood or cross-laminated timber.

Combined Lateral and Withdrawal Loads

When the load applied to a wood screw is at an angle, α, to the wood surface, the wood screw is subjected to combined lateral and withdrawal loading, and the design value is determined by the Hankinson formula given by NDS Sec. 12.4.1 as

$$Z'_\alpha = \frac{W' p_t Z'}{W' p_t \cos^2 \alpha + Z' \sin^2 \alpha}$$

Example 6.17

A 7g steel strap is secured to a select structural Douglas fir-larch collector with ten 14g $\times$ 3 in wood screws. Edge and end distances and spacing are sufficient to prevent splitting of the wood. Determine the maximum tensile force, T, due to wind load that can be resisted by the connection.

Solution

The reference design value for single shear is tabulated in NDS Table 12M as

$$Z = 202 \text{ lbf}$$
$$C_M = 1.0, \quad C_t = 1.0$$

The penetration of the screw shank plus the threaded length is

$$p = L - t_s$$
$$= 3 \text{ in} - 0.179 \text{ in}$$
$$= 2.821 \text{ in}$$

This is greater than $10D$, and from NDS Table 12M, the penetration depth factor is

$$C_d = 1.0$$

ASD Method

From Table 6.5, the load duration factor is

$$C_D = 1.60$$

The allowable lateral design value for 10 screws is

$$T = nZC_D C_d C_M C_t$$
$$= (10)(202 \text{ lbf})(1.60)(1.0)(1.0)(1.0)$$
$$= 3232 \text{ lbf}$$

LRFD Method

From Table 6.4, the time effect factor for wind load is

$$\lambda = 1.0$$

The format conversion factor given in NDS Table 11.3.1 is

$$K_F = 3.32$$

The resistance factor given in NDS Table 11.3.1 is

$$\phi = 0.65$$

The strength level lateral design value for the 10 screws is

$$T = nZC_M C_t C_d K_F \lambda \phi$$
$$= (10)(202 \text{ lbf})(1.0)(1.0)(1.0)(3.32)(1.0)(0.65)$$
$$= 4359 \text{ lbf}$$

Connections with Nails and Spikes

Installation

NDS specifications apply to common nails and spikes, box nails, sinker nails, and threaded hardened-steel nails. The tabulated nominal design values apply to nailed connections with or without pre-bored holes. As specified in NDS Sec. 12.1.6.2, pre-bored holes may be used to prevent the splitting of wood. For wood with a specific gravity greater than 0.6, the hole diameter may not exceed 90% of the diameter of the nail. For wood with a specific gravity less than or equal to 0.6, the hole diameter may not exceed 75% of the diameter of the nail.

As shown in Fig. 6.9 and as specified in NDS Sec. 12.1.6.3, toe nails are driven into the member at an angle of approximately 30° with the point of penetration approximately one-third the length of the nail from the member end. The design values for lateral loads are

$$t_s = \frac{L}{3}$$

$$p = L \cos 30° - \frac{L}{3}$$

Figure 6.9 *Toe-Nailed Connection*

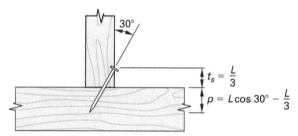

Lateral Design Values in Side Grain

Recommended edge distances, end distances, and spacing are tabulated in NDS Comm. Table C12.1.6.6 for wood and steel side plates with and without pre-bored holes. Nails and spikes are not subject to the group action factor, C_g, or the geometry factor, C_Δ.

As specified in NDS Table 12N, for full design values to be applicable, the depth of penetration must not be less than

$$p = 10D$$

The minimum allowable penetration is $6D$. When the penetration is between $6D$ and $10D$, the reference design value is multiplied by the penetration factor, which is defined in NDS Table 12N as

$$C_d = \frac{p}{10D}$$
$$\leq 1.0$$

Wood

Reference design values for nails and spikes used in diaphragm construction are multiplied by the diaphragm factor $C_{di} = 1.1$.

Reference lateral design values for nails and spikes used in toe-nailed connections are multiplied by the toe-nail factor $C_{tn} = 0.83$.

Reference withdrawal design values for nails and spikes used in toe-nailed connections are multiplied by the toe-nail factor $C_{tn} = 0.67$. The wet service factor does not apply. The penetration of the toe-nail in a withdrawal connection is

$$p = L - L/3 \cos 30°$$

Reference design values for single shear connections for two sawn lumber members of identical species are tabulated in NDS Table 12N.

Reference design values for single shear connections for a sawn lumber member with steel side plates are tabulated in NDS Table 12P.

The reference double shear value for a three-member sawn lumber connection is twice the lesser of the nominal design value for each shear plane. The minimum penetration into the side member must be six times the connector diameter, or when the side member is at least $\frac{3}{8}$ in thick and 0.148 in diameter or smaller nails extend at least three diameters beyond the side member the nails must be clinched.

Example 6.18

A 4×8 select structural Douglas fir-larch collector is secured to the Douglas fir-larch top plate of a shear wall with a 12-gage steel strap, as shown. Fourteen 16d common nails $3\frac{1}{2}$ in long are on each side of the strap. Edge and end distances and spacing are sufficient to prevent splitting of the wood. $C_M = C_t = 1.0$. Determine the maximum tensile force due to wind load that can be resisted by the nails.

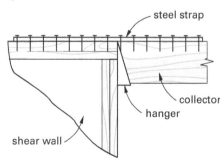

Solution

From NDS Table 12P, the reference single shear design value for a 16d common nail in a Douglas fir-larch member with a 12-gage side plate is

$$Z = 149 \text{ lbf}$$

As specified in NDS Table 12P, for full design values to be applicable, the depth of penetration must not be less than

$$\begin{aligned}
p_{\text{full}} &= 10D = (10)(0.162 \text{ in}) \\
&= 1.62 \text{ in}
\end{aligned}$$

The actual penetration of the nails is

$$\begin{aligned}
p &= L - t_s \\
&= 3\frac{1}{2} \text{ in} - 0.105 \text{ in} \\
&= 3.395 \text{ in} \\
&> p_{\text{full}}
\end{aligned}$$

Since $p > p_{\text{full}}$, the penetration depth factor is given by

$$C_d = 1.0$$

ASD Method

From Table 6.5, the load duration factor for wind load is

$$C_D = 1.60$$

The allowable lateral design value for fourteen 16d nails is

$$\begin{aligned}
T &= nZC_DC_MC_dC_t \\
&= (14)(149 \text{ lbf})(1.60)(1.0)(1.0)(1.0) \\
&= 3338 \text{ lbf}
\end{aligned}$$

LRFD Method

From Table 6.4, the time effect factor for wind load is

$$\lambda = 1.0$$

The format conversion factor given in NDS Table 11.3.1 is

$$K_F = 3.32$$

The resistance factor given in NDS Table 11.3.1 is

$$\phi = 0.65$$

The strength level lateral design value for the 14 nails is

$$\begin{aligned}
T &= nZC_MC_tC_dK_F\lambda\phi \\
&= (14)(149 \text{ lbf})(1.0)(1.0)(1.0)(3.32)(1.0)(0.65) \\
&= 4502 \text{ lbf}
\end{aligned}$$

Wood

Withdrawal Design Values in Side Grain

Reference withdrawal design values for nails and spikes used in toe-nailed connections are multiplied by the toe-nail factor $C_{tn} = 0.67$. The wet service factor does not apply. The penetration of the toe-nail in a withdrawal connection is

$$p = L - L/3 \cos 30°$$

Reference withdrawal values, W, in lbf/in of penetration for smooth shank bright or galvanized nails are given in NDS Table 12.2C. Reference withdrawal values may also be calculated using NDS Eq. 12.2-3, which is

$$W = 1380 G^{5/2} D$$

The withdrawal design resistance, R_W, in lbf is given by

$$R_W = W p_t$$

Reference withdrawal values, W, in lbf/in of penetration for smooth shank stainless steel nails, are given in NDS Table 12.2D. Reference withdrawal values may also be calculated using NDS Eq. 12.2-4, which is

$$W = 465 G^{3/2} D$$

The withdrawal design resistance, R_W, in lbf is given by

$$R_W = W p_t$$

Reference withdrawal values, W, in lbf/in of ring shank penetration for roof sheathing or post-frame ring shank nails are given in NDS Table 12.2E. Reference withdrawal values may also be calculated using NDS Eq. 12.2-5, which is

$$W = 1800 G^2 D$$

For uncoated carbon steel roof sheathing or post-frame ring shank nails, reference withdrawal values are given by NDS Sec. 12.2.3.2 as

$$W = (1.25)(1800 G^2 D)$$

The withdrawal design resistance, R_W, in lbf is given by multiplying the reference withdrawal value, W, by the ring shank penetration, TL, given in NDS Table L5 to give

$$R_W = W(TL)$$

The adjustment factors applicable to withdrawal values are given in NDS Table 11.3.1 and are load duration factor (ASD only), temperature factor, wet service factor, format conversion factor (LRFD only), resistance factor (LRFD only), and time effect factor (LRFD only).

Nails and spikes must not be used in withdrawal from the end grain of wood or cross-laminated timber.

Combined Lateral and Withdrawal Loads

When the load applied to a nail or spike is at an angle, α, to the wood surface, the nail or spike is subjected to combined lateral and withdrawal loading, and in accordance with the NDS Sec. 12.4.2, the design value is determined by the interaction equation

$$Z'_\alpha = \frac{W' p_t Z'}{W' p_t \cos\alpha + Z' \sin\alpha}$$

Example 6.19

Determine the lateral design value for the 3 in long 10d common wire nail in the toe-nailed connection shown. Loading applied to the connection is due to wind load, and all members are Douglas fir-larch.

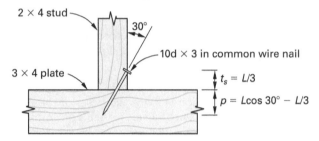

Solution

As specified in NDS Sec. 12.1.6.3, toe nails are driven at an angle of 30° from the face of the member, with the point of penetration one-third the length of the nail from the member end. In accordance with NDS Comm. Sec. C12.1.5, the side member thickness is taken to be equal to this end distance and

$$t_s = \frac{L}{3} = \frac{3 \text{ in}}{3}$$
$$= 1 \text{ in}$$

The nominal design value for single shear is tabulated in NDS Table 12N as

$$Z = 118 \text{ lbf}$$
$$C_M = 1.0, \quad C_t = 1.0$$

The applicable adjustment factors for the nail are as follows.

- The penetration of the nail into the sill plate, in accordance with NDS Comm. Sec. C12.1.6, is taken as the vertically projected length of the nail in the member and

$$p = L \cos 30° - \frac{L}{3}$$
$$= (3 \text{ in})(0.866) - \frac{3 \text{ in}}{3}$$
$$= 1.60 \text{ in}$$

This is greater than $10D$, and from NDS Table 12N, the penetration depth factor is

$$C_d = 1.0$$

- C_{tn} = toe-nail factor from NDS Sec. 12.5.4.

$$C_{tn} = 0.83$$

ASD Method

From Table 6.5, the load duration factor is

$$C_D = 1.60$$

The allowable lateral design value for the nail is

$$
\begin{aligned}
Z' &= ZC_D C_d C_{tn} C_M C_t \\
&= (118 \text{ lbf})(1.60)(1.0)(0.83)(1.0)(1.0) \\
&= 157 \text{ lbf}
\end{aligned}
$$

LRFD Method

From Table 6.4, the time effect factor for wind load is

$$\lambda = 1.0$$

The format conversion factor given in NDS Table 11.3.1 is

$$K_F = 3.32$$

The resistance factor given in NDS Table 11.3.1 is

$$\phi = 0.65$$

The strength level lateral design value for the nail is

$$
\begin{aligned}
T &= ZC_M C_t C_d C_{tn} K_F \lambda \phi \\
&= (118 \text{ lbf})(1.0)(1.0)(1.0)(0.83)(3.32)(1.0)(0.65) \\
&= 211 \text{ lbf}
\end{aligned}
$$

Fastener Head Pull-Through Design Values

For fasteners with round heads, the reference pull-through values, W_H, in lbf for wood side members are given in NDS Table 12.2F. Reference pull-through values may also be calculated using NDS Eq. 12.2-6a and Eq. 12.2-6b, which are

$$
\begin{aligned}
W_H &= 690\pi D_H G^2 t_{ns} \quad [t_{ns} \leq 2.5 D_H] \\
&= 1725\pi D_H^2 G^2 \quad [t_{ns} > 2.5 D_H]
\end{aligned}
$$

The adjustment factors applicable to pull-through values are given in NDS Table 11.3.1 and are load duration factor (ASD only), wet service factor, temperature

factor, format conversion factor (LRFD only), resistance factor (LRFD only), and time effect factor (LRFD only).

Fastener Uplift Capacity

Fasteners connecting roof sheathing to the structural framework must resist the uplift wind forces as shown in Fig. 6.10.

Figure 6.10 Fastener Uplift Capacity

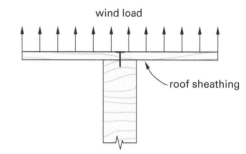

The uplift capacity of the connection is the lesser of the fastener withdrawal design value and the fastener head pull-through design value.

Example 6.20

Determine the uplift capacity of the roofing connection shown in Fig. 6.10. The fastener consists of a Dash No. 04 galvanized roof sheathing ring shank nail, and the roof sheathing is $^7/_{16}$ in thick Structural 1 plywood. The framing consists of a 3×8 select structural Douglas fir-larch visually graded member. Normal temperature and dry service conditions are applicable.

Solution

The ring shank nail properties of the Dash No. 04 nail are obtained from NDS Table L6 and are

$$
\begin{aligned}
\text{diameter, } D &= 0.120 \text{ in} \\
\text{length of threaded shank, } TL &= 1.5 \text{ in} \\
\text{head diameter, } H &= 0.281 \text{ in}
\end{aligned}
$$

The specific gravity, G, of the Structural 1 plywood is obtained from NDS Table 12.3.3B and is 0.5.

The specific gravity, G, of the select structural Douglas fir-larch visually graded member is obtained from NDS Table 12.3.3A and is 0.5.

From the problem statement,

$$
\begin{aligned}
C_M &= 1.0 \quad \text{[wet service factor]} \\
C_t &= 1.0 \quad \text{[temperature factor]}
\end{aligned}
$$

Reference withdrawal value of the ring shank nail, W, is obtained from NDS Table 12.2E and is 54 lbf/in.

Reference pull-through value of the ring shank nail, W_H, is obtained from NDS Table 12.2F and is 67 lbf.

ASD Method

The load duration factor for wind loading is obtained from NDS Table 2.3.2 as

$$C_D = 1.6 \quad \text{[load duration factor]}$$

The adjusted allowable withdrawal value is

$$W' = WC_D C_M C_t$$
$$= \left(54 \, \frac{\text{lbf}}{\text{in}}\right)(1.6)(1.0)(1.0)$$
$$= 86.4 \, \text{lbf/in}$$

The adjusted allowable withdrawal design resistance is

$$R'_W = W'(TL)$$
$$= \left(86.4 \, \frac{\text{lbf}}{\text{in}}\right)(1.5 \, \text{in})$$
$$= 129.6 \, \text{lbf}$$

The adjusted allowable pull-through value is

$$W'_H = W_H C_D C_M C_t$$
$$= (67 \, \text{lbf})(1.6)(1.0)(1.0)$$
$$= 107.2 \, \text{lbf} \quad \text{[governs]}$$
$$< \text{adjusted allowable withdrawal design resistance}$$

LRFD Method

From NDS Table N3.3, the applicable time effect factor is

$$\lambda = 1.0$$

From NDS Table 11.3.1, the applicable resistance factor is

$$\phi = 0.65$$

From NDS Table 11.3.1, the applicable format conversion factor is

$$K_F = 3.32$$

The adjusted factored design withdrawal value is

$$W' = WC_M C_t K_F \lambda \phi$$
$$= \left(54 \, \frac{\text{lbf}}{\text{in}}\right)(1.0)(1.0)(3.32)(1.0)(0.65)$$
$$= 116.5 \, \text{lbf/in}$$

The adjusted factored design withdrawal resistance is

$$R'_W = W'(TL)$$
$$= \left(116.5 \, \frac{\text{lbf}}{\text{in}}\right)(1.5 \, \text{in})$$
$$= 174.8 \, \text{lbf}$$

The adjusted factored design pull-through resistance value is

$$W'_H = W_H C_M C_t K_F \lambda \phi$$
$$= (67 \, \text{lbf})(1.0)(1.0)(3.32)(1.0)(0.65)$$
$$= 144.6 \, \text{lbf} \quad \text{[governs]}$$
$$< \text{adjusted factored design withdrawal resistance}$$

12. REFERENCES

1. American Wood Council. *National Design Specification for Wood Construction.* Leesburg, VA: American Wood Council, 2018.

2. American Wood Council. *National Design Specification Supplement: Design Values for Wood Construction.* Leesburg, VA: American Wood Council, 2018.

3. International Code Council. *International Building Code.* Falls Church, VA: International Code Council, 2018.

4. American Wood Preservers' Association. *Book of Standards.* Selma, AL, 2011.

5. American Plywood Association. *Glued Laminated Beam Design Tables.* Tacoma, WA: Engineered Wood Systems, 2016.

6. Western Wood Products Association. *Western Lumber Span Tables.* Portland, OR: Western Wood Products Association.

Wood

7 Reinforced Masonry Design

1. CONSTRUCTION DETAILS

Requirements for the placement of reinforcement are given in TMS 402[1] Sec. 6.1. Reinforcing bars must be securely supported to prevent displacement during grout placement and be completely embedded in grout. Grout proportioning and mixing must comply with ASTM C476 and have a slump between 8 in and 11 in. Small cells or cavities require grout with a higher slump than larger cells or cavities. Grout is classified in ASTM C476 as either fine or coarse. When suitable, coarse grout is preferable as it shrinks less and requires less cement than fine grout. Fine grout is preferable for small cells to ensure that the grout adequately fills the confined space. The least clear dimension for grouting between wythes, the minimum cell dimensions when grouting hollow units, and the appropriate type of grout to use are given by TMS 402 Table 3.2.1. Obstructions into the grout space and the diameter of horizontal reinforcing bars must be considered when determining the minimum dimension.

Grout may be placed by either pumping or pouring. In accordance with TMS 602[1] Sec. 3.5D, grout may be placed in one continuous operation, or lift, not exceeding 12.67 ft in height, provided that

- the masonry has cured for a minimum of four hours to minimize potential displacement of the units

- the grout slump is maintained between 10 in and 11 in to ensure the grout will flow into and completely fill all spaces

- there are no intermediate reinforced bond beams between the top and bottom of the lift

Grout pours not exceeding 1 ft in height may be consolidated by puddling; otherwise, mechanical vibration is required to fill the grout space completely. After consolidation, water loss and settlement of the grout occurs, and reconsolidation is necessary to eliminate any voids formed in the grout. After reconsolidation, additional lifts may be placed, provided that the total height of the pour does not exceed the value given in TMS 402 Table 3.2.1.

2. ASD AND SD METHODS

Masonry structures may be designed using the provisions for either the allowable stress design of masonry (ASD) method or the strength design of masonry (SD) method.

The ASD method is the traditional method of designing masonry structures, and it is based on elastic theory to calculate the stresses produced in the member. Factored loads using ASD load combinations are applied to the member, and the stresses produced in the member must not exceed the specified allowable stress. TMS 402 Chap. 8 details the ASD method. IBC[2] Sec. 2107.1 adopts TMS 402 Chap. 1 through Chap. 8 except as modified by IBC Sec. 2107.2 through 2107.3.

In the SD method, factored loads using SD load combinations are applied to the member to determine the required ultimate strength. This required strength must not exceed the design strength, which is calculated as the member nominal strength multiplied by a resistance factor, ϕ. TMS 402 Chap. 9 details the SD method. IBC Sec. 2108.1 adopts the SD method of TMS 402 Chap. 9 except as modified by IBC Sec. 2108.2 through 2108.3.

3. LOAD COMBINATIONS

Nomenclature

D	dead load	lbf or kips
E	earthquake load	lbf or kips
E_m	modulus of elasticity of masonry in compression	lbf/in^2
E_s	modulus of elasticity of steel reinforcement	lbf/in^2
f'_m	specified masonry compressive strength	lbf/in^2
F_b	allowable compressive stress in masonry due to flexure	lbf/in^2

F_s	allowable tensile stress in reinforcement due to flexure	lbf/in^2
H	load due to lateral pressure	lbf or kips
L	live load	lbf or kips
L_r	roof live load	lbf or kips
Q	load effect due to service load	lbf or kips
R	rain load	lbf or kips
S	snow load	lbf or kips
W	wind load	lbf or kips

Symbols

ϕ	strength reduction factor	–
γ	load factor	–

ASD Required Strength

The required strength of a member consists of the most critical combination of factored loads applied to the member. Factored loads consist of nominal, or service, loads multiplied by the appropriate ASD load factors. In accordance with TMS 402 Sec. 4.1.2, load combinations must be as specified in the applicable building code. The required strength, $\sum \gamma Q$, is defined by nine load combinations given in IBC Sec. 1605.3.1. The seismic and wind loads specified in the IBC are at the strength design level, in contrast to other loads that are at the allowable stress design level. In the ASD load combinations, the load factor for seismic loads is 0.7, and the load factor for wind loads is 0.6 to reduce them to allowable stress design-level values. The combinations, with uncommon load conditions (self-straining loads and fluid pressure) omitted, are as follows.

$$\sum \gamma Q = D \qquad \text{[IBC 16-8]}$$

$$\sum \gamma Q = D + H + L \qquad \text{[IBC 16-9]}$$

$$\sum \gamma Q = D + H + (L_r \text{ or } S \text{ or } R) \qquad \text{[IBC 16-10]}$$

$$\sum \gamma Q = D + H + 0.75L \\ +0.75(L_r \text{ or } S \text{ or } R) \qquad \text{[IBC 16-11]}$$

$$\sum \gamma Q = D + H + (0.6W \text{ or } 0.7E) \qquad \text{[IBC 16-12]}$$

$$\sum \gamma Q = D + H + 0.75(0.6W) + 0.75L \\ +0.75(L_r \text{ or } S \text{ or } R) \qquad \text{[IBC 16-13]}$$

$$\sum \gamma Q = D + H + 0.75(0.7E) \\ +0.75L + 0.75S \qquad \text{[IBC 16-14]}$$

$$\sum \gamma Q = 0.6D + 0.6W + H \qquad \text{[IBC 16-15]}$$

$$\sum \gamma Q = 0.6D + 0.7E + H \qquad \text{[IBC 16-16]}$$

In accordance with TMS 402 Comm. Sec. 8.1.1, allowable stresses may not be increased by one-third for wind or for seismic load combinations.

SD Required Strength

The required ultimate strength of a member consists of the most critical combination of factored loads applied to the member. Factored loads consist of nominal, or service, loads multiplied by the appropriate SD load factors. In accordance with TMS 402 Sec. 4.1.2, load combinations must be as specified in the applicable building code. The required strength, $\sum \gamma Q$, is defined by seven combinations given in IBC Sec. 1605.2. The combinations, with uncommon load conditions (self-straining loads and fluid pressure) omitted, are as follows.

$$\sum \gamma Q = 1.4D \qquad \text{[IBC 16-1]}$$

$$\sum \gamma Q = 1.2D + 1.6(L + H) \\ +0.5(L_r \text{ or } S \text{ or } R) \qquad \text{[IBC 16-2]}$$

$$\sum \gamma Q = 1.2D + 1.6(L_r \text{ or } S \text{ or } R) \\ +1.6H + (f_1 L \text{ or } 0.5W) \qquad \text{[IBC 16-3]}$$

$$\sum \gamma Q = 1.2D + 1.0W + f_1 L + 1.6H \\ +0.5(L_r \text{ or } S \text{ or } R) \qquad \text{[IBC 16-4]}$$

$$\sum \gamma Q = 1.2D \pm 1.0E + f_1 L + 1.6H \\ +f_2 S \qquad \text{[IBC 16-5]}$$

$$\sum \gamma Q = 0.9D + 1.0W + 1.6H \qquad \text{[IBC 16-6]}$$

$$\sum \gamma Q = 0.9D + 1.0E + 1.6H \qquad \text{[IBC 16-7]}$$

For IBC Eq. 16-3, Eq. 16-4, and Eq. 16-5, use $f_1 = 1.0$ for garages, places of public assembly, and areas where $L > 100 \ lbf/ft^2$. Use $f_1 = 0.5$ for all other live loads. For IBC Eq. 16-5, use $f_2 = 0.7$ for roof configurations that do not shed snow, and use 0.2 for other roof configurations.

Where the effect of H resists the primary variable load effect, a load factor of 0.9 must be included with H where H is permanent. H must be set to zero for all other conditions.

ASD Allowable Stresses

Allowable stresses are determined by applying safety factors to material nominal stresses. Allowable tensile stress in bar reinforcement is given in TMS 402 Sec. 8.3.3 as

$$F_s = 32{,}000 \ lbf/in^2 \quad \text{[for grade 60 reinforcement]}$$

$$F_s = 20{,}000 \ lbf/in^2 \quad \begin{bmatrix} \text{for grade 40 or grade 50} \\ \text{reinforcement} \end{bmatrix}$$

Allowable compressive stress in masonry due to flexure is given in TMS 402 Sec. 4.2.2 as

$$F_b = 0.45 f'_m$$

The modulus of elasticity of steel reinforcement is given by TMS 402 Table 4.2.2 as

$$E_s = 29,000,000 \text{ lbf/in}^2$$

The modulus of elasticity of concrete masonry is based on the chord modulus of elasticity. It is between 0.05 and 0.33 of the maximum compressive strength of the masonry prism. Alternatively, from TMS 402 Table 4.2.2, the modulus of elasticity may be derived from the equation

$$E_m = 900f'_m$$

SD Design Strength

The design strength of a member consists of the nominal, or theoretical ultimate, strength of the member multiplied by the appropriate strength reduction factor, ϕ. The design strength must equal or exceed the required strength, which consists of combinations of nominal loads multiplied by the appropriate load factors. The reduction factors for reinforced masonry are defined in TMS 402 Sec. 9.1.4 as

$$\phi = 0.90 \begin{bmatrix} \text{flexure, axial load, or combinations} \\ \text{of flexure and axial load} \end{bmatrix}$$

$$\phi = 0.80 \; [\text{shear and shear friction}]$$

$$\phi = 0.50 \begin{bmatrix} \text{anchor bolts, strength governed by} \\ \text{masonry breakout, crushing, or pryout} \end{bmatrix}$$

$$\phi = 0.90 \begin{bmatrix} \text{anchor bolts, strength governed} \\ \text{by anchor bolt steel} \end{bmatrix}$$

$$\phi = 0.65 \begin{bmatrix} \text{anchor bolts, strength governed} \\ \text{by anchor pullout} \end{bmatrix}$$

$$\phi = 0.60 \; [\text{bearing on masonry surfaces}]$$

4. MASONRY BEAMS IN FLEXURE

Nomenclature

a	depth of equivalent rectangular stress block	in
$a_{\max}$	maximum allowable value of a	in
A_s	area of tension reinforcement	in^2
b	width of beam	in
c	distance from extreme compression fiber to neutral axis	in
c_c	cover	in
$c_{\max}$	maximum allowable value of c	in
C	compressive force	lbf
d	effective depth, distance from extreme compression fiber to centroid of tension reinforcement	in
d_b	diameter of reinforcement	in

E_m	modulus of elasticity of masonry in compression	lbf/in^2
E_s	modulus of elasticity of steel reinforcement	lbf/in^2
f_b	calculated compressive stress in masonry due to flexure	lbf/in^2
f'_m	specified masonry compressive strength	lbf/in^2
f_r	modulus of rupture of masonry	lbf/in^2
f_s	calculated stress in reinforcement	lbf/in^2
f_y	yield strength of reinforcement	lbf/in^2
F_b	allowable compressive stress in masonry due to flexure	lbf/in^2
F_s	allowable stress in reinforcement	lbf/in^2
h	overall dimension of member	in
j	lever-arm factor, ratio of distance between centroid of flexural compression forces and centroid of tensile forces to effective depth, $1 - k/3$ (ASD)	–
j	lever-arm factor, ratio of distance between centroid of flexural compression forces and centroid of tensile forces to effective depth, $(d - a/2)/d$ (SD method)	–
k	neutral axis depth factor, $\sqrt{2\rho n + (\rho n)^2} - \rho n$	–
K	lesser of masonry cover, clear spacing between adjacent reinforcement, or $9d_b$	in
K_u	design moment factor, $M_u/b_w d^2$	lbf/in^2
l	clear span length of beam	ft
l_c	distance between points of lateral support	ft
l_d	development length of straight reinforcement	in
l_e	effective span	ft
l_e	equivalent development length provided by a standard hook	in
l_s	distance between centers of supports	ft
M	applied moment	ft-kips or in-lbf
M_{cr}	nominal cracking moment strength of a member	ft-kips
$M_{\max}$	maximum design flexural strength	ft-kips
M_n	nominal flexural strength of a member	ft-kips
n	modular ratio, E_s/E_m	–
s_c	clear spacing of reinforcement	in
S_n	section modulus of the beam	in^3
T	tensile force	lbf

Symbols

γ	reinforcement size factor	–
ϵ_m	strain in masonry	–

ϵ_{mu} maximum usable compressive strain of masonry, 0.0035 for clay masonry and 0.0025 for concrete masonry —

ϵ_s strain in reinforcement —

ϵ_y strain at yield in tension reinforcement —

ρ tension reinforcement ratio, A_s/b_wd —

ρ_{max} maximum tension reinforcement ratio in a rectangular beam with tension reinforcement only —

ρ_{min} minimum allowable reinforcement ratio —

ϕ strength reduction factor —

Reinforcement Requirements

Reinforcement sizes in masonry members are limited in order to control bond stresses in the bars, reduce congestion in the cells, and facilitate grout consolidation. The size of reinforcement is specified by TMS 402 Sec. 6.1.2 and Sec. 9.3.3. For allowable stress design, the reinforcement diameter must not exceed the lesser of

- one-eighth of the nominal member thickness

- one-half of the least clear dimension of the cell, bond beam, or collar joint in which it is placed

- no. 11 bar

For allowable stress design, the area of vertical reinforcement must not exceed 6% of the area of the grout space.

For strength design, the reinforcement diameter must not exceed the lesser of

- one-eighth of the nominal member thickness

- one-quarter of the least clear dimension of the cell, bond beam, or collar joint in which the bar is placed

- no. 9 bar

For strength design, the area of reinforcing bars placed in a cell or in a course of hollow unit construction must not exceed 4% of the area of the cell area.

For both allowable stress design and strength design, reinforcement is required by TMS 402 Sec. 6.1.4.1 to have the following protective cover.

- 2 in for bars larger than no. 5 and $1\frac{1}{2}$ in for no. 5 bars or smaller, where the masonry face is exposed to earth or weather

- $1\frac{1}{2}$ in for all bars not exposed to earth or weather

In accordance with TMS 402 Sec. 6.1.3, the clear distance between parallel bars must not be less than the nominal diameter of the bars, with a minimum distance of 1 in. In columns and pilasters, the clear distance between vertical bars must not be less than 1.5 times the nominal bar diameter, with a minimum distance of $1\frac{1}{2}$ in. The thickness of grout between the reinforcement

and the masonry unit must be a minimum of $\frac{1}{4}$ in for fine grout, or $\frac{1}{2}$ in for coarse grout. Not more than two reinforcing bars may be bundled (ASD). TMS 402 Sec. 9.3.4.2.2.1 specifies that not more than two bar sizes may be used in a beam and that variations in bar size cannot exceed one bar size (SD).

In accordance with TMS 402 Sec. 9.3.3.3, for strength design reinforcing bars must not be bundled.

Example 7.1

A nominal 8 in beam, not exposed to the weather, is shown. Determine the maximum permissible reinforcement bar size.

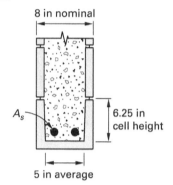

Solution

ASD Method

From TMS 402 Sec. 6.1.2, the reinforcement bar size must not exceed

$$d_b = \text{no. 11 bar} \quad \text{[TMS 402 Sec. 6.1.2.1]}$$

Or,

$$d_b = \text{one-half least dimension of cell}$$
$$= \frac{5 \text{ in}}{2} \quad \text{[TMS 402 Sec. 6.1.2.2]}$$
$$= 2.5 \text{ in}$$

Or,

$$d_b = \text{one-eighth nominal member thickness}$$
$$= \frac{8 \text{ in}}{8} \quad \text{[TMS 402 Sec. 6.1.2.5]}$$
$$= 1 \text{ in} \quad \text{[governs]}$$

To conform to the governing bar size of 1 in, two no. 8 bars may be used. This provides a clear spacing between bars of 2 in and the thickness of grout between the reinforcement and the masonry unit of $\frac{1}{2}$ in, which satisfies TMS 402 Sec. 6.1.3.

The cover provided to the no. 8 bars is

$$c_c = \frac{b - s_c - 2d_b}{2}$$
$$= \frac{7.63 \text{ in} - 2 \text{ in} - (2)(1 \text{ in})}{2}$$
$$= 1.82 \text{ in}$$
$$> 1\tfrac{1}{2} \text{ in} \qquad [\text{satisfies TMS 402 Sec. 6.1.4.1}]$$

SD Method

From TMS 402 Sec. 9.3.3, the reinforcement bar size must not exceed

$$d_b = \text{no. 9 bar} \quad [\text{TMS 402 Sec. 9.3.3.1}]$$

Or,

$$d_b = \text{one-quarter least dimension of cell}$$
$$= \frac{5 \text{ in}}{4} \quad [\text{TMS 402 SEC. 9.3.3.1}]$$
$$= 1.25 \text{ in}$$

Or,

$$d_b = \text{one-eighth nominal member thickness}$$
$$= \frac{8 \text{ in}}{8} \quad [\text{TMS 402 Sec. 6.1.2.5}]$$
$$= 1 \text{ in} \quad [\text{governs}]$$

From TMS 402 Sec 9.3.3.1, the maximum area of reinforcing bars permitted is

$$A_s = 4\% \text{ cell area}$$
$$= (0.04)(5 \text{ in average cell width})(6.25 \text{ in cell height})$$
$$= 1.25 \text{ in}^2 \quad [\text{governs}]$$

To conform to the governing reinforcing bar area of 1.25 in^2, two no. 7 bars that give an area of 1.20 in^2 are satisfactory. This provides a clear spacing between bars of 2 in and the thickness of grout between the reinforcement and the masonry unit of $\tfrac{5}{8}$ in, which satisfies TMS 402 Sec. 6.1.3.

The cover provided to the no. 7 bars is

$$c_c = \frac{b - s_c - 2d_b}{2}$$
$$= \frac{7.63 \text{ in} - 2 \text{ in} - (2)(0.875 \text{ in})}{2}$$
$$= 1.94 \text{ in}$$
$$> 1\tfrac{1}{2} \text{ in} \qquad [\text{satisfies TMS 402 Sec. 6.1.4.1}]$$

Dimensional Limitations

The maximum permitted unbraced length on the compression side of a masonry beam is given by TMS 402 Sec. 5.2.1.2 as

$$l_c = 32b$$
$$\leq \frac{120b^2}{d}$$

The minimum permitted bearing length of a masonry beam is given by TMS 402 Sec. 5.2.1.3 as

$$b_r = 4 \text{ in}$$

TMS 402 Sec. 9.3.4.2.4 for strength design requires all beams to be solid grouted.

Development Length of Bar Reinforcement

The basic development length of compression and tension reinforcement for both ASD and SD is given by TMS 402 Sec. 6.1.5.1.1 Eq. 6-1 as

$$l_d = \frac{0.13d_b^2 f_y \gamma}{K\sqrt{f_m'}} \quad [\text{TMS 402 6-1}]$$
$$\geq 12 \text{ in}$$

In accordance with IBC Sec. 2108.2, for strength design, l_d need not exceed $72d_b$.

The development length is increased by 50% for epoxy-coated bars.

K is the lesser of masonry cover, clear spacing of reinforcement, and nine times the bar diameter, d_b. γ is 1.0 for no. 3 through no. 5 bars; 1.3 for no. 6 through no. 7 bars; and 1.5 for no. 8 and larger bars.

The equivalent development length of a standard hook in tension is specified in TMS 402 Sec. 6.1.5.1.3 as

$$l_e = 13d_b \quad [\text{TMS 402 Eq. 6-2}]$$

Example 7.2

The nominal 8 in beam shown has a masonry compressive strength of 1500 lbf/in^2 and reinforcement consisting of two grade 60 no. 7 bars. The clear distance between bars is 2 in. Determine the required development length for straight bars and for bars provided with a standard hook.

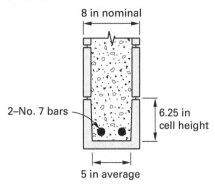

Solution

ASD and SD Method

The development parameter, K, is the lesser of the masonry cover, the clear spacing of reinforcement, and $9d_b$.

For the masonry cover,

$$K = \frac{b - s_c - 2d_b}{2}$$
$$= \frac{7.63 \text{ in} - 2 \text{ in} - (2)(0.875 \text{ in})}{2}$$
$$= 1.94 \text{ in} \quad \text{[governs]}$$

For the clear spacing of reinforcement,

$$K = s_c = 2 \text{ in}$$

For the bar diameter,

$$K = 9d_b$$
$$= (9)(0.875 \text{ in})$$
$$= 7.88 \text{ in}$$

The reinforcement size factor for a no. 7 bar is

$$\gamma = 1.3$$

The required development length for a straight bar is given by TMS 402 Eq. 6-1 as

$$l_d = \frac{0.13d_b^2 f_y \gamma}{K\sqrt{f_m'}}$$
$$= \frac{(0.13)(0.875 \text{ in})^2 \left(60{,}000 \ \dfrac{\text{lbf}}{\text{in}^2}\right)(1.3)}{(1.94 \text{ in})\sqrt{1500 \ \dfrac{\text{lbf}}{\text{in}^2}}}$$
$$= 103 \text{ in}$$

The equivalent development length provided by a standard hook is given by TMS 402 Sec. 6.1.5.1.3 as

$$l_e = 13d_b$$
$$= (13)(0.875 \text{ in})$$
$$= 11 \text{ in}$$

The required development length of the no. 7 bars, for bars provided with a standard hook, is

$$l_d = 103 \text{ in} - 11 \text{ in}$$
$$= 92 \text{ in}$$

Splicing of Bar Reinforcement

Splicing of bar reinforcement is used to provide the transfer of force between the bars being spliced. Splicing is accomplished by means of lap splices, welded splices, or mechanical connections.

Lap splices

In accordance with TMS 402 Sec. 6.1.6.1.1.1, the length of a lap splice in tension and compression, for both allowable stress design and strength design, must equal the development length determined by TMS 402 Sec. 6.1.5.1.1 using TMS 402 Eq. 6-1, which is

$$l_d = \frac{0.13d_b^2 f_y \gamma}{K\sqrt{f_m'}}$$
$$\geq 12 \text{ in}$$

K = lesser of masonry cover, clear spacing of reinforcement, and $9d_b$

$\gamma = 1.0$ for no. 3 through no. 5 bars, 1.3 for no. 6 through no. 7 bars, and 1.5 for no. 8 and larger bars

Spliced bars are required by TMS 402 Sec. 6.1.6.1.1.3 to be spaced apart a distance of

$$s_c \leq \text{one-fifth the required length of lap}$$
$$\leq 8 \text{ in}$$

Reduction in lap splice

For transverse confinement reinforcement provided at the splice, TMS 402 Sec. 6.1.6.1.1.2 permits the splice length to be reduced. As shown in Fig 7.1, reinforcement consisting of a no. 3 or larger bar is placed transversely within the lap, with at least one bar 8 in or less from each end of the lap. The lap length determined by TMS 402 Eq. 6-1 is then reduced by multiplying by the confinement factor ξ, given by TMS 402 Eq. 6-4 as

$$\xi = 1.0 - \frac{2.3A_{sc}}{d_b^{2.5}}$$

$$\frac{2.3A_{sc}}{d_b^{2.5}} \leq 1.0$$

A_{sc} = area of the transverse bars at each end of the lap splice

$\leq 0.35 \text{ in}^2$

Figure 7.1 *Transverse Reinforcement for Lap Splices*

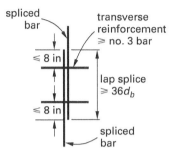

The reduced lap splice length must not be less than $36d_b$. The clear space between transverse bars and lapped bars must not exceed 1½ in, and the transverse bars must be fully developed in grouted masonry.

Alternative IBC provisions

IBC Sec. 2107.2.1 provides an alternative to TMS 402 Sec. 6.1.6.1.1 for the determination of lap splices. This alternative is applicable only to allowable stress design. The minimum length of lap splices for reinforcing bars in tension or compression, l_d, is given by IBC Eq. 21-1 as

$$l_d = 0.002d_b f_s$$
$$\geq 12 \text{ in}$$
$$\geq 40d_b$$

In regions of moment where the calculated tensile stress in the reinforcement, f_s, is greater than 80% of the allowable steel tension stress, F_s, the lap length of splices is increased by 50% of the minimum required length, but need not be greater than $72d_b$. Other equivalent means of stress transfer to accomplish the same 50% increase should be permitted. Where epoxy-coated bars are used, lap length is increased by 50%.

Welded splices of bar reinforcement

IBC Sec. 2107.3 modifies TMS 402 Sec. 6.1.6.1 and requires welding to conform to AWS D1.4[3] with welded splices of ASTM A706[4] steel reinforcement. Reinforcement larger than no. 9 must be spliced using mechanical connections in accordance with TMS 402 Sec. 6.1.6.1.3.

IBC Sec. 2108.3 modifies TMS 402 Sec. 6.1.6.1.2 and requires a welded splice to have the bars butted and welded to develop not less than 125% of the yield strength, f_y, of the bar in tension or compression, as required. For strength design, welded splices are not permitted in plastic hinge zones of intermediate or special reinforced walls.

Mechanical splices of bar reinforcement

TMS 402 Sec 6.1.6.1.3 requires the connected bars to develop not less than 125% of the yield strength, f_y, of the bar in tension or compression, as required. IBC Sec. 2108.3 modifies TMS 402 Sec. 6.1.6.1.3 and requires for strength design that mechanical splices are classified as Type 1 or 2 in accordance with ACI 318[5] Sec. 18.2.7.1. Type 1 mechanical splices must not be used within a plastic hinge zone or within a beam-column joint of intermediate or special reinforced masonry shear walls. Type 2 mechanical splices are permitted in any location within a member.

Example 7.3

The retaining wall shown has a reinforced concrete base and a nominal 8 in solid grouted concrete masonry stem. The calculated stress, f_s, in the no. 5 bars at the base of the stem is 30,000 lbf/in². Use IBC Eq. 21-1 to determine the splice length required with the no. 5 dowel bar in the base.

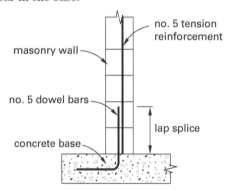

Solution

The minimum splice length is given by IBC Eq. 21-1 as

$$l_d = 0.002d_b f_s$$
$$= (0.002)(0.625 \text{ in})\left(30,000 \ \frac{\text{lbf}}{\text{in}^2}\right)$$
$$= 37.5 \text{ in}$$
$$\geq 12 \text{ in} \quad \text{[satisfactory]}$$
$$\geq 40d_b \quad \text{[satisfactory]}$$

Find 80% of the allowable steel tension stress.

$$0.8F_s = (0.8)\left(32,000 \ \frac{\text{lbf}}{\text{in}^2}\right)$$
$$= 25,600 \text{ lbf/in}^2$$
$$< f_s$$

Hence, the splice length must be increased by 50%, and the required splice length is

$$l_d = (1.5)(37.5 \text{ in})$$
$$= 56 \text{ in}$$

Effective Span Length of Masonry Beams

The effective span length of a simply supported beam is defined in TMS 402 Sec. 5.2.1.1 and is illustrated in Fig. 7.2.

For beams not built integrally with the support, the effective span is defined as the clear span plus the depth of member, as shown in Fig. 7.2(b).

$$l_e = l + h$$

However, the effective span need not exceed the distance between centers of supports, as shown in Fig. 7.2(a).

$$l_e = l_s = l + b_s$$

For a continuous beam, TMS 402 Sec. 5.2.1.1.2 defines the effective span as the distance between centers of supports, as shown in Fig. 7.2(c).

$$l_e = l_s = l + b_s$$

As shown in Fig. 7.2(d), the effective span length of beams built integrally with supports is customarily taken as equal to the clear span.

$$l_e = l$$

Deep Beams

In TMS 402 Sec. 2.2, a deep beam is defined as a beam that has an effective span-to-depth ratio, l_e/d_v, less than 3 for a continuous span and less than 2 for a simple span where d_v is the actual depth of masonry in the direction of shear considered. Normal beam theory does not apply to deep beams, as the assumption of a linear strain distribution across the depth of the section is not applicable. TMS 402 Sec. 5.2.2 requires that deep beams are designed using deep beam theory. Alternatively, TMS 402 Sec. 5.2.2.2 permits the assumption of an internal lever arm, z, for the determination of the required horizontal flexural reinforcement. The area of flexural reinforcement required is given by

$$A_s = \frac{M}{zF_s}$$

Values of z depend on the l_e/d_v ratio and are given in TMS 402 Table 5.2.2.2, as shown in Table 7.1.

Table 7.1 *Internal Lever Arm*

span condition	ratio	internal lever arm, z
simply supported	$1 \le l_e/d_v < 2$	$0.2(l_e + 2d_v)$
	$l_e/d_v < 1$	$0.6\, l_e$
continuous	$1 \le l_e/d_v < 3$	$0.2(l_e + 1.5d_v)$
	$l_e/d_v < 1$	$0.5 l_e$

The flexural reinforcement is positioned in the lower half of the beam depth, d_v, at a maximum spacing not exceeding $d_v/5$ nor 16 in.

The requirements for vertical and horizontal shear reinforcement are given in TMS 402 Sec. 5.2.2.4 and Sec. 5.2.2.5 as follows.

- Vertical shear reinforcement to resist applied loads is designed in accordance with TMS 402 Sec. 8.3.5 (ASD) or TMS 402 Sec. 9.3.4 (SD).

- The minimum area of vertical shear reinforcement is $0.0007bd_v$.

- The minimum area of horizontal shear reinforcement is one-half the area of the required vertical reinforcement, and this area may include the horizontal flexural reinforcement.

- Maximum spacing of vertical and horizontal shear reinforcement is not to exceed $d_v/5$ nor 16 in.

- The total area of horizontal and vertical reinforcement must not be less than $0.001bd_v$.

As shown in TMS 402 Fig. CC-5.2-2, it is unnecessary to consider the full depth of masonry above an opening. An arbitrary shallower depth may be selected that does not meet the deep beam provisions. This shallower beam is then designed conventionally.

Beams with Tension Reinforcement Only

ASD Method

For grade 40 or grade 50 reinforcement, the allowable tensile stress and the allowable compressive stress are given by TMS 402 Sec. 8.3.3 as

$$F_s = 20,000 \text{ lbf/in}^2$$

For grade 60 reinforcement, the allowable tensile stress and the allowable compressive stress are given by TMS 402 Sec. 8.3.3 as

$$F_s = 32,000 \text{ lbf/in}^2$$

Figure 7.2 *Effective Span Length*

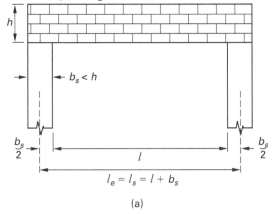

$$l_e = l_s = l + b_s$$

(a)

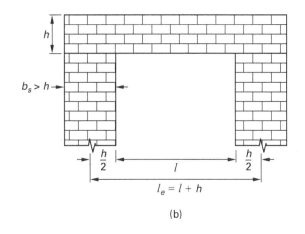

$$l_e = l + h$$

(b)

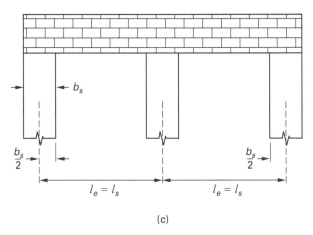

(c)

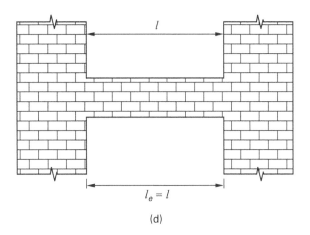

(d)

The allowable compressive stress in masonry due to flexure is given by TMS 402 Sec. 8.3.4.2.2 as

$$F_b = 0.45 f'_m$$

The elastic design method, illustrated in Fig. 7.3, is used to calculate the stresses in a masonry beam under the action of the applied service loads, and to ensure that these stresses do not exceed allowable values.

The basic assumptions adopted[6] in the elastic, or allowable stress, design method are as follows.

- The strain distribution over the depth of the member is linear, as shown in Fig. 7.3.

- Stresses in the masonry and in the reinforcement are proportional to the induced strain.

- Tensile stress in the masonry is neglected.

The modular ratio is defined as

$$n = \frac{E_s}{E_m}$$

The tension reinforcement ratio is

$$\rho = \frac{A_s}{bd}$$

From the strain diagram, with a neutral axis depth of kd, as shown in Fig. 7.3,

$$\frac{\epsilon_s}{\epsilon_m} = \frac{d - kd}{kd}$$

$$= \frac{1 - k}{k}$$

$$\frac{f_s E_m}{f_b E_s} = \frac{1 - k}{k}$$

$$\frac{f_s}{f_b} = \frac{n(1 - k)}{k}$$

A triangular compressive stress distribution is formed in the masonry, above the neutral axis, and the line of action of the compressive force, C, acts at a depth of $kd/3$ with

$$C = \frac{f_b k d b}{2}$$

The tensile force developed in the reinforcement is

$$T = f_s A_s$$
$$= f_s \rho b d$$

Equating tensile and compressive forces acting on the section gives

$$f_s \rho b d = \frac{f_b k d b}{2}$$
$$\frac{f_s}{f_b} = \frac{k}{2\rho}$$
$$= \frac{n(1-k)}{k}$$
$$k = \sqrt{2\rho n + (\rho n)^2} - \rho n$$

The lever arm of the internal resisting moment is obtained from Fig. 7.3 as

$$jd = d - \frac{kd}{3}$$

Figure 7.3 *Elastic Design of Reinforced Masonry Beam*

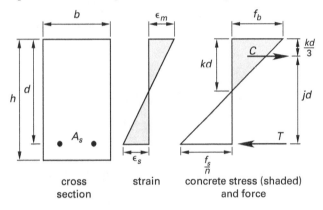

The lever arm factor is

$$j = 1 - \frac{k}{3}$$

The resisting moment of the masonry is

$$M_m = Cjd$$
$$= \frac{f_b j k b d^2}{2}$$

The resisting moment of the reinforcement is

$$M_s = Tjd$$
$$= f_s j \rho b d^2$$

For an allowable masonry stress of F_b and an allowable reinforcement stress of F_s, the allowable moment capacity of the section, M_R, is the lesser of M_m and M_s as follows.

$$M_m = \frac{F_b j k b d^2}{2}$$
$$M_s = F_s j \rho b d^2$$

To facilitate the determination of k, App. 2.B tabulates values of k against ρn.

The stress in the reinforcement due to an applied moment, M, is

$$f_s = \frac{M}{A_s j d}$$

The stress in the masonry due to an applied moment, M, is

$$f_b = \frac{2M}{jkbd^2}$$

ASD Design Procedure

The ASD design procedure consists of the following steps.

step 1: Assume beam dimensions and masonry strength.

step 2: Assume that $j = 0.9$.

step 3: Calculate $A_s = M/(F_s jd)$.

step 4: Select bar size and number required.

step 5: Calculate ρ and ρn.

step 6: Determine k from the $\rho n/k$ table in App. 2.B.

step 7: Calculate j.

step 8: Calculate M_m.

step 9: If $M_m < M$, increase beam size or f'_m.

step 10: Calculate M_s.

step 11: If $M_s < M$, increase A_s.

step 12: If both M_m and M_s are greater than M, the beam is satisfactory.

ASD Analysis Procedure

The ASD analysis procedure consists of the following steps.

step 1: Calculate ρ and ρn.

step 2: Determine k from the $\rho n/k$ table in App. 2.B.

step 3: Calculate j.

step 4: Calculate f_b.

step 5: Calculate f_s.

SD Method

The strength design method[7], illustrated in Fig. 7.4 for a beam reinforced in tension, is used to calculate the design strength of a concrete masonry beam under the action of the applied factored loads, to ensure that the design strength is greater than the most critical load combination. In accordance with TMS 402 Sec. 9.3.2, during the loading to failure of a reinforced masonry beam, the strain distribution over the depth of the beam is linear. TMS 402 Sec. 9.3.2 provisions also state that an equivalent rectangular stress distribution has a depth of 0.8 times the depth to the neutral axis and a stress of 0.8 times the specified masonry compressive strength; that at failure, the tension reinforcement has yielded; and that the maximum compressive strain in the masonry is 0.0035 for clay masonry and 0.0025 for concrete masonry. The tensile strength of the masonry is neglected.

The depth of the equivalent rectangular stress block is obtained by equating the tensile and compressive forces acting on the section.

$$
\begin{aligned}
a &= \frac{A_s f_y}{0.80 b f_m'} \\
&= \frac{1.25 \rho d f_y}{f_m'} \\
&= 0.80c
\end{aligned}
$$

From Fig. 7.4, the nominal flexural strength of the member is derived as

$$
\begin{aligned}
M_n &= A_s f_y \left(d - \frac{a}{2} \right) \\
&= A_s f_y \left(d - \frac{A_s f_y}{(2)(0.80) b f_m'} \right) \\
&= A_s f_y d \left(1 - \frac{0.625 A_s f_y}{b d f_m'} \right) \\
&= \rho f_y b d^2 \left(1 - \frac{0.625 \rho f_y}{f_m'} \right)
\end{aligned}
$$

In addition, the expression for nominal strength is

$$
\begin{aligned}
M_n &= \frac{M_u}{\phi} \\
&= \frac{M_u}{0.9} \\
&= \rho f_y b d^2 - \frac{0.625 \rho^2 b d^2 f_y^2}{f_m'} \\
0 &= \rho^2 - \rho \left(\frac{f_m'}{0.625 f_y} \right) + \frac{f_m' M_u}{0.563 b d^2 f_y^2}
\end{aligned}
$$

The reinforcement ratio required to provide a given factored moment, M_u, is then

$$
\begin{aligned}
\rho &= 0.80 f_m' \left(\frac{1 - \sqrt{1 - \dfrac{M_u}{0.36 b d^2 f_m'}}}{f_y} \right) \\
&= 0.80 f_m' \left(\frac{1 - \sqrt{1 - \dfrac{K_u}{0.36 f_m'}}}{f_y} \right)
\end{aligned}
$$

The design moment factor is

$$
K_u = \frac{M_u}{b d^2}
$$

Minimum Reinforcement Area

To prevent brittle failure of a lightly reinforced beam, TMS 402 Sec. 9.3.4.2.2.2 requires the nominal flexural strength of a beam to be not less than 1.3 times the nominal cracking moment strength of the beam. This provision is imposed so that there is a minimum area of reinforcement in a beam to ensure compliance. The required nominal moment is

$$
M_n \geq 1.3 M_{cr}
$$

The nominal cracking moment of the beam is

$$
M_{cr} = f_r S_n
$$

S_n is the section modulus of the beam, and f_r is the modulus of rupture of the masonry. The modulus of rupture is given in TMS 402 Table 9.1.9.2.

Figure 7.4 *Strength Design of a Reinforced Masonry Beam*

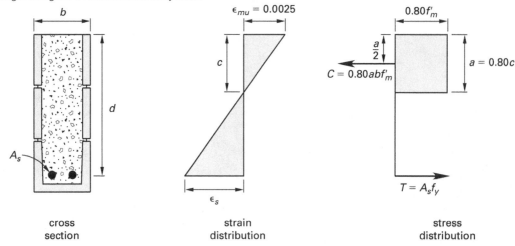

cross section   strain distribution   stress distribution

Example 7.4

The nominal 8 in beam shown has a masonry compressive strength of 1500 lbf/in². Its reinforcement consists of two grade 60 no. 5 bars with an effective depth of 21 in. The cells are solid grouted, and the masonry is laid in running bond with type S mortar cement. Use the SD method to determine if the minimum reinforcement meets TMS 402 requirements.

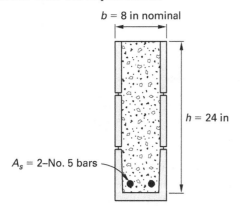

$b = 8$ in nominal

$h = 24$ in

$A_s = 2$–No. 5 bars

Solution

Use the SD method only.

The section modulus of the beam is

$$S_n = \frac{bh^2}{6}$$
$$= \frac{(7.63 \text{ in})(24 \text{ in})^2}{6}$$
$$= 732 \text{ in}^3$$

The modulus of rupture of the solid grouted masonry, for tension parallel to bed joints and type S mortar, is given by TMS 402 Table 9.1.9.2 as

$$f_r = 267 \text{ lbf/in}^2$$

The cracking moment of the beam is

$$M_{cr} = f_r S_n$$
$$= \frac{\left(267 \ \dfrac{\text{lbf}}{\text{in}^2}\right)(732 \text{ in}^3)}{\left(12 \ \dfrac{\text{in}}{\text{ft}}\right)\left(1000 \ \dfrac{\text{lbf}}{\text{kip}}\right)}$$
$$= 16.29 \text{ ft-kips}$$
$$1.3M_{cr} = (1.3)(16.29 \text{ ft-kips})$$
$$= 21.18 \text{ ft-kips}$$

The area of tension reinforcement is given as two no. 5 bars, which provide an area of

$$A_s = 0.62 \text{ in}^2$$

The depth of the equivalent rectangular stress block is obtained by equating tensile and compressive forces acting on the section. The depth is given by

$$a = \frac{A_s f_y}{0.80 b f'_m}$$
$$= \frac{(0.62 \text{ in}^2)\left(60{,}000 \ \dfrac{\text{lbf}}{\text{in}^2}\right)}{(0.80)(7.63 \text{ in})\left(1500 \ \dfrac{\text{lbf}}{\text{in}^2}\right)}$$
$$= 4.06 \text{ in}$$

The lever arm of the compressive and tensile forces is given by

$$jd = d - \frac{a}{2}$$
$$= 21 \text{ in} - \frac{4.06 \text{ in}}{2}$$
$$= 18.97 \text{ in}$$

The nominal flexural strength of the beam is given by

$$M_n = A_s j d f_y$$

$$= \frac{(0.62 \text{ in}^2)(18.97 \text{ in})\left(60 \ \dfrac{\text{kips}}{\text{in}^2}\right)}{12 \ \dfrac{\text{in}}{\text{ft}}}$$

$$= 58.81 \text{ ft-kips}$$

$$> 1.3M_{cr} \quad \text{[satisfies TMS 402 Sec. 9.3.4.2.2.2]}$$

Maximum Reinforcement Ratio

In order to provide adequate ductile response in a beam, TMS 402 Sec. 9.3.3.2.1 limits the maximum reinforcement ratio, $\rho_{\max}$, in accordance with a prescribed strain distribution. The area of tensile reinforcement must not exceed the area required to develop a strain of $1.5\epsilon_y$ in the extreme tensile reinforcement, simultaneously with a maximum masonry compressive strain of 0.0025 for concrete masonry or 0.0035 for clay masonry. This is applicable when the structure is designed using a value for the response modification factor, R, greater than 1.5, as defined by ASCE/SEI7[8] Table 12.2-1. Since the yield strain for grade 60 reinforcement is 0.00207, as shown in Fig. 7.5, this requirement effectively limits the depth of the neutral axis for reinforced concrete masonry to a minimum of

$$c = d\left(\frac{\epsilon_{mu}}{\epsilon_{mu} + 1.5\epsilon_y}\right) = d\left(\frac{0.0025}{0.0025 + (1.5)(0.00207)}\right)$$

$$= 0.446d$$

The depth of the equivalent rectangular stress block for concrete masonry is limited to

$$a = 0.8c = 0.8(0.446d) = 0.357d$$

When calculating the maximum reinforcement area, it is assumed that the stress in the reinforcement equals the product of the modulus of elasticity of the steel and the strain in the reinforcement, and that the stress is not greater than f_y. The maximum reinforcement area corresponding to the depth of the equivalent rectangular stress block is given by

$$A_{\max} = \frac{0.80abf'_m}{f_y} = \frac{0.286bdf'_m}{f_y}$$

The corresponding maximum reinforcement ratio is given by

$$\rho_{\max} = \frac{A_{\max}}{bd} = \frac{0.286f'_m}{f_y}$$

Example 7.5

The nominal 8 in beam shown has a masonry compressive strength of 1500 lbf/in² and reinforcement consisting of two grade 60 no. 5 bars. Use the SD method to determine if the maximum reinforcement meets TMS 402 requirements.

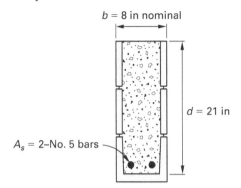

Solution

Use the SD method only.

The maximum allowable reinforcement ratio is

$$\rho_{\max} = \frac{0.286f'_m}{f_y}$$

$$= \frac{(0.286)\left(1500 \ \dfrac{\text{lbf}}{\text{in}^2}\right)}{60{,}000 \ \dfrac{\text{lbf}}{\text{in}^2}}$$

$$= 0.0072$$

The actual reinforcement ratio provided is

$$\rho = \frac{A_s}{bd} = \frac{(2)(0.31 \text{ in}^2)}{(7.63 \text{ in})(21 \text{ in})} = 0.0039$$

$$< \rho_{\max} \quad \text{[satisfies TMS 402 Sec. 9.3.3.2.1]}$$

SD Design Procedure

The SD design procedure consists of the following steps.

step 1: Assume beam dimensions and masonry strength.

step 2: Calculate $K_u = M_u/bd^2$.

step 3: Calculate the reinforcement ratio.

$$\rho = 0.80f'_m\left(\frac{1 - \sqrt{1 - \dfrac{K_u}{0.36f'_m}}}{f_y}\right)$$

step 4: Select the bar size and number required.

Figure 7.5 *Maximum Reinforcement in Concrete Masonry Beams*

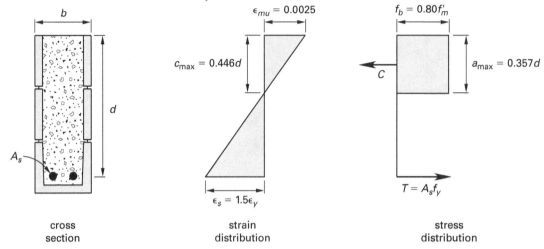

cross
section

strain
distribution

stress
distribution

step 5: Check that the beam complies with the maximum reinforcement requirements of TMS 402 Sec. 9.3.3.2. Increase beam size or f'_m if necessary.

step 6: Check that $M_n \geq 1.3M_{cr}$. Increase beam size or f'_m if necessary.

SD Analysis Procedure

The SD analysis procedure consists of the following steps.

step 1: Calculate stress block depth.

$$a = \frac{A_s f_y}{0.80 b f'_m}$$

step 2: Calculate nominal strength.

$$M_n = A_s f_y(d - a/2)$$

step 3: Calculate design strength, ϕM_n.

Example 7.6

The 8 in solid grouted, concrete block masonry beam shown in *Illustration for Ex. 7.6* is simply supported over an effective span of 12 ft. The masonry has a compressive strength of 1500 lbf/in² and a modulus of elasticity of 1,000,000 lbf/in². Reinforcement consists of four no. 6 grade 60 bars. The effective depth is 45 in, the overall depth is 48 in, and the beam is laterally braced at both ends. The 20 kips concentrated loads are floor live loads. The self-weight of the beam is 69 lbf/ft². Determine whether the beam is adequate.

Solution

The beam self-weight is

$$w = \left(69 \ \frac{\text{lbf}}{\text{ft}^2}\right)\left(\frac{48 \ \text{in}}{12 \ \frac{\text{in}}{\text{ft}}}\right) = 276 \ \text{lbf/ft}$$

At midspan, the bending moment produced by this self-weight is

$$M_s = \frac{wl^2}{8} = \frac{\left(276 \ \frac{\text{lbf}}{\text{ft}}\right)(12 \ \text{ft})^2}{(8)\left(1000 \ \frac{\text{lbf}}{\text{kip}}\right)} = 4.97 \ \text{ft-kips}$$

At midspan, the bending moment produced by the concentrated loads is

$$M_c = Wa = \frac{(20 \ \text{kips})(12 \ \text{ft} - 7 \ \text{ft})}{2} = 50 \ \text{ft-kips}$$

ASD Method

At midspan, the total ASD moment is given by IBC Eq. 16-9 as

$$\begin{aligned} M_y &= 1.0M_s + 1.0M_c \\ &= (1.0)(4.97 \ \text{ft-kips}) + (1.0)(50 \ \text{ft-kips}) \\ &= 54.97 \ \text{ft-kips} \end{aligned}$$

The allowable stresses, in accordance with TMS 402 Sec. 8.3.3 and Sec. 8.3.4.2.2, are

$$F_b = 0.45f'_m = (0.45)\left(1500 \ \frac{\text{lbf}}{\text{in}^2}\right) = 675 \ \text{lbf/in}^2$$

$$F_s = 32{,}000 \ \text{lbf/in}^2$$

Illustration for Ex. 7.6

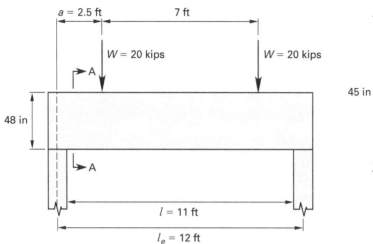

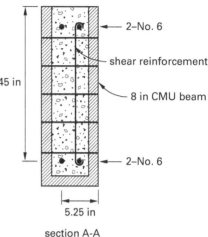

section A-A

The relevant parameters of the beam are

$$E_m = 1{,}000{,}000 \text{ lbf/in}^2$$

$$E_s = 29{,}000{,}000 \text{ lbf/in}^2$$

$$b = 7.63 \text{ in}$$

$$d = 45 \text{ in}$$

$$l_e = 12 \text{ ft}$$

$$A_s = (2)(0.44 \text{ in}^2) = 0.88 \text{ in}^2$$

$$\frac{l_e}{b} = \frac{(12 \text{ ft})\left(12 \dfrac{\text{in}}{\text{ft}}\right)}{7.63 \text{ in}} = 18.9$$

$$< 32 \quad \begin{bmatrix} \text{satisfies TMS 402} \\ \text{Sec. 5.2.1.2(a)} \end{bmatrix}$$

$$\frac{120b^2}{d} = \frac{(120 \text{ ft})(7.63 \text{ in})^2}{45 \text{ in}} = 155 \text{ in}$$

$$> l_e \quad \begin{bmatrix} \text{satisfies TMS 402} \\ \text{Sec. 5.2.1.2(b)} \end{bmatrix}$$

$$n = \frac{E_s}{E_m} = \frac{29{,}000{,}000 \dfrac{\text{lbf}}{\text{in}^2}}{1{,}000{,}000 \dfrac{\text{lbf}}{\text{in}^2}} = 29$$

$$\rho = \frac{A_s}{bd} = \frac{0.88 \text{ in}^2}{(7.63 \text{ in})(45 \text{ in})} = 0.00256$$

$$\rho n = (0.00256)(29) = 0.0743$$

From App. 2.B, the beam parameters and stresses, in accordance with TMS 402 Sec. 8.3.2, are

$$k = \sqrt{2\rho n + (\rho n)^2} - \rho n$$

$$= \sqrt{(2)(0.0743) + (0.0743)^2} - 0.0743$$

$$= 0.318$$

$$j = 1 - \frac{k}{3} = 1 - \frac{0.318}{3} = 0.894$$

$$f_b = \frac{2M_y}{jkbd^2} = \frac{(2)(54.97 \text{ ft-kips})\left(12 \dfrac{\text{in}}{\text{ft}}\right)\left(1000 \dfrac{\text{lbf}}{\text{kip}}\right)}{(0.894)(0.318)(7.63 \text{ in})(45 \text{ in})^2}$$

$$= 300 \text{ lbf/in}^2$$

$$< F_b \quad \text{[satisfactory]}$$

$$f_s = \frac{M_y}{jdA_s} = \frac{(54.97 \text{ ft-kips})\left(12 \dfrac{\text{in}}{\text{ft}}\right)\left(1000 \dfrac{\text{lbf}}{\text{kip}}\right)}{(0.894)(45 \text{ in})(0.88 \text{ in}^2)}$$

$$= 18{,}632 \text{ lbf/in}^2$$

$$< F_s \quad \text{[satisfactory]}$$

The beam is adequate.

SD Method

The total factored moment at midspan is given by IBC Eq. 16-2 as

$$M_u = 1.2M_s + 1.6M_c$$

$$= (1.2)(4.97 \text{ ft-kips}) + (1.6)(50 \text{ ft-kips})$$

$$= 85.96 \text{ ft-kips}$$

The stress block depth is

$$a = \frac{A_s f_y}{0.80 b f'_m} = \frac{(0.88 \text{ in}^2)\left(60 \frac{\text{kips}}{\text{in}^2}\right)\left(1000 \frac{\text{lbf}}{\text{kip}}\right)}{(0.80)(7.63 \text{ in})\left(1500 \frac{\text{lbf}}{\text{in}^2}\right)}$$

$$= 5.77 \text{ in}$$

The nominal strength is

$$M_n = A_s f_y \left(d - \frac{a}{2}\right)$$

$$= (0.88 \text{ in}^2)\left(60 \frac{\text{kips}}{\text{in}^2}\right)\left(\frac{45 \text{ in} - \dfrac{5.77 \text{ in}}{2}}{12 \frac{\text{in}}{\text{ft}}}\right)$$

$$= 185 \text{ ft-kips}$$

The design strength is

$$\phi M_n = (0.9)(185 \text{ ft-kips}) = 167 \text{ ft-kips}$$
$$> M_u \quad [\text{satisfactory}]$$

The beam is adequate.

Biaxial Bending

Using the ASD method, the combined stresses produced by biaxial bending must not exceed the allowable values. Using the SD method, the interaction equation is used to determine the adequacy of a member.

Example 7.7

The masonry beam described in Ex. 7.6, in addition to the vertical loads indicated, is subjected to a lateral force of 140 lbf/ft due to wind. Determine whether the beam is adequate.

Solution

At midspan, the bending moment produced by the wind load is

$$M_x = \frac{q l_e^2}{8} = \frac{\left(140 \frac{\text{lbf}}{\text{ft}}\right)(12 \text{ ft})^2}{(8)\left(1000 \frac{\text{lbf}}{\text{kip}}\right)} = 2.52 \text{ ft-kips}$$

The relevant parameters of the beam in the transverse direction are

$$b = 48 \text{ in}$$
$$d = 5.25 \text{ in}$$
$$l_e = 12 \text{ ft}$$
$$A_s = 0.88 \text{ in}^2$$
$$n = 29$$

$$\rho = \frac{A_s}{bd} = \frac{0.88 \text{ in}^2}{(48 \text{ in})(5.25 \text{ in})} = 0.00349$$

$$\rho n = (0.00349)(29) = 0.101$$

ASD Method

From App. 2.B, the beam parameters and stresses caused by the wind load, in accordance with TMS 402 Sec. 8.3.2, are

$$k = \sqrt{2\rho n + (\rho n)^2} - \rho n$$

$$= \sqrt{(2)(0.101) + (0.101)^2} - 0.101$$

$$= 0.360$$

$$j = 1 - \frac{k}{3} = 1 - \frac{0.360}{3} = 0.88$$

$$f_{bw} = \frac{2M_x}{jkbd^2} = \frac{(2)(2.52 \text{ ft-kips})\left(12 \frac{\text{in}}{\text{ft}}\right)\left(1000 \frac{\text{lbf}}{\text{kip}}\right)}{(0.88)(0.360)(48 \text{ in})(5.25 \text{ in})^2}$$

$$= 144 \text{ lbf/in}^2$$

$$f_{sw} = \frac{M_x}{jdA_s} = \frac{(2.52 \text{ ft-kips})\left(12 \frac{\text{in}}{\text{ft}}\right)\left(1000 \frac{\text{lbf}}{\text{kip}}\right)}{(0.88)(5.25 \text{ in})(0.88 \text{ in}^2)}$$

$$= 7438 \text{ lbf/in}^2$$

From Ex. 7.6, the allowable stresses are

$$F_b = 675 \text{ lbf/in}^2$$
$$F_s = 32{,}000 \text{ lbf/in}^2$$

The bending moment produced by the beam self-weight is

$$M_s = 4.97 \text{ ft-kips}$$

Reinforced Masonry

The stresses produced in the masonry and in the reinforcement by the beam self-weight are

$$f_{bs} = \frac{2M_s}{jkbd^2}$$

$$= \frac{(2)(4.97 \text{ ft-kips})\left(12 \frac{\text{in}}{\text{ft}}\right)\left(1000 \frac{\text{lbf}}{\text{kip}}\right)}{(0.894)(0.318)(7.63 \text{ in})(45 \text{ in})^2}$$

$$= 27 \text{ lbf/in}^2$$

$$f_{ss} = \frac{M_s}{jdA_s}$$

$$= \frac{(4.97 \text{ ft-kips})\left(12 \frac{\text{in}}{\text{ft}}\right)\left(1000 \frac{\text{lbf}}{\text{kip}}\right)}{(0.894)(45 \text{ in})(0.88 \text{ in}^2)}$$

$$= 1684 \text{ lbf/in}^2$$

The bending moment produced by the concentrated loads is

$$M_c = 50 \text{ ft-kips}$$

The stresses produced in the masonry and in the reinforcement by the live load are

$$f_{bc} = \frac{2M_c}{jkbd^2}$$

$$= \frac{(2)(50 \text{ ft-kips})\left(12 \frac{\text{in}}{\text{ft}}\right)\left(1000 \frac{\text{lbf}}{\text{kip}}\right)}{(0.894)(0.318)(7.63 \text{ in})(45 \text{ in})^2}$$

$$= 273 \text{ lbf/in}^2$$

$$f_{sc} = \frac{M_c}{jdA_s}$$

$$= \frac{(50 \text{ ft-kips})\left(12 \frac{\text{in}}{\text{ft}}\right)\left(1000 \frac{\text{lbf}}{\text{kip}}\right)}{(0.894)(45 \text{ in})(0.88 \text{ in}^2)}$$

$$= 16,948 \text{ lbf/in}^2$$

Applying IBC Eq. 16-12 gives the combined stresses caused by the beam self-weight and the wind load as

$$f_b = 1.0f_{bs} + 0.6f_{bw}$$

$$= (1.0)\left(27 \frac{\text{lbf}}{\text{in}^2}\right) + (0.6)\left(144 \frac{\text{lbf}}{\text{in}^2}\right)$$

$$= 113 \text{ lbf/in}^2$$

$$< F_b \quad [\text{satisfactory}]$$

$$f_s = 1.0f_{ss} + 0.6f_{sw}$$

$$= (1.0)\left(1684 \frac{\text{lbf}}{\text{in}^2}\right) + (0.6)\left(7438 \frac{\text{lbf}}{\text{in}^2}\right)$$

$$= 6147 \text{ lbf/in}^2$$

$$< F_s \quad [\text{satisfactory}]$$

Applying IBC Eq. 16-13 gives the combined stresses caused by the beam self-weight, the wind load, and the live load as

$$f_b = 1.0f_{bs} + 0.75(0.6)f_{bw} + 0.75f_{bc}$$

$$= (1.0)\left(27 \frac{\text{lbf}}{\text{in}^2}\right) + (0.75)(0.6)\left(144 \frac{\text{lbf}}{\text{in}^2}\right)$$

$$+ (0.75)\left(273 \frac{\text{lbf}}{\text{in}^2}\right)$$

$$= 297 \text{ lbf/in}^2$$

$$< F_b \quad [\text{satisfactory}]$$

$$f_s = 1.0f_{ss} + 0.75(0.6)f_{sw} + 0.75f_{sc}$$

$$= (1.0)\left(1684 \frac{\text{lbf}}{\text{in}^2}\right) + (0.75)(0.6)\left(7438 \frac{\text{lbf}}{\text{in}^2}\right)$$

$$+ (0.75)\left(16,948 \frac{\text{lbf}}{\text{in}^2}\right)$$

$$= 17,742 \text{ lbf/in}^2$$

$$< F_s \quad [\text{satisfactory}]$$

The beam is adequate.

SD Method

The service bending moments are from the ASD solution for Ex. 7.7.

The bending moment produced by the beam self-weight is

$$M_s = 4.97 \text{ ft-kips}$$

The bending moment produced by the live load is

$$M_c = 50 \text{ ft-kips}$$

Reinforced Masonry

The bending moment produced by the wind load is

$$M_x = 2.52 \text{ ft-kips}$$

IBC Eq. 16-4 governs. Applying this load combination to the vertical loads gives the combined factored moment caused by the beam self-weight and the live load as

$$
\begin{aligned}
M_u &= 1.2M_s + 0.5M_c \\
&= (1.2)(4.97 \text{ ft-kips}) + (0.5)(50 \text{ ft-kips}) \\
&= 31.0 \text{ ft-kips}
\end{aligned}
$$

The design strength for vertical loads from Ex. 7.6 is

$$\phi M_{\hat{y}} = 167 \text{ ft-kips}$$

Applying IBC Eq. 16-4 to the wind load gives the factored moment as

$$
\begin{aligned}
M_w &= 1.0M_x \\
&= (1.0)(2.52 \text{ ft-kips}) \\
&= 2.52 \text{ ft-kips}
\end{aligned}
$$

For lateral loads, the stress block depth is

$$
\begin{aligned}
a &= \frac{A_s f_y}{0.80 b f'_m} \\
&= \frac{(0.88 \text{ in}^2)\left(60 \, \frac{\text{kips}}{\text{in}^2}\right)\left(1000 \, \frac{\text{lbf}}{\text{kip}}\right)}{(0.80)(48 \text{ in})\left(1500 \, \frac{\text{lbf}}{\text{in}^2}\right)} \\
&= 0.92 \text{ in}
\end{aligned}
$$

The nominal strength in the lateral direction is

$$
\begin{aligned}
M_n &= A_s f_y\left(d - \frac{a}{2}\right) \\
&= \frac{(0.88 \text{ in}^2)\left(60 \, \frac{\text{kips}}{\text{in}^2}\right)\left(5.25 \text{ in} - \dfrac{0.92 \text{ in}}{2}\right)}{12 \, \dfrac{\text{in}}{\text{ft}}} \\
&= 21.1 \text{ ft-kips}
\end{aligned}
$$

The design strength is

$$
\begin{aligned}
\phi M_{nx} &= (0.9)(21.1 \text{ ft-kips}) \\
&= 19 \text{ ft-kips}
\end{aligned}
$$

Applying the interaction equation gives

$$
\begin{aligned}
\frac{M_u}{\phi M_y} + \frac{M_{\hat{x}}}{\phi M_{nx}} &= \frac{31.0 \text{ ft-kips}}{167 \text{ ft-kips}} + \frac{2.52 \text{ ft-kips}}{19 \text{ ft-kips}} \\
&= 0.32 \\
&< 1.0 \quad \text{[satisfactory]}
\end{aligned}
$$

The beam is adequate.

5. BEAMS IN SHEAR

Nomenclature

A_n	net area of cross-section	in^2
A_{nv}	net shear area	in^2
A_v	area of shear reinforcement	in^2
d	distance from extreme compression fiber to centroid of tension reinforcement	in
d_v	actual depth of masonry in the direction of shear considered	in
f_v	calculated shear stress in masonry	in
F_s	allowable stress in reinforcement	lbf/in^2
F_v	allowable shear stress	lbf/in^2
F_{vm}	allowable shear stress resisted by the masonry	lbf/in^2
F_{vs}	allowable shear stress resisted by the shear reinforcement	lbf/in^2
M	moment at the section under consideration	ft-kips
M_u	factored moment	ft-kips
P	axial force on a beam	kips
P_u	factored axial force	kips
s	spacing of shear reinforcement	in
V	shear force at the section under consideration	kips
V_n	nominal shear strength	kips
V_{nm}	nominal shear strength provided by the masonry	kips
V_{ns}	nominal shear strength provided by the shear reinforcement	kips
V_u	factored shear force	kips

Shear Reinforcement

ASD Method

When the shear stress on a member, f_v, exceeds the allowable shear stress of the masonry, F_v, TMS 402 Comm. Sec. 8.3.5 requires the provision of shear reinforcement to resist the residual shear. Stirrup requirements are given in TMS 402 Sec. 6.1.7.2, Sec. 8.3.5.2.1, and Sec. 8.3.5.4, and are summarized as follows.

- Stirrups must be effectively anchored at each end in order to develop the shear reinforcement as specified in TMS 402 Sec. 6.1.7.2.1.

- Stirrups must be placed as close to the compression face of the beam as permitted by cover requirements, due to the extent of flexural tension cracks.

- Spacing of stirrups must not exceed one-half the beam depth or 48 in to ensure that every potential shear crack is crossed by at least one stirrup.

- For noncantilever beams, the maximum design shear may be calculated at a distance of $d/2$ from the face of the support. This location of the critical section is applicable provided that no concentrated load occurs between the face of the support and a distance $d/2$ from the face, and that the support reaction introduces compression into the end regions of the beam.

SD Method

When the factored shear force on a section, V_u, exceeds the design shear strength of the masonry, V_{nm}, TMS 402 Sec. 9.3.4.1.2 requires the provision of shear reinforcement to resist the excess. As specified in TMS 402 Sec. 9.3.4.2.3 and shown in Fig. 7.6, shear reinforcement must comply with the following requirements.

- Shear reinforcement should consist of a single bar with a standard 180° hook at each end to reduce congestion in the member.

- Shear reinforcement must be hooked round longitudinal reinforcement at each end so as to develop the shear reinforcement.

- The first shear reinforcing bar must be located no more than one-fourth the beam depth from the beam end so as to intersect any diagonal crack formed at the support.

- The minimum area of shear reinforcement must be $0.0007bd_v$.

- The spacing of shear reinforcing bars must not exceed one-half the beam depth or 48 in to improve ductility.

- As specified in TMS 402 Sec. 9.3.4.2.4, beams must be fully grouted.

Figure 7.6 *Shear Reinforcement*

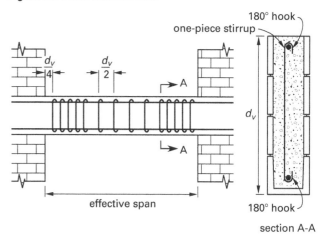

section A-A

Design for Shear in Beams

ASD Method

In accordance with TMS 402 Sec. 8.3.5.1, the shear stress in a *solid* grouted masonry beam is

$$f_v = \frac{V}{bd_v} = \frac{V}{A_{nv}}$$

The allowable shear stress in a beam without shear reinforcement is given by TMS 402 Eq. 8-26 as

$$F_{vm} = \frac{1}{2}\left(\left(4.0 - 1.75\left(\frac{M}{Vd_v}\right)\right)\sqrt{f'_m} + 0.25\left(\frac{P}{A_n}\right)\right)$$

When $f_v > F_{vm}$, shear reinforcement is provided to carry the residual shear stress. The area of shear reinforcement required is given by TMS 402 Eq. 8-27 as

$$F_{vs} = 0.5\left(\frac{A_v F_s d_v}{A_{nv}s}\right)$$

For a *solid* grouted masonry beam, the net shear area of the beam is

$$A_{nv} = bd_v$$

For a grouted beam in accordance with TMS 402 Eq. 8-22, the shear stress resisted by the masonry and the shear stress resisted by the shear reinforcement are additive to give a combined allowable shear stress of

$$F_v = (F_{vm} + F_{vs})\gamma_g$$

The grouted shear factor is given by TMS 402 Sec. 8.3.5.1.2 as

$$\gamma_g = 0.75 \text{ for partially grouted shear walls}$$
$$= 1.0 \text{ otherwise}$$

The allowable shear stress, when $M/(Vd_v) \leq 0.25$, is limited by TMS 402 Eq. 8-23 to

$$F_v \leq \left(3\sqrt{f_m'} \right) \gamma_g$$

The allowable shear stress, when $M/(Vd_v) \geq 1.0$, is limited by TMS 402 Eq. 8-24 to

$$F_v \leq \left(2\sqrt{f_m'} \right) \gamma_g$$

To simplify the procedure, TMS 402 Comm. Sec. 8.3.5.1.2 permits $M/(Vd_v)$ to be 1.0 in TMS 402 Eq. 8-26 and Sec. 8.3.5.1.2.

SD Method

The nominal shear strength of a beam without shear reinforcement is given by TMS 402 Eq. 9-20 as

$$V_{nm} = \left(4.0 - 1.75 \left(\frac{M_u}{V_u d_v} \right) \right) A_{nv} \sqrt{f_m'} + 0.25 P_u$$

The nominal shear strength provided by shear reinforcement is given by TMS 402 Eq. 9-21 as

$$V_{ns} = 0.5 \left(\frac{A_v}{s} \right) f_y d_v$$

For a solid grouted masonry beam, the net shear area of the beam is

$$A_{nv} = b d_v$$

For a grouted masonry beam in accordance with TMS 402 Eq. 9-17, the nominal shear strength provided by the masonry and the nominal shear strength provided by the shear reinforcement are additive to give a combined nominal shear strength of

$$V_n = (V_{nm} + V_{ns}) \gamma_g$$

The grouted shear factor is given by TMS 402 Sec. 9.3.4.1.2 as

$$\gamma_g = 0.75 \text{ for partially grouted shear walls}$$
$$= 1.0 \text{ otherwise}$$

For a grouted masonry beam the nominal shear strength, when $M_u/(V_u d_v) \leq 0.25$, is limited by TMS 402 Eq. 9-18 to

$$V_n \leq \left(6 A_{nv} \sqrt{f_m'} \right) \gamma_g$$

For a grouted masonry beam, the nominal shear strength when $M_u/(V_u d_v) \geq 1.0$ is limited by TMS 402 Eq. 9-19 to

$$V_n \leq \left(4 A_{nv} \sqrt{f_m'} \right) \gamma_g$$

To simplify the procedure, TMS 402 Comm. Sec. 9.3.4.1.2 permits $M_u/(V_u d_v)$ to be 1.0 in TMS 402 Eq. 9-20, Sec. 9.3.4.1.2, and Sec. 9.3.4.1.2.1.

Example 7.8

For the solid grouted masonry beam of Ex. 7.6, determine whether the shear reinforcement provided is adequate. Shear reinforcement consists of no. 4 bars at 16 in centers (ASD) and no. 5 bars at 24 in centers (SD).

Solution

ASD Method

The maximum permitted shear stress for a solid grouted beam, assuming $M/Vd_v = 1$, is given by TMS 402 Eq. 8-24 as

$$F_v = 2\sqrt{f_m'}$$
$$= 2\sqrt{1500 \ \frac{\text{lbf}}{\text{in}^2}}$$
$$= 77.5 \ \text{lbf/in}^2$$

The shear force at a distance of $d/2$ from each support is given by

$$V = \frac{w(l-d)}{2} + W$$
$$= \frac{\left(276 \ \dfrac{\text{lbf}}{\text{ft}} \right) \left(11 \ \text{ft} - \left(\dfrac{45 \ \text{in}}{12 \ \dfrac{\text{in}}{\text{ft}}} \right) \right)}{(2)\left(1000 \ \dfrac{\text{lbf}}{\text{kip}} \right)} + 20 \ \text{kips}$$
$$= 21.0 \ \text{kips}$$

The shear stress at a distance of $d/2$ from each support is given by TMS 402 Eq. 8-21 as

$$f_v = \frac{V}{A_{nv}} = \frac{V}{bd_v} = \frac{(21.0 \text{ kips})\left(1000 \frac{\text{lbf}}{\text{kip}}\right)}{(7.63 \text{ in})(48 \text{ in})}$$

$$= 57.3 \text{ lbf/in}^2$$

$$< 77.5 \text{ lbf/in}^2 \quad \begin{bmatrix} \text{satisfies TMS 402} \\ \text{Sec. 8.3.5.1.2} \end{bmatrix}$$

The allowable shear stress in a beam without shear reinforcement is given by TMS 402 Eq. 8-26. Since $P = 0 \text{ lbf/in}^2$,

$$F_{vm} = \frac{1}{2}\left[\left(4.0 - 1.75\left(\frac{M}{Vd_v}\right)\right)\sqrt{f'_m} + 0.25\left(\frac{P}{A_n}\right)\right]$$

$$= \left(\frac{1}{2}\right)(4.0 - (1.75)(1.0))\sqrt{1500 \frac{\text{lbf}}{\text{in}^2}} + 0 \frac{\text{lbf}}{\text{in}^2}$$

$$= 43.6 \text{ lbf/in}^2$$

$$< f_v = 57.3 \text{ lbf/in}^2 \quad \text{[shear reinforcement is required]}$$

The reinforcement provided consists of a no. 4 bar (0.20 in^2) at 16 in centers. This provides an allowable shear stress, as specified by TMS 402 Eq. 8-27, of

$$F_{vs} = 0.5\left(\frac{A_v F_s d_v}{A_{nv} s}\right)$$

$$= \frac{(0.5)(0.20 \text{ in}^2)\left(32{,}000 \frac{\text{lbf}}{\text{in}^2}\right)(48 \text{ in})}{(7.63 \text{ in})(48 \text{ in})(16 \text{ in})}$$

$$= 26.2 \text{ lbf/in}^2$$

The combined allowable shear stress is given by TMS 402 Eq. 8-22 as

$$F_v = F_{vm} + F_{vs} = 43.6 \frac{\text{lbf}}{\text{in}^2} + 26.2 \frac{\text{lbf}}{\text{in}^2}$$

$$= 69.8 \text{ lbf/in}^2$$

$$> f_v = 57.3 \text{ lbf/in}^2 \quad \text{[satisfactory]}$$

The shear reinforcement provided is adequate.

SD Method

The total factored shear force at a distance of $d/2$ from each support is given by IBC Eq. 16-2 as

$$V_u = 1.2V_s + 1.6V_c$$

$$= (1.2)(1.0 \text{ kips}) + (1.6)(20 \text{ kips})$$

$$= 33.2 \text{ kips}$$

For a solid grouted beam, if $M_u/(V_u d_v) = 1.0$, the maximum nominal shear capacity permitted is limited by TMS 402 Eq. 9-19 to

$$V_n \leq 4A_{nv}\sqrt{f'_m} = \frac{(4)(7.63 \text{ in})(48 \text{ in})\sqrt{1500 \frac{\text{lbf}}{\text{in}^2}}}{1000 \frac{\text{lbf}}{\text{kip}}}$$

$$= 56.7 \text{ kips}$$

The maximum design shear capacity permitted is

$$\phi V_n = (0.8)(56.7 \text{ kips}) = 45.4 \text{ kips}$$

$$> V_u = 33.2 \text{ kips} \quad \text{[satisfactory]}$$

The nominal shear capacity of the beam without shear reinforcement is given by TMS 402 Eq. 9-20. Since $P_u = 0 \text{ lbf}$,

$$V_{nm} = \left(4.0 - 1.75\left(\frac{M_u}{V_u d_v}\right)\right)A_{nv}\sqrt{f'_m} + 0.25P_u$$

$$= \frac{(4.0 - (1.75)(1.0))(7.63 \text{ in})(48 \text{ in})}{\times \sqrt{1500 \frac{\text{lbf}}{\text{in}^2}} + 0 \text{ lbf}}{1000 \frac{\text{lbf}}{\text{kip}}}$$

$$= 31.9 \text{ kips}$$

$$\phi V_{nm} = (0.8)(31.9 \text{ kips}) = 25.5 \text{ kips}$$

$$< V_u = 33.2 \text{ kips} \quad \text{[shear reinforcement is required]}$$

The minimum permitted area of shear reinforcement is specified in TMS 402 Sec. 9.3.4.2.3(c) as

$$A_v = 0.0007bd_v$$

$$= (0.0007)(7.63 \text{ in})(48 \text{ in})$$

$$= 0.26 \text{ in}^2$$

For a no. 5 bar,

$$A_v = 0.31 \text{ in}^2 > 0.26 \text{ in}^2 \quad \text{[satisfactory]}$$

The maximum permitted spacing of shear reinforcement is specified in TMS 402 Sec. 9.3.4.2.3(e) as

$$s_{\max} = 0.5d_v$$

$$= (0.5)(48 \text{ in})$$

$$= 24 \text{ in}$$

Reinforced Masonry

For a spacing of 24 in,

$$s = 24 \text{ in} = s_{\max} \quad \text{[satisfactory]}$$

The shear reinforcement provided is 0.31 in^2 at 24 in centers. This provides a nominal shear capacity, as specified by TMS 402 Eq. 9-21, of

$$V_{ns} = 0.5 \left(\frac{A_v}{s} \right) f_y d_v$$

$$= \frac{(0.5) \left(\dfrac{0.31 \text{ in}^2}{24 \text{ in}} \right) \left(60{,}000 \ \dfrac{\text{lbf}}{\text{in}^2} \right) (48 \text{ in})}{1000 \ \dfrac{\text{lbf}}{\text{kip}}}$$

$$= 18.6 \text{ kips}$$

The combined nominal shear capacity is given by TMS 402 Eq. 9-21 as

$$V_n = V_{nm} + V_{ns} = 31.9 \text{ kips} + 18.6 \text{ kips} = 50.5 \text{ kips}$$

The design shear capacity is

$$\phi V_n = (0.8)(50.5)$$
$$= 40.4 \text{ kips}$$
$$> V_u = 33.2 \text{ kips} \quad \text{[satisfactory]}$$

The shear reinforcement provided is adequate.

6. DESIGN OF MASONRY COLUMNS

Nomenclature

a	distance between column reinforcement	in
A	area of reinforcement in compression	in^2
$A_{\max}$	maximum area of the tension reinforcement that will satisfy TMS 402 Sec. 9.3.3.5.1	in^2
A_n	net effective area of column	in^2
A_s	area of reinforcement	in^2
A_s'	area of reinforcement in tension	in^2
A_{st}	area of laterally tied longitudinal reinforcement	in^2
A_t	transformed area of column, $A_n(1 + (2n - 1)\rho)$	in^2
A_{ts}	transformed area of reinforcement, $A_s(2n - 1)$	in^2
b	width of section	in
c	depth of neutral axis	in
C_m	force in masonry stress block	kips
C_s	force is compression steel	kips
d	effective depth of tension reinforcement	in

d'	depth of compression reinforcement	in
E	modulus of elasticity	lbf/in^2
f_a	calculated compressive stress due to axial load only	lbf/in^2
f_m	calculated stress in the masonry	lbf/in^2
f_s	stress in tension reinforcement	lbf/in^2
f_s'	stress in compression reinforcement	lbf/in^2
F_a	allowable compressive stress due to axial load only	lbf/in^2
h	effective height of column	in
I_n	net effective moment of inertia of column	in^4
I_t	transformed moment of inertia of column	in^4
M	bending moment	ft-kips
M_n	nominal flexural strength of a member	ft-kips
n	modular ratio, E_s/E_m	–
P	axial load	kips
P_a	allowable compressive force due to axial load only	kips
P_m	allowable compressive force on the masonry due to axial load only	kips
P_n	nominal axial strength	kips
P_s	allowable compressive force on the reinforcement due to axial load only	kips
P_u	factored axial load	kips
Q_E	effect of horizontal seismic forces	kips
r	radius of gyration	in
R	response modification factor	–
s	center-to-center spacing of items	in
S_t	transformed section modulus of column	in^3
t	nominal thickness of member	in
T	force in tension steel	kips

Symbols

ϵ_s	strain in tension reinforcement	–
ϵ_s'	strain in compression reinforcement	–
ϵ_y	strain at yield in tension reinforcement	–
ρ	ratio of tension reinforcement	–

Dimensional Limitations

Dimensional requirements for columns, as shown in Fig. 7.7, are specified in TMS 402 Sec. 5.3.1 as follows.

- The minimum nominal column width is 8 in.

- Columns must be fully grouted.

- The distance between lateral supports is limited to 99 multiplied by the least radius of gyration, r.

Figure 7.7 *Column Dimensions*

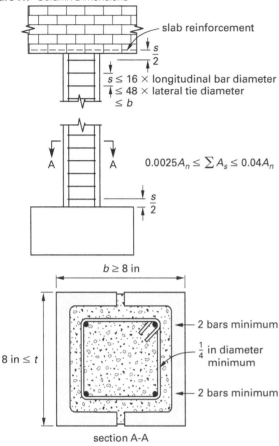

slab reinforcement

$s \le 16 \times$ longitudinal bar diameter
$\le 48 \times$ lateral tie diameter
$\le b$

$0.0025A_n \le \sum A_s \le 0.04A_n$

$b \ge 8$ in

8 in $\le t$

2 bars minimum

$\frac{1}{4}$ in diameter minimum

2 bars minimum

section A-A

- The area of longitudinal reinforcement is limited to a maximum of 4% and a minimum of 0.25% of the net column area of cross section.

- At least four longitudinal reinforcing bars must be provided, one in each corner of the column.

- Lateral ties must not have a diameter of less than $\frac{1}{4}$ in.

- Lateral ties must be placed at a spacing not exceeding the lesser of 16 longitudinal bar diameters, 48 lateral tie diameters, or the least cross-sectional dimension of the column.

- Lateral ties must be arranged so that every corner and alternate longitudinal bar has support provided by the corner of a lateral tie, and no bar is farther than 6 in clear on each side from a supported bar.

- Lateral ties must be located not more than one-half the lateral tie spacing above the top of footing or slab in any story. Lateral ties must be placed not more than one-half the lateral tie spacing below the horizontal reinforcement in the beam or slab reinforcement above.

Example 7.9

The nominal 16 in square, solid grouted concrete block masonry column shown is reinforced with four no. 4 grade 60 bars. Determine the required size and spacing of lateral ties.

Illustration for Ex. 7.10

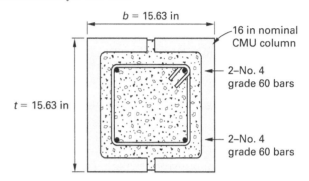

$b = 15.63$ in

16 in nominal CMU column

2–No. 4 grade 60 bars

$t = 15.63$ in

2–No. 4 grade 60 bars

Solution

As specified by TMS 402 Sec. 5.3.1, lateral ties for the confinement of longitudinal reinforcement must not have a diameter of less than $\frac{1}{4}$ in. Using no. 3 bars for the ties, the spacing must not exceed the lesser of the following.

$$s = 48d_{\text{lateral}} = (48)(0.375 \text{ in})$$
$$= 18 \text{ in}$$
$$s = 16 \text{ in} \quad [\text{least cross-sectional column dimension}]$$
$$s = 16d_{\text{longitudinal}} = (16)(0.5 \text{ in})$$
$$= 8 \text{ in} \quad [\text{governs}]$$

Axial Compression in Columns

ASD Method

The allowable compressive stress in an axially loaded reinforced masonry column is given by

$$F_a = \frac{P_a}{A_n}$$

For columns having an h/r ratio not greater than 99, the allowable axial load is given by TMS 402 Sec. 8.3.4.2.1 as

$$P_a = (0.25f_m'A_n + 0.65A_{st}F_s)\left(1 - \left(\frac{h}{140r}\right)^2\right)$$
[TMS 402 8-18]

For columns having an h/r ratio greater than 99, the allowable axial load is given by TMS 402 Sec. 8.3.4.2.1 as

$$P_a = (0.25f_m'A_n + 0.65A_{st}F_s)\left(\frac{70r}{h}\right)^2 \quad [\text{TMS 402 8-19}]$$

Reinforced Masonry

To allow for accidental eccentricities, TMS 402 Sec. 8.3.4.3 requires that a column be designed for a minimum eccentricity equal to 0.1 times each side dimension. When actual eccentricity exceeds the minimum eccentricity, the actual eccentricity should be used.

For columns with lateral reinforcement, the allowable steel stress is given by TMS 402 Sec. 8.3.3 as

$$F_s = 20{,}000 \ \text{lbf/in}^2 \quad \text{[grade 40 or grade 50 reinforcement]}$$

$$F_s = 32{,}000 \ \text{lbf/in}^2 \quad \text{[grade 60 reinforcement]}$$

SD Method

The nominal allowable axial compressive strength in an axially loaded reinforced masonry column, as given by TMS 402 Sec. 9.3.4.1.1, allows for slenderness effects and accidental eccentricity of the applied load. The nominal axial compressive strength must not exceed TMS 402 Eq. 9-15 or Eq. 9-16, as appropriate. For members having an h/r ratio not greater than 99, TMS 402 Eq. 9-15 applies and is given by

$$P_n = 0.80 \Big(0.80 f'_m (A_n - A_{st}) + f_y A_{st} \Big) \left[1 - \left(\frac{h}{140r} \right)^2 \right]$$

For members having an h/r ratio greater than 99, TMS 402 Eq. 9-16 applies and is given by

$$P_n = 0.80 \Big(0.80 f'_m (A_n - A_{st}) + f_y A_{st} \Big) \left(\frac{70r}{h} \right)^2$$

Example 7.10

The nominal 16 in square, solid grouted concrete block masonry column shown has a specified strength of 1500 lbf/in², a modulus of elasticity of 1,000,000 lbf/in², and is reinforced with four no. 4 grade 60 bars. The column has a height of 15 ft and is pinned at each end. Neglecting accidental eccentricity, determine the available column strength.

Illustration for Ex. 7.10

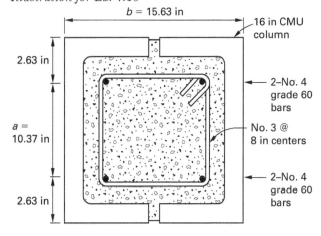

Solution

The relevant properties of the column are

$$
\begin{aligned}
b &= \text{effective column width} \\
&= 15.63 \ \text{in} \\
h &= \text{effective column height} \\
&= 15 \ \text{ft} \\
A_{st} &= \text{reinforcement area} \\
&= (4)(0.20 \ \text{in}^2) \\
&= 0.80 \ \text{in}^2 \\
A_n &= \text{effective column area} \\
&= b^2 \\
&= (15.63 \ \text{in})^2 \\
&= 244 \ \text{in}^2 \\
\rho &= \frac{A_{st}}{A_n} \\
&= \frac{0.80 \ \text{in}^2}{244 \ \text{in}^2} \\
&= 0.0033 \\
&< 0.04 \\
&> 0.0025 \quad \begin{bmatrix} \text{satisfies TMS 402} \\ \text{Sec. 5.3.1} \end{bmatrix}
\end{aligned}
$$

The radius of gyration of the column is

$$
\begin{aligned}
r &= \sqrt{\frac{I_n}{A_n}} \\
&= \sqrt{\frac{\left(\dfrac{1}{12} \right)(15.63 \ \text{in})(15.63 \ \text{in})^3}{244 \ \text{in}^2}} \\
&= 4.51 \ \text{in}
\end{aligned}
$$

The slenderness ratio of the column is

$$
\begin{aligned}
\frac{h}{r} &= \frac{(15 \ \text{ft})\left(12 \ \dfrac{\text{in}}{\text{ft}} \right)}{4.51 \ \text{in}} \\
&= 39.9 \\
&< 99 \quad \begin{bmatrix} \text{TMS 402 Eq. 8-18} \\ \text{is applicable.} \end{bmatrix}
\end{aligned}
$$

ASD Method

The allowable steel stress is given by TMS 402 Sec. 8.3.3 as

$$F_s = 32,000 \text{ lbf/in}^2 \quad \text{[for grade 60 bars]}$$

The allowable column load is given by

$$P_a = \left(0.25 f'_m A_n + 0.65 A_{st} F_s\right)\left(1 - \left(\frac{h}{140r}\right)^2\right)$$

$$= \frac{\left(\begin{array}{c} (0.25)\left(1500 \dfrac{\text{lbf}}{\text{in}^2}\right)(244 \text{ in}^2) \\[2mm] + (0.65)(0.80 \text{ in}^2)\left(32,000 \dfrac{\text{lbf}}{\text{in}^2}\right) \end{array} \right)}{1000 \dfrac{\text{lbf}}{\text{kip}}}$$

$$\times \left(1 - \left(\frac{(15 \text{ ft})\left(12 \dfrac{\text{in}}{\text{ft}}\right)}{(140)(4.51 \text{ in})}\right)^2\right)$$

$$= (91.5 \text{ kips} + 16.6 \text{ kips})(0.919)$$
$$= 84.1 \text{ kips} + 15.3 \text{ kips}$$
$$= P_m + P_s \left[\begin{array}{c}\text{allowable loads on the} \\ \text{masonry and reinforcement}\end{array}\right]$$
$$= 99.4 \text{ kips}$$

The allowable column strength is 99.4 kips.

SD Method

The slenderness ratio of the column is

$$\frac{h}{r} = 39.9$$

$$< 99 \quad \text{[TMS 402 Eq. 9-15 is applicable]}$$

From TMS 402 Eq. 9-15, the nominal axial strength is

$$P_n = 0.80\left(0.80 f'_m (A_n - A_{st}) + f_y A_{st}\right)\left(1 - \left(\frac{h}{140r}\right)^2\right)$$

$$= \frac{(0.80)\left(\begin{array}{c} (0.80)\left(1500 \dfrac{\text{lbf}}{\text{in}^2}\right)(244 \text{ in}^2 - 0.80 \text{ in}^2) \\[2mm] + \left(60,000 \dfrac{\text{lbf}}{\text{in}^2}\right)(0.80 \text{ in}^2) \end{array} \right)}{1000 \dfrac{\text{lbf}}{\text{kip}}}$$

$$\times \left(1 - \left(\frac{(15 \text{ ft})\left(12 \dfrac{\text{in}}{\text{ft}}\right)}{(140)(4.51 \text{ in})}\right)^2\right)$$

$$= (0.80)(291.8 \text{ kips} + 48 \text{ kips})(0.919)$$
$$= 250 \text{ kips}$$

The design column strength is

$$\phi P_n = (0.9)(250 \text{ kips})$$
$$= 225 \text{ kips}$$

Combined Compression and Flexure

ASD Method

The allowable compressive stress in masonry due to combined axial load and flexure is given by TMS 402 Sec. 8.3.4.2.2 as

$$F_b = 0.45 f'_m$$

In addition, the calculated compressive stress due to the axial load cannot exceed the allowable values given in TMS 402 Sec. 8.2.4.1. For columns having an h/r ratio not greater than 99, this is

$$F_a = (0.25 f'_m)\left(1.0 - \left(\frac{h}{140r}\right)^2\right) \quad \text{[TMS 402 8-13]}$$

For columns having an h/r ratio greater than 99, the allowable value is

$$F_a = (0.25 f'_m)\left(\frac{70r}{h}\right)^2 \quad \text{[TMS 402 8-14]}$$

When the axial load on the column causes a compressive stress larger than the tensile stress produced by the applied bending moment, the section is uncracked and stresses may be calculated by using the transformed section properties.[6] To allow for creep in the masonry,[9] the transformed reinforcement area is taken as $A_s(2n - 1)$, and the resultant stresses at the extreme fibers of the section, as shown in Fig. 7.8, are given by

$$f_m = f_a \pm f_b$$
$$= \frac{P}{A_t} \pm \frac{M}{S_t}$$

Stress in the reinforcement is equal to $2n$ times the stress in the adjacent masonry.

When the applied moment produces cracking in the section, the principle of superposition is no longer applicable. To determine the stresses on the section, the strain distribution over the section is estimated, and forces are determined as shown in Fig. 7.9. Internal forces on the section are compared with the applied loads, and the procedure is repeated until external and internal forces balance.

Reinforced Masonry

Figure 7.8 *Uncracked Section Properties*

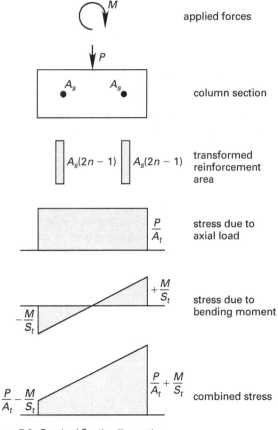

applied forces

column section

transformed reinforcement area

stress due to axial load

stress due to bending moment

combined stress

Figure 7.9 *Cracked Section Properties*

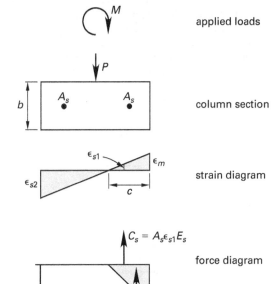

applied loads

column section

strain diagram

force diagram

SD Method

As specified in TMS 402 Sec. 9.1.2, columns must be designed for the maximum design moment accompanying the axial load. As in the case of flexural members, the requirements of TMS 402 Sec. 9.3.2 are also applied to columns, and are shown in Fig. 7.10. The design assumptions are as follows.

- Stress in the reinforcement in the compression zone is based on a linear strain distribution.

- Stress in the reinforcement below the yield strain is taken as E_s multiplied by strain. Stress above the yield strain is taken as f_y.

- The extreme compressive fiber strain is 0.0025 for concrete masonry and 0.0035 for clay masonry.

- A stress of $0.8f'_m$ and an equivalent rectangular stress distribution are assumed in the masonry with a depth of 0.8 times the depth to the neutral axis.

- The tensile strength of the masonry is neglected.

Figure 7.10 *Combined Compression and Flexure*

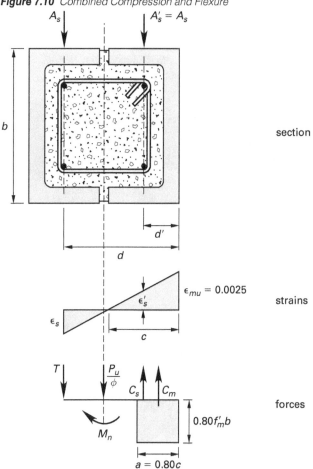

To analyze a given column section, the neutral axis depth, c, is assumed. Using the notation in Fig. 7.10 and equating compressive and tensile forces acting on the section gives

$$\frac{P_u}{\phi} = C_m + C_s - T$$

$$= 0.64cbf'_m + A'_s\epsilon'_s E_s - A_s f_y$$

$$= 0.64cbf'_m + A'_s f'_s - A_s f_y \quad [f'_s \leq f_y]$$

The neutral axis depth is adjusted until this equation is balanced. The nominal flexural strength is then determined by summing moments about the mid-depth of the section.

$$M_n = C_m\left(\frac{b}{2} - \frac{a}{2}\right) + C_s\left(\frac{b}{2} - d'\right) + T\left(d - \frac{b}{2}\right)$$

Example 7.11

The masonry column described in Ex. 7.10 is subjected to factored forces of $P = 75$ kips and $M = 10$ ft-kips (ASD), or $P_u = 76$ kips and $M_u = 50$ ft-kips (SD). The axial load includes the column weight. Determine whether the column is adequate.

Solution

ASD Method

From Ex. 7.10 and TMS 402 Eq. 8-13, the allowable compressive stress in the masonry due to axial load is

$$F_a = \frac{P_m}{A_n} = \frac{(84.1 \text{ kips})\left(1000 \frac{\text{lbf}}{\text{kip}}\right)}{244 \text{ in}^2}$$

$$= 345 \text{ lbf/in}^2$$

As shown in Ex. 7.10, the distance between reinforcement bars is

$$a = 10.37 \text{ in}$$

The modular ratio is

$$n = \frac{E_s}{E_m} = \frac{29,000,000 \frac{\text{lbf}}{\text{in}^2}}{1,000,000 \frac{\text{lbf}}{\text{in}^2}} = 29$$

If the section is not cracked, the transformed area of two no. 4 bars is

$$A_{st} = A_s(2n - 1) = \left((2)(0.2 \text{ in}^2)\right)\left((2)(29) - 1\right)$$

$$= 22.8 \text{ in}^2$$

The transformed area of the column is

$$A_t = A_n\left(1 + (2n - 1)\rho\right)$$

$$= (244 \text{ in}^2)\left(1 + ((2)(29) - 1)(0.0033)\right)$$

$$= 290 \text{ in}^2$$

The transformed moment of inertia of the column is

$$I_t = \frac{A_{st}a^2}{2} + \frac{b^4}{12}$$

$$= \frac{(22.8 \text{ in}^2)(10.37 \text{ in})^2}{2} + \frac{(15.63)^4}{12}$$

$$= 6199 \text{ in}^4$$

The stress in the masonry due to the axial load is

$$f_a = \frac{P}{A_t} = \frac{(75 \text{ kips})\left(1000 \frac{\text{lbf}}{\text{kip}}\right)}{290 \text{ in}^2}$$

$$= 259 \text{ lbf/in}^2$$

$$< F_a \quad \begin{bmatrix} \text{TMS 402 Sec. 8.3.4.2.2} \\ \text{is satisfied.} \end{bmatrix}$$

The stresses in the extreme fibers of the column due to the applied moment are

$$f_b = \frac{Mb}{2I_t} = \pm\frac{(10 \text{ ft-kips})(15.63 \text{ in})\left(12 \frac{\text{in}}{\text{ft}}\right)\times\left(1000 \frac{\text{lbf}}{\text{kip}}\right)}{(2)(6199 \text{ in}^4)}$$

$$= \pm 151 \text{ lbf/in}^2$$

$$< f_a \quad [\text{The section is uncracked.}]$$

From Ex. 7.6, the allowable stresses due to the bending moment are

$$F_b = 675 \text{ lbf/in}^2$$

$$F_s = 32,000 \text{ lbf/in}^2$$

The maximum stress in the masonry due to combined axial and flexural load is

$$f_m = f_a + f_b = 259 \frac{\text{lbf}}{\text{in}^2} + 151 \frac{\text{lbf}}{\text{in}^2}$$

$$= 410 \text{ lbf/in}^2$$

$$< F_b \text{ lbf/in}^2 \quad [\text{satisfactory}]$$

The maximum compressive stress in the reinforcement due to combined axial and flexural load is given by

$$
\begin{aligned}
f_s &= 2n\left(f_a + \frac{f_b a}{b}\right) \\[2mm]
&= (2)(29)\left(259 \ \frac{\text{lbf}}{\text{in}^2} + \frac{\left(151 \ \dfrac{\text{lbf}}{\text{in}^2}\right)(10.37 \ \text{in})}{15.63 \ \text{in}}\right) \\[2mm]
&= 20{,}832 \ \text{lbf/in}^2 \\
&< F_s \quad [\text{satisfactory}]
\end{aligned}
$$

The column is adequate.

SD Method

The required nominal axial strength is given by TMS 402 Sec. 9.1.4.4 as

$$
\begin{aligned}
P_n &= \frac{P_u}{\phi} = \frac{76 \ \text{kips}}{0.90} \\
&= 84.4 \ \text{kips}
\end{aligned}
$$

From *Illustration for Ex. 7.10*, the depth of the compression reinforcement and the effective depth of the tension reinforcement are, respectively,

$$
\begin{aligned}
d' &= 2.63 \ \text{in} \\
d &= 2.63 \ \text{in} + 10.37 \ \text{in} \\
&= 13.0 \ \text{in}
\end{aligned}
$$

Assuming the depth of the neutral axis is 6.0 in, the depth of the equivalent rectangular stress block is

$$
\begin{aligned}
a &= 0.80c \\
&= (0.80)(6.0 \ \text{in}) \\
&= 4.80 \ \text{in}
\end{aligned}
$$

The strain in the tension steel is

$$
\begin{aligned}
\epsilon_s &= \frac{\epsilon_{mu}(d - c)}{c} \\[2mm]
&= \frac{(0.0025)(13.0 \ \text{in} - 6.0 \ \text{in})}{6.0 \ \text{in}} \\[2mm]
&= 0.00292 \\
&> \epsilon_y = 0.00207
\end{aligned}
$$

The stress in the tension reinforcement is

$$
f_s = f_y = 60 \ \text{kips/in}^2
$$

The strain in the compression steel is given by

$$
\begin{aligned}
\epsilon'_s &= \frac{\epsilon_{mu}(c - d')}{c} \\[2mm]
&= \frac{(0.0025)(6.0 \ \text{in} - 2.63 \ \text{in})}{6.0 \ \text{in}} \\[2mm]
&= 0.00140
\end{aligned}
$$

The stress in the compression steel is

$$
\begin{aligned}
f'_s &= \epsilon'_s E_s \\[2mm]
&= (0.00140)\left(\frac{29{,}000{,}000 \ \dfrac{\text{lbf}}{\text{in}^2}}{1000 \ \dfrac{\text{lbf}}{\text{kip}}}\right) \\[2mm]
&= 40.6 \ \text{kips/in}^2
\end{aligned}
$$

The compressive strength of the masonry block is

$$
\begin{aligned}
C_m &= 0.64cbf'_m \\[2mm]
&= \frac{(0.64)(6.0 \ \text{in})(15.63 \ \text{in})\left(1500 \ \dfrac{\text{lbf}}{\text{in}^2}\right)}{1000 \ \dfrac{\text{lbf}}{\text{kip}}} \\[2mm]
&= 90.0 \ \text{kips}
\end{aligned}
$$

The compressive strength of the steel is

$$
\begin{aligned}
C_s &= A'_s \epsilon'_s E_s = A'_s f'_s \\[2mm]
&= ((2)(0.2 \ \text{in}^2))\left(40.6 \ \frac{\text{kips}}{\text{in}^2}\right) \\[2mm]
&= 16.2 \ \text{kips}
\end{aligned}
$$

The tensile strength of the steel is

$$
\begin{aligned}
T = A_s f_y &= ((2)(0.2 \ \text{in}^2))\left(60 \ \frac{\text{kips}}{\text{in}^2}\right) \\[2mm]
&= 24 \ \text{kips}
\end{aligned}
$$

The nominal axial load capacity for this strain condition is

$$
\begin{aligned}
\frac{P_u}{\phi} &= C_m + C_s - T \\[2mm]
&= 90.0 \ \text{kips} + 16.2 \ \text{kips} - 24 \ \text{kips} \\
&= 82.2 \ \text{kips} \\
&\approx P_n = 84.44 \ \text{kips} \quad [\text{satisfactory}]
\end{aligned}
$$

The nominal flexural strength is given by

$$M_n = C_m\left(\frac{b}{2} - \frac{a}{2}\right) + C_s\left(\frac{b}{2} - d'\right) + T\left(d - \frac{b}{2}\right)$$

$$(90.0 \text{ kips})\left(\dfrac{15.63 \text{ in}}{2} - \dfrac{4.80 \text{ in}}{2}\right)$$

$$+ (16.2 \text{ kips})\left(\dfrac{15.63 \text{ in}}{2} - 2.63 \text{ in}\right)$$

$$= \dfrac{+ (24 \text{ kips})\left(13.0 \text{ in} - \dfrac{15.63 \text{ in}}{2}\right)}{12 \, \dfrac{\text{in}}{\text{ft}}}$$

$$= 58.0 \text{ ft-kips}$$

The design moment strength is given by TMS 402 Sec. 9.1.4.4 as

$$\phi M_n = (0.9)(58.0 \text{ ft-kips}) = 52.2 \text{ ft-kips}$$
$$> M_u = 50 \text{ ft-kips}$$

The column is adequate.

Maximum Reinforcement Ratio for Columns

As in the case of flexural members, the requirements of TMS 402 Sec. 9.3.3.2.1 are also applied to columns. These requirements are shown in Fig. 7.11. When the structure is designed using a value for the response modification factor, R, greater than 1.5, as defined by ASCE/SEI7 Table 12.2-1, the maximum reinforcement ratio is determined using the following design assumptions.

- Strain in the extreme tension reinforcement is 1.5 times the strain associated with the reinforcement yield stress, f_y.

- Flexural strength is calculated by assuming that

$$f_s = \epsilon_s E_s \le f_y$$

- Axial loads, P, are included in the analysis with $P = D + 0.75L + 0.525Q_E$.

Figure 7.11 *Maximum Reinforcement Requirements for Columns*

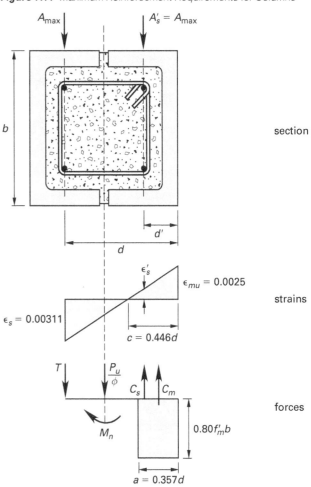

From the strain distribution shown in Fig. 7.11, the neutral axis depth is

$$c = 0.446d$$

The depth of the equivalent rectangular stress block is

$$a = 0.357d$$

The force in the equivalent rectangular stress block is

$$C_m = 0.80abf'_m$$
$$= 0.286bdf'_m$$

The force in the reinforcing bars on the tension side of the column is given by TMS 402 Sec. 9.3.3.2.1 as

$$T = A_{\max}f_y$$

The force in the reinforcing bars on the compression side of the column is given by TMS 402 Sec. 9.3.3.2.1 as

$$C_s = A_s' \epsilon_s' E_s$$
$$= A_{max} \epsilon_s' E_s$$
$$= A_{max} f_s' \quad [f_s' \leq f_y]$$

Equating compressive and tensile forces acting on the section gives

$$\frac{P_u}{\phi} = C_m + C_s - T$$
$$= 0.286 bd f_m' + A_{max} f_s' - A_{max} f_y$$
$$= (D + 0.75L + 0.525 Q_E)/\phi \quad [\text{TMS 402 Sec. 9.3.3.2.1(d)}]$$

The maximum area of the tension reinforcement that will satisfy TMS 402 Sec. 9.3.3.2.1 is

$$A_{max} = \frac{0.286 bd f_m' - \dfrac{P_u}{\phi}}{f_y - f_s'}$$

Example 7.12

The nominal 16 in square, solid grouted concrete block masonry column described in Ex. 7.10 has a specified strength of 1500 lbf/in². It is reinforced with four no. 4 grade 60 bars. The column has a height of 15 ft. It is pinned at each end. The axial force is $P_u = 65$ kips. Determine whether the reinforcement area provided satisfies TMS 402 Sec. 9.3.3.2.1. Use the SD method.

Solution

Use the SD method only.

The relevant dimensions are obtained from Ex. 7.10 as

$$d = 13.0 \text{ in}$$
$$c = 0.446d$$
$$= (0.446)(13.0 \text{ in})$$
$$= 5.80 \text{ in}$$
$$\epsilon_s' = \frac{\epsilon_{mu}(c - d')}{c}$$
$$= (0.0025)\left(\frac{5.80 \text{ in} - 2.63 \text{ in}}{5.80 \text{ in}}\right)$$
$$= 0.00137$$
$$f_s' = \epsilon_s' E_s$$
$$= (0.00137)\left(29{,}000 \ \frac{\text{kips}}{\text{in}^2}\right)$$
$$= 39.7 \text{ kips/in}^2$$

The maximum area of tension reinforcement that will satisfy TMS 402 Sec. 9.3.3.2.1 is

$$A_{max} = \frac{0.286 bd f_m' - \dfrac{P_u}{\phi}}{f_y - f_s'}$$

$$= \frac{(0.286)(15.63 \text{ in})(13.0 \text{ in})\left(\dfrac{1500 \ \frac{\text{lbf}}{\text{in}^2}}{1000 \ \frac{\text{lbf}}{\text{kip}}}\right) - \dfrac{65 \text{ kips}}{0.9}}{60 \ \dfrac{\text{kips}}{\text{in}^2} - 39.7 \ \dfrac{\text{kips}}{\text{in}^2}}$$

$$= 0.74 \text{ in}^2$$
$$> 0.40 \text{ in}^2 \text{ provided} \quad [\text{satisfactory}]$$

7. DESIGN OF SHEAR WALLS

Nomenclature

A_g	gross cross-sectional area of masonry	in²
A_g	gross cross-sectional area of the wall using specified dimensions	in²
A_{max}	maximum area of reinforcement	in²
A_n	net effective area of shear wall	in²
A_s	area of reinforcement	in²
A_{sh}	area of horizontal reinforcement in shear wall	in²
A_{sv}	area of vertical reinforcement in shear wall	in²
A_v	area of shear reinforcement	in²
b	width of section	in
c	depth of neutral axis	in
C_m	force in masonry stress block	kips
C_s	force in compression steel	kips
d	effective depth of tension reinforcement	in
d_v	actual depth of masonry in direction of shear	in
f_m'	specified masonry compressive strength	lbf/in²
f_s	stress in reinforcement	lbf/in²
f_y	yield strength of reinforcement	lbf/in²
h	height of masonry shear wall	in
M	moment occurring simultaneously with V at the section under consideration	in-lbf
M_n	nominal bending moment strength	ft-kips
M_u	factored bending moment	ft-kips
P	unfactored axial load	kips
P_n	nominal axial strength	kips
P_u	factored axial load	kips
s	spacing of reinforcement	in
T	force in tension steel	kips
V	shear force	lbf

V_n	total nominal shear strength of shear wall	kips
V_{nm}	shear strength provided by masonry	kips
V_{ns}	shear strength provided by shear reinforcement	kips
V_u	factored shear force	kips

Symbols

ρ_{max}	maximum reinforcement ratio	–
γ_g	grouted shear wall factor	–

Shear Wall Types

Several types of shear walls are classified in TMS 402 Sec. 2.2, and the determination of which type to adopt depends on the seismic design category of the structure and the materials used. Design of shear walls must comply with TMS 402 Sec. 7.3.2. The different types of reinforced concrete masonry shear walls are described as follows.

- *Ordinary plain (unreinforced) masonry shear walls* are shear walls designed to resist lateral forces without reinforcement, or where stresses in the reinforcement, if present, are neglected. This type of wall may be used only in seismic design categories A and B.

- *Detailed plain (unreinforced) masonry shear walls* are shear walls with specific minimum reinforcement and connection requirements that are designed to resist lateral forces. Stresses in the reinforcement are neglected. This type of wall may be used only in seismic design categories A and B. The reinforcement requirements are specified in TMS 402 Sec. 7.3.2.3.1 as horizontal and vertical reinforcement of at least no. 4 bars at a maximum spacing of 120 in. Additional reinforcement is required at wall openings and corners.

- *Ordinary reinforced masonry shear walls* are shear walls having the minimum reinforcement as specified in TMS 402 Sec. 7.3.2.3.1 and that are designed to resist lateral forces while considering the stresses in the reinforcement. This type of wall may be used only in seismic design categories A, B, and C. The maximum permitted height in seismic design category C is 160 ft.

- *Intermediate reinforced masonry shear walls* are shear walls having the minimum reinforcement as specified in TMS 402 Sec. 7.3.2.3.1, with the exception that the spacing of vertical reinforcement is limited to a maximum of 48 in. Walls are designed to resist lateral forces while considering the stresses in the reinforcement. This type of wall may be used only in seismic design categories A, B, and C. There is no limitation on height in seismic design category C.

- *Special reinforced masonry shear walls* are shear walls having the minimum reinforcement as specified in TMS 402 Sec. 7.3.2.6 and that are designed to resist lateral forces while considering the stresses in the reinforcement. This type of wall must be used in seismic design categories D, E, and F. When used in bearing wall or building frame systems, the maximum permitted height in seismic design categories D and E is 160 ft, and in seismic design category F is 100 ft.

Special Reinforced Shear Wall Reinforcement Requirements

In accordance with TMS 402 Sec. 7.3.2.6.1.2, special reinforced shear walls designed to resist seismic forces, using the allowable stress method, must be designed to resist 1.5 times the seismic forces calculated by IBC Chap. 16. The 1.5 multiplier need not be applied to the overturning moment.

The reinforcement requirements for a shear wall depend on the shear force in the wall and on the seismic design category assigned to the structure. In accordance with TMS 402 Sec. 7.3.2.6, shear reinforcement in special reinforced shear walls should be anchored around vertical reinforcement with a standard hook. The reinforcement requirements for a special reinforced wall are given in TMS 402 Sec. 7.3.2.6 and are shown in Fig. 7.12.

The reinforcement requirements for a special reinforced shear wall in stack bond masonry are given in TMS 402 Sec. 7.3.2.6 and are shown in Fig. 7.13.

For special reinforced shear walls designed using the strength design method, TMS 402 Sec. 7.3.2.6.1.1 specifies that the design shear strength of the wall, ϕV_n, must

Figure 7.12 *Reinforcement Details for Special Reinforced Shear Wall Laid in Running Bond*

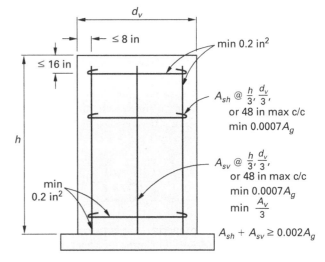

A_v = area of required shear reinforcement

Reinforced Masonry

Figure 7.13 *Reinforcement Details for Stack Bond Special Reinforced Shear Wall*

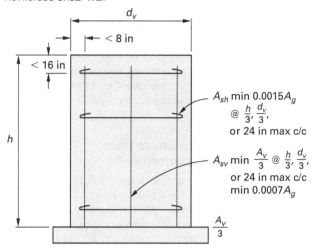

A_v = area of required shear reinforcement

exceed the shear corresponding to 1.25 times the nominal flexural strength, M_n, except that the nominal shear strength, V_n, need not exceed the value given by

$$V_n = 2.5 V_u$$

Design for Shear in Shear Walls

ASD Method

In accordance with TMS 402 Eq. 8-21, the shear stress in a masonry shear wall is determined using

$$f_v = \frac{V}{A_{nv}}$$

The allowable shear stress resisted by the masonry in a shear wall is given by TMS 402 Eq. 8-26 as

$$F_{vm} = \frac{1}{2}\left(\left(4.0 - 1.75\left(\frac{M}{Vd_v}\right)\right)\sqrt{f'_m}\right) + 0.25\frac{P}{A_n}$$

For a special reinforced masonry shear wall, degradation of the masonry shear strength may occur in plastic hinge regions. The allowable shear stress resisted by the masonry in a shear wall is then given by TMS 402 Eq. 8-25 as

$$F_{vm} = \frac{1}{4}\left(\left(4.0 - 1.75\left(\frac{M}{Vd_v}\right)\right)\sqrt{f'_m}\right) + 0.25\frac{P}{A_n}$$

When $f_v > F_{vm}$, shear reinforcement is provided to carry the residual shear stress. The area of shear reinforcement required is given by TMS 402 Eq. 8-27 as

$$F_{vs} = 0.5\left(\frac{A_v F_s d_v}{A_{nv} s}\right)$$

For a solid grouted masonry shear wall, the net shear area of the wall is

$$A_{nv} = bd_v$$

For a shear wall in accordance with TMS 402 Eq. 8-22, the shear stress resisted by the masonry and the shear stress resisted by the shear reinforcement are additive to give a combined allowable shear stress of

$$F_v = (F_{vm} + F_{vs})\gamma_g$$

The grouted shear factor is given by TMS 402 Sec. 8.3.5.1.2 as

$$\gamma_g = 0.75 \text{ for partially grouted shear walls}$$
$$= 1.0 \text{ otherwise}$$

The allowable shear stress, when $M/(Vd_v) \leq 0.25$, is limited by TMS 402 Eq. 8-23 to

$$F_v \leq \left(3\sqrt{f'_m}\right)\gamma_g$$

The allowable shear stress when $M/(Vd_v) \geq 1.0$ is limited by TMS 402 Eq. 8-24 to

$$F_v \leq \left(2\sqrt{f'_m}\right)\gamma_g$$

To simplify the procedure, TMS 402 Comm. Sec. 8.3.5.1.2 permits $M/(Vd_v)$ to be 1.0 in TMS 402 Eq. 8.25 and Eq. 8.26 and Sec. 8.3.5.1.2.

In accordance with TMS 402 Sec. 8.3.5.2, shear reinforcement is provided parallel to the direction of the applied shear force with a spacing not exceeding $d/2$ or 48 in. In accordance with TMS 402 Sec. 8.3.5.2.2, reinforcement is provided perpendicular to the shear reinforcement with an area not less than one-third the area of the shear reinforcement and with a spacing not exceeding 8 ft.

SD Method

The nominal shear strength of a shear wall without shear reinforcement is given by TMS 402 Eq. 9-20 as

$$V_{nm} = \left(4.0 - 1.75\left(\frac{M_u}{V_u d_v}\right)\right)A_{nv}\sqrt{f'_m} + 0.25P_u$$

The nominal shear strength provided by shear reinforcement is given by TMS 402 Eq. 9-21 as

$$V_{ns} = 0.5\left(\frac{A_v}{s}\right)f_y d_v$$

For a solid grouted masonry shear wall, the net shear area is

$$A_{nv} = bd_v$$

d_v is the actual depth of the wall in the direction of the applied shear and is equal to the length of shear wall, l_w.

For a masonry shear wall in accordance with TMS 402 Eq. 9-17, the nominal shear strength provided by the masonry and the nominal shear strength provided by the shear reinforcement are additive to give a combined nominal shear strength of

$$V_n = (V_{nm} + V_{ns})\gamma_g$$

The grouted shear factor is given by TMS 402 Sec. 9.3.4.1.2 as

$\gamma_g = 0.75$ for partially grouted shear walls

$\quad = 1.0$ otherwise

The nominal shear strength when $M_u/(V_u d_v) \leq 0.25$ is limited by TMS 402 Eq. 9-18 to

$$V_n \leq \left(6 A_{nv}\sqrt{f'_m}\right)\gamma_g$$

The nominal shear strength when $M_u/(V_u d_v) \geq 1.0$ is limited by TMS 402 Eq. 9-19 to

$$V_n \leq \left(4 A_{nv}\sqrt{f'_m}\right)\gamma_g$$

To simplify the procedure, TMS 402 Comm. Sec. 9.3.4.1.2 permits $M_u/(V_u d_v)$ to be 1.0 in TMS 402 Eq. 9-20 and Sec. 9.3.4.1.2.

Design for Flexure in Shear Walls

ASD Method

When flexural reinforcement is concentrated at the ends of a shear wall and axial loads are comparatively light, the shear wall may be designed in the same manner as a beam in bending. When these suppositions are not valid, it is necessary to analyze the wall using basic principles. The compressive resistance of steel reinforcement is neglected, in accordance with TMS 402 Sec. 8.3.3.3, unless lateral tie reinforcement is provided.

The maximum flexural reinforcement ratio for special reinforced masonry shear walls with $M/(V d_v) \geq 1.0$ and with $P > 0.05 f'_m A_n$ is given by TMS 402 Eq. 8-20 as

$$\rho_{\max} = \frac{n f'_m}{2 f_y \left(n + \dfrac{f_y}{f'_m}\right)}$$

SD Method

When flexural reinforcement is concentrated at the ends of a shear wall and axial loads are comparatively light, the shear wall may be designed in the same manner as a beam in bending. When these conditions are not valid, it is necessary to analyze the wall using basic principles.

To prevent brittle failure of a lightly reinforced shear wall, TMS 402 Sec. 9.3.4.2.2.2 requires the nominal flexural strength of the shear wall to be greater than 1.3 times the nominal cracking moment strength. The required nominal moment is

$$M_n \geq 1.3 M_{cr}$$

In order to provide adequate ductile response in a shear wall, TMS 402 Sec. 9.3.3.2.1 through TMS 402 Sec. 9.3.3.2.4 limit the maximum reinforcement ratio in accordance with a prescribed strain distribution. The masonry compressive strain is defined as 0.0025 for concrete masonry. The tensile strain, in the extreme tension reinforcement, depends on the seismic design category, the shear wall type, the response modification factor, R, and the value of $M_u/(V d_v)$. The tensile strain is limited to the following values.

- $\epsilon_s = 1.5\epsilon_y$ for $R \geq 1.5$ and $M_u/(V_u d_v) \leq 1.0$

- ϵ_s is not limited for $R \leq 1.5$ and $M_u/(V_u d_v) \leq 1.0$

- $\epsilon_s = 3.0\epsilon_y$ for an intermediate reinforced masonry wall with $M_u/(V_u d_v) \geq 1.0$

- $\epsilon_s = 4.0\epsilon_y$ for a special reinforced masonry wall with $M_u/(V_u d_v) \geq 1.0$

The response modification factor, R, is defined in ASCE/SEI7[8] Table 12.2-1. For masonry shear walls used in bearing wall structural systems, the values of R are given in Table 7.2.

Table 7.2 *Response Modification Factor*

masonry shear wall type	R
special reinforced	5.0
intermediate reinforced	3.5
ordinary reinforced	2.0
detailed plain	2.0
ordinary plain	1.5

The limit on maximum tensile reinforcement ratio for shear walls is waived if special boundary elements are provided to the shear wall in compliance with TMS 402 Sec. 9.3.6.6. Boundary elements consisting of vertical reinforcement with lateral restraint increase the strain capacity of the compressive stress block. Special boundary elements need not be provided in shear walls meeting the following conditions.

The factored axial load does not exceed the value $P_u \le 0.10A_g f'_m$ for geometrically symmetrical wall sections, or $P_u \le 0.05A_g f'_m$ for geometrically unsymmetrical wall sections. In addition, one of the following conditions must apply.

$$\frac{M_u}{V_u d_v} \le 1.0$$

$$V_u \le 3A_{nv}\sqrt{f'_m} \quad \left[\text{when } \frac{M_u}{V_u d_v} \le 3.0\right]$$

For walls bent in single curvature, TMS 402 Sec. 9.3.6.6.3(a) requires walls designed by the strength design method to be provided with boundary elements where the depth of the neutral axis is

$$c \ge \frac{h_w l_w}{600 C_d \delta_{ne}}$$

h_w = height of wall

l_w = length of wall

c = neutral axis depth calculated using the strength level load combination of IBC Eq. 16-5, which is $1.2(D+F) + 1.0E + f_1 L + 1.6H + f_2 S$

$\quad = \dfrac{A_s f_y + P_u}{0.64 f'_m t}$

C_d = deflection amplification factor tabulated in ASCE/SEI Table 12.2-1

δ_{ne} = elastic deflection at the top of the wall due to code-prescribed seismic forces

$C_d\delta_{ne}$ = inelastic wall drift defined in ASCE/SEI Sec. 12.8.6

The minimum height of the special confinement reinforcement is given by TMS 402 Sec. 9.3.6.6.3(b) as the larger of l_w or $M_u/4V_u$. The special boundary element must extend horizontally from the extreme compression fiber a distance given by TMS 402 Sec. 9.3.6.6.5(a) as the larger of $(c - 0.1l_w)$ or $c/2$. TMS 402 Sec. 9.3.6.6.5 requires testing to verify the strain capacity of the element.

Example 7.13

The nominal 8 in solid grouted concrete block masonry shear wall shown in the illustration has a specified strength of 1500 lbf/in^2 and a modulus of elasticity of 1,000,000 lbf/in^2. An in-plane wind load of 32 kips acts at the top of the wall, as shown, and this is the governing shear load. The wall is located in a structure assigned to seismic design category C and is laid in running bond. Determine the reinforcement required in the wall. Axial load may be neglected.

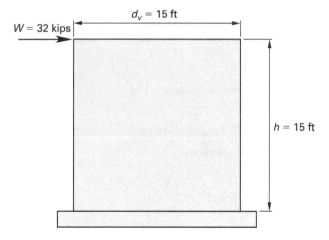

Solution

ASD Method

The wall is located in seismic design category C. An ordinary reinforced masonry wall may be used since $h < 160$ ft.

and Sec. 8.3.4.2.2, are

$$F_b = 0.45f'_m$$
$$= (0.45)\left(1500 \ \frac{\text{lbf}}{\text{in}^2}\right)$$
$$= 675 \ \text{lbf/in}^2$$
$$F_s = 32,000 \ \text{lbf/in}^2$$

Using IBC Eq. 16-15, the factored load is

$$V = 0.6W$$
$$= (0.6)(32 \ \text{kips})$$
$$= 19.2 \ \text{kips}$$

The bending moment, produced by the wind load, at the base of the wall is

$$M = Vh$$
$$= (19.2 \ \text{kips})(15 \ \text{ft})$$
$$= 288 \ \text{ft-kips}$$

Use $M_u/(Vd_v) = 1.0$ per TMS 402 Comm. Sec. 8.3.5.1.2. Assuming that two no. 6 reinforcing bars are located 4 in from each end of the wall, the relevant parameters of the wall are

$$b = 7.63 \text{ in}$$

$$d = (15 \text{ ft})\left(12 \frac{\text{in}}{\text{ft}}\right) - 4 \text{ in} = 176 \text{ in}$$

$$A_s = (2)(0.44 \text{ in}^2) = 0.88 \text{ in}^2$$

$$n = \frac{E_s}{E_m}$$

$$= \frac{29{,}000{,}000 \dfrac{\text{lbf}}{\text{in}^2}}{1{,}000{,}000 \dfrac{\text{lbf}}{\text{in}^2}}$$

$$= 29$$

$$\rho = \frac{A_s}{bd}$$

$$= \frac{0.88 \text{ in}^2}{(7.63 \text{ in})(176 \text{ in})}$$

$$= 0.000655$$

$$\rho n = (0.000655)(29)$$

$$= 0.0190$$

From App. 2.B, the wall stresses caused by the wind load, in accordance with TMS 402 Sec. 8.3.2, are

$$k = \sqrt{2\rho n + (\rho n)^2} - \rho n$$

$$= \sqrt{(2)(0.0190) + (0.0190)^2} - 0.0190$$

$$= 0.177$$

$$j = 1 - \frac{k}{3}$$

$$= 1 - \frac{0.177}{3}$$

$$= 0.941$$

$$f_b = \frac{2M}{jkbd^2}$$

$$= \frac{(2)(288 \text{ ft-kips})\left(12 \dfrac{\text{in}}{\text{ft}}\right)\left(1000 \dfrac{\text{lbf}}{\text{kip}}\right)}{(0.941)(0.177)(7.63 \text{ in})(176 \text{ in})^2}$$

$$= 176 \text{ lbf/in}^2$$

$$< 675 \text{ lbf/in}^2 \quad \text{[satisfactory]}$$

$$f_s = \frac{M}{jdA_s}$$

$$= \frac{(288 \text{ ft-kips})\left(12 \dfrac{\text{in}}{\text{ft}}\right)\left(1000 \dfrac{\text{lbf}}{\text{kip}}\right)}{(0.941)(176 \text{ in})(0.88 \text{ in}^2)}$$

$$= 23{,}713 \text{ lbf/in}^2$$

$$< 32{,}000 \text{ lbf/in}^2 \quad \text{[satisfactory]}$$

The flexural reinforcement provided is adequate.

The shear stress in the masonry wall is given by TMS 402 Eq. 8-21 as

$$f_v = \frac{V}{A_{nv}} = \frac{(19.2 \text{ kips})\left(1000 \dfrac{\text{lbf}}{\text{kip}}\right)}{(7.63 \text{ in})(15 \text{ ft})\left(12 \dfrac{\text{in}}{\text{ft}}\right)}$$

$$= 14 \text{ lbf/in}^2$$

The allowable stress is obtained by applying TMS 402 Sec. 8.3.5.1.3.

The allowable shear stress in an ordinary reinforced shear wall without shear reinforcement is given by TMS 402 Eq. 8-26 as

$$F_{vm} = \frac{1}{2}\left(\left(4.0 - 1.75\frac{M}{Vd}\right)\sqrt{f'_m}\right) + 0.25\left(\frac{P}{A_n}\right)$$

$$= \left(\frac{1}{2}\right)\left((4.0 - (1.75)(1.0))\sqrt{1500 \frac{\text{lbf}}{\text{in}^2}}\right) + 0 \frac{\text{lbf}}{\text{in}^2}$$

$$= 43.6 \text{ lbf/in}^2$$

$$> f_v = 14 \text{ lbf/in}^2 \quad \text{[satisfactory]}$$

Masonry takes all the shear force, and nominal reinforcement is required as described in TMS 402 Sec. 7.3.2.4.

SD Method

The structure is assigned to seismic design category C and an ordinary reinforced masonry wall may be used since $h < 100 \text{ ft}$.

Using IBC Eq. 16-6, the factored load is

$$V_u = 1.0 \ W$$

$$= (1.0)(32 \text{ kips})$$

$$= 32 \text{ kips}$$

The bending moment produced by the wind load at the base of the wall is

$$M_u = V_u h$$

$$= (32 \text{ kips})(15 \text{ ft})$$

$$= 480 \text{ kips-ft}$$

Assuming two no. 6 reinforcing bars are located 4 in from each end of the wall, the relevant parameters of the wall are

$$b = 7.63 \text{ in}$$

$$d = (15 \text{ ft})\left(12 \frac{\text{in}}{\text{ft}}\right) - 4 \text{ in} = 176 \text{ in}$$

$$A_s = (2)(0.44 \text{ in}^2) = 0.88 \text{ in}^2$$

The stress block depth is

$$a = \frac{A_s f_y}{0.80 b f'_m}$$

$$= \frac{(0.88 \text{ in}^2)\left(60 \; \frac{\text{kips}}{\text{in}^2}\right)\left(1000 \; \frac{\text{lbf}}{\text{kip}}\right)}{(0.80)(7.63 \text{ in})\left(1500 \; \frac{\text{lbf}}{\text{in}^2}\right)}$$

$$= 5.77 \text{ in}$$

The nominal strength is

$$M_n = A_s f_y \left(d - \frac{a}{2}\right)$$

$$= \frac{(0.88 \text{ in}^2)\left(60 \; \frac{\text{kips}}{\text{in}^2}\right)\left(176 \text{ in} - \frac{5.77 \text{ in}}{2}\right)}{12 \; \frac{\text{in}}{\text{ft}}}$$

$$= 762 \text{ ft-kips}$$

The design strength is

$$\phi M_n = (0.9)(762 \text{ ft-kips})$$
$$= 686 \text{ ft-kips}$$
$$> M_u \quad [\text{satisfactory}]$$

The flexural reinforcement provided is adequate.

TMS 402 Comm. Sec. 9.3.4.1.2 permits the adoption of $M_u/V_u d_v = 1.0$. The nominal shear strength of a shear wall without shear reinforcement is given by TMS 402 Eq. 9-20. Since $P_u = 0$ lbf,

$$V_{nm} = \left(4.0 - 1.75\left(\frac{M_u}{V_u d_v}\right)\right) A_{nv}\sqrt{f'_m} + 0.25 P_u$$

$$= \frac{(4.0 - (1.75)(1.0))(7.63 \text{ in})(15 \text{ ft})}{1000 \; \frac{\text{lbf}}{\text{kip}}}$$
$$\times \sqrt{1500 \; \frac{\text{lbf}}{\text{in}^2}}\left(12 \; \frac{\text{in}}{\text{ft}}\right) + 0 \text{ lbf}$$

$$= 120 \text{ kips}$$
$$\phi V_{nm} = (0.8)(120 \text{ kips})$$
$$= 96 \text{ kips}$$
$$> V_u = 32 \text{ kips} \quad [\text{satisfactory}]$$

The masonry takes all the shear force, and nominal reinforcement is required as detailed in TMS 402 Sec. 7.3.2.4.

Design for Axial Compression in Shear Walls

ASD Method

The vertical reinforcement in a shear wall is usually not laterally restrained. Hence, in accordance with TMS 402 Sec. 8.3.2(d), the vertical reinforcement cannot contribute to the axial strength of the wall. Then, the allowable axial compressive stress in the wall is given by TMS 402 Eq. 8-13 and Eq. 8-14, which are

$$F_a = 0.25 f'_m \left(1 - \left(\frac{h}{140r}\right)^2\right) \quad [\text{for } h/r \leq 99]$$

$$F_a = 0.25 f'_m \left(\frac{70r}{h}\right)^2 \quad [\text{for } h/r > 99]$$

SD Method

The vertical reinforcement in a shear wall is usually not laterally restrained. Hence, in accordance with TMS 402 Sec. 9.3.2(e), the longitudinal reinforcement cannot contribute to the axial strength of the wall. Then, the axial design strength of the wall is given by TMS 402 Eq. 9-11 and Eq. 9-12, which are

$$\phi P_n = (0.80)(0.90)(0.80) f'_m A_n \left(1 - \left(\frac{h}{140r}\right)^2\right) \quad [\text{for } h/r \leq 99]$$

$$\phi P_n = (0.80)(0.90)(0.80) f'_m A_n \left(\frac{70r}{h}\right)^2 \quad [\text{for } h/r > 99]$$

Example 7.14

The nominal 8 in solid grouted concrete block masonry shear wall described in Ex. 7.13 has a specified strength of 1500 lbf/in². Determine the available wall axial strength.

Solution

The relevant properties of the shear wall are

nominal thickness of the wall, $t = 7.63$ in

length of shear wall, $l = 15$ ft

height of shear wall, $h = 15$ ft

The gross cross-sectional area of the wall is

$$A_g = lt$$
$$= (15 \text{ ft})\left(12 \; \frac{\text{in}}{\text{ft}}\right)(7.63 \text{ in})$$
$$= 1373 \text{ in}^2$$

The radius of gyration of the wall is

$$r = 0.289t$$
$$= (0.289)(7.63 \text{ in})$$
$$= 2.21 \text{ in}$$

The slenderness ratio of the wall is

$$\frac{h}{r} = \frac{(15 \text{ ft})\left(12 \dfrac{\text{in}}{\text{ft}}\right)}{(2.21 \text{ in})}$$
$$= 81.5$$

ASD Method

Since $h/r < 99$, TMS 402 Eq. 8-13 applies, and the allowable axial compressive stress in the wall is

$$F_a = 0.25 f_m' \left[1 - \left(\frac{h}{140r}\right)^2\right]$$
$$= (0.25)\left(1500 \frac{\text{lbf}}{\text{in}^2}\right)\left(1 - \left(\frac{81.5}{140}\right)^2\right)$$
$$= 248 \text{ lbf/in}^2$$

The allowable wall axial strength is

$$P_a = F_a A_g$$
$$= \frac{\left(248 \dfrac{\text{lbf}}{\text{in}^2}\right)(1373 \text{ in}^2)}{1000 \dfrac{\text{lbf}}{\text{kip}}}$$
$$= 341 \text{ kips}$$

SD Method

Since $h/r < 99$, the design axial strength of the wall is given by TMS 402 Eq. 9-11, which is

$$\phi P_n = (0.80)(0.90)(0.80) f_m' A_n \left(1 - \left(\frac{h}{140r}\right)^2\right)$$
$$= \frac{\left(\begin{array}{l}(0.80)(0.90)(0.80)\left(1500 \dfrac{\text{lbf}}{\text{in}^2}\right)(1373 \text{ in}^2) \\ \times\left(1 - \left(\dfrac{81.5}{140}\right)^2\right)\end{array}\right)}{1000 \dfrac{\text{lbf}}{\text{kip}}}$$
$$= 784 \text{ kips}$$

Shear Friction Design Requirements

Nomenclature

A_{nc}	net cross-sectional area between the neutral axis and the compressive face	in^2
A_{nv}	net shear area	in^2
A_{sp}	cross-sectional area of reinforcement within the net shear area	in^2
d_v	depth of member in direction of shear	in^2
F_f	allowable shear friction stress	lbf/in^2
F_s	allowable tensile stress in reinforcement	lbf/in^2
f_m'	specified compressive strength of masonry	lbf/in^2
f_y	yield strength of reinforcement	lbf/in^2
M	moment	ft-kips
P	axial load	kips
P_u	factored axial load	kips
V	shear	kips
V_u	factored shear	kips
V_{nf}	nominal shear friction strength	kips

Symbols

μ	coefficient of friction	–

$\mu = 1.0$ for masonry on concrete with an unfinished surface, or masonry on concrete with a finished surface that has been intentionally roughened

$\mu = 0.7$ for all other conditions

Squat shear walls with an aspect ratio less than one and with low axial compressive load are vulnerable to sliding shear at the base. A low-rise structure with long shear walls usually has considerable overstrength in flexure and shear with base sliding as the governing failure mechanism. Resistance to sliding is provided by friction at the interface, dowel action of reinforcement crossing the interface, and the shear strength of the reinforcement.

ASD Method

The allowable shear friction stress at the interface is given by TMS 402 Eq. 8-28 and Eq. 8-29, which are

$$F_f = \frac{\mu(A_{sp} F_s + P)}{A_{nv}} \quad [\text{where } M/V d_v \leq 0.5]$$

$$F_f = \frac{0.65(0.6 A_{sp} F_s + P)}{A_{nv}} \quad [\text{where } M/V d_v \geq 1.0]$$

For values of $M_u/V_u d_v$ between 0.5 and 1.0, the value of F_f is linearly interpolated between the values given by TMS 402 Eq. 8-28 and Eq. 8-29.

SD Method

The nominal shear friction strength at the interface is given by TMS 402 Eq. 9-33 and Eq. 9-34, which are

$$V_{nf} = \mu(A_{sp}f_y + P_u) \quad [\text{where } M_u/V_u d_v \leq 0.5]$$
$$V_{nf} = 0.42f'_m A_{nc} \quad [\text{where } M_u/V_u d_v \geq 1.0]$$

For values of $M_u/V_u d_v$ between 0.5 and 1.0, the value of V_{nf} is linearly interpolated between the values given by TMS 402 Eq. 9-33 and Eq. 9-34.

Example 7.15

The nominal 8 in solid grouted concrete block masonry shear wall shown in the illustration has a specified strength of 1500 lbf/in². The vertical dead load on the wall from the roof diaphragm is $p_d = 1$ kip/ft, and the weight of the shear wall is $\gamma_w = 69$ lbf/ft². The shear wall is laid on a concrete foundation with an unfinished surface. An in-plane wind load of $W = 60$ kips acts at the top of the wall, as shown. Eight #4 vertical dowel bars are provided and are anchored above and below the slip plane to develop the yield strength of the bars. Determine if the sliding shear capacity is adequate.

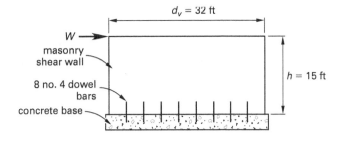

Solution

The total weight of the wall is

$$P_w = \gamma_w h d_v$$
$$= \frac{\left(69 \, \dfrac{\text{lbf}}{\text{ft}^2}\right)(15 \text{ ft})(32 \text{ ft})}{1000 \, \dfrac{\text{lbf}}{\text{kip}}}$$
$$= 33.12 \text{ kips}$$

The total dead load from the roof diaphragm is

$$P_d = p_d d_v$$
$$= \left(1.0 \, \frac{\text{kip}}{\text{ft}}\right)(32 \text{ ft})$$
$$= 32 \text{ kips}$$

The total dead load at the base of the wall is

$$D = P_w + P_d$$
$$= 33.12 \text{ kips} + 32.0 \text{ kips}$$
$$= 65.12 \text{ kips}$$

The net shear area of the wall is

$$A_{nv} = bd_v$$
$$= (7.63 \text{ in})(32 \text{ ft})\left(12 \, \frac{\text{in}}{\text{ft}}\right)$$
$$= 2930 \text{ in}^2$$

The area of dowels provided at the base is

$$A_{sp} = (8 \text{ bars})(0.2 \text{ in}^2)$$
$$= 1.6 \text{ in}^2$$

ASD Method

From IBC Eq. 16-15, the appropriate shear force acting at the base of the wall is

$$V = 0.6W$$
$$= (0.6)(60 \text{ kips})$$
$$= 36 \text{ kips}$$

The shear stress at the base of the wall is

$$f_v = \frac{V}{A_{nv}}$$
$$= \frac{(36 \text{ kips})\left(1000 \, \dfrac{\text{lbf}}{\text{kip}}\right)}{2930 \text{ in}^2}$$
$$= 12.29 \text{ lbf/in}^2$$

From IBC Eq. 16-15, the appropriate axial load acting on the wall is

$$P = 0.6D = (0.6)(65.12 \text{ kips})$$
$$= 39.1 \text{ kips}$$

$$\frac{M}{Vd_v} = \frac{Vh}{Vd_v} = \frac{h}{d_v}$$

$$= \frac{15 \text{ ft}}{32 \text{ ft}}$$

$$= 0.47$$

$$< 0.5 \qquad [\text{TMS 402 Eq. 8-28 is applicable}]$$

For masonry on concrete with an unfinished surface, the coefficient of friction is given by TMS 402 Sec. 8.3.6 as $\mu = 1.0$.

From TMS 402 Eq. 8-28, the allowable shear friction stress is

$$F_f = \frac{\mu(A_{sp}F_s + P)}{A_{nv}}$$

$$= \frac{(1.0)\left((1.6 \text{ in}^2)\left(32 \frac{\text{kips}}{\text{in}^2}\right) + 39.1 \text{ kips}\right)\left(1000 \frac{\text{lbf}}{\text{kip}}\right)}{2930 \text{ in}^2}$$

$$= 30.8 \text{ lbf/in}^2$$

$$> f_v \qquad [\text{satisfactory}]$$

SD Method

From IBC Eq. 16-6, the appropriate factored shear force acting at the base of the wall is

$$V_u = W = 60 \text{ kips}$$

From IBC Eq. 16-6, the appropriate factored axial load acting on the wall is

$$P_u = 0.9D$$

$$= (0.9)(65.12 \text{ kips})$$

$$= 58.6 \text{ kips}$$

$$\frac{M_u}{V_u d_v} = \frac{V_u h}{V_u d_v} = \frac{h}{d_v}$$

$$= \frac{15 \text{ ft}}{32 \text{ ft}}$$

$$= 0.47$$

$$< 0.5 \qquad [\text{TMS 402 Eq. 9-33 is applicable}]$$

For masonry on concrete with an unfinished surface, the coefficient of friction is given by TMS 402 Sec. 9.3.6.5 as $\mu = 1.0$.

From TMS 402 Eq. 9-33, the design shear friction strength is

$$\phi V_{nf} = \phi\mu(A_{sp}f_y + P_u)$$

$$= (0.8)(1.0)\left((1.6 \text{ in}^2)\left(60 \frac{\text{kips}}{\text{in}^2}\right) + 58.6 \text{ kips}\right)$$

$$= 123 \text{ kips}$$

$$> V_u \qquad [\text{satisfactory}]$$

8. DESIGN OF SLENDER WALLS

Nomenclature

a	depth of equivalent rectangular stress block	in
A_g	gross cross-sectional area of member	in^2
$A_{\max}$	maximum area of reinforcement that will satisfy TMS 402 Sec. 9.3.3.2.1	in^2
A_n	net cross-sectional area of member	in^2
A_s	cross-sectional area of reinforcing steel	in^2
A_{se}	equivalent area of reinforcing steel	in^2
b	width of section	in
c	depth of neutral axis	in
C_m	force in masonry stress block	lbf
d	effective depth	in
D	dead load or related internal moments or forces	–
D_w	wall dead load	lbf/ft
e_u	eccentricity of applied axial load	in
E_s	modulus of elasticity of reinforcement	lbf/in^2
E_m	modulus of elasticity of masonry	lbf/in^2
f_a	compressive stress due to axial load only	lbf/in^2
f_m	stress in masonry	lbf/in^2
f'_m	specified masonry compressive strength	lbf/in^2
f_r	modulus of rupture of masonry	lbf/in^2
f_s	stress in reinforcement	lbf/in^2
f_y	yield strength of reinforcement	lbf/in^2
h	wall height	in
I_{cr}	moment of inertia of cracked transformed section about the neutral axis	in^4
I_{eff}	effective moment of inertia	in^4
I_g	moment of inertia of gross wall section	in^4
I_n	moment of inertia of net cross-sectional area	in^4
j	lever arm factor, $1 - k/3$	–
k	neutral axis depth factor	–
L	live load or related internal moments or forces	–
M_{cr}	cracking moment	in-lbf
M_n	nominal bending moment strength	in-lbf

M_{ser}	service moment at midheight of wall including P-delta effects	in-lbf
M_u	factored bending moment at midheight of wall, including P-delta effects	in-lbf
M_{u0}	factored moment from first-order analysis	in-lbf
n	modular ratio	–
P	nonfactored axial load	lbf
P_d	nonfactored dead load from tributary floor or roof loads	lbf
P_e	Euler buckling load	lbf
P_f	nonfactored load from tributary floor or roof loads, sum of P_d and P_r	lbf
P_r	nonfactored live load from tributary floor or roof loads	lbf
P_u	sum of P_{uw} and P_{uf}	lbf
P_{uf}	factored load from tributary floor or roof loads	lbf
P_{uw}	factored weight of wall tributary to section considered	lbf
P_w	nonfactored weight of wall tributary to section considered	lbf
Q_E	the effect of horizontal seismic forces	–
S_n	section modulus of net wall section	in^3
t	nominal thickness of wall	in
T	tensile force on section	lbf
w	nonfactored lateral load	lbf/ft^2
w_u	factored lateral load	lbf/ft^2
W	wind load or related internal moments or forces	–

Symbols

γ_w	weight of masonry wall	lbf/ft^2
δ_s	deflection at midheight of wall due to service loads and including P-delta effects	in
δ_u	deflection at midheight of wall due to factored loads and including P-delta effects	in
ϵ_{mu}	maximum usable compressive strain of masonry	–
ϵ_s	strain in reinforcement	–
ϵ_{su}	maximum strain in reinforcement	–
ϵ_y	strain at yield in tension reinforcement	–
ρ_e	equivalent tension reinforcement ratio, A_{se}/bd	–
ρ_{max}	maximum reinforcement ratio that will satisfy TMS 402 Sec. 9.3.3.2.1	–
ϕ	strength reduction factor	–

Design Basis

The design of slender masonry walls with out-of-plane loads is covered in TMS 402 Sec. 9.3.5. The principal requirements are as follows.[10]

- The P-delta effect caused by the wall deflection is considered in the design analysis.

- In order to minimize residual deflections, the maximum allowable deflection of the wall due to service loads is limited by TMS 402 Eq. 9-32 to

$$\delta_s = 0.007h$$

- To minimize P-delta effects, limits are imposed by TMS 402 Sec. 9.3.5.4.2 on the factored axial stress, and the iterative design procedure of TMS 402 Sec. 9.3.5.4.2 is permitted when

$$h/t > 30, \text{ and } f_a \leq 0.05 f'_m$$

or

$$h/t \leq 30, \text{ and } f_a \leq 0.20 f'_m$$

$$f_a = P_u/A_g$$

When $f_a > 0.2 f'_m$, a second order analysis or the moment magnifier method of TMS 402 Sec. 9.3.5.4.3 must be used.

- The restriction that $f_a \leq 0.05 f'_m$ does not apply to the moment magnifier method.

- The maximum reinforcement ratio is limited by TMS 402 Sec. 9.3.3.2.1 to ensure ductility and prevent brittle compression failure.

- The design of the wall must follow strength design procedures in accordance with TMS 402 Sec. 9.3.5.2.

- The generally accepted minimum wall thickness is given by TMS 402 App. A Sec. A.6.2 as

  $t_{min} = 6$ in for one-story buildings

  $t_{min} = 8$ in for buildings with more than one story

  $t_{min} = 8$ in for shear walls

Strength Design Method

Figure 7.14 illustrates the design assumptions of TMS 402 Sec. 9.3.2 for a concrete masonry wall with a factored axial load of P_u and with the reinforcement area A_s located in the center of the wall. The effective depth of the section is

$$d = \frac{t}{2}$$

Allowing for the axial load on the wall, the equivalent reinforcement area is

$$A_{se} = \frac{\dfrac{P_u}{\phi} + A_s f_y}{f_y}$$

The stress in the reinforcement is assumed to be equal to the yield strength, f_y, and the equivalent tensile force on the section is

$$T = A_{se}f_y$$
$$= \frac{P_u}{\phi} + A_sf_y$$

Figure 7.14 *Flexural Capacity of a Slender Concrete Masonry Wall*

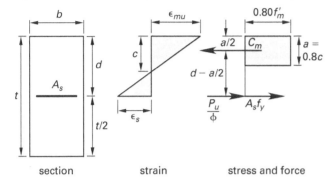

The strain in the extreme compression fiber is 0.25% for concrete masonry and 0.35% for clay masonry, and the depth of the neutral axis is c. A rectangular compression stress block is assumed with a magnitude of $0.80f_m'$ and a depth of

$$a = 0.80c$$

The compression force on the stress block acts at mid-depth of the stress block, and is given by

$$C_m = 0.80abf_m'$$

The depth of the stress block is obtained by equating forces on the section and is given by TMS 402 Comm. Sec. 9.3.5.2 as

$$a = \frac{\dfrac{P_u}{\phi} + A_sf_y}{0.80bf_m'}$$

The nominal moment capacity of the section is obtained by taking moments about the line of action of the compression force and is given by TMS 402 Comm. Sec 9.3.5.2 as

$$M_n = \left(\frac{P_u}{\phi} + A_sf_y\right)\left(d - \frac{a}{2}\right)$$

The design moment capacity is given by TMS 402 Sec. 9.1.4 as

$$\phi M_n = 0.9\left(\frac{P_u}{\phi} + A_sf_y\right)\left(d - \frac{a}{2}\right)$$

Example 7.16

A nominal 8 in solid grouted, concrete block masonry wall is shown. It has a specified strength of 1500 lbf/in² and is reinforced longitudinally with no. 4 grade 60 bars at 16 in centers placed centrally in the wall. The wall has an effective height of $h = 20$ ft and is simply supported at the top and bottom. The unfactored roof live load is $P_r = 200$ lbf/ft, and the unfactored roof dead load is $P_d = 400$ lbf/ft. The masonry wall has a weight of $\gamma_w = 69$ lbf/ft². Wind load governs and has a value of $W = 30$ lbf/ft². Determine the design flexural strength of the wall.

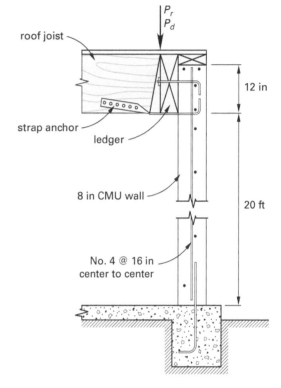

Solution

In determining the moment demand on the wall, the governing load combination is IBC Eq. 16-6, which is

$$Q_u = 0.9D + 1.0W$$

The same load factors will be used to determine the design flexural capacity of the wall.

The critical section of the wall is at midheight of the wall between supports. Use a wall width of $b = 1$ ft. The weight of wall tributary to the critical section is

$$P_w = \gamma_w\left(\frac{h}{2} + 1 \text{ ft}\right)b$$
$$= \left(69 \ \frac{\text{lbf}}{\text{ft}^2}\right)\left(\frac{20 \text{ ft}}{2} + 1 \text{ ft}\right)(1 \text{ ft})$$
$$= 759 \text{ lbf}$$

Reinforced Masonry

The dead load from the roof over a 1 ft width is given as

$$P_d = 400 \text{ lbf}$$
$$= P_f \quad \begin{bmatrix} \text{since roof live load is not} \\ \text{included in IBC Eq. 16-6} \end{bmatrix}$$

The total nonfactored gravity load at the critical section is

$$P = P_w + P_f = \frac{759 \text{ lbf} + 400 \text{ lbf}}{1000 \dfrac{\text{lbf}}{\text{kip}}}$$
$$= 1.16 \text{ kips}$$

The total factored axial load at the critical section of the wall is

$$P_u = 0.9(P_w + P_f)$$
$$= \frac{(0.9)(759 \text{ lbf} + 400 \text{ lbf})}{1000 \dfrac{\text{lbf}}{\text{kip}}}$$
$$= 1.04 \text{ kips}$$
$$\frac{P_u}{\phi} = \frac{1.04 \text{ kips}}{0.9} = 1.16 \text{ kips}$$

The reinforcement area over a 1 ft width of the wall is

$$A_s = \frac{(1 \text{ ft})(0.20 \text{ in}^2)\left(12 \dfrac{\text{in}}{\text{ft}}\right)}{16 \text{ in}}$$
$$= 0.15 \text{ in}^2$$

For strength level loads, the equivalent reinforcement area is given by

$$A_{se} = \frac{\dfrac{P_u}{\phi} + A_s f_y}{f_y}$$
$$= \frac{1.16 \text{ kips} + (0.15 \text{ in}^2)\left(60 \dfrac{\text{kips}}{\text{in}^2}\right)}{60 \dfrac{\text{kips}}{\text{in}^2}}$$
$$= 0.169 \text{ in}^2$$

The depth of the rectangular stress block is

$$a = \frac{A_{se} f_y}{0.80 f'_m b}$$
$$= \frac{(0.169 \text{ in}^2)\left(60 \dfrac{\text{kips}}{\text{in}^2}\right)\left(1000 \dfrac{\text{lbf}}{\text{kip}}\right)}{(0.80)\left(1500 \dfrac{\text{lbf}}{\text{in}^2}\right)(1 \text{ ft})\left(12 \dfrac{\text{in}}{\text{ft}}\right)}$$
$$= 0.70 \text{ in}$$

The nominal moment strength is

$$M_n = A_{se} f_y \left(d - \frac{a}{2}\right)$$
$$= \frac{(0.169 \text{ in}^2)\left(60 \dfrac{\text{kips}}{\text{in}^2}\right)\left(\dfrac{7.63 \text{ in}}{2} - \dfrac{0.70 \text{ in}}{2}\right)}{12 \dfrac{\text{in}}{\text{ft}}}$$
$$= 2.93 \text{ ft-kips}$$

The design moment strength is

$$\phi M_n = (0.9)(2.93 \text{ ft-kips})$$
$$= 2.64 \text{ ft-kips}$$

Flexural Demand on a Slender Masonry Wall

Two design methods are presented in TMS 402. These are the P-delta iterative method of TMS 402 Sec. 9.3.5.4.2 and the moment magnifier method of TMS 402 Sec. 9.3.5.4.3. Slender masonry walls are designed for the factored applied loads, which must include the axial loads and eccentricities, the P-delta effects caused by the vertical loads and the lateral deflection of the wall, and the lateral loads. The design method of TMS 402 Sec. 9.3.5 assumes the wall is simply supported at the top and bottom and is uniformly laterally loaded with the load w_u to produce the lateral deflection δ_u at the midheight of the wall. The critical section then occurs at the midheight of the wall. The factored moment at the critical section is derived from the free-body diagram in Fig. 7.15 to give TMS 402 Eq. 9-23, which is

$$M_u = \frac{w_u h^2}{8} + \frac{P_{uf} e_u}{2} + P_u \delta_u$$

Reinforced Masonry

Figure 7.15 *Analysis of a Slender Concrete Masonry Wall*

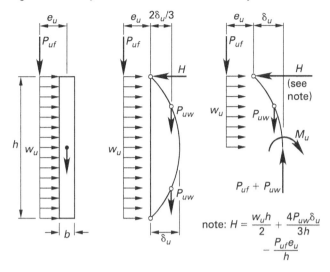

note: $H = \dfrac{w_u h}{2} + \dfrac{4 P_{uw} \delta_u}{3h} - \dfrac{P_{uf} e_u}{h}$

TMS 402 Iterative Procedure

The deflection at midheight of the wall due to the factored loads, and including P-delta effects, is given by TMS 402 Eq. 9-25 and Eq. 9-26. When the moment demand is less than the cracking moment, TMS 402 Eq. 9-25 governs, which is

$$\delta_u = \frac{5 M_u h^2}{48 E_m I_n}$$

When the moment demand is greater than the cracking moment, TMS 402 Eq. 9-26 governs, which is

$$\delta_u = \frac{5 M_{cr} h^2}{48 E_m I_n} + \frac{5 h^2 (M_u - M_{cr})}{48 E_m I_{cr}}$$

Calculation of the lateral deflection requires the determination of several additional factors, including the following.

The modulus of elasticity of the reinforcement is given by TMS 402 Table 4.2.2 as

$$E_s = 29{,}000 \text{ kips/in}^2$$

The modulus of elasticity of the concrete masonry is given by TMS 402 Table 4.2.2 as

$$E_m = 900 f_m'$$

The modular ratio is given by TMS 402 Sec. 2.1 as

$$n = \frac{E_s}{E_m}$$

The modulus of rupture for out-of-plane forces on a fully grouted masonry wall with type S mortar is given in TMS 402 Table 9.1.9.2 as

$$f_r = 163 \text{ lbf/in}^2$$

The moment of inertia of a fully grouted wall prior to cracking is

$$I_n = \frac{bt^3}{12}$$

For centered reinforcement the section modulus of the cross-sectional area of the wall prior to cracking is

$$S_n = \frac{bt^2}{6}$$

The depth to the neutral axis is given by TMS 402 Eq. 9-31 as

$$c = \frac{A_s f_y + P_u}{0.64 f_m' b}$$

As shown in Fig. 7.16, the effective reinforcement depth is

$$d = \frac{t}{2}$$

The distance of a reinforcing bar from the neutral axis is $d - c$.

Figure 7.16 *Transformed Section of a Slender Concrete Masonry Wall with Centered Reinforcement*

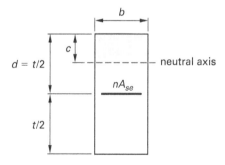

For centered reinforcement, allowing for the axial load on the wall, the moment of inertia of the cracked transformed section about the neutral axis is given by TMS 402 Eq. 9-30 as

$$I_{cr} = \frac{bc^3}{3} + n\left(A_s + \frac{P_u}{f_y}\right)(d - c)^2$$

Since the moment demand depends on the lateral deflection, and the lateral deflection depends on the moment demand, an iterative process is required until the values for M_u and δ_u converge.

TMS 402 Moment Magnifier Procedure

The first-order strength level moment at midheight of the wall can be obtained from TMS 402 Eq. 9-23 by omitting the P-delta component to give

$$M_{u,0} = \frac{w_u h^2}{8} + \frac{P_{uf} e_u}{2}$$

The second-order moment can then be obtained by multiplying the first-order moment by the moment multiplier, ψ, as shown by TMS 402 Eq. 9-27 as

$$M_u = \psi M_{u,0}$$

The moment magnifier is given by TMS 402 Eq. 9-28 as

$$\psi = \frac{1}{1 - \dfrac{P_u}{P_e}}$$

In this equation, P_e is the Euler buckling load, which is given by TMS 402 Eq. 9-29 as

$$P_e = \frac{\pi^2 E_m I_{\text{eff}}}{h^2}$$

I_{eff} is the effective moment of inertia, which is

$$I_{\text{eff}} = 0.75 I_n \quad [\text{for } M_u < M_{cr}]$$
$$I_{\text{eff}} = I_{cr} \quad [\text{for } M_u \geq M_{cr}]$$

Example 7.17

The nominal 8 in solid grouted, concrete block masonry wall described in Ex. 7.16, has type S mortar and a specified strength of 1500 lbf/in². It is reinforced longitudinally with no. 4 grade 60 bars at 16 in centers placed centrally in the wall. The wall has an effective height of $h = 20$ ft and is simply supported at the top and bottom. The unfactored roof live load is $P_r = 200$ lbf/ft. The unfactored roof dead load is $P_d = 400$ lbf/ft. The masonry wall has a

weight of $\gamma_w = 69$ lbf/ft². Wind load governs and has a value of $W = 30$ lbf/ft². Determine if the flexural capacity of the wall is adequate.

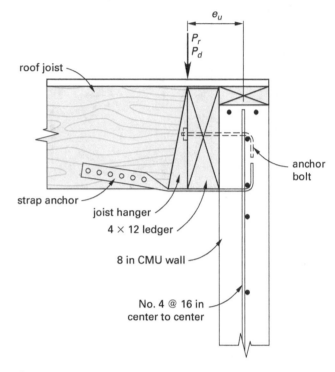

Solution

Using the Iterative Design Procedure of TMS 402 Sec. 9.3.5.4.2

From Ex. 7.16, the nominal flexural capacity for a 1 ft width of wall is

$$M_n = 2.93 \text{ ft-kips}$$

The design moment strength is

$$\phi M_n = 2.64 \text{ ft-kips}$$

The factored tributary roof dead load over a 1 ft width using IBC Eq. 16-6 is

$$P_{uf} = 0.9 P_d \quad \begin{bmatrix} \text{since roof live load} \\ \text{is not included in} \\ \text{IBC Eq. 16-6} \end{bmatrix}$$

$$= (0.9) \left(\frac{400 \ \dfrac{\text{lbf}}{\text{ft}}}{1000 \ \dfrac{\text{lbf}}{\text{kip}}} \right) (1 \text{ ft})$$

$$= 0.36 \text{ kip}$$

The total factored axial load at the critical section of the wall is

$$P_u = 1.04 \text{ kips}$$

Using IBC Eq. 16-6, the factored wind load over a 1 ft width is

$$w_u = W$$

$$= \left(\frac{30 \ \frac{\text{lbf}}{\text{ft}^2}}{1000 \ \frac{\text{lbf}}{\text{kip}}} \right) (1 \ \text{ft})$$

$$= 0.03 \ \text{kip/ft}$$

From the illustration, the eccentricity of the tributary roof dead load is

$$e_u = \text{ledger width} + \text{half wall thickness}$$

$$= 3.5 \ \text{in} + \frac{7.63}{2} \ \text{in}$$

$$= 7.32 \ \text{in}$$

The modulus of rupture for out-of-plane forces on a fully grouted masonry wall, with type S mortar, is given by TMS 402 Table 9.1.9.2 as

$$f_r = 163 \ \text{lbf/in}^2$$

For centered reinforcement the section modulus of the gross wall section is

$$S_n = \frac{bt^2}{6}$$

$$= \frac{(1 \ \text{ft})(7.63 \ \text{in})^2 \left(12 \ \frac{\text{in}}{\text{ft}} \right)}{6}$$

$$= 116.4 \ \text{in}^3$$

The gross wall area is

$$A_g = (7.63 \ \text{in})(12 \ \text{in})$$

$$= 91.56 \ \text{in}^2$$

Including the effects of axial load, the nominal cracking moment strength is

$$M_{cr} = S_n \left(f_r + \frac{P_u}{A_g} \right)$$

$$= \frac{(116.4 \ \text{in}^3) \left(163 \ \frac{\text{lbf}}{\text{in}^2} + \frac{(1.04 \ \text{kips}) \left(1000 \ \frac{\text{lbf}}{\text{kip}} \right)}{91.56 \ \text{in}^2} \right)}{\left(12 \ \frac{\text{in}}{\text{ft}} \right) \left(1000 \ \frac{\text{lbf}}{\text{kip}} \right)}$$

$$= 1.69 \ \text{ft-kips}$$

The nominal wall moment is

$$M_n = 2.93 \ \text{ft-kips}$$

$$\frac{M_n}{M_{cr}} = \frac{2.93 \ \text{ft-kips}}{1.69 \ \text{ft-kips}}$$

$$= 1.73$$

$$> 1.3 \quad \begin{bmatrix} \text{complies with TMS 402} \\ \text{Sec. 9.3.4.2.2.2} \end{bmatrix}$$

Assume a deflection at midheight due to factored loads of

$$\delta_{u1} = 0.20 \ \text{in}$$

The applied strength level moment at midheight of the wall is given by TMS 402 Eq. 9-23 as

$$M_{u1} = \frac{w_u h^2}{8} + \frac{P_{uf} e_u}{2} + P_u \delta_{u1}$$

$$= \frac{\left(0.03 \ \frac{\text{kip}}{\text{ft}} \right)(20 \ \text{ft})^2}{8} + \frac{(0.36 \ \text{kip})(7.32 \ \text{in})}{(2) \left(12 \ \frac{\text{in}}{\text{ft}} \right)}$$

$$+ \frac{(1.04 \ \text{kips})(0.20 \ \text{in})}{12 \ \frac{\text{in}}{\text{ft}}}$$

$$= 1.63 \ \text{ft-kips}$$

$$< \phi M_n \quad \text{[satisfactory]}$$

$$< M_{cr} \quad \text{[TMS 402 Eq. 9-25 applies]}$$

The deflection corresponding to the factored moment is determined in accordance with TMS 402 Sec. 9.3.5.4.2. The moment of inertia of the gross wall section is

$$I_n = \frac{bt^3}{12}$$

$$= \frac{(1 \ \text{ft})(7.63 \ \text{in})^3 \left(12 \ \frac{\text{in}}{\text{ft}} \right)}{12}$$

$$= 444 \ \text{in}^4$$

The modulus of elasticity of reinforcement is given by TMS 402 Table 4.2.2 as

$$E_s = 29,000 \ \text{kips/in}^2$$

The modulus of elasticity of concrete masonry is given by TMS 402 Table 4.2.2 as

$$E_m = 900f'_m$$

$$= (900) \left(\frac{1500 \, \frac{\text{lbf}}{\text{in}^2}}{1000 \, \frac{\text{lbf}}{\text{kip}}} \right)$$

$$= 1350 \text{ kips/in}^2$$

The modular ratio is

$$n = \frac{E_s}{E_m} = \frac{29{,}000 \, \frac{\text{kips}}{\text{in}^2}}{1350 \, \frac{\text{kips}}{\text{in}^2}}$$

$$= 21.5$$

The distance from the extreme compression fiber to the neutral axis is given by TMS 402 Eq. 9-35 as

$$c = \frac{A_s f_y + P_u}{0.64 f'_m b}$$

$$= \frac{\left[(0.15 \text{ in}^2) \left(60 \, \frac{\text{kips}}{\text{in}^2} \right) + 1.04 \text{ kips} \right] \left(1000 \, \frac{\text{lbf}}{\text{kip}} \right)}{(0.64) \left(1500 \, \frac{\text{lbf}}{\text{in}^2} \right) (12 \text{ in})}$$

$$= 0.87 \text{ in}$$

Allowing for the axial load, the moment of inertia of the cracked transformed section about the neutral axis is given by TMS 402 Eq. 9-30 as

$$I_{cr} = \frac{bc^3}{3} + n \left(A_s + \frac{P_u}{f_y} \right) (d - c)^2$$

$$= \frac{(1 \text{ ft})(0.87 \text{ in})^3 \left(12 \, \frac{\text{in}}{\text{ft}} \right)}{3}$$

$$+ (21.5) \left(0.15 \text{ in}^2 + \frac{1.04 \text{ kips}}{60 \, \frac{\text{kips}}{\text{in}^2}} \right)$$

$$\times \left(\frac{7.63 \text{ in}}{2} - 0.87 \text{ in} \right)^2$$

$$= 33.8 \text{ in}^4$$

Since $M_{u1} < M_{cr}$, the midheight deflection corresponding to the factored moment is derived from TMS 402 Eq. 9-25 as

$$\delta_u = \frac{5 M_u h^2}{48 E_m I_n}$$

$$= \frac{(5)(1.63 \text{ ft-kips}) \left(12 \, \frac{\text{in}}{\text{ft}} \right) \left((20 \text{ ft}) \left(12 \, \frac{\text{in}}{\text{ft}} \right) \right)^2}{(48) \left(1350 \, \frac{\text{kips}}{\text{in}^2} \right) (444 \text{ in}^4)}$$

$$= 0.20 \text{ in}$$

$$\left[\begin{array}{c} \text{This is equal to} \\ \text{the assumed deflection; further} \\ \text{iterations are unnecessary.} \end{array} \right]$$

So,

$$M_u = 1.63 \text{ ft-kips}$$

$$< \phi M_n \quad [\text{satisfactory}]$$

The flexural capacity is adequate.

Using the Moment Magnification Process of TMS 402 Sec. 9.3.5.4.3

The first-order strength level moment at midheight of the wall is

$$M_{u,0} = \frac{w_u h^2}{8} + \frac{P_{uf} e_u}{2}$$

$$= \frac{\left(0.03 \, \frac{\text{kip}}{\text{ft}} \right) (20 \text{ ft})^2}{8} + \frac{(0.36 \text{ kip}) \left(\frac{7.32 \text{ in}}{12 \, \frac{\text{in}}{\text{ft}}} \right)}{2}$$

$$= 1.61 \text{ ft-kips}$$

The moment of inertia of the net wall section is

$$I_n = 444 \text{ in}^4$$

M_u is less than M_{cr}, so the effective moment of inertia is

$$I_{\text{eff}} = 0.75 I_n = (0.75)(444 \text{ in}^4) = 333 \text{ in}^4$$

The Euler buckling load is

$$P_e = \frac{\pi^2 E_m I_{\text{eff}}}{h^2} = \frac{\pi^2 \left(1350 \, \frac{\text{kips}}{\text{in}^2} \right) (333 \text{ in}^4)}{\left((20 \text{ ft}) \left(12 \, \frac{\text{in}}{\text{ft}} \right) \right)^2}$$

$$= 77 \text{ kips}$$

The moment magnifier is

$$\psi = \frac{1}{1 - \dfrac{P_u}{P_e}} = \frac{1}{1 - \dfrac{1.04 \text{ kips}}{77 \text{ kips}}} = 1.014$$

The second-order moment is

$$\begin{aligned} M_u &= \psi M_{u,0} = (1.014)(1.61 \text{ ft-kips}) \\ &= 1.63 \text{ ft-kips} \\ &< \phi M_n \quad \text{[satisfactory]} \end{aligned}$$

The flexural capacity is adequate.

Maximum Reinforcement Limit for a Slender Masonry Wall

The amount of tensile reinforcement allowed in slender masonry walls is limited so as to provide adequate ductility in the wall by ensuring that the tensile reinforcement yields prior to the masonry compressive zone crushing. To achieve this, TMS 402 Sec. 9.3.3.2.1 stipulates the following conditions, as shown in Fig. 7.17.

- Strain in the extreme tension reinforcement is 1.5 times the strain associated with the reinforcement yield stress, f_y, and for grade 60 reinforcement is

$$\begin{aligned} \epsilon_{su} &= (1.5)(0.00207) \\ &= 0.00311 \end{aligned}$$

- Maximum strain in the extreme masonry compression fiber, for concrete masonry, is

$$\epsilon_{mu} = 0.00250$$

- Unfactored gravity axial loads are included in the analysis using the combination

$$P_u = D + 0.75L + 0.525Q_E$$

Figure 7.17 *Maximum Reinforcement Requirements for a Slender Concrete Masonry Wall*

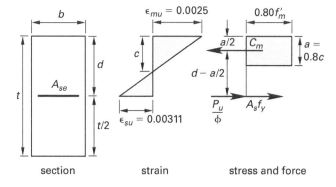

section   strain   stress and force

From the strain distribution shown in Fig. 7.17, the neutral axis depth is obtained as

$$\begin{aligned} c &= \frac{0.0025d}{0.00561} \\ &= 0.446d \end{aligned}$$

The depth of the equivalent rectangular stress block is

$$\begin{aligned} a &= 0.80c = (0.80)(0.446d) \\ &= 0.357d \end{aligned}$$

The force in the equivalent rectangular stress block is

$$\begin{aligned} C_m &= 0.80abf'_m = (0.80)(0.357d)\,bf'_m \\ &= 0.286bdf'_m \end{aligned}$$

The force in the reinforcing bars is given by TMS 402 Sec. 9.3.3.2.1 as

$$T = A_{\max}f_y$$

Equating compressive and tensile forces acting on the section gives

$$\begin{aligned} \frac{P_u}{\phi} &= C_m - T \\ &= 0.286bdf'_m - A_{\max}f_y \end{aligned}$$

So, the maximum area of the tension reinforcement that will satisfy TMS 402 Sec. 9.3.3.2.1 is

$$A_{\max} = \frac{0.286bdf'_m - P_n}{f_y}$$

Example 7.18

The nominal 8 in solid grouted, concrete block masonry wall, with type S mortar and described in Ex. 7.16, has a specified strength of 1500 lbf/in². It is reinforced longitudinally with no. 4 grade 60 bars at 16 in centers placed centrally in the wall. The wall has an effective height of $h = 20$ ft and is simply supported at the top and bottom. The unfactored roof live load is $P_r = 200$ lbf/ft. The unfactored roof dead load is $P_d = 400$ lbf/ft. The masonry wall has a weight of $\gamma_w = 69$ lbf/ft². Wind load governs and has a value of $W = 30$ lbf/ft². Determine if the reinforcement area provided satisfies TMS 402 Sec. 9.3.3.2.1.

Solution

From Ex. 7.16, the reinforcement area over a 1 ft width of the wall is

$$A_s = 0.15 \text{ in}^2$$

The tributary roof dead load on a 1 ft width of wall is

$$P_d = 0.40 \text{ kip}$$

The tributary roof live load on a 1 ft width of wall is

$$L = 0 \text{ kips}$$

The weight of wall tributary to the critical section is

$$P_w = 0.759 \text{ kip}$$

The specified gravity axial load combination is

$$
\begin{aligned}
P_u &= D + 0.75L + 0.525 Q_E \\
&= P_d + P_w + 0 \text{ kips} + 0 \text{ kips} \\
&= 0.40 \text{ kip} + 0.759 \text{ kip} + 0 \text{ kips} + 0 \text{ kips} \\
&= 1.16 \text{ kips} \\
P_n &= \frac{P_u}{\phi} \\
&= \frac{1.16 \text{ kips}}{0.9} \\
&= 1.29 \text{ kips}
\end{aligned}
$$

The maximum area of the tension reinforcement that will satisfy TMS 402 Sec. 9.3.3.2.1 is

$$
\begin{aligned}
A_{\max} &= \frac{0.286 b d f'_m - P_n}{f_y} \\
&= \frac{(0.286)(12 \text{ in})(3.82 \text{ in})\left(1.5 \dfrac{\text{kips}}{\text{in}^2}\right) - 1.29 \text{ kips}}{60 \dfrac{\text{kips}}{\text{in}^2}} \\
&= 0.31 \text{ in}^2 \\
&> A_s \text{ provided} \quad [\text{satisfactory}]
\end{aligned}
$$

Lateral Deflection of a Slender Masonry Wall Under Service Loads

The maximum permissible deflection at midheight of the wall due to service level vertical and lateral loads, and including P-delta effects, is given by TMS 402 Eq. 9-32 as

$$\delta_s = 0.007 h$$

When the applied service moment is less than the cracking moment, the service deflection is given by TMS 402 Eq. 9-25 after replacing M_u with M_{ser} and δ_u with δ_s.

$$\delta_s = \frac{5 M_{\text{ser}} h^2}{48 E_m I_n}$$

When the applied service moment, M_{ser}, exceeds the cracking moment, M_{cr}, the service deflection is given by TMS 402 Eq. 9-26 after replacing M_u with M_{ser} and δ_u with δ_s.

$$\delta_s = \frac{5 M_{cr} h^2}{48 E_m I_n} + \frac{5(M_{\text{ser}} - M_{cr}) h^2}{48 E_m I_{cr}}$$

The service moment at midheight of wall, including P-delta effects, is

$$M_{\text{ser}} = \frac{w h^2}{8} + \frac{P_f e}{2} + P \delta_s$$

Example 7.19

The nominal 8 in solid grouted, concrete block masonry wall described in Ex. 7.16 has type S mortar and a specified strength of 1500 lbf/in². It is reinforced longitudinally with no. 4 grade 60 bars at 16 in centers placed centrally in the wall. The wall has an effective height of $h = 20$ ft and is simply supported at the top and bottom. The unfactored roof live load is $P_r = 200$ lbf/ft. The unfactored roof dead load is $P_d = 400$ lbf/ft. The masonry wall has a weight of $\gamma_w = 69$ lbf/ft². Wind load governs and has a factored value of $w_u = 30$ lbf/ft². Determine if the midheight deflection of the wall under service level loads is within the permissible limits.

Solution

The maximum permissible deflection at midheight of the wall due to service level vertical and lateral loads, and including P-delta effects, is given by TMS 402 Eq. 9-32 as

$$
\begin{aligned}
\delta_s &= 0.007 h \\
&= (0.007)(240 \text{ in}) \\
&= 1.68 \text{ in}
\end{aligned}
$$

From Ex. 7.17, the midheight deflection produced by the factored loads is

$$
\begin{aligned}
\delta_u &= 0.203 \text{ in} \\
&< \delta_s
\end{aligned}
$$

So, the deflection under service loads is less than that permissible.

9. DESIGN OF ANCHOR BOLTS

Nomenclature

A_b	effective cross-sectional area of an anchor bolt	in²
A_o	overlap of projected areas	in²
A_{pt}	projected area of tensile breakout surface	in²
A_{pv}	projected area of shear breakout surface	in²

b_a	applied tensile force on an anchor bolt	lbf
b_{au}	factored tensile force on an anchor bolt	kips
b_v	applied shear force on an anchor bolt	lbf
b_{vu}	factored shear forced on an anchor bolt	kips
B_a	allowable axial load on an anchor bolt	lbf
B_{ab}	allowable axial strength in tension of an anchor bolt when governed by masonry breakout	lbf
B_{an}	nominal axial capacity in tension of an anchor bolt	kips
B_{ap}	allowable axial strength in tension of an anchor bolt when governed by anchor pullout	lbf
B_{as}	allowable axial strength in tension of an anchor bolt when governed by steel yielding	lbf
B_v	allowable shear load on an anchor bolt	lbf
B_{vb}	allowable strength in shear of an anchor bolt when governed by masonry breakout	lbf
B_{vc}	allowable strength in shear of an anchor bolt when governed by masonry crushing	lbf
B_{vn}	nominal capacity in shear of an anchor bolt	kips
B_{vpry}	allowable strength in shear of an anchor bolt when governed by anchor pryout	lbf
B_{vs}	allowable strength in shear of an anchor bolt when governed by steel yielding	lbf
d_b	nominal bolt diameter	in
e_b	projected leg extension of a bent-bar anchor measured from inside edge of anchor at bend to farthest point of anchor in the plane of the hook	in
f_y	yield strength of the anchor bolt	kips/in^2
l_b	effective embedment depth of anchor bolt	in
l_{be}	anchor bolt edge distance measured from edge of masonry to center of the cross section of anchor bolt	in
r	radius of projected area	in
s	bolt spacing	in
T	tension force	kips
V	shear force	kips

Symbols

θ	half the angle subtended by the chord at the intersection of overlapping projected areas	degree

Placement Details

Details of headed and bent-bar anchor bolts are given in TMS 402 Sec. 6.3 and are illustrated in Fig. 7.18.

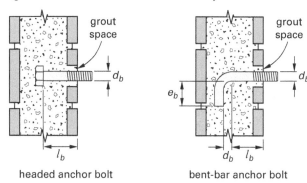

Figure 7.18 *Anchor Bolts in Concrete Masonry*

The dimensional limitations specified for anchor bolts in TMS 602 Sec. 3.4D are as follows.

- Anchor bolts in excess of $\frac{1}{4}$ in diameter must be embedded in grout.

- A minimum of either $\frac{1}{2}$ in coarse grout or $\frac{1}{4}$ in fine grout is required between bolts and the masonry units.

- Anchor bolts of $\frac{1}{4}$ in diameter or less must be placed in grout or mortar bed joints that are at least $\frac{1}{2}$ in thick.

- The clear distance between parallel anchor bolts must not be less than the diameter of the bolt or less than 1 in.

- The minimum permissible effective embedment length is specified in TMS 402 Sec. 6.3.6 as $l_b = 4d_b \geq 2$ in.

- For a headed anchor bolt, the effective embedment length is the length of embedment measured from the masonry surface to the bearing surface of the anchor head.

- For a bent-bar anchor bolt, the effective embedment length is the length of embedment measured from the masonry surface to the bearing surface of the bent end, minus one anchor bolt diameter.

Seismic Design Requirements

In accordance with ASCE 7 Sec. 13.4.2.2, the design of masonry anchors should be governed by ductile failure. To ensure this, either of the following two design approaches may be adopted.

- The strength of the connection must be governed by the ductile yielding of the support or component that the anchor is connecting to the structure.

- The connection must be designed to resist the load combinations of ASCE 7 Sec. 12.4.3, including the overstrength factor Ω_0 given in Table 13.5-1 and Table 13.6-1.

Reinforced Masonry

Anchor Bolt in Tension

Anchor bolts may fail under tensile forces by[11]

- tensile yielding of the steel anchor

- masonry tensile breakout

- straightening of the hook followed by pullout from the masonry in the case of bent-bar anchors

In accordance with TMS 402 Comm. Sec. 6.3.2, masonry breakout of an anchor that is solidly grouted in masonry occurs by the pullout of a conically shaped section of masonry. As shown in Fig. 7.19, the failure surface slopes at 45°. The projected area of the cone on the masonry surface is a circle with a radius equal to the embedment length of the bolt, and with an area of

$$A_{pt} = \pi l_b^2$$

Figure 7.19 *Masonry Breakout of Anchor Bolt in Tension*

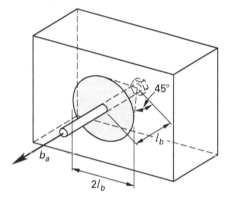

As indicated in TMS 402 Comm. Sec. 6.3.2, if the projected areas of anchor bolts overlap, an adjustment must be made so as not to overstress the masonry. When bolts are spaced at less than twice the embedment length apart, the projected areas of the bolts overlap, and the anchors must be treated as an anchor group. As specified in TMS 402 Sec. 6.3.2, the combined projected area of the group is reduced by the overlapping areas. In effect, the projected areas are adjusted so that no area is included more than once. For two anchor bolts, the overlapping area is shown in Fig. 7.20, and is given by

$$A_o = \left(\frac{\pi 2\theta}{180°} - \sin 2\theta \right) r^2$$

The angle subtended at the center of the projected area by the chord of the intersecting circles is 2θ, where

$$\theta = \arccos \frac{s}{2r}$$

$$r = l_b$$

Figure 7.20 *Overlap of Projected Areas*

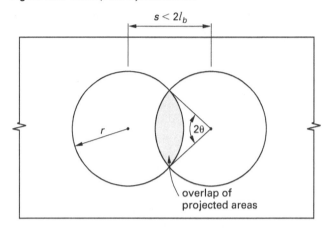

Similarly, when a bolt is located less than the embedment length from the edge of a member, that portion of the projected area falling outside the masonry member must be deducted from the calculated area. As shown in Fig. 7.21, where the anchor bolt edge distance is l_{be}, the projected area falling outside the masonry wall is

$$A_o = \frac{\left(\dfrac{\pi 2\theta}{180°} - \sin 2\theta \right) r^2}{2}$$

The angle subtended at the center of the projected area by the chord formed by the member edge is 2θ, where

$$\theta = \arccos \frac{l_{be}}{r}$$

Figure 7.21 *Projected Area Extends Beyond Wall Edge*

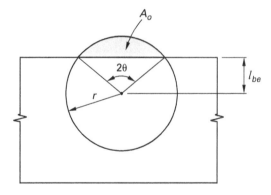

Bent-Bar Anchor Bolts

ASD Method

For tensile strength governed by the tensile yielding of a bent-bar steel anchor, the allowable strength is given by TMS 402 Eq. 8-5 as

$$B_{as} = 0.6 A_b f_y$$

For tensile strength governed by masonry breakout, the allowable strength is given by TMS 402 Eq. 8-3 as

$$B_{ab} = 1.25 A_{pt}\sqrt{f_m'}$$

For bent-bar anchors, when the tensile strength is governed by straightening the hook followed by pullout from the masonry, the allowable strength is given by TMS 402 Eq. 8-4 as

$$B_{ap} = 0.6 f_m' e_b d_b + 120\pi(l_b + e_b + d_b)d_b$$

The allowable strength, B_a, for bent-bar anchor bolts is the smallest value given by TMS 402 Eq. 8-3, Eq. 8-4, or Eq. 8-5.

SD Method

For tensile strength governed by the tensile yielding of a bent-bar steel anchor, the design capacity is given by TMS 402 Eq. 9-5 as

$$\phi B_{ans} = \phi A_b f_y$$
$$\phi = 0.9$$

For tensile strength governed by masonry breakout, the design capacity is given by TMS 402 Eq. 9-3 as

$$\phi B_{anb} = \phi(4 A_{pt})\sqrt{f_m'}$$
$$\phi = 0.5$$

For bent-bar anchors, when the tensile strength is governed by straightening the hook followed by pullout from the masonry, the design capacity is given by TMS 402 Eq. 9-4 as

$$\phi B_{anp} = \phi\big(1.5 f_m' e_b d_b + 300\pi\big(l_b + e_b + d_b\big)d_b\big)$$
$$\phi = 0.65$$

The design capacity, ϕB_{an}, is the smallest value given by TMS 402 Eq. 9-3, Eq. 9-4, and Eq. 9-5.

Headed Anchor Bolts

ASD Method

For tensile strength governed by the tensile yielding of a headed anchor bolt, the allowable strength is given by TMS 402 Eq. 8-2 as

$$B_{as} = 0.65 A_b f_y$$

For tensile strength governed by masonry breakout, the allowable strength is given by TMS 402 Eq. 8-1 as

$$B_{ab} = 1.25 A_{pt}\sqrt{f_m'}$$

The allowable strength, B_a, for headed anchor bolts is the smaller value given by TMS 402 Eq. 8-1 or Eq. 8-2.

SD Method

For tensile strength governed by the tensile yielding of a headed anchor bolt, the design capacity is given by TMS 402 Eq. 9-2 as

$$\phi B_{ans} = \phi A_b f_y$$
$$\phi = 0.9$$

For tensile strength governed by masonry breakout, the design capacity is given by TMS 402 Eq. 9-1 as

$$\phi B_{anb} = \phi(4 A_{pt})\sqrt{f_m'}$$
$$\phi = 0.5$$

The design capacity, B_{an}, is the smaller value given by TMS 402 Eq. 9-1 or Eq. 9-2.

Example 7.20

A glulam crosstie between diaphragm chords is supported at one end in a steel beam bucket as shown in the illustration. The bucket is attached to a solid grouted concrete block masonry wall with a masonry compressive strength of 1500 lbf/in². The four headed anchor bolts are ³⁄₄ in diameter ASTM A 307 type C, with an effective minimum specified yield strength of 36 kips/in² and an effective cross-sectional area of 0.334 in². The effective embedment length of the anchor bolts is 6 in. The ASD governing load combination gives a tension force on the beam bucket of $T = 13$ kips and a shear force of $V = 1.5$ kips. The SD governing load combination gives a tension force of $T_u = 21.7$ kips and a shear force of $V_u = 2.25$ kips. Determine if the bolts are adequate for the tension force on the bucket. The tensile forces are due to wind loads.

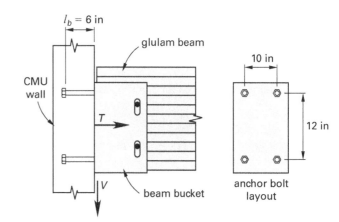

Solution

Masonry breakout

The effective embedment length of an anchor bolt measured from the surface of the masonry to the bearing surface of the bolt head is given as

$$l_b = 6 \text{ in}$$
$$> 4d_b \quad [\text{satisfies TMS 402 Sec. 6.3.6}]$$
$$> 2 \text{ in} \quad [\text{satisfies TMS 402 Sec. 6.3.6}]$$

The projected area of one bolt, before considering overlapping areas, is

$$A_{pt} = \pi l_b^2 = \pi(6 \text{ in})^2 = 113 \text{ in}^2$$

The vertical spacing of the bolts is

$$s = 12 \text{ in} = 2l_b$$

The projected areas of the bolts do not overlap in the vertical direction. The horizontal spacing of the bolts is

$$s = 10 \text{ in} < 2l_b$$

The projected areas of the bolts overlap in the horizontal direction. The angle subtended at the center of the projected area by the chord of the intersecting circles is

$$2\theta = 2 \arccos \frac{s}{2r}$$
$$= 2 \arccos \frac{10 \text{ in}}{(2)(6 \text{ in})}$$
$$= 67.1°$$

For the top two anchor bolts, the overlapping area is

$$A_o = \left(\frac{\pi 2\theta}{180°} - \sin 2\theta \right) r^2$$
$$= \left(\frac{\pi(67.1°)}{180°} - \sin 67.1° \right)(6 \text{ in})^2$$
$$= 9 \text{ in}^2$$

The reduced projected area for the top two bolts is

$$A'_{pt} = 2A_{pt} - A_o$$
$$= (2)(113 \text{ in}^2) - 9 \text{ in}^2$$
$$= 217 \text{ in}^2$$

For all four bolts the reduced projected area is

$$A''_{pt} = (2)(217 \text{ in}^2)$$
$$= 434 \text{ in}^2$$

ASD Method

For tensile strength governed by masonry breakout, the allowable strength of the four bolts is given by TMS 402 Eq. 8-1 as

$$B_{ab} = 1.25 A''_{pt} \sqrt{f'_m}$$
$$= \frac{(1.25)(434 \text{ in}^2) \sqrt{1500 \, \dfrac{\text{lbf}}{\text{in}^2}}}{1000 \, \dfrac{\text{lbf}}{\text{kip}}}$$
$$= 21.0 \text{ kips}$$
$$> T \quad [\text{satisfactory}]$$

For tensile strength governed by the tensile yielding of a steel anchor, the allowable strength of each anchor bolt is given by TMS 402 Eq. 8-2 as

$$B_{as} = 0.65 A_b f_y$$
$$= (0.65)(0.334 \text{ in}^2)\left(36 \, \frac{\text{kips}}{\text{in}^2} \right)$$
$$= 7.82 \text{ kips}$$

The allowable strength of the four bolts is

$$4B_{as} = (4)(7.82 \text{ kips})$$
$$= 31.3 \text{ kips} > B_{ab} = 21.0 \text{ kips} \quad [\text{governs}]$$

The bolts are adequate for the tension force on the bucket. The allowable tension force on one bolt in the absence of shear force is

$$B_{as} = \frac{21.0 \text{ kips}}{4} = 5.25 \text{ kips}$$

SD Method

For tensile strength governed by the tensile yielding of a steel anchor, the design capacity is given by TMS 402 Eq. 9.2 as

$$\phi B_{ans} = \phi A_b f_y = (0.9)(0.334 \text{ in}^2)\left(36 \, \frac{\text{kips}}{\text{in}^2} \right)$$
$$= 10.8 \text{ kips}$$

The design capacity of the four bolts is

$$4\phi B_{ans} = (4)(10.8 \text{ kips})$$
$$= 43.2 \text{ kips} > T_u \quad [\text{satisfactory}]$$

Reinforced Masonry

For tensile strength governed by masonry breakout, the design capacity of the four bolts is given by TMS 402 Eq. 9-1 as

$$4\phi B_{anb} = \phi(4A_{pt})\sqrt{f'_m}$$

$$= \frac{(0.5)(4)(434 \text{ in}^2)\sqrt{1500 \dfrac{\text{lbf}}{\text{in}^2}}}{1000 \dfrac{\text{lbf}}{\text{kip}}}$$

$$= 33.6 \text{ kips} \quad [\text{governs}]$$
$$< 4\phi B_{ans}$$
$$> T_u \quad [\text{satisfactory}]$$

The bolts are adequate for the tension force on the bucket. The design capacity of one bolt in the absence of shear force is

$$\phi B_{an} = \frac{33.6 \text{ kips}}{4}$$
$$= 8.40 \text{ kips}$$

Anchor Bolt in Shear

Anchor bolts may fail under shear forces by

- steel anchor shear yielding
- masonry shear breakout
- masonry shear crushing
- anchor bolt shear pryout

The four modes of failure are shown in Fig. 7.22.

Figure 7.22 *Shear Failure Modes for Anchor Bolts*

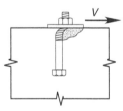

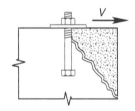
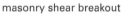

steel anchor shear yielding   masonry shear breakout

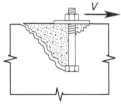
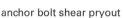
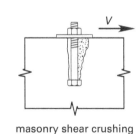

anchor bolt shear pryout   masonry shear crushing

Anchors near an edge in solidly grouted masonry fail in shear by the breakout of one-half of a conically shaped section of masonry. (See Fig. 7.23.) The failure surface slopes at 45°. The bolt's projected area is a semicircle of radius equal to the anchor bolt edge distance with an area given by TMS 402 Eq. 6-6 of

$$A_{pv} = \frac{\pi l_{be}^2}{2}$$

As indicated in TMS 402 Sec. 6.3.3, if the projected area of anchor bolts overlap or portions of the projected area fall outside the masonry member, an adjustment must be made to the projected area.

Figure 7.23 *Masonry Breakout in Shear*

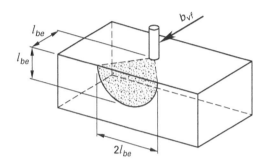

ASD Method

For shear strength governed by masonry breakout, the allowable strength is given by TMS 402 Eq. 8-6 as

$$B_{vb} = 1.25 A_{pv}\sqrt{f'_m}$$

For shear strength governed by masonry crushing, the allowable strength is given by TMS 402 Eq. 8-7 as

$$B_{vc} = 580\sqrt[4]{f'_m A_b}$$

For shear strength governed by masonry pryout, the allowable strength is given by TMS 402 Eq. 8-8 as

$$B_{vpry} = 2.5 A_{pt}\sqrt{f'_m}$$
$$= 2B_{ab}$$

For shear strength governed by the shear yielding of a steel anchor, the allowable strength is given by TMS 402 Eq. 8-9 as

$$B_{vs} = 0.36 A_b f_y$$

The allowable strength is the smallest value given by TMS 402 Eq. 8-6, Eq. 8-7, Eq. 8-8, and Eq. 8-9.

SD Method

For shear strength governed by masonry breakout, the design capacity is given by TMS 402 Eq. 9-6 as

$$\phi B_{vnb} = \phi(4A_{pv}\sqrt{f_m'})$$
$$\phi = 0.5$$

For shear strength governed by masonry crushing, the design capacity is given by TMS 402 Eq. 9-7 as

$$\phi B_{vnc} = \phi(1750\sqrt[4]{f_m' A_b})$$
$$\phi = 0.5$$

For shear strength governed by masonry pryout, the design capacity is given by TMS 402 Eq. 9-8 as

$$\phi B_{vnpry} = \phi(8A_{pt}\sqrt{f_m'})$$
$$= \phi(2B_{anb})$$
$$\phi = 0.5$$

The design strength of the anchor bolt in shear yielding is given by TMS 402 Eq. 9-9 as

$$\phi B_{vns} = \phi(0.6A_b f_y)$$
$$\phi = 0.9$$

Example 7.21

As indicated in Ex. 7.20, a glulam crosstie between diaphragm chords is supported at one end in a steel beam bucket. The bucket is attached to a solid grouted concrete block masonry wall with a masonry compressive strength of 1500 lbf/in². The four headed anchor bolts are ¾ in diameter ASTM A 307 type C, with an effective minimum specified yield strength of 36 kips/in² and an effective cross-sectional area of 0.334 in². The effective embedment length of the anchor bolts is 6 in, and the threaded portion of the bolts is not located within the shear plane. The ASD governing load combination gives a tension force on the beam bucket of $T = 13$ kips and a shear force of $V = 1.5$ kips. The SD governing load combination gives a tensile force of $T_u = 21.7$ kips and a shear force of $V_u = 2.25$ kips. The bolts are far from a free edge of the wall and masonry breakout in shear cannot occur. Determine if the bolts are adequate for the shear force on the bucket.

Solution

From Ex. 7.20, the reduced projected area for the top two bolts is

$$A_{pt}' = 217 \text{ in}^2$$

For all four bolts, the reduced projected area is

$$A_{pt}'' = (2)(217 \text{ in}^2)$$
$$= 434 \text{ in}^2$$

ASD Method

Anchor yielding

For shear strength governed by the shear yielding of the steel anchor, the allowable strength of each anchor bolt is given by TMS 402 Eq. 8-9 as

$$B_{vs} = 0.36A_b f_y$$
$$= (0.36)(0.334 \text{ in}^2)\left(36 \frac{\text{kips}}{\text{in}^2}\right)$$
$$= 4.33 \text{ kips}$$

The allowable strength of the four bolts is

$$4B_{vs} = (4)(4.33 \text{ kips})$$
$$= 17.3 \text{ kips}$$

Masonry pryout

For shear strength governed by masonry pryout, the allowable strength of all four bolts is given by TMS 402 Eq. 8-8 as

$$4B_{vpry} = 2.5A_{pt}''\sqrt{f_m'}$$
$$= \frac{(2.5)(434 \text{ in}^2)\sqrt{1500 \frac{\text{lbf}}{\text{in}^2}}}{1000 \frac{\text{lbf}}{\text{kip}}}$$
$$= 42.0 \text{ kips}$$

Masonry crushing

For shear strength governed by masonry crushing, the allowable strength of each anchor bolt is given by TMS 402 Eq. 8-7 as

$$B_{vc} = 580\sqrt[4]{f_m' A_b}$$
$$= \frac{(580)\sqrt[4]{\left(1500 \frac{\text{lbf}}{\text{in}^2}\right)(0.334 \text{ in}^2)}}{1000 \frac{\text{lbf}}{\text{kip}}}$$
$$= 2.74 \text{ kips} \quad \text{[governs]}$$

The allowable strength of the four bolts is

$$4B_{vc} = (4)(2.74 \text{ kips})$$
$$= 11.0 \text{ kips}$$
$$< 4B_{vpry}$$
$$< 4B_{vs}$$
$$> V \quad [\text{satisfactory}]$$

The bolts are adequate for the shear force on the bucket. The allowable shear force on one bolt in the absence of tensile force is

$$B_v = \frac{11.0 \text{ kips}}{4} = 2.75 \text{ kips}$$

SD Method

Anchor yielding

The design strength of each anchor bolt in shear yielding is given by TMS 402 Eq. 9-9 as

$$\phi B_{vns} = \phi(0.6 A_b f_y)$$
$$= (0.9)(0.6)(0.334 \text{ in}^2)\left(36 \frac{\text{kips}}{\text{in}^2}\right)$$
$$= 6.49 \text{ kips}$$

The design strength of the four bolts is

$$4\phi B_{vns} = (4)(6.49 \text{ kips})$$
$$= 26.0 \text{ kips}$$

Masonry pryout

For shear strength governed by masonry pryout, the design capacity for all four bolts is given by TMS 402 Eq. 9-8 as

$$\phi B_{vnpry} = \phi(2 B_{anb})$$
$$= (2)(33.6 \text{ kips})$$
$$= 67.2 \text{ kips}$$

Masonry crushing

For shear strength governed by masonry crushing, the design capacity of each anchor bolt is given by TMS 402 Eq. 9-7 as

$$\phi B_{vnc} = \phi(1750 \sqrt[4]{f'_m A_b})$$
$$= \frac{(0.5)(1750)\sqrt[4]{\left(1500 \frac{\text{lbf}}{\text{in}^2}\right)(0.334 \text{ in}^2)}}{1000 \frac{\text{lbf}}{\text{kip}}}$$
$$= 4.14 \text{ kips} \quad [\text{governs}]$$

The design shear strength of the four bolts is

$$4\phi B_{vnc} = (4)(4.14 \text{ kips})$$
$$= 16.6 \text{ kips}$$
$$> V_u \quad [\text{satisfactory}]$$

The bolts are adequate for the shear force on the bucket. The design capacity of one bolt in the absence of tensile force is

$$\phi B_{vnc} = 4.14 \text{ kips}$$

Anchor Bolts in Combined Tension and Shear

ASD Method

The allowable axial load on an anchor bolt, B_a, is the smaller value given by TMS 402 Eq. 8-1 or Eq. 8-2.

The allowable shear load on an anchor bolt, B_v, is the smallest value given by TMS 402 Eq. 8-6, Eq. 8-7, Eq. 8-8, or Eq. 8-9.

The applied axial load on an anchor bolt is b_a.

The applied shear load on an anchor bolt is b_v.

Anchor bolts subjected to combined axial load and shear load must comply with TMS 402 Eq. 8-10, which is

$$\left(\frac{b_a}{B_a}\right)^{5/3} + \left(\frac{b_v}{B_v}\right)^{5/3} \leq 1$$

In addition, the allowable strength in shear and tension must each exceed the applied loads.

SD Method

The nominal axial strength of an anchor bolt, B_{an}, is the smaller value given by TMS 402 Eq. 9-1 or Eq. 9-2.

The nominal shear strength of an anchor bolt, B_{vn}, is the smallest value given by TMS 402 Eq. 9-6, Eq. 9-7, Eq. 9-8, or Eq. 9-9.

The strength level axial load on an anchor bolt is b_{au}.

The strength level shear load on an anchor bolt is b_{vu}.

Anchor bolts subjected to combined axial load and shear load must comply with TMS 402 Eq. 9-10, which is

$$\left(\frac{b_{au}}{\phi B_{an}}\right)^{5/3} + \left(\frac{b_{vu}}{\phi B_{vn}}\right)^{5/3} \leq 1$$

In addition, the design capacity in shear and tension must each exceed the strength level applied loads.

Reinforced Masonry

Example 7.22

As given in Ex. 7.20, a glulam crosstie between diaphragm chords is supported at one end in a steel beam bucket. The bucket is attached to a solid grouted concrete block masonry wall with a masonry compressive strength of 1500 lbf/in². The four headed anchor bolts are ³/₄ in diameter ASTM A 307 type C, with a minimum specified yield strength of 36 kips/in² and an effective cross-sectional area of 0.334 in². The effective embedment length of the anchor bolts is 6 in, and the threaded portion of the bolts is not located within the shear plane. The ASD governing load combination gives a tension force on the beam bucket of $T = 13$ kips and a shear force of $V = 1.5$ kips. The SD governing load combination gives a tension force of $T_u = 21.7$ kips and a shear force of $V_u = 2.25$ kips. Determine if the bolts are adequate for the combined tension and shear forces.

Solution

The applied tension force on one bolt is

$$b_a = \frac{T}{4}$$
$$= \frac{13 \text{ kips}}{4}$$
$$= 3.25 \text{ kips}$$

The applied shear force on one bolt is

$$b_v = \frac{V}{4}$$
$$= \frac{1.5 \text{ kips}}{4}$$
$$= 0.38 \text{ kip}$$

For combined tension and shear, TMS 402 Eq. 8-10 must be satisfied.

$$\left(\frac{b_a}{B_a}\right)^{5/3} + \left(\frac{b_v}{B_v}\right)^{5/3} \leq 1$$

$$\left(\frac{3.25 \text{ kips}}{5.25 \text{ kips}}\right)^{5/3} + \left(\frac{0.38 \text{ kip}}{2.75 \text{ kips}}\right)^{5/3} = 0.49$$
$$< 1.0 \quad \text{[satisfactory]}$$

SD Method

The factored tension force on one bolt is

$$b_{af} = \frac{T_u}{4}$$
$$= \frac{21.7 \text{ kips}}{4}$$
$$= 5.43 \text{ kips}$$

The factored shear force on one bolt is

$$b_{vf} = \frac{V_u}{4}$$
$$= \frac{2.25 \text{ kips}}{4}$$
$$= 0.56 \text{ kips}$$

For combined tension and shear, anchor bolts must comply with TMS 402 Eq. 9-10, which is

$$\left(\frac{b_{au}}{\phi B_{an}}\right)^{5/3} + \left(\frac{b_{vu}}{\phi B_{vn}}\right)^{5/3} \leq 1$$

$$\left(\frac{5.43 \text{ kips}}{8.41 \text{ kips}}\right)^{5/3} + \left(\frac{0.56 \text{ kip}}{4.14 \text{ kips}}\right)^{5/3} = 0.52$$
$$< 1.0 \quad \text{[satisfactory]}$$

10. DESIGN OF PRESTRESSED MASONRY

Nomenclature

A_n	net cross-sectional area of masonry	in²
A_{ps}	area of prestressing steel	in²
A_s	area of nonprestressed reinforcement	in²
b	width of section	in
C_m	force in masonry stress block	lbf
d	distance from extreme compression fiber to centroid of prestressing tendon or reinforcement	in
d_t	tendon diameter	in
D	dead load or related internal moments or forces	lbf
e	eccentricity of axial load	in
E_m	modulus of elasticity of masonry	lbf/in²
E_{mi}	modulus of elasticity of masonry at transfer	lbf/in²
E_{ps}	modulus of elasticity of prestressing steel	lbf/in²
f_a	calculated compressive stress in masonry due to axial load	lbf/in²
f_{ai}	calculated compressive stress in masonry at transfer due to axial load	lbf/in²
f_b	calculated compressive stress in masonry due to flexure	lbf/in²
f_{bi}	calculated compressive stress in masonry at transfer due to flexure	lbf/in²
f'_m	specified compressive strength of masonry	lbf/in²
f'_{mi}	specified compressive strength of masonry at time of prestress transfer	lbf/in²
f_{pj}	stress in prestressing tendon due to jacking force	lbf/in²
f_{ps}	stress in prestressing tendon at nominal strength	lbf/in²

f_{psi}	initial stress in prestressing tendon	lbf/in^2
f_{pu}	specified tensile strength of prestressing tendons	lbf/in^2
f_{py}	specified yield strength of prestressing tendons	lbf/in^2
f_{se}	effective stress in prestressing tendon after allowance for all prestress losses	lbf/in^2
f_y	specified yield stress of nonprestressed reinforcement	lbf/in^2
F_a	allowable compressive stress in masonry due to axial load	lbf/in^2
F_{ai}	allowable compressive stress in masonry at transfer due to axial load	lbf/in^2
F_b	allowable compressive stress in masonry due to flexure	lbf/in^2
F_{bi}	allowable compressive stress in masonry at transfer due to flexure	lbf/in^2
F_t	allowable tensile stress in masonry due to flexure	lbf/in^2
h	effective height of wall	ft
I_n	moment of inertia of net cross-sectional area	in^4
l_p	clear span of the prestressed member in the direction of the prestressing tendon	ft
M	maximum moment at the section under consideration	ft-kips
M_n	nominal flexural strength	ft-kips
M_u	factored moment at the section under consideration	ft-kips
P	nonfactored axial load	lbf
P_d	nonfactored dead load from tributary floor or roof loads	lbf
P_e	Euler critical load	lbf
P_f	nonfactored load from tributary floor or roof loads, sum of P_d and P_r	lbf
P_i	axial load at transfer	lbf
P_{ps}	prestressing tendon force at time and location relevant for design	kips
P_{psi}	prestressing tendon force at transfer	kips
P_{psj}	prestressing force at the jack	kips
P_r	nonfactored live load from tributary floor or roof loads	lbf
P_u	factored axial load	lbf
P_w	nonfactored weight of wall tributary to section considered	lbf
Q	load combination	–
r	radius of gyration	in
S_n	section modulus of net cross-sectional area	in^3
t	nominal thickness of member	in
t_f	face-shell thickness	in
w	wind load	lbf/ft^2
W	wind load or related internal moments or forces	lbf

Symbols

| γ_w | weight of wall | lbf/ft^2 |
| ϕ | strength reduction factor | – |

General Considerations

A masonry wall is prestressed by tensioning a prestressing tendon that is typically located in the mid-plane of the wall. This produces compression in the member, which increases its flexural tensile capacity and improves its resistance to lateral loading.[12] Prestressed masonry members must be designed for three distinct design stages.

- the *transfer design stage*, when a prestressing force is applied to the wall and immediate prestress losses occur due to elastic deformation of the masonry and seating of the anchorage

- the *serviceability design stage*, when all time-dependent prestress losses have occurred as a result of creep and shrinkage of the masonry and relaxation of the tendons

- the *strength design stage*, when the design capacity of a wall is checked to ensure that it is not less than the demand produced by the factored loads

By comparison with traditional reinforced masonry construction, prestressed masonry has the advantage of not requiring grouting of the cores in which the prestressing tendons are placed. This results in

- a reduced wall weight, requiring smaller footings and reduced seismic forces

- the elimination of grout, requiring less material and labor

- a reduction in construction times

Construction Details

Prestressing tendons must be of high tensile steel as required by TMS 602 Sec. 2.4B. Tendons may consist of wire, rods with threaded ends, or strand. Rods may be connected during construction using couplers. Strand is provided in the full required length. A typical wall using rods with threaded ends is shown in Fig. 7.24.

Walls are usually post-tensioned and ungrouted, as shown in Fig. 7.24. Walls subject to a moist and corrosive environment require corrosion protection for the tendon. Corrosion protection may be provided by coating the tendon with a corrosion-inhibiting material and enclosing the tendon in a continuous plastic sheath. Alternatively, tendons may be galvanized threaded high-tensile steel rods.

Reinforced Masonry

Figure 7.24 *Prestressed Masonry Details*

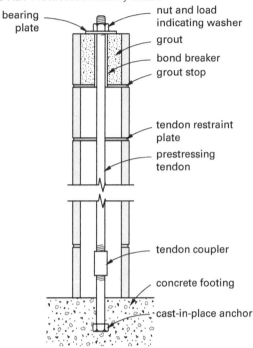

The prestressing force is transferred to the masonry by means of anchorages at the foot of the wall and at the top of the wall. As specified in TMS 402 Sec. 10.8.1, tendons may be anchored by

- mechanical devices bearing directly on masonry or placed inside an end block of concrete or fully grouted masonry

- bond in reinforced concrete end blocks or members

An example of an anchor bolt embedded in the concrete foundation is shown in Fig. 7.24. The prestressing rod is attached to the anchor bolt by a threaded coupler. An inspection port is provided at the location of the coupler to enable the rod to be connected to the coupler after the block wall has been constructed to partial or full height. In accordance with TMS 402 Sec. 10.8.2, couplers must develop 95% of the specified tensile strength of the prestressing rods. When a moment connection is not required at the base of the wall, anchor bolts may be located in the first block course of masonry.

The tendons are normally laterally restrained so as to deflect laterally with the wall. The advantage of laterally restraining the tendons is that the prestressing force does not contribute to the elastic instability of the wall, as lateral displacement of the wall is counter-balanced by an equal and opposite restraint from the prestressing tendons. So, the wall cannot buckle under its own prestressing force. In addition, the ultimate strength of the wall is increased by ensuring that the wall and the tendon deform together. TMS 402 Comm. Sec. 10.4.2 stipulates that three restraints along the length of a tendon are sufficient to provide adequate restraint. As shown in Fig. 7.24, restraint is provided by a steel plate with a

central hole. The sides of the plate are embedded in the joint between courses. Alternatively, a cell can be filled with grout, and a bond breaker is applied to the tendon to allow it to move freely within the cell.

A mechanical anchorage is used at the top of the wall shown in Fig. 7.24. The topmost cell is filled with grout and a steel bearing plate is bedded on top. The size of the bearing plate is determined from the requirement of TMS 402 Sec. 10.8.4.2, which stipulates that bearing stresses must not exceed 50% of the masonry compressive strength at the time of transfer. Reinforcement is required in the grouted cell, as specified in TMS 402 Sec. 10.8.3, to resist the bursting and tensile forces produced when the tendon is stressed. The tendon may be stressed by means of a hydraulic jack or by tightening a nut with a standard wrench. In the former method, the load is measured by a load cell or a calibrated gauge on the jack. In the latter method, the force produced by tightening a nut may be measured with a direct tension indicator washer (a washer with dimples formed on its top face). A hardened steel washer is placed on top of the indicator washer and as the nut is tightened, the dimples are compressed. The required force in the tendon has been produced when the gap between the two washers reaches a specified amount as measured by a feeler gauge. TMS 602 Sec. 3.6B requires that the elongation of the tendons be measured and compared with the elongation anticipated for the applied prestressing force. If the discrepancy between the two methods exceeds 7% for post-tensioned walls, the cause must be determined and corrected.

Masonry blocks are usually laid in running bond in face-shell mortar bedding, using type S mortar to provide early strength gain. The small percentage of prestressing steel in the wall results in a highly ductile system with the neutral axis entirely within the face shell. Properties of concrete masonry units with face-shell mortar bedding are given in the *Masonry Designers' Guide*[13] and in Table 7.3.

Table 7.3 *Properties of CMU Walls with Face-Shell Mortar Bedding and Unit Density of 110 lbf/ft[3]*

nominal wall thickness	6 in	8 in	10 in	12 in
face-shell thickness	1.00	1.25	1.38	1.5
area, A_n (in^2/ft)	24	30	33	36
moment of inertia, I_n (in^4/ft)	130	309	567	929
section modulus, S_n (in^3/ft)	46.3	81	118	160
radius of gyration, r (in/ft)	2.33	3.21	4.14	5.08
weight of wall, γ_w (lbf/ft^2)	28.5	37	45	51

Transfer Design Stage

The transfer design stage occurs immediately after a prestressing force is applied to the wall. The prestressing force is usually applied to the wall three days after construction of the wall when the masonry strength is f'_{mi}.

Since the masonry will continue to gain strength, TMS 402 Sec. 10.4.1.2 permits a 20% overstress in the customary allowable axial and flexural stresses, and in the customary allowable axial force. The allowable compressive stress at transfer in a member having an h/r ratio not greater than 99 and that is subjected to axial load and flexure is given by TMS 402 Eq. 8-13 as

$$F_{ai} = 1.2(0.25f'_{mi})\left[1 - \left(\frac{h}{140r}\right)^2\right]$$

For members having an h/r ratio greater than 99, the allowable compressive stress at transfer is given by TMS 402 Eq. 8-14 as

$$F_{ai} = 1.2(0.25f'_{mi})\left(\frac{70r}{h}\right)^2$$

The allowable compressive stress at transfer in a member due to flexure is given by TMS 402 Eq. 8-15 as

$$F_{bi} = \frac{1.2f'_{mi}}{3}$$

The axial load at transfer is due entirely to the self-weight of the wall and the prestressing force at transfer, since additional floor or roof loads are not present and the axial force in the masonry is

$$P_i = P_w + P_{psi}$$

The axial stress in the masonry at transfer is

$$f_{ai} = \frac{P_i}{A_n}$$

In the event that the full lateral load occurs at transfer, the bending compressive stress is

$$f_{bi} = \frac{M}{S_n}$$

The resulting compressive stress due to combined axial load and flexure must satisfy the interaction expression of TMS 402 Eq. 8-11, which is

$$\frac{f_{ai}}{F_{ai}} + \frac{f_{bi}}{F_{bi}} \leq 1.0$$

In applying this expression, the governing load combination is IBC Eq. 16-12, which is

$$Q = D + 0.6W$$

The allowable tensile stress, F_t, due to flexure is given by TMS 402 Table 8.2.4.2 and, in accordance with TMS 402 Sec. 8.2.4.2, is

$$f_{bi} - f_{ai} \leq F_t$$

In applying this expression, the governing load combination is IBC Eq. 16-15, which is

$$Q = 0.6D + 0.6W$$

To ensure elastic stability at transfer, the axial force is limited by TMS 402 Eq. 8-12 and TMS 402 Sec. 10.4.1.2 to

$$P_i \leq \frac{1.2P_{ei}}{4}$$

The Euler critical load, P_{ei}, is given by TMS 402 Sec. 10.4.2.2 and TMS 402 Eq. 8-16 as

$$P_{ei} = \frac{\pi^2 E_{mi} I_n \left(1 - \dfrac{0.577e}{r}\right)^3}{h^2}$$

Using laterally restrained tendons, the wall cannot buckle under its own prestressing force and, in accordance with TMS 402 Sec. 10.4.2.2, for stability calculations, P_{psi} is not considered in the determination of P_i. At transfer, additional floor or roof loads are not present, and the axial force in the masonry is

$$P_i = P_w$$

The stress in a prestressing tendon due to the jacking force is limited by TMS 402 Sec. 10.3.1 to

$$f_{pj} \leq 0.94f_{py}$$
$$\leq 0.80f_{pu}$$

Immediately after transfer, when initial loss of prestress has occurred, the stress in a post-tensioned tendon at anchorages and couplers is limited by TMS 402 Sec. 10.3.3 to

$$f_{psi} \leq 0.78f_{py}$$
$$\leq 0.70f_{pu}$$

The initial loss of prestress in a tendon at transfer due to elastic deformation of the masonry and seating of the anchorage is given by TMS 402 Comm. Sec. 10.3.4 as 5% to 10%.

Reinforced Masonry

Example 7.23

A nominal 8 in ungrouted concrete block masonry wall with face-shell mortar bedding of type S portland cement/lime mortar has a specified strength of $f'_m = 2500$ lbf/in² at 28 days, and a specified strength of $f'_{mi} = 2000$ lbf/in² at transfer. The wall is post-tensioned with $^7/_{16}$ in diameter steel rods at 16 in centers placed centrally in the wall, and the rods are laterally restrained. The wall has an effective height of $h = 20$ ft and is simply supported at the top and bottom. The roof live load is 200 lbf/ft and the roof dead load is 400 lbf/ft applied to the wall without eccentricity. The masonry wall has a weight of $\gamma_w = 37$ lbf/ft². The properties of the steel rod are $A_{ps} = 0.142$ in², $f_{py} = 100$ kips/in², and $f_{pu} = 122$ kips/in². Wind load governs and has a value of $w = 25$ lbf/ft². Assume the loss of prestress at transfer is 5% and the final loss of prestress is 30%. Determine if the wall under transfer level loads is within the permissible limits. Ignore P-delta effects.

Solution

The calculations are based on a 1 ft length of wall. The relevant wall section properties are obtained from Table 7.3 as

$$A_n = 30 \text{ in}^2$$
$$I_n = 309 \text{ in}^4$$
$$S_n = 81 \text{ in}^3$$
$$r = 3.21 \text{ in}$$

The weight of the wall above midheight is

$$P_w = \frac{\gamma_w h}{2}$$
$$= \frac{\left(37 \dfrac{\text{lbf}}{\text{ft}^2}\right)(20 \text{ ft})(1 \text{ ft})}{2}$$
$$= 370 \text{ lbf}$$

Prestress details

The equivalent area of prestressing tendon per foot of wall is

$$A_{ps} = \frac{(0.142 \text{ in}^2)(1 \text{ ft})(12 \text{ in})}{16 \text{ in}}$$
$$= 0.107 \text{ in}^2$$

The stress in a prestressing tendon due to the jacking force is limited by TMS 402 Sec. 10.3.1 to

$$f_{pj} \leq 0.80 f_{pu}$$
$$= (0.80)\left(122 \frac{\text{kips}}{\text{in}^2}\right)$$
$$= 97.6 \text{ kips/in}^2$$
$$\leq 0.94 f_{py}$$
$$= (0.94)\left(100 \frac{\text{kips}}{\text{in}^2}\right)$$
$$= 94.0 \text{ kips/in}^2 \quad \text{[governs]}$$

After 5% initial losses, the stress in the prestressing tendon is

$$f_{psi} = 0.95 f_{pj}$$
$$= (0.95)\left(94.0 \frac{\text{kips}}{\text{in}^2}\right)$$
$$= 89.3 \text{ kips/in}^2$$

The stress in a post-tensioned tendon after transfer is limited by TMS 402 Sec. 10.3.3 to

$$f_{psi} \leq 0.70 f_{pu}$$
$$= (0.70)\left(122 \frac{\text{kips}}{\text{in}^2}\right)$$
$$= 85.4 \text{ kips/in}^2$$
$$\leq 0.78 f_{py}$$
$$= (0.78)\left(100 \frac{\text{kips}}{\text{in}^2}\right)$$
$$= 78.0 \text{ kips/in}^2 \quad \text{[governs]}$$
$$< 89.3 \text{ kips/in}^2$$

Therefore, the initial jacking force per foot of wall is restricted to

$$P_{psj} = 1.05 f_{psi} A_{ps}$$
$$= (1.05)\left(78.0 \frac{\text{kips}}{\text{in}^2}\right)(0.107 \text{ in}^2)$$
$$= 8.76 \text{ kips}$$

After 5% initial losses, the prestressing tendon force per foot of wall is

$$P_{psi} = 0.95 P_{psj}$$
$$= (0.95)(8.76 \text{ kips})$$
$$= 8.32 \text{ kips}$$

Compressive stress check

The slenderness ratio of the wall is

$$\frac{h}{r} = \frac{(20 \text{ ft})\left(12 \frac{\text{in}}{\text{ft}}\right)}{3.21 \text{ in}}$$
$$= 74.8$$

After allowing for the 20% increase in allowable stress at transfer, the allowable compressive stress in a member having an h/r ratio not greater than 99 and subjected to axial load is given by TMS 402 Eq. 8-13 as

$$F_{ai} = 1.2(0.25f'_{mi})\left[1 - \left(\frac{h}{140r}\right)^2\right]$$
$$= 1.2(0.25f'_{mi})\left(1 - \left(\frac{\left(\frac{h}{r}\right)}{140}\right)^2\right)$$
$$= (1.2)(0.25)\left(2000 \frac{\text{lbf}}{\text{in}^2}\right)\left(1 - \left(\frac{74.8}{140}\right)^2\right)$$
$$= 429 \text{ lbf/in}^2$$

After allowing for the 20% increase in stress at transfer, the allowable compressive stress in a member due to flexure is given by TMS 402 Eq. 8-15 as

$$F_{bi} = \frac{1.2f'_{mi}}{3}$$
$$= \frac{(1.2)\left(2000 \frac{\text{lbf}}{\text{in}^2}\right)}{3}$$
$$= 800 \text{ lbf/in}^2$$

The compressive stress due to combined axial load and flexure must be checked using IBC Eq. 16-12, which is

$$Q = D + 0.6W$$

The total axial load at midheight of the wall at transfer is

$$P_i = P_w + P_{psi}$$
$$= 370 \text{ lbf} + (8.32 \text{ kips})\left(1000 \frac{\text{lbf}}{\text{kip}}\right)$$
$$= 8690 \text{ lbf}$$

The compressive stress produced in the wall by the wall self-weight and the effective prestressing force after initial losses at transfer is

$$f_{ai} = \frac{P_i}{A_n}$$
$$= \frac{8690 \text{ lbf}}{30 \text{ in}^2}$$
$$= 290 \text{ lbf/in}^2$$

Assuming the full lateral load occurs at transfer, the bending moment acting on the wall is

$$M = \frac{0.6wh^2}{8}$$
$$= \frac{(0.6)\left(25 \frac{\text{lbf}}{\text{ft}}\right)(20 \text{ ft})^2}{8}$$
$$= 750 \text{ ft-lbf}$$

The flexural stress produced in the wall by applied lateral load is

$$f_{bi} = \frac{M}{S_n}$$
$$= \frac{(750 \text{ ft-lbf})\left(12 \frac{\text{in}}{\text{ft}}\right)}{81 \text{ in}^3}$$
$$= 111 \text{ lbf/in}^2$$

The combined stresses must satisfy the interaction equation TMS 402 Eq. 8-11, which is

$$\frac{f_{ai}}{F_{ai}} + \frac{f_{bi}}{F_{bi}} \le 1.0$$
$$\frac{290 \frac{\text{lbf}}{\text{in}^2}}{429 \frac{\text{lbf}}{\text{in}^2}} + \frac{111 \frac{\text{lbf}}{\text{in}^2}}{800 \frac{\text{lbf}}{\text{in}^2}} = 0.82$$
$$\le 1.0 \quad \text{[satisfactory]}$$

Tensile stress check

Per TMS 402 Table 8.2.4.2, the allowable tensile stress for ungrouted units laid in type S portland cement/lime mortar normal to bed joints is

$$F_t = 33 \text{ lbf/in}^2$$

Reinforced Masonry

The tensile stress due to combined axial load and flexure must be checked using IBC Eq. 16-15, which is

$$Q = 0.6D + 0.6W$$
$$P_i = 0.6P_w + P_{psi}$$
$$= (0.6)(370 \text{ lbf})$$
$$+ (8.32 \text{ kips})\left(1000 \frac{\text{lbf}}{\text{kip}}\right)$$
$$= 8542 \text{ lbf}$$

The compressive stress produced in the wall by the wall self-weight and the effective prestressing force after initial losses at transfer is

$$f_{ai} = \frac{P_i}{A_n}$$
$$= \frac{8542 \text{ lbf}}{30 \text{ in}^2}$$
$$= 285 \text{ lbf/in}^2$$

The combined stresses must satisfy the equation

$$f_{bi} - f_{ai} \le F_t$$
$$111 \frac{\text{lbf}}{\text{in}^2} - 285 \frac{\text{lbf}}{\text{in}^2} = -174 \frac{\text{lbf}}{\text{in}^2}$$
$$< 33 \text{ lbf/in}^2 \quad \text{[satisfactory]}$$

Stability check

The modulus of elasticity of the concrete masonry at transfer is given by TMS 402 Table 4.2.2 as

$$E_{mi} = 900f'_{mi}$$
$$= (900)\left(\frac{2000 \frac{\text{lbf}}{\text{in}^2}}{1000 \frac{\text{lbf}}{\text{kip}}}\right)$$
$$= 1800 \text{ kips/in}^2$$

For laterally restrained tendons, the Euler critical load is given by TMS 402 Eq. 8-16 as

$$P_{ei} = \frac{\pi^2 E_{mi} I_n \left(1 - \frac{0.577e}{r}\right)^3}{h^2}$$
$$= \frac{\pi^2 \left(1800 \frac{\text{kips}}{\text{in}^2}\right)(309 \text{ in}^4)\left(1 - \frac{(0.577)(0 \text{ in})}{3.21 \text{ in}}\right)^3}{\left((20 \text{ ft})\left(12 \frac{\text{in}}{\text{ft}}\right)\right)^2}$$
$$= 95 \text{ kips}$$

To ensure elastic stability, the axial load on a wall, not including prestressing force, is limited by TMS 402 Eq. 8-12 and TMS 402 Sec. 10.4.1.2 to a maximum value of

$$P_i = \frac{1.2P_{ei}}{4}$$
$$= \frac{(1.2)(95 \text{ kips})}{4}$$
$$= 28.5 \text{ kips}$$
$$> P_w \quad \text{[satisfactory]}$$

The wall is adequate at transfer.

Serviceability Design Stage

At the serviceability design stage, all time-dependent prestress losses have occurred as a result of creep and shrinkage of the masonry and relaxation of the tendons. The total loss of prestress after long-term service for concrete masonry is given by TMS 402 Comm. Sec. 10.3.4 as 30% to 35%.

The design procedure for the serviceability design stage is similar to that used for the transfer design stage.

Example 7.24

The nominal 8 in ungrouted concrete block masonry wall described in Ex. 7.23 has face-shell mortar bedding of type S portland cement/lime mortar, a specified strength of $f'_m = 2500 \text{ lbf/in}^2$ at 28 days, and a specified strength of $f'_{mI} = 2000 \text{ lbf/in}^2$ at transfer. The wall is post-tensioned with $7/16$ in diameter steel rods at 16 in centers placed centrally in the wall, and the rods are laterally restrained. The wall has an effective height of $h = 20 \text{ ft}$ and is simply supported at the top and bottom. The roof live load is 200 lbf/ft, and the roof dead load is 400 lbf/ft applied to the wall without eccentricity. The wall has a weight of $\gamma_w = 37 \text{ lbf/ft}^2$. The properties of the steel rod are $A_{ps} = 0.142 \text{ in}^2$, $f_{py} = 100 \text{ kips/in}^2$, $f_{pu} = 122 \text{ kips/in}^2$. Wind load governs and has a value of $w = 25 \text{ lbf/ft}^2$. Assume the loss of prestress at transfer is 5% and the final loss of prestress is 30%. Determine if the wall under service level loads is within the permissible limits. Ignore P-delta effects.

Solution

The following calculations are based on a 1 ft length of wall. The relevant details from Ex. 7.23 are

$$A_n = 30 \text{ in}^2$$
$$I_n = 309 \text{ in}^4$$
$$S_n = 81 \text{ in}^3$$
$$r = 3.21 \text{ in}$$

The weight of the wall above midheight is

$$P_w = 370 \text{ lbf}$$

The roof dead load is

$$P_d = 400 \text{ lbf}$$

Roof live load is not included in IBC Eq. 16-12 and Eq. 16-15, so

$$P_f = P_d + P_r$$
$$= 400 \text{ lbf} + 0 \text{ lbf}$$
$$= 400 \text{ lbf}$$

Prestress details

The initial jacking force per foot of wall is

$$P_{psj} = 8.76 \text{ kips}$$

After 30% total losses, the prestressing tendon force per foot of wall is

$$P_{ps} = 0.70 P_{psj} = (0.70)(8.76 \text{ kips})$$
$$= 6.13 \text{ kips}$$

Compressive stress check

The slenderness ratio of the wall is

$$\frac{h}{r} = 74.8$$

The allowable compressive stress in a member having an h/r ratio not greater than 99 and subjected to combined axial load and flexure is given by TMS 402 Eq. 8-13 as

$$F_a = (0.25 f'_m)\left(1 - \left(\frac{h}{140r}\right)^2\right)$$
$$= (0.25)\left(2500 \frac{\text{lbf}}{\text{in}^2}\right)\left(1 - \left(\frac{74.8}{140}\right)^2\right)$$
$$= 447 \text{ lbf/in}^2$$

The allowable compressive stress in a member due to flexure is given by TMS 402 Eq. 8-15 as

$$F_b = \frac{f'_m}{3}$$
$$= \frac{2500 \frac{\text{lbf}}{\text{in}^2}}{3}$$
$$= 833 \text{ lbf/in}^2$$

The compressive stress due to combined axial load and flexure must be checked using IBC Eq. 16-12, which is

$$Q = D + 0.6W$$

The total axial load at midheight of the wall under service loads is

$$P = P_f + P_w + P_{ps}$$
$$= 400 \text{ lbf} + 370 \text{ lbf} + (6.13 \text{ kips})\left(1000 \frac{\text{lbf}}{\text{kip}}\right)$$
$$= 6900 \text{ lbf}$$

The compressive stress produced in the wall by the wall self-weight, roof dead load, and effective prestressing force after all losses is

$$f_a = \frac{P}{A_n}$$
$$= \frac{6900 \text{ lbf}}{30 \text{ in}^2}$$
$$= 230 \text{ lbf/in}^2$$

The bending moment produced on the wall by the lateral wind load is

$$M = \frac{0.6wh^2}{8}$$
$$= \frac{(0.6)\left(25 \frac{\text{lbf}}{\text{ft}}\right)(20 \text{ ft})^2}{8}$$
$$= 750 \text{ ft-lbf}$$

The flexural stress produced in the wall by applied lateral load is

$$f_b = \frac{M}{S_n} = \frac{(750 \text{ ft-lbf})\left(12 \frac{\text{in}}{\text{ft}}\right)}{81 \text{ in}^3}$$
$$= 111 \text{ lbf/in}^2$$

The combined stresses must satisfy the interaction equation TMS 402 Eq. 8-11, which is

$$\frac{f_a}{F_a} + \frac{f_b}{F_b} \leq 1.0$$

$$\frac{230 \frac{\text{lbf}}{\text{in}^2}}{447 \frac{\text{lbf}}{\text{in}^2}} + \frac{111 \frac{\text{lbf}}{\text{in}^2}}{833 \frac{\text{lbf}}{\text{in}^2}} = 0.65$$

$$\leq 1.0 \quad \text{[satisfactory]}$$

Tensile stress check

The allowable tensile stress for ungrouted units laid in type S portland cement/lime mortar normal to bed joints is given by TMS 402 Table 8.2.4.2 as

$$F_t = 33 \ \text{lbf/in}^2$$

The tensile stress due to combined axial load and flexure must be checked using IBC Eq. 16-15, which is

$$Q = 0.6D + 0.6W$$
$$P = 0.6(P_f + P_w) + P_{ps}$$
$$= (0.6)(400 \ \text{lbf} + 370 \ \text{lbf}) + (6.13 \ \text{kips})\left(1000 \ \frac{\text{lbf}}{\text{kip}}\right)$$
$$= 6592 \ \text{lbf}$$

The compressive stress produced in the wall by the wall self-weight and the effective prestressing force after total losses at transfer is

$$f_a = \frac{P}{A_n}$$
$$= \frac{6592 \ \text{lbf}}{30 \ \text{in}^2}$$
$$= 220 \ \text{lbf/in}^2$$

The combined stresses must satisfy the equation

$$f_b - f_a \le F_t$$
$$111 \ \frac{\text{lbf}}{\text{in}^2} - 220 \ \frac{\text{lbf}}{\text{in}^2} = -109 \ \text{lbf/in}^2$$
$$< 33 \ \text{lbf/in}^2 \quad \text{[satisfactory]}$$

Stability check

The modulus of elasticity of the concrete masonry at the service load stage is given by TMS 402 Table 4.2.2 as

$$E_m = 900f'_m$$
$$= (900)\left(\frac{2500 \ \dfrac{\text{lbf}}{\text{in}^2}}{1000 \ \dfrac{\text{lbf}}{\text{kip}}}\right)$$
$$= 2250 \ \text{kips/in}^2$$

For laterally restrained tendons, the Euler critical load is given by TMS 402 Eq. 8-19 as

$$P_e = \frac{\pi^2 E_m I_n \left(1 - \dfrac{0.577e}{r}\right)^3}{h^2}$$
$$= \frac{\pi^2 \left(2250 \ \dfrac{\text{kips}}{\text{in}^2}\right)(309 \ \text{in}^4)\left(1 - \dfrac{(0.577)(0 \ \text{in})}{3.21 \ \text{in}}\right)^3}{\left((20 \ \text{ft})\left(12 \ \dfrac{\text{in}}{\text{ft}}\right)\right)^2}$$
$$= 119 \ \text{kips}$$

To ensure elastic stability, the axial load on a wall, not including prestressing force, is limited by TMS 402 Eq. 8-12 to a maximum value of

$$P = \frac{P_e}{4}$$
$$= \frac{119 \ \text{kips}}{4}$$
$$= 30 \ \text{kips}$$
$$> P_f + P_w \quad \text{[satisfactory]}$$

The wall is adequate under service level loads.

Strength Design Stage

The strength design method, illustrated in Fig. 7.25 for a prestressed concrete masonry wall, is used to calculate the design strength of the wall under the action of the applied factored loads so as to ensure that the design strength is not less than the most critical load combination. Figure 7.25 shows the face shell of the concrete masonry unit and the centrally placed and laterally restrained prestressing tendon. The forces and stresses developed at ultimate load are indicated.

Figure 7.25 *Strength Design of Prestressed Masonry*

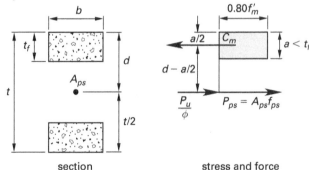

section   stress and force

For a centrally located, laterally restrained or unrestrained, unbonded prestressing tendon, the stress in the tendon at nominal load has been determined empirically, and it is given by TMS 402 Eq. 10-3 as

$$f_{ps} = f_{se} + \cfrac{0.03\left(\cfrac{E_{ps}d}{l_p}\right)\left(1 - 1.56\left(\cfrac{A_{ps}f_{se} + \cfrac{P_u}{\phi}}{f'_m bd}\right)\right)}{1 + 0.0468\left(\cfrac{E_{ps}A_{ps}}{f'_m bl_p}\right)}$$

$$\geq f_{se}$$
$$\leq f_{py}$$

The force in the prestressing tendon at nominal load is

$$P_{ps} = f_{ps}A_{ps}$$

A rectangular compression stress block is assumed in the face shell on the compression face of the wall with a magnitude of $0.80f'_m$ and a depth of a. The compression force on the stress block acts at mid-depth of the stress block and is given by

$$C_m = 0.80abf'_m$$

Stress in the face shell on the tension face of the wall is ignored. Allowing for the factored axial load, P_u, on the wall, the depth of the stress block is obtained by equating forces on the section to give

$$a = \frac{f_{ps}A_{ps} + \dfrac{P_u}{\phi}}{0.80f'_m b}$$

When the wall also contains concentrically placed, bonded, nonprestressed reinforcement in grouted cores, the depth of the stress block is given by TMS 402 Eq. 10-1 as

$$a = \frac{f_{ps}A_{ps} + f_y A_s + \dfrac{P_u}{\phi}}{0.80f'_m b}$$

The nominal flexural capacity of the section is obtained by taking moments about the line of action of the compression force to give TMS 402 Eq. 10-2 which is

$$M_n = \left(f_{ps}A_{ps} + f_y A_s + \frac{P_u}{\phi}\right)\left(d - \frac{a}{2}\right)$$

The design flexural capacity is given by TMS 402 Sec. 10.4.3.3 as

$$\phi M_n = 0.8 M_n$$

To ensure a ductile failure in concrete block walls subject to out-of-plane loading, TMS 402 Sec. 10.4.3.6 specifies a maximum depth for the stress block of

$$a = 0.36d$$

In addition, the depth of the stress block may not exceed the thickness of the face shell.

Example 7.25

The nominal 8 in ungrouted concrete block masonry wall described in Ex. 7.23 has face-shell mortar bedding of type S portland cement/lime mortar, a specified strength of $f'_m = 2500 \ \text{lbf/in}^2$ at 28 days, and a specified strength of $f'_{mi} = 2000 \ \text{lbf/in}^2$ at transfer. The wall is post-tensioned with $^7/_{16}$ in diameter steel rods at 16 in centers placed centrally in the wall, and the rods are laterally restrained. The wall has an effective height of $h = 20$ ft and is simply supported at the top and bottom. The roof live load is 200 lbf/ft, and the roof dead load is 400 lbf/ft applied to the wall without eccentricity. The masonry wall has a weight of $\gamma_w = 37 \ \text{lbf/in}^2$. The properties of the steel rod are $E_{ps} = 29{,}000 \ \text{kips/in}^2$, $A_{ps} = 0.142 \ \text{in}^2$, $f_{py} = 100 \ \text{kips/in}^2$, $f_{pu} = 122 \ \text{kips/in}^2$. Wind load governs and has a value of $w = 25 \ \text{lbf/ft}^2$. Assume the loss of prestress at transfer is 5% and the final loss of prestress is 30%. Determine if the wall design flexural capacity is adequate. Ignore P-delta effects.

Solution

The following calculations are based on a 1 ft length of wall and on information obtained from Ex. 7.23 and Ex. 7.24.

The weight of the wall above midheight plus the roof dead load is

$$P = 770 \ \text{lbf}$$

In determining the moment demand on the wall, the governing load combination is IBC Eq. 16-6, which is

$$Q_u = 0.9D + 1.0W$$

The factored axial load on the wall is

$$P_u = 0.9P = (0.9)(770 \ \text{lbf}) = 693 \ \text{lbf}$$

$$\frac{P_u}{\phi} = \frac{693 \ \text{lbf}}{0.8} = 866 \ \text{lbf}$$

The bending moment produced on the wall by the factored lateral wind load is

$$M_u = \frac{wh^2}{8} = \frac{\left(25 \ \dfrac{\text{lbf}}{\text{ft}}\right)(20 \ \text{ft})^2}{8}$$
$$= 1250 \ \text{ft-lbf}$$

The equivalent area of prestressing tendon per foot of wall is

$$A_{ps} = 0.107 \ \text{in}^2$$

$$d = \frac{7.63 \ \text{in}}{2} = 3.815 \ \text{in}$$

After 30% total losses, the prestressing tendon force per foot of wall is

$$P_{ps} = 6.13 \ \text{kips}$$

After 30% total losses, the prestressing tendon stress per foot of wall is

$$f_{se} = \frac{P_{ps}}{A_{ps}} = \frac{6.13 \ \text{kips}}{0.107 \ \text{in}^2} = 57.3 \ \text{kips/in}^2$$

The stress in the tendon at nominal load is given by TMS 402 Eq. 10-3 as

$$f_{ps} = f_{se} + \frac{0.03\left(\dfrac{E_{ps}d}{l_p}\right)\left(1 - 1.56\left(\dfrac{A_{ps}f_{se} + \dfrac{P_u}{\phi}}{f'_m bd}\right)\right)}{1 + 0.0468\left(\dfrac{E_{ps}A_{ps}}{f'_m bl_p}\right)}$$

$$f_{se} = \left(57.3 \ \frac{\text{kips}}{\text{in}^2}\right)\left(1000 \ \frac{\text{lbf}}{\text{kip}}\right) = 57{,}300 \ \text{lbf/in}^2$$

$$\frac{E_{ps}d}{l_p} = \frac{\left(29{,}000{,}000 \ \dfrac{\text{lbf}}{\text{in}^2}\right)(3.815 \ \text{in})}{(20 \ \text{ft})\left(12 \ \dfrac{\text{in}}{\text{ft}}\right)} = 460{,}979 \ \text{lbf/in}^2$$

$$\frac{A_{ps}f_{se} + \dfrac{P_u}{\phi}}{f'_m bd} = \frac{\left[(0.107 \ \text{in}^2)\left(57{,}300 \ \dfrac{\text{lbf}}{\text{in}^2}\right) + (866 \ \text{lbf})\right]}{\left(2500 \ \dfrac{\text{lbf}}{\text{in}^2}\right)(12 \ \text{in})(3.815 \ \text{in})}$$
$$= 0.061$$

$$\frac{E_{ps}A_{ps}}{f'_m bl_p} = \frac{\left(29{,}000{,}000 \ \dfrac{\text{lbf}}{\text{in}^2}\right)(0.107 \ \text{in}^2)}{\left(2500 \ \dfrac{\text{lbf}}{\text{in}^2}\right)(12 \ \text{in})(20 \ \text{ft})\left(12 \ \dfrac{\text{in}}{\text{ft}}\right)} = 0.43$$

$$f_{ps} = 57{,}300 \ \frac{\text{lbf}}{\text{in}^2} + \left[\frac{(0.03)\left(460{,}979 \ \dfrac{\text{lbf}}{\text{in}^2}\right)(1 - (1.56)(0.061))}{1 + (0.0468)(0.43)}\right]$$

$$= 69{,}570 \ \text{lbf/in}^2$$
$$> f_{se}$$
$$< f_{py}$$

The depth of the stress block is given by TMS 402 Eq. 10-1 as

$$a = \frac{f_{ps}A_{ps} + f_y A_s + \dfrac{P_u}{\phi}}{0.80f'_m b}$$

$$= \frac{\left(69{,}570 \ \dfrac{\text{lbf}}{\text{in}^2}\right)(0.107 \ \text{in}^2) + \left(0 \ \dfrac{\text{lbf}}{\text{in}^2}\right)(0 \ \text{in}^2) + 866 \ \text{lbf}}{(0.80)\left(2500 \ \dfrac{\text{lbf}}{\text{in}^2}\right)(12 \ \text{in})}$$

$$= 0.35 \ \text{in} \quad \text{[satisfactory]}$$
$$< t_f$$
$$< 0.36d \quad \text{[satisfies TMS 402 Sec. 10.4.3.6]}$$

The nominal moment is given by TMS 402 Eq. 10-2 as

$$M_n = \left(f_{ps}A_{ps} + f_y A_s + \frac{P_u}{\phi}\right)\left(d - \frac{a}{2}\right)$$

$$= \frac{\left[\left(69{,}570 \ \dfrac{\text{lbf}}{\text{in}^2}\right)(0.107 \ \text{in}^2) + \left(0 \ \dfrac{\text{lbf}}{\text{in}^2}\right)(0 \ \text{in}^2) + 866 \ \text{lbf}\right] \times \left(\dfrac{7.63 \ \text{in}}{2} - \dfrac{0.35 \ \text{in}}{2}\right)}{12 \ \dfrac{\text{in}}{\text{ft}}}$$

$$= 2520 \ \text{ft-lbf}$$

The design flexural strength for a prestressed member is given by TMS 402 Sec. 10.4.3.3 as

$$\phi M_n = 0.8 M_n$$
$$= (0.8)(2520 \text{ ft-lbf})$$
$$= 2016 \text{ ft-lbf}$$
$$> M_u \quad [\text{satisfactory}]$$

The design moment strength is adequate.

11. QUALITY ASSURANCE, TESTING, AND INSPECTION

A quality assurance program is used to ensure that the constructed masonry is in compliance with the construction documents. The quality assurance plan specifies the required level of quality for the work, and testing and inspection determines the acceptability of the final construction. For masonry construction, these issues are addressed in IBC Sec. 110, Sec. 1704.5, Sec. 1705.4, and Sec. 2105, as well as TMS 402 Sec. 3.1 and TMS 602 Sec. 1.6.

A quality assurance plan, in conformance with IBC requirements, is developed by the engineer of record and incorporated into the contract documents. An approved testing agency is appointed to sample and test the materials used on the project and to determine compliance with the contract documents. The test results are reported to the engineer of record and to the contractor. An approved inspection agency is appointed to perform the inspection and evaluation of the construction as required by the contract documents. The inspection reports are provided to the engineer of record and to the contractor. Both the testing agency and the inspection agency are required to draw any observed deficiencies to the attention of the engineer of record, the building official, and the contractor.

Types of Inspections

Inspections may be of two types, periodic or continuous. *Periodic inspection* is the part-time or intermittent observation of work requiring periodic inspection by an inspector who is present at selected stages of construction and at the completion of the work. *Continuous inspection* is the full-time observation of work requiring continuous inspection by an inspector who is continuously present in the area where the work is being performed. In addition, inspections may be classified as preliminary, standard, special, and final. *Standard inspections* are the basic inspection requirements specified in IBC Sec. 110 and are applicable to all projects. *Special inspections* are required by IBC Sec. 1704.2 for the installation of critical components and connections. These inspections are performed by special inspectors, with special expertise in the specific work involved, to ensure compliance with the contract documents. A *preliminary inspection* of the work site may be requested by

the building official prior to the issuance of a building permit. The *final inspection* is made after all work required by the building permit is completed.

Level of Inspection

The level of inspection required by the IBC depends on the design process used for the project and on the risk category assigned to the completed structure. The project may be designed using either an empirical approach or by an engineered design procedure. Engineered masonry comprises projects designed by allowable stress design, strength design, and prestressed design as specified in Chap. 8 through Chap. 9 in the TMS 402. Empirically designed masonry is specified in App. A.1 through App. A.9 of TMS 402. Prescriptive designed projects comprise veneer and glass unit masonry, as specified in TMS 402 Chap. 12 and Chap. 13 and masonry partition walls in Chap. 14. Empirical design is, of necessity, more conservative than engineered design.

Risk Category

The *risk category* is a designation used to determine the structural requirements of a building based on the nature of its occupancy. IBC Table 1604.5 lists risk categories, and an abbreviated listing is given in Table 7.4.

Table 7.4 *Risk Category of Buildings*

risk category	nature of occupancy
I	low hazard structures
II	standard occupancy structures
III	assembly structures
IV	essential structures or structures housing hazardous materials

- Category I structures are facilities that represent a low hazard to human life in the event of failure. These include agricultural facilities, temporary facilities, and minor storage facilities.

- Category II structures are all facilities except those listed in risk categories I, III, and IV.

- Category III structures are facilities that represent a substantial hazard to human life in the event of failure because of their high occupant load. These include facilities where more than 300 people congregate in one area, school or daycare facilities with a capacity exceeding 250, colleges and universities with a capacity exceeding 500, healthcare facilities with a capacity of 50 or more resident patients but not having surgery or emergency treatment facilities, correctional centers, power stations, and water treatment facilities. Also included are facilities not included in risk category IV containing sufficient quantities of toxic or explosive substances to be dangerous to the public if released.

Reinforced Masonry

- Category IV structures are facilities housing essential equipment that is required for post-disaster recovery. These include healthcare facilities with surgery or emergency treatment facilities; fire, rescue, ambulance and police stations; emergency vehicle garages; emergency shelters; emergency response centers; air-traffic control centers; structures having critical national defense functions; water storage facilities and pump structures required to maintain water pressure for fire suppression; and buildings housing equipment and utilities required as emergency backup facilities for risk category IV structures. Also included are facilities containing highly toxic materials where the quantity of the material exceeds the maximum allowable quantities of IBC Table 307.1(2).

In accordance with IBC Sec. 1705.4, special inspections and tests of masonry construction must comply with the quality assurance program requirements of TMS 402 and TMS 602. TMS 402 Sec. 3.1 specifies three levels of quality assurance, depending on the risk category assigned to the structure and the design method used to prepare the contract plans. A level 1 quality assurance program is the least stringent program, and a level 3 quality assurance program is the most stringent. Three categories of design methods are specified in TMS 402, as follows

- engineered design methods in accordance with TMS 402 Part 3 (allowable stress design, strength design, prestressed design, and autoclaved aerated concrete masonry design), TMS 402 App. B (masonry infill design), and TMS 402 App. C (limit design)

- prescriptive design methods in accordance with TMS 402 Part 4 (glass unit masonry, masonry veneer, and masonry partition walls)

- empirical design methods in accordance with TMS 402 App. A (glass unit masonry, masonry veneer, and masonry partition walls)

Table 7.5 indicates the applicability of each level of quality assurance program to the design method and risk category of a structure.

Table 7.5 *Masonry Quality Assurance Levels*

risk category	empirically designed methods	prescriptive design methods	engineered design method
I, II, or III	level 1	level 1	level 2
IV	not permitted	level 2	level 3

TMS 602 Table 3 specifies the verification requirements corresponding to the quality assurance level assigned to the structure. This is shown in a modified form in Table 7.6.

Table 7.6 *Masonry Verification Requirements*

minimum verification	required for quality assurance*		
	level 1	level 2	level 3
prior to construction, verification of compliance of submittals	R	R	R
prior to construction, verification of f'_m and f'_{AAC}, except where specifically exempted by the Code	NR	R	R
during construction, verification of slump flow and Visual Stability Index (VSI) when self-consolidating grout is delivered to the project site	NR	R	R
during construction, verification of f'_m and f'_{AAC} for every 5000 ft² (465 m²)	NR	NR	R
during construction, verification of proportions of materials as delivered to the project site for premixed or preblended mortar, prestressing grout, and grout other than self-consolidating grout	NR	NR	R

*R = required; NR = not required

Adapted from TMS 602-16 Table 3, Minimum Verification Requirements.

TMS 602 Table 4 specifies the special inspection requirements corresponding to the quality assurance level assigned to the structure. This is shown in a modified form in Table 7.7.

Structural Observation

Structural observation is required by IBC Sec. 1704.6 for some structures. Structural observation consists of the visual observation of the structural system, usually by the engineer of record, to ensure general conformance to the contract documents. Structural observation, when required, is additional to normal inspection procedures. IBC Sec. 1704.6 requires structural observation for a structure when

- the structure is classified as risk category IV

- the height of the structure is greater than 75 ft above the lowest level of fire department vehicle access

- the structure is assigned to seismic design category D, E, or F and is classified as risk category III or IV

- the structure is assigned to seismic design category E, is classified as risk category I or II, and is greater than two stories above grade plane

- such observation is stipulated by the registered design professional responsible for the structural design

- such observation is specifically required by the building official

Table 7.7 *Masonry Special Inspection Requirements*

inspection task	level 1	level 2	level 3
1. As masonry construction begins, verify that the following are in compliance.			
a. proportions of site-prepared mortar	NR	P	P
b. grade and size of prestressing tendons and anchorages	NR	P	P
c. grade, type, and size of reinforcement, connectors, anchor bolts, and prestressing tendons and anchorages	NR	P	P
d. prestressing technique	NR	P	P
e. properties of thin-bed mortar for AAC masonry	NR	C^b/P^c	C
f. sample panel construction	NR	P	C
2. Prior to grouting, verify that the following are in compliance.			
a. grout space	NR	P	C
b. placement of prestressing tendons and anchorages	NR	P	P
c. placement of reinforcement, connectors, and anchor bolts	NR	P	C
d. proportions of site-prepared grout and prestressing grout for bonded tendons	NR	P	P
3. Verify compliance of the following during construction.			
a. materials and procedures with the approved submittals	NR	P	P
b. placement of masonry units and mortar joint construction	NR	P	P
c. size and location of structural members	NR	P	P
d. type, size, and location of anchors, including other details of anchorage of masonry to structural members, frames, or other construction	NR	P	C
e. welding of reinforcement	NR	C	C
f. preparations, construction, and protection of masonry during cold weather (temperature below 40°F (4.4°C)) or hot weather (temperature above 90°F (32.2°C))	NR	P	P
g. application and measurement of prestressing force	NR	C	C
h. placement of grout and prestressing grout for bonded tendons is in compliance	NR	C	C
i. placement of AAC masonry units and construction of thin-bed mortar joints	NR	C^b/P^c	C
4. Observe preparation of grout specimens, mortar specimens, and/or prisms.	NR	P	C

^{a}frequency of inspection, which may be continuous during the listed task or periodically; NR = not required; P = periodic; C = continuous

^{b}required for the first 5000 ft^2 (465 m^2) of AAC masonry

^{c}required after the first 5000 ft^2 (465 m^2) of AAC masonry

Adapted from TMS 602-16 Table 4, Minimum Special Inspection Requirements.

IBC Sec. 1704.6.3 requires structural observation for structures located where the basic wind speed exceeds 130 mph and the structure is classified as risk category III or IV.

12. REFERENCES

1. TMS 402/602. *Building Code Requirements* and *Specification for Masonry Structures.* Longmont, CO: The Masonry Society, 2016.

2. International Code Council. *2018 International Building Code.* Falls Church, VA: International Code Council, 2018.

3. AWS D1.4/D1.4M-17: *Structural Welding Code-Reinforcing Steel Including Metal Inserts and Connections in Reinforced Concrete Construction.* Miami, FL: American Welding Society, 2017.

4. ASTM A706/A706M-15: *Specification for Low-alloy Steel Deformed and Plain Bars for Concrete Reinforcement.* West Conshohocken, PA: American Society for Testing Materials, 2015.

5. ACI 318-14: *Building Code Requirements for Structural Concrete.* Farmington Hills, MI: American Concrete Institute, 2014.

6. Brandow, Gregg E., Chukwuma Ekwueme, and Gary C. Hart. *2009 Design of Reinforced Masonry Structures.* Citrus Heights, CA: Concrete Masonry Association of California and Nevada, 2011.

7. Kubischta, M. "Comparison of the 1997 UBC and the 2002 TMS 402 Code." *Masonry Chronicles* (Spring 2003).

8. ASCE/SEI7-16. *Minimum Design Loads and Associated Criteria for Buildings and Other Structures.* Reston, VA: American Society of Civil Engineers, 2016.

9. Kubischta, M. "In-Plane Loads on Masonry Walls." *Masonry Chronicles* (Fall 2003).

10. Huang, H. and Chukwuma Ekwueme. "The Effects of Axial Load on the Strength Design of Slender Out-of-Plane Concrete Masonry Walls." *Masonry Chronicles* (Summer 2007).

11. Ekwueme, Chukwuma. "Design of Anchor Bolts in Concrete Masonry." *Masonry Chronicles* (Winter 2009–2010).

12. Durning, T.A. "Prestressed Masonry." *Structural Engineer* (June 2000).

13. The Masonry Society. *Masonry Designers' Guide,* Longmont, CO: The Masonry Society, 2015.

Lateral Forces

PART 1: LATERAL FORCE-RESISTING SYSTEMS

Nomenclature

b_s	width of shear wall	ft
B	width of diaphragm	ft
C	compression chord force in diaphragm or shear wall	kips
C_o	shear capacity adjustment factor	–
E	modulus of elasticity	kips/in^2
h	height of shear wall	ft
I	moment of inertia	in^4
L	length of diaphragm, shear wall or wall segment, or member	ft
M_D	bending moment in a diaphragm	ft-kips
s_{ij}	stiffness factor at end i of member ij	–
t	uniform uplift force in perforated shear wall	lbf/ft
T	tensile chord force in diaphragm or shear wall, or hold-down force on shear wall	kips
v	induced unit shear	lbf/ft

V	shear force	lbf
w	distributed load	lbf/ft

Symbols

$\sum L_i$	sum of perforated shear wall segment lengths	ft
ϕ_D	resistance factor	–

1. INTRODUCTION

The *International Building Code* (IBC)[1] adopts by reference the American Society of Civil Engineers' *Minimum Design Loads and Associated Criteria for Buildings and Other Structures* (ASCE/SEI7[2]) for many of its code requirements. In accordance with IBC Sec. 1613.1,

> Every structure, and portion thereof, including nonstructural components that are permanently attached to structures and their supports and attachments, shall be designed and constructed to resist the effects of earthquake motions in accordance with Chapters 11, 12, 13, 15, 17 and 18 of ASCE/SEI7-16, as applicable.

The seismic design criteria of ASCE/SEI7 are primarily based on FEMA P-750[3], the 2009 edition of the *NEHRP Recommended Seismic Provisions for New Buildings and Other Structures.*

In accordance with IBC Sec. 1613.1, the following buildings are exempt from seismic design requirements.

- detached one- and two-family dwellings, assigned to seismic design category A, B, or C or located where the mapped short-period spectral response acceleration, S_S, is less than $0.4g$

- the seismic force-resisting systems of wood framed buildings that conform to the provisions of IBC Sec. 2308

- agricultural storage structures intended only for incidental human occupancy

- structures that require special consideration of their response characteristics and environment that are not addressed by the IBC or ASCE/SEI7, and for which other regulations provide seismic criteria, such as vehicular bridges, electrical transmission towers, hydraulic structures, buried utility lines and their appurtenances, and nuclear reactors

References within ASCE/SEI7 to Ch. 14 do not apply, except as specifically required.

2. BASIC COMPONENTS

The basic function of a *lateral force-resisting system* is to transfer the lateral forces acting on the structure to the foundation. Vertical and horizontal resisting components are used to provide a continuous and competent load path from the top of the structure to the foundation.

For the single-story, light-framed steel structure shown in Fig. 8.1, the steel roof deck forms the horizontal resisting component or diaphragm.[4] The lateral wind or seismic force acting on the south wall of the structure delivers a line load of w (in units of lbf/ft) to the roof diaphragm. The diaphragm acts as a deep beam with end reactions of

$$V = \frac{wL}{2}$$

These reactions produce a maximum unit shear at the ends of the diaphragm of

$$v = \frac{V}{B} = \frac{wL}{2B}$$

Figure 8.1 *Lateral Force-Resisting Components*

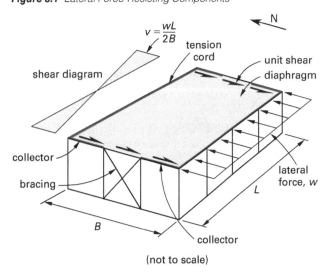

(not to scale)

The edges of the diaphragm normal to the direction of the lateral force are known as *chords* and act in a manner similar to the flanges of a steel beam to resist the bending moment produced by the lateral force. As shown in Fig. 8.1, one chord is in tension and the other in compression. The force in a chord is given by the equation

$$T = C = \frac{M_D}{B} = \frac{wL^2}{8B}$$

As shown in Fig. 8.1, the diaphragm must be designed with a capacity to resist the maximum shear at the ends, but may be designed for a smaller shear nearer midspan. Diaphragms are typically constructed from concrete,[4] composite and non-composite formed steel deck,[5] or wood structural panels.[6] Alternatively, in place of a diaphragm, horizontal bracing may be used to resist the lateral load.

The diaphragm shear force is transferred by the collector elements, or *drag struts*, to the vertical resisting components. In Fig. 8.1, these are the bracing elements in the end walls. The bracing elements transfer the lateral force to the structure's foundation.

Structural Irregularities

The structure illustrated in Fig. 8.1 is defined as a *regular building*. A regular building has the following characteristics.

- a continuous load path that is provided to transfer the applied lateral forces to the foundation

- a plan shape that is symmetrical and vertical lateral force-resisting elements that have similar strengths so as to minimize torsion

- vertical lateral force-resisting elements that are located so as to provide the maximum torsional capacity

- uniformly distributed mass, stiffness, and strength to minimize stress concentrations.

- no geometric irregularities, discontinuities, or reentrant corners that will produce stress concentrations

Buildings that have structural irregularities are known as *non-regular buildings*. *Structural irregularities* produce stress concentrations and increased torsional effects that may cause collapse of the building. In order to discourage the use of irregular features in a design and encourage regularity and redundancy, ASCE/SEI7 Chap. 12 imposes penalties on non-regular structures.

3. STRUCTURAL SYSTEMS

The *lateral force-resisting systems* listed in ASCE/SEI7 Table 12.2-1 consist of the following nine types.

- bearing wall systems

- moment-resisting frame systems

- building frame systems with shear walls

- building frame systems with braced frames

- dual systems with special moment frames

- dual systems with intermediate moment frames

- shear wall-frame interactive system with ordinary reinforced concrete moment frames and ordinary reinforced concrete shear walls

- cantilever column systems

- steel systems not specifically detailed for seismic resistance

Bearing Wall Systems

A *bearing wall system*, as shown in Fig. 8.2, has shear walls that are designed to support most of the gravity load of the building and to resist all lateral forces. The shear walls receive the shear force from the floor and roof diaphragms and transfer the lateral force to the building foundation. The system lacks redundancy, since a failure of the lateral capacity of the walls will also produce collapse of the gravity load carrying capacity. Therefore, the system has a comparatively low value of the response modification coefficient, R, and a correspondingly high lateral design force. The system has considerable stiffness and low inelastic deformation, which results in negligible damage to architectural features and nonstructural elements in the event of an earthquake. Shear walls must extend from the roof to the foundation without offsets or extensive openings, and this restricts architectural design freedom. In seismic design categories D, E, and F, concrete and masonry shear walls must be specially reinforced and detailed, and they are restricted in height. Wood framed residential and small commercial and industrial structures are also constructed using this system, with shear walls sheathed with wood structural panels.

Figure 8.2 Bearing Wall System

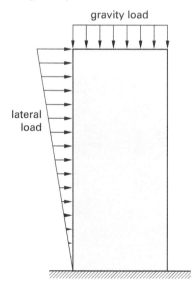

Moment-Resisting Frame Systems

A *moment-resisting frame system*, as shown in Fig. 8.3, provides support for both gravity and lateral loads by flexural action induced by rigid connections at the beam/column joints. In seismic design categories D, E, and F, joints must be capable of developing an interstory drift angle of 0.04 rad at a flexural strength of 80% of the plastic moment capacity of the beam. For steel special moment-resisting frames, details of prequalified joints that meet this requirement have been published.[7,8] Special reinforced concrete moment-resisting frames are detailed to ensure that large inelastic displacements can occur without impairing the structural integrity of the frame.[9,10]

Figure 8.3 Moment-Resisting Frame System

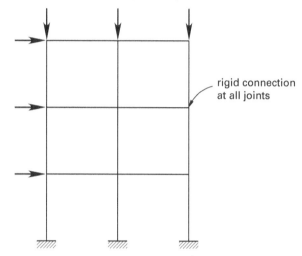

rigid connection
at all joints

In moment-resisting frame systems, no diagonal bracing or shear walls are necessary, and the open bays allow for great freedom with architectural designs. No restrictions are placed on the height of special moment-resisting frames. The system has considerable flexibility and high inelastic deformation capacity, which may result in significant damage to architectural features and nonstructural elements in the event of an earthquake. Therefore, the design of moment-resisting frames is often governed by code requirements to limit interstory drift rather than to provide minimum strength. The system is highly redundant and has a high value of the response modification coefficient, R, and a correspondingly low lateral design force.

A uniform distribution of stiffness, strength, and mass is required over the height of the frame in order to prevent P-delta instability. In addition, a strong-column/weak-beam design approach is required to ensure that column flexural strength exceeds the beam flexural strength at each joint. Inelastic drift must be uniformly distributed over the height of the frame, and inelastic deformations are concentrated at the ends of the beams. The resulting collapse mechanism is shown in Fig. 8.4.

Figure 8.4 *Strong-Column/Weak-Beam Frame Collapse Mechanism*

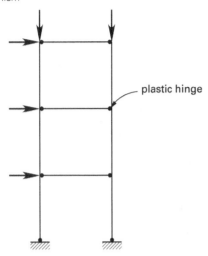

Where the stiffness of the columns in a story is reduced, a *soft story irregularity* is produced. Large inelastic drift occurs in the story, and plastic hinges are formed at the top and bottom of the columns in the soft story. *P*-delta effects cause a *story mechanism* to form, and the collapse mechanism is shown in Fig. 8.5.

Figure 8.5 *Story Mechanism*

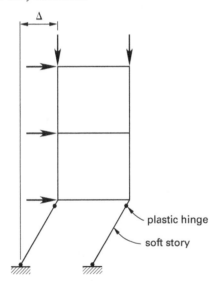

Several methods are available for determining an approximate estimate of the forces in a rigid frame subjected to lateral loads.[11] The simplest of these is the *portal method*. The portal method assumes the following.

- A point of inflection occurs at the mid-height of each column.

- A point of inflection occurs at the midpoint of each girder.

- The shear and moment in each interior column is twice that of an exterior column.

- The axial force in an interior column is zero.

An accurate determination of the member forces in a rigid frame, with axial effects neglected, may be obtained using the *moment distribution procedure*. In the case of lateral loads applied to a *symmetrical single bay frame*, as shown in Fig. 8.6, the skew symmetry may be used to simplify the process.[12] Allowing for skew symmetry, the modified stiffness of each column is EI/L, and the modified stiffness of each beam is $6EI/L$. The carryover factor in the columns is -1, and there is no carryover between girder ends. The initial fixed-end moments are obtained by imposing unit virtual sway displacement on each story in turn.

Figure 8.6 *Sway Distribution for Single Bay Frame*

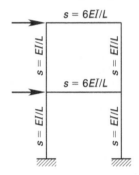

Multi-bay frames of the type shown in Fig. 8.7, where the relative EI/L values are shown ringed, satisfy the principle of multiples. Using the *principle of multiples*, the original frame is replaced by two substitute frames, with the ratio of applied load to member stiffness the same in each frame. Joint rotations and sway displacements are the same in the substitute frames and the original frame, and member forces in the substitute frames can be added to give the corresponding member force in the original frame.

Figure 8.7 *Principle of Multiples*

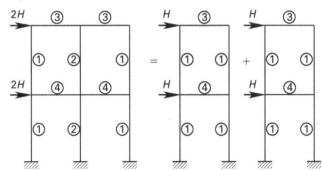

Example 8.1

Using the portal method, determine the bending moments and support reactions produced in the two-story rigid frame shown.

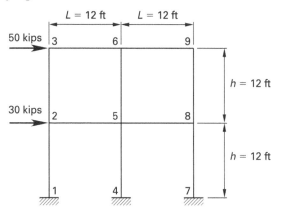

Solution

Equating the reactions at the column bases with the applied horizontal forces gives

$$H_1 + H_4 + H_7 = H_2 + H_3$$
$$= 30 \text{ kips} + 50 \text{ kips}$$
$$= 80 \text{ kips}$$

Since the shear in an interior column is twice that of an exterior column,

$$H_1 + H_4 + H_7 = 0.5H_4 + H_4 + 0.5H_4$$
$$= 2H_4$$
$$= 80 \text{ kips}$$
$$H_4 = 40 \text{ kips}$$
$$H_1 = H_7 = 0.5H_4$$
$$= (0.5)(40 \text{ kips})$$
$$= 20 \text{ kips}$$

A point of inflection occurs at the midpoint of each column, and the moment in each interior column is twice that of an exterior column. So,

$$M_{12} = M_{21} = M_{78} = M_{87}$$
$$= \frac{H_1 h}{2}$$
$$= \frac{(20 \text{ kips})(12 \text{ ft})}{2}$$
$$= 120 \text{ ft-kips}$$
$$M_{45} = M_{54} = 2M_{12}$$
$$= (2)(120 \text{ ft-kips})$$
$$= 240 \text{ ft-kips}$$

Since the axial force in the central column is zero, taking moments about support 1 gives

$$\sum M = V_7(2L) + M_{12} + M_{45} + M_{78}$$
$$- H_2 h - H_3(2h) = 0$$
$$V_7 = \frac{H_2 h + H_3(2h) - M_{12} - M_{45} - M_{78}}{2L}$$
$$= \frac{\begin{array}{c}(30 \text{ kips})(12 \text{ ft}) + (50 \text{ kips})(2)(12 \text{ ft}) \\ -120 \text{ ft-kips} - 240 \text{ ft-kips} \\ -120 \text{ ft-kips}\end{array}}{(2)(12 \text{ ft})}$$
$$= 45 \text{ kips} \quad [45 \text{ kips up}]$$
$$V_1 = -V_7 = -45 \text{ kips} \quad [45 \text{ kips down}]$$

Imposing unit virtual sway displacement on the top story gives

$$M_{23} + M_{56} + M_{89} + M_{32} + M_{65} + M_{98}$$
$$= \text{moment of } H_3 \text{ about node 2}$$
$$= H_3 h$$
$$= (50 \text{ kips})(12 \text{ ft})$$
$$= 600 \text{ ft-kips}$$

Each of the two interior moments, M_{56} and M_{65}, is twice as large as each of the four exterior moments, so

$$M_{23} + M_{56} + M_{89} + M_{32} + M_{65} + M_{98}$$
$$= 600 \text{ ft-kips}$$
$$= M_{23} + 2M_{23} + M_{23} + M_{23}$$
$$\qquad + 2M_{23} + M_{23}$$
$$= 8M_{23}$$
$$M_{23} = \frac{600 \text{ ft-kips}}{8}$$
$$= 75 \text{ ft-kips}$$
$$= M_{89} = M_{32} = M_{98}$$
$$2M_{23} = (2)(75 \text{ ft-kips})$$
$$= 150 \text{ ft-kips}$$
$$= M_{56} = M_{65}$$

Equating moments at node 2 gives

$$M_{25} = M_{21} + M_{23}$$
$$= 120 \text{ ft-kips} + 75 \text{ ft-kips}$$
$$= 195 \text{ ft-kips}$$
$$= M_{52} = M_{58} = M_{85}$$

Equating moments at node 3 gives

$$M_{36} = M_{32}$$
$$= 75 \text{ ft-kips}$$
$$= M_{63} = M_{69} = M_{96}$$

The member forces (kips) are shown in the following illustration with moments (ft-kips) drawn on the tension side of the members.

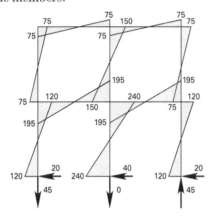

Example 8.2

Using the principle of multiples, determine the bending moments and support reactions produced in the two-story rigid frame analyzed in Ex. 8.1. The relative EI/L values for columns 45 and 56 are twice the values for each of the other members.

Solution

The frame satisfies the principle of multiples, and each of the two substitute frames is identical with the single-bay frame shown.

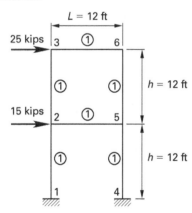

The skew symmetry in the substitute frame allows automatically for sidesway. For all members,

$$\frac{EI}{L} = k$$

Allowing for skew symmetry, the modified stiffness of the members is

$$s_{21} = s_{23} = s_{32} = k$$
$$s_{25} = s_{36} = 6k$$

The distribution factors at joint 2 are

$$d_{25} = \frac{s_{25}}{s_{25} + s_{21} + s_{23}}$$
$$= \frac{6k}{8k}$$
$$= \frac{3}{4}$$
$$d_{21} = d_{23} = \frac{1}{8}$$

The distribution factors at joint 3 are

$$d_{36} = \frac{s_{36}}{s_{36} + s_{32}}$$
$$= \frac{6k}{7k}$$
$$= \frac{6}{7}$$
$$d_{32} = \frac{1}{7}$$

Allowing for skew symmetry, the carry-over factor in the columns is -1, and there is no carry-over between girder ends. The initial fixed-end moments are obtained by imposing unit virtual sway displacement on each story.

For the upper story,

$$M_{F23} = M_{F32} = \frac{-V_3 h_{23}}{4}$$
$$= \frac{-(25 \text{ kips})(12 \text{ ft})\left(12 \frac{\text{in}}{\text{ft}}\right)}{4}$$
$$= -900 \text{ in-kips}$$

For the lower story,

$$M_{F12} = M_{F21} = \frac{-(V_3 + V_2)h_{12}}{4}$$
$$= \frac{-(40 \text{ kips})(12 \text{ ft})\left(12 \frac{\text{in}}{\text{ft}}\right)}{4}$$
$$= -1440 \text{ in-kips}$$

The distribution of moments is given in the following table, with distribution occurring in the left half of the substitute frame only.

member	12	21	25	23	32	36
distribution factor	0	1/8	3/4	1/8	1/7	6/7
fixed-end moments	−1440	−1440		−900	−900	
distribution		293	1754	293	129	771
carry-over	−293			−129	−293	
distribution		16	97	16	42	251
carry-over	−16			−42	−16	
distribution		5	32	5	2	14
carry-over	−5			−2	−5	
distribution			2		1	4
final moments (in-kips)	−1754	−1126	1885	−759	−1040	1040
final moments (ft-kips)	−146	−94	157	−63	−87	87

The final bending moments (in ft-kips) produced in the members are shown in the following illustration with the moments drawn on the tension sides of the members.

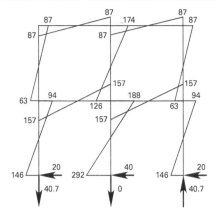

Building Frame Systems with Shear Walls

A *building frame system with shear walls*, as shown in Fig. 8.8, consists of two independent systems: one to provide support for gravity loads, the other to resist lateral loads. The shear walls resist all lateral loads, and a column and beam frame supports most gravity loads. Since the gravity frame does not contribute to lateral resistance, it does not require special ductile detailing, and simple shear connections are used to connect the beams and columns. However, the gravity frame is required to satisfy deformation compatibility requirements. Since the gravity and lateral-force systems are independent, the building frame system is more redundant than the bearing wall system. It has a higher value of the response modification coefficient, R, and a correspondingly lower lateral design force. The system has considerable stiffness and low inelastic deformation resulting in

negligible damage to architectural features and nonstructural elements in the event of an earthquake. Shear walls must extend from the roof to the foundation without offsets or extensive openings, and this restricts architectural design freedom.

Figure 8.8 *Building Frame System with Shear Walls*

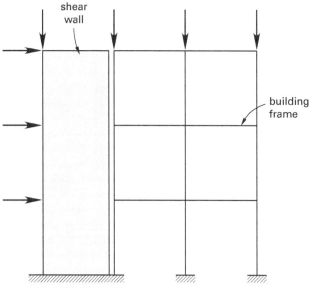

In seismic design categories D, E, and F, concrete and masonry shear walls must be specially reinforced and detailed, and they are restricted in height. Steel special plate shear walls are also used in a building frame system, and they are restricted in height in seismic design categories D, E, and F. Steel plate shear walls consist of a steel frame stiffened by thin steel plates. Lateral loads are resisted by buckling of the plate mobilizing diagonal tension-field action. Wood structural panel shear walls in wood framed small commercial and industrial structures may also be used in a building frame system and are restricted to a height of 65 ft in seismic design categories D, E, and F.

IBC Sec. 2306.3 requires wood structural panel shear walls be designed and constructed in accordance with the American Wood Council *Special Design Provisions for Wind and Seismic* (SDPWS[13]). The construction details of a typical *plywood sheathed shear wall* are shown in Fig. 8.9. The shear capacity of a shear wall depends on the thickness and grade of the plywood sheathing, width of framing members, support of the panel edges, and nail spacing and penetration. Nominal shear capacities are given in SDPWS Table 4.3A for seismic and wind loading of walls with plywood sheathing on one side, all panel edges blocked, and 2 in nominal framing of Douglas fir-larch or southern pine.

Figure 8.9 *Shear Wall Details*

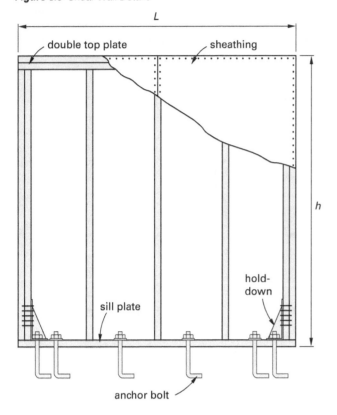

The nominal unit shear capacities tabulated in SDPWS Table 4.3A specifically relate to shear walls with plywood panels attached to one side of the wall using 6d, 8d, or 10d common or galvanized box nails. The table provides *nominal unit shear capacity* values for seismic design in column A and values for wind design are provided in column B. The corresponding ASD *allowable unit shear capacity* values are obtained by dividing the tabulated nominal unit shear capacity values by the ASD reduction factor of 2.0. The LRFD *factored unit resistance* values are obtained by multiplying the tabulated nominal unit shear capacity values by the resistance factor, ϕ, of 0.80.

SDPWS Table 4.3B, Table 4.3C, and Table 4.3D provide nominal unit shear capacities for shear walls sheathed with wood structural panels applied over gypsum wallboard; gypsum and Portland cement plaster; and lumber, respectively.

IBC Table 2306.3(1) tabulates allowable unit shear capacity values for shear walls sheathed with plywood panels using *staple fasteners*. IBC Table 2306.3(2) and Table 2306.3(3) provide allowable unit shear capacities for shear walls sheathed with fiberboard and gypsum board, respectively, using staple fasteners.

To control the stiffness of shear walls, SDPWS Table 4.3.4 imposes a maximum aspect ratio (b_s/h) on wood structural panel blocked shear walls of 3.5/1. For wood structural panels with aspect ratios greater than 2/1, the nominal shear values in SDPWS Table 4.3A are

multiplied by the aspect ratio factor $1.25 - 0.125\ h/b_s$. The shear wall height, h, is defined in SDPWS Sec. 2.3 as the clear height from the top of the foundation to the bottom of the diaphragm framing above. The shear wall width is defined in SDPWS Sec. 4.3.4.3 as the segment width of a perforated shear wall and in SDPWS Sec. 4.3.4.4 as the overall horizontal sheathed dimension of the wall. For unblocked wood structural panel shear walls, the maximum aspect ratio permitted is 2/1.

To resist the uplift due to overturning moments on the shear wall, a *hold-down* must be provided at each end of each wall. Transfer of the lateral force to the foundation is achieved with anchor bolts at a maximum spacing of 6 ft. A minimum of two bolts is required.

Since the shear wall is considered non-rigid, gravity loads along the top of the wall do not provide a restoring moment to the wall. Therefore, the tension force, T, in the hold-down, which is also the force in the end posts, or chords, is given by SDPWS Eq. 4.3-7 as

$$T = C = vh$$

C is the compression force; v is the induced unit shear; and h is the shear wall height.

To accommodate the bolts or screws in the hold-down, a double end post is usually provided. Similarly, to provide continuity for the top plate and to provide overlapping at intersections, a double top plate is normally used. IBC Sec. 2308.3.1 requires sill plates to be anchored to the foundation with not less than $\frac{1}{2}$ in diameter steel bolts or approved anchors spaced not more than 6 ft apart. Bolts must be embedded at least 7 in into concrete or masonry, and there must be a minimum of two bolts or anchor straps per wall, with one bolt or anchor strap located not more than 12 in, or less than 4 in, from the end of each wall. As shown in Fig. 8.10, to minimize the potential for cross grain bending in the sill plate, SDPWS Sec. 4.3.6.4.3 requires a steel plate washer under each nut not less than 0.229 in × 3 in × 3 in in size. The plate washer must extend to within $\frac{1}{2}$ in of the edge of the sill plate on the sheathed side when the required nominal unit shear capacity exceeds 400 lbf/ft for wind or seismic. Standard cut washers may be used when anchor bolts are designed to resist shear only and the following requirements are met.

- The shear wall is designed as an individual full-height wall segment with required uplift anchorage at shear wall ends sized to resist overturning, neglecting the dead load stabilizing moment.

- The shear wall aspect ratio does not exceed 2/1.

- The nominal unit shear capacity of the shear wall does not exceed 980 lbf/ft for seismic, or 1370 lbf/ft for wind.

Figure 8.10 *Anchor Bolt Detail*

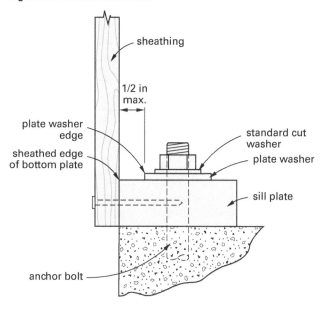

Adapted with permission from *Special Design Provisions for Wind and Seismic with Commentary*, copyright © 2015, by the American Wood Council.

In accordance with SDPWS Sec. 4.3.7, the maximum stud spacing is 24 in. At intermediate framing members, the maximum permitted nail spacing is 6 in, except that 12 in is permitted when stud spacing is less than 24 in, or panel thickness is $^{7}/_{16}$ in or more. Nails along intermediate framing members must be the same as nails specified for panel edge nailing. The width of the nailed face of framing members and blocking is required to be 2 in nominal or greater at adjoining panel edges, except that a 3 in nominal or greater width at adjoining panel edges and staggered nailing at all panel edges are required where any of the following conditions exist.

- Nail spacing of 2 in or less at adjoining panel edges is specified.

- 10d common nails having penetration into framing members and blocking of more than $1^{1}/_{2}$ in are specified at 3 in on center or less at adjoining panel edges.

- The nominal unit shear capacity on either side of the shear wall exceeds 700 lbf/ft in seismic design categories D, E, or F.

In accordance with SDPWS Sec. 4.3.3, for walls sheathed with plywood of equal shear capacity on opposite sides of the wall, the shear capacity of the wall may be taken as twice the value permitted for one side. For shear walls sheathed with dissimilar materials on opposite sides, the combined shear capacity may be taken as the maximum value given by twice the smaller capacity or equal to the larger capacity. Summing capacities of dissimilar materials applied to shear walls in the same wall line is not permitted.

Example 8.3

The shear wall shown in Fig. 8.9 has dimensions of $L = 8$ ft and $h = 8$ ft. The sheathing is $^{15}/_{32}$ in Structural I and the stud spacing is 16 in. The dead load on the wall from the roof diaphragm is $w = 80$ lbf/ft, and the strength level wind load transmitted by the diaphragm is $V = 11.6$ kips. Determine the nailing requirements for the shear wall, the force in the hold-downs, and the number of $^{5}/_{8}$ in diameter A307 anchor bolts required in the 4 in × 3 in Douglas fir-larch sill plate. Neglect the self-weight of the shear wall.

Solution

The strength level unit shear acting on the shear wall is

$$v = \frac{V}{L}$$

$$= \frac{(11.6 \text{ kips}) \left(1000 \ \dfrac{\text{lbf}}{\text{kip}} \right)}{8 \text{ ft}}$$

$$= 1450 \text{ lbf/ft}$$

The required spacing of 10d common nails with $1^{1}/_{2}$ in penetration is obtained from SDPWS Table 4.3A. The required nail spacing is

all panel edges	3 in
intermediate framing members	12 in
capacity provided	$(0.8) \left(1860 \ \dfrac{\text{lbf}}{\text{ft}} \right) = 1488$ lbf/ft
capacity required	1450 lbf/ft

From SDPWS Sec. 4.3.7, the framing at adjoining panel edges must be nominal 3 in, and staggered nailing at all panel edges is required.

The service level value of the unit shear is given by IBC Eq. 16-15 as

$$v' = 0.6v$$

$$= (0.6) \left(1450 \ \dfrac{\text{lbf}}{\text{ft}} \right)$$

$$= 870 \text{ lbf/ft}$$

Lateral Forces

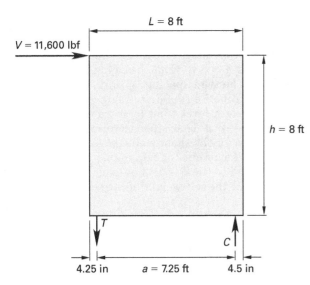

SDPWS Eq. 4.3-7 may be used to obtain the hold-down force. This ignores the vertical load on the wall and the distance of the hold-down anchor rod from the end of the wall. SDPWS Eq. 4.3-7 gives the hold-down force as

$$T = v'h = \left(870 \; \frac{\text{lbf}}{\text{ft}}\right)(8 \text{ ft}) = 6960 \text{ lbf}$$

The allowable parallel-to-grain load, $Z_{\parallel}$, on a $\frac{5}{8}$ in diameter bolt in the $2\frac{1}{2}$ in thick Douglas fir-larch sill plate is obtained from NDS[14] Table 12E as

$$Z_{\parallel} = 1180 \text{ lbf}$$

The load duration factor, C_D, for wind load is given by NDS Table 2.3.2 as

$$C_D = 1.6$$

Therefore, the number, N, of $\frac{5}{8}$ in diameter anchor bolts required is

$$N = \frac{0.6V}{C_D Z_{\parallel}} = \frac{(0.6)(11,600 \text{ lbf})}{(1.6)(1180 \text{ lbf})}$$
$$= 3.69 \quad [\text{use 4 bolts}]$$

Shear Walls with Openings

A shear wall with openings has been designed traditionally by considering each full height segment of the wall as an individual shear wall. As shown in Fig. 8.11, this results in a wall with a single opening being designed as two separate shear walls requiring a total of four hold-downs, one at either end of the two shear walls. Sheathing above and below the opening is not considered to contribute to the overall shear capacity of the wall, and the shear capacity of the wall is calculated as the sum of the capacities of the individual segments.

Figure 8.11 Segmented Shear Wall

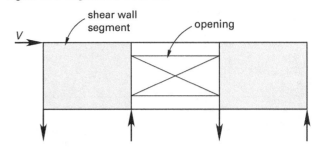

An alternative design method is the *perforated shear wall method*.[15] The shear capacity of a perforated shear wall is calculated as a percentage of the capacity of the wall without openings, and the method is specified in SDPWS Sec. 4.3.3.5. As shown in Fig. 8.12, the advantage of the method is that only two hold-downs are necessary, one at either end of the wall. Sheathed areas above and below openings are not designed for force transfer and are considered to provide only local restraint at their ends. The shear capacity of a perforated wall depends on the maximum opening height and on the percentage of full height sheathing.

Figure 8.12 Perforated Shear Wall

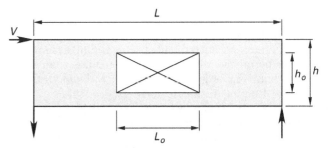

SDPWS Sec. 4.3.5.3 requires the following for perforated walls.

- A segment without openings must be located at each end of the perforated shear wall.

- The aspect ratio limitations of SDPWS Sec. 4.3.4.1 apply.

- The required nominal unit shear capacity for a single sided wall is limited to a maximum of 1740 lbf/ft for seismic or 2435 lbf/ft for wind.

- Where out-of-plane offsets occur, portions of the wall on each side of the offset must be considered separate perforated shear walls.

- Collectors for shear transfer must be provided through the full length of the wall.

- A perforated shear wall must have uniform top-of-wall and bottom-of-wall elevations.

- The height must not exceed 20 ft.

The design shear capacity, V, of a perforated shear wall is obtained from SDPWS Eq. 4.3-9 as

$$V = v C_o \sum L_i$$

The variables are defined as

> v = allowable unit shear capacity in a segmented shear wall (lbf/ft)
>
> C_o = shear capacity adjustment factor given in SDPWS Table 4.3.3.5
>
> $\sum L_i$ = sum of perforated shear wall segment lengths (ft)

Values of C_o are given in Table 8.1.

Table 8.1 *Capacity Adjustment Factors*

wall height, h	maximum opening height				
	$h/3$	$h/2$	$2h/3$	$5h/6$	h
8'-0" wall	2'-8"	4'-0"	5'-4"	6'-8"	8'-0"
10'-0" wall	3'-4"	5'-0"	6'-8"	8'-4"	10'-0"
full-height sheathing (%)	shear capacity adjustment factor				
10	1.00	0.69	0.53	0.43	0.36
20	1.00	0.71	0.56	0.45	0.38
40	1.00	0.77	0.63	0.53	0.45
60	1.00	0.83	0.71	0.63	0.56
80	1.00	0.91	0.83	0.77	0.71
100	1.00	1.00	1.00	1.00	1.00

Adapted with permission from *Special Design Provisions for Wind and Seismic with Commentary*, copyright © 2015, by the American Wood Council.

The force in a hold-down, which is also the force in the end posts, is given by SDPWS Eq. 4.3-8 as

$$T = C$$
$$= \frac{Vh}{C_o \sum L_i}$$

The unit shear force in a perforated shear wall is given by SDPWS Eq. 4.3-9 as

$$v_{max} = \frac{V}{C_o \sum L_i}$$

Anchor bolts, in addition to resisting the horizontal shear force $v_{max} \Sigma L_i$, must also resist a uniformly distributed uplift force. This force is given by SDPWS Sec. 4.3.6.4.2.1 as

$$t = v_{max}$$

Example 8.4

The perforated shear wall shown in Fig. 8.12 has an overall length, L, of 24 ft and a height, h, of 8 ft. The sheathing is $^{15}/_{32}$ in Structural I and the stud spacing is 16 in. The centrally placed opening has dimensions of $L_o = 8$ ft and $h_o = 4$ ft. The dead load on the wall from the roof diaphragm is $w = 80$ lbf/ft, and the strength level wind load transmitted by the diaphragm is $V = 11.6$ kips. Determine the nailing requirements for the shear wall, the force in the hold-downs, and the number of $^5/_8$ in diameter A307 anchor bolts required in the 4 in × 3 in Douglas fir-larch sill plate. Neglect the self-weight of the shear wall.

Solution

The aspect ratio of each segment is

$$a = \frac{L - L_o}{2h}$$
$$= \frac{24 \text{ ft} - 8 \text{ ft}}{(2)(8 \text{ ft})}$$
$$= 1.0 \quad [\text{complies with SDPWS Table 4.3.4}]$$

The percentage of full height sheathing is

$$\frac{\sum L_i}{L} = \frac{8 \text{ ft} + 8 \text{ ft}}{24} \times 100\%$$
$$= 67\%$$

The maximum opening height ratio is

$$\frac{h_o}{h} = \frac{4 \text{ ft}}{8 \text{ ft}}$$
$$= 0.5$$

From SDPWS Table 4.3.3.5, the shear capacity adjustment factor is

$$C_o = 0.86$$

The strength level unit shear acting on the shear wall is

$$v = \frac{V}{L - L_o}$$
$$= \frac{(11.6 \text{ kips})\left(1000 \dfrac{\text{lbf}}{\text{kip}}\right)}{24 \text{ ft} - 8 \text{ ft}}$$
$$= 725 \text{ lbf/ft}$$

Allowing for the shear capacity adjustment factor, the equivalent unit shear for a perforated shear wall is

$$v' = \frac{v}{C_o}$$
$$= \frac{725 \; \frac{\text{lbf}}{\text{ft}}}{0.86}$$
$$= 843 \; \text{lbf/ft}$$

The required spacing of 10d common nails with $1\frac{1}{2}$ in penetration is obtained from SDPWS Table 4.3A as

all panel edges	4 in
intermediate framing members	12 in
capacity provided	$(0.8)(1430 \; \text{lbf/ft}) = 1144 \; \text{lbf/ft}$
capacity required	843 lbf/ft [satisfactory]

The allowable parallel-to-grain load on a $\frac{5}{8}$ in diameter bolt in the $2\frac{1}{2}$ in thick Douglas fir-larch sill plate is obtained from NDS Table 12E as

$$Z_{\parallel} = 1180 \; \text{lbf}$$

The load duration factor for wind load is given by NDS Table 2.3.2 as

$$C_D = 1.6$$

Converting to service level values using IBC Eq. 16-15, the number of $\frac{5}{8}$ in diameter bolts required is

$$N = \frac{0.6V}{C_D Z_{\parallel}}$$
$$= \frac{(0.6)(11.6 \; \text{kips})\left(1000 \; \frac{\text{lbf}}{\text{kip}}\right)}{(1.6)(1180 \; \text{lbf})}$$
$$= 3.7 \quad [\text{use 4 bolts}]$$

Two bolts in each segment must be provided. The strength level shear force on each bolt is

$$P_v = \frac{V}{4}$$
$$= \frac{(11.6 \; \text{kips})\left(1000 \; \frac{\text{lbf}}{\text{kip}}\right)}{4 \; \text{bolts}}$$
$$= 2900 \; \text{lbf/bolt}$$

The available strength level shear force on a $\frac{5}{8}$ in diameter A307 bolt is given by *AISC Manual*[16] Table 7-1 as

$$\phi r_n = 6230 \; \text{lbf}$$
$$> P_v \quad [\text{satisfactory}]$$

The uniformly distributed design uplift anchorage force on the sill plate of each full height, perforated shear wall segment is given by SDPWS Sec. 4.3.6.4.2.1 as

$$t = v_{\max}$$
$$= \frac{V}{C_o \sum L_i}$$
$$= \frac{(11.6 \; \text{kips})\left(1000 \; \frac{\text{lbf}}{\text{kip}}\right)}{(0.86)(8 \; \text{ft} + 8 \; \text{ft})}$$
$$= 843 \; \text{lbf/ft}$$

The net uplift on the shear wall is given by IBC Eq. 16-6 as

$$u = 0.9w + t$$
$$= (0.9)\left(-80 \; \frac{\text{lbf}}{\text{ft}}\right) + 843 \; \frac{\text{lbf}}{\text{ft}}$$
$$= 771 \; \text{lbf/ft}$$

The strength level tensile force in each bolt is

$$P_t = \frac{u \sum L_i}{4 \; \text{bolts}} = \frac{\left(771 \; \frac{\text{lbf}}{\text{ft}}\right)(8 \; \text{ft} + 8 \; \text{ft})}{4 \; \text{bolts}}$$
$$= 3084 \; \text{lbf/bolt}$$

The available strength level tensile force on a $\frac{5}{8}$ in diameter A307 bolt is given by *AISC Manual* Table 7-2 as

$$\phi r_n = 10{,}400 \; \text{lbf}$$
$$> P_t \quad [\text{satisfactory}]$$

The anchor bolts provided are adequate for both shear and uplift. The force in a hold-down is given by SDPWS Eq. 4.3-8 as

$$T = C$$
$$= \frac{Vh}{C_o \sum L_i}$$
$$= v_{\max} h$$
$$= \left(843 \; \frac{\text{lbf}}{\text{ft}}\right)(8 \; \text{ft})$$
$$= 6744 \; \text{lbf} \quad [\text{at strength level}]$$
$$= 4046 \; \text{lbf} \quad [\text{at service level}]$$

Building Frame System with Braced Frames

A *building frame system with braced frames*, as shown in Fig. 8.13, consists of a braced frame that resists lateral loads by truss action and an independent column and beam frame that supports most gravity loads. Since the gravity frame does not contribute to lateral resistance, it does not require special ductile detailing, and simple shear connections are used to connect the beams and columns. However, the gravity frame is required to satisfy deformation compatibility requirements. A braced frame is more ductile than a shear wall and provides more architectural freedom than a shear wall.

Figure 8.13 Building Frame System with Braced Frame

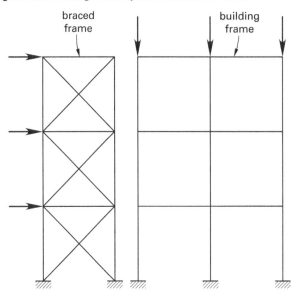

There are four general types of braced frames.

- steel special concentrically braced frames

- steel ordinary concentrically braced frames

- buckling-restrained braced frames

- steel eccentrically braced frames

In *steel special concentrically braced frames*, the center-line of all the frame members, beams, columns, and braces are coincident. Therefore, applied lateral loads are resisted by axial forces in the members and flexure is eliminated. Many different configurations are possible, and several are shown in Fig. 8.14. To ensure ductile behavior in the brace, limits are placed on the slenderness and compactness of the brace. Also, to ensure that the full strength of the brace can be developed, brace connections must be designed to resist the yield strength of the brace.

Figure 8.14 Steel Special Concentrically Braced Frames

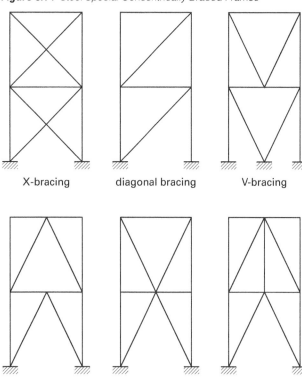

In *V-bracing* and *inverted V-bracing systems*, the intersecting beam must be designed to carry all gravity loads without support from the braces. In addition, the beam must be designed for the unbalanced force produced in the event of buckling of the compression brace and yielding of the tensile brace. In *X-bracing* and *zipper bracing systems*, this unbalanced force on the beam is eliminated. The K-bracing shown in Fig. 8.15 is not permitted for steel special concentrically braced frames.

A building frame system with steel special concentrically braced frames is assigned the same value of the response modification coefficient, R, as a building frame system with special reinforced concrete shear walls and is subject to the same height restrictions in seismic design categories D, E, and F. A building frame system with steel special concentrically braced frames is generally less expensive than a special moment-resisting frame system because it avoids the high cost of the special rigid joints in moment-resisting frames.

Steel ordinary concentrically braced frames are similar to steel special concentrically braced frames in configuration and may utilize all of the systems indicated in Fig. 8.14. However, the value of the response modification coefficient, R, is much lower and, consequently, structures are designed for a higher seismic force. Therefore, the system remains essentially elastic under a seismic event and does not require the special detailing necessary for the steel special concentrically braced frame system. K-bracing, shown in Fig. 8.15, is not allowed in steel ordinary concentrically braced building frames because of the unbalanced force produced in the column in the event of buckling of the compression brace.

Figure 8.15 *K-Bracing System*

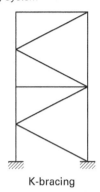

K-bracing

As specified in ASCE/SEI7 Table 12.2-1, no limitation is imposed on the height of steel ordinary concentrically braced frames in seismic design categories A, B, and C. The maximum height permitted in seismic design categories D and E is 35 ft, with exceptions permitted for single-story buildings up to a height of 60 ft when the dead load of the roof does not exceed 20 lbf/ft^2, and for penthouse structures.

The system is not permitted in seismic design category F, except for single-story storage warehouse buildings less than or equal to 45 ft high.

Buckling-restrained braced frames are a special type of concentrically braced frame. In this type of frame, special braces are used that consist of a steel core surrounded by a casing of steel and mortar to prevent the core from buckling. Therefore, compression yielding of the core can occur, as well as tensile yielding. This ensures significantly better ductility than a steel special concentrically braced frame.

The system has a higher value of the response modification coefficient, R, than a steel special concentrically braced frame and a correspondingly lower lateral design force. The buckling-restrained braced frame is subject to the same height restrictions as a steel special concentrically braced frame in seismic design categories D, E, and F.

In an *eccentrically braced frame*, as shown in Fig. 8.16, one end of the brace is connected to the beam so as to form a short link between the brace and the column, or between two opposing braces. The link acts as a structural fuse by providing the inelastic behavior necessary to absorb the input seismic energy while the other framing elements remain elastic. The eccentrically braced frame provides lateral stiffness equivalent to that of a concentrically braced frame, and ductility equivalent to that of a special moment-resisting frame.

Figure 8.16 *Steel Eccentrically Braced Frames*

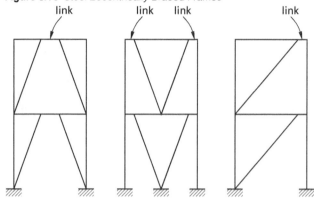

The larger spaces between braces give greater freedom in architectural design than there is with concentrically braced frames. The increased stiffness of the system, compared with special moment-resisting frames, limits damage to architectural features and nonstructural elements in the event of an earthquake. The eccentrically braced frame is subject to the same height restrictions as a steel special concentrically braced frame in seismic design categories D, E, and F.

The system has a higher value of the response modification coefficient, R, than a steel special concentrically braced frame, and a correspondingly lower lateral design force. However, the additional design and detailing requirements for the link beams is generally more expensive than a steel special concentrically braced frame configuration because of the high cost of the link beam.

Dual System with Moment-Resisting Frames

As shown in Fig. 8.17, a *dual system* has a secondary lateral support system coupled with the primary non-gravity-load-bearing lateral support system. Shear walls or braced frames provide the primary lateral support system with a special or intermediate moment-resisting frame providing primary support for gravity loads, and act as a backup for the lateral support system. The moment-resisting frame must provide resistance to at least 25% of the seismic forces.

The special moment-resisting frame system has a high value of the response modification coefficient, R, and no restrictions are placed on height. The intermediate moment-resisting frame system has a lower value of the response modification coefficient, R, and restrictions are placed on the height of systems in seismic design categories D, E, and F.

Shear Wall-Frame Interactive System

A shear wall-frame interactive system with ordinary reinforced concrete moment frames and ordinary reinforced concrete shear walls is a dual system and it is used for structures assigned to seismic design categories A and B. The shear walls and frames are designed to resist lateral forces in proportion to their rigidities considering the interaction between shear walls and frames on all levels. In accordance

Figure 8.17 *Dual System with Moment-Resisting Frames*

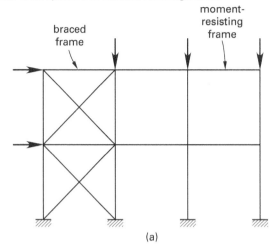

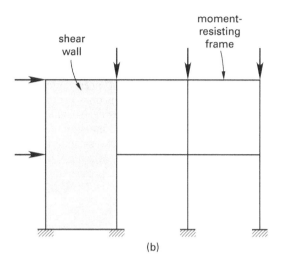

(a)   (b)

with ASCE/SEI7 Sec. 12.2.5.8, the shear strength of the shear wall must be at least 75% of the design story shear at each story. The frames of the shear wall-frame interactive system must be capable of resisting at least 25% of the design story shear at each story.

Structures utilizing this design system are located in zones of low seismicity, so it is not necessary to use the special seismic detailing required for a dual system with moment-resisting frames. For structures assigned to seismic design category A, ACI Sec. 18.2.1.6 specifies that both ordinary reinforced concrete moment frames and ordinary reinforced concrete shear walls need not comply with ACI Chap. 18. Ordinary reinforced concrete shear walls assigned to seismic design category B do not need to comply with Chap. 18 requirements. For ordinary reinforced concrete moment frames assigned to seismic design category B, compliance requirements are given in ACI Sec. 18.3. These requirements are that beams must have at least two longitudinal bars continuous along both the top and bottom faces, and that columns with a clear height less than or equal to five times the column width must be designed for shear in accordance with ACI Sec. 18.3.3.

Cantilever Column Systems

As shown in Fig. 8.18, a *cantilevered column system* consists of a structure supported on columns cantilevering from their base. This system lacks redundancy, as the inelastic behavior necessary to absorb the input seismic energy is concentrated at the base of the columns, and this produces a sidesway collapse mechanism. Additionally, the excessive flexibility of the system leads to excessive drift and consequent *P*-delta instability. The system has a low value for the response modification coefficient, *R*, and is restricted in height in all seismic design categories.

Figure 8.18 *Cantilever Column System*

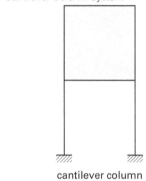

cantilever column

Steel Systems Not Specifically Detailed for Seismic Resistance

AISC 360[17] provides criteria for the design of structural steel buildings. It is specifically intended for low-seismic applications where design is based on a seismic response modification coefficient, *R*, less than 3. AISC 341[18] provides criteria for the design of structural steel buildings and is specifically intended for high-seismic applications where design is based on a seismic response modification coefficient greater than 3. In accordance with IBC Sec. 2205.2.2.1, steel building structures assigned to seismic design categories D, E, or F, must be designed and detailed as specified by AISC 341. In accordance with IBC Sec. 2205.2.2.2, AISC 341 may also be used to design and detail steel building structures assigned to seismic design categories B or C. When AISC 341 is used, the seismic loads are computed using the response modification coefficient given in ASCE/SEI7 Table 12.2-1.

In accordance with IBC Sec. 2205.2.1.1, steel building structures assigned to seismic design categories B or C, with the exception of cantilever column systems, may also be designed and detailed as specified by AISC 360. When AISC 360 is used, the seismic loads are computed using a response modification coefficient of 3, which ensures a nominally elastic response to the applied loads. This may often result in a structure that is less

Lateral Forces

expensive to build. For seismic design category A, special detailing is not required, and steel building structures may be designed and detailed as specified by AISC 360.

When wind effects exceed seismic effects, the building elements must still be detailed in accordance with AISC 341 provisions. These provisions provide the design requirements for structural steel seismic force-resisting systems to sustain the large inelastic deformations necessary to dissipate the seismic-induced demand. The AISC's *Seismic Design Manual*[19] provides guidance on the application of the provisions to the design of structural steel seismic force-resisting systems.

4. DIAPHRAGMS

Nomenclature

A_x	amplification factor	–
B	width of diaphragm	ft
F	force	kips
J	polar moment of inertia $= \Sigma r_i^2 R_i$	ft-kips
r_i	distance of wall i from the center of rigidity	ft
R	rigidity	kips/ft
R_i	rigidity (stiffness) of wall i	kips/ft
T	torsion	ft-kips
V	shear	kips

Symbols

δ_{avg}	average of displacements at extreme points of the structure at level x computed assuming $A_x = 1$	in
δ_{max}	maximum displacement at level x computed assuming $A_x = 1$	in
δ_{MDD}	deflection of diaphragm	in
Δ_{ADVE}	average story drift below the diaphragm	in

In accordance with ASCE/SEI7 Sec. 12.3.1, diaphragms are classified as flexible, semirigid, or rigid.

Flexible Diaphragms

A *flexible diaphragm*, as defined in ASCE/SEI7 Sec. 12.3.1.3, is a diaphragm whose lateral deflection under a lateral load is more than twice the average story drift of the adjoining vertical elements of the lateral force-resisting system. This is illustrated in Fig. 8.19. To qualify as a flexible diaphragm, the diaphragm must satisfy the following requirement.

$$\delta_{\text{MDD}} > 2\Delta_{\text{ADVE}} \quad \text{[ASCE/SEI7 12.3-1]}$$

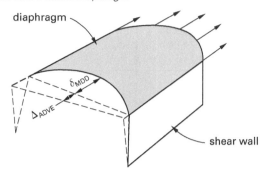

Figure 8.19 *Flexible Diaphragm*

When subjected to a transverse force, a flexible diaphragm undergoes lateral displacement without rotation, and a lateral force is distributed to the vertical seismic-load-resisting elements based on tributary areas.

In accordance with ASCE/SEI7 Sec. 12.3.1.1, the following types of diaphragms may be considered flexible.

- untopped steel decking or wood structural panels supported by vertical elements of steel or composite braced frames, or by concrete, masonry, steel, or composite shear walls

- untopped steel decking or wood structural panels in one- and two-family residential buildings

In addition, in structures of light-frame construction, diaphragms of untopped steel decking or wood structural panels are considered flexible, provided all of the following conditions are met.

- Toppings of concrete or similar materials are not placed over wood structural panel diaphragms, except for nonstructural toppings no greater than 1.5 in thick.

- Each line of the lateral force-resisting system complies with the allowable story drift of ASCE/SEI7 Table 12.12-1.

Semirigid Diaphragm

Diaphragms not satisfying the requirements of ASCE/SEI7 Sec. 12.3.1.1, Sec. 12.3.1.2, or Sec. 12.3.1.3 are considered semirigid. Analysis of a structure with semirigid diaphragms must include consideration of the actual stiffness of the diaphragm.

Rigid Diaphragm

Diaphragms of concrete slabs or of concrete filled metal decks with span-to-depth ratios of 3 or less in structures that have no horizontal irregularities are considered by ASCE/SEI7 Sec. 12.3.1.2 to be rigid.

How a rigid diaphragm distributes lateral force to the vertical seismic-load-resisting elements depends on two things: first, the relative rigidity of these elements, and second, the torsional displacements produced by the rigid-body rotation of the diaphragm. To calculate the torsional displacements, the center of rigidity and the center of mass for the structure must both be known. The *center of rigidity* is the point about which a structure rotates when subjected to a torsional moment. The *center of mass* is the point through which the lateral force, V, acts. For the structure shown in Fig. 8.20(a), the torsional moment acting on the diaphragm is defined as $T = Ve$, where e is the eccentricity of the center of mass with respect to the center of rigidity.

Figure 8.20 *Rigid Diaphragm*

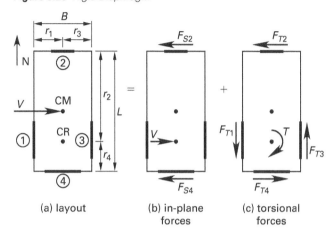

(a) layout   (b) in-plane forces   (c) torsional forces

The displacement of the building consists of an east-west translation and a clockwise rotation about the center of rigidity. As shown in Fig. 8.20(b), the translation produces in-plane forces in vertical elements 2 and 4 proportional to their relative translational stiffness. The in-plane shear force, F_{Si}, in wall i is

$$F_{Si} = \frac{VR}{\sum R_i}$$

No forces are produced in vertical elements 1 and 3 by this translation, and $\sum R_i = R_2 + R_4$. The clockwise rotation produces forces in all four walls, proportional to their torsional stiffnesses, as shown in Fig. 8.20(c). The torsional shear force in wall i is

$$F_{Ti} = \frac{Tr_iR_i}{J}$$

The total force in wall i is

$$F_i = F_{Si} + F_{Ti}$$

In a perfectly symmetric building, the centers of mass and rigidity coincide, and torsion is not produced. However, the centers of mass and rigidity may not be calculated accurately because of uncertainties in determining the mass and stiffness distribution in the building. In addition, torsional components of the ground motion may also cause torsion to develop. Therefore, accidental eccentricity may in fact exist even in a nominally

symmetric structure. Torsion resulting from this accidental eccentricity is referred to as *accidental torsion*. To account for accidental torsion, ASCE/SEI7 Sec. 12.8.4.2 specifies that the center of mass is assumed displaced each way from its actual location by a distance equal to 5% of the building dimension perpendicular to the direction of the applied force.

When a building that is assigned to a seismic design category C through F has a torsional irregularity as defined in ASCE/SEI7 Table 12.3-1 (horizontal structural irregularity type 1a or 1b), the accidental torsion is amplified as specified in ASCE/SEI7 Sec. 12.8.4.3. The *amplification factor* is given by ASCE/SEI7 Eq. 12.8-14 as

$$A_x = \left(\frac{\delta_{\max}}{1.2\delta_{\text{avg}}} \right)^2$$

Accidental torsion is applied to all structures to determine if a horizontal structural irregularity exists, as defined in ASCE/SEI7 Table 12.3-1. In accordance with ASCE/SEI7 Sec. 12.8.4.2, accidental torsion need not be considered when determining the seismic forces in the structure or in the determination of story drift, with the following exceptions.

- for a structure assigned to seismic design category C through F with type 1a or type 1b horizontal structural irregularity

- for a structure assigned to seismic design category B with type 1b horizontal structural irregularity

In accordance with ASCE/SEI7 Sec. 12.3.3.1, structures assigned to seismic design category E or F with torsional irregularity type 1b are not permitted.

Example 8.5

The single-story building shown in Fig. 8.20 has a rigid roof diaphragm that is acted on by an east-west force of 40 kips. Determine the force produced on shear wall 2 if the center of gravity is located at the center of the building. The building dimensions and the relative wall rigidities are as follows.

$$L = 80 \text{ ft}$$
$$B = 40 \text{ ft}$$
$$R_4 = 3R$$
$$R_1 = R_2 = R_3 = R$$

Neglect accidental eccentricity.

Solution

From the problem statement, the distance from wall 4 to the center of mass is

$$x = \frac{80 \text{ ft}}{2} = 40 \text{ ft}$$

Due to symmetry, the center of rigidity is located midway between wall 1 and wall 3.

$$r_1 = r_3$$
$$= 20 \text{ ft}$$

The distance from wall 4 to the center of rigidity is

$$r_4 = \frac{R_2 L}{R_2 + R_4}$$
$$= \frac{(1R)(80 \text{ ft})}{1R + 3R}$$
$$= 20 \text{ ft}$$

The distance of the center of rigidity from wall 2 is

$$r_2 = L - r_4 = 80 \text{ ft} - 20 \text{ ft}$$
$$= 60 \text{ ft}$$

The sum of the values of $r^2 R$ for the four walls is

$$J = \sum r_i^2 R_i$$
$$= r_1^2 R_1 + r_2^2 R_2 + r_3^2 R_3 + r_4^2 R_4$$
$$= (20 \text{ ft})^2(1R) + (60 \text{ ft})^2(1R) + (20 \text{ ft})^2(1R)$$
$$\quad + (20 \text{ ft})^2(3R)$$
$$= (5600 \text{ ft}^2)R$$

For a seismic load in the east-west direction, the eccentricity is

$$e = x - r_4 = 40 \text{ ft} - 20 \text{ ft}$$
$$= 20 \text{ ft}$$

The torsional moment acting about the center of rigidity is

$$T = Ve = (40 \text{ kips})(20 \text{ ft})$$
$$= 800 \text{ ft-kips}$$

The torsional shear force in wall 2 is

$$F_{T2} = \frac{T r_2 R_2}{J} = \frac{(800 \text{ ft-kips})(60 \text{ ft})(1R)}{(5600 \text{ ft}^2)R}$$
$$= 8.57 \text{ kips}$$

The in-plane shear force in wall 2 is

$$F_{S2} = \frac{V R_2}{R_2 + R_4}$$
$$= \frac{(40 \text{ kips})(1R)}{1R + 3R}$$
$$= 10 \text{ kips}$$

The total force in wall 2 is

$$F_2 = F_{S2} + F_{T2}$$
$$= 10 \text{ kips} + 8.57 \text{ kips}$$
$$= 18.57 \text{ kips}$$

Subdiaphragms[20]

Nomenclature

a_p	component amplification factor from ASCE/SEI7 Table 13.5-1 is 2.5 for an unbraced parapet	–
F_p	force on diaphragm, force on wall	lbf
h	height of roof above the base	ft
I_e	occupancy importance factor	–
I_p	component importance factor given in ASCE/SEI7 Sec. 13.1.3	–
k_a	amplification factor for diaphragm flexibility	–
L_f	span of a flexible diaphragm	ft
R_p	component response modification factor from ASCE/SEI7 Table 13.5-1 is 2.5 for an unbraced parapet	–
S_{DS}	design response acceleration at a period of 0.2 second	–
W_p	weight of the wall tributary to the anchor	lbf
z	height of point of attachment of parapet above the base $= h$	ft

In seismic design categories C through F, ASCE/SEI7 Sec. 12.11.2.2.1 requires that continuous ties be provided across the complete depth of the diaphragm. This is to transfer the diaphragm anchorage forces across the depth of the diaphragm and to prevent the walls and diaphragm from separating. To reduce the number of full depth ties required, *subdiaphragms* and added chords are used to span between the full depth ties. The maximum permitted length-to-width ratio of the subdiaphragm is 2.5 to 1.

A typical arrangement of subdiaphragms and crossties is shown in Fig. 8.21. IBC Sec. 1604.8.2 requires all structural walls to be anchored to the diaphragm to prevent separation of the walls from the diaphragm. Wall anchors are provided at the ends of the subdiaphragm ties, and must have a capacity to resist the horizontal force. For walls of structures assigned to seismic design category A, this force is given in ASCE/SEI7 Sec. 1.4.5 as

$$F_p = 0.2 W_p$$
$$\geq 5 \text{ lbf/ft}^2$$

The subdiaphragm ties transfer the anchor force to the subdiaphragm chords which, in turn, transfer the anchor force to the continuous crossties. In accordance with ASCE/SEI7 Sec. 12.11.2.2.3 and Sec. 12.11.2.2.4, the continuous ties must be in addition to plywood sheathing or metal deck that is considered ineffective in providing the ties.

Figure 8.21 *Subdiaphragms and Crossties*

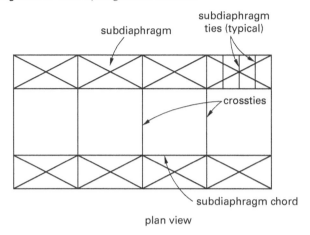

plan view

Lateral Design Force on Walls and Parapets

The out-of-plane seismic force on a wall is specified in ASCE/SEI7 Sec. 12.11.1 as

$$F_p = 0.4S_{DS}I_eW_p$$
$$\geq 0.1W_p$$

ASCE/SEI7 Sec. 13.3.1 requires parapets in seismic design category B through F to be designed as architectural components using

$$F_p = \frac{0.4a_pS_{DS}W_p}{\dfrac{R_p}{I_p}}\left(1 + 2\frac{z}{h}\right) \quad \text{[ASCE/SEI7 13.3-1]}$$
$$\leq 1.6S_{DS}I_pW_p \quad \text{[ASCE/SEI7 13.3-2]}$$
$$\geq 0.3S_{DS}I_pW_p \quad \text{[ASCE/SEI7 13.3-3]}$$

The poor seismic performance and the lack of redundancy of parapets can create a safety hazard to the public. Therefore, parapets are designed for a higher design load than walls, and a high value of 2.5 is assigned to the component amplification factor, a_p. The lateral force is considered uniformly distributed over the height of the parapet.

Anchorage of Structural Walls to Flexible Diaphragms

During past earthquakes, a major cause of failure has been the separation of flexible diaphragms from concrete and masonry supporting walls. This separation is due to the diaphragm flexibility amplifying out-of-plane accelerations. To prevent separation from occurring, supporting walls must be securely anchored to the

subdiaphragm ties. Where the wall anchor spacing exceeds 4 ft, in accordance with ASCE/SEI7 Sec. 12.11.2.1, the wall must be designed to span between anchors. In accordance with ASCE/SEI7 Sec. 12.11.2.2, steel elements in the anchorage system are required to resist 1.4 times the calculated anchorage force.

For buildings assigned to seismic design categories B through F, ASCE/SEI7 Sec. 12.11.2.1 requires anchors to be designed for the force as

$$F_p = 0.4S_{DS}k_aI_eW_p \quad \text{[ASCE/SEI7 12.11-1]}$$
$$\geq 0.2k_aI_eW_p$$
$$k_a = 1.0 + \frac{L_f}{100} \quad \text{[ASCE/SEI7 12.11-2]}$$

k_a is the amplification factor for diaphragm flexibility. L_f is the span, in feet, of a flexible diaphragm measured between vertical elements. These vertical elements provide lateral support to the diaphragm in the direction considered. For rigid diaphragms, $L_f = 0$.

Anchorage of Structural Walls to Rigid Diaphragms

For buildings with rigid diaphragms assigned to seismic design categories B through F, ASCE/SEI7 Sec. 12.11.2.1 requires anchors that are not at roof level to be designed for the force.

$$F_p = 0.4S_{DS}I_eW_p\left(\frac{1 + \dfrac{2z}{h}}{3}\right)$$
$$\geq 0.2I_eW_p$$

Anchorage force for rigid diaphragms at roof level is determined from ASCE/SEI7 Eq. 12.11-1 with $k_a = 1.0$, which gives

$$F_p = 0.4S_{DS}I_eW_p$$
$$\geq 0.2I_eW_p$$

Example 8.6

For the north-south direction, design a suitable subdiaphragm arrangement for the roof diaphragm shown in the illustration. The blocked plywood diaphragm is flexible, and the concrete walls have a weight, w, of 75 lbf/ft². The building is assigned to seismic design category D with an importance factor of $I_e = 1.0$ and a design response acceleration, S_{DS}, of $1.0g$.

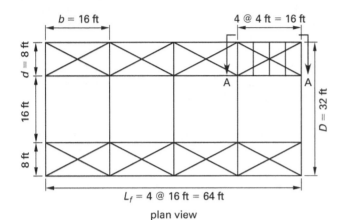

plan view

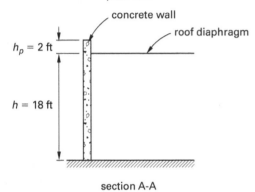

section A-A

Solution

For the layout shown in the illustration, the aspect ratio of the whole diaphragm is

$$\frac{B}{D} = \frac{64 \text{ ft}}{32 \text{ ft}}$$
$$= 2$$
$$< 4 \quad \text{[conforms to SDPWS Sec. 4.2.4]}$$

The aspect ratio of the subdiaphragm is

$$\frac{b}{d} = \frac{16 \text{ ft}}{8 \text{ ft}}$$
$$= 2$$
$$< 2.5 \quad \text{[conforms to ASCE/SEI7 Sec. 12.11.2.2.1]}$$

Provide wall anchors at each subdiaphragm tie. Spacing, s, is 4 ft on center. The equivalent area of wall tributary to each anchor is obtained by taking moments about the base of the wall.

$$A = \frac{s(h + h_p)^2}{2h}$$
$$= \frac{(4 \text{ ft})(20 \text{ ft})^2}{(2)(18 \text{ ft})}$$
$$= 44.44 \text{ ft}^2$$

The weight of wall tributary to each anchor is

$$W_p = wA$$
$$= \left(75 \frac{\text{lbf}}{\text{ft}^2}\right)(44.44 \text{ ft}^2)$$
$$= 3333 \text{ lbf}$$

The span of the flexible diaphragm is

$$L_f = 64 \text{ ft}$$

The amplification factor for diaphragm flexibility is

$$k_a = 1.0 + \frac{L_f}{100}$$
$$= 1.0 + \frac{64 \text{ ft}}{100 \text{ ft}}$$
$$= 1.64$$

For seismic design category D, the seismic lateral force on an anchor is given by ASCE/SEI7 Eq. 12.11-1 as

$$F_p = 0.4S_{DS}k_aI_eW_p$$
$$= (0.4)(1.0)(1.64)(1.0)(3333 \text{ lbf})$$
$$= 2186 \text{ lbf} \quad \text{[governs]}$$

The minimum permissible force on an anchor is

$$F_p = 0.2k_aI_eW_p$$
$$= (0.2)(1.64)(1.0)(3333 \text{ lbf})$$
$$= 1093 \text{ lbf}$$
$$< 2186 \text{ lbf}$$

The required seismic design force for the anchors is

$$F_p = 2186 \text{ lbf}$$

Therefore, the pull-out force, p, along the wall is

$$p = \frac{F_p}{s} = \frac{2186 \text{ lbf}}{4 \text{ ft}}$$
$$= 547 \text{ lbf/ft}$$

The unit shear stress at the subdiaphragm boundary is

$$v = \frac{pb}{2d}$$
$$= \frac{\left(547 \dfrac{\text{lbf}}{\text{ft}}\right)(16 \text{ ft})}{(2)(8 \text{ ft})}$$
$$= 547 \text{ lbf/ft}$$

The subdiaphragm chord force is

$$P_c = \frac{pb^2}{8d}$$
$$= \frac{\left(547 \dfrac{\text{lbf}}{\text{ft}}\right)(16 \text{ ft})^2}{(8)(8 \text{ ft})}$$
$$= 2188 \text{ lbf}$$

The force in the crossties is

$$P_t = pb$$
$$= \left(547 \dfrac{\text{lbf}}{\text{ft}}\right)(16 \text{ ft})$$
$$= 8752 \text{ lbf}$$

PART 2: SEISMIC DESIGN

Nomenclature

C_d	deflection amplification factor from ASCE/SEI7 Table 12.2-1	–
C_s	seismic response coefficient specified in ASCE/SEI7 Sec. 12.8.1	–
C_u	coefficient for upper limit on calculated period from ASCE/SEI7 Table 12.8-1	–
D	dead load applied to a structural element	lbf or kips
E	calculated seismic load on an element of a structure resulting from both horizontal and vertical earthquake induced forces as given by ASCE/SEI7 Eq. 12.4-1 and Eq. 12.4-2	lbf or kips
f_i	design seismic lateral force at level i	lbf or kips
F_a	short-period site coefficient	–
F_p	force on diaphragm	lbf or kips
F_v	long-period site coefficient	–
F_x	design seismic lateral force at level x as specified in ASCE/SEI7 Sec. 12.8.3	lbf or kips
g	gravitational acceleration, 32.2	ft/sec²
h_i	height above the base to level i	ft
h_n	height of the roof above the base, not including the height of penthouses or parapets	ft
h_{sx}	story height below level x	ft
h_x	height above the base to level x	ft
I_e	seismic importance factor	–
k	distribution exponent given in ASCE/SEI7 Sec. 12.8.3	–
M_P	primary moment	ft-kips
M_S	secondary moment	ft-kips
N	number of stories	–
P_x	total unfactored vertical design load at and above level x	lbf or kips
Q_E	effect of horizontal seismic forces	lbf or kips
R	response modification coefficient for a specific structural system from ASCE/SEI7 Table 12.2-1	–
S_1	maximum considered response acceleration for a period of 1.0 sec	–
S_a	design spectral response acceleration	–
S_{DS}	design spectral response acceleration at a period of 0.2 sec	–
S_{D1}	design spectral response acceleration at a period of 1.0 sec	–
S_{MS}	modified spectral response acceleration at a period of 0.2 sec	–
S_{M1}	modified spectral response acceleration at a period of 1.0 sec	–
S_S	maximum considered response acceleration for a period of 0.2 sec	–
T	fundamental period of vibration, defined in ASCE/SEI7 Sec. 12.8.2	sec
T_a	approximate fundamental period of vibration determined using ASCE/SEI7 Sec. 12.8.2.1	sec
T_L	long-period transition period	sec
T_0	defined in ASCE/SEI7 Sec. 11.4.5 as $0.2S_{D1}/S_{DS}$	–
T_S	defined in ASCE/SEI7 Sec. 11.4.5 as S_{D1}/S_{DS}	–
V	total seismic base shear	lbf or kips
V_x	total shear force at level x	lbf or kips
V_S	design base shear	–
V_Y	base shear at formation of the collapse mechanism	–
w_i	seismic dead load located at level i	lbf or kips
w_x	seismic dead load located at level x	lbf or kips
W	wind load applied to a structural element	lbf or kips
W	effective seismic weight defined in ASCE/SEI7 Sec. 12.7.2	lbf or kips
$\sum F_i$	total shear force at level i	lbf or kips
$\sum w_i$	total seismic dead load at level i and above	lbf or kips

Symbols

β	ratio of shear demand to shear capacity for the story between levels x and x-1 as defined in ASCE/SEI7 Sec. 12.8.7	–
δ_x	amplified horizontal deflection at level x, defined in ASCE/SEI7 Sec. 12.8.6	in
δ_{xe}	horizontal deflection at level x, determined by an elastic analysis, as defined in ASCE/SEI7 Sec. 12.8.6	in
Δ	design story drift, occurring simultaneously with the story shear V_x, defined in ASCE/SEI7 Sec. 12.8.6, and calculated using the amplification factor C_d	in
Δ_a	allowable story drift, defined in ASCE/SEI7 Table 12.12-1	in
Ω_0	overstrength factor tabulated in ASCE/SEI7 Table 12.2-1	–
θ	stability coefficient defined in ASCE/SEI7 Sec. 12.8.7	–

5. EQUIVALENT LATERAL FORCE PROCEDURE

Determination of the seismic response of a structure depends on several factors, including ground motion parameters, site classification, site coefficient, adjusted response acceleration, design spectral response acceleration, importance factor, seismic design category, classification of the structural system, response modification coefficient, deflection amplification factor, overstrength factor, effective seismic weight, fundamental period of vibration, and seismic response coefficient. A summary of these factors follows.

Ground Motion Parameters

Ground motion parameters defined in ASCE/SEI7 Sec. 11.4.2 are values of the maximum considered ground acceleration that may be experienced at a specific location. As defined in ASCE/SEI7 Sec. 11.2, these are the most severe earthquake effects considered by the code. The parameters are risk-adjusted to provide a uniform risk with a 1% probability of collapse in 50 years. Two values of the ground acceleration are required, and these are designated S_S and S_1. S_S represents the 5% damped, maximum considered earthquake spectral response acceleration for a period of 0.2 sec and is applicable to short-period structures. S_1 represents the 5% damped, maximum considered earthquake spectral response acceleration for a period of 1 sec and is applicable to tall structures with longer periods. Values of the ground accelerations S_S and S_1 are mapped in ASCE/SEI7 Fig. 22-1 through Fig. 22-8. The parameters are given as a percentage of the acceleration due to gravity.

Site Classification Characteristics

Site classification is defined in ASCE/SEI7 Sec. 11.4.3 and ASCE/SEI7 Table 20.3-1. Six different soil types are specified and range from site class A, which consists of hard rock, through site class F, which consists of peat, highly plastic clay, or collapsible soil. The soil profile may be determined on site from the average shear wave velocity in the top 100 ft of material. Alternatively, for site classification types C, D, or E, the classification may be made by measuring the standard penetration resistance or undrained shear strength of the material. An abbreviated listing of the site classifications is provided in Table 8.2.

Table 8.2 Site Classification Definitions

site classification	soil profile name	shear wave velocity (ft/sec)
A	hard rock	> 5000
B	rock	2500 to 5000
C	soft rock	1200 to 2500
D	stiff soil	600 to 1200
E	soft clay soil	< 600
F	–	–

Soil classification type A has the effect of reducing the ground response by 20%. Soil classification type E is defined as soft soil and has the effect of increasing the long period ground response by up to 350%. When soil parameters are unknown, in accordance with ASCE/SEI7 Sec. 11.4.3, soil classification type D may be assumed unless the building official determines that soil classification types E or F are likely to be present at the site.

Site Coefficients

Site coefficients are amplification factors applied to the maximum considered ground acceleration and are a function of the site classification. F_a is the short-period or acceleration-based amplification factor and is tabulated in ASCE/SEI7 Table 11.4-1. F_v is the long-period or velocity-based amplification factor and is tabulated in ASCE/SEI7 Table 11.4-2. ASCE/SEI7 Table 11.4-1 and Table 11.4-2 are combined and reproduced in Table 8.3. Linear interpolation may be used to obtain intermediate values.

For situations in which site investigations reveal competent rock conditions with moderate fracturing and weathering consistent with site class B, but site-specific velocity measurements are not made, the site coefficients F_a and F_v are taken as unity (1.0).

Where site class D is selected as the default site class in accordance with ASCE/SEI7 Sec. 11.4.3, the value of F_a must not be less than 1.2.

Lateral Forces

Table 8.3 Site Coefficients F_a corresponding to S_S, and F_v corresponding to S_1

site classification	response acceleration, S_S						response acceleration, S_1					
	≤ 0.25	0.50	0.75	1.00	1.25	≥ 1.50	≤ 0.1	0.2	0.3	0.4	0.5	≥ 0.6
A	0.8	0.8	0.8	0.8	0.8	0.8	0.8	0.8	0.8	0.8	0.8	0.8
B	0.9	0.9	0.9	0.9	0.9	0.9	0.8	0.8	0.8	0.8	0.8	0.8
C	1.3	1.3	1.2	1.2	1.2	1.2	1.5	1.5	1.5	1.5	1.5	1.4
D	1.6	1.4	1.2	1.1	1.0	1.0	2.4	2.2[c]	2.0[c]	1.9[c]	1.8[c]	1.7[c]
E	2.4	1.7	1.3	(a)	(a)	(a)	4.2	(b)	(b)	(b)	(b)	(b)
F	(d)	(d)	(d)	(d)	(d)	(d)	(d)	(d)	(d)	(d)	(d)	(d)

Notes: [a]A ground motion hazard analysis is required unless the site coefficient F_a is taken as equal to that of site class C ($F_a = 1.2$).

[b]A ground motion hazard analysis is required unless T is less than or equal to T_S and the ELF procedure is used for design.

[c]A ground motion hazard analysis is required unless the value of the seismic response coefficient C_s is conservatively calculated using ASCE/SEI7 Eq. 12.8-2 for $T \leq 1.5T_S$ and taken as equal to 1.5 times the value computed in accordance with either ASCE/SEI7 Eq. 12.8-3 for $T_L \geq T > 1.5T_S$ or ASCE/SEI7 Eq. 12.8-4 for $T > T_L$.

[d]A site response analysis is required unless any of the exceptions to ASCE/SEI7 Sec. 20.3.1 are applicable.

Adapted with permission from *Minimum Design Loads and Associated Criteria for Buildings and Other Structures*, copyright © 2016, by the American Society of Civil Engineers.

Adjusted Response Accelerations

The maximum considered ground accelerations must be adjusted by the site coefficients to allow for the site clas-sification effects. ASCE/SEI7 Sec. 11.4.4 defines the modified spectral response accelerations at short peri-ods and at a period of 1 sec as

$$S_{MS} = F_a S_S \quad \text{[ASCE/SEI7 11.4-1]}$$
$$S_{M1} = F_v S_1 \quad \text{[ASCE/SEI7 11.4-2]}$$

Design Spectral Response Acceleration Parameters

The relevant design parameters are defined in ASCE/SEI7 Sec. 11.4.5 and are given by

S_{DS} = 5% damped design spectral response acceleration for a period of 0.2 sec

$$= \frac{2S_{MS}}{3} \quad \text{[ASCE/SEI7 11.4-3]}$$

S_{D1} = 5% damped design spectral response acceleration for a period of 1 sec

$$= \frac{2S_{M1}}{3} \quad \text{[ASCE/SEI7 11.4-4]}$$

Example 8.7

The two-story, reinforced concrete, moment-resisting frame shown is located on a site with a soil profile of stiff soil having a shear wave velocity of 600 ft/sec. The 5% damped, maximum considered earthquake spectral response accelerations are obtained from the ASCE/

SEI7 standard. They are $S_S = 1.5g$ and $S_1 = 0.7g$. Determine the 5% damped design spectral response accelerations S_{DS} and S_{D1}.

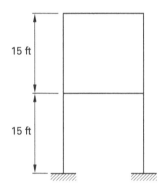

Solution

From ASCE/SEI7 Table 20.3-1 or from Table 8.2, the applicable site classification for stiff soil with a shear wave velocity of 600 ft/sec is site classification D. The site coefficients for this site classification and for the given values of the 5% damped, maximum considered earthquake spectral response accelerations are obtained from ASCE/SEI7 Table 11.4-1 and Table 11.4-2 or from Table 8.3 as

$$F_a = 1.0$$
$$F_v = 1.7$$

The adjusted spectral response accelerations are given by ASCE/SEI7 Sec. 11.4.4 as

$$S_{MS} = F_a S_S$$
$$= (1.0)(1.5g)$$
$$= 1.5g$$
$$S_{M1} = F_v S_1$$
$$= (1.7)(0.7g)$$
$$= 1.19g$$

The 5% damped design spectral response accelerations are given by ASCE/SEI7 Sec. 11.4.5 as

$$S_{DS} = \frac{2S_{MS}}{3}$$
$$= \frac{(2)(1.5g)}{3}$$
$$= 1.0g$$
$$S_{D1} = \frac{2S_{M1}}{3}$$
$$= \frac{(2)(1.19g)}{3}$$
$$= 0.79g$$

Risk Category and Importance Factors

In accordance with ASCE/SEI7 Table 1.5-1, each structure is assigned to a risk category, depending on the nature of its *occupancy*, with the corresponding *importance factor* indicated in ASCE/SEI7 Table 1.5-2. Table 8.4 lists the risk categories, occupancy type, and seismic importance factors.

Category IV structures are those housing essential facilities that are required for post-earthquake recovery. Also included in category IV are structures containing substantial quantities of highly toxic substances that would endanger the safety of the public if released. Essential facilities are defined in IBC Table 1604.5 as hospitals, fire and police stations, emergency response centers, and buildings housing utilities and equipment required for these facilities. In order to ensure that category IV facilities remain functional after an upper level earthquake, an importance factor, I_e, of 1.5 is assigned to these facilities. This has the effect of increasing the design seismic forces by 50%, and raises the seismic level at which inelastic behavior occurs and the level at which the operation of essential facilities is compromised.

Table 8.4 *Occupancies and Importance Factors*

risk category	occupancy type	importance factor, I_e
I	low hazard structures	1.00
II	standard occupancy structures	1.00
III	assembly structures	1.25
IV	essential or hazardous structures	1.50

Adapted with permission from *Minimum Design Loads and Associated Criteria for Buildings and Other Structures*, copyright © 2016, by the American Society of Civil Engineers.

For situations in which site investigations reveal competent rock conditions with moderate fracturing and weathering consistent with site class B, but site-specific velocity measurements are not made, the site coefficients F_a and F_v are taken as unity (1.0).

Where site class D is selected as the default site class in accordance ASCE/SEI7 Sec. 11.4.4, the value of F_a must not be less than 1.2.

Category III structures are facilities that, if they failed, would become a substantial public hazard because of their high occupant load. These facilities are buildings where more than 300 people congregate in one area, schools with a capacity exceeding 250, colleges with a capacity exceeding 500, health care facilities with a capacity of 50 or more that do not have emergency treatment facilities, jails, and power stations. Also included are facilities containing explosive or toxic substances in a quantity exceeding the exempt amounts in IBC Table 307.1(1). These structures are allocated a seismic importance factor, I_e, of 1.25.

Category II structures comprise standard occupancy structures and are allocated an importance factor, I_e, of 1.00. Standard occupancy structures consist of residential, commercial, and office buildings.

Category I structures comprise low-hazard structures and are allocated an importance factor, I_e, of 1.00. Low-hazard structures consist of agricultural facilities, temporary facilities, and minor storage facilities.

Determination of Seismic Design Category

Structures are assigned to a *seismic design category* A through F based on their risk category and the design spectral response coefficients S_{DS} and S_{D1}. The seismic design category is defined in ASCE/SEI7 Sec. 11.6 and ASCE/SEI7 Table 11.6-1 and Table 11.6-2, and establishes the design and detailing requirements necessary in a structure. The seismic design category is determined twice, first as a function of S_{DS} using ASCE/SEI7 Table 11.6-1, and then as a function of S_{D1} using ASCE/SEI7 Table 11.6-2. The most severe seismic design category governs. ASCE/SEI7 Table 11.6-1 and ASCE/SEI7 Table 11.6-2 are combined and reproduced in Table 8.5.

Table 8.5 Seismic Design Categories

		risk category	
S_{DS}	S_{D1}	I, II, or III	IV
$S_{DS} < 0.167g$	$S_{D1} < 0.067g$	A	A
$0.167g \leq S_{DS} < 0.33g$	$0.067g \leq S_{D1} < 0.133g$	B	C
$0.33g \leq S_{DS} < 0.50g$	$0.133g \leq S_{D1} < 0.20g$	C	D
$0.50g \leq S_{DS}$	$0.20g \leq S_{D1}$	D	D
MCE$_R$ acceleration at 1.0 sec period, $S_1 \geq 0.75g$		E	F

Adapted with permission from *Minimum Design Loads and Associated Criteria for Buildings and Other Structures*, copyright © 2016, by the American Society of Civil Engineers.

Six seismic design categories are defined, categories A through F, and these establish the design and detailing requirements necessary in a structure. Seismic design category A is applicable to structures in locations where anticipated ground movements are minimal. ASCE/ SEI7 Sec. 11.7 specifies requirements to ensure the integrity of the structure in the event of a minor earthquake. Seismic design category B is applicable to structures in risk categories I, II, and III in regions of moderate seismicity. Seismic design category C is applicable to category IV structures in regions of moderate seismicity as well as structures in risk categories I, II, and III in regions of somewhat more severe seismicity. The use of some structural systems is restricted in this design category. Plain concrete and masonry structures are not permitted. Seismic design category D includes structures in risk categories I, II, III, and IV in regions of high seismicity, but not located close to a major active fault, as well as risk category IV structures in regions with less severe seismicity. In this design category some types of structural systems must be designed by dynamic analysis methods. Seismic design category E includes structures in risk categories I, II, and III located close to a major active fault. Seismic design category F includes risk category IV structures located close to a major active fault. In this design category restrictions are imposed on the use of structural systems and analysis methods.

Example 8.8

The two-story, reinforced concrete, moment-resisting frame analyzed in Ex. 8.7 is used as a residential building. Determine the applicable risk category, importance factor, and seismic design category.

Solution

The 5% damped design spectral response accelerations are obtained from Ex. 8.7 as

$$S_{DS} = 1.0g$$
$$S_{D1} = 0.79g$$

A residential building is classified as a standard occupancy structure. The applicable risk category is obtained from ASCE/SEI7 Table 1.5-1 or Table 8.4.

The risk category is II. For risk category II, the seismic importance factor is obtained from ASCE/SEI7 Table 1.5-2 or Table 8.4 as

$$I_e = 1.00$$

The design spectral response acceleration at short periods is

$$S_{DS} = 1.0g$$
$$> 0.50g$$

For a risk category of II, the seismic design category for this acceleration is obtained from ASCE/SEI7 Table 11.6-1 or Table 8.5. The seismic design category is D.

The design spectral response acceleration at a period of 1 sec is

$$S_{D1} = 0.79g$$
$$> 0.20g$$

For a risk category of II, the seismic design category for this acceleration is obtained from ASCE/SEI7 Table 11.6-2 or Table 8.5. The category is D. Therefore, the seismic design category for this building is D.

Classification of the Structural System

ASCE/SEI7 Sec. 12.2.1 and ASCE/SEI7 Table 12.2-1 detail eight major categories of building types characterized by the method used to resist the lateral force. These categories consist of bearing walls, building frames, moment-resisting frames, dual systems with a special moment-resisting frame, dual systems with a reinforced concrete intermediate moment frame or a steel ordinary moment frame, shear wall-frame interactive, inverted pendulum structures, and steel systems not detailed for seismic resistance.

A *bearing wall system* consists of shear walls that provide support for the gravity loads and resist all lateral loads. A *building frame system* consists of shear walls or braced frames that resist all lateral loads, and a separate framework that provides support for gravity loads. *Moment-resisting frames* provide support for both lateral and gravity loads by flexural action. In a *dual system*, nonbearing walls or braced frames supply the primary resistance to lateral loads, with a moment frame providing primary support for gravity loads plus additional resistance to lateral loads. A *shear wall-frame interactive system* is a dual system using ordinary reinforced concrete moment frames and ordinary reinforced concrete shear walls. A *cantilevered column structure* consists of a building supported on column elements to produce an inverted pendulum structure. A *steel system* not detailed for seismic resistance is designed using AISC 360.

Lateral Forces

Response Modification Coefficient

The structure *response modification coefficient, R,* is a measure of the ability of a specific structural system to resist lateral loads without collapse. ASCE/SEI7 Table 12.2-1 lists the different structural framing systems, with the height limitations, response modification coefficients, and deflection amplification factors for each. An abbreviated listing of structural systems, response modification coefficients, overstrength factors, and deflection amplification factors is provided in Table 8.6.

Deflection Amplification Factor

The *deflection amplification factor* is tabulated in ASCE/SEI7 Table 12.2-1 and in Table 8.6 and is given by

$$C_d = \frac{\delta_x}{\delta_{xe}}$$

After allowing for the risk importance factor, ASCE/SEI7 Eq. 12.8-15 gives the value of the actual displacement as

$$\delta_x = \frac{C_d \delta_{xe}}{I_e}$$

Overstrength Factor

The *overstrength factor* is a measure of the actual strength of a structure compared to the design seismic force. Values of the overstrength factor for various building systems are tabulated in ASCE/SEI7 Table 12.2-1 and shown in Table 8.6. The overstrength factor is given by

$$\Omega_0 = \frac{V_Y}{V_S}$$

The system overstrength is produced by the following factors: conservative design methods, system redundancy, material overstrength, oversized members, application of load factors, and drift limitations controlling design.

Effective Seismic Weight

The *effective seismic weight, W,* as specified in ASCE/SEI7 Sec. 12.7.2, is the total dead load of the structure and the part of the service load that may be expected to be attached to the building. The effective seismic weight consists of the following.

- 25% of the reduced floor live load for storage and warehouse occupancies, except for public parking structures and storage loads adding not more than 5% to the seismic weight

- a minimum allowance of 10 lbf/ft² for moveable partitions or the actual weight if larger

- flat roof snow loads exceeding 30 lbf/ft², which are reduced by 80%

- the total weight of permanent equipment and fittings

- weight of landscaping and other materials at roof gardens

Roof and floor live loads, except as noted above, are not included in the value of W.

Fundamental Period of Vibration

ASCE/SEI7 Sec. 12.8.2 and Sec. 12.8.2.1 provide three methods for determining the *fundamental period* of a structure. These three methods are the two approximate methods given in ASCE/SEI7 Sec. 12.8.2.1 and the properly substantiated analysis method, such as the Rayleigh method, given in ASCE/SEI7 Sec. 12.8.2.

From ASCE/SEI7 Eq. 12.8-7, the approximate fundamental period is given by

$$T_a = 0.028 h_n^{0.8} \text{ for steel moment-resisting} \\ \text{frames}$$

$$T_a = 0.016 h_n^{0.9} \text{ for reinforced concrete} \\ \text{moment-resisting frames systems}$$

$$T_a = 0.030 h_n^{0.75} \text{ for eccentrically braced} \\ \text{steel frames}$$

$$T_a = 0.020 h_n^{0.75} \text{ for all other structural} \\ \text{systems}$$

Alternatively, for moment-resisting frames not exceeding 12 stories in height and with a story height not less than 10 ft, the approximate fundamental period may be determined by ASCE/SEI7 Eq. 12.8-8 as

$$T_a = 0.1N$$

Example 8.9

A two-story, reinforced concrete, moment-resisting frame is shown in the illustration. Calculate the natural period of vibration T_a.

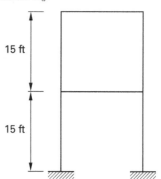

Table 8.6 *Seismic Parameters and Building Height*

structural system	R	Ω_0	C_d	system and height limitations seismic design category				
				B	C	D	E	F
bearing wall								
light-framed walls sheathed with wood shear panels	6.5	3.0	4.0	NL	NL	65	65	65
special reinforced concrete shear walls	5.0	2.5	5.0	NL	NL	160	160	100
special reinforced masonry shear walls	5.0	2.5	3.5	NL	NL	160	160	100
building frame								
steel eccentrically braced frame	8.0	2.0	4.0	NL	NL	160	160	100
steel special concentrically braced frames	6.0	2.0	5.0	NL	NL	160	160	100
steel ordinary concentrically braced frames	3.25	2.0	3.25	NL	NL	35	35	NP
light frame walls sheathed with wood shear panels	7.0	2.5	4.5	NL	NL	65	65	65
buckling-restrained braced frame	8.0	2.5	5.0	NL	NL	160	160	100
steel special plate shear walls	7.0	2.0	6.0	NL	NL	160	160	100
special reinforced concrete shear walls	6.0	2.5	5.0	NL	NL	160	160	100
special reinforced masonry shear walls	5.5	2.5	4.0	NL	NL	160	160	100
moment-resisting frame								
steel or concrete special moment frames	8.0	3.0	5.5	NL	NL	NL	NL	NL
steel special truss moment frames	7.0	3.0	5.5	NL	NL	160	100	NP
steel intermediate moment frames	4.5	3.0	4.0	NL	NL	35	NP	NP
steel ordinary moment frames	3.5	3.0	3.0	NL	NL	NP	NP	NP
intermediate moment frames of reinforced concrete	5.0	3.0	4.5	NL	NL	NP	NP	NP
ordinary moment frames of reinforced concrete	3.0	3.0	2.5	NL	NP	NP	NP	NP
dual system with special moment-resisting frames								
steel eccentrically braced frames	8.0	2.5	4.0	NL	NL	NL	NL	NL
steel special concentrically braced frames	7.0	2.5	5.5	NL	NL	NL	NL	NL
buckling-restrained braced frames	8.0	2.5	5.0	NL	NL	NL	NL	NL
steel special plate shear walls	8.0	2.5	6.5	NL	NL	NL	NL	NL
special reinforced concrete shear walls	7.0	2.5	5.5	NL	NL	NL	NL	NL
special reinforced masonry shear walls	5.5	3.0	5.0	NL	NL	NL	NL	NL
dual system with intermediate moment frames								
steel special concentrically braced frames	6.0	2.5	5.0	NL	NL	35	NP	NP
steel and concrete composite special concentrically braced frames	5.5	2.5	4.5	NL	NL	160	100	NP
special reinforced concrete shear walls	6.5	2.5	5.0	NL	NL	160	100	100
intermediate reinforced masonry shear walls	3.5	3.0	3.0	NL	NL	NP	NP	NP
shear wall-frame interactive system with ordinary reinforced concrete moment frames and ordinary reinforced shear walls	4.5	2.5	4	NL	NP	NP	NP	NP
cantilevered column								
steel special cantilever column systems	2.5	1.25	2.5	35	35	35	35	35
steel ordinary cantilever column systems	1.25	1.25	1.25	35	35	NP	NP	NP
special reinforced concrete moment frames	2.5	1.25	2.5	35	35	35	35	35
steel systems not specifically detailed for seismic resistance, excluding cantilever column systems	3.0	3.0	3.0	NL	NL	NP	NP	NP

Note: NL = not limited and NP = not permitted.

Adapted with permission from *Minimum Design Loads and Associated Criteria for Buildings and Other Structures*, Table 12.2-1, copyright © 2016, by the American Society of Civil Engineers.

Solution

The number of stories is

$$N = 2$$
$$< 12$$

Then, for a moment-resisting frame, ASCE/SEI7 Eq. 12.8-8 specifies a value for a building period of

$$T_a = 0.1N$$
$$= \left(0.1 \frac{\text{sec}}{\text{story}}\right)(2 \text{ stories})$$
$$= 0.20 \text{ sec}$$

Example 8.10

For the two-story, reinforced concrete, moment-resisting frame analyzed in Ex. 8.9, calculate the fundamental period of vibration, T_a, by using ASCE/SEI7 Eq. 12.8-7.

Solution

For a reinforced concrete frame, the fundamental period is given by ASCE/SEI7 Eq. 12.8-7 as

$$T_a = (0.016)(30 \text{ ft})^{0.9}$$
$$= 0.342 \text{ sec}$$

Rayleigh Procedure

ASCE/SEI7 Sec. 12.8.2 permits the fundamental period to be determined by a "properly substantiated analysis." The *Rayleigh procedure* is an acceptable method, and the fundamental period is given by

$$T = 2\pi \sqrt{\frac{\sum w_i \delta_i^2}{g \sum f_i \delta_i}}$$
$$= 0.32 \sqrt{\frac{\sum w_i \delta_i^2}{\sum f_i \delta_i}}$$

The terms in this expression are illustrated in Fig. 8.22, where δ_i represents the elastic displacements due to a lateral force distribution f_i increasing approximately uniformly with height.

Figure 8.22 *Application of the Rayleigh Procedure*

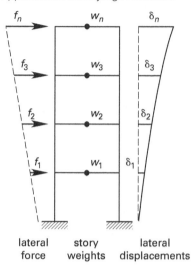

lateral force   story weights   lateral displacements

To allow for a possible underestimation of the stiffness of the structure, ASCE/SEI7 Sec. 12.8.2 specifies that the value of the natural period determined by this method may not exceed the value of

$$T = C_u T_a$$

When the calculated value, T, is less than $C_u T_a$, the lower value of T must be used.

Values of C_u are given in ASCE/SEI7 Table 12.8-1 and are shown in Table 8.7.

Table 8.7 *Coefficient for Upper Limit on the Calculated Period*

S_{D1}	≥ 0.40	0.30	0.20	0.15	≤ 0.10
C_u	1.4	1.4	1.5	1.6	1.7

Example 8.11

Using ASCE/SEI7 Sec. 12.8.2, determine the fundamental period of vibration of the two-story frame of Ex. 8.9, which is located in an area with a value for S_{D1} exceeding 0.4. The force system shown in the illustration may be used; the effective seismic weight at each level and the total stiffness of each story are indicated.

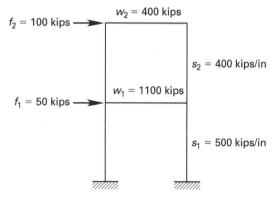

Solution

For the force system indicated, the displacements at each level are given by

$$\delta_1 = \frac{f_2 + f_1}{s_1} = \frac{100 \text{ kips} + 50 \text{ kips}}{500 \; \dfrac{\text{kips}}{\text{in}}}$$

$$= 0.30 \text{ in}$$

$$\delta_2 = \frac{f_2}{s_2} + \delta_1 = \frac{100 \text{ kips}}{400 \; \dfrac{\text{kips}}{\text{in}}} + 0.30 \text{ in}$$

$$= 0.55 \text{ in}$$

The natural period is given by Rayleigh's procedure as

$$T = 0.32 \sqrt{\frac{\sum w_i \delta_i^2}{\sum f_i \delta_i}}$$

The relevant values are given in the following table.

Rayleigh's Procedure

level	w_i (kips)	f_i (kips)	δ_i (in)	$w_i \delta_i^2$ (kips-in^2)	$f_i \delta_i$ (in-kips)
2	400	100	0.55	121	55
1	1100	50	0.30	99	15
total	1500	–	–	220	70

$$T = \left(0.32 \; \frac{\text{sec}}{\sqrt{\text{in}}} \right) \sqrt{\frac{220 \text{ kips-in}^2}{70 \text{ in-kips}}}$$

$$= 0.567 \text{ sec}$$

In an area with a value for $S_{D1} > 0.4$, the value of the coefficient for the upper limit on the calculated period is obtained from ASCE/SEI7 Table 12.8-1 or Table 8.7 as

$$C_u = 1.4$$

The fundamental period, in accordance with ASCE/SEI7 Sec. 12.8.2, is limited to

$$T = 1.4 T_a$$
$$= (1.4)(0.2 \text{ sec})$$
$$= 0.28 \text{ sec}$$
$$< 0.567 \text{ sec}$$

Use the value of

$$T = 0.28 \text{ sec}$$

Alternatively, the value obtained for T_a in Ex. 8.10 may be used to give

$$T = (1.4)(0.342 \text{ sec})$$
$$= 0.479 \text{ sec}$$
$$< 0.567 \text{ sec}$$

General Procedure Response Spectrum

The *general procedure response spectrum* is defined in ASCE/SEI7 Sec. 11.4.5 and shown in ASCE/SEI7 Fig. 11.4-1. The response spectrum is reproduced in Fig. 8.23.

Figure 8.23 *Construction of ASCE/SEI7 Response Spectra*

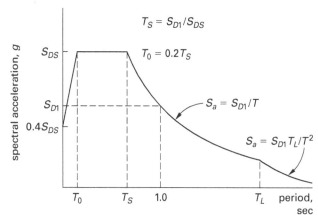

The response spectrum is constructed as shown using the following functions.

$$T_S = \frac{S_{D1}}{S_{DS}}$$

$$T_0 = \frac{0.2 S_{D1}}{S_{DS}}$$

For periods less than or equal to T_0, the design spectral response acceleration is given by ASCE/SEI7 Eq. 11.4-5 as

$$S_a = \frac{0.6 S_{DS} T}{T_0} + 0.4 S_{DS}$$

At $T = 0$,

$$S_a = 0.4 S_{DS}$$

For periods greater than or equal to T_0 and less than or equal to T_S, the design response acceleration is given by

$$S_a = S_{DS}$$

For periods greater than T_S and less than or equal to T_L, the design response acceleration is given by ASCE/SEI7 Eq. 11.4-6 as

$$S_a = \frac{S_{D1}}{T}$$

For periods greater than T_L, the design spectral response acceleration is given by ASCE/SEI7 Eq. 11.4-7 as

$$S_a = \frac{S_{D1}T_L}{T^2}$$

Values of T_L range from 4 sec to 6 sec, and are mapped in ASCE/SEI7 Fig. 22-14 through Fig. 22-17.

Seismic Response Coefficient

The *seismic response coefficient*, C_s, given in ASCE/SEI7 Sec. 12.8.1.1 represents the code design spectrum and is given by ASCE/SEI7 Eq. 12.8-3 for values of T not greater than T_L as

$$C_s = \frac{S_{D1}I_e}{RT}$$

For values of T greater than T_L, the seismic response coefficient is given by ASCE/SEI7 Eq. 12.8-4 as

$$C_s = \frac{S_{D1}T_LI_e}{RT^2}$$

The maximum value of the seismic response coefficient is given by ASCE/SEI7 Eq. 12.8-2 as

$$C_s = \frac{S_{DS}I_e}{R}$$

This latter expression controls for shorter periods up to approximately 1 sec. For longer periods, the expression provides conservative values.

In accordance with ASCE/SEI7 Eq. 12.8-5, the value of the seismic response coefficient must not be taken less than

$$C_s = 0.044S_{DS}I_e \geq 0.01$$

For those structures for which the 1 sec spectral response value is $S_1 \geq 0.6g$, the minimum value of the seismic response coefficient is given by ASCE/SEI7 Eq. 12.8-6 as

$$C_s - \frac{0.5S_1I_e}{R}$$

In accordance with ASCE/SEI7 Sec. 12.8.1.3, the value of C_s may be calculated using a value of $S_{DS} = 1.0$, but not less than $0.7(2S_{MS}/3)$, provided that all of the following criteria are met.

- The structure is regular with a maximum of five stories. Each mezzanine level is considered a story for the purposes of this limit.

- The fundamental period $T \leq 0.5$ sec as determined by ASCE/SEI7 Sec. 12.8.2.

- The redundancy factor $\rho = 1.0$ as determined by ASCE/SEI7 Sec. 12.3.4.2.

- The structure is assigned to risk category I or II.

- The structure is not located on a site defined as site class E or F.

Example 8.12

The two-story, reinforced concrete, special moment-resisting frame of Ex. 8.11 is used for a residential building. Calculate the seismic response coefficient by using the alternative value for the fundamental period determined in Ex. 8.11.

Solution

From previous examples, the relevant parameters are

$$S_{DS} = 1.0g$$
$$S_{D1} = 0.79g$$
$$I_e = 1.00$$
$$T = 0.479 \text{ sec}$$

From Fig. 8.23,

$$T_S = \frac{S_{D1}}{S_{DS}} = \frac{0.79 \text{ sec}}{1.0} = 0.79 \text{ sec}$$

The value of the response modification coefficient for a special moment-resisting frame is obtained from Table 8.6 as

$$R = 8.0$$

The seismic response coefficient is given in ASCE/SEI7 Sec. 12.8.1.1 as

$$\begin{aligned} C_s &= \frac{S_{D1}I_e}{RT} \\ &= \frac{(0.79)(1.00)}{(8.0)(0.479 \text{ sec})} \\ &= 0.206 \end{aligned}$$

The maximum value of the seismic response coefficient is

$$C_s = \frac{S_{DS}I_e}{R}$$
$$= \frac{(1.0)(1.00)}{8.0}$$
$$= 0.125 \quad [\text{governs}]$$

This follows since $T < T_S$ and the response of the moment-resisting frame lies in the flat topped segment of the response curve.

Seismic Base Shear

ASCE/SEI7 Eq. 12.8-1 specifies the *seismic base shear* as

$$V = C_s W$$

Example 8.13

Calculate the seismic base shear for the two-story, reinforced concrete, moment-resisting frame of Ex. 8.12.

Solution

The value of the effective seismic weight was derived in Ex. 8.11 as

$$W = 1500 \text{ kips}$$

The value of the seismic response coefficient was derived in Ex. 8.12 as

$$C_s = 0.125$$

The base shear is given by ASCE/SEI7 Eq. 12.8-1 as

$$V = C_s W$$
$$= (0.125)(1500 \text{ kips})$$
$$= 188 \text{ kips}$$

Building Configuration Requirements

The static lateral force procedure is applicable to structures that satisfy prescribed conditions of regularity, occupancy, location, and height. A regular structure has mass, stiffness, and strength uniformly distributed over the height of the structure and is without irregular features that will produce stress concentrations. Vertical irregularities are defined in ASCE/SEI7 Table 12.3-2 and horizontal irregularities in ASCE/SEI7 Table 12.3-1. As defined in ASCE/SEI7 Table 12.6-1, the equivalent lateral force method may be used in the design of a structure when a structure is assigned to seismic design category B or C. Additionally, the equivalent lateral force method may be used when a structure is assigned to seismic design category D, E, or F and conforms to one of the following conditions.

- light-frame construction
- a risk category of I or II and does not exceed two stories in height
- height not exceeding 160 ft and is a regular building
- height not exceeding 160 ft and has neither horizontal irregularities 1a (torsional) or 1b (extreme torsional), nor vertical irregularities 1a (soft story), 1b (extreme soft story), 2 (mass), or 3 (geometric)
- height exceeding 160 ft, a fundamental period $T < 3.5 T_S$, and has no structural irregularities

All other structures not described above that are assigned to seismic design categories D, E, and F need a modal analysis.

Redundancy Factor

To improve the seismic performance of structures in seismic design categories D, E, and F, redundancy is incorporated in the structures by providing multiple load-resisting paths. The *redundancy factor*, ρ, is a factor that penalizes structures with relatively few lateral load-resisting elements and is specified in ASCE/SEI7 Sec. 12.3.4. The redundancy factor is further defined in this text in Sec. 8.12.

6. VERTICAL DISTRIBUTION OF SEISMIC FORCES

The distribution of base shear over the height of a building is obtained from ASCE/SEI7 Sec. 12.8.3, and the design lateral force at level x is given by

$$F_x = \frac{V w_x h_x^k}{\sum w_i h_i^k}$$

The terms in this expression are illustrated in Fig. 8.24, where h_i represents the height above the base to any level i, h_x represents the height above the base to a specific level x, and $\sum w_i h_i^k$ represents the summation over the whole structure of the product of w_i and h_i^k. To allow for higher mode effects in long period buildings, when T has a value of 2.5 sec or more, the distribution exponent k is given by

$$k = 2$$

Figure 8.24 *Vertical Force Distribution*

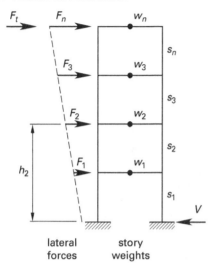

lateral
forces

story
weights

When T has a value not exceeding 0.5 sec, the distribution exponent is

$$k = 1$$

For intermediate values of T, a linear variation of k may be assumed.

Example 8.14

Determine the vertical force distribution for the two-story reinforced concrete, moment-resisting frame of Ex. 8.12.

Solution

The fundamental period was derived in Ex. 8.11 as

$$T = 0.479 \text{ sec}$$
$$< 0.5 \text{ sec}$$

The value of the distribution exponent factor is obtained from ASCE/SEI7 Sec. 12.8.3 as

$$k = 1.0$$

In accordance with ASCE/SEI7 Sec. 12.8.3, the expression for F_x reduces to

$$F_x = \frac{V w_x h_x}{\sum w_i h_i}$$

The effective seismic weights located at levels 1 and 2 are obtained from Ex. 8.11, and the relevant values are given in the following table.

Vertical Force Distribution

level	w_x (kips)	h_x (ft)	$w_x h_x$ (ft-kips)	F_x (kips)
2	400	30	12,000	79
1	1100	15	16,500	109
total	1500	–	28,500	188

From Ex. 8.13, the base shear is given by

$$V = 188 \text{ kips}$$

The design lateral force at level x is

$$F_x = \frac{V w_x h_x}{\sum w_i h_i}$$
$$= \frac{(188 \text{ kips}) w_x h_x}{28,500 \text{ ft-kips}}$$
$$= 0.00660 w_x h_x$$

The values of F_x are given in the previous table.

7. DIAPHRAGM LOADS

With the exception of precast concrete diaphragms, the load acting on a horizontal diaphragm is given by ASCE/SEI7 Sec. 12.10.1.1 as

$$F_{px} = \frac{w_{px} \sum F_i}{\sum w_i} \qquad \text{[ASCE/SEI7 12.10-1]}$$
$$\geq 0.2 S_{DS} I w_{px}$$
$$\leq 0.4 S_{DS} I w_{px}$$

The terms in the expression are illustrated in Fig. 8.25, where $\sum F_i$ represents the total shear force at level i, $\sum w_i$ represents the total seismic weight at level i and above, and w_{px} represents the seismic weight tributary to the diaphragm at level x, not including walls parallel to the direction of the seismic load.

Figure 8.25 *Diaphragm Loads*

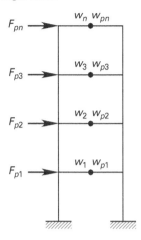

For a single-story structure, the expression reduces to

$$F_{px} = \frac{V w_{px}}{W} = C_s w_{px}$$

For a multistory structure, at the second-floor level,

$$\frac{\sum F_i}{\sum w_i} = \frac{V}{W} = C_s$$

Example 8.15

Determine the diaphragm loads for the regular two-story reinforced concrete, moment-resisting frame of Ex. 8.14. The effective seismic weight tributary to the diaphragm at roof level is 300 kips and at the second-floor level is 600 kips.

Solution

From Ex. 8.7, the design response coefficient is

$$S_{DS} = 1.0g$$

The diaphragm load is given by ASCE/SEI7 Eq. 12.10-1 as

$$F_{px} = \frac{w_{px} \sum F_i}{\sum w_i}$$

level	w_i (kips)	$\sum w_i$ (kips)	F_i (kips)	$\sum F_i$ (kips)	$\dfrac{\sum F_i}{\sum w_i}$
2	400	400	79	79	0.198
1	1100	1500	109	188	0.125

level	max	min	w_{px} (kips)	F_{px} (kips)
2	0.40	0.20	300	60
1	0.40	0.20	600	120

The maximum value for the diaphragm load is given by ASCE/SEI7 Sec. 12.10.1.1 as

$$\begin{aligned} F_{p(\text{max})} &= 0.4 S_{DS} I_e w_{px} \\ &= (0.4)(1.0)(1.00) w_{px} \\ &= 0.40 w_{px} \end{aligned}$$

The minimum value for the diaphragm load is given by ASCE/SEI7 Sec. 12.10.1.1 as

$$\begin{aligned} F_{p(\text{min})} &= 0.2 S_{DS} I_e w_{px} \\ &= (0.2)(1.0)(1.00) w_{px} \\ &= 0.20 w_{px} \quad \text{[governs at both levels]} \end{aligned}$$

The relevant values are given in the previous table.

8. COLLECTOR DESIGN

A *collector*, also known as a *drag strut*, is defined in IBC Sec. 202 as a horizontal diaphragm element, parallel and in line with the applied force, that collects and transfers diaphragm shear forces to the vertical elements of the lateral force-resisting system. Collectors are required where shear walls or braced frames terminate along the boundary of a diaphragm.

Collectors provide deformation compatibility at the termination of shear walls and braced frames.

With the exception of structures with precast concrete diaphragms, collector elements and their connections in structures assigned to seismic design categories C, D, E, or F are designed in accordance with ASCE/SEI7 Sec. 12.10.2. The collector design forces are the maximum of the three following conditions.

- Forces resulting from application at each level of the design lateral force, F_x, calculated from ASCE/SEI7 Eq. 12.8-11 and Eq. 12.8-12, which give

$$F_x = \frac{V w_x h_x^k}{\sum w_i h_i^k}$$

The value of F_x is determined using the overstrength requirements of ASCE/SEI7 Sec. 12.4.3.

Lateral Forces

- Forces resulting from the application at each level of the diaphragm design force, F_{px}, calculated from ASCE/SEI7 Eq. 12.10-1 as

$$F_{px} = \frac{w_{px} \sum F_i}{\sum w_i}$$

The value of F_{px} is determined using the overstrength requirements of ASCE/SEI7 Sec. 12.4.3.

- Forces resulting from the application at each level of the load combinations of ASCE/SEI7 Sec. 2.3.6, with the minimum diaphragm design force, F_{px}, given by ASCE/SEI7 Eq. 12.10-2 as

$$F_{px} = 0.2 S_{DS} I_e w_{px}$$

The following exception is permitted by ASCE/SEI7 Sec. 12.10.2.1.

- For structures braced entirely by wood light-frame shear walls, collector elements and their connections need only be designed for forces resulting from the application at each level of the diaphragm design force, F_{px}, calculated from ASCE/SEI7 Eq. 12.10-1 as

$$F_{px} = \frac{w_{px} \sum F_i}{\sum w_i}$$

The value of F_{px} is determined using load combinations of ASCE/SEI7 Sec. 2.3.6.

The following increase in forces is required by ASCE/SEI7 Sec. 12.3.3.4.

- For structures assigned to seismic design category D, E, or F and having a horizontal structural irregularity of type 1, 2, 3, or 4 or a vertical structural irregularity of type 4, the design force determined from ASCE/SEI7 Sec. 12.10.1.1 is increased by 25% for collectors and their connections. Where the design force is calculated using the seismic load effects, including the overstrength factor of ASCE/SEI7 Sec. 2.3.6, the 25% increase is not applied.

Example 8.16

A single-story building has a flexible roof diaphragm and wood structural panel shear walls, and has a north-south seismic force acting on the building, as shown in *Illustration for Ex. 8.16*. Determine the force in collector 34.

Solution

The structure is braced entirely by wood structural panel shear walls, and the exception permitted by ASCE/SEI7 Sec. 12.10.2.1 is applicable. ASCE/SEI7 Eq. 12.10-1 applies and for a single-story structure. This equation reduces to the expression

$$F_{\text{roof}} = C_s w_{\text{roof}}$$

As shown in the illustration part (a), shear wall 45 terminates at the diaphragm boundary, and drag strut 34 is required to anchor the diaphragm to the shear wall. The diaphragm is flexible, and shear wall 45 effectively subdivides the diaphragm into two simply supported segments 12543 and 3476. These may be treated as two independent simply supported beams, 31 and 36, as shown in the illustration part (b). The equivalent beam reactions, which represent the shear force at the boundaries of each diaphragm, are

$$V_{31} = \frac{\left(1 \, \frac{\text{kip}}{\text{ft}} \right)(40 \, \text{ft})}{2}$$
$$= 20 \, \text{kips}$$

$$V_{36} = \frac{\left(0.5 \, \frac{\text{kip}}{\text{ft}} \right)(40 \, \text{ft})}{2}$$
$$= 10 \, \text{kips}$$

The unit shears at the boundaries of each diaphragm are

$$v_{31} = \frac{V_{31}}{L_{35}} = \frac{20 \, \text{kips}}{80 \, \text{ft}}$$
$$= 0.25 \, \text{kip/ft}$$

$$v_{36} = \frac{V_{36}}{L_{34}} = \frac{10 \, \text{kips}}{40 \, \text{ft}}$$
$$= 0.25 \, \text{kip/ft}$$

The unit shears acting on the diaphragms are shown in the illustration part (c).

The shear wall 45 resists a shear force of

$$V_W = V_{31} + V_{36}$$
$$= 20 \, \text{kips} + 10 \, \text{kips}$$
$$= 30 \, \text{kips}$$

Illustration for Ex. 8.16

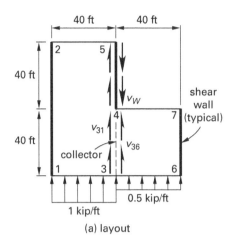

(a) layout

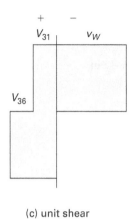

(c) unit shear

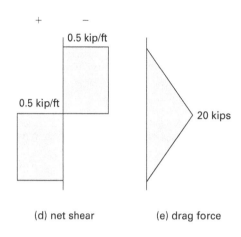

(d) net shear   (e) drag force

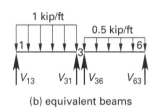

(b) equivalent beams

The unit shear in the shear wall is

$$v_W = \frac{V_W}{L_{45}}$$
$$= \frac{30 \text{ kips}}{40 \text{ ft}}$$
$$= 0.75 \text{ kip/ft}$$

The unit shear acting on the shear wall is shown in the illustration part (c).

The net shear along the diaphragm boundary at 54 is

$$v_{54} = v_{31} - v_W$$
$$= 0.25 \frac{\text{kip}}{\text{ft}} - 0.75 \frac{\text{kip}}{\text{ft}}$$
$$= -0.50 \text{ kip/ft}$$

The net shear along the diaphragm interface at 34 is

$$v_{34} = v_{31} + v_{36}$$
$$= 0.25 \frac{\text{kip}}{\text{ft}} + 0.25 \frac{\text{kip}}{\text{ft}}$$
$$= 0.50 \text{ kip/ft}$$

The net shears are plotted as shown in the illustration part (d).

The maximum drag force occurs at the end of the collector, at the connection to the shear wall. It is given by

$$F = v_{34}L_{34}$$
$$= \left(0.50 \frac{\text{kip}}{\text{ft}}\right)(40 \text{ ft})$$
$$= 20 \text{ kips}$$

The drag force diagram is shown in the illustration part (e).

9. STORY DRIFT

Story drift, Δ, is defined in ASCE/SEI7 Sec. 12.8.6 as the difference of the inelastic deflections at the centers of mass at the top and bottom of the story under consideration. Where centers of mass do not align vertically, it is permitted to compute the deflection at the bottom of the story based on the vertical projection of the center of mass at the top of the story. Where allowable stress design is used, the story drift is computed using strength level seismic forces without reduction for allowable stress design.

The maximum allowable story drift, Δ_a, is given in ASCE/SEI7 Table 12.12-1 and is shown in Table 8.8.

Lateral Forces

Table 8.8 *Maximum Allowable Story Drift, Δ_a*

building type	risk category		
	I or II	III	IV
one-story buildings with fittings designed to accomodate drift	no limit	no limit	no limit
buildings other than masonry buildings of four stories or less with fittings designed to accommodate drift	$0.025h_{sx}$	$0.020h_{sx}$	$0.015h_{sx}$
masonry cantilever shear wall buildings	$0.010h_{sx}$	$0.010h_{sx}$	$0.010h_{sx}$
other masonry shear wall buildings	$0.007h_{sx}$	$0.007h_{sx}$	$0.007h_{sx}$
all other buildings	$0.020h_{sx}$	$0.015h_{sx}$	$0.010h_{sx}$

To allow for inelastic deformations, drift is determined by using the deflection amplification factor C_d defined in Table 8.6. The amplified deflection at level x is given by ASCE/SEI7 Eq. 12.8-15 as

$$\delta_x = \frac{C_d \delta_{xe}}{I_e}$$

The term δ_{xe} represents the horizontal deflection at level x, determined by an elastic analysis using the code-prescribed design level forces. In accordance with ASCE/SEI7 Sec. 12.8.7, P-delta effects need not be included in the calculation of drift when the stability coefficient θ does not exceed 0.10.

Using the nomenclature from Fig. 8.24, the elastic lateral displacement in the bottom story is

$$\delta_{1e} = \frac{F_n + F_3 + F_2 + F_1}{s_1}$$

The elastic lateral displacement in the second story is

$$\delta_{2e} = \frac{F_n + F_3 + F_2}{s_2} + \delta_{1e}$$

The elastic lateral displacement in the third story is

$$\delta_{3e} = \frac{F_n + F_3}{s_3} + \delta_{2e}$$

The elastic lateral displacement in the top story is

$$\delta_{ne} = \frac{F_n}{s_n} + \delta_{3e}$$

The drift in the bottom story is

$$\Delta_1 = \frac{C_d \delta_{1e}}{I_e}$$

The drift in the second story is

$$\Delta_2 = \frac{C_d(\delta_{2e} - \delta_{1e})}{I_e}$$

The drift in the third story is

$$\Delta_3 = \frac{C_d(\delta_{3e} - \delta_{2e})}{I_e}$$

The drift in the top story is

$$\Delta_n = \frac{C_d(\delta_{ne} - \delta_{3e})}{I_e}$$

For the calculation of drift, in accordance with ASCE/SEI7 Sec. 12.8.6.2, the full value of T, the fundamental period determined by using the Rayleigh procedure, may be used to determine the seismic base shear.

The redundancy factor, ρ, is not used.

Example 8.17

Determine the drift in the bottom story of the two-story, reinforced concrete, special moment-resisting frame of Ex. 8.14, which is used as a residential building. The relevant details are shown. Fittings are designed to accommodate drift.

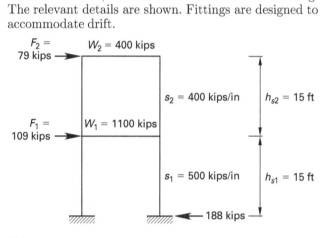

Solution

From Ex. 8.11, the fundamental period obtained by using the Rayleigh procedure is

$$T = 0.567 \text{ sec}$$

This value does not need to be reduced.

The seismic response coefficient is given in ASCE/SEI7 Sec. 12.8.1.1 as

$$C_s = \frac{S_{D1}I_e}{RT} = \frac{(0.79)(1.00)}{(8.0)(0.567 \text{ sec})} = 0.174$$

The maximum value of the seismic design coefficient is

$$C_s = \frac{S_{DS}I_e}{R} = \frac{(1.0)(1.00)}{8.0} = 0.125 \quad [\text{governs}]$$

The seismic base shear is given by ASCE/SEI7 Sec. 12.8.1 and is identical with the value calculated in Ex. 8.13 as

$$V = 188 \text{ kips}$$

In addition, the lateral forces are identical with the values calculated in Ex. 8.14.

For a moment-resisting frame, the amplification factor is obtained from ASCE/SEI7 Table 12.2-1 or Table 8.6 as

$$C_d = 5.5$$

From the lateral forces determined in Ex. 8.14, the drift in the bottom story, allowing for $I_e = 1.0$, is

$$\begin{aligned} \Delta_1 &= C_d \delta_{xe} \\ &= \frac{C_d(F_2 + F_1)}{s_1} \\ &= \frac{(5.5)(79 \text{ kips} + 109 \text{ kips})}{500 \dfrac{\text{kips}}{\text{in}}} \\ &= 2.07 \text{ in} \end{aligned}$$

In accordance with ASCE/SEI7 Table 12.12-1, the maximum allowable drift for a two-story structure in risk category II is

$$\begin{aligned} \Delta_a &= 0.025h_{s1} = (0.025)(15 \text{ ft})\left(12 \frac{\text{in}}{\text{ft}}\right) \\ &= 4.50 \text{ in} \\ &> 2.07 \text{ in} \end{aligned}$$

The drift is acceptable.

10. *P*-DELTA EFFECTS

The *P-delta effects* are calculated by using the design level seismic forces and elastic displacements determined in accordance with ASCE/SEI7 Sec. 12.8.1. *P*-delta effects in a given story are a result of the secondary moments caused by the eccentricity of the gravity loads above that story. The *secondary moment in a story* is defined as the product of the total dead load, floor live load, and snow load above the story multiplied by the elastic drift of that story. The *primary moment in a story* is defined as the seismic shear in the story multiplied by the height of the story.

The ratio of the secondary moment to primary moment is termed the *stability coefficient* and is given by ASCE/SEI7 Sec. 12.8.7 as

$$\begin{aligned} \theta &= \frac{P_x \Delta I_e}{V_x h_{sx} C_d} \\ &= \frac{P_x(\delta_{xe} - \delta_{(x-1)e})}{V_x h_{sx}} \end{aligned}$$

The stability coefficient must not exceed the value

$$\begin{aligned} \theta_{\max} &= \frac{0.5}{\beta C_d} \\ &\leq 0.25 \end{aligned}$$

If $\theta > \theta_{\max}$, the structure is unstable and must be redesigned.

The term β is the ratio of the shear demand to the shear capacity in a story and may conservatively be considered equal to 1.0. If the stability coefficient in any story exceeds 0.1, the effects of the secondary moments must be included in the analysis of the whole structure. The revised story drift, allowing for *P*-delta effects, is obtained as the product of the calculated drift and the factor $1/(1 - \theta)$.

As shown in Fig. 8.26, with the designated lateral forces and story drift and with the combined dead load plus floor live load indicated by W_1 and the combined dead load plus roof snow load indicated by W_2, the primary moment in the second story of the frame is

$$M_{P2} = F_2 h_{s2}$$

Figure 8.26 *P-delta Effects*

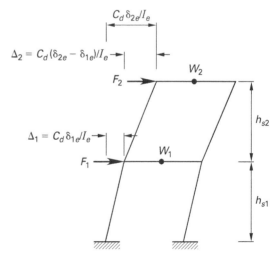

The secondary moment in the second story is

$$M_{S2} = \frac{P_2 \Delta_2 I_e}{C_d}$$
$$= W_2(\delta_{2e} - \delta_{1e})$$

The stability coefficient in the second story is

$$\theta_2 = \frac{M_{S2}}{M_{P2}}$$

The primary moment in the first story of the frame is

$$M_{P1} = (F_1 + F_2)h_{s1}$$

The secondary moment in the first story is

$$M_{S1} = \frac{P_1 \Delta_1 I_e}{C_d}$$
$$= (W_2 + W_1)\delta_{1e}$$

The stability coefficient in the first story is

$$\theta_1 = \frac{M_{S1}}{M_{P1}}$$

Example 8.18

Determine the stability coefficient for the bottom story of the two-story reinforced concrete, moment-resisting frame of Ex. 8.17.

Solution

The drift in the bottom story is derived in Ex. 8.17 as

$$\Delta_1 = 2.07 \text{ in}$$

The primary moment in the bottom story is

$$M_{P1} = (F_2 + F_1)h_{s1}$$
$$= (79 \text{ kips} + 109 \text{ kips})(15 \text{ ft})\left(12 \ \frac{\text{in}}{\text{ft}}\right)$$
$$= 33{,}840 \text{ in-kips}$$

The secondary moment is

$$M_{S1} = \frac{(W_2 + W_1)\Delta_1 I_e}{C_d}$$
$$= \frac{(1500 \text{ kips})(2.07 \text{ in})(1.0)}{5.5}$$
$$= 565 \text{ in-kips}$$

The stability coefficient is

$$\theta_1 = \frac{M_{S1}}{M_{P1}}$$
$$= \frac{565 \text{ in-kips}}{33{,}840 \text{ in-kips}}$$
$$= 0.017$$
$$< 0.1 \quad \text{[Secondary moments need not be considered.]}$$

11. SIMPLIFIED LATERAL FORCE PROCEDURE

For some low-rise structures, ASCE/SEI7 Sec. 12.14 permits an alternative, conservative design method. The simplified method is applicable to a structure in which the following 12 limitations are met.

1. The structure does not exceed three stories in height.

2. The structure is assigned to risk category I or II.

3. The structure is located at a site with a soil profile of site class A through D.

4. The structure's lateral force-resisting system is either a bearing wall system or a building frame system.

5. The structure must have at least two lines of lateral resistance in each of two major axis directions. (See Fig. 8.27.)

6. At least one line of resistance must be provided on each side of the center of mass (CM) in each direction. (See Fig. 8.27.)

Figure 8.27 *Lines of Lateral Resistance*

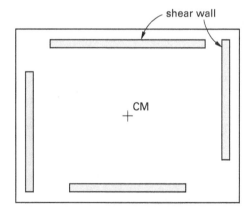

7. A structure with flexible diaphragms, overhangs beyond the outside line of shear walls, or braced frames must satisfy ASCE/SEI7 Eq. 12.14-2. (See Fig. 8.28.) The distance, a, perpendicular to the forces being considered from the extreme edge of the diaphragm to the line of vertical resistance closest to that edge must not exceed one-fifth the depth, d, of the diaphragm parallel

to the forces being considered at the line of vertical resistance closest to the edge.

$$a \le \frac{d}{5} \quad \text{[ASCE/SEI7 12.14-2]}$$

For a cast-in-place concrete diaphragm, $a \le d/3$

Figure 8.28 *Flexible Diaphragm Overhang*

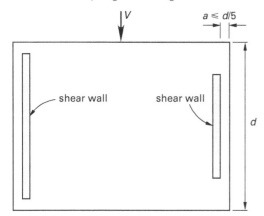

8. The distance between the center of rigidity and the center of mass in each story must not exceed 10% of the length of the diaphragm parallel to the eccentricity. See Fig. 8.29.

Figure 8.29 *Torsion Check for Nonflexible Diaphragms*

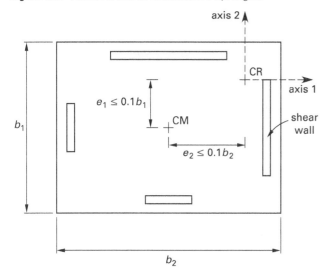

9. For buildings with a nonflexible diaphragm, forces in the vertical elements are determined as if the diaphragm is flexible and, in addition,

> (a) for buildings with two lines of resistance in a given direction, the distance between the two lines is at least 50% of the diaphragm length perpendicular to the two lines

> (b) for buildings with more than two lines of resistance in a given direction, the distance between the two most extreme lines of resistance in that direction is at least 60 percent of

the diaphragm length perpendicular to the lines

> (c) for buildings with two or more lines of resistance closer together than one-half the length of the longer of the walls, as shown in Fig. 8.30, the walls may be replaced by a single wall at the centroid of the group for the initial distribution of forces, and the resultant force to the group is then distributed to the members of the group based on their relative stiffnesses

Figure 8.30 *Closely Spaced Walls*

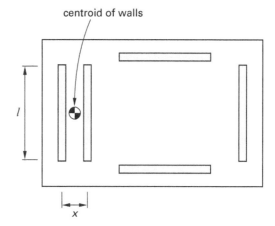

10. Lines of resistance of the structure's lateral force-resisting system must be oriented at angles of no more than 15° from alignment with the major orthogonal horizontal axes of the building.

11. The simplified design procedure must be used for each major orthogonal horizontal axis direction of the structure.

12. System irregularities caused by in plane or out of plane offsets of lateral force-resisting elements are not permitted, except in two-story structures of light frame construction provided that the upper wall is designed for a factor of safety of 2.5 against overturning.

13. The lateral load resistance of any story must not be less than 80% of the story above it.

When using the simplified design procedure, ASCE/SEI7 Sec. 12.14.3.2.1 states that the overstrength factor is

$$\Omega_0 = 2.5$$

When the simplified design procedure is used, ASCE/SEI7 Sec. 12.14.8.5 specifies that structural drift need not be calculated. If a drift value is required for design of cladding or to determine building separation, it may be assumed to be 1% of building height.

Lateral Forces

Example 8.19

The two-story structure shown is an office building located on a site with a soil classification type D. The structure consists of a bearing wall system with the reinforced concrete shear walls continuous through both stories, and the roof and floor diaphragms are rigid. Determine whether the simplified lateral force procedure is applicable.

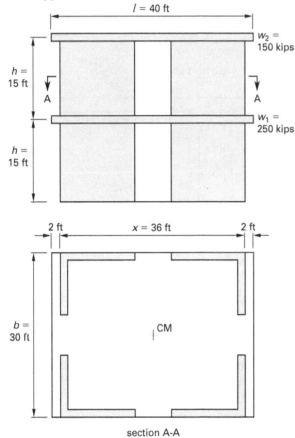

section A-A

Solution

The following hold true for the office building and are satisfactory.

- The risk category for an office building is II.
- It is situated on a site with a soil classification type D.
- It is two stories in height, which is less than three stories.
- The seismic force-resisting system is a bearing wall system.
- It has two lines of lateral resistance in each of the two major axis directions.
- One line of resistance is provided on each side of the center of mass in each direction.

- The distance between the two lines of resistance in each direction exceeds 50% of the diaphragm length perpendicular to the two lines.

- The lines of resistance of the lateral force-resisting system are parallel to the major orthogonal horizontal axes of the building.

- There are no irregularities caused by in-plane or out-of-plane offsets of lateral force-resisting elements.

- The lateral load resistance is identical in both stories.

Simplified Determination of Seismic Base Shear

The simplified seismic base shear is given by ASCE/SEI7 Eq. 12.14-12 as

$$V = \left(\frac{F S_{DS}}{R} \right) W$$

The design spectral response acceleration at short periods is given by ASCE/SEI7 Sec. 12.14.8.1 as

$$S_{DS} = \frac{2 F_a S_S}{3}$$

The 5% damped, maximum considered earthquake spectral response acceleration, for a period of 0.2 sec for structures founded on rock, is given by ASCE/SEI7 Sec. 11.4.2 with the limitation

$$S_S \leq 1.5g$$

The short-period site coefficient is obtained from ASCE/SEI7 Table 11.4-1, or may be taken as

$$F_a = 1.0 \quad \text{[rock sites]}$$
$$F_a = 1.4 \quad \text{[soil sites]}$$

ASCE/SEI7 Sec. 12.14.8.1 defines a rock site as having the height of the soil between the rock surface and the bottom of the building's foundations no greater than 10 ft.

The modification factor for building type is

$$F = 1.0 \quad \text{[one-story buildings]}$$
$$F = 1.1 \quad \text{[two-story buildings]}$$
$$F = 1.2 \quad \text{[three-story buildings]}$$

The effective seismic weight, W, as specified in ASCE/SEI7 Sec. 12.14.8.1, includes the total dead load of the structure plus the following loads.

- 25% of the floor live load for storage and warehouse occupancies, except for public parking structures and storage loads adding not more than 5% to the seismic weight

- a minimum allowance of 10 lbf/ft² for moveable partitions or the actual weight if greater

- flat roof snow loads exceeding 30 lbf/ft², which are reduced by 80%

- weight of landscaping and other materials at roof gardens

- the total weight of permanent equipment and fittings

The response modification factor, R, and the limitations on the use of the various lateral force-resisting systems are given in ASCE/SEI7 Table 12.14-1 and are summarized in Table 8.9.

Table 8.9 Design Factors for Simplified Lateral Force Procedure

structural system	response modification coefficient, R	limitations		
		seismic design category		
		B	C	D, E
bearing wall				
special reinforced concrete shear walls	5	P	P	P
ordinary reinforced concrete shear walls	4	P	P	NP
special reinforced masonry shear walls	5	P	P	P
intermediate reinforced masonry shear walls	3.5	P	P	NP
ordinary reinforced masonry shear walls	2	P	NP	NP
light-framed walls with wood structural panels	6.5	P	P	P
building frame				
steel eccentrically braced frame	8	P	P	P
steel special concentrically braced frames	6	P	P	P
steel ordinary concentrically braced frames	3.25	P	P	P
special reinforced concrete shear walls	6	P	P	P
ordinary reinforced concrete shear walls	5	P	P	NP
steel and concrete composite special concentrically braced frames	5	P	P	P
special reinforced masonry shear walls	5.5	P	P	P
intermediate reinforced masonry shear walls	4	P	P	NP
light-framed walls with wood structural panels	7	P	P	P

P = permitted, NP = not permitted

Adapted with permission from *Minimum Design Loads and Associated Criteria for Buildings and Other Structures*, copyright © 2016, by the American Society of Civil Engineers.

Example 8.20

Determine the seismic base shear using the simplified procedure for the two-story special reinforced concrete bearing wall structure of Ex. 8.19. The 5% damped, maximum considered earthquake spectral response acceleration, for a period of 0.2 sec, is $S_S = 1.2g$. The effective seismic weight of the building is 400 kips.

Solution

The 5% damped, maximum considered earthquake spectral response acceleration for a period of 0.2 sec is given as

$$S_S = 1.2g$$

The short-period site coefficient for a soil site is given by ASCE/SEI7 Sec. 12.14.8.1 as

$$F_a = 1.4$$

The design spectral response acceleration at short periods is given by ASCE/SEI7 Sec. 12.14.8.1 as

$$S_{DS} = \frac{2F_a S_S}{3} = \frac{(2)(1.4)(1.2g)}{3}$$
$$= 1.12g$$

For a bearing wall structure with special reinforced concrete shear walls, the response modification factor is obtained from ASCE/SEI7 Table 12.14-1, or from Table 8.9, as

$$R = 5$$

The modification factor for a two-story building is

$$F = 1.1$$

The effective seismic weight of the building is given as

$$W = 400 \text{ kips}$$

Therefore, the simplified base shear is given by ASCE/SEI7 Eq. 12.14-12 as

$$V = \left(\frac{FS_{DS}}{R}\right) W$$
$$= \left(\frac{(1.1)(1.12g)}{5}\right)(400 \text{ kips})$$
$$= 98.6 \text{ kips}$$

Lateral Forces

Simplified Vertical Distribution of Base Shear

When the simplified procedure is used to determine the seismic base shear, the forces at each level may be determined from ASCE/SEI7 Sec. 12.14.8.2 as

$$F_x = \frac{w_x V}{W} \qquad \text{[ASCE/SEI7 12.14-13]}$$

Example 8.21

Determine the vertical force distribution by using the simplified procedure for the two-story, reinforced concrete, bearing wall structure of Ex. 8.20.

Solution

From Ex. 8.20, the following values are obtained.

$$V = 98.6 \text{ kips}$$
$$W = 400 \text{ kips}$$

The forces at each level are determined from ASCE/SEI7 Eq. 12.14-13 as

$$\begin{aligned} F_x &= \frac{w_x V}{W} \\ &= \frac{w_x(98.6 \text{ kips})}{400 \text{ kips}} \\ &= 0.247 w_x \end{aligned}$$

The values of F_x are given in the following table.

level	w_x (kips)	F_x (kips)
2	150	37.05
1	250	61.75
total	400	98.80

Simplified Determination of Drift

In accordance with ASCE/SEI7 Sec. 12.14.8.5, when the simplified procedure is used to determine the seismic base shear, the design story drift in any story must be taken as

$$\Delta_x = 0.01 h_{sx}$$

Example 8.22

Using the simplified procedure, determine the drift in the bottom story of the two-story, reinforced concrete, bearing wall structure of Ex. 8.20.

Solution

The design story drift in the bottom story is given by

$$\begin{aligned} \Delta_1 &= 0.01 h_{s1} \\ &= (0.01)(15 \text{ ft})\left(12 \, \frac{\text{in}}{\text{ft}}\right) \\ &= 1.80 \text{ in} \end{aligned}$$

12. SEISMIC LOAD ON AN ELEMENT OF A STRUCTURE

The *seismic load*, E, is a function of both horizontal and vertical earthquake-induced forces and is given by ASCE/SEI7 Sec. 12.4.2 as

$$E = \rho Q_E + 0.2 S_{DS} D$$

The term Q_E is the lateral force produced by the calculated base shear V. The term $0.2 S_{DS} D$ is the vertical force due to the effects of vertical acceleration. The redundancy factor, ρ, is a factor that penalizes structures with relatively few lateral load-resisting elements and is defined by ASCE/SEI7 Sec. 12.3.4.

The seismic performance of a structure is enhanced by providing multiple lateral load-resisting elements. Therefore, the yield of one element will not result in an unstable condition that may lead to collapse of the structure. For this situation, the value of the redundancy factor is $\rho = 1.0$. When the yield of an element will result in an unstable condition, the value of the redundancy factor is $\rho = 1.3$.

In accordance with ASCE/SEI7 Sec. 12.3.4.1, the value of the redundancy factor is equal to 1.0 for the following.

- structures assigned to seismic design categories B and C

- drift calculations and P-delta effects

- design of nonstructural components

- design of nonbuilding structures that are not similar to buildings

- design of collector elements, splices, and their connections for which load combinations with overstrength factors are used

- design of members and connections for which load-combinations with overstrength factors are required

- diaphragm loads that are determined using ASCE/SEI7 Eq. 12.10-1, 12.10-2, or 12.10-3

- structures with damping systems designed in accordance with ASCE/SEI7 Chap. 18

- design of structural walls for out-of-plane forces, including their anchorages

For structures assigned to seismic design category D and having an extreme torsional irregularity, ρ is taken as 1.3. For other structures assigned to seismic design category D, and for structures assigned to seismic design categories E or F, ρ is taken as 1.3 unless one of the conditions given in ASCE/SEI7 Sec. 12.3.4.2 is met when the value of the redundancy factor is permitted to be taken as 1.0.

In accordance with ASCE/SEI7 Sec. 12.3.4.2a, the value of the redundancy factor may be taken equal to 1.0, provided that each story resisting more than 35% of the base shear complies with the following.

- For a braced frame, the removal of an individual brace, or connection thereto, does not result in more than a 33% reduction in story strength, nor create an extreme torsional irregularity (horizontal structural irregularity type 1b).

- For a moment frame, loss of moment resistance at the beam to column connections at both ends of a single beam does not result in more than a 33% reduction in story strength, nor create an extreme torsional irregularity (horizontal structural irregularity type 1b).

- For a shear wall or a wall-pier system with a height to length ratio greater than 1.0, removal of a wall or pier, or collector connections thereto, does not result in more than a 33% reduction in story strength, nor create an extreme torsional irregularity (horizontal structural irregularity type 1b).

- For a cantilever column, loss of moment resistance at the base connections of any single cantilever column does not result in more than a 33% reduction in story strength, nor create an extreme torsional irregularity (horizontal structural irregularity type 1b).

- There are no requirements for all other structural systems.

In accordance with ASCE/SEI7 Sec. 12.3.4.2b, the value of the redundancy factor may be taken equal to 1.0 provided that the building is regular in plan at all levels with not less than two bays of lateral load-resisting perimeter framing on each side of the building in each orthogonal direction at each story resisting more than 35% of the base shear. The number of bays for a shear wall is calculated as the length of the shear wall divided by the story height. For light-framed construction, the number of bays for a shear wall is calculated as twice the length of the shear wall divided by the story height.

Example 8.23

Determine the redundancy factor for the moment-resisting framed structure shown. The stiffness of all frames is identical. The roof diaphragm is flexible. The structure is assigned to seismic design category D.

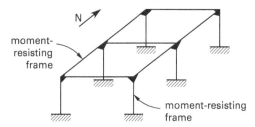

Solution

The building is regular in plan, and this complies with ASCE/SEI7 Sec. 12.3.4.2b.

In the north-south direction, two bays of moment-resisting perimeter frames are provided on each side of the building. The frames on each perimeter resist 50% of the base shear. This complies with ASCE/SEI7 Sec. 12.3.4.2b.

In the east-west direction, only one bay of moment-resisting perimeter frames is provided on each side of the building. This does not comply with ASCE/SEI7 Sec. 12.3.4.2a, and it is necessary to check compliance with ASCE/SEI7 Sec. 12.3.4.2a.

In the north-south direction, removing one frame results in a reduction of shear strength of

$$\frac{1}{4} = 0.25$$
$$< 0.33$$

This complies with ASCE/SEI7 Sec. 12.3.4.2a.

In the east-west direction, three moment resisting frames are provided and removing one frame results in a reduction of shear strength of

$$\frac{1}{3} = 0.33$$

This complies with ASCE/SEI7 Sec. 12.3.4.2a.

The redundancy factor is given by ASCE/SEI7 Sec. 12.3.4 as

$$\rho = 1.0$$

Example 8.24

Determine the redundancy factor for the structure shown. The stiffness of all braced frames in the north-south direction is identical, and the stiffness of all moment-resisting frames in the east-west direction is identical. The roof diaphragm is flexible. The structure is assigned to seismic design category D.

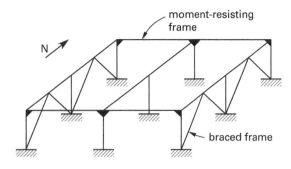

moment-resisting frame

N

braced frame

Solution

The building is regular in plan and this complies with ASCE/SEI7 Sec. 12.3.4.2b.

In the north-south direction, two bays of chevron bracing are provided on each side of the building. The frames on each perimeter resist 50% of the base shear. This complies with ASCE/SEI7 Sec. 12.3.4.2b.

Similarly, in the east-west direction, two bays of moment-resisting frames are provided on each side of the building. The frames on each perimeter resist 50% of the base shear. This complies with ASCE/SEI7 Sec. 12.3.4.2b.

The redundancy factor is given by ASCE/SEI7 Sec. 12.3.4 as

$$\rho = 1.0$$

PART 3: WIND DESIGN

Nomenclature

a	width of pressure coefficient zone	ft
A	effective wind area	ft^2
A_g	gross area of wall in which A_o is identified	ft^2
A_{gi}	sum of gross surface areas of building envelope (walls and roof), excluding A_g	ft^2
A_o	total area of openings in a wall receiving positive external pressure	ft^2
A_{oi}	sum of areas of all openings in building envelope (walls and roof), excluding A_o	ft^2
b_e	effective joist spacing	ft
B	horizontal dimension of building measured normal to wind direction	ft
C_{net}	net pressure coefficient based on $K_d((GC_p) - (GC_{pi}))$ per IBC Table 1609.6.2	–
C_p	external pressure coefficient	–
C_{pi}	internal pressure coefficient	–
G	gust effect factor	–
(GC_p)	product of gust effect factor and external pressure coefficient	–
(GC_{pf})	product of gust effect factor and equivalent external pressure coefficient	–

	for determining wind loads in MWFRS of low-rise buildings	
(GC_{pi})	product of internal pressure coefficient and gust effect factor	–
h	mean roof or eave height	ft
K_1, K_2, K_3	multipliers from ASCE/SEI7 Fig. 26.8-1 used to obtain K_{zt}	–
K_d	wind directionality factor given in ASCE/SEI7 Table 26.6-1	–
K_e	ground elevation factor	–
K_h	velocity pressure exposure coefficient evaluated at height $z = h$	–
K_z	velocity pressure exposure coefficient evaluated at height z	–
K_{zt}	topographic factor as defined in ASCE/SEI7 Sec. 26.8.1	–
l	joist spacing	ft
L	horizontal dimension of a building measured parallel to wind direction	ft
MWFRS	main wind force-resisting system	–
p	design pressure for determining wind loads	lbf/ft^2
P_{net}	design wind pressure for determining wind loads on buildings, building components, or cladding	lbf/ft^2
q_h	velocity pressure evaluated at height $z = h$	lbf/ft^2
q_s	wind stagnation pressure	lbf/ft^2
q_z	velocity pressure evaluated at height z above ground	lbf/ft^2
s	joist spacing	–
V	basic wind speed	mi/hr
w	distributed load	lbf/ft
z	height above ground	ft

Symbols

γ	exposure adjustment factor	–
λ	adjustment factor for height and exposure	–
θ	angle of plane of roof from horizontal	degree

13. WIND LOADS

Design Procedures

Two basic wind design procedures, the directional procedure and the envelope procedure, are defined in ASCE/SEI7 Sec. 26.2. The *directional procedure* determines the wind loads on buildings for specific wind directions. It uses external pressure coefficients based on wind tunnel testing of prototypical building models for the corresponding direction of wind. The *envelope procedure* determines the wind load cases on buildings.

Lateral Forces

It uses pseudo-external pressure coefficients derived from wind tunnel testing of prototypical building models successively rotated through 360°. These pseudo-pressure cases produce key structural actions (uplift, horizontal shear, bending moments, etc.) that envelop the actions' maximum values among all possible wind directions.

Building Characteristics

The following terms are defined in ASCE/SEI7 Sec. 26.2 and are used to designate building types and components.

- The *main wind-force resisting system* is an assemblage of structural elements assigned to provide support and stability for the overall structure. The system generally receives wind loading from more than one surface.

- *Components* and *cladding* are elements of the building envelope that do not qualify as part of the main wind-force resisting system. *Building cladding* receives wind loading directly. Examples of cladding include wall and roof sheathing, windows, and doors. *Building components* receive wind loading from the cladding and transfer the load to the main wind force-resisting system. Components include purlins, studs, girts, fasteners, and roof trusses. Some elements, such as roof trusses and sheathing, may also form part of the main wind force-resisting system and must be designed for both conditions.

- *Low-rise buildings* are enclosed or partially enclosed buildings that comply with the following conditions: the mean roof height, h, is less than or equal to 60 ft; and the mean roof height also does not exceed the least horizontal dimension.

- *Simple diaphragm buildings* are buildings in which both windward and leeward wind loads are transmitted by roof and vertically spanning wall assemblies, through continuous floor and roof diaphragms, to the main wind-force resisting system.

- *Rigid buildings* have a fundamental frequency greater than or equal to 1 Hz. Most buildings with a height-to-minimum-width ratio of less than 4 may be considered rigid.

- *Regular-shaped buildings* have no unusual geometrical irregularity in spatial form.

Wind Load Determination

In accordance with IBC Sec. 1609.1.1, wind loads on buildings are determined by the provisions of ASCE/SEI7. Several different procedures for determining wind loads on the main wind-force resisting system of buildings are specified as follows.

- The analytical directional design method of ASCE/SEI7 Sec. 27.3 is applicable to enclosed, partially enclosed, and open buildings of all heights and roof geometry.

- The simplified method of ASCE/SEI7 Sec. 27.5 is based on the analytical method of ASCE/SEI7 Chap. 27, Part 1 and is applicable to enclosed, simple diaphragm buildings of any roof geometry with a height not exceeding 160 ft.

- The envelope design method of ASCE/SEI7 Sec. 28.3 is applicable to enclosed, partially enclosed, and open low-rise buildings having a flat, gable, or hip roof with a height not exceeding 60 ft.

- The simplified method of ASCE/SEI7 Sec. 28.5 is based on the envelope procedure of ASCE/SEI7 Chap. 28, Part 1 and is applicable to enclosed, simple diaphragm low-rise buildings having a flat, gable, or hip roof with a height not exceeding 60 ft.

- The wind tunnel procedure of ASCE/SEI7 Chap. 31 may be used for any structure.

Several different procedures for determining wind loads on components and cladding of buildings are specified as follows.

- The analytical envelope design method of ASCE/SEI7 Sec. 30.3 is applicable to enclosed and partially enclosed low-rise buildings.

- The simplified envelope design method of ASCE/SEI7 Sec. 30.4 is applicable to enclosed low-rise buildings.

- The analytical directional design method of ASCE/SEI7 Sec. 30.5 is applicable to enclosed and partially enclosed buildings with $h > 60$ ft.

- The simplified directional design method of ASCE/SEI7 Sec. 30.6 is applicable to enclosed buildings with $h \leq 160$ ft.

- The analytical directional design method of ASCE/SEI7 Sec. 30.7 is applicable to open buildings of all heights.

Wind loads on parapets are covered in ASCE/SEI7 Sec. 30.8.

The following parameters must be known in order to determine a structure's wind load.

- exposure category of the site

- basic wind speed at the location of the structure

- velocity pressure exposure coefficient

- topographic effects

Lateral Forces

- directionality factor

- wind importance factor

- wind velocity pressure

Surface Roughness Category

Ground surface roughness is defined in ASCE/SEI7 Sec. 26.7.2. (See Table 8.10.)

Table 8.10 Surface Roughness Categories

surface roughness category	applicable ground surface roughness
B	urban and suburban areas, wooded areas, or other terrain with numerous closely spaced obstructions the size of single-family dwellings or larger
C	open terrain containing scattered obstructions that have heights generally less than 30 ft; includes flat open country and grasslands
D	flat, unobstructed areas and water surfaces; includes smooth mud flats, salt flats, and unbroken ice

Site Exposure Category

A site's exposure category accounts for the effect of terrain roughness on wind speed and is described in ASCE/SEI7 Sec. C26.7.3. Table 8.11 lists the three exposure categories defined in ASCE/SEI7 Sec. 26.7.3.

Table 8.11 Site Exposure Categories

site exposure category	applicable site exposure
B	applies (1) for building height $h \leq 30$ ft and where surface roughness B prevails in the upwind direction for more than 1500 ft, or (2) for building height $h > 30$ ft, and where surface roughness B prevails in the upwind direction for more than the greater of 2600 ft or $20h$
C	applies in all cases where exposures B or D do not apply
D	applies (1) where surface roughness D prevails in the upwind direction for more than the greater of 5000 ft or $20h$, or (2) where surface roughness immediately upwind of the site is B or C for a distance not exceeding the greater of 600 ft or $20h$ is followed by surface roughness D for more than the greater of 5000 ft or $20h$

Basic Wind Speed

Basic wind speed, V, is used to determine the design wind loads on buildings and other structures, and it is determined from ASCE/SEI7 Fig. 26.5-1 and Fig. 26.5-2. Wind is assumed to come from any horizontal direction. Basic wind speed values are based on the 3 sec gust wind speed (in miles per hour). They are adjusted to a reference height of 33 ft for exposure category C. Drag effects slow down

wind flow close to the ground, and wind speed increases with height above ground level until the gradient height is reached and the speed becomes constant. The gradient heights, z_g, for different exposure conditions are given in ASCE/SEI7 Table 26.11-1.

The wind speed is given at the strength level that gives a load factor of 1.0 for wind loads in the strength design load combinations. ASCE/SEI7 uses four wind speed maps for buildings with different risk categories. This ensures that high-risk facilities are designed for higher loads so as to reduce possible structural damage. Details of the different risk categories and corresponding wind speed maps are given in Table 8.12 for the contiguous United States.

Table 8.12 Risk Category and Wind Speed Maps

risk category	nature of occupancy	ASCE/SEI7 wind speed maps
I	low hazard structures	26.5-1A
II	standard occupancy structures	26.5-1B
III	assembly structures	26.5-1C
IV	essential or hazardous structures	26.5-1D

Similarly, ASCE/SEI7 Fig. 26.5-2A through 26.5-2D give basic wind speeds for risk categories I through IV buildings for Hawaii.

Velocity Pressure Exposure Coefficients

The velocity pressure exposure coefficient, K_z, reflects the change in wind speed with height and exposure category. K_z values for different exposure conditions are given in ASCE/SEI7 Table 26.10.1. These are tabulated for a limited number of heights in Table 8.13 for main wind force-resisting systems for the purpose of determining overall wind loads on the building.

Table 8.13 Velocity Pressure Exposure Coefficients for Main Wind Force-Resisting Systems

exposure	height above ground level (ft)					
	0–15	20	25	30	40	50
B[a]	0.57	0.62	0.66	0.70	0.76	0.81
B[b]	0.70	0.70	0.70	0.70	0.76	0.81
C	0.85	0.90	0.94	0.98	1.04	1.09
D	1.03	1.08	1.12	1.16	1.22	1.27

[a]directional procedure
[b]envelope procedure

Adapted with permission from *Minimum Design Loads and Associated Criteria for Buildings and Other Structures*, copyright © 2016, by the American Society of Civil Engineers.

Minimum Design Wind Loads

The minimum design wind loads for an enclosed or partially enclosed building are given in ASCE/SEI7 Sec. 27.1.5 and are shown in ASCE/SEI7 Fig. C27.1-1. The loading consists of an external pressure of 16 lbf/ft^2 on wall areas and 8 lbf/ft^2 on roof areas projected onto a

Lateral Forces

vertical plane normal to the wind direction. Details are shown in Fig. 8.31. The minimum loads are to be applied as a separate load case in addition to the normal load cases specified.

Figure 8.31 *Minimum Design Wind Loads*

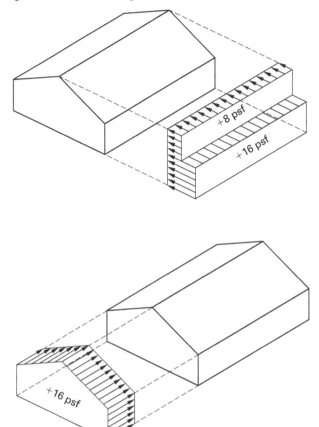

Topographic Effects

Increased wind speed effects are produced at isolated hills, ridges, and escarpments constituting abrupt changes in the general topography. To account for this, the velocity pressure exposure coefficient is multiplied by the *topographic factor*, K_{zt}. The topographic factor is a function of the following criteria.

- slope of the hill

- distance of the building from the crest

- height of the building above the local ground surface

These three criteria are represented by the topographic multipliers K_1, K_2, and K_3, and are tabulated in ASCE/SEI7 Fig. 26.8-1. The topographic factor is given by ASCE/SEI7 Eq. 26.8-1 as

$$K_{zt} = (1 + K_1 K_2 K_3)^2$$

When the topographic effect need not be considered, it is given by

$$K_{zt} = 1.0$$

Directionality Factor

The *wind directionality factor*, K_d, is determined from ASCE/SEI7 Table 26.6-1. For buildings, it is given as 0.85. The directionality factor accounts for the reduced probability of the following.

- extreme winds occurring in any specific direction

- the peak pressure coefficient occurring for a specific wind direction

Wind Velocity Pressure

The *wind velocity pressure*, q_z (in units of $\mathrm{lbf/ft^2}$), is evaluated at a height, z, using ASCE/SEI7 Eq. 26.10-1. (This formula is not dimensionally consistent.)

$$q_{z,\mathrm{lbf/ft^2}} = 0.00256 K_z K_{zt} K_d K_e V_{\mathrm{mi/hr}}^2$$

The constant 0.00256 reflects the mass density of air at a temperature of 59°F and a pressure of 29.92 in of mercury. This value should be used unless sufficient data is available to justify a different value. The velocity pressure varies as the value of the velocity pressure exposure coefficient varies with the height above ground level.

Example 8.25

The steel framed factory shown is located in a suburban area 25 miles inland from the coast of Southern California. The factory is not located in a special wind region. Determine the wind velocity pressure at roof height for the main wind force-resisting system.

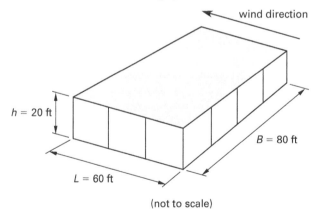

(not to scale)

Solution

For a factory building, the risk category is II. From Table 8.12, wind speed map ASCE/SEI7 Fig. 26.5-1B applies.

From ASCE/SEI7 Fig. 26.5-1B, the basic wind speed is

$$V = 95 \text{ mi/hr}$$

The height-to-minimum-width ratio is

$$\frac{h}{L} = \frac{20 \text{ ft}}{60 \text{ ft}} = 0.33$$
$$< 1 \quad \text{[low-rise building]}$$
$$< 4 \quad \text{[rigid building]}$$

The eaves roof height is

$$h = 20 \text{ ft}$$
$$< 60 \text{ ft} \quad \text{[low-rise building]}$$

So, the building qualifies as a low-rise, rigid building. For a building designed using ASCE/SEI7 Fig. 28.3-1 values from ASCE/SEI7 Table 26.10-1 are applicable for the velocity pressure exposure coefficients.

For a suburban area, the exposure is category B and the relevant parameters are obtained as follows.

K_z = velocity pressure exposure coefficient from ASCE/SEI7 Table 26.10-1 for a height of 20 ft for the main wind force-resisting system and exposure category B

= 0.70

K_{zt} = topographic factor from ASCE/SEI7 Fig. 26.8-1

= 1.0

K_d = wind directionality factor from ASCE/SEI7 Table 26.6-1

= 0.85

The wind velocity pressure, q_h, at a roof height of 20 ft above the ground, is given by ASCE/SEI7 Eq. 26.10-1.

$$q_{h,\text{lbf/ft}^2} = 0.00256 K_z K_{zt} K_d K_e V_{\text{mi/hr}}^2$$
$$= (0.00256)(0.70)(1.0)(0.85)(1.0)\left(95 \ \frac{\text{mi}}{\text{hr}}\right)^2$$
$$= 13.75 \text{ lbf/ft}^2$$

14. DESIGN WIND PRESSURE

In order to determine the design wind pressures on a structure, it is necessary to convert the wind velocity pressure to *design wind pressure*. To do so, the following items must be determined.

- rigidity of the structure
- gust effect factor
- enclosure classification

Rigidity of the Structure

A *flexible structure* is defined in ASCE/SEI7 Sec. 26.2 as a slender structure with a fundamental frequency less than 1 Hz. Therefore, a *rigid structure* is a structure with a fundamental frequency greater than or equal to 1 Hz. A flexible structure exhibits a significant dynamic resonant response to wind gusts. According to ASCE/SEI7 Sec. 26.11.2, a low-rise building may be considered rigid. Where necessary, the fundamental frequency may be determined using the procedures given in ASCE/SEI7 Sec. 26.11.3.

Gust Effect Factor

For a rigid structure, the *gust effect factor* is taken as 0.85. Alternatively, the gust effect factor may be calculated using the procedure given in ASCE/SEI7 Sec. 26.11.4. The gust effect factor accounts for along-wind loading effects (that is, loading effects in the direction of the wind) caused by dynamic amplification in flexible structures and for wind turbulence-structure interaction.

Enclosure Classification

A structure's *enclosure classification* is determined in accordance with ASCE/SEI7 Sec. 26.2, and is based on the number of openings in its building envelope. ASCE/SEI7 Sec. 26.2 defines *openings* as apertures or holes in the building envelope that allow air to flow through the building envelope. The *building envelope* is defined as cladding, roofing, exterior walls, glazing, door assemblies, window assemblies, skylight assemblies, and other components enclosing the structure or building. Per ASCE/SEI7 Sec. 26.2, there are three types of building enclosures: enclosed, partially enclosed, and open.

A *partially enclosed building* is defined as satisfying both of the following conditions.

- *condition 1:* The total area of openings in a wall that receives positive external pressure must exceed the sum of the areas of openings in the balance of the building envelope (walls and roof) by more than 10%.

- *condition 2:* The total area of openings in a wall that receives positive external pressure exceeds 4 ft^2 or 1% of the area of that wall, whichever is smaller, and the percentage of openings in the balance of the building envelope does not exceed 20%.

These conditions are expressed by the following equations.

- $A_o > 1.10 A_{oi}$

- $A_o > 4 \text{ ft}^2$ or $> 0.01 A_g$, whichever is smaller, and $A_{oi}/A_{gi} \leq 0.20$

An *open building* is defined as a building where each wall is at least 80% open. This condition is expressed for each wall by

$$A_o \geq 0.8 A_g$$

An *enclosed building* is defined as a building that receives positive external pressure and has a total wall opening area of less than or equal to 4 ft² or 1% of the area of the wall, whichever is smaller. These conditions are expressed by the following equations.

$$A_o < 0.01 A_g$$

but not more than

$$A_o = 4 \text{ ft}^2$$

A *partially open building* is defined as a building that does not comply with the requirements for open, partially enclosed, or enclosed buildings.

Ground Elevation Factor

Air density and air pressure decrease with increasing altitude. The ground elevation factor, K_e, accounts for this and is determined from ASCE/SEI 7 Table 26.9-1. It is permitted to take K_e as 1.0 for all elevations.

Wind-Borne Debris Region

ASCE/SEI7 Sec. 26.12.3.1 states that if a building is in a wind-borne debris region, then all glazing in the lower 60 ft of the building must be assumed to be openings unless the glazing is impact resistant or protected with an impact-resistant covering. The same requirement applies to glazing that is less than 30 ft above any aggregate surface roof located within 1500 ft of the building. Glazing in risk category I buildings may be unprotected.

A *wind-borne debris region* is defined in ASCE/SEI7 Sec. 26.12.3.1 as being

- within 1 mi of the coastal mean high water line and where the basic wind speed is at least 130 mph

- any region where the basic wind speed is at least 140 mph

Internal Pressure

A structure's *internal pressure* (i.e., the pressure that is produced in a structure by wind) depends on the size and location of openings in the external walls of the structure. Figure 8.32 shows an opening in the windward wall of a structure, which produces an internal pressure, and an opening in the leeward wall of a structure, which produces an internal suction.

Figure 8.32 Building Openings

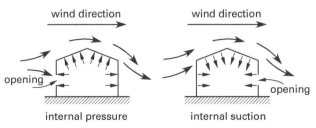

15. LOW-RISE REGULAR BUILDING, MAIN WIND FORCE-RESISTING SYSTEM

Nomenclature

(GC_{pf})	product of the equivalent external pressure coefficient and gust effect factor as given in ASCE/SEI7 Fig. 28.3-1	–
(GC_{pi})	product of the internal pressure coefficient and gust effect factor as given in ASCE/SEI7 Table 26.13-1	–
q_h	wind velocity pressure at mean roof height h for the applicable exposure category	lbf/ft²

Envelope Design Method for MWFRS

The analytical envelope procedure (Part 1) defined in ASCE/SEI7 Sec. 28.3 is applicable to a low-rise structure that meets both of the following requirements.

- The structure must be regular-shaped as defined in ASCE/SEI7 Sec. 26.2, without any irregularities such as projections or indentations.

- Either the structure does not have response characteristics that make it subject to across-wind loading, vortex shedding, or instability due to galloping or flutter; or it does not have a site location for which channeling effects or buffeting in the wake of upwind obstructions warrant special consideration.

For low-rise buildings, a simplification is introduced into the analytical procedure by combining the gust effect factor with the pressure coefficient and treating the combination as a single factor. The design wind pressure, p, on the main system is given by ASCE/SEI7 Eq. 28.3-1 as

$$p = q_h \big((GC_{pf}) - (GC_{pi}) \big)$$

Internal Pressure Coefficients

For the envelope procedure of ASCE/SEI7 Chap. 28, Part 1, the gust effect factor is combined with the internal pressure coefficient and denoted by (GC_{pi}). Values of (GC_{pi}) are tabulated in ASCE/SEI7 Table 26.13-1 for the three different building enclosure classifications and are given in Table 8.14.

Table 8.14 Values of Internal Pressure Coefficients

enclosure classification	(GC_{pi})
open buildings	0.00
partially open buildings	±0.18
partially enclosed buildings	±0.55
enclosed buildings	±0.18

The pressure acting on internal surfaces is obtained from the second term of ASCE/SEI7 Eq. 28.3-1 as

$$p_i = \pm q_h (GC_{pi})$$

Lateral Forces

Pressures act normal to wall and roof surfaces. They are positive when acting toward the surface and negative when acting away from the surface. The conditions that produce internal suction and internal pressure are shown in Fig. 8.32. Both cases must be considered for any building and added algebraically to external pressures to determine the most critical loading condition.

Design Wind Load Cases

In the envelope procedure, a building is designed for all wind directions by considering each corner of the building as the windward corner. At each corner, wind acting in both the transverse and longitudinal directions must be considered, giving eight basic load cases as illustrated in ASCE/SEI7 Fig. 28.3-1. Two of these load cases are shown in Fig. 8.33. Load case A covers wind in the transverse direction, and load case B covers wind in the longitudinal direction.

Figure 8.33 *Load Case A and Load Case B*

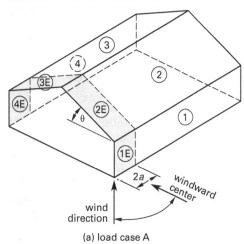

(a) load case A

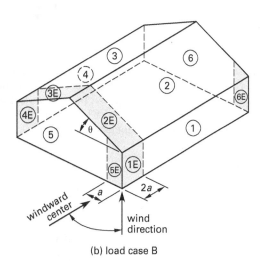

(b) load case B

Adapted with permission from *Minimum Design Loads and Associated Criteria for Buildings and Other Structures*, copyright © 2016, by the American Society of Civil Engineers.

For each of the eight basic load cases, both positive and negative internal pressure must be considered, resulting in a total of 16 combinations. When the building is symmetrical about one axis, only two corners need to be investigated. If the building is doubly symmetrical, only one corner needs to be investigated. When torsion must be considered, each of these load cases is also modified as indicated in ASCE/SEI7 Fig. 28.3-1, Note 5. Torsion need not be considered if the building is one story with a height less than 30 ft, or if the building is no more than two stories and either designed with flexible diaphragms or framed with light frame construction.

Local turbulence at building corners and at the roof eaves produces an increase in pressure in these areas. To allow for the differences in pressure, the method subdivides the building surface into distinct zones. As shown in Fig. 8.33, eight zones are designated for transverse load, and twelve zones are designated for longitudinal wind. External pressure coefficients are tabulated for each zone.

External Pressure Coefficients

For the envelope procedure described in ASCE/SEI7 Chap. 28, Part 1, the gust effect factor is combined with the external pressure coefficients and denoted by (GC_{pf}). Values of (GC_{pf}) are tabulated in ASCE/SEI7 Fig. 28.3-1 and given in Table 8.15 and Table 8.16. Values are given for two separate loading conditions—case A for wind acting transversely and case B for wind acting longitudinally to the building.

The pressure acting on external surfaces is obtained from the first term of ASCE/SEI7 Eq. 28.3-1 as

$$p_e = q_h(GC_{pf})$$

On each wall and roof surface external pressure coefficients are given for two zones: an end zone and an interior zone. The end zone width is given by ASCE/SEI7 Fig. 28.3-1 as either a or $2a$, where a is the lesser of the following.

$$a = (0.1)(\text{least horizontal dimension})$$
$$a = 0.4h$$

However, a cannot be less than either of the following.

$$a = (0.04)(\text{least horizontal dimension})$$
$$a = 3 \text{ ft}$$

Pressures act normal to wall and roof surfaces and are positive when acting toward the surface. They are negative when acting away from the surface.

Table 8.15 External Pressure Coefficients for Load Case A

roof angle, θ (degrees)	load case A building surface							
	1	2	3	4	1E	2E	3E	4E
0–5	0.40	−0.69	−0.37	−0.29	0.61	−1.07	−0.53	−0.43
20	0.53	−0.69	−0.48	−0.43	0.80	−1.07	−0.69	−0.64
30–45	0.56	0.21	−0.43	−0.37	0.69	0.27	−0.53	−0.48
90	0.56	0.56	−0.37	−0.37	0.69	0.69	−0.48	−0.48

Adapted with permission from *Minimum Design Loads and Associated Criteria for Buildings and Other Structures*, Fig. 28.3-1, copyright © 2016, by the American Society of Civil Engineers.

Table 8.16 External Pressure Coefficients for Load Case B

roof angle, θ (degrees)	load case B building surface											
	1	2	3	4	5	6	1E	2E	3E	4E	5E	6E
0–90	−0.45	−0.69	−0.37	−0.45	0.40	−0.29	−0.48	−1.07	−0.53	−0.48	0.61	−0.43

Adapted with permission from *Minimum Design Loads and Associated Criteria for Buildings and Other Structures*, Fig. 28.3-1, copyright © 2016, by the American Society of Civil Engineers.

Envelope Procedure

The following information is needed to determine wind loads using the envelope procedure.

- risk category I, II, III, or IV from Table 8.12 [ASCE/SEI7 Table 1.5-1]

- basic wind speed, V, for the applicable risk category from ASCE/SEI7 Fig. 26.5-1, and Fig. 26.5-2

- exposure category B, C, or D from Table 8.11 [ASCE/SEI7 Sec. 26.7]

- velocity pressure exposure coefficients, K_z, for the applicable exposure category from Table 8.13 [ASCE/SEI7 Table 26.10-1]

- topographic factor, K_{zt}, from ASCE/SEI7 Fig. 26.8-1

- ground elevation factor, K_e, from ASCE/SEI7 Table 26.9-1

- directionality factor, K_d, from ASCE/SEI7 Table 26.6-1

- enclosure classification from ASCE/SEI7 Sec. 26.12

- internal pressure coefficient, (GC_{pi}), from Table 8.14 [ASCE/SEI7 Table 26.13-1]

- wind velocity pressure, q_h, from ASCE/SEI7 Eq. 26.10-1

- external pressure coefficient, (GC_{pf}), from Table 8.15 and Table 8.16 [ASCE/SEI7 Fig. 28.3-1]

- internal wind pressure, $p_i = q_h(GC_{pi})$, from ASCE/SEI7 Eq. 28.3-1

- external wind pressure, $p_e = q_h(GC_{pf})$, from ASCE/SEI7 Eq. 28.3-1

- combined internal and external wind pressures, p, from ASCE/SEI7 Eq. 28.3-1

- minimum applicable design wind loads from Fig. 8.31 [ASCE/SEI7 Sec. 28.3.4]

The wind pressures are applied to each building corner in turn, as shown in ASCE/SEI7 Fig. 28.3-1. Torsional effects are evaluated if necessary.

Example 8.26

The steel framed factory analyzed in Ex. 8.25 may be considered a closed structure. Determine the design wind pressure for the main wind force-resisting system at the ends of the 80 ft side, for transverse wind load.

Solution

From Ex. 8.25, the velocity pressure at mean roof height is obtained as

$$q_h = 18.43 \text{ lbf/ft}^2$$

The height-to-minimum-width ratio is

$$\frac{h}{L} = \frac{20 \text{ ft}}{60 \text{ ft}} = 0.33$$
$$< 1$$

The mean roof height is

$$h = 20 \text{ ft}$$
$$< 60 \text{ ft}$$

Lateral Forces

So, the building qualifies as a low-rise, regular building as defined by ASCE/SEI7 Sec. 26.2, and the low-rise building analytical method of ASCE/SEI7 Sec. 28.3 is applicable. Values of (GC_{pf}) may be obtained from Table 8.15.

For a one-story building with a height less than 30 ft, ASCE/SEI7 Fig. 28.3-1 specifies that torsional load cases may be neglected. To determine the design wind pressure at the ends of the building, the pressures on surfaces 1E, 2E, 3E, and 4E must be determined for load case A. For an enclosed building, the product of the internal pressure coefficient and gust effect factor is given by Table 8.14 as

$$(GC_{pi}) = \pm 0.18$$

For surface 1E, the product of the equivalent external pressure coefficient and gust effect factor is given by Table 8.15 as

$$(GC_{pf}) = 0.61$$

The design wind pressure is given by ASCE/SEI7 Eq. 28.3-1 as

$$p = q_h\big((GC_{pf}) - (GC_{pi})\big)$$
$$= \left(18.43 \; \frac{\text{lbf}}{\text{ft}^2}\right)(0.61 - (\pm 0.18))$$
$$-14.56 \; \text{lbf/ft}^2 \quad \begin{bmatrix} \text{for negative internal} \\ \text{pressure (suction)} \end{bmatrix}$$
$$= 7.93 \; \text{lbf/ft}^2 \quad \text{[for positive internal pressure]}$$

For surface 2E, the product of the equivalent external pressure coefficient and gust effect factor is given by ASCE/SEI7 Fig 28.3-1 as

$$(GC_{pf}) = -1.07$$

The design wind pressure is given by ASCE/SEI7 Eq. 28.3-1 as

$$p = q_h\big((GC_{pf}) - (GC_{pi})\big)$$
$$= \left(18.43 \; \frac{\text{lbf}}{\text{ft}^2}\right)(-1.07 - (\pm 0.18))$$
$$= -16.40 \; \text{lbf/ft}^2 \quad \begin{bmatrix} \text{for negative internal} \\ \text{pressure (suction)} \end{bmatrix}$$
$$= -23.04 \; \text{lbf/ft}^2 \quad \text{[for positive internal pressure]}$$

For surface 3E, the product of the equivalent external pressure coefficient and gust effect factor is given by ASCE/SEI7 Fig. 28.3-1 as

$$(GC_{pf}) = -0.53$$

The design wind pressure is given by ASCE/SEI7 Eq. 28.3-1 as

$$p = q_h\big((GC_{pf}) - (GC_{pi})\big)$$
$$= \left(18.43 \; \frac{\text{lbf}}{\text{ft}^2}\right)(-0.53 - (\pm 0.18))$$
$$= -6.45 \; \text{lbf/ft}^2 \quad \begin{bmatrix} \text{for negative internal} \\ \text{pressure (suction)} \end{bmatrix}$$
$$= -13.09 \; \text{lbf/ft}^2 \quad \text{[for positive internal pressure]}$$

For surface 4E, the product of the equivalent external pressure coefficient and gust effect factor is given by ASCE/SEI7 Fig. 28.3-1 as

$$(GC_{pf}) = -0.43$$

The design wind pressure is given by ASCE/SEI7 Eq. 28.4-1 as

$$p = q_h\big((GC_{pf}) - (GC_{pi})\big)$$
$$= \left(18.43 \; \frac{\text{lbf}}{\text{ft}^2}\right)(-0.43 - (\pm 0.18))$$
$$= -4.61 \; \text{lbf/ft}^2 \quad \begin{bmatrix} \text{for negative internal} \\ \text{pressure (suction)} \end{bmatrix}$$
$$= -11.24 \; \text{lbf/ft}^2 \quad \begin{bmatrix} \text{for positive internal} \\ \text{pressure} \end{bmatrix}$$

The wind pressure diagrams for both internal suction and internal pressure are shown.

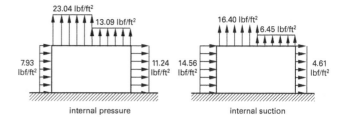

16. LOW-RISE REGULAR BUILDING, COMPONENTS AND CLADDING

Nomenclature

b_e	effective tributary width, $\geq l/3$	ft
(GC_p)	product of the equivalent external pressure coefficient and gust effect factor, as given in ASCE/SEI7 Fig. 30.3-1 through Fig. 30.3-7	–
(GC_{pi})	product of the internal pressure coefficient and gust effect	–
l	element span length	ft
q_h	wind velocity pressure at mean roof height, h	lbf/ft^2

Envelope Design Procedure for Components and Cladding

The envelope design procedure of ASCE/SEI7 Sec. 30.3 is applicable to enclosed and partially enclosed buildings which are low-rise buildings as defined in ASCE/SEI7 Sec. 26.2; with $h \leq 60$ ft; with flat roofs, gable roofs, multispan gable roofs, hip roofs, monoslope roofs, stepped roofs, or sawtooth roofs.

This procedure is simplified by combining the gust effect factor with the pressure coefficient and treating the combination as a single factor. Then the design wind pressure, p, on the main system is given by ASCE/SEI7 Eq. 30.3-1 as

$$p = q_h\big((GC_p) - (GC_{pi})\big)$$

Velocity Pressure Exposure Coefficients

The velocity pressure exposure coefficient, K_z, reflects the change in wind speed with height and exposure category and is tabulated for different exposure categories given in ASCE/SEI7 Table 26.10-1. Table 8.17 lists the velocity pressure exposure coefficients for a limited number of heights.

Table 8.17 Velocity Pressure Exposure Coefficients for Components and Cladding Systems

exposure category	height above ground level (ft)					
	0–15	20	25	30	40	50
B	0.70	0.70	0.70	0.70	0.76	0.81
C	0.85	0.90	0.94	0.98	1.04	1.09
D	1.03	1.08	1.12	1.16	1.22	1.27

Adapted with permission from *Minimum Design Loads and Associated Criteria for Buildings and Other Structures*, copyright © 2016, by the American Society of Civil Engineers.

Minimum Design Wind Loads

The minimum design wind loads for an enclosed or partially enclosed building are given in ASCE/SEI7 Sec. 30.2.2. The loading consists of a net pressure of 16 lbf/ft^2 applied in either direction normal to the surface.

Internal Pressure Coefficients

For the envelope procedure of ASCE/SEI7 Sec. 30.3, the gust effect factor is combined with the internal pressure coefficient and denoted by (GC_{pi}). Refer to Table 8.14 for (GC_{pi}) values for enclosed and partially enclosed buildings.

The pressure acting on internal surfaces is obtained from the second term of ASCE/SEI7 Eq. 30.3-1 as

$$p_i = \pm q_h(GC_{pi})$$

Pressures act normal to wall and roof surfaces and are positive when acting towards the surface and negative when acting away from the surface. The conditions that produce internal suction and internal pressure are shown in Fig. 8.32. Both cases must be considered for any building and added algebraically to external pressures to determine the most critical loading condition.

Effective Wind Area

Because of local turbulence, which may occur over small areas and at ridges and corners of buildings, components and cladding are designed for higher wind pressures than the main wind force-resisting system. An effective wind area is used to determine the external pressure coefficient. This is defined in ASCE/SEI7 Sec. 26.2 as the span length multiplied by an effective tributary width and is given by

$$A = b_e l$$
$$b_e \geq l/3$$

l is the element span length and b_e is the effective tributary width.

For cladding fasteners, the effective wind area must not be greater than the area that is tributary to an individual fastener.

In accordance with ASCE/SEI7 Sec. 30.2.3, component and cladding elements with tributary areas greater than 700 ft^2 may be designed using provisions for the main wind force-resisting system.

External Pressure Coefficients

For the envelope procedure of ASCE/SEI7 Sec. 30.3, the gust effect factor is combined with the external pressure

Lateral Forces

coefficients and denoted by (GC_p). The pressure acting on external surfaces is obtained from the first term of ASCE/SEI7 Eq. 30.3-1 as

$$p_e = \pm q_h(GC_p)$$

Values of (GC_p) for walls are tabulated in ASCE/SEI7 Fig. 30.3-1 and for roofs in ASCE/SEI7 Fig. 30.3-2 through Fig. 30.3-7. The values of (GC_p) are a function of the effective area attributed to the element considered. Because turbulence at wall corners and at roof eaves produces a large increase in pressure, the building surface is divided into distinct zones as shown in Fig. 8.34.

Figure 8.34 *Components and Cladding External Pressure Zones for $\theta \leq 7°$*

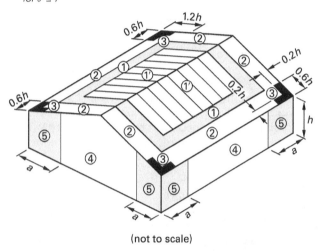

(not to scale)

For $\theta \leq 7°$, walls are divided into two zones, and roofs are divided into four zones, with a different wind pressure coefficient assigned to each. For the walls, the end zone width is given by ASCE/SEI7 Fig. 30.3-1 as a where a is the lesser of

$$a = (0.1)(\text{least horizontal dimension})$$
$$a = 0.4h$$

However, a cannot be less than either of the following.

$$a = (0.04)(\text{least horizontal dimension})$$
$$a = 3 \text{ ft}$$

When the least horizontal dimension of the building is greater than 300 ft and the roof angle, θ, is less than or equal to 7°, the end zone width is given by the lesser of

$$a = (0.1)(\text{least horizontal dimension})$$
$$\text{or}$$
$$a = 0.8h$$

For $\theta \leq 10°$, h is the eave height; otherwise, h is the mean roof height.

For a roof with $\theta \leq 7°$

The two edge zone widths, as given by ASCE/SEI7 Fig. 30.3-2A, are each $0.6h$. The four corner zones are each $0.2h$ wide and extend a distance of $0.6h$ along the eave and ridge.

For $\theta \leq 10°$, h is the eave height; otherwise, h is the mean roof height.

Pressures act normal to wall and roof surfaces and are positive when acting toward the surface, and negative when acting away from the surface. In accordance with ASCE/SEI7 Fig. 30.3-1, Note 5, the values of (GC_p) must be reduced by 10% for the walls of buildings with a roof slope of 10° or less.

Envelope Procedure for Components and Cladding

The following information is needed to determine wind loads for components and cladding using the envelope procedure.

- risk category I, II, III, or IV from Table 8.12 [ASCE/SEI7 Table 1.5-1]

- basic wind speed, V, for the applicable risk category from ASCE/SEI7 Sec. 26.5

- exposure category B, C, or D from Table 8.11 [ASCE/SEI7 Sec. 26.7]

- velocity pressure exposure coefficients, K_z, for the applicable exposure category from Table 8.17 [ASCE/SEI7 Table 26.10-1]

- topographic factor, K_{zt}, from ASCE/SEI7 Fig. 26.8-1

- ground elevation factor, K_e, from ASCE/SEI7 Table 26.9-1

- directionality factor, K_d, from ASCE/SEI7 Table 26.6-1

- enclosure classification from ASCE/SEI7 Sec. 26.12

- internal pressure coefficient, (GC_{pi}), from Table 8.14 [ASCE/SEI7 Table 26.13-1]

- wind velocity pressure, q_h, from ASCE/SEI7 Eq. 26.10-1

- external pressure coefficient, (GC_p), from ASCE/SEI7 Fig. 30.3-1 through Fig. 30.3-7

- internal wind pressure, $p_i = q_h(GC_{pi})$, from ASCE/SEI7 Eq. 30.3-1

- external wind pressure, $p_e = q_h(GC_p)$, from ASCE/SEI7 Eq. 30.3-1

- combined internal and external wind pressures, p, from ASCE/SEI7 Eq. 30.3-1

- minimum design wind loads from ASCE/SEI7 Sec. 30.2.2 (16 lbf/ft^2)

Example 8.27

The roof framing of the factory analyzed in Ex. 8.25 consists of open web joists spaced at 5 ft centers that span 20 ft parallel to the long side of the building. For wind flowing normal to the 80 ft long side of the building, determine the design wind pressure acting on a roof joist in zone 2. The building is enclosed.

Solution

From Ex. 8.25, the velocity pressure at eaves roof height using case 1 values for K_z is

$$q_h = 18.43 \text{ lbf/ft}^2$$

The product of the internal pressure coefficient and the gust effect factor for an enclosed building is given by Table 8.14 as

$$(GC_{pi}) = \pm 0.18$$

The roof slope is $\theta = 0°$ and the eave roof height is

$$h = 20 \text{ ft}$$
$$< 60 \text{ ft}$$

Therefore, the low-rise building method of ASCE/SEI7 Sec. 30.3 is applicable.

The effective tributary width of a roof joist is defined in ASCE/SEI7 Sec. 26.2 as the larger of the following.

$$b_e = \text{joist spacing}$$
$$= 5 \text{ ft}$$

Or,

$$b_e = \frac{l}{3}$$
$$= \frac{20 \text{ ft}}{3}$$
$$= 6.67 \text{ ft} \quad \text{[governs]}$$

The effective wind area attributed to the roof joist is

$$A = b_e l$$
$$= (6.67 \text{ ft})(20 \text{ ft})$$
$$= 133 \text{ ft}^2$$

The negative external pressure coefficient for roof zone 2, for an effective wind area of 133 ft^2, is obtained from ASCE/SEI7 Fig. 30.3-2A as

$$(GC_p) = -1.7$$

The negative design wind pressure on a roof joist for zone 2 is obtained from ASCE/SEI7 Eq. 30.3-1 as

$$p = q_h\big((GC_p) - (GC_{pi})\big)$$
$$= \left(18.43 \, \frac{\text{lbf}}{\text{ft}^2}\right)(-1.7 - 0.18)$$
$$= -34.65 \text{ lbf/ft}^2$$

The upward load on the roof joist over zone 2 is

$$w = ps$$
$$= \left(-34.65 \, \frac{\text{lbf}}{\text{ft}^2}\right)(5 \text{ ft})$$
$$= -173 \text{ lbf/ft}$$

The positive external pressure coefficient for roof zone 2, for an effective wind area of 133 ft^2, is obtained from ASCE/SEI7 Fig. 30.3-2A as

$$(GC_p) = 0.2$$

The positive design wind pressure on a roof joist for zone 2 is obtained from ASCE/SEI7 Eq. 30.3-1 as

$$p = q_h\big((GC_p) - (GC_{pi})\big)$$
$$= \left(18.43 \, \frac{\text{lbf}}{\text{ft}^2}\right)\big(0.2 - (-0.18)\big)$$
$$= 7.00 \text{ lbf/ft}^2$$
$$\text{Use 16 lbf/ft}^2 \text{ minimum.}$$

The downward load on the roof joist over zone 2 is

$$w = ps$$
$$= \left(16.00 \, \frac{\text{lbf}}{\text{ft}^2}\right)(5 \text{ ft})$$
$$= 80.00 \text{ lbf/ft}$$

The wind loading acting on the roof joist is shown.

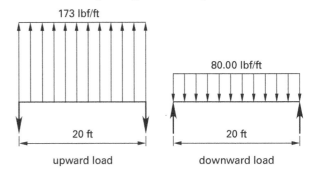

Lateral Forces

17. REFERENCES

1. International Code Council. *2018 International Building Code.* Falls Church, VA: International Code Council, 2018.

2. American Society of Civil Engineers. *Minimum Design Loads and Associated Criteria for Buildings and Other Structures.* Reston, VA: American Society of Civil Engineers, 2016.

3. Building Seismic Safety Council of the National Institute of Building Sciences. *NEHRP Recommended Seismic Provisions for New Buildings and Other Structures.* Washington, DC: Building Seismic Safety Council, 2015.

4. National Council of Structural Engineers Associations. *Guide to the Design of Diaphragms, Chords and Collectors.* Chicago, IL: National Council of Structural Engineers Associations, 2006.

5. Steel Deck Institute. *Diaphragm Design Manual.* Fox River Grove, IL: SDI, 2004.

6. APA–The Engineered Wood Association. *Diaphragms and Shear Walls: Design/Construction Guide.* Tacoma, WA: APA–The Engineered Wood Association, 2007.

7. American Institute of Steel Construction. *Prequalified Connections for Special and Intermediate Steel Moment Frames for Seismic Applications.* Chicago, IL: American Institute of Steel Construction, 2016.

8. Hamburger, Ronald O. et al. *Seismic Design of Steel Special Moment Frames: A Guide for Practicing Engineers*, NEHRP Seismic Design Technical Brief no. 2, vol. 6. Gaithersburg, MD: National Institute of Standards and Technology, 2009.

9. American Concrete Institute. *Building Code Requirements for Structural Concrete and Commentary.* Farmington Hills, MI: American Concrete Institute, 2014.

10. Moehle, Jack P., John D. Hooper, and Chris D. Lubke. *Seismic Design of Reinforced Concrete Special Moment Frames: A Guide for Practicing Engineers.* NEHRP Seismic Design Technical Brief no. 1, vol. 8. Gaithersburg, MD: National Institute of Standards and Technology, 2008.

11. Grinter, Linton E. *Theory of Modern Steel Structures, Vol. 1, Statically Determinate Structures.* New York, NY: Macmillan, 1949.

12. Williams, Alan. *Structural Analysis in Theory and Practice.* Burlington, MA: Elsevier/International Codes Council, 2009.

13. American Wood Council. *Special Design Provisions for Wind and Seismic.* Leesburg, VA. American Wood Council, 2015.

14. American Wood Council. *National Design Specification for Wood Construction.* Leesburg, VA. American Wood Council, 2018.

15. Line, P. *Perforated Shear Wall Design.* Washington, DC: American Wood Council, 2002.

16. American Institute of Steel Construction. *Steel Construction Manual,* Fifteenth ed. Chicago, IL: American Institute of Steel Construction, 2017.

17. American Institute of Steel Construction. *Specification for Structural Steel Buildings.* Chicago, IL: American Institute of Steel Construction, 2016.

18. American Institute of Steel Construction. *Seismic Provisions for Structural Steel Buildings.* Chicago, IL: American Institute of Steel Construction, 2016.

19. American Institute of Steel Construction. *Seismic Design Manual,* Third ed., Chicago, IL: American Institute of Steel Construction, 2018.

20. Lawson, J. Tilt-Up Panel Subdiaphragm Example. Structural Engineers Association of California Design Seminar, 1992.

Bridge Design

1. DESIGN LOADS

Nomenclature

a	tabulated force coefficient for distributed loads	–
A	area of beam	in^2
b	width of beam	in
B_s	width of equivalent strip	in
d	depth of beam	in
d_e	depth to the resultant of the tensile force	in
d_v	distance between the resultants of the tensile and compressive forces due to flexure	in
DC	dead load of components and attachments	kips
DW	dead load of wearing surface and utilities	kips
e_g	distance between the centers of gravity of the beam and the deck slab	in
E	width of equivalent strip for a slab bridge	in
EQ	earthquake load	lbf or kips
g	distribution factor for moment or shear	–
I	dynamic factor	–
I	moment of inertia of beam	in^4
IM	dynamic load allowance	–
k_s	stiffness of equivalent strip	in^4
K_g	longitudinal stiffness parameter	in^4
L	span length	ft
L_1	modified span length	ft
LL	vehicular live load	kips
m	multiple presence factor	–
M	bending moment	ft-kips
M_D	dead load moment	ft-kips
M_L	live load moment	ft-kips
M_s	service load moment	ft-kips
M_u	factored design moment	ft-kips

n	ratio of the modulus of elasticity of the beam and the deck slab	–
N_b	number of beams	–
N_L	number of traffic lanes	–
P	wheel load	kips
Q	factored force effect	–
R_n	nominal resistance capacity	ft-kips
S	spacing of beams	ft
t_s	deck slab thickness	in
V	shear force	kips
w	distributed load	kips/ft
w	roadway width between curbs	ft
w_D	dead load	kips/ft
w_L	lane width	ft
W	concentrated load	kips
W	edge-to-edge width of the bridge	ft
W_1	modified edge-to-edge width of the bridge	ft
x	distance from the center line of a stringer to the face of the stringer	in

Symbols

γ	load factor	–
η	load modifier	–
ϕ	strength reduction factor	–

Design Lanes

A bridge deck is divided into design lanes as defined in AASHTO[1] Sec. 3.6. For deck widths between 20 ft and 24 ft, two design lanes are specified, each equal to one-half the deck width. For all other deck widths, design lanes are defined as being 12 ft wide, with fractional parts of a lane discounted, and the number of design lanes is given by

$$N_L = \text{INT}\left(\frac{w}{12}\right)$$

INT is the integer part of the ratio.

Design lanes are positioned on the deck to produce the maximum effect. The determination of the number of design lanes is illustrated in Fig. 9.1.

Figure 9.1 *Design Traffic Lanes*

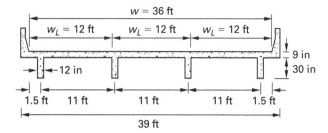

Example 9.1

For the bridge deck shown in the illustration, determine the number of design lanes.

Solution

From AASHTO Sec. 3.6, the number of design lanes is given by

$$N_L = \frac{w}{12} = \frac{36 \text{ ft}}{12 \, \dfrac{\text{ft}}{\text{lane}}} = 3 \text{ lanes}$$

Live Loads

The vehicular live loading for bridges is designated HL-93 and is specified in AASHTO Sec. 3.6.1.2. The loading consists of the more critical of the following two loading types.

- a design lane load combined with a design truck
- a design lane load combined with a design tandem

As shown in Fig. 9.2, the design lane load consists of a load of 0.64 kip/ft uniformly distributed in the longitudinal direction. In considering the design lane load in the design of continuous spans, as many spans should be loaded with the 0.64 kip/ft uniform load as is necessary

to produce the maximum effect. The design lane load is placed longitudinally only on those portions of the spans of a bridge to give the most critical effect. Transversely, the design lane load is uniformly distributed over a 10 ft width. The 10 ft loaded width is placed in the design lane to give the most critical effect without encroaching on the adjacent lane. A dynamic load allowance is not applied to the design lane load.

Figure 9.2 *Design Lane Load*

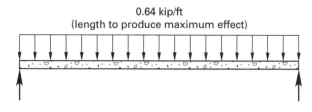

As shown in Fig. 9.3, the design truck load consists of three axles—the lead axle of 8 kips and the two following axles of 32 kips. The spacing between the two 32 kip axles is varied between 14 ft and 30 ft to produce the most critical effect. The transverse spacing of the wheels is 6 ft. Transversely, the design truck is positioned in a lane, as specified in AASHTO Sec. 3.6.1.3.1, so that the center of any wheel load is not closer than

- 1 ft from the face of a curb for the design of a deck overhang
- 2 ft from the edge of the design lane for all other components

Figure 9.3 *Design Truck Load*

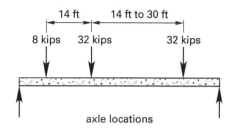

axle locations

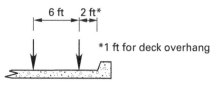

wheel locations

A dynamic load allowance is applied to the design truck load.

As shown in Fig. 9.4, the design tandem load consists of a pair of 25 kip axles spaced 4 ft apart. The transverse spacing of the wheels is 6 ft. Transversely, the design tandem is positioned in a lane in the same manner as the design truck. A dynamic load allowance is applied to the design tandem load.

Figure 9.4 Design Tandem Load

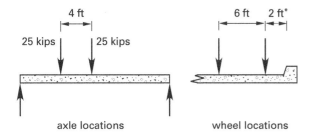

axle locations   wheel locations

*1 ft for deck overhang

In accordance with AASHTO Sec. 3.6.1.1.2, the number of loaded lanes is selected to produce the most critical effect. As specified in AASHTO Sec. 3.6.1.2.1, each lane under consideration must be occupied by either the design truck or tandem combined with the lane load. Figure 9.5 shows the location of the design truck and the design lane load to produce the maximum positive moment at midspan of the end span of a three-span continuous deck. The design lane load is placed on both end spans so as to produce the maximum effect. The design truck is placed with its central axle at the midspan of the end span.

Figure 9.5 Design Truck Positioned for Maximum Positive Moment at Point 5

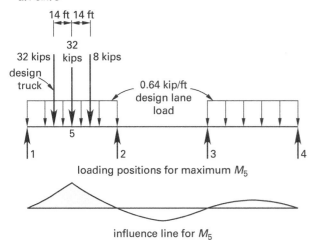

loading positions for maximum M_5

influence line for M_5

Similarly, Fig. 9.6 shows the design tandem positioned to produce the maximum positive moment at midspan of the end span of a three-span continuous deck.

Figure 9.6 Design Tandem Positioned for Maximum Positive Moment at Point 5

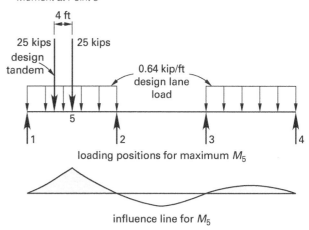

loading positions for maximum M_5

influence line for M_5

When several lanes are loaded, the force effect determined is multiplied by a multiple presence factor to account for the probability of simultaneous lane occupation by the full HL-93 design live load. AASHTO Table 3.6.1.1.2-1 gives the multiple presence factor, m, as

- 1.2 for one loaded lane

- 1.0 for two loaded lanes

- 0.85 for three loaded lanes

- 0.65 for more than three loaded lanes

For the determination of maximum negative moments in a continuous deck, AASHTO Sec. 3.6.1.3.1 specifies that two design trucks may be located in each lane with a minimum distance of 50 ft between the lead axle of one truck and the rear axle of the other truck. The distance between the 32 kip axles of each truck is 14 ft. The two design trucks are placed in adjacent spans to produce maximum force effects. Axles that do not contribute to the negative moment are neglected. The truck loading is combined with the design lane load using patch loading to produce the maximum effect. The total combined moment is multiplied by a reduction factor of 0.9 to give the design moment. The same procedure is used to determine the reaction at interior piers.

Similarly, as shown in Fig. 9.7, two design tandems may be applied, spaced 26–40 ft apart, and combined with the design lane load. This represents the loading caused by "low-boy" type vehicles weighing in excess of 110 kips. The total combined moment obtained is the required design moment without multiplying by a reduction factor.

For continuous spans, influence lines may be used to determine the maximum effect, and these are available[2,3] for standard cases. For nonstandard situations, several methods[4,5] may be used to determine the required influence lines.

Figure 9.7 *Design Lane Load and Two Design Tandems Positioned for Maximum Moment at Support 3*

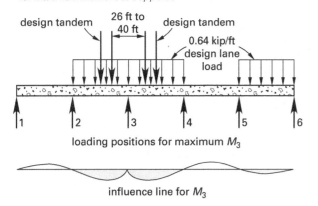

Example 9.2

The four-span bridge shown in the illustration has the superstructure analyzed in Ex. 9.1. Determine the maximum moment at support 2 produced by loading one design lane with the design lane load combined with the design truck. Neglect the multiple presence factor and dynamic load allowance.

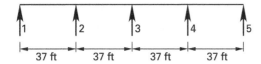

Solution

The locations of the design lane load to produce the maximum moment at support 2 are obtained[2] as shown in the illustration. Span 34 is not loaded.

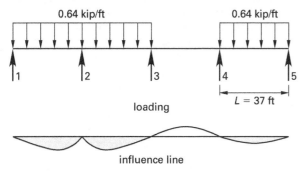

The bending moment at support 2, produced by one design lane load, is given by[2]

$$M_2 = awL^2$$
$$= (0.1205)\left(0.64 \ \frac{\text{kip}}{\text{ft}}\right)(37 \ \text{ft})^2$$
$$= 106 \ \text{ft-kips}$$

The axle locations of the two design trucks to produce the maximum moment at support 2 are obtained[2] as shown in the illustration. The 8 kip lead axle on each truck is neglected.

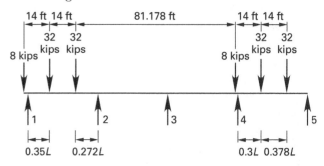

The bending moment at support 2, produced by the two design trucks, is given by[2]

$$M_2 = \sum \gamma WL$$
$$= \binom{0.083 + 0.090}{+ 0.0064 + 0.0051}(32 \ \text{kips})(37 \ \text{ft})$$
$$= 218 \ \text{ft-kips}$$

The combined moment produced by the design lane load and the two standard trucks is

$$M_2 = 106 \ \text{ft-kips} + 218 \ \text{ft-kips}$$
$$= 324 \ \text{ft-kips}$$

AASHTO Sec. 3.6.1.3.1 gives the design moment as

$$M_2 = (0.9)(324 \ \text{ft-kips})$$
$$= 292 \ \text{ft-kips}$$

Dynamic Load Allowance

In accordance with AASHTO Sec. 3.6.2, an allowance for dynamic effects is applied to the static axle loads of the design truck and the design tandem. AASHTO Table 3.6.2.1-1 gives the dynamic load allowance, *IM*, as

- 75% for deck joints for all limit states

- 15% for all other components for fatigue and fracture limit states

- 33% for all other components for all other limit states

The dynamic factor to be applied to the static load is

$$I = 1 + \frac{IM}{100}$$

The dynamic factor is not applied to

- the design lane load
- pedestrian loads
- centrifugal forces and braking forces
- retaining walls not subject to vertical loads from the superstructure
- foundation components that are entirely below ground level
- wood structures

The dynamic load allowance for culverts and other buried structures is given by AASHTO Eq. 3.6.2.2-1 as

$$IM = 33(1.0 - 0.125D_E)$$
$$\geq 0\%$$

D_E, the minimum depth of earth cover over the structure, is in ft.

Example 9.3

For the four-span bridge of Ex. 9.2, determine the maximum moment at support 2 produced by loading one design lane with the design lane load combined with the design truck. Include the effect of the dynamic load allowance.

Solution

The dynamic load allowance for the support moment is given by AASHTO Table 3.6.2.1-1 as

$$IM = 33\%$$

The dynamic factor is derived as

$$I = 1 + \frac{IM}{100}$$
$$= 1 + \frac{33}{100}$$
$$= 1.33$$

The static moment at support 2 caused by the two design trucks is given by Ex. 9.2 as

$$M_2 = 218 \text{ ft-kips}$$

The moment at support 2 caused by the two design trucks, including the dynamic load allowance, is

$$M_2 = (1.33)(218 \text{ ft-kips})$$
$$= 290 \text{ ft-kips}$$

The combined moment produced by the design lane load and the two standard trucks, including the dynamic load allowance, is

$$M_2 = 106 \text{ ft-kips} + 290 \text{ ft-kips}$$
$$= 396 \text{ ft-kips}$$

The final design moment is given by AASHTO Sec. 3.6.1.3.1 as

$$M_2 = (0.9)(396 \text{ ft-kips})$$
$$= 356 \text{ ft-kips}$$

Lateral Distribution of Loads

In the calculation of bending moments for T-beam bridges, permanent loads of and on the deck may be distributed uniformly to all beams.

The distribution of live load depends on the torsional stiffness of the bridge deck system and, if necessary, may be determined by several[6,7,8,9] analytical methods. In accordance with AASHTO Sec. 4.6.2.2, however, the distribution of live load may be calculated by empirical expressions, depending on the superstructure type and the stringer spacing.

The types of superstructure for which the distribution factor method may be used are illustrated in AASHTO Table 4.6.2.2.1-1. The method may be applied provided that the following conditions are met.

- A single lane of live loading is analyzed.
- Multiple lanes of live loading producing approximately the same force effect per lane are analyzed.
- The deck width is constant.
- The number of beams is not less than four (with some exceptions).
- Beams are parallel and have approximately the same stiffness.
- The roadway part of the overhang does not exceed 3 ft (with some exceptions).
- The curvature of the superstructure is less than the limit specified in AASHTO Sec. 4.6.1.2.4.

Additional requirements are specified for each specific superstructure illustrated in AASHTO Table 4.6.2.2.1-1, and these are listed in AASHTO Table 4.6.2.2.2b-1.

A monolithic T-beam superstructure is listed as case (e) in AASHTO Table 4.6.2.2.1-1. For this type of bridge, the limitation on the beam spacing is

$$3.5 \text{ ft} \leq S \leq 16.0 \text{ ft}$$

The limitation on the deck slab thickness is

$$4.5 \text{ in} \le t_s \le 12.0 \text{ in}$$

The limitation on the superstructure span is

$$20 \text{ ft} \le L \le 240 \text{ ft}$$

The limitation on the number of beams is

$$N_b \ge 4$$

The limitation on the longitudinal stiffness parameter is

$$10,000 \text{ in}^4 \le K_g \le 7,000,000 \text{ in}^4$$

The longitudinal stiffness parameter is defined by AASHTO Eq. 4.6.2.2.1-1 as

$$K_g = n(I + Ae_g^2)$$

The ratio of the modulus of elasticity of the beam and the deck slab is defined by AASHTO Eq. 4.6.2.2.1-2 as

$$n = \frac{E_B}{E_D}$$

The moment of inertia of the beam is

$$I = \frac{bd^3}{12}$$

The area of the beam is

$$A = bd$$

The distance between the centers of gravity of the beam and the deck slab is

$$e_g = \frac{t_s + d}{2}$$

When these conditions are complied with, AASHTO Table 4.6.2.2.2b-1 gives the distribution factor for moment for one design lane loaded as

$$g_1 = 0.06 + \left(\frac{S}{14}\right)^{0.4}\left(\frac{S}{L}\right)^{0.3}\left(\frac{K_g}{12.0Lt_s^3}\right)^{0.1}$$

When two or more design lanes are loaded, the distribution factor for moment is

$$g_m = 0.075 + \left(\frac{S}{9.5}\right)^{0.6}\left(\frac{S}{L}\right)^{0.2}\left(\frac{K_g}{12.0Lt_s^3}\right)^{0.1}$$

In accordance with AASHTO Sec. 3.6.1.1.2, the multiple presence factors specified in AASHTO Table 3.6.1.1.2-1 are not applicable as these factors are already incorporated in the distribution factors. The dynamic load allowance must be applied to that portion of the bending moment produced by design trucks and design tandems.

These distribution factors are not applicable for the determination of bending moments in exterior beams. For exterior beams, with one lane loaded, and for interior beams in decks with less than four beams, the lever-rule method specified in AASHTO Sec. C4.6.2.2.1 may be used. For these analyses, both the multiple presence factor and the dynamic load allowance must be applied. Irrespective of the calculated moment, an exterior beam must have a carrying capacity not less than that of an interior beam.

Example 9.4

For the four-span concrete T-beam bridge of Ex. 9.2, determine the maximum live load moment for design of an interior beam at support 2. The ratio of the modulus of elasticity of the beam and the deck slab is $n = 1.0$.

Solution

From Ex. 9.3, the maximum live load moment produced at support 2 by loading one design lane with two design trucks, plus dynamic load allowance, and the design lane load is

$$M_2 = 356 \text{ ft-kips}$$

The ratio of the modulus of elasticity of the beam and the deck slab is given as

$$n = \frac{E_B}{E_D} = 1.0$$

The moment of inertia of the beam is

$$I = \frac{bd^3}{12} = \frac{(12 \text{ in})(30 \text{ in})^3}{12}$$
$$= 27,000 \text{ in}^4$$

The area of the beam is

$$A = bd = (12 \text{ in})(30 \text{ in})$$
$$= 360 \text{ in}^2$$

The distance between the centers of gravity of the beam and the deck slab is

$$e_g = \frac{t_s + d}{2} = \frac{9 \text{ in} + 30 \text{ in}}{2}$$
$$= 19.5 \text{ in}$$

The longitudinal stiffness parameter of the deck is defined by AASHTO Eq. 4.6.2.2.1-1 as

$$K_g = n(I + A e_g^2)$$
$$= (1.0)(27{,}000 \text{ in}^4 + (360 \text{ in}^2)(19.5 \text{ in})^2)$$
$$= 163{,}890 \text{ in}^4 \quad \begin{bmatrix} \text{complies with} \\ \text{AASHTO Table 4.6.2.2b-1} \end{bmatrix}$$
$$> 10{,}000 \text{ in}^4$$
$$< 7{,}000{,}000 \text{ in}^4$$

The beam spacing is

$$S = 11 \text{ ft} \quad \text{[complies with AASHTO Table 4.6.2.2.2b-1]}$$
$$> 3.5 \text{ ft}$$
$$< 16.0 \text{ ft}$$

The deck slab thickness is

$$t_s = 9 \text{ in} \quad \text{[complies with AASHTO Table 4.6.2.2.2b-1]}$$
$$> 4.5 \text{ in}$$
$$< 12.0 \text{ in}$$

The superstructure span is

$$L = 37 \text{ ft} \quad \text{[complies with AASHTO Table 4.6.2.2.2b-1]}$$
$$> 20 \text{ ft}$$
$$< 240 \text{ ft}$$

The number of beams in the deck is

$$N_b = 4 \quad \text{[complies with AASHTO Table 4.6.2.2.2b-1]}$$

Thus, the configuration of the deck completely conforms to the requirements of AASHTO Table 4.6.2.2.2b-1.

With one lane loaded, AASHTO Table 4.6.2.2.2b-1 gives the distribution factor for moment as

$$g_1 = 0.06 + \left(\frac{S}{14}\right)^{0.4}\left(\frac{S}{L}\right)^{0.3}\left(\frac{K_g}{12.0 L t_s^3}\right)^{0.1}$$
$$= 0.06 + \left(\frac{11 \text{ ft}}{14}\right)^{0.4}\left(\frac{11 \text{ ft}}{37 \text{ ft}}\right)^{0.3}$$
$$\times \left(\frac{163{,}890 \text{ in}^4}{(12.0)(37 \text{ ft})(9 \text{ in})^3}\right)^{0.1}$$
$$= 0.650$$

With two lanes loaded, as shown in the illustration, AASHTO Table 4.6.2.2.2b-1 gives the distribution factor for moment as

$$g_m = 0.075 + \left(\frac{S}{9.5}\right)^{0.6}\left(\frac{S}{L}\right)^{0.2}\left(\frac{K_g}{12.0 L t_s^3}\right)^{0.1}$$
$$= 0.075 + \left(\frac{11 \text{ ft}}{9.5}\right)^{0.6}\left(\frac{11 \text{ ft}}{37 \text{ ft}}\right)^{0.2}$$
$$\times \left(\frac{163{,}890 \text{ in}^4}{(12.0)(37 \text{ ft})(9 \text{ in})^3}\right)^{0.1}$$
$$= 0.875 \quad \text{[governs]}$$

The live load moment for the design of an interior beam at support 2 is

$$M_L = g_m M_2$$
$$= (0.875)(356 \text{ ft-kips})$$
$$= 312 \text{ ft-kips}$$

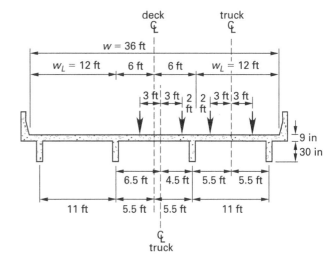

Shear Determination

The distribution factor method is also used to calculate design shear in interior beams. The distribution factors and the range of applicability are listed in AASHTO Table 4.6.2.2.3a-1. For a monolithic T-beam superstructure, the limitations on beam spacing, slab thickness, span length, and number of beams are identical with those for determining the distribution factor for moment. There is no requirement specified for the longitudinal stiffness parameter.

When these conditions are complied with, AASHTO Table 4.6.2.2.3a-1 gives the distribution factor for shear for one design lane loaded as

$$g_1 = 0.36 + \frac{S}{25}$$

When two or more design lanes are loaded, the distribution factor for shear is

$$g_m = 0.2 + \frac{S}{12} - \left(\frac{S}{35}\right)^{2.0}$$

In accordance with AASHTO Sec. 3.6.1.1.2, the multiple presence factors that are specified in AASHTO Table 3.6.1.1.2-1 are not applicable as these factors are already incorporated in the distribution factors. The dynamic load allowance must be applied to that portion of the shear produced by design trucks and design tandems.

These distribution factors are not applicable for the determination of shear in exterior beams. For exterior beams, with one lane loaded, and for interior beams in decks with less than four beams, the lever rule method specified in AASHTO Sec. C4.6.2.2.1 may be used. For these analyses, both the multiple presence factor and the dynamic load allowance must be applied. Irrespective of the calculated shear, an exterior beam must not have less resistance than an interior beam.

The application of the lever rule is illustrated in Fig. 9.8 for the determination of the distribution factor for shear in the exterior girder of a T-beam superstructure. For one lane loaded, the center of one wheel of an axle of the design truck or the design tandem is located 2 ft from the edge of the design lane as specified in AASHTO Sec. 3.6.1.3.1. A notional hinge is introduced into the deck slab at the position of beam 2 and moments are taken about this hinge. The reaction at beam 1, in terms of one wheel load, is

$$V_1 = \frac{P(4.5 \text{ ft} + 10.5 \text{ ft})}{11 \text{ ft}}$$
$$= 1.364P$$

The distribution factor for shear for one lane loaded with one axle is

$$g_1 = \frac{V_1}{2P}$$
$$= \frac{1.364P}{2P}$$
$$= 0.682$$

Applying the multiple presence factor for one lane loaded gives a distribution factor of

$$g = 1.2g_1$$
$$= (1.2)(0.682)$$
$$= 0.818$$

Figure 9.8 *Lever Rule for Shear in an Exterior Girder*

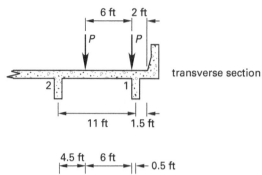

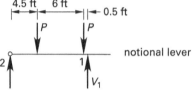

Example 9.5

For the four-span concrete T-beam bridge of Ex. 9.2, determine the live load shear, V_{23}, for design of an interior beam. Use the design truck load combined with the design lane load.

Solution

The influence line for V_{23} is shown in the illustration.

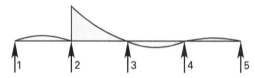

To produce the maximum value of V_{23}, the truck is positioned as shown in the illustration.

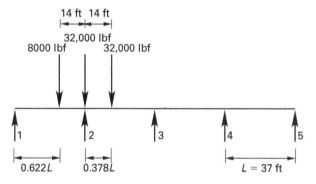

The shear at end 2 of span 23 produced by one standard truck is given by[2]

$$V_{23} = \sum \gamma W$$
$$= \frac{(0.128)(8000 \text{ lbf})}{1000 \, \frac{\text{lbf}}{\text{kip}}} + \frac{(1.635)(32{,}000 \text{ lbf})}{1000 \, \frac{\text{lbf}}{\text{kip}}}$$
$$= 53.34 \text{ kips}$$

As determined in Ex. 9.4, the superstructure dimensions are within the allowable range of applicability.

AASHTO Table 4.6.2.2.3a-1 gives the distribution factor for shear for one design lane loaded as

$$g_1 = 0.36 + \frac{S}{25}$$
$$= 0.36 + \frac{11 \text{ ft}}{25 \text{ ft}}$$
$$= 0.800$$

When two or more design lanes are loaded, the distribution factor for shear is

$$g_m = 0.2 + \frac{S}{12} - \left(\frac{S}{35}\right)^{2.0}$$
$$= 0.2 + \frac{11 \text{ ft}}{12 \text{ ft}} - \left(\frac{11 \text{ ft}}{35 \text{ ft}}\right)^{2.0}$$
$$= 1.018 \quad \text{[governs]}$$

The dynamic load allowance for the shear is given by AASHTO Table 3.6.2.1-1 as

$$IM = 33\%$$

This is applied to the static axle load of the design truck to give the dynamic factor

$$I = 1 + \frac{IM}{100} = 1 + \frac{33}{100}$$
$$= 1.33$$

The shear at support 2 caused by the design truck, including the dynamic load allowance, is

$$V_{23} = (1.33)(53.34 \text{ kips})$$
$$= 70.94 \text{ kips}$$

The design lane load is positioned in spans 12, 23, and 45 in order to produce the maximum shear at support 2. The shear produced by one design lane load is given by[2]

$$V_{23} = awL$$
$$= (0.6027)\left(0.64 \frac{\text{kip}}{\text{ft}}\right)(37 \text{ ft})$$
$$= 14.27 \text{ kips}$$

The combined shear produced by one lane of the design lane load and the standard truck, including the dynamic load allowance, is

$$V_{23} = 70.94 \text{ kips} + 14.27 \text{ kips}$$
$$= 85.21 \text{ kips}$$

The live load shear, produced by two loaded lanes for the design of an interior beam at support 2, is

$$V_L = g_m V_{23}$$
$$= (1.018)(85.21 \text{ kips})$$
$$= 86.74 \text{ kips}$$

Design of Concrete Deck Slabs

The bending moments, caused by wheel loads, in concrete deck slabs supported by longitudinal stringers and transverse girders may be obtained by the methods proposed by Westergaard[10] and Pucher.[11] AASHTO Sec. 4.6.2.1 provides an equivalent strip method for the design of concrete deck slabs.

This method consists of dividing the deck into strips perpendicular to the supporting stringers and transverse girders. The principal features of the method are as follows.

- The extreme positive moment in any deck panel must be applied to all positive moment regions.

- The extreme negative moment over any supporting component must be applied to all negative moment regions.

- Where the deck slab spans primarily in the transverse direction, only the axles of the design truck or the design tandem should be applied to the deck slab.

- Where the deck slab spans primarily in the longitudinal direction, and the span does not exceed 15 ft, only the axles of the design truck or the design tandem should be applied to the deck slab.

- Where the deck slab spans primarily in the longitudinal direction, and the span exceeds 15 ft, the design truck combined with the design lane load, or the design tandem combined with the design lane load, should be applied to the deck slab and the provisions of AASHTO Sec. 4.6.2.3 apply.

- Where the deck slab spans primarily in the longitudinal direction, the width of the equivalent strip supporting an axle load must not be taken greater than 40 in for open grids.

- Where the deck slab spans primarily in the transverse direction, the equivalent strip is not subject to width limits.

- Where the spacing of supporting components in the secondary direction exceeds 1.5 times the spacing in the primary direction, all of the wheel loads may be considered to be applied to the primary strip. Distribution reinforcement that complies with AASHTO Sec. 9.7.3.2 may be applied in the secondary direction.

- Where the spacing of supporting components in the secondary direction is less than 1.5 times the spacing in the primary direction, the deck should be modeled as a system of intersecting strips.

- Wheel loads are distributed to the intersecting strips in proportion to their stiffnesses.

- The stiffness of a strip is specified as $k_s = EI_s/S^3$.

- Strips are treated as simply supported or continuous beams as appropriate with a span length equal to the center-to-center distance between the supporting components.

- Wheel loads may be modeled as concentrated loads or as patch loads whose length along the span is equal to the length of the tire contact area plus the depth of the deck slab.

- Both the multiple presence factor and the dynamic load allowance must be applied to the bending moments calculated.

- In lieu of determining the width of the equivalent strip, the moments may be obtained directly from AASHTO Table A4-1 and these values include an allowance for both the multiple presence factor and the dynamic load allowance.

AASHTO Table 4.6.2.1.3-1 defines the width of an equivalent strip. For cast-in-place deck slabs, the width, in inches, of both longitudinal and transverse strips for calculating positive moment is

$$B_s = 26 \text{ in} + 6.6 S_{\text{ft}}$$

The width of both longitudinal and transverse strips for calculating negative moment is

$$B_s = 48 \text{ in} + 3.0 S_{\text{ft}}$$

Example 9.6

For the four-span concrete T-beam bridge of Ex. 9.2, determine the maximum negative live load moment in the slab and the width of the equivalent strip. The layout of longitudinal and transverse girders is shown in the illustration.

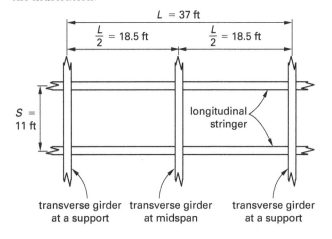

Solution

The aspect ratio of the slab is

$$AR = \frac{\dfrac{L}{2}}{S} = \frac{18.5 \text{ ft}}{11 \text{ ft}}$$
$$= 1.68$$
$$> 1.5$$

Therefore, all of the wheel loads may be considered to be applied to the primary strip in the transverse direction. Distribution reinforcement that complies with AASHTO Sec. 9.7.3.2 may be applied in the secondary direction. Since the deck slab spans primarily in the transverse direction, only the axles of the design vehicle should be applied to the deck slab.

The width, in inches, of the transverse strip, which is used for calculating negative moment, is given by AASHTO Table 4.6.2.1.3-1 as

$$B_s = 48 \text{ in} + 3.0 S_{\text{ft}}$$
$$= \frac{48 \text{ in} + (3.0)(11 \text{ ft})}{12 \dfrac{\text{in}}{\text{ft}}}$$
$$= 6.75 \text{ ft}$$

The required moment may be determined from AASHTO Table A4-1. The span length of the transverse strip is

$$S = 11 \text{ ft}$$

The distance from the center line of a longitudinal stringer to the face of the stringer is

$$x = 6 \text{ in}$$

Therefore, from AASHTO Table A4-1, the maximum negative bending moment is

$$M_s = 7.38 \text{ ft-kips/ft}$$

Design of Slab-Type Bridges

The bending moments and shears in concrete slab-type decks, caused by axle loads, may be obtained by an equivalent strip method that is defined in AASHTO Sec. 4.6.2.3. The equivalent width of a longitudinal strip with two lines of wheels in one lane is given by AASHTO Eq. 4.6.2.3-1 as

$$E = 10.0 + 5.0(L_1 W_1)^{0.5}$$

The modified span length, L_1, is equal to the lesser of the actual span length or 60 ft. The modified edge-to-edge width of the bridge, W_1, is equal to the lesser of the actual width, W, or 30 ft for single-lane loading, or 60 ft for multi-lane loading.

The equivalent width of a longitudinal strip with more than one lane loaded is given by AASHTO Eq. 4.6.2.3-2 as

$$E = 84.0 + 1.44(L_1W_1)^{0.5}$$
$$\leq 12.0\,W/N_L$$

The number of design lanes, N_L, is determined as specified in AASHTO Sec. 3.6.1.1.1.

An allowance for the multiple presence factor is included in the equivalent strip width. The dynamic load allowance must be applied to the bending moments calculated.

Example 9.7

A prestressed concrete slab bridge has a simply supported span of $L = 37$ ft. The overall width of the bridge is $W = 39$ ft, and the distance between curbs is $w = 36$ ft. Determine the width of the equivalent strip.

Solution

From AASHTO Sec. 3.6, the number of design lanes is

$$N_L = \frac{w}{12}$$
$$= \frac{36 \text{ ft}}{12 \dfrac{\text{ft}}{\text{lane}}}$$
$$= 3 \text{ lanes}$$

For one design lane loaded, the modified span length is equal to the lesser of the actual span length or 60 ft, and

$$L_1 = L = 37 \text{ ft}$$

The modified edge-to-edge width of the bridge is equal to the lesser of the actual width or 30 ft, and

$$W_1 = 30 \text{ ft}$$

The equivalent width of a longitudinal strip is given by AASHTO Eq. 4.6.2.3-1 as

$$E = 10.0 + 5.0(L_1W_1)^{0.5}$$
$$= 10.0 \text{ in} + (5.0)\big((37 \text{ ft})(30 \text{ ft})\big)^{0.5}$$
$$= 176.6 \text{ in}$$

For more than one design lane loaded, the modified span length is equal to the lesser of the actual span length or 60 ft, and

$$L_1 = L = 37 \text{ ft}$$

The modified edge-to-edge width of the bridge is equal to the lesser of the actual width or 60 ft, and

$$W_1 = W = 39 \text{ ft}$$

The equivalent width of a longitudinal strip is given by AASHTO Eq. 4.6.2.3-2 as

$$E = 84.0 + 1.44(L_1W_1)^{0.5}$$
$$= 84.0 \text{ in} + (1.44)\big((37 \text{ ft})(39 \text{ ft})\big)^{0.5}$$
$$= 138.7 \text{ in}$$
$$\frac{12.0\,W}{N_L} = \frac{(12.0)(39 \text{ ft})}{3 \text{ lanes}}$$
$$= 156 \text{ in}$$
$$> E \quad [E = 138.7 \text{ in}]$$

The equivalent width for more than one design lane loaded governs, and

$$E = 138.7 \text{ in}$$

Combinations of Loads

The load and resistance factor design method presented in AASHTO Sec. 1.3.2, defines four limit states: the service limit state, the fatigue and fracture limit state, the strength limit state, and the extreme event limit state.

The *service limit state* governs the design of the structure under regular service conditions to ensure satisfactory stresses, deformations, and crack widths. Four service limit states are defined, with service I limit state comprising the load combination relating to the normal operational use of the bridge with a 55 mph wind, and all loads taken at their nominal values.

The *fatigue limit state* governs the design of the structure loaded with a single design truck for a given number of stress range cycles. The *fracture limit state* is defined as a set of material toughness requirements given in the AASHTO Materials Specifications.

The *strength limit state* ensures the structure's strength and structural integrity under the various load combinations imposed on the bridge during its design life. Five strength limit states are defined, with strength I limit state comprising the load combination relating to the normal vehicular use of the bridge without wind.

The *extreme event limit state* ensures the survival of the structure during a major earthquake or flood, or when subject to collision from a vessel, vehicle, or ice flow.

Two extreme limit states are defined, with extreme event I limit state comprising the load combination that includes earthquake.

The factored load is influenced by the ductility of the components, the redundancy of the structure, and the operational importance of the bridge based on social or defense requirements. It is preferable for components to exhibit ductile behavior, as this provides warning of impending failure by large inelastic deformations. Brittle components are undesirable because failure occurs suddenly, with little or no warning, when the elastic limit is exceeded. For the strength limit state, the load modifier for ductility is given by AASHTO Sec. 1.3.3 as

$$\eta_D = 1.05 \quad \text{[nonductile components]}$$
$$= 1.00 \quad \text{[conventional designs and details]}$$
$$\geq 0.95 \quad \left[\begin{array}{c} \text{components with} \\ \text{ductility-enhancing features} \end{array} \right]$$

For all other limit states, the load modifier for ductility is given by AASHTO Sec. 1.3.3 as

$$\eta_D = 1.00$$

The component redundancy classification is based on the contribution of the component to the bridge safety. Major components, whose failure will cause collapse of the structure, are designated as failure-critical, and the associated structural system is designated nonredundant. Alternatively, components whose failure will not cause collapse of the structure are designated as nonfailure-critical, and the associated structural system is designated redundant. For the strength limit state, the load modifier for redundancy is given by AASHTO Sec. 1.3.4 as

$$\eta_R = 1.05 \quad \text{[nonredundant components]}$$
$$= 1.00 \quad \text{[conventional levels of redundancy]}$$
$$\geq 0.95 \quad \text{[exceptional levels of redundancy]}$$

For all other limit states, the load modifier for redundancy is given by AASHTO Sec. 1.3.4 as

$$\eta_R = 1.00$$

A bridge may be declared to be of operational importance based on survival or security reasons. For the strength limit state, the load modifier for operational importance is given by AASHTO Sec. 1.3.5 as

$$\eta_I = 1.05 \quad \text{[for important bridges]}$$
$$= 1.00 \quad \text{[for typical bridges]}$$
$$\geq 0.95 \quad \text{[for relatively less important bridges]}$$

For all other limit states, the load modifier for importance is given by AASHTO Sec. 1.3.5 as

$$\eta_I = 1.00$$

For loads where a maximum value is appropriate, the combined load modifier relating to ductility, redundancy, and operational importance is given by AASHTO Eq. 1.3.2.1-2 as

$$\eta_i = \eta_D \eta_R \eta_I$$
$$\geq 0.95$$

For loads where a minimum value is appropriate, the combined load modifier is given by AASHTO Eq. 1.3.2.1-3 as

$$\eta_i = \frac{1}{\eta_D \eta_R \eta_I}$$
$$\leq 1.0$$

The load factors applicable to permanent loads are listed in AASHTO Table 3.4.1-2 and are summarized in Table 9.1.

Table 9.1 Load Factors for Permanent Loads

type of load	load factor, γ_p	
	max	min
components and attachments, DC	1.25	0.90
wearing surfaces and utilities, DW	1.5	0.65

The actual value of permanent loads may be less than or more than the nominal value, and both possibilities must be considered by using the maximum and minimum values given for the load factor.

Load combinations and load factors are listed in AASHTO Table 3.4.1-1, and those applicable to gravity and earthquake loads are summarized in Table 9.2.

Table 9.2 Load Factors and Load Combinations

load combination limit state	DC and DW	LL and IM	EQ
strength I	γ_p	1.75	–
extreme event I	1.00	γ_{EQ}	1.00
service I	1.00	1.00	–
fatigue I	–	1.75	–
fatigue II	–	0.80	–

The value of the load factor γ_p for the dead load of components and wearing surfaces is obtained from Table 9.1. The load factor γ_{EQ} in extreme event limit state I has traditionally been taken as 0.0. However, partial live load should be considered, and a reasonable value for the load factor is

$$\gamma_{EQ} = 0.50$$

The total factored force effect is given by AASHTO Eq. 3.4.1-1 as

$$Q = \sum \eta_i \gamma_i Q_i$$
$$\leq \phi R_n$$

Both positive and negative extremes must be considered for each load combination. For permanent loads, the load factor that produces the more critical effect is selected from Table 9.1. In strength I limit state, when the permanent loads produce a positive effect and the live loads produce a negative effect, the appropriate total factored force effect is

$$Q = 0.9DC + 0.65DW + 1.75(LL + IM)$$

In strength I limit state, when both the permanent loads and the live loads produce a negative effect, the appropriate total factored force effect is

$$Q = 1.25DC + 1.50DW + 1.75(LL + IM)$$

Example 9.8

For the four-span concrete T-beam bridge of Ex. 9.2, determine the strength I factored moment for design of an interior beam at support 2. Each concrete parapet has a weight of 0.5 kip/ft, and the parapets are constructed after the deck slab has cured. Assume a unit weight of concrete of 0.15 kip/ft³.

Solution

The dead load acting on an interior beam consists of the beam self-weight, plus the applicable portion of the deck slab, plus the applicable portion of the two parapets. The dead load of a beam is

$$w_B = \left(0.15 \, \frac{\text{kip}}{\text{ft}^3}\right)(2.5 \text{ ft})(1 \text{ ft})$$
$$= 0.375 \text{ kip/ft}$$

The dead load of the applicable portion of the deck slab is determined by dividing the total dead load by four beams.

$$w_S = \frac{\left(0.15 \, \frac{\text{kip}}{\text{ft}^3}\right)(39 \text{ ft})(0.75 \text{ ft})}{4 \text{ beams}}$$
$$= 1.097 \text{ kips/ft}$$

The weights of the two concrete parapets may be distributed equally to the four beams. Then, the applicable weight distributed to an interior beam is

$$w_P = \frac{\left(0.5 \, \frac{\text{kip}}{\text{ft}}\right)(2 \text{ parapets})}{4 \text{ beams}}$$
$$= 0.25 \text{ kip/ft}$$

The total dead load supported by an interior beam is

$$w_D = w_B + w_S + w_P$$
$$= 0.375 \, \frac{\text{kip}}{\text{ft}} + 1.097 \, \frac{\text{kips}}{\text{ft}} + 0.25 \, \frac{\text{kip}}{\text{ft}}$$
$$= 1.722 \text{ kips/ft}$$

The bending moment produced in an interior beam at support 2 by the uniformly distributed dead load is given by[2]

$$M_D = aw_D L^2$$
$$= (0.1071)\left(1.722 \, \frac{\text{kips}}{\text{ft}}\right)(37 \text{ ft})^2$$
$$= 252 \text{ ft-kips}$$

The live load bending moment plus impact at support 2 is obtained from Ex. 9.4 as

$$M_L = 312 \text{ ft-kips}$$

The factored design moment for a strength I limit state is given by AASHTO Eq. 3.4.1-1 and by AASHTO Table 3.4.1-1 as

$$M_u = \eta_i(\gamma_p M_D + \gamma_{LL+IM} M_L)$$
$$= 1.0(1.25 M_D + 1.75 M_L)$$
$$= (1.0)\big((1.25)(252 \text{ ft-kips}) + (1.75)(312 \text{ ft-kips})\big)$$
$$= 861 \text{ ft-kips}$$

Critical Section for Shear

AASHTO Sec. 5.7.3.2 specifies that when the support reaction produces a compressive stress in a reinforced concrete beam, the critical section for shear is located at a distance from the support equal to the depth, d_v. The depth, d_v, is defined in AASHTO Sec. 5.7.2.8 as the distance between the resultants of the tensile and compressive forces due to flexure. The effective depth, d_e, is the depth to the resultant of the tensile force.

$$d_v \geq 0.9 d_e$$
$$\geq 0.72h$$

Example 9.9

For the four-span concrete T-beam bridge of Ex. 9.2, determine the factored shear force, V_{23}, for design of an interior beam at support 2. The depth $d_v = 31.4$ in.

Solution

The live load shear force, including impact, on an interior beam at support 2 is obtained from Ex. 9.5 as

$$V_L = 86.74 \text{ kips}$$

The dead load supported by an interior beam is obtained from Ex. 9.8 as

$$w_D = 1.722 \text{ kips/ft}$$

The dead load shear at the support of an interior beam is given by[2]

$$
\begin{aligned}
V_s &= a w_D L \\
&= (0.536)\left(1.722 \ \frac{\text{kips}}{\text{ft}}\right)(37 \text{ ft}) \\
&= 34.15 \text{ kips}
\end{aligned}
$$

In accordance with AASHTO Sec. 5.7.3.2, the design shear for a distributed load may be determined at a distance d_v from the support and is given by

$$
\begin{aligned}
V_D &= V_s - w_D d_v \\
&= 34.15 \text{ kips} - \frac{\left(1.722 \ \dfrac{\text{kips}}{\text{ft}}\right)(31.4 \text{ in})}{12 \ \dfrac{\text{in}}{\text{ft}}} \\
&= 29.64 \text{ kips}
\end{aligned}
$$

The factored design shear for strength I limit state is given by AASHTO Eq. 3.4.1-1 and AASHTO Table 3.4.1-1 as

$$
\begin{aligned}
V_{23} &= \eta_i(\gamma_p V_D + \gamma_{LL+IM} V_L) \\
&= 1.0(1.25 V_D + 1.75 V_L) \\
&= (1.0)\big((1.25)(29.64 \text{ kips}) + (1.75)(86.74 \text{ kips})\big) \\
&= 189 \text{ kips}
\end{aligned}
$$

Service Limit State

The service limit state governs stresses, deformations, and crack widths under regular service conditions. In accordance with AASHTO Sec. 3.4.1, the service I limit state comprises the load combination relating to the normal operational use of a bridge with a 55 mph wind and all loads taken at their nominal values.

Example 9.10

For the four-span concrete T-beam bridge of Ex. 9.1, determine the service I design moment for an interior beam at support 2. Each concrete parapet has a weight of 0.5 kip/ft, and the parapets are constructed after the deck slab has cured. Wind effects may be neglected.

Solution

From Ex. 9.8, the bending moment at support 2 produced by the uniformly distributed dead load is

$$M_D = 252 \text{ ft-kips}$$

The live load bending moment plus impact at support 2 is obtained from Ex. 9.8 as

$$M_L = 312 \text{ ft-kips}$$

The service I design moment is given by AASHTO Sec. 3.4.1 as

$$
\begin{aligned}
M_s &= \eta_i(\gamma_p M_D + \gamma_{LL+IM} M_L) \\
&= 1.0(1.0 M_D + 1.0 M_L) \\
&= 252 \text{ ft-kips} + 312 \text{ ft-kips} \\
&= 564 \text{ ft-kips}
\end{aligned}
$$

2. REINFORCED CONCRETE DESIGN

Design for Flexure

Nomenclature

a	depth of equivalent rectangular stress block	in
A_s	area of tension reinforcement	in²
b	width of compression face of member	in
$\bar{c}$	distance from extreme tension fiber to centroid of tension reinforcement	in
d	distance from extreme compression fiber to centroid of tension reinforcement	in
d_b	diameter of bar	in
d_c	thickness of concrete cover measured from extreme tension fiber to center of nearest bar	in
d_l	distance from extreme compression fiber to centroid of extreme tension element	in
f_{ct}	splitting tensile stress	kips/in²
f_c'	compressive strength of concrete	kips/in²
f_{min}	minimum stress in reinforcement	kips/in²
f_r	modulus of rupture of concrete	kips/in²
f_{ss}	calculated stress in tension reinforcement at service loads	kips/in²

f_y	yield strength of reinforcement	$kips/in^2$
h	overall dimension of member	in
h_f	flange depth	in
h_{min}	recommended minimum depth of superstructure	ft
I_g	moment of inertia of gross concrete section	in^4
K_u	design moment factor	lbf/in^2
l_a	lever arm for elastic design	in
L	span	ft
M_{cr}	cracking moment	ft-kips
M_{dnc}	bending moment due to noncomposite dead load acting on the precast section	ft-kips
M_D	dead load moment	ft-kips
M_{max}	maximum moment	ft-kips
M_{min}	minimum design flexural strength	ft-kips
M_{mr}	maximum moment range	ft-kips
M_n	nominal flexural strength of a member	ft-kips
M_u	factored moment on the member	ft-kips
s	spacing of reinforcement	in
S_c	section modulus of the composite section referred to the bottom fiber	in^3
S_{nc}	section modulus of the noncomposite section referred to the bottom fiber	in^3
w_c	unit weight of concrete	$kips/ft^3$

Symbols

β_1	compression zone factor	–
β_s	ratio of flexural strain at the extreme tension face to the strain of the centroid of the reinforcement layer nearest to the tension face	–
γ	influence line coefficient, load factor	–
γ_1	flexural cracking variability factor	–
γ_2	prestress variability factor	–
γ_3	ratio of specified minimum yield strength to ultimate tensile strength of the reinforcement	–
γ_e	exposure factor	–
λ	concrete density modification factor	–
ρ	ratio of tension reinforcement	–
ϕ	strength reduction factor	–
ω	tension reinforcement index	–
Δf	live load stress range	$kips/in^2$
$(\Delta F)_{TH}$	constant-amplitude fatigue threshold	$kips/in^2$

Strength Design Method

The procedure specified in AASHTO Sec. 5.6 is similar to the procedure adopted in the ACI[12] building code. In addition, stresses at service load must be limited to ensure satisfactory performance under service load conditions, and the requirements for deflection, cracking moment, flexural cracking, skin reinforcement, and fatigue must be satisfied.

Load Factor Design

When the depth of the equivalent stress block is not greater than the flange depth of a reinforced concrete T-beam, the section may be designed as a rectangular beam. The resistance factor for a tension-controlled reinforced concrete section is given by AASHTO Sec. 5.5.4.2 as

$$\phi = 0.90$$

Example 9.11

For the four-span concrete T-beam bridge of Ex. 9.1, determine the tensile reinforcement required in an interior beam in the end span 12. The concrete strength is 4 kips/in^2, and the reinforcement consists of no. 9 grade 60 bars. Assume that the strength I factored moment is $M_u = 1216$ ft-kips.

Solution

The effective compression flange width is given by AASHTO Sec. 4.6.2.6.1 as the tributary width, which is

$$b = S$$
$$= (11 \text{ ft})\left(12 \frac{\text{in}}{\text{ft}}\right)$$
$$= 132 \text{ in}$$

The factored design moment is given as

$$M_u = 1216 \text{ ft-kips}$$

Assuming that the stress block lies within the flange and the effective depth, d, is 34.6 in, the required tension reinforcement is determined from the principles of AASHTO Sec. 5.6. The design moment factor is

$$K_u = \frac{M_u}{bd^2}$$
$$= \frac{(1216 \text{ ft-kips})\left(12 \frac{\text{in}}{\text{ft}}\right)\left(1000 \frac{\text{lbf}}{\text{kip}}\right)}{(132 \text{ in})(34.6 \text{ in})^2}$$
$$= 92.3 \text{ lbf/in}^2$$

$$\frac{K_u}{f_c'} = \frac{92.3 \frac{\text{lbf}}{\text{in}^2}}{4000 \frac{\text{lbf}}{\text{in}^2}}$$
$$= 0.0231$$

From App. 2.A, the corresponding tension reinforcement index is

$$\omega = 0.026$$
$$< 0.319\beta_1$$
$$= (0.319)(0.85)$$
$$= 0.271$$

Hence, the section is tension controlled, and $\phi = 0.9$.

The required reinforcement ratio is

$$\rho = \frac{\omega f_c'}{f_y} = \frac{(0.026)\left(4\ \dfrac{\text{kips}}{\text{in}^2}\right)}{60\ \dfrac{\text{kips}}{\text{in}^2}}$$
$$= 0.00173$$

The reinforcement area required is

$$A_s = \rho b d$$
$$= (0.00173)(132\ \text{in})(34.6\ \text{in})$$
$$= 7.90\ \text{in}^2$$

Using eight no. 9 bars as shown in the illustration, the reinforcement area provided is

$$A_s = 8\ \text{in}^2$$
$$> 7.90\ \text{in}^2 \quad [\text{satisfactory}]$$
$$\phi M_n = \frac{M_u(8\ \text{in}^2)}{7.90\ \text{in}^2} = \frac{(1216\ \text{ft-kips})(8\ \text{in}^2)}{7.90\ \text{in}^2}$$
$$= 1231\ \text{ft-kips}$$

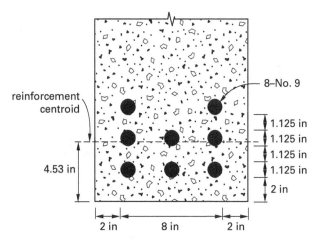

The height of the centroid of the tensile reinforcement is

$$\bar{c} = \frac{(3)(2.563\ \text{in} + 4.813\ \text{in}) + (2)(7.063\ \text{in})}{8}$$
$$= 4.53\ \text{in}$$

The effective depth is

$$d = h - \bar{c}$$
$$= 39\ \text{in} - 4.53\ \text{in}$$
$$= 34.47\ \text{in}$$
$$\approx 34.6\ \text{in} \quad [\text{assumed value of } d \text{ satisfactory}]$$

The stress block depth is

$$a = \frac{A_s f_y}{0.85 b f_c'}$$
$$= \frac{(8\ \text{in}^2)\left(60{,}000\ \dfrac{\text{lbf}}{\text{in}^2}\right)}{(0.85)(132\ \text{in})\left(4000\ \dfrac{\text{lbf}}{\text{in}^2}\right)}$$
$$= 1.07\ \text{in}$$
$$< h_f \quad \left[\begin{array}{c}\text{The stress block is contained}\\ \text{within the flange.}\end{array}\right]$$

Deflection Requirements

Deflections due to service live load plus impact are limited by AASHTO Sec. 2.5.2.6.2 to

$$\delta_{\max} = \frac{L}{800}$$

To achieve these limits, AASHTO Table 2.5.2.6.3-1 provides expressions for the determination of minimum superstructure depths. These are summarized in Table 9.3.

Table 9.3 *Recommended Minimum Depths*

superstructure type	minimum depth (ft)	
	simple spans	continuous spans
slabs spanning in direction of traffic	$(1.2)(S+10)/30$	$(S+10)/30 \geq 0.54$
T-beams	$0.070L$	$0.065L$
box girders	$0.060L$	$0.055L$

Actual deflections may be calculated in accordance with AASHTO Sec. 5.6.3.5.2, with the modulus of elasticity of normal weight concrete with $f_c' \leq \text{kips/in}^2$ given by AASHTO Eq. C5.4.2.4-3 as

$$E_c = 1820\sqrt{f_c'}$$

In determining deflections, the effective moment of inertia may be taken as the moment of inertia of the gross concrete section.

Example 9.12

Determine whether the deflection under live load of the four-span concrete T-beam bridge of Ex. 9.1 is satisfactory.

Solution

The recommended minimum depth of the T-beam superstructure, in accordance with AASHTO Table 2.5.2.6.3-1, is

$$h_{\min} = 0.065L$$
$$= (0.065)(37 \text{ ft})$$
$$= 2.4 \text{ ft}$$

The depth provided is

$$h = 3.25 \text{ ft}$$
$$> 2.4 \text{ ft} \quad [\text{satisfactory}]$$

Cracking Moment Requirements

The *cracking moment* is the moment that when applied to a reinforced concrete member, will produce cracking in the tension face of the member. In determining the cracking moment, AASHTO Sec. 5.6.3.5.2 allows the use of the gross section properties neglecting reinforcement. In the case of T-beam construction, it is appropriate to include the full width of the flange, tributary to the web, in determining the gross moment of inertia, I_g. The modulus of rupture of normal weight concrete is given by AASHTO Sec. 5.4.2.6 as

$$f_r = 0.24\lambda\sqrt{f'_c}$$

λ = concrete density modification factor = 1.0 for normal weight concrete; otherwise,

$$\lambda = \frac{4.7f_{ct}}{\sqrt{f'_c}} \le 1.0$$

or

$$0.75 \le \lambda = 7.5w_c \le 1.0 \quad [\text{where } f_{ct} \text{ is not specified}]$$

For a noncomposite reinforced concrete member, $\gamma_2 = 0$, $S_c/S_{nc} = 1$, and M_{dnc} is not applicable. When the neutral axis of the section is a distance $\bar{y}$ from the tension face, the cracking moment is given by the reduced version of AASHTO Eq. 5.6.3.3-1 as

$$M_{cr} = \frac{\gamma_1\gamma_3 f_r I_g}{\bar{y}}$$

The applicable factors for the cracking moment are as follows.

γ_1 = flexural cracking variability factor

$\quad = 1.2$ [precast segmental structures]

$\quad = 1.6$ [other concrete structures]

γ_3 = ratio of specified minimum yield strength to ultimate tensile strength of the reinforcement

$\quad = 0.67$ [A615, grade 60 reinforcement]

$\quad = 0.75$ [A615, grade 75 reinforcement]

To prevent sudden tensile failure of a flexural member, AASHTO Sec. 5.6.3.3 requires the member to have a moment capacity at least equal to the lesser of

$$\phi M_n = M_{cr}$$
$$\phi M_n = 1.33M_u$$

Example 9.13

Determine whether the interior beam in the end span 12 of the four-span concrete T-beam bridge of Ex. 9.11 complies with AASHTO Sec. 5.6.3.3. The bridge has A615, grade 60 reinforcement and normal weight concrete.

Solution

The gross moment of inertia of an interior beam is obtained as shown in the following table.

part	A (in^2)	y (in)	I (in^4)	Ay (in^3)	Ay^2 (in^4)
beams	360	15.0	27,000	5400	81,000
flange	1188	34.5	8019	40,986	1,414,017
total	1548	–	35,019	46,386	1,495,017

The height of the neutral axis of the section is

$$\bar{y} = \frac{\sum Ay}{\sum A} = \frac{46,386 \text{ in}^3}{1548 \text{ in}^2}$$
$$= 30 \text{ in}$$
$$I_g = \sum I + \sum Ay^2 + \bar{y}^2 \sum A - 2\bar{y}\sum Ay$$
$$= 35,019 \text{ in}^4 + 1,495,017 \text{ in}^4$$
$$\quad + (1548 \text{ in}^2)(30 \text{ in})^2 - (60 \text{ in})(46,386 \text{ in}^3)$$
$$= 140,074 \text{ in}^4$$

The concrete density modification factor is $\lambda = 1.0$, and the modulus of rupture of the concrete is given by AASHTO Sec. 5.4.2.6 as

$$f_r = 0.24\sqrt{f_c'} = 0.24\sqrt{4\ \frac{\text{kips}}{\text{in}^2}}$$
$$= 0.48\ \text{kip/in}^2$$

For a nonsegmental concrete structure,

$$\gamma_1 = 1.6$$

For a nonprestressed concrete structure with A615, grade 60 reinforcement,

$$\gamma_3 = 0.67$$

Therefore,

$$\gamma_1\gamma_3 = 1.072$$

The cracking moment of an interior beam is given by the reduced version of AASHTO Eq. 5.6.3.3-1 as

$$M_{cr} = \frac{\gamma_1\gamma_3 f_r I_g}{\bar{y}}$$
$$= \frac{(1.072)\left(0.48\ \frac{\text{kip}}{\text{in}^2}\right)(140{,}074\ \text{in}^4)}{(30\ \text{in})\left(12\ \frac{\text{in}}{\text{ft}}\right)}$$
$$= 200\ \text{ft-kips}$$

From Ex. 9.11, the factored applied moment is

$$M_u = 1216\ \text{ft-kips}$$
$$1.33 M_u = (1.33)(1216\ \text{ft-kips})$$
$$= 1617\ \text{ft-kips}$$
$$> M_{cr}\quad [M_{cr}\ \text{governs}]$$

From Ex. 9.11, the design strength of an interior beam is

$$\phi M_n = 1231\ \text{ft-kips}$$
$$> M_{cr}\quad [\text{satisfactory}]$$

The beam complies with AASHTO Sec. 5.6.3.3.

Control of Flexural Cracking

To control flexural cracking of the concrete, the size and arrangement of tension reinforcement must be adjusted.

Two exposure conditions are defined in AASHTO Sec. 5.6.7. Class 1 exposure condition applies when cracks can be tolerated because of reduced concern for

appearance or corrosion. Class 2 exposure condition applies when there is greater concern for appearance or corrosion.

The anticipated crack width depends on the following factors.

- the spacing, s, of reinforcement in the layer closest to the tension face
- the tensile stress, f_{ss}, in reinforcement at the service limit state $\leq 0.6 f_y$
- the thickness of concrete cover, d_c, measured from the extreme tension fiber to center of reinforcement in the layer closest to the tension face

The exposure factor is defined as

$$\gamma_e = 1.00\quad [\text{class 1 exposure conditions}]$$
$$\gamma_e = 0.75\quad [\text{class 2 exposure conditions}]$$

The ratio of flexural strain at the extreme tension face to the strain at the centroid of the reinforcement layer nearest to the tension face is defined as

$$\beta_s = 1 + \frac{d_c}{0.7(h - d_c)}$$

The required spacing of reinforcement in the layer closest to the tension face is given by AASHTO Eq. 5.6.7-1 as

$$s \leq \frac{700\gamma_e}{\beta_s f_{ss}} - 2 d_c$$

Example 9.14

For an interior beam in the end span 12 of the four-span concrete T-beam bridge of Ex. 9.1, determine the allowable spacing of tension reinforcement. Assume that the service I moment is $M_s = 639$ ft-kips.

Solution

The concrete cover measured to the center of the reinforcing bar closest to the tension face of the member is obtained from Ex. 9.11 as

$$d_c = 2.56\ \text{in}$$

The lever arm for elastic design may conservatively be taken as

$$l_a = d - \frac{h_f}{2} = 34.47\ \text{in} - 4.5\ \text{in}$$
$$= 29.97\ \text{in}$$

The maximum service dead plus live load moment in an interior beam in the end span 12 is given as

$$M_s = 639 \text{ ft-kips}$$

The stress in the reinforcement is given by

$$
\begin{aligned}
f_{ss} &= \frac{M_s}{l_a A_s} \\
&= \frac{(639 \text{ ft-kips})\left(12 \dfrac{\text{in}}{\text{ft}}\right)}{(29.97 \text{ in})(8 \text{ in}^2)} \\
&= 31.98 \text{ kips/in}^2 \\
&< 0.60 f_y = 36 \text{ kips/in}^2
\end{aligned}
$$

The exposure factor for class 1 exposure conditions is given by AASHTO Sec. 5.6.7 as

$$\gamma_e = 1.00$$

The ratio of flexural strain at the extreme tension face to the strain at the centroid of the reinforcement layer nearest to the tension face is

$$
\begin{aligned}
\beta_s &= 1 + \frac{d_c}{0.7(h - d_c)} \\
&= 1 + \frac{2.56 \text{ in}}{(0.7)(39 \text{ in} - 2.56 \text{ in})} \\
&= 1.10
\end{aligned}
$$

The required spacing of reinforcement in the layer closest to the tension face is given by AASHTO Eq. 5.6.7-1 as

$$
\begin{aligned}
s &\le \frac{700\gamma_e}{\beta_s f_{ss}} - 2d_c \\
&= \frac{(700)(1.00)}{(1.10)\left(31.98 \dfrac{\text{kips}}{\text{in}^2}\right)} - (2)(2.56 \text{ in}) \\
&= 14.8 \text{ in}
\end{aligned}
$$

The spacing provided is

$$s = 2.3 \text{ in} \quad [\text{satisfactory}]$$

Longitudinal Skin Reinforcement

Longitudinal skin reinforcement is required to control cracking in the side faces of members where d_l exceeds 3 ft. In accordance with AASHTO Sec. 5.6.7, skin reinforcement must be provided over a distance of $d_l/2$ nearest the flexural tension reinforcement, and the area in each face, per foot of height, must be not less than

$$
\begin{aligned}
A_{s(\min)} &= (0.012)(d_l - 30) \quad [\text{in}^2/\text{ft}] \\
&\le A_s/4
\end{aligned}
$$

The spacing of this reinforcement must not exceed the lesser of

$$s = \frac{d_l}{6}$$

or

$$s = 12 \text{ in}$$

Example 9.15

Determine the skin reinforcement required, in an interior beam, in the end span 12 of the four-span concrete T-beam bridge of Ex. 9.1.

Solution

The distance from the extreme compression fiber to the centroid of the extreme tension element is

$$
\begin{aligned}
d_l &= 39 \text{ in} - 2.56 \text{ in} \\
&= 36.4 \text{ in} \\
&\approx 36 \text{ in}
\end{aligned}
$$

Hence, in accordance with AASHTO Sec. 5.6.7, skin reinforcement is not required.

Fatigue Limits

Fatigue stress limits are defined in AASHTO Sec. 5.5.3 and depend on the stress in the reinforcement and the range of stress resulting from the fatigue I load combination. The constant-amplitude fatigue threshold, $(\Delta F)_{\text{TH}}$, for straight reinforcement, is defined by AASHTO Sec. 5.5.3.2 as

$$(\Delta F)_{\text{TH}} = 26 - \frac{0.22 f_{\min}}{f_y} \quad [\text{AASHTO 5.5.3.2-1}]$$

$f_{\min}$ is the minimum live-load stress (in kips/in^2) resulting from the fatigue I load combination combined with the more severe stress from either the unfactored permanent loads, or the unfactored permanent loads, shrinkage, and creep-induced external loads. The minimum stress is positive if tension, negative if compression.

The factored live load stress range due to the passage of the fatigue I vehicle is

$$\gamma(\Delta f) \le (\Delta F)_{\text{TH}}$$

Stress levels are determined at the fatigue I limit state load. In accordance with AASHTO Table 3.4.1-1, this consists of 150% of the design vehicle live load including dynamic load allowance. As specified in AASHTO Sec. 3.6.1.4, the design vehicle consists of a single design truck with a constant spacing of 30 ft between the 32 kip axles. In accordance with AASHTO Sec. 3.6.1.3.1, axles that do not contribute to the maximum force under consideration are neglected.

Example 9.16

Determine whether the fatigue stress limits, in an interior beam, at the midspan of span 12 of the four-span concrete T-beam bridge of Ex. 9.1 and 9.2 are satisfactory. Assume a value of $g = 0.65$ for the load distribution to the beam. Ignore the effects of creep and shrinkage.

Solution

The maximum moment at the midspan of an interior beam in span 12 caused by the design truck, plus the dynamic load allowance, is derived as

$$M_{\max} = IWL\gamma g$$
$$= (1.15)(32 \text{ kips})(37 \text{ ft})(0.1998)(0.65)$$
$$= 177 \text{ ft-kips}$$

150% of the maximum moment is

$$M_{1.5,\max} = (1.5)(177 \text{ ft-kips})$$
$$= 266 \text{ ft-kips}$$

The influence line for the bending moment at midspan of beam 12 is shown in the illustration.

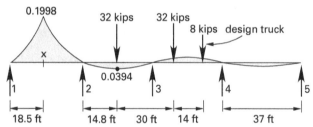

The location of the design truck to produce the minimum live load moment at point x is shown in the illustration. In accordance with AASHTO Sec. 3.6.1.3.1, the leading axle and the first 32 kips axle are ignored. Hence, the minimum moment caused by the design truck plus the dynamic load allowance is derived as

$$M_{\min} = IWL\gamma g$$
$$= (1.15)(32 \text{ kips})(37 \text{ ft})(-0.0394)(0.65)$$
$$= -35 \text{ ft-kips}$$

150% of the minimum moment is

$$M_{1.5,\min} = (1.5)(-35 \text{ ft-kips})$$
$$= -53 \text{ ft-kips}$$

The maximum moment range is

$$M_{mr} = M_{1.5,\max} - M_{1.5,\min}$$
$$= 266 \text{ ft-kips} - (-53 \text{ ft-kips})$$
$$= 319 \text{ ft-kips}$$

The lever arm for elastic design is obtained from Ex. 9.14 as

$$l_a = 29.97 \text{ in}$$

The actual factored stress range is

$$\gamma(\Delta f) = \frac{M_{mr}}{l_a A_s} = \frac{(319 \text{ ft-kips})\left(12 \dfrac{\text{in}}{\text{ft}}\right)}{(29.97 \text{ in})(8 \text{ in}^2)}$$
$$= 15.97 \text{ kips/in}^2$$

From Ex. 9.8, the dead load acting on beam 12 is $w = 1.722 \text{ kips/ft}$.

The permanent dead load moment at section x is derived as

$$M_D = wL^2\gamma$$
$$= \left(1.722 \dfrac{\text{kips}}{\text{ft}}\right)(37 \text{ ft})^2(0.0714)$$
$$= 168 \text{ ft-kips}$$

The minimum live load moment resulting from the fatigue load combined with the permanent dead load moment is

$$M = M_D + M_{1.5,\min}$$
$$= 168 \text{ ft-kips} - 53 \text{ ft-kips}$$
$$= 115 \text{ ft-kips}$$

The corresponding stress is

$$f_{\min} = \frac{M}{l_a A_s}$$
$$= \frac{(115 \text{ ft-kips})\left(12 \dfrac{\text{in}}{\text{ft}}\right)}{(29.97 \text{ in})(8 \text{ in}^2)}$$
$$= 5.76 \text{ kips/in}^2$$

The constant-amplitude fatigue threshold is given by AASHTO Eq. 5.5.3.2-1 as

$$(\Delta F)_{\text{TH}} = 26 - \frac{22f_{\min}}{f_y}$$

$$= 26 - \frac{(22)\left(5.76 \ \dfrac{\text{kips}}{\text{in}^2}\right)}{60 \ \dfrac{\text{kips}}{\text{in}^2}}$$

$$= 23.89 \ \text{kips/in}^2$$

$$> \gamma(\Delta f)$$

The fatigue stress limits are satisfactory.

Design for Shear

Nomenclature

A_s	area of tension reinforcement	in^2
A_v	area of shear reinforcement perpendicular to flexural tension reinforcement	in^2
b_v	web width	in
d_e	effective depth from extreme compression fiber to the centroid of the tensile force in the tensile reinforcement	in
d_v	effective shear depth	in
f_c'	specified compressive strength of concrete	kips/in^2
f_y	specified yield strength of reinforcing bars	kips/in^2
h	overall depth of member	–
M_n	nominal flexural resistance	ft-kips
s	spacing of transverse reinforcement	in
v_u	average factored shear stress	kips/in^2
V_c	nominal shear strength provided by concrete	kips
V_s	nominal shear strength provided by shear reinforcement	kips
V_u	factored shear force at section	kips

Symbols

β	factor relating effect of longitudinal strain on the shear capacity of concrete	–
θ	angle of inclination of diagonal compressive stress	degree
ϕ	resistance factor	–

Design Methods

Three design methods are described in the *AASHTO LRFD Bridge Design Specifications.* For members in which the strain distribution is nonlinear, AASHTO Sec. 5.8.2 specifies the use of a strut-and-tie model. This method is applicable to pile caps and deep footings, and

to members with abrupt changes in cross section. The traditional sectional model is applicable where engineering beam theory is valid, as is the case for typical bridge girders and slabs. The sectional model is specified in AASHTO Sec. 5.7.3. For nonprestressed concrete sections, not subjected to axial tension and with the minimum area of transverse reinforcement specified in AASHTO Eq. 5.7.2.5-1, a simplified procedure is permissible as specified in AASHTO Sec. 5.7.3.4.1.

Simplified Design Method

In the simplified method, the value for the longitudinal strain factor is taken as $\beta = 2.0$, and the value of the angle of inclination of diagonal compressive stress is taken as $\theta = 45°$. This produces the simplified expressions for nominal concrete shear capacity, V_c, and nominal stirrup shear capacity, V_s, used in the procedure. This is similar to the approach adopted in the ACI[12] building code.

The nominal shear capacity of the concrete section for $\beta = 2.0$ is given by AASHTO Sec. 5.7.3.3 as

$$V_c = 0.0632 b_v d_v \lambda \sqrt{f_c'} \quad \text{[AASHTO 5.7.3.3-3]}$$

The effective shear depth, d_v, is taken as the distance between the resultants of the tensile and compressive forces due to flexure.

The effective shear depth for a reinforced concrete beam is given by AASHTO Eq. C5.7.2.8-1 as

$$d_v = \frac{M_n}{A_s f_y}$$

The effective shear depth need not be taken to be less than the greater of $0.9d_e$ or $0.72h$. The effective web width, b_v, is taken as the minimum web width between the resultants of the tensile and compressive forces due to flexure.

The nominal shear capacity of vertical stirrups with $\theta = 45°$ is given by AASHTO Sec. 5.7.3.3 as

$$V_s = \frac{A_v f_y d_v}{s} \quad \text{[AASHTO C5.7.3.3-1]}$$

The shear stress on the concrete is calculated by AASHTO Eq. 5.7.2.8-1 as

$$v_u = \frac{V_u}{\phi b_v d_v}$$

For a value of v_u less than $0.125f_c'$, AASHTO Sec. 5.7.2.6 limits the spacing of transverse reinforcement to the lesser of $0.8d_v$, or 24 in. When the value of v_u is not less than $0.125f_c'$, the spacing is reduced to the lesser of $0.4d_v$, or 12 in.

A minimum area of transverse reinforcement is required to control diagonal cracking and is specified by AASHTO Eq. 5.7.2.5-1 as

$$A_v = \frac{0.0316\lambda\sqrt{f_c'}\,b_v s}{f_y}$$

The combined nominal shear resistance of the concrete section and the shear reinforcement is given by AASHTO Sec. 5.7.3.3 as the lesser of

$$V_n = V_c + V_s$$
$$V_n = 0.25f_c' b_v d_v$$

The combined shear capacity of the concrete section and the shear reinforcement is

$$\phi V_n = \phi V_c + \phi V_s$$
$$\geq V_u$$

The resistance factor for shear and torsion is given by AASHTO Sec. 5.5.4.2 as

$$\phi = 0.90 \quad \text{[normal weight concrete]}$$
$$\phi = 0.90 \quad \text{[lightweight concrete]}$$

Example 9.17

For the four-span concrete T-beam bridge of Ex. 9.1, determine the shear reinforcement required in an interior beam at end 2 of span 23. The concrete strength is 4 kips/in^2, and the shear reinforcement consists of no. 4 grade 60 bars. The depth $d_v = 31.4$ in.

Solution

From Ex. 9.9, the factored shear at a distance d_v from the support is

$$V_{23} = 189 \text{ kips}$$

For normal weight concrete, $\lambda = 1.0$.

The shear strength provided by the concrete is given by

$$\phi V_c = 0.0632\phi b_v d_v \sqrt{f_c'}$$
$$= (0.0632)(0.90)(12 \text{ in})(31.4 \text{ in})\sqrt{4 \frac{\text{kips}}{\text{in}^2}}$$
$$= 42.87 \text{ kips}$$
$$< V_{23}$$

The factored shear force exceeds the shear strength of the concrete, and the shear strength required from shear reinforcement is given by

$$\phi V_s = V_{23} - \phi V_c = 189 \text{ kips} - 42.87 \text{ kips}$$
$$= 146.13 \text{ kips}$$

The shear stress is given by AASHTO Eq. 5.7.2.8-1 as

$$v_u = \frac{V_{23}}{\phi b_v d_v} = \frac{189 \text{ kips}}{(0.9)(12 \text{ in})(31.4 \text{ in})}$$
$$= 0.56 \text{ kips/in}^2$$
$$> 0.125f_c'$$

Therefore, stirrups are required at a maximum spacing of 12 in. The area of shear reinforcement required is given by AASHTO Eq. C5.7.3.3-1 as

$$\frac{A_v}{s} = \frac{\phi V_s}{\phi d_v f_y} = \frac{(146.13 \text{ kips})\left(12 \frac{\text{in}}{\text{ft}}\right)}{(0.90)(31.4 \text{ in})\left(60 \frac{\text{kips}}{\text{in}^2}\right)}$$
$$= 1.03 \text{ in}^2/\text{ft}$$

Shear reinforcement consisting of two arms of no. 4 bars at 4 in spacing provides a reinforcement area of

$$\frac{A_v}{s} = 1.2 \frac{\text{in}^2}{\text{ft}}$$
$$> 1.03 \text{ in}^2/\text{ft} \quad \text{[satisfactory]}$$

3. PRESTRESSED CONCRETE DESIGN

Design for Flexure

Nomenclature

a	depth of equivalent rectangular stress block	in
A	area of concrete section	in^2
A_c	area of composite section	in^2
A_{ps}	area of prestressing steel	in^2
A_s	area of nonprestressed tension reinforcement	in^2
b	width of compression face of member	in
c	distance from the extreme compression fiber to the neutral axis	in
d_p	distance from extreme compression fiber to centroid of prestressing tendons	in
d_s	distance from the extreme compression fiber to the centroid of nonprestressed tensile reinforcement	in

e	eccentricity of prestressing force	in
f_{be}	bottom fiber stress at service load after allowance for all prestress losses	kips/in^2
f_{bi}	bottom fiber stress immediately after prestress transfer and before time-dependent prestress losses	kips/in^2
f_c'	specified compressive strength of concrete	kips/in^2
f_{ci}'	compressive strength of concrete at time of prestress transfer	kips/in^2
f_{cpe}	bottom fiber stress due only to effective prestressing force after allowance for all prestress losses	kips/in^2
f_{pbt}	allowable stress in prestressing steel immediately prior to prestress transfer	kips/in^2
f_{pe}	effective stress in prestressing steel after allowance for all prestress losses	kips/in^2
f_{pj}	stress in the prestressing steel at jacking	kips/in^2
f_{ps}	stress in prestressing steel at ultimate load	kips/in^2
f_{pt}	stress in the prestressing steel immediately after transfer	kips/in^2
f_{pu}	specified tensile strength of prestressing steel	kips/in^2
f_{py}	specified yield strength of prestressing steel	kips/in^2
f_r	modulus of rupture of concrete	kips/in^2
f_s	stress in the tension reinforcement at nominal flexural resistance	kips/in^2
f_{te}	top fiber stress at service loads after allowance for all prestress losses	kips/in^2
f_{ti}	top fiber stress immediately after prestress transfer and before time-dependent prestress losses	kips/in^2
f_y	specified yield strength of reinforcing bars	kips/in^2
h_f	compression flange thickness	in
k	prestressing steel factor	–
L	span length	ft
M_{cr}	cracking moment	in-kips
M_D	bending moment due to superimposed dead load	in-kips
M_{DC}	bending moment due to superimposed dead load on composite section	in-kips
M_{dnc}	bending moment due to noncomposite dead load acting on the precast section, $M_g + M_S$	in-kips
M_g	bending moment due to self-weight of girder	in-kips
M_L	bending moment due to superimposed live load	in-kips
M_n	nominal flexural strength	in-kips
M_r	factored flexural resistance	in-kips
M_S	bending moment due to weight of deck slab	in-kips
M_u	factored moment	in-kips
P_e	force in prestressing steel at service loads after allowance for all losses	kips

P_i	force in prestressing steel immediately after prestress transfer	kips
S_b	section modulus of the concrete section referred to the bottom fiber	in^3
S_c	section modulus of the composite section referred to the bottom fiber	in^3
S_{ci}	section modulus of the composite section referred to the interface of girder and slab	in^3
S_{nc}	section modulus of the noncomposite section referred to the bottom fiber	in^3
S_t	section modulus of the concrete section referred to the top fiber	in^3
w	distributed load	kips/ft
y_b	height of centroid of the concrete section	in
y_s	height of centroid of the prestressing steel	in

Symbols

α_1	stress block factor	–
β_1	stress block factor	–
γ_1	flexural cracking variability factor	–
γ_2	prestress variability factor	–
γ_3	ratio of specified minimum yield strength to ultimate tensile strength of the reinforcement	–
ϵ_{cu}	failure strain of concrete in compression	–
ϵ_t	net tensile strain in extreme tension steel at nominal resistance	–
ϕ	strength reduction factor	–
ϕ_w	reduction factor for slender members	–

Conditions at Transfer

The allowable stresses in the concrete at transfer, in other than segmentally constructed bridges, are specified in AASHTO Sec. 5.9.2.3 and AASHTO Table 5.9.2.3.1b-1 and are

$$f_{ti} \geq -0.0948\lambda\sqrt{f_{ci}'} \quad \text{[without bonded reinforcement]}$$

$$\geq -0.2 \text{ kips/in}^2 \quad \text{[without bonded reinforcement]}$$

$$f_{ti} \geq -0.24\lambda\sqrt{f_{ci}'} \quad \text{[with bonded reinforcement]}$$

$$f_{bi} \leq 0.65f_{ci}' \quad \text{[pretensioned members]}$$

$$f_{bi} \leq 0.65f_{ci}' \quad \text{[post-tensioned members]}$$

In accordance with AASHTO Table 5.9.2.2-1, the maximum allowable stress in pretensioned tendons immediately prior to transfer is

$$f_{pbt} = 0.75f_{pu} \quad \text{[low-relaxation strand]}$$

$$f_{pbt} = 0.70f_{pu} \quad \text{[high-strength bars]}$$

The maximum allowable stress in post-tensioned low-relaxation strand immediately after transfer is

$$f_{pt} = 0.70 f_{pu} \quad \text{[at the anchorage]}$$
$$f_{pt} = 0.74 f_{pu} \quad \text{[elsewhere]}$$

The maximum allowable stress at jacking is

$$f_{pj} = 0.90 f_{py}$$

Example 9.18

The post-tensioned girder with grouted tendons shown in the illustration is simply supported over a span of 100 ft and has the following properties.

A	S_t	S_b	y_b	f'_{ci}
800 in^2	14,700 in^3	15,600 in^3	37.8 in	4500 lbf/in^2

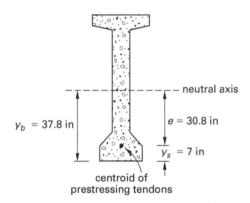

The normal weight concrete strength at transfer is $f'_{ci} = 4.5$ kips/in^2. The prestressing force immediately after transfer is 1000 kips, and the centroid of the tendons is 7 in above the bottom of the beam. Determine the actual and allowable stresses in the girder at midspan immediately after transfer if only prestressed reinforcement is provided.

Solution

For normal weight concrete, $\lambda = 1.0$.

At midspan, the minimum allowable tensile stress in the top fiber without bonded reinforcement is given by AASHTO Table 9.5.2.3.1b-1 as

$$f_{ti} = -0.0948\sqrt{f'_{ci}}$$
$$= -0.0948\sqrt{4.5 \ \frac{\text{kips}}{\text{in}^2}}$$
$$= -0.201 \ \text{kip/in}^2$$

Use the minimum allowable value of

$$f_{ti} = -0.200 \ \text{kip/in}^2$$

At midspan, the maximum allowable compressive stress in the bottom fiber is given by AASHTO Sec. 5.9.2.3 as

$$f_{bi} = 0.65 f'_{ci}$$
$$= (0.65)\left(4.5 \ \frac{\text{kips}}{\text{in}^2}\right)$$
$$= 2.93 \ \text{kips/in}^2$$

At midspan, the self-weight moment is

$$M_g = \frac{wL^2}{8}$$
$$= \frac{\left(150 \ \dfrac{\text{lbf}}{\text{ft}^3}\right)\left(\dfrac{800 \ \text{in}^2}{\left(12 \ \dfrac{\text{in}}{\text{ft}}\right)^2}\right)(100 \ \text{ft})^2\left(12 \ \dfrac{\text{in}}{\text{ft}}\right)}{(8)\left(1000 \ \dfrac{\text{lbf}}{\text{kip}}\right)}$$
$$= 12{,}500 \ \text{in-kips}$$

At midspan, the eccentricity of the prestressing force is

$$e = y_b - y_s$$
$$= 37.8 \ \text{in} - 7 \ \text{in}$$
$$= 30.8 \ \text{in}$$

At midspan, the actual stress in the top fiber is given by

$$f_{ti} = \frac{P_i}{A} - \frac{P_i e}{S_t} + \frac{M_g}{S_t}$$
$$= \frac{1000 \ \text{kips}}{800 \ \text{in}^2} - \frac{(1000 \ \text{kips})(30.8 \ \text{in})}{14{,}700 \ \text{in}^3}$$
$$\quad + \frac{12{,}500 \ \text{in-kips}}{14{,}700 \ \text{in}^3}$$
$$= +0.005 \ \text{kip/in}^2$$
$$> -0.20 \ \text{kip/in}^2 \quad \text{[satisfactory]}$$

At midspan, the actual compressive stress in the bottom fiber is given by

$$
\begin{aligned}
f_{bi} &= \frac{P_i}{A} + \frac{P_i e}{S_b} - \frac{M_g}{S_b} \\
&= \frac{1000 \text{ kips}}{800 \text{ in}^2} + \frac{(1000 \text{ kips})(30.8 \text{ in})}{15{,}600 \text{ in}^3} \\
&\quad - \frac{12{,}500 \text{ in-kips}}{15{,}600 \text{ in}^3} \\
&= 2.42 \text{ kips/in}^2 \\
&< 2.93 \text{ kips/in}^2 \quad \text{[satisfactory]}
\end{aligned}
$$

Service Load Conditions

The allowable stresses in the concrete at service limit state, in other than segmentally constructed bridges, after all prestressing losses have occurred are specified in AASHTO Table 5.9.2.3.2a-1 and 5.9.2.3.2b-1 as

$$f_{te} \leq 0.45 f_c' \quad \text{[for permanent load]}$$

$$f_{te} \leq 0.60 \phi_w f_c' \quad \text{[for permanent + transient loads]}$$

$$f_{be} \geq -0.19\lambda\sqrt{f_c'} \quad \left[\begin{array}{l}\text{with bonded prestressing tendons} \\ \text{or reinforcement and moderate exposure}\end{array}\right]$$

$$\geq -0.60 \text{ kips/in}^2$$

$$f_{be} \geq -0.0948\lambda\sqrt{f_c'} \quad \left[\begin{array}{c}\text{with bonded prestressing} \\ \text{tendons or reinforcement} \\ \text{and severe exposure}\end{array}\right]$$

$$\geq -0.30 \text{ kips/in}^2$$

$$f_{be} \geq 0 \quad \text{[with unbonded prestressing tendons]}$$

In accordance with AASHTO Table 5.9.2.2-1, the maximum allowable stress in the tendons after all losses is

$$f_{pe} = 0.80 f_{py}$$

Example 9.19

The post-tensioned girder of Ex. 9.18 forms part of a composite deck, as shown in the illustration, with girders located at 8 ft centers. The resulting composite section properties, allowing for the different moduli of elasticity of the girder and the slab, are tabulated as follows.

A_c	S_{ci}	S_c	f_c' (girder)
1250 in^2	45,400 in^3	21,200 in^3	6000 lbf/in^2

The concrete strength of the normal weight girder at 28 days is $f_c' = 6$ kips/in^2.

The prestressing force after all losses is 800 kips, and the losses occur before the deck slab is cast. The bending moment, M_{DC}, due to dead load imposed on the composite section is 3000 in-kips. The bending moment, M_L, due to live load plus impact is 16,250 in-kips. Neglect the dead load of wearing surface and utilities. Determine the actual and allowable stresses in the girder at midspan if the girder is subject to moderate exposure. Bonded reinforcement is provided at the bottom of the girder. Assume $\phi_w = 1.0$.

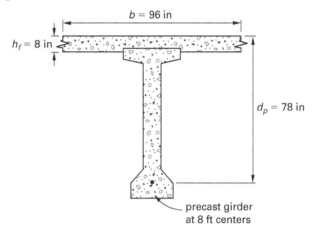

Solution

For normal weight concrete, $\lambda = 1.0$.

At midspan, the minimum allowable tensile stress in the bottom fiber with bonded reinforcement is given by AASHTO Table 9.5.2.3.2b-1 as

$$
\begin{aligned}
f_{be} &= -0.19\sqrt{f_c'} \\
&= -0.19\sqrt{6 \, \frac{\text{kips}}{\text{in}^2}} \\
&= -0.465 \text{ kip/in}^2
\end{aligned}
$$

At midspan, the maximum allowable compressive stress in the top fiber is given by AASHTO Table 9.5.2.3.2a-1 as

$$
\begin{aligned}
f_{te} &= 0.45 f_c' \quad \text{[for permanent loads]} \\
&= (0.45)\left(6 \, \frac{\text{kips}}{\text{in}^2}\right) \\
&= 2.7 \text{ kips/in}^2
\end{aligned}
$$

$$
\begin{aligned}
f_{te} &= 0.60 \phi_w f_c' \quad \text{[for permanent and transient loads]} \\
&= (0.60)(1.0)\left(6 \, \frac{\text{kips}}{\text{in}^2}\right) \\
&= 3.6 \text{ kips/in}^2
\end{aligned}
$$

From Ex. 9.18, the midspan moment due to the self-weight of the girder is

$$M_g = 12{,}500 \text{ in-kips}$$

The resulting stresses in the girder are

$$f_{Gt} = \frac{M_g}{S_t} = \frac{12{,}500 \text{ in-kips}}{14{,}700 \text{ in}^3}$$

$$= 0.850 \text{ kip/in}^2$$

$$f_{Gb} = -\frac{M_g}{S_b} = -\frac{12{,}500 \text{ in-kips}}{15{,}600 \text{ in}^3}$$

$$= -0.801 \text{ kip/in}^2$$

At midspan, the moment due to the weight of the deck slab is

$$M_S = \frac{wL^2}{8} = \frac{\left(150 \dfrac{\text{lbf}}{\text{ft}^3}\right)\left(\dfrac{8 \text{ in}}{12 \dfrac{\text{in}}{\text{ft}}}\right)(8 \text{ ft})(100 \text{ ft})^2\left(12 \dfrac{\text{in}}{\text{ft}}\right)}{(8)\left(1000 \dfrac{\text{lbf}}{\text{kip}}\right)}$$

$$= 12{,}000 \text{ in-kips}$$

The resulting stresses in the girder are

$$f_{St} = \frac{M_S}{S_t} = \frac{12{,}000 \text{ in-kips}}{14{,}700 \text{ in}^3}$$

$$= 0.816 \text{ kip/in}^2$$

$$f_{Sb} = -\frac{M_S}{S_b} = -\frac{12{,}000 \text{ in-kips}}{15{,}600 \text{ in}^3}$$

$$= -0.769 \text{ kip/in}^2$$

The resulting stresses in the girder due to the dead load imposed on the composite section are

$$f_{Dt} = \frac{M_{DC}}{S_{ci}} = \frac{3000 \text{ in-kips}}{45{,}400 \text{ in}^3}$$

$$= 0.066 \text{ kip/in}^2$$

$$f_{Db} = -\frac{M_{DC}}{S_c} = -\frac{3000 \text{ in-kips}}{21{,}200 \text{ in}^3}$$

$$= -0.142 \text{ kip/in}^2$$

The resulting stresses in the girder due to the live load imposed on the composite section are

$$f_{Lt} = \frac{M_L}{S_{ci}} = \frac{16{,}250 \text{ in-kips}}{45{,}400 \text{ in}^3}$$

$$= 0.358 \text{ kip/in}^2$$

$$f_{Lb} = -\frac{M_L}{S_c} = -\frac{16{,}250 \text{ in-kips}}{21{,}200 \text{ in}^3}$$

$$= -0.767 \text{ kip/in}^2$$

The stresses in the girder due to the effective prestressing force after all losses are

$$f_{Pt} = P_e\left(\frac{1}{A} - \frac{e}{S_t}\right)$$

$$= (800 \text{ kips})\left(\frac{1}{800 \text{ in}^2} - \frac{30.8 \text{ in}}{14{,}700 \text{ in}^3}\right)$$

$$= -0.676 \text{ kip/in}^2$$

$$f_{Pb} = P_e\left(\frac{1}{A} + \frac{e}{S_b}\right)$$

$$= (800 \text{ kips})\left(\frac{1}{800 \text{ in}^2} + \frac{30.8 \text{ in}}{15{,}600 \text{ in}^3}\right)$$

$$= 2.579 \text{ kips/in}^2$$

The final bottom fiber stress in the girder due to all permanent and transient loads is

$$f_{be} = f_{Gb} + f_{Sb} + f_{Db} + f_{Lb} + f_{Pb}$$

$$= -0.801 \frac{\text{kip}}{\text{in}^2} + \left(-0.769 \frac{\text{kip}}{\text{in}^2}\right) + \left(-0.142 \frac{\text{kip}}{\text{in}^2}\right)$$

$$+ \left(-0.767 \frac{\text{kip}}{\text{in}^2}\right) + 2.579 \frac{\text{kips}}{\text{in}^2}$$

$$= 0.1 \text{ kip/in}^2$$

$$> -0.465 \text{ kip/in}^2 \quad \text{[satisfactory]}$$

The final top fiber stress in the girder due to all permanent and transient loads is

$$f_{te} = f_{Gt} + f_{St} + f_{Dt} + f_{Lt} + f_{Pt}$$

$$= 0.850 \frac{\text{kip}}{\text{in}^2} + 0.816 \frac{\text{kip}}{\text{in}^2} + 0.066 \frac{\text{kip}}{\text{in}^2}$$

$$+ 0.358 \frac{\text{kip}}{\text{in}^2} + \left(-0.676 \frac{\text{kip}}{\text{in}^2}\right)$$

$$= 1.414 \text{ kips/in}^2$$

$$< 3.600 \text{ kips/in}^2 \quad \text{[satisfactory]}$$

The final top fiber stress in the girder due to sustained loads is

$$f_t = f_{Gt} + f_{St} + f_{Dt} + f_{Pt}$$
$$= 0.850 \ \frac{\text{kip}}{\text{in}^2} + 0.816 \ \frac{\text{kip}}{\text{in}^2} + 0.066 \ \frac{\text{kip}}{\text{in}^2}$$
$$+ \left(-0.676 \ \frac{\text{kip}}{\text{in}^2} \right)$$
$$= 1.056 \ \text{kips/in}^2$$
$$< 2.700 \ \text{kips/in}^2 \quad \text{[satisfactory]}$$

Ultimate Load Conditions

Provided that the effective prestress in the tendons after losses, f_{pe}, is not less than half the tensile strength of the tendons, f_{pu}, the stress in bonded tendons at ultimate load is given by AASHTO Eq. 5.6.3.1.1-1 as

$$f_{ps} = f_{pu} \left(1 - \frac{kc}{d_p} \right)$$

This expression is based on the assumption that all of the prestressing steel is concentrated at a distance d_p from the extreme compression fiber. If this assumption is not justified, a method based on strain compatibility must be used.

The prestressing steel factor, k, is given by AASHTO Table C5.6.3.1.1-1 as

- 0.48 for type 2 high-strength bars with $f_{py}/f_{pu} = 0.80$

- 0.38 for stress-relieved strands and type 1 high-strength bars with $f_{py}/f_{pu} = 0.85$

- 0.28 for low-relaxation strands with $f_{py}/f_{pu} = 0.90$

As specified in AASHTO Sec. 5.6.2.1, a rectangular stress block is assumed in the concrete at ultimate flexural load in a beam. The intensity of the stress is given by $\alpha_1 f'_c$, and the depth of the stress block, a, is given by $\beta_1 c$, where c is the neutral axis depth.

The stress block factor, α_1, given in AASHTO Sec. 5.6.2.2 is

- 0.85 for $f'_c \le 10 \ \text{kips/in}^2$

- $0.85 - (f'_c - 10)/50$ for $10 \ \text{kips/in}^2 < f'_c \le 15 \ \text{kips/in}^2$

- 0.75 for $f'_c > 15 \ \text{kips/in}^2$

The stress block factor, β_1, given in AASHTO Sec. 5.6.2.1 is

- 0.85 for $f'_c \le 4 \ \text{kips/in}^2$

- $0.85 - (f'_c - 4)/20$ for $4 \ \text{kips/in}^2 < f'_c \le 8 \ \text{kips/in}^2$

- 0.65 for $f'_c > 8 \ \text{kips/in}^2$

For a rectangular section, with nonprestressed tension reinforcement and without compression reinforcement, the distance from the extreme compression fiber to the neutral axis is given by AASHTO Eq. 5.6.3.1.1-4 as

$$c = \frac{A_{ps}f_{pu} + A_s f_s}{\alpha_1 f'_c \beta_1 b + \dfrac{k A_{ps} f_{pu}}{d_p}}$$

The previous expression is also applicable to a flanged section with the neutral axis within the flange.

The nominal flexural strength of a rectangular section, without nonprestressed compression reinforcement, is given by AASHTO Eq. 5.6.3.2.2-1 as

$$M_n = A_{ps} f_{ps} \left(d_p - \frac{a}{2} \right) + A_s f_s \left(d_s - \frac{a}{2} \right)$$

As specified in AASHTO Sec. 5.6.2.1, f_y may replace f_s when the resulting ratio c/d_s does not exceed $0.003/(0.003 + \epsilon_{cl})$. If c/d_s exceeds this limit, strain compatibility should be used to determine the stress in the nonprestressed tension reinforcement.

The compression-controlled strain limit, ϵ_{cl}, is the net tensile strain in the reinforcement when the concrete in compression reaches a strain limit of 0.003. For grade 60 reinforcement, $\epsilon_{cl} = 0.002$ and the limiting ratio, c/d_s, is 0.6. For grade 100 reinforcement, $\epsilon_{cl} = 0.004$ and the limiting ratio, c/d_s, is 0.43. The limiting ratio, c/d_s, for other grades of reinforcement may be determined by linear interpolation. For all prestressed reinforcement, the value of $\epsilon_{cl} = 0.002$.

The tension-controlled strain limit, ϵ_{tl}, is the net tensile strain in the reinforcement when the concrete in compression reaches a strain limit of 0.003. For nonprestressed reinforcement with $f_y \le 75 \ \text{kips/in}^2$, $\epsilon_{tl} = 0.005$ and the limiting ratio, c/d_s, is 0.38. For nonprestressed reinforcement with $f_y = 100 \ \text{kips/in}^2$, $\epsilon_{tl} = 0.008$ and the limiting ratio, c/d_s, is 0.27. The limiting ratio, c/d_s, for other grades of reinforcement may be determined by linear interpolation. For all prestressed reinforcement, the value of $\epsilon_{tl} = 0.005$.

The factored flexural resistance is given by AASHTO Eq. 5.6.3.2.1-1 as

$$M_r = \phi M_n$$

The resistance factor for a tension-controlled prestressed concrete section with bonded tendons is given by AASHTO Sec. 5.5.4.2 as

$$\phi = 1.0$$

Example 9.20

The area of the low-relaxation strand in the post-tensioned girder of Ex. 9.19 is 5.36 in^2, and the strand has a specified tensile strength of 270 kips/in^2. The 28-day compressive strength of the deck slab is 3 kips/in^2. Determine the maximum factored moment at midspan and the design flexural capacity of the composite section.

Solution

From Ex. 9.19, the total dead load moment (neglecting wearing surface and utilities) on the composite section is

$$M_D = M_g + M_S + M_{DC}$$
$$= 12{,}500 \text{ in-kips} + 12{,}000 \text{ in-kips}$$
$$+ 3000 \text{ in-kips}$$
$$= 27{,}500 \text{ in-kips}$$

The live load moment plus impact is

$$M_L = 16{,}250 \text{ in-kips}$$

The strength I limit state moment is given by AASHTO Eq. 3.4.1-1 as

$$M_u = \gamma_p M_D + \gamma_{LL+IM} M_L$$
$$= (1.25)(27{,}500 \text{ in-kips})$$
$$+ (1.75)(16{,}250 \text{ in-kips})$$
$$= 62{,}813 \text{ in-kips}$$

The effective prestress in the tendons after all losses is obtained from Ex. 9.19 as

$$f_{pe} = \frac{P_e}{A_{ps}}$$
$$= \frac{800 \text{ kips}}{5.36 \text{ in}^2}$$
$$= 149 \text{ kips/in}^2$$
$$> 0.5 f_{pu}$$

Therefore, AASHTO Sec. 5.6.3.1 is applicable.

The compression zone factor for 3 kips/in^2 concrete is

$$\beta_1 = 0.85$$

The prestressing steel factor is given by AASHTO Table C5.6.3.1.1-1 as

$$k = 0.28 \quad \text{[for low-relaxation strand]}$$

Assuming that the neutral axis lies within the flange, for a section without nonprestressed tension reinforcement, the distance from the extreme compression fiber to the neutral axis is given by AASHTO Eq. 5.6.3.1.1-4 as

$$c = \frac{A_{ps} f_{pu}}{\alpha_1 f'_c \beta_1 b + \dfrac{k A_{ps} f_{pu}}{d_p}}$$

$$= \frac{(5.36 \text{ in}^2)\left(270 \, \dfrac{\text{kips}}{\text{in}^2}\right)}{(0.85)\left(3 \, \dfrac{\text{kips}}{\text{in}^2}\right)(0.85)(96 \text{ in})}$$
$$+ \frac{(0.28)(5.36 \text{ in}^2)\left(270 \, \dfrac{\text{kips}}{\text{in}^2}\right)}{78 \text{ in}}$$

$$= 6.79 \text{ in}$$
$$< 8 \text{ in}$$

Therefore, the neutral axis does lie within the flange.

The depth of the equivalent rectangular stress block is given by AASHTO Sec. 5.6.2 as

$$a = \beta_1 c = (0.85)(6.79 \text{ in})$$
$$= 5.8 \text{ in}$$

The stress in bonded tendons at ultimate load is given by AASHTO Eq. 5.6.3.1.1-1 as

$$f_{ps} = f_{pu}\left(1 - \frac{kc}{d_p}\right)$$
$$= \left(270 \, \frac{\text{kips}}{\text{in}^2}\right)\left(1 - \frac{(0.28)(6.79 \text{ in})}{78 \text{ in}}\right)$$
$$= 263.42 \text{ kips/in}^2$$

The nominal flexural strength of the section is given by AASHTO Eq. 5.6.3.2.2-1 as

$$M_n = A_{ps} f_{ps}\left(d_p - \frac{a}{2}\right)$$
$$= (5.36 \text{ in}^2)\left(263.42 \, \frac{\text{kips}}{\text{in}^2}\right)\left(78 \text{ in} - \frac{5.8 \text{ in}}{2}\right)$$
$$= 106{,}036 \text{ in-kips}$$

The strain in the prestressing tendons at the nominal flexural strength is

$$\epsilon_t = \epsilon_{cu}\left(\frac{d_p - c}{c}\right)$$

$$= (0.003)\left(\frac{78 \text{ in} - 6.79 \text{ in}}{6.79 \text{ in}}\right)$$

$$= 0.031$$

$$> 0.005$$

Therefore, from AASHTO Sec. 5.6.2.1, the section is tension controlled and the resistance factor is given by AASHTO Sec. 5.5.4.2 as

$$\phi = 1.0$$

The factored flexural resistance is

$$M_r = \phi M_n = (1.0)(106{,}036 \text{ in-kips})$$

$$= 106{,}036 \text{ in-kips}$$

$$> M_u \quad \text{[satisfactory]}$$

$$> 1.33 \, M_u \quad \text{[satisfies AASHTO Sec. 5.6.3.3]}$$

Cracking Moment

The *cracking moment* is the external moment that, when applied to the member after all losses have occurred, will cause cracking in the bottom fiber. This cracking occurs when the stress in the bottom fiber exceeds the modulus of rupture, which is defined in AASHTO Sec. 5.4.2.6 as

$$f_r = 0.24\lambda\sqrt{f_c'}$$

For a composite section, the cracking moment is defined in AASHTO Eq. 5.6.3.3-1 as

$$M_{cr} = \gamma_3\left(S_c(\gamma_2 f_{cpe} + \gamma_1 f_r) - M_{dnc}\left(\frac{S_c}{S_{nc}} - 1\right)\right)$$

$$\geq S_c f_r$$

The applicable factors for the cracking moment are as follows.

γ_1 = flexural cracking variability factor
  = 1.2 [precast segmental structures]
  = 1.6 [other concrete structures]
γ_2 = prestress variability factor
  = 1.1 [bonded tendons]
  = 1.0 [unbonded tendons]
γ_3 = ratio of specified minimum yield strength to ultimate tensile strength of the reinforcement
  = 0.67 [A615, grade 60 reinforcement]
  = 0.75 [A615, grade 75 reinforcement]
  = 1.00 [prestressing steel]

For noncomposite beams, S_{nc} is substituted for S_c in the previous expression. To prevent sudden tensile failure, AASHTO Sec. 5.6.3.3 requires that a beam have a moment capacity at least equal to the lesser of

$$\phi M_n = M_{cr}$$

$$\phi M_n = 1.33 M_u$$

Example 9.21

Determine the cracking moment of the composite section of Ex. 9.19.

Solution

From Ex. 9.19, the bottom fiber stress due only to the effective prestressing force after allowance for all prestress losses is

$$f_{cpe} = 2.579 \text{ kips/in}^2$$

In addition, the bending moment due to the noncomposite dead load acting on the precast section is given by

$$M_{dnc} = M_g + M_S = 12{,}500 \text{ in-kips} + 12{,}000 \text{ in-kips}$$

$$= 24{,}500 \text{ in-kips}$$

The modulus of rupture for normal weight concrete is given by AASHTO Sec. 5.4.2.6 as

$$f_r = 0.24\lambda\sqrt{f_c'} = 0.24\sqrt{6 \, \frac{\text{kips}}{\text{in}^2}} = 0.588 \text{ kip/in}^2$$

From Ex. 9.18, the section modulus of the noncomposite section referred to the bottom fiber is

$$S_{nc} = S_b = 15{,}600 \text{ in}^3$$

The cracking moment is given by AASHTO Eq. 5.6.3.3-1 as

$$M_{cr} = \gamma_3\left(S_c(\gamma_2 f_{cpe} + \gamma_1 f_r) - M_{dnc}\left(\frac{S_c}{S_{nc}} - 1\right)\right)$$

$$= (1.0)\left(\begin{array}{l}(21{,}200 \text{ in}^3)\left(\begin{array}{l}(1.1)\left(2.579 \, \dfrac{\text{kips}}{\text{in}^2}\right) \\ + (1.6)\left(0.588 \, \dfrac{\text{kip}}{\text{in}^2}\right)\end{array}\right) \\ - (24{,}500 \text{ in-kips})\left(\dfrac{21{,}200 \text{ in}^3}{15{,}600 \text{ in}^3} - 1\right)\end{array}\right)$$

$$= 71{,}292 \text{ in-kips}$$

$$< \phi M_n \quad \left[\begin{array}{l}\text{satisfies AASHTO} \\ \text{Sec. 5.6.3.3}\end{array}\right]$$

Design for Shear

Nomenclature

A_v	area of shear reinforcement	in²
b_v	web width	in
d_e	effective depth from the extreme compression fiber to the centroid of the tensile force in the tensile reinforcement	in
d_p	distance from the extreme compression fiber to the centroid of the prestressing tendons	in
d_v	effective shear depth	in
f_c'	specified compressive strength of concrete	kips/in²
f_{cpe}	compressive stress in the concrete, due to the final prestressing force only, at the bottom fiber of the section	lbf/in²
f_d	tensile stress at bottom fiber of precast member due to unfactored dead load acting on the precast member	lbf/in²
f_{pc}	compressive stress in the concrete, due to the final prestressing force and applied loads resisted by precast member, at the centroid of the composite section	lbf/in²
f_r	modulus of rupture	kips/in²
f_y	specified yield strength of reinforcing bars	kips/in²
g	drape of the prestressing cable	in
h	depth of section	in
M_{cre}	moment causing flexural cracking at section due to externally applied loads	in-lbf or in-kips
M_d	moment due to unfactored dead load	in-kips
M_{dnc}	total unfactored dead load moment acting on the precast member	in-kips
M_L	bending moment due to live load	in-kips
M_{max}	maximum factored moment at sectiondue to externally applied loads	in-lbf or in-kips
M_u	factored moment at the section due tototal factored loads	in-kips
s	longitudinal spacing of shear reinforcement	in
S	section modulus at the centroid of the composite section	in³
S_c	section modulus at the bottom of the composite member	in³
S_{nc}	section modulus at the bottom of the precast member	in³
v_u	average factored shear stress on the concrete	kips/in²
V_c	nominal shear strength provided by concrete	kips
V_{ci}	nominal shear strength provided by concrete when diagonal cracking results from combined shear and moment	kips
V_{cw}	nominal shear strength provided by concrete when diagonal cracking results from excessive principal tensile stress in the web	kips
V_d	shear force at section due to unfactored dead load	kips
V_i	factored shear force at section due to externally applied loads occurring simultaneously with M_{max}	kips
V_L	unfactored shear due to live load	kips
V_p	vertical component of effective prestress force at section	kips
V_s	nominal shear strength provided by shear reinforcement	kips
V_u	factored shear force at section	kips
y_s	height of cable above beam soffit	in
x	horizontal distance	ft
z	half the beam length	ft

Symbols

α	angle of inclination of transverse reinforcement to the longitudinal axis	degree
β	factor indicating ability of diagonally cracked concrete to transmit tension and shear	–
ε_s	net tensile strain at the centroid of the tension reinforcement	–
γ_c	specific weight of concrete	kips/ft³
θ	angle of inclination of diagonal compressive stresses	degree
ϕ	strength reduction factor	–

Ultimate Load Design for Shear

The general procedure of AASHTO Sec. 5.7.3.4.2 is permissible for the design of prestressed concrete sections in shear. The nominal shear capacity of the concrete is given by AASHTO Eq. 5.7.3.3-3 as

$$V_c = 0.0316\beta\lambda\sqrt{f_c'}\,b_v d_v$$

The nominal shear capacity of vertical ($\alpha = 90°$) shear reinforcement is given by AASHTO Eq. C5.7.3.3-1 as

$$V_s = \frac{A_v f_y d_v \cot\theta}{s}$$

The effective shear depth, d_v, is taken as the distance between the resultants of the tensile and compressive forces due to flexure. The effective shear depth is given by AASHTO Eq. C5.7.2.8-1 as

$$d_v = \frac{M_n}{A_s f_y + A_{ps} f_{ps}}$$

The effective shear depth need not be taken to be less than the greater of $0.9d_e$ or $0.72h$. The effective web width, b_v, is taken as the minimum web width between the resultants of the tensile and compressive forces due to flexure.

In order to determine the values of the parameters β and θ, the applicable value of ϵ_s must first be determined. The term ϵ_s is defined in AASHTO Sec. 5.7.3.4.2 as the net tensile strain in the section at the centroid of the tension reinforcement, and is given by AASHTO Eq. 5.7.3.4.2-4 as

$$\epsilon_s = \frac{\dfrac{|M_u|}{d_v} + 0.5N_u + |V_u - V_p| - A_{ps}f_{po}}{E_sA_s + E_pA_{ps}}$$

In this equation,

$N_u = 0$ where there is no axial force applied to the member

$M_u \geq (V_u - V_p)d_v$

$f_{po} \approx 0.7f_{pu}$

$E_s = 29{,}000 \text{ kips/in}^2$ for steel reinforcement

$E_p = 28{,}500 \text{ kips/in}^2$ for strand, $30{,}000 \text{ kips/in}^2$ for bars

When the calculated value ϵ_s is negative, in accordance with AASHTO Sec. 5.7.3.4.2, ϵ_s is taken as zero.

For members containing not less than the minimum amount of transverse reinforcement specified by AASHTO Eq. 5.7.2.5-1, the value of the parameter β, the factor relating to the ability of diagonally cracked concrete to transmit tension, is given by AASHTO Eq. 5.7.3.4.2-1 as

$$\beta = \frac{4.8}{(1 + 750\epsilon_s)}$$

The minimum area of transverse reinforcement is specified by AASHTO Eq. 5.7.2.5-1 as

$$A_v \geq 0.0316\lambda\sqrt{f_c'}\,\frac{b_vs}{f_y}$$

The parameter θ, the angle of inclination of diagonal compressive stresses, is given by AASHTO Eq. 5.7.3.4.2-3 as

$$\theta = 29 + 3500\epsilon_s$$

The combined nominal shear resistance is given by AASHTO Sec. 5.7.3.3 as the lesser of

$$V_n = V_c + V_s + V_p$$
$$V_n = 0.25f_c'b_vd_v + V_p$$

The combined shear capacity is

$$\phi V_n \geq V_u$$

The resistance factor for prestressed concrete members with bonded tendons is given by AASHTO Sec. 5.5.4.2 as

$$\phi = 0.90$$

The shear stress on the concrete is calculated by AASHTO Eq. 5.7.2.8-1 as

$$v_u = \frac{V_u - \phi V_p}{\phi b_v d_v}$$

For a value of v_u less than $0.125f_c'$, AASHTO Sec. 5.7.2.6 limits the spacing of transverse reinforcement to the lesser of $0.8d_v$ or 24 in. When the value of v_u is not less than $0.125f_c'$, the spacing is reduced to the lesser of $0.4d_v$ or 12 in.

When the support reaction produces a compressive stress in the member, AASHTO Sec. 5.7.3.2 specifies that the critical section for shear may be taken at a distance from the internal face of the support equal to the effective shear depth, d_v.

Example 9.22

The tendon centroid of the post-tensioned girder of Ex. 9.18 is parabolic in shape, as shown in *Illustration for Ex. 9.22*. At section A-A, the unfactored shear and moment due to live load plus impact are $V_L = 58$ kips and $M_L = 2340$ in-kips. The moment of inertia of the precast section is $I = 589{,}680 \text{ in}^4$. Determine the required spacing of no. 3 grade 60 stirrups.

Solution

The equation of the parabolic cable profile is

$$y = \frac{gx^2}{z^2}$$

At section A-A, the rise of the cable is given by

$$y = \frac{(30 \text{ in})(600 \text{ in} - 61.2 \text{ in})^2}{(600 \text{ in})^2}$$
$$= 24.2 \text{ in}$$

Illustration for Ex. 9.22

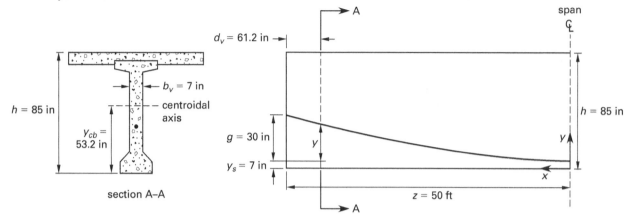

section A–A

At section A-A, the effective depth of the prestressing cable referred to the composite section is

$$d_p = h - y - y_s$$
$$= 85 \text{ in} - 24.2 \text{ in} - 7 \text{ in}$$
$$= 53.8 \text{ in}$$
$$= d_e \quad \begin{bmatrix} \text{effective depth from the extreme compression} \\ \text{fiber to the centroid of the tensile force} \end{bmatrix}$$

From Ex. 9.20, the depth of the stress block at midspan of the composite section is

$$a = 5.8 \text{ in}$$

The value of a may be conservatively taken as the stress block depth at section A-A and the effective shear depth is

$$d_v = d_e - \frac{a}{2} = 53.8 \text{ in} - \frac{5.8 \text{ in}}{2}$$
$$= 50.9 \text{ in}$$

The effective shear depth need not be taken to be less than the greater of

$$0.9d_e = (0.9)(53.8 \text{ in})$$
$$= 48.4 \text{ in}$$
$$0.72h = (0.72)(85 \text{ in})$$
$$= 61.2 \text{ in} \quad [\text{governs}]$$

Therefore, as specified by AASHTO Sec. 5.7.3.2, the critical section for shear is located a distance of 61.2 in from the support.

At section A-A, the cable eccentricity referred to the precast section is given by

$$e = y_b - y_s - y$$
$$= 37.8 \text{ in} - 7 \text{ in} - 24.2 \text{ in}$$
$$= 6.6 \text{ in}$$

At section A-A, the slope of the cable is given by

$$\frac{dy}{dx} = \frac{2gx}{z^2} = \frac{(2)(30 \text{ in})(538.8 \text{ in})}{(600 \text{ in})^2}$$
$$= 0.0898$$

The vertical component of the final effective prestressing force at section A-A is

$$V_P = P_e\left(\frac{dy}{dx}\right) = (800 \text{ kips})(0.0898)$$
$$= 72 \text{ kips}$$

The centroid of the composite section is at a height of

$$y_{cb} = 53.2 \text{ in}$$

At the beam midspan, the total dead load moment is obtained from Ex. 9.20 as

$$M_{D,\text{comp}} = 27,500 \text{ in-kips}$$

At section A-A, the total dead load moment is

$$M_D = M_{D,\text{comp}}\left(1 - \left(\frac{x}{z}\right)^2\right)$$
$$= (27,500 \text{ in-kips})\left(1 - \left(\frac{538.8 \text{ in}}{600 \text{ in}}\right)^2\right)$$
$$= 5324 \text{ in-kips}$$

At section A-A, the moment due to live load plus impact is given as

$$M_L = 2340 \text{ in-kips}$$

At section A-A, the strength I limit state moment is given by AASHTO Eq. 3.4.1-1 as

$$M_u = \gamma_P M_D + \gamma_{LL+IM} M_L$$
$$= (1.25)(5324 \text{ in-kips}) + (1.75)(2340 \text{ in-kips})$$
$$= 10{,}750 \text{ in-kips}$$

$$(V_u - V_p)d_v = (205 \text{ kips} - 72 \text{ kips})(61.2 \text{ in})$$
$$= 8140 \text{ in-kips}$$
$$< M_u \quad [\text{as required by AASHTO Sec. 5.7.3.4.2}]$$

The properties of the low-relaxation strand are obtained from Ex. 9.20 and are

$$A_{ps} = 5.36 \text{ in}^2$$
$$f_{pu} = 270 \text{ kips/in}^2$$
$$f_{po} \approx 0.7 f_{pu}$$
$$= (0.7)\left(270 \frac{\text{kips}}{\text{in}^2}\right)$$
$$= 189 \text{ kips/in}^2$$
$$E_p = 28{,}500 \text{ kips/in}^2$$

The net tensile strain in the section at the centroid of the tension reinforcement is given by AASHTO Eq. 5.7.3.4.2-4 as

$$\epsilon_s = \frac{\left(\dfrac{|M_u|}{d_v} + 0.5 N_u + |V_u - V_p| - A_{ps}f_{po}\right)}{E_s A_s + E_p A_{ps}}$$

$$= \frac{\left(\begin{array}{c}\dfrac{10{,}750 \text{ in-kips}}{61.2 \text{ in}} + (205 \text{ kips} - 72 \text{ kips}) \\ -(5.36 \text{ in}^2)\left(189 \frac{\text{kips}}{\text{in}^2}\right)\end{array}\right)}{\left(28{,}500 \frac{\text{kips}}{\text{in}^2}\right)(5.36 \text{ in}^2)}$$

$$= -0.0046$$

Since this value is negative, AASHTO Sec. 5.7.3.4.2 requires that ϵ_s be taken as zero.

The parameter β, the factor relating to the ability of diagonally cracked concrete to transmit tension, is given by AASHTO Eq. 5.7.3.4.2-1 as

$$\beta = \frac{4.8}{(1 + 750\epsilon_s)}$$
$$= 4.8$$

The nominal shear capacity of the concrete is given by AASHTO Eq. 5.7.3.3-3 as

$$V_c = 0.0316\beta\lambda\sqrt{f_c'}\, b_v d_v$$
$$= (0.0316)(4.8)(1.0)\left(6 \frac{\text{kips}}{\text{in}^2}\right)^{0.5}(7 \text{ in})(61.2 \text{ in})$$
$$= 159 \text{ kips}$$

The design shear capacity is

$$\phi V_c = (0.9)(159 \text{ kips})$$
$$= 143 \text{ kips}$$

From Ex. 9.18, the cross-sectional area of the girder is $A_G = 800 \text{ in}^2$. At section A-A, the shear force due to the girder self-weight is

$$V_G = w_G x = \gamma_c A_G x$$
$$= \frac{\left(0.150 \frac{\text{kip}}{\text{ft}^3}\right)\left(\dfrac{800 \text{ in}^2}{\left(12 \frac{\text{in}}{\text{ft}}\right)^2}\right)(538.8 \text{ in})}{12 \frac{\text{in}}{\text{ft}}}$$
$$= 37.4 \text{ kips}$$

The cross-sectional area of the slab is $A_F = (8 \text{ in})(96 \text{ in}) = 768 \text{ in}^2$. At section A-A, the shear force due to the self-weight of the slab is

$$V_S = w_S x = \gamma_c A_F x$$
$$= \frac{\left(0.150 \frac{\text{kip}}{\text{ft}^3}\right)\left(\dfrac{768 \text{ in}^2}{\left(12 \frac{\text{in}}{\text{ft}}\right)^2}\right)(538.8 \text{ in})}{12 \frac{\text{in}}{\text{ft}}}$$
$$= 36.0 \text{ kips}$$

At section A-A, the shear force due to the dead load imposed on the composite section is

$$V_{DC} = w_{DC} x = \frac{\left(0.20 \frac{\text{kip}}{\text{ft}}\right)(538.8 \text{ in})}{12 \frac{\text{in}}{\text{ft}}}$$
$$= 9.0 \text{ kips}$$

At section A-A, the total dead load shear force on the composite section is

$$V_D = V_G + V_S + V_{DC}$$
$$= 37.4 \text{ kips} + 36.0 \text{ kips} + 9.0 \text{ kips}$$
$$= 82.4 \text{ kips}$$

The live load shear plus impact is given as

$$V_L = 58 \text{ kips}$$

The strength I limit state shear force is given by AASHTO Eq. 3.4.1-1 as

$$V_u = \gamma_p V_D + \gamma_{LL+IM} V_L$$
$$= (1.25)(82.4 \text{ kips}) + (1.75)(58 \text{ kips})$$
$$= 205 \text{ kips}$$
$$> \phi V_c = 143 \text{ kips}$$

The factored shear force exceeds the shear strength of the concrete, and the shear strength required from shear reinforcement is

$$\phi V_s = V_u - \phi V_c$$
$$= 205 \text{ kips} - 143 \text{ kips}$$
$$= 62 \text{ kips}$$

The parameter θ, the angle of inclination of diagonal compressive stresses, is given by AASHTO Eq. 5.7.3.4.2-3 as

$$\theta = 29 + 3500\epsilon_s$$
$$= 29°$$

The area of shear reinforcement required is given by AASHTO Eq. C5.7.3.3-1 as

$$\frac{A_v}{s} = \frac{V_s}{f_y d_v \cot\theta}$$
$$= \frac{62 \text{ kips}}{(0.9)\left(60 \dfrac{\text{kips}}{\text{in}^2}\right)(61.2 \text{ in})(1.80)}$$
$$= 0.0104 \text{ in}^2/\text{in}$$

Providing no. 3 stirrups necessitates a spacing of

$$s = \frac{0.22}{0.0104}$$
$$= 21 \text{ in}$$

The minimum area of shear reinforcement is required and is given by AASHTO Eq. 5.7.2.5-1 as

$$\frac{A_v}{s} = 0.0316\lambda\left(\frac{\sqrt{f_c'}\, b_v}{f_y}\right)$$
$$= (0.0316)(1.0)\left(\frac{\left(\sqrt{6 \dfrac{\text{kips}}{\text{in}^2}}\right)(7 \text{ in})}{60 \dfrac{\text{kips}}{\text{in}^2}}\right)$$
$$= 0.0090 \text{ in}^2/\text{in} \quad \text{[satisfactory]}$$

This is less than the area provided.

The shear stress on the concrete is calculated by AASHTO Eq. 5.7.2.8-1 as

$$v_u = \frac{V_u - \phi V_p}{\phi b_v d_v}$$
$$= \frac{205 \text{ kips} - (0.9)(72 \text{ kips})}{(0.9)(7 \text{ in})(61.2 \text{ in})}$$
$$= 0.36 \text{ kip/in}^2$$
$$< 0.125 f_c' \quad (0.75 \text{ kip/in}^2)$$

Therefore, AASHTO Sec. 5.7.2.6 limits the spacing of transverse reinforcement to the lesser of

$$s = 0.8 d_v$$
$$= (0.8)(61.2 \text{ in})$$
$$= 49 \text{ in}$$
$$s = 24 \text{ in} \quad \text{[governs]}$$

The 21 in spacing provided is satisfactory.

Prestress Losses

Nomenclature

A_g	gross area of section	in^2
A_{ps}	area of prestressing steel	in^2
E_{ci}	modulus of elasticity of concrete at time of initial prestress	kips/in^2
E_p	modulus of elasticity of prestressing steel	kips/in^2
f_{cgp}	compressive stress at centroid of prestressing steel due to prestress and self-weight of girder at transfer	kips/in^2
f_{pj}	stress in the prestressing steel at jacking	kips/in^2
g	drape of prestressing steel	in
I_g	moment of inertia of precast girder	in^4

K	wobble friction coefficient per foot of prestressing tendon	–
l_{px}	distance from free end of cable to section under consideration	ft
M_g	bending moment due to self-weight of precast member	in-kips
n_i	modular ratio at transfer	–
N	number of identical prestressing tendons	–
P_{ES}	loss of prestress force due to elastic shortening	kips
P_i	force in prestressing steel immediately after transfer	kips
P_o	force in prestressing steel at anchorage	kips
R	radius of curvature of tendon profile	ft

Symbols

α	angular change of tendon profile from jacking end to any point x	radians
Δf_{pES}	loss of prestress due to elastic shortening	kips/in²
Δf_{pF}	prestress loss due to friction	kips/in²
Δf_{pLT}	long-term prestress loss due to creep and shrinkage of concrete and relaxation of steel	kips/in²
μ	curvature friction coefficient	–

Friction Losses

AASHTO Sec. 5.9.3.2.2 determines friction losses from the equation

$$\Delta f_{pF} = f_{pj}\left(1 - e^{-(Kx + \mu\alpha)}\right)$$

Values of the wobble and curvature friction coefficients are given in AASHTO Table 5.9.3.2.2b-1 and for prestressing strand are

$$K = 0.0002$$
$$\mu = 0.15 \text{ to } 0.25$$

Example 9.23

The beam of Ex. 9.22 is post-tensioned with low-relaxation strands with a total area of 5.36 in², a yield strength of 243 kips/in², and a tensile strength of 270 kips/in². The strands are located in 4 cables. The centroid of the prestressing steel is parabolic in shape and is stressed simultaneously from both ends with a jacking force, P_o, of 1036 kips. The value of the wobble friction coefficient is 0.0002/ft, and the curvature friction coefficient is 0.25. Determine the force in the prestressing steel at midspan of the member before elastic losses.

Solution

The nominal radius of the profile of the prestressing steel is

$$R = \frac{z^2}{2g} = \frac{(50 \text{ ft})^2}{(2)(2.5 \text{ ft})}$$
$$= 500 \text{ ft}$$

The length along the curve from the jacking end to midspan is

$$l_{px} = z + \frac{g^2}{3z} = 50 \text{ ft} + \frac{(2.5 \text{ ft})^2}{(3)(50 \text{ ft})}$$
$$= 50.04 \text{ ft}$$

The angular change of the cable profile over this length is

$$\alpha = \frac{l_{px}}{R} = \frac{50.04 \text{ ft}}{500 \text{ ft}}$$
$$= 0.100 \text{ radians}$$
$$(Kl_{px} + \mu\alpha) = (0.0002)(50.04 \text{ ft})$$
$$+ (0.25)(0.100 \text{ radians})$$
$$= 0.035$$

The cable force at midspan is given by

$$P_x = P_o e^{-(Kl_{px} + \mu\alpha)}$$
$$= (1036 \text{ kips})\left(e^{-0.035}\right)$$
$$= 1000 \text{ kips}$$

Elastic Shortening

Losses occur due to the elastic shortening of the concrete. The concrete stress at the level of the centroid of the prestressing steel after elastic shortening is

$$f_{cgp} = P_i\left(\frac{1}{A_g} + \frac{e^2}{I_g}\right) - \frac{eM_g}{I_g}$$

AASHTO Sec. C5.4.2.4 specifies that the modulus of elasticity of normal weight concrete at transfer is

$$E_{ci} = 1820\sqrt{f'_{ci}}$$

The modulus of elasticity of prestressing strand is given as

$$E_p = 28{,}500 \text{ kips/in}^2$$

The modular ratio at transfer is

$$n_i = \frac{E_p}{E_{ci}}$$

For a pretensioned member, the loss of prestress due to elastic shortening is given by AASHTO Sec. 5.9.3.2.3a as

$$\Delta f_{pES} = n_i f_{cgp}$$

For a post-tensioned member, the loss of prestress is given by AASHTO Sec. 5.9.3.2.3b.

$$\Delta f_{pES} = \frac{N-1}{2N} n_i f_{cgp}$$

Example 9.24

The post-tensioned beam of Ex. 9.23 has a concrete strength at transfer of 4.5 kips/in². The initial force at midspan, after friction losses and before allowance for elastic shortening, is 1000 kips. The moment of inertia of the girder is 589,680 in⁴. Determine the loss of prestress due to elastic shortening.

Solution

From AASHTO Eq. C5.4.2.4-1, the modulus of elasticity of the concrete at transfer is

$$E_{ci} = 1820\sqrt{f'_{ci}}$$
$$= 1820\sqrt{4.5\ \frac{\text{kips}}{\text{in}^2}}$$
$$= 3861\ \text{kips/in}^2$$

The modular ratio at transfer is

$$n_i = \frac{E_s}{E_{ci}} = \frac{28{,}500\ \dfrac{\text{kips}}{\text{in}^2}}{3861\ \dfrac{\text{kips}}{\text{in}^2}}$$
$$= 7.38$$

Assuming a 3% loss due to elastic shortening, the initial prestressing force at midspan is

$$P_i = 970\ \text{kips}$$

The compressive stress at the centroid of the prestressing steel immediately after transfer is

$$f_{cgp} = P_i \left(\frac{1}{A_g} + \frac{e^2}{I_g} \right) - \frac{e M_g}{I_g}$$
$$= (970\ \text{kips}) \left(\frac{1}{800\ \text{in}^2} + \frac{(30.8\ \text{in})^2}{589{,}680\ \text{in}^4} \right)$$
$$\quad - \frac{(30.8\ \text{in})(12{,}500\ \text{in-kips})}{589{,}680\ \text{in}^4}$$
$$= 2.120\ \text{kips/in}^2$$

For a post-tensioned member, the loss of prestress is given by AASHTO Sec. 5.9.3.2.3b as

$$\Delta f_{pES} = \frac{N-1}{2N} n_i f_{cgp}$$
$$= \frac{(3)(7.38)\left(2.120\ \dfrac{\text{kips}}{\text{in}^2} \right)}{(2)(4)}$$
$$= 5.87\ \text{kips/in}^2$$

The loss of prestressing force is

$$P_{ES} = A_{ps}\Delta f_{pES} = (5.36\ \text{in}^2)\left(5.87\ \frac{\text{kips}}{\text{in}^2} \right)$$
$$= 31.45\ \text{kips}$$
$$\approx 30\ \text{kips} \quad \begin{bmatrix} \text{assumed value is} \\ \text{sufficiently accurate} \end{bmatrix}$$

Estimated Time-Dependent Losses

An estimate of time-dependent losses for pretensioned members of usual design, with normal loading, using normal weight concrete, and exposed to average exposure conditions may be obtained from AASHTO Sec. 5.9.3.3.

The *long-term prestress loss*, Δf_{pLT}, due to creep of concrete, shrinkage of concrete, and relaxation of steel may be estimated using AASHTO Eq. 5.9.3.3-1.

$$\Delta f_{pLT} = \frac{10.0\gamma_h\gamma_{st}f_{pi}A_{ps}}{A_g} + 12.0\gamma_h\gamma_{st} + \Delta f_{pR}$$

Where,

γ_h = correction factor for relative humidity of the ambient air
$= 1.7 - 0.01H$

γ_{st} = correction factor for specified concrete strength at time of transfer
$= 5/\left(1 + f'_{ci} \right)$

γ_{pi} = prestressing steel stress immediately prior to transfer (kips/in²)

H = the average annual ambient relative humidity (%)

Δf_{pR} = an estimate of relaxation loss taken as 2.4 kips/in² for low relaxation strand; 10.0 kips/in² for stress relieved strand; and in accordance with manufacturers' recommendations for other types of strand (kips/in²)

The first term in AASHTO Eq. 5.9.3.3-1 corresponds to creep losses, the second term to shrinkage losses, and the third to relaxation losses. The elastic losses at transfer must be added to the time-dependent losses to determine the total losses.

Example 9.25

A precast pretensioned girder with composite deck has the following properties.

$$A_g = 800 \text{ in}^2$$

$$A_{ps} = 5.36 \text{ in}^2$$

$$f'_{ci} = 4 \text{ kips/in}^2$$

$$f_{pi} = 190 \text{ kips/in}^2$$

The girder is pretensioned with low-relaxation strand and the average annual ambient relative humidity is 70%. Determine the time-dependent prestress losses.

Solution

The correction factor for relative humidity of the ambient air is

$$\gamma_h = 1.7 - 0.01H$$
$$= 1.7 - (0.01)(70\%)$$
$$= 1.0$$

The correction factor for specified concrete strength at the time of transfer is

$$\gamma_{st} = \frac{5}{1 + f'_{ci}} = \frac{5 \dfrac{\text{kips}}{\text{in}^2}}{1 \dfrac{\text{kip}}{\text{in}^2} + 4 \dfrac{\text{kips}}{\text{in}^2}}$$
$$= 1.0$$

The long-term prestress loss is given by AASHTO Eq. 5.9.3.3-1 as

$$\Delta f_{pLT} = 10.0\gamma_h\gamma_{st}f_{pi}\frac{A_{ps}}{A_g} + 12.0\gamma_h\gamma_{st} + \Delta f_{pR}$$

$$= (10.0)(1.0)(1.0)\left(190 \frac{\text{kips}}{\text{in}^2}\right)\left(\frac{5.36 \text{ in}^2}{800 \text{ in}^2}\right)$$

$$+ \left(12.0 \frac{\text{kips}}{\text{in}^2}\right)(1.0)(1.0) + 2.4 \frac{\text{kips}}{\text{in}^2}$$

$$= 27.1 \text{ kips/in}^2$$

4. STRUCTURAL STEEL DESIGN

Design for Flexure

Nomenclature

A_{rb}	area of bottom layer of longitudinal reinforcement within the effective concrete deck width	in^2
A_{rt}	area of top layer of longitudinal reinforcement within the effective concrete deck width	in^2
A_s	cross-sectional area of structural steel	in^2
b_c	width of the compression flange of the steel beam	in
b_s	effective concrete flange width	in or ft
b_t	width of the tension flange of the steel beam	in
C	compressive force in slab at ultimate load	kips
d	depth of steel beam	in
d_c	distance from the plastic neutral axis to the midthickness of the compression flange used to compute the plastic moment	in
d_{rb}	distance from the plastic neutral axis to the centerline of the bottom layer of longitudinal concrete deck reinforcement used to compute the plastic moment	in
d_{rt}	distance from the plastic neutral axis to the centerline of the top layer of longitudinal concrete deck reinforcement used to compute the plastic moment	in
d_t	distance from the plastic neutral axis to the midthickness of the tension flange used to compute the plastic moment	in
d_w	distance from the plastic neutral axis to the midthickness of the web used to compute the plastic moment	in
D	depth of the web of the steel beam	in
D_p	distance from the top of slab to the plastic neutral axis	in
D_t	total depth of composite section	in
f'_c	specified compressive strength of the concrete	kips/in^2
F_y	specified minimum yield strength of the structural steel section	kips/in^2
F_{yc}	specified yield strength of the compression flange of the steel beam	kips/in^2
F_{yrb}	specified yield strength of the bottom layer of longitudinal deck reinforcement	kips/in^2
F_{yrt}	specified yield strength of the top layer of longitudinal deck reinforcement	kips/in^2
F_{yt}	specified yield strength of the tension flange of the steel beam	kips/in^2
F_{yw}	specified yield strength of the web of the steel beam	kips/in^2
L	span length	ft

M_n	nominal flexural resistance of the composite beam	in-kips or ft-kips
M_p	full plastic moment of the member	in-kips or ft-kips
M_u	moment due to factored loads	in-kips or ft-kips
P_c	plastic force in the compression flange of the steel beam	kips
P_{rb}	plastic force in the bottom layer of longitudinal deck reinforcement	kips
P_{rt}	plastic force in the top layer of longitudinal deck reinforcement	kips
P_s	plastic force in the full depth of the concrete deck	kips
P_t	plastic force in the tension flange of the steel beam	kips
P_w	plastic force in the web of the steel beam	kips
S	beam spacing	ft or in
t_s	slab thickness	in
y	moment arm between centroids of tensile force and compressive force	in

Symbols

γ	compression zone factor	–
ϕ_f	resistance factor for flexure	–

Strength Design Method

The strength design of a composite member is detailed in AASHTO Sec. 6.10.7 and AASHTO App. D6. As shown in Fig. 9.9, for positive bending with the plastic neutral axis (PNA) within the concrete slab, AASHTO Table D6.1-1 case V provides expressions for the depth of the plastic neutral axis and for the plastic moment of resistance. The depth of the plastic neutral axis is

$$D_p = t_s \left(\frac{P_{rb} + P_c + P_w + P_t - P_{rt}}{P_s} \right)$$

Figure 9.9 Determination of M_p

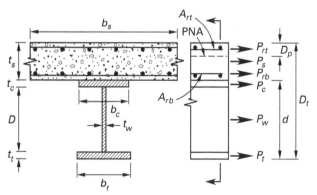

The plastic moment of resistance is

$$M_p = \frac{D_p^2 P_s}{2t_s} + P_{rb}d_{rb} + P_c d_c + P_w d_w + P_t d_t + P_{rt}d_{rt}$$

The plastic force in the bottom layer of longitudinal deck reinforcement is given by AASHTO Table D6.1-1 as

$$P_{rb} = F_{yrb}A_{rb}$$

The plastic force in the top layer of longitudinal deck reinforcement is given by AASHTO Table D6.1-1 as

$$P_{rt} = F_{yrt}A_{rt}$$

The plastic force in the top flange of the steel beam is given by AASHTO Table D6.1-1 as

$$P_c = F_{yc}b_c t_c$$

The plastic force in the bottom flange of the steel beam is given by AASHTO Table D6.1-1 as

$$P_t = F_{yt}b_t t_t$$

The plastic force in the web of the steel beam is given by AASHTO Table D6.1-1 as

$$P_w = F_{yw}Dt_w$$

The plastic compressive force in the full depth of the concrete deck is given by AASHTO Table D6.1-1 as

$$P_s = 0.85 f_c' b_s t_s$$

In the derivation of the expressions, concrete in tension is neglected. The plastic force in the portion of the concrete slab that is in compression is based on a magnitude of the compressive stress equal to $0.85 f_c'$.

The forces in the longitudinal reinforcement in the slab may be conservatively neglected. The plastic moment of resistance of the composite section may then be determined as shown in Fig. 9.10.

Figure 9.10 *Fully Composite Beam Ultimate Strength*

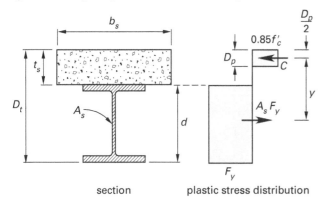

For positive bending where the top flange of the steel beam is encased in concrete or anchored to the deck slab by shear connectors, the flange is considered continuously braced. Lateral bending stresses are then considered equal to zero. For this condition, at the strength limit state, the moment due to factored loads must satisfy the expression

$$M_u \leq \phi_f M_n$$

The resistance factor for flexure is given by AASHTO Sec. 6.5.4.2 as

$$\phi_f = 1.0$$

For $D_p \leq 0.1 D_t$ the nominal flexural resistance of the section is given by AASHTO Eq. 6.10.7.1.2-1 as

$$M_n = M_p$$

For $D_p > 0.1 D_t$ the nominal flexural resistance of the section is given by AASHTO Eq. 6.10.7.1.2-2 as

$$M_n = M_p \left(1.07 - \frac{0.7 D_p}{D_t} \right)$$

For the composite beam shown in Fig. 9.10, the effective width of the concrete slab is given by AASHTO Sec. 4.6.2.6 1 as the tributary width, which is

$$b_s = S$$
$$= \phi M_n$$

Figure 9.10 shows conditions at the strength limit state when the depth of the compression zone at the ultimate load is less than the depth of the slab. From AASHTO Sec. D6.1, the depth of the stress block is given by

$$D_p = \frac{F_y A_s}{0.85 f'_c b_s}$$

The distance between the centroids of the compressive force in the slab and the tensile force in the girder is

$$y = \frac{d}{2} + t_s - \frac{D_p}{2}$$

The plastic moment capacity in bending is given by

$$M_p = F_y A_s y$$

Example 9.26

The simply supported composite beam shown in the illustration consists of an 8 in concrete slab cast on W36 × 194 grade A50 steel beams with adequate shear connection. The beams are spaced at 8 ft centers and span 100 ft; the slab consists of 4.5 kips/in² normal weight concrete. Neglect the dead load of wearing surface and utilities. The bending moment, M_{DC}, due to dead load imposed on the composite section is 250 ft-kips. The bending moment, M_L, due to live load plus impact is 1354 ft-kips. Determine the maximum factored applied moment and the maximum flexural strength of the composite section in bending.

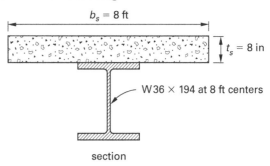

Solution

The W36 × 194 girder has the properties tabulated as follows.

A_s	d	t_w	t_f	D	I
57 in²	36.5 in	0.765 in	1.26 in	33.98 in	12,100 in⁴

The effective width of the concrete slab is given by AASHTO Sec. 4.6.2.6.1 as the tributary width, which is

$$b_s = S = (8 \text{ ft}) \left(12 \, \frac{\text{in}}{\text{ft}} \right)$$
$$= 96 \text{ in}$$

Assuming that the compression zone is contained within the slab, the depth of the stress block is given by AASHTO Sec. D6.1 as

$$D_p = \frac{F_y A_s}{0.85 f_c' b_s}$$

$$= \frac{\left(50 \dfrac{\text{kips}}{\text{in}^2}\right)(57 \text{ in}^2)}{(0.85)\left(4.5 \dfrac{\text{kips}}{\text{in}^2}\right)(96 \text{ in})}$$

$$= 7.76 \text{ in}$$

Hence, the compression zone is located within the slab. The distance between the centroids of the compressive force in the slab and the tensile force in the girder is

$$y = \frac{d}{2} + t_s - \frac{D_p}{2}$$

$$= \frac{36.5 \text{ in}}{2} + 8 \text{ in} - \frac{7.76 \text{ in}}{2}$$

$$= 22.37 \text{ in}$$

The plastic moment capacity in bending is given by

$$M_p = F_y A_s y$$

$$= \frac{\left(50 \dfrac{\text{kips}}{\text{in}^2}\right)(57 \text{ in}^2)(22.37 \text{ in})}{12 \dfrac{\text{in}}{\text{ft}}}$$

$$= 5313 \text{ ft-kips}$$

$$\frac{D_p}{D_t} = \frac{7.76 \text{ in}}{44.5 \text{ in}} = 0.174$$

$$> 0.1$$

Therefore, the nominal flexural resistance of the section is given by AASHTO Eq. 6.10.7.1.2-2 as

$$M_n = \left(1.07 - \frac{0.7 D_p}{D_t}\right) M_p$$

$$= \left(1.07 - (0.7)\left(\frac{7.76 \text{ in}}{44.5 \text{ in}}\right)\right)(5313 \text{ ft-kips})$$

$$= 5038 \text{ ft-kips}$$

$$= \phi_f M_n$$

At midspan, the self-weight moment of the steel beam is

$$M_g = \frac{wL^2}{8} = \frac{\left(0.194 \dfrac{\text{kip}}{\text{ft}}\right)(100 \text{ ft})^2}{8}$$

$$= 243 \text{ ft-kips}$$

The moment due to the weight of the deck slab is

$$M_s = \frac{wL^2}{8} = \frac{\left(150 \dfrac{\text{lbf}}{\text{ft}^3}\right)\left(\dfrac{8 \text{ in}}{12 \dfrac{\text{in}}{\text{ft}}}\right)(8 \text{ ft})(100 \text{ ft})^2}{(8)\left(1000 \dfrac{\text{lbf}}{\text{kip}}\right)}$$

$$= 1000 \text{ ft-kips}$$

The total dead load moment on the composite section is

$$M_D = M_g + M_S + M_{DC}$$

$$= 243 \text{ ft-kips} + 1000 \text{ ft-kips} + 250 \text{ ft-kips}$$

$$= 1493 \text{ ft-kips}$$

The live load moment plus impact is

$$M_L = 1354 \text{ ft-kips}$$

The factored applied moment is given by AASHTO Eq. 3.4.1-1 as

$$M_u = \gamma_p M_D + \gamma_{LL+IM} M_L$$

$$= (1.25)(1493 \text{ ft-kips}) + (1.75)(1354 \text{ ft-kips})$$

$$= 4236 \text{ ft-kips}$$

$$< \phi_f M_n \quad [\text{satisfactory}]$$

Design for Shear

Nomenclature

C	web buckling coefficient	–
D	depth of the web of the steel beam	in
E	modulus of elasticity of the steel beam	kips/in²
t_w	web thickness	in
V_n	nominal shear strength	kips
V_p	shear yielding strength of the web	kips
V_r	factored shear resistance	kips
V_u	factored applied shear force	kips

Symbols

γ	compression zone factor	–
ϕ_v	resistance factor for shear	–

Strength Design Method

AASHTO Sec. 6.10.9.2 defines the nominal shear strength of a girder with unstiffened web as

$$V_n = C V_p \quad [\text{AASHTO 6.10.9.2-1}]$$

The plastic shear force of the web is given by

$$V_p = 0.58 F_{yw} D t_w \qquad \text{[AASHTO 6.10.9.2-2]}$$

For values of $D/t_w \le 1.12\sqrt{5E/F_{yw}}$, the web buckling coefficient is defined by AASHTO Eq. 6.10.9.3.2-4 as

$$C = 1.0$$

The factored shear resistance is given by AASHTO Eq. 6.12.1.2.3a-1 as

$$V_r = \phi_v V_n$$

The resistance factor for shear is given by AASHTO Sec. 6.5.4.2 as

$$\phi_v = 1.0$$

Example 9.27

The simply supported composite beam of Ex. 9.26 is subjected to a support reaction, V_L, of 58 kips due to live load plus impact and a support reaction, V_{DC}, of 10 kips due to dead load imposed on the composite section. Determine whether the section is adequate.

Solution

The ratio of web height to web thickness is

$$\begin{aligned}
\frac{D}{t_w} &= \frac{33.98 \text{ in}}{0.765 \text{ in}} \\
&= 44.42
\end{aligned}$$

$$\begin{aligned}
1.12\sqrt{\frac{5E}{F_{yw}}} &= (1.12)\sqrt{\frac{(5)\left(29{,}000 \ \dfrac{\text{kips}}{\text{in}^2}\right)}{50 \ \dfrac{\text{kips}}{\text{in}^2}}} \\
&= 64.62 \\
&> \frac{D}{t_w}
\end{aligned}$$

Hence, from AASHTO Sec. 6.10.9.3.2, the web buckling coefficient for the composite section is

$$C = 1.0$$

The nominal shear strength of the composite section is given by AASHTO Sec. 6.10.9.2 as

$$\begin{aligned}
V_n = C V_p &= 0.58 F_{yw} D t_w \\
&= (0.58)\left(50 \ \frac{\text{kips}}{\text{in}^2}\right)(33.97 \text{ in})(0.765 \text{ in}) \\
&= 754 \text{ kips}
\end{aligned}$$

The factored shear resistance is

$$\begin{aligned}
V_r = \phi_v V_n &\\
&= (1.0)(754 \text{ kips}) \\
&= 754 \text{ kips}
\end{aligned}$$

The support reaction due to the self-weight of the girder is

$$\begin{aligned}
V_g = \frac{w_g L}{2} &= \frac{\left(0.194 \ \dfrac{\text{kip}}{\text{ft}}\right)(100 \text{ ft})}{2} \\
&= 9.7 \text{ kips}
\end{aligned}$$

The support reaction due to the weight of the slab is

$$\begin{aligned}
V_S = \frac{w_S L}{2} &= \left(0.150 \ \frac{\text{kip}}{\text{ft}^3}\right)\left(\frac{64 \text{ in}}{12 \ \dfrac{\text{in}}{\text{ft}}}\right)(50 \text{ ft}) \\
&= 40 \text{ kips}
\end{aligned}$$

The total dead load support reaction is

$$\begin{aligned}
V_D = V_g + V_S + V_{DC} &= 9.7 \text{ kips} + 40 \text{ kips} + 10 \text{ kips} \\
&= 59.7 \text{ kips}
\end{aligned}$$

The live load reaction plus impact is

$$V_L = 58 \text{ kips}$$

The factored applied reaction is given by AASHTO Eq. 3.4.1-1 as

$$\begin{aligned}
V_u = \gamma_p V_D + \gamma_{LL+IM} V_L &\\
&= (1.25)(59.7 \text{ kips}) + (1.75)(58 \text{ kips}) \\
&= 176 \text{ kips} \\
&< V_r \quad \text{[The section is adequate.]}
\end{aligned}$$

Shear Connection

Nomenclature

A_{ct}	transformed area of concrete slab	in^2
A_s	area of steel beam	in^2
A_{sc}	cross-sectional area of stud shear connector	in^2
b_s	effective width of the concrete deck	in
b_t	transformed width of concrete slab	in
d	diameter of stud shear connector	in
E_c	modulus of elasticity of concrete	kips/in^2
f_c'	specified 28-day compressive strength of concrete	kips/in^2
F_u	specified tensile strength of the steel beam	kips/in^2
F_y	specified yield strength of the steel beam	kips/in^2
H	stud height	in

I	impact factor	–
I	moment of inertia of transformed composite section	in^4
n	modular ratio	–
n	number of shear connectors between point of maximum positive moment and point of zero moment	–
n	number of shear connectors in a cross section	–
N	number of cycles	–
p	connector spacing	in
P	total shear force at interface at ultimate limit state	kips
Q	moment of transformed compressive concrete area about neutral axis	in^3
V_f	range of shear force due to live load plus impact	kips
V_{sr}	range of horizontal shear at interface	kips/in
y'	distance from slab center to neutral axis of transformed composite	in
Z_r	allowable range of shear for a welded stud	lbf

Symbols

α	stress cycle factor	–
γ	load factor	–
ϕ	reduction factor	–

General

Shear connectors are designed for fatigue and are checked for ultimate strength. Fatigue stresses are caused by the range of shear produced on the connector by live load plus impact and are calculated by using elastic design principles. The ultimate strength of the connectors must be adequate to develop the lesser of the strength of the steel girder or the ultimate strength of the concrete slab. The pitch of the connectors is determined from the fatigue limit state. The total number of connectors required is determined from the strength limit state.

Fatigue Strength

The elastic design properties of the composite section are determined by using the transformed width of the concrete slab. As shown in Fig. 9.11, the transformed width is given by

$$b_t = \frac{b_s}{n}$$

For transient loads, the modular ratio, n, for short-term loads is used.

The transformed area of the concrete slab is

$$A_{ct} = b_t t_s$$

The statical moment of the transformed concrete area about the neutral axis of the composite section is

$$Q = y' A_{ct}$$

For a straight girder, the range of horizontal shear at the interface is given by AASHTO Eq. 6.10.10.1.2-3 as

$$V_{sr} = \frac{V_f Q}{I}$$

The range of shear force, V_f, due to live load plus impact is the difference between the maximum and minimum applied shear under the fatigue load combination. In accordance with AASHTO Sec. 6.10.10.1.1, the ratio of height to diameter, H/d, of a welded stud must not be less than 4. The allowable range of shear for a welded stud is given by AASHTO Sec. 6.10.10.2.

For a projected 75-year single-lane average daily truck traffic greater than or equal to 960 trucks per day, the fatigue I load combination is used. The fatigue shear resistance for infinite life is taken as

$$Z_r = 5.5d^2 \quad \text{[AASHTO 6.10.10.2-1]}$$

For other loading conditions, the fatigue II load combination is used and the fatigue shear resistance for finite life is taken as

$$Z_r = \alpha d^2 \quad \text{[AASHTO 6.10.10.2-2]}$$

$$\alpha = 34.5 - 4.28 \log N \quad \text{[AASHTO 6.10.10.2-3]}$$

The required connector pitch is given by AASHTO Eq. 6.10.10.1.2-1 as

$$p = \frac{n Z_r}{V_{sr}}$$
$$\leq 24 \text{ in} \quad \text{[for web depth} < 24 \text{ in]}$$
$$\leq 48 \text{ in} \quad \text{[for web depth} \geq 24 \text{ in]}$$
$$\geq 6d$$

Example 9.28

For the composite beam of Ex. 9.27, determine the required spacing of $\frac{3}{4}$ in diameter stud shear connectors at the support. The fatigue I limit state applies, and the maximum shear force at the support due to the fatigue I vehicle is $V = 36$ kips, including impact.

Solution

The compressive strength of the normal weight concrete in the slab is given as

$$f_c' = 4.5 \text{ kips/in}^2$$

Figure 9.11 *Composite Section Properties*

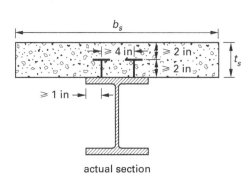 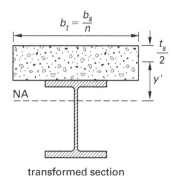

actual section   transformed section

AASHTO Sec. C5.4.2.4 specifies that the modulus of elasticity of normal weight concrete in the slab is

$$E_c = 1820\sqrt{f'_c}$$
$$= 1820\sqrt{4.5\ \frac{\text{kips}}{\text{in}^2}}$$
$$= 3861\ \text{kips/in}^2$$

The corresponding modular ratio is

$$n = \frac{E_s}{E_c}$$
$$= \frac{29{,}000\ \dfrac{\text{kips}}{\text{in}^2}}{3861\ \dfrac{\text{kips}}{\text{in}^2}}$$
$$= 7.51$$

Round to

$$n = 8$$

The transformed area of the concrete slab is

$$A_{ct} = \frac{b_s t_s}{n} = \frac{(96\ \text{in})(8\ \text{in})}{8}$$
$$= 96\ \text{in}^2$$

The moment of inertia of the transformed section is derived as shown in the following table.

part	A (in²)	y (in)	I (in⁴)	Ay (in³)	Ay^2 (in⁴)
girder	57	18.25	12,100	1040	18,985
slab	96	40.49	512	3887	157,385
total	153	–	12,612	4927	176,370

The height of the neutral axis of the transformed section is

$$\bar{y} = \frac{\sum Ay}{\sum A} = \frac{4927\ \text{in}^3}{153\ \text{in}^2}$$
$$= 32.2\ \text{in}$$

The moment of inertia of the transformed section is

$$I = \sum I + \sum Ay^2 + \bar{y}^2 \sum A - 2\bar{y}\sum Ay$$
$$= 12{,}612\ \text{in}^4 + 176{,}370\ \text{in}^4 + (32.2\ \text{in})^2(153\ \text{in}^2)$$
$$\quad - (64.4\ \text{in})(4927\ \text{in}^3)$$
$$= 30{,}320\ \text{in}^4$$

The statical moment of the transformed slab about the neutral axis is

$$Q = y'A_{ct} = \left(d + \frac{t_s}{2} - \bar{y}\right)A_{ct}$$
$$= \left(36.5\ \text{in} + \frac{8\ \text{in}}{2} - 32.2\ \text{in}\right)(96\ \text{in}^2)$$
$$= 797\ \text{in}^3$$
$$\frac{Q}{I} = \frac{797\ \text{in}^3}{30{,}320\ \text{in}^4}$$
$$= 0.0263\ \text{in}^{-1}$$

The transformed section is shown.

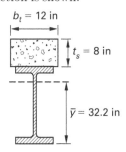

transformed composite section

The fatigue shear resistance for infinite life for a ¾ in diameter welded stud subjected to fatigue I loading is given by AASHTO Eq. 6.10.10.2-1 as

$$Z_r = 5.5d^2 = \left(5.5\ \frac{\text{kips}}{\text{in}^2}\right)(0.75\ \text{in})^2$$
$$= 3.09\ \text{kips}$$

From AASHTO Table 3.4.1-1, the maximum factored shear force at the support due to the fatigue I vehicle plus impact is

$$V_{\max} = \gamma V = (1.75)(36\ \text{kips})$$
$$= 63\ \text{kips}$$

The minimum shear force at the support is

$$V_{\min} = 0\ \text{kips}$$

The range of shear force at the support for the fatigue I limit state is

$$V_f = V_{\max} - V_{\min}$$
$$= 63\ \text{kips} - 0\ \text{kips}$$
$$= 63\ \text{kips}$$

The range of horizontal shear at the support at the interface is given by AASHTO Eq. 6.10.10.1.2-3 as

$$V_{sr} = \frac{V_f Q}{I} = (63\ \text{kips})(0.0263\ \text{in}^{-1})$$
$$= 1.66\ \text{kips/in}$$

With three studs per row, the required stud spacing is

$$p = \frac{nZ_r}{V_{sr}} = \frac{(3)(3.09\ \text{kips})}{1.66\ \dfrac{\text{kips}}{\text{in}}}$$
$$= 5.6\ \text{in}$$

Ultimate Strength

To provide adequate connection at the interface at ultimate load, the number of connectors required on each side of the point of maximum moment is given by AASHTO Sec. 6.10.10.4.1 as

$$n = \frac{P}{\phi_{sc} Q_n}$$

For a straight rolled section girder, the total shear force at the interface at the ultimate limit state is given in AASHTO Sec. 6.10.10.4.2 as the lesser of

$$P = A_s F_y \quad [\text{girder governs}]$$
$$P = 0.85 f_c' b_s t_s \quad [\text{deck slab governs}]$$

The resistance factor for a shear connector is given by AASHTO Sec. 6.5.4.2 as

$$\phi_{sc} = 0.85$$

The nominal strength of a welded stud shear connector is given by AASHTO Eq. 6.10.10.4.3-1 as

$$Q_n = 0.5 A_{sc}\sqrt{f_c' E_c}$$
$$\leq A_{sc} F_u$$

The modulus of elasticity of normal weight concrete is given by AASHTO Sec. C5.4.2.4 as

$$E_c = 1820\sqrt{f_c'}$$

Example 9.29

For the composite beam of Ex. 9.27, determine the required number of ¾ in diameter stud shear connectors to provide adequate connection at the ultimate load. The shear connectors have a tensile strength of $F_u = 60\ \text{kips/in}^2$.

Solution

The compressive strength of the concrete slab is $4.5\ \text{kips/in}^2$, and the corresponding modulus of elasticity is given by AASHTO Sec. 5.4.2.4 as

$$E_c = 1820\sqrt{f_c'}$$
$$= 1820\sqrt{4.5\ \frac{\text{kips}}{\text{in}^2}}$$
$$= 3861\ \text{kips/in}^2$$

The ultimate strength of a ¾ in diameter welded stud shear connector is given by AASHTO Eq. 6.10.10.4.3-1 as

$$Q_n = 0.5 A_{sc}\sqrt{f_c' E_c}$$
$$= (0.5)(0.44\ \text{in}^2)\sqrt{\left(4.5\ \frac{\text{kips}}{\text{in}^2}\right)\left(3861\ \frac{\text{kips}}{\text{in}^2}\right)}$$
$$= 29\ \text{kips}$$

Maximum strength is

$$Q_{n(\max)} = F_u A_{sc}$$
$$= (60)(0.44)$$
$$= 26.40 \text{ kips} \quad [\text{governs}]$$

The nominal shear force at the interface is governed by the girder, and the total number of connectors required on the girder is

$$2n = \frac{2A_s F_y}{\phi_{sc} Q_n} = \frac{(2)(57 \text{ in}^2)\left(50 \frac{\text{kips}}{\text{in}^2}\right)}{(0.85)(26.40 \text{ kips})}$$
$$= 254$$

5. WOOD STRUCTURES

Basic Design Values and Adjustment Factors

Nomenclature

a	species parameter for volume factor	–
A	ratio of F_{bE} to F_b	–
b	breadth of rectangular bending member	in
C_d	deck factor	–
C_F	size factor for sawn lumber	–
C_{fu}	flat use factor	–
C_i	incising factor	–
C_{KF}	format conversion factor	–
C_L	beam stability factor	–
C_M	wet service factor	–
C_V	volume factor for structural glued laminated timber	–
C_λ	time effect factor	–
d	depth of member	in
E, E_o	adjusted and reference modulus of elasticity	kips/in^2
F_b	reference bending design value multiplied by all applicable adjustment factors	kips/in^2
F_{bE}	critical buckling design value for bending members	kips/in^2
F_{bo}	reference bending design value	kips/in^2
F_v	adjusted design value of wood in shear	kips/in^2
F_{vo}	reference design value of wood in shear	kips/in^2
K_{bE}	Euler buckling coefficient for beams	–
L	span length of bending member	ft or in
L_e	effective bending member length	ft or in
L_u	laterally unsupported bending member length	ft or in
M_n	nominal flexural resistance	in-kips
M_r	factored flexural resistance	in-kips
R_B	slenderness ratio of bending member	–
S	section modulus	in^3
V_n	nominal shear resistance	kips
V_r	factored shear resistance	kips

Symbols

ϕ	resistance factor	–

Reference Design Values

The reference design values for sawn lumber are given in AASHTO Table 8.4.1.1.4-1, Table 8.4.1.1.4-2, and Table 8.4.1.1.4-3. The reference design values for glued laminated timber are given in AASHTO Table 8.4.1.2.3-1 and Table 8.4.1.2.3-2. These tabulated design values are applicable to normal conditions of use as defined in AASHTO Sec. C8.4.1. For other conditions of use, the tabulated values are multiplied by adjustment factors, specified in AASHTO Sec. 8.4.4, to determine the corresponding adjusted design values. AASHTO Sec. 8.4.4.1 specifies the adjusted design value in bending as

$$F_b = F_{bo} C_{KF} C_M (C_F \text{ or } C_V) C_{fu} C_i C_d C_\lambda$$

[AASHTO 8.4.4.1-1]

Adjustment Factors

The *time effect factor*, C_λ, given in AASHTO Sec. 8.4.4.9 is applicable to all reference design values with the exception of the modulus of elasticity. Values of the time effect factor are given in Table 9.4 [AASHTO Table 8.4.4.9-1].

Table 9.4 *Time Effect Factor, C_λ*

limit state	C_λ
strength I	0.8
strength II	1.0
strength III	1.0
strength IV (permanent)	0.6
extreme event I	1.0

The *wet service factor*, C_M, given in AASHTO Table 8.4.4.3-1, is applicable to sawn lumber when the moisture content exceeds 19%. Values of the wet service factor are given in Table 9.5. For a moisture content $\leq 19\%$, $C_M = 1.0$.

Bridge Design

Table 9.5 *Wet Service Factor, C_M, for Sawn Lumber*

design function	$F_{bo}C_F > 1.15$	F_{vo}
members not exceeding 4 in thickness	0.85	0.97
members exceeding 4 in thickness	1.00	1.00

When the moisture content of a glued laminated member exceeds 16%, the adjustment factor given in AASHTO Table 8.4.4.3-2 is applicable. Values of the wet service factor are given in Table 9.6. For a moisture content $\leq$ 16%, $C_M = 1.0$.

Table 9.6 *Wet Service Factor, C_M, for Glued Laminated Members*

design function	F_{bo}	F_{vo}
wet service factor	0.80	0.875

The *beam stability factor* is applicable to the tabulated bending reference design value for sawn lumber and glued laminated members. For glued laminated members, C_L is not applied simultaneously with the volume factor, C_V, and the lesser of these two values is applicable.

The beam stability factor is given by AASHTO Sec. 8.6.2 as

$$C_L = \frac{1.0 + A}{1.9} - \sqrt{\frac{(1.0 + A)^2}{3.61} - \frac{A}{0.95}}$$

[AASHTO 8.6.2-2]

The variables are defined as

$$A = \frac{F_{bE}}{F_b}$$

[AASHTO 8.6.2-3]

F_b = reference bending design value multiplied by all applicable adjustment factors

$$= F_{bo}C_{KF}C_M C_F C_{fu}C_i C_d C_\lambda \begin{bmatrix} C_F \text{ applies only} \\ \text{to visually graded} \\ \text{sawn lumber} \end{bmatrix}$$

F_{bE} = critical buckling design value

$$= \frac{K_{bE}E}{R_B^2}$$

[AASHTO 8.6.2-4]

K_{bE} = Euler buckling coefficient
$= 1.10$ [for glued laminated timber]
$= 0.76$ [for visually graded lumber]

E = adjusted modulus of elasticity
$= E_o C_M C_i$

[AASHTO 8.4.4.1-6]

R_B = slenderness ratio

$$= \sqrt{\frac{L_e d}{b^2}}$$

[AASHTO 8.6.2-5]

≤ 50

The term L_e is the effective length of a bending member and is defined in AASHTO Sec. 8.6.2 and tabulated in Table 9.7.

Table 9.7 *Effective Length, L_e*

member dimensions	L_e
$\dfrac{L_u}{d} < 7$	$2.06L_u$
$7 \leq \dfrac{L_u}{d} \leq 14.3$	$1.63L_u + 3d$
$\dfrac{L_u}{d} > 14.3$	$1.84L_u$

In accordance with AASHTO Sec. 8.6.2, $C_L = 1.0$ when either of the following two conditions apply.

- $\dfrac{d}{b} \leq 1.0$

- the compression edge is continuously restrained and rotation is prevented at supports

The *size factor*, C_F, is taken as 1.0 for mechanically graded lumber and structural glued laminated timber.

For visually graded dimension lumber, except southern pine and mixed southern pine species, values of the size factor are given in AASHTO Table 8.4.4.4-1.

For southern pine and mixed southern pine species dimension lumber wider than 12 in, the size factor for bending, compression, and tension parallel to grain is $C_F = 0.9$. Reference design values for all other sizes of southern pine and mixed southern pine species dimension lumber are size adjusted, and no further adjustment is required.

For sawn beams and stringers with loads applied to the wide face, values of the size factor are given in AASHTO Table 8.4.4.4-2. For sawn beams and stringers with loads applied to the narrow face, and posts and timbers with loads applied to either face, the size factor is

$$C_F = 1.0 \text{ where } d \leq 12 \text{ in}$$

Where $d > 12$ in,

$$C_F = \left(\frac{12}{d}\right)^{1/9}$$

[AASHTO 8.4.4.4-2]

The *volume factor*, C_V, is applicable to the reference design value for bending of glued laminated members and is not applied simultaneously with the beam stability factor, C_L; the lesser of these two factors is applicable.

The volume factor is defined in AASHTO Sec. 8.4.4.5 as

$$C_V = \left(\frac{1291.5}{bdL} \right)^a \quad \text{[AASHTO 8.4.4.5-1]}$$
$$\leq 1.0$$

The volume factor is applied only when the depth, width, or length of the member exceeds 12 in, 5.125 in, or 21 ft, respectively.

The variables are defined as

$L =$ length of beam between points
of zero moment, ft
$b =$ beam width, in
$d =$ beam depth, in
$a = 0.05$ [for southern pine]
$= 0.10$ [for all other species]

The *flat-use factor*, C_{fu}, is applicable to dimension lumber with the load applied to the wide face. Values of the flat-use factor are given in AASHTO Table 8.4.4.6-1. Flat-use factors for glued laminated members, with the load applied parallel to the wide faces of the laminations, are given in AASHTO Table 8.4.4.6-2. The flat-use factor is not applied to dimension lumber graded as decking, as the design values already incorporate the appropriate factor.

Values of the *incising factor*, C_i, for a prescribed incising pattern are given in AASHTO Table 8.4.4.7-1. The prescribed pattern consists of incisions parallel to the grain a maximum depth of 0.4 in, a maximum length of 3/8 in, and a density of incisions of up to 1100/ft^2.

The *deck factor*, C_d, is applied to mechanically laminated decks. Values of the deck factor for stressed wood, spike-laminated, and nail-laminated decks are given in AASHTO Table 8.4.4.8-1.

The format conversion factor, C_{KF}, is used to ensure that load and resistance factor design will result in the same size members as allowable stress design. AASHTO Sec. 8.4.4.2 gives the format conversion factor for all loading conditions except compression perpendicular to the grain, as

$$C_{KF} = \frac{2.5}{\phi}$$

For compression perpendicular to the grain,

$$C_{KF} = \frac{2.1}{\phi}$$

Resistance factors are tabulated in AASHTO Sec. 8.5.2.2 and these include

$$\phi = 0.85 \text{ for flexure}$$
$$\phi = 0.75 \text{ for shear}$$

The factored flexural resistance is given by AASHTO Eq. 8.6.1-1 as

$$M_r = \phi M_n$$

The nominal flexural resistance of a rectangular member is given by AASHTO Eq. 8.6.2-1 as

$$M_n = F_b S C_L$$

For shear, the corresponding values are

$$V_r = \phi V_n \quad \text{[AASHTO 8.7-1]}$$
$$V_n = \frac{F_v b d}{1.5} \quad \text{[AASHTO 8.7-2]}$$

Example 9.30

The bridge superstructure shown in the illustration is simply supported over a span of 35 ft, and rotation at the supports is prevented. The moisture content exceeds 16%. Determine the nominal resistance values in bending and shear for the strength I limit state for the glued laminated girders, with combination 24F-V4 Douglas fir.

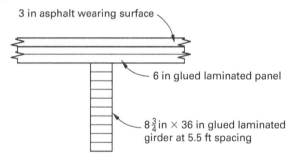

3 in asphalt wearing surface

6 in glued laminated panel

$8\frac{3}{4}$ in $\times$ 36 in glued laminated girder at 5.5 ft spacing

Solution

The reference design bending stress for a Douglas fir 24F-V4 glued laminated stringer is obtained from AASHTO Table 8.4.1.2.3-1 and is

$$F_{bo} = 2.4 \text{ kips/in}^2$$

The applicable adjustment factors for bending stress are as follows.

C_λ = time effect factor for the strength I limit state from Table 8.4.4.9-1
   = 0.8

C_M = wet service factor from AASHTO Table 8.4.4.3-2
   = 0.80

C_L = beam stability factor from AASHTO Sec. 8.6.2
   = 1.0 $\begin{bmatrix} \text{compression face of the girder} \\ \text{fully supported laterally} \end{bmatrix}$

C_V = volume factor given by AASHTO Eq. 8.4.4.5-1
   $= \left(\dfrac{1291.5}{bdL}\right)^a$
   $= \left(\dfrac{1291.5 \text{ in}^2\text{-ft}}{(8.75 \text{ in})(36.0 \text{ in})(35 \text{ ft})}\right)^{0.10}$
   = 0.81   [governs]
   $< C_L$

In accordance with AASHTO Eq. 8.4.4.1-1, the adjusted design bending stress is

$$F_b = F_{bo}C_\lambda C_M C_V C_{KF}$$
$$= \left(2.4 \text{ } \frac{\text{kips}}{\text{in}^2}\right)(0.8)(0.80)(0.81)\left(\frac{2.5}{0.85}\right)$$
$$= 3.66 \text{ kips/in}^2$$

The nominal flexural resistance is given by AASHTO Eq. 8.6.2-1 as

$$M_n = F_b S C_L$$
$$= (3.66 \text{ kips})(1890 \text{ in}^3)(1.0)$$
$$= 6917 \text{ in-kips}$$

The reference design shear stress for a Douglas fir 24F-V4 glued laminated stringer is obtained from AASHTO Table 8.4.1.2.3-1 and is

$$F_{vo} = 0.265 \text{ kip/in}^2$$

The applicable adjustment factors for shear stress are as follows.

C_λ = time effect factor for the strength I limit state from AASHTO Table 8.4.4.9-1
   = 0.8

C_M = wet service factor from AASHTO Table 8.4.4.3-2
   = 0.875

In accordance with AASHTO Eq. 8.4.4.1-2, the adjusted design shear stress is

$$F_v = F_{vo}C_\lambda C_M C_{KF}$$
$$= \left(0.265 \text{ } \frac{\text{kip}}{\text{in}^2}\right)(0.8)(0.875)\left(\frac{2.5}{0.75}\right)$$
$$= 0.62 \text{ kip/in}^2$$

Design Requirements for Flexure

AASHTO Sec. 8.6.2 specifies that when the depth of a beam does not exceed its breadth or when continuous lateral restraint is provided to the compression edge of a beam with the ends restrained against rotation, the beam stability factor $C_L = 1.0$. For other situations, the value of C_L is calculated in accordance with AASHTO Sec. 8.6.2, and the effective span length, L_e, is determined in accordance with AASHTO Sec. 8.6.2. For visually graded sawn lumber, both the stability factor, C_L, and the size factor, C_F, must be considered concurrently. For glued laminated members, both the stability factor, C_L, and the volume factor, C_V, must be determined. Only the lesser of these two factors is applicable in determining the allowable design value in bending. In accordance with AASHTO Sec. 3.6.2.3, the dynamic load allowance need not be considered in timber structures.

Design Requirements for Shear

In accordance with AASHTO Sec. 8.7, shear should be investigated at a distance from the support equal to the depth of the beam. The governing *shear force, V, for vehicle loads* is determined by placing the live load to produce the maximum shear at a distance from the support given by the lesser of

- $3d$
- $L/4$

In accordance with AASHTO Sec. 3.6.2.3, the dynamic load allowance need not be considered in timber structures.

6. SEISMIC DESIGN

Nomenclature

A_S	peak seismic ground acceleration coefficient modified by zero period site factor from AASHTO 3.10.4.2	–
C_{sm}	seismic response coefficient specified in AASHTO Sec. 3.10.4.2	–
F_a	site factor for short-period range of acceleration response spectrum from AASHTO Sec. 3.10.3.2	–
F_{pga}	site factor at zero-period on acceleration response spectrum from AASHTO Sec. 3.10.3.2	–

Bridge Design

F_v	site factor for long-period range of acceleration response spectrum from AASHTO Sec. 3.10.3.2	–
g	gravitational acceleration, 32.2	ft/sec^2
IC	importance category	–
K	total lateral stiffness of bridge	lbf/in or kips/in
L	length of bridge deck	ft
N	minimum support length for girders	in
$p_e(x)$	intensity of the equivalent static seismic loading used to calculate the period in AASHTO Sec. C4.7.4.3.2b	lbf/in or kips/in
p_o	assumed uniform loading used to calculate the period in AASHTO Sec. C.4.7.4.3.2b	lbf/ft or kips/ft
PGA	peak seismic ground acceleration coefficient on rock (site class B) from AASHTO Sec. 3.10.2.1	–
R	response modification factor from AASHTO Sec. 3.10.7.1	–
S_1	horizontal response spectral acceleration coefficient at 1.0 sec period on rock (site class B) from AASHTO Sec. 3.10.2.1	–
S_{D1}	horizontal response spectral acceleration coefficient at 1.0 sec modified by long-period site factor from AASHTO Sec. 3.10.4.2	–
S_{DS}	horizontal response spectral acceleration coefficient at 0.2 sec period modified by short-period site factor from AASHTO Sec. 3.10.4.2	–
S_S	horizontal response spectral acceleration coefficient at 0.2 sec period on rock (site class B) from AASHTO Sec. 3.10.2.1	–
SC	site class from AASHTO Sec. 3.10.3.1	–
SPZ	seismic performance zone	–
T_0	reference period used to define shape of acceleration response spectrum from AASHTO Sec. 3.10.4.2	–
T_m	fundamental period of vibration, defined in AASHTO Sec. C4.7.4.3.2b	sec
T_S	corner period at which acceleration response spectrum changes from being independent of period to being inversely proportional to period from AASHTO Sec. 3.10.4.2	sec
$v_s(x)$	static displacement profile resulting from applied load p_o used in AASHTO Sec. C4.7.4.3.2b	in
$w(x)$	dead weight of bridge superstructure and tributary substructure per unit length	lbf/ft or kips/ft
W	total weight of bridge superstructure and tributary substructure	lbf or kips

Symbols

α	coefficient used to calculate period of the bridge in AASHTO Sec. C4.7.4.3.2b	ft^2
β	coefficient used to calculate period of the bridge in AASHTO Sec. C4.7.4.3.2b	ft-kips
γ	coefficient used to calculate period of the bridge in AASHTO Sec. C4.7.4.3.2b	ft^2-kips

Analysis Procedures

To determine the seismic response of the structure, several factors must be considered. These factors include the ground motion parameters, site class, fundamental period, and response modification factors. Selection of the design procedure depends on the type of bridge, the importance category, and the seismic zone. Four analysis procedures[13,14] are presented in AASHTO Sec. 4.7.4.3.1 and are shown in Table 9.8.

Table 9.8 Analysis Procedures

procedure	method
UL	uniform load elastic
SM	single-mode elastic
MM	multimode elastic
TH	time history

The uniform load and single-mode procedures both assume that the seismic response of a bridge can be represented by a single mode of vibration and are suitable for hand computation. The multimode and time history procedures account for higher modes of vibration and require analysis by computer.

Acceleration Coefficients

The acceleration coefficients PGA, S_S, and S_1 are defined in AASHTO Sec. 3.10.4.2, and they are shown in AASHTO Fig. 3.10.2.1-1 to Fig. 3.10.2.1-21. These are an estimate of the site-dependent design ground acceleration expressed as a percentage of the gravity constant, g. The acceleration coefficients correspond to ground acceleration values with a recurrence interval of 1000 yr, which gives a 7% probability of being exceeded in a 75 yr period. This is termed the design earthquake.

Example 9.31

The two-span bridge shown in the illustration is located at 33.70° north and −117.50° west on a non-essential route. The central circular column is fixed at the top and bottom. The soil profile at the site consists of a stiff soil with a shear wave velocity of 700 ft/sec. The relevant criteria are (a) column moment of inertia $I_c = 60$ ft^4, (b) column modulus of elasticity $E_c = 450{,}000$ kips/ft^2, (c) column height $h_c = 30$ ft, (d) weight of the superstructure and tributary substructure $w = 10$ kips/ft, (e) superstructure moment of inertia $I_s = 4000$ ft^4, and (f) superstructure modulus of elasticity $E_s = 450{,}000$ kips/ft^2. Determine the applicable acceleration coefficients.

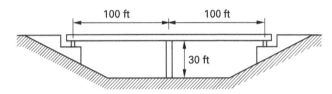

Solution

From AASHTO Fig. 3.10.2.1-4 to Fig. 3.10.2.1-6, the applicable acceleration coefficients are

$$PGA = 0.61$$
$$S_S = 1.45$$
$$S_1 = 0.52$$

Importance Category

The importance category is defined in AASHTO Sec. 3.10.5, and three categories are specified: critical bridges, essential bridges, and other bridges. The importance category of a bridge is determined on the basis of social and security requirements. An importance category of *critical* is assigned to bridges that must remain functional immediately after a 2500-year return period earthquake. An importance category of *essential* is assigned to bridges that must remain functional immediately after the design earthquake. An importance category of *other* is assigned to nonessential bridges.

Example 9.32

Determine the importance category for the bridge of Ex. 9.31.

Solution

From AASHTO Sec. 3.10.5, for a bridge on a nonessential route, the importance category is

$$IC = other$$

Site Class

Six soil profile types are identified in AASHTO Table 3.10.3.1-1. Table 9.9 gives a summary of the soil profile types.

Table 9.9 Site Classes

site class	soil profile name	shear wave velocity
A	hard rock	>5000
B	rock	2500–5000
C	soft rock	1200–2500
D	stiff soil	600–1200
E	soft soil	<600
F	*	*

* consists of peat or high plasticity clay requiring a site-specific geotechnical investigation.

Example 9.33

Determine the site class for the bridge of Ex. 9.31.

Solution

From AASHTO Sec. 3.10.3.1 and Table 9.9, the relevant site class for a stiff soil is D.

Site Factors

Site factors are amplification factors applied to the ground accelerations and are a function of the site class. Site factor F_{pga} corresponds to PGA, F_a corresponds to S_S, and F_v corresponds to S_1. Site Class B is the reference site category and has a site factor of 1.0. The site factors generally increase as the soil profile becomes softer (in going from site class A to E). The factors also decrease as the ground motion level increases due to the nonlinear behavior of the soil. Site factors F_{pga}, F_a, and F_v are specified in AASHTO Table 3.10.3.2-1, Table 3.10.3.2-2, and Table 3.10.3.2-3 and are summarized in Table 9.10. Linear interpolation may be used to obtain intermediate values.

Example 9.34

Determine the site factors for the bridge of Ex. 9.31.

Solution

From Ex. 9.31, the ground motion parameters are PGA $= 0.61g$, $S_S = 1.45g$, and $S_1 = 0.52g$. From Ex. 9.33, the site class at the location of the bridge is SC $= D$. From Table 9.10 the site factors are

$$F_{pga} = 1.0$$
$$F_a = 1.0$$
$$F_v = 1.5$$

Table 9.10 Site Factors (F_{pga} corresponding to PGA; F_a corresponding to S_S; F_v corresponding to S_1)

site class	ground acceleration, PGA					ground acceleration, S_S					ground acceleration, S_1				
	≤ 0.1	0.2	0.3	0.4	≥ 0.5	≤ 0.25	0.50	0.75	1.00	≥ 1.25	≤ 0.1	0.2	0.3	0.4	≥ 0.5
A	0.8	0.8	0.8	0.8	0.8	0.8	0.8	0.8	0.8	0.8	0.8	0.8	0.8	0.8	0.8
B	1.0	1.0	1.0	1.0	1.0	1.0	1.0	1.0	1.0	1.0	1.0	1.0	1.0	1.0	1.0
C	1.2	1.2	1.1	1.0	1.0	1.2	1.2	1.1	1.0	1.0	1.7	1.6	1.5	1.4	1.3
D	1.6	1.4	1.2	1.1	1.0	1.6	1.4	1.2	1.1	1.0	2.4	2.0	1.8	1.6	1.5
E	2.5	1.7	1.2	0.9	0.9	2.5	1.7	1.2	0.9	0.9	3.5	3.2	2.8	2.4	2.4
F	(*)	(*)	(*)	(*)	(*)	(*)	(*)	(*)	(*)	(*)	(*)	(*)	(*)	(*)	(*)

Note: (*) Site-specific geotechnical investigation and dynamic site response analysis is required.

Adjusted Response Parameters

As specified in AASHTO Sec. 3.10.4.2, the ground motion parameters are modified by the site factors to allow for the site class effects. Therefore, the adjusted response parameters are

$$A_S = F_{pga}(\text{PGA})$$
$$S_{DS} = F_a S_S$$
$$S_{D1} = F_v S_1$$

Example 9.35

Determine the adjusted response parameters for the bridge of Ex. 9.31.

Solution

From Ex. 9.31, the ground motion parameters are PGA $= 0.61g$, $S_S = 1.45g$, and $S_1 = 0.52g$. From Ex. 9.34, the site factors are $F_{pga} = 1.0$, $F_a = 1.0$, and $F_v = 1.5$. From AASHTO Sec. 3.10.4.2, the adjusted response parameters are

$$
\begin{aligned}
A_s &= F_{pga}(\text{PGA}) \\
&= (1.0)(0.61g) \\
&= 0.61g \\
S_{DS} &= F_a S_S \\
&= (1.0)(1.45g) \\
&= 1.45g \\
S_{D1} &= F_v S_1 = (1.5)(0.52g) \\
&= (1.5)(0.52g) \\
&= 0.78g
\end{aligned}
$$

Design Response Spectrum

The adopted design response spectrum is given by AASHTO Fig. 3.10.4.1-1 and is shown in Fig. 9.12. The spectrum for a specific location is a graph of the elastic seismic response coefficient, C_{sm}, over a range of periods of vibration, T_m.

Figure 9.12 Design Response Spectrum

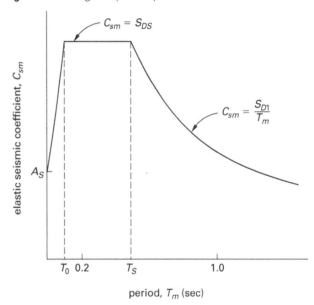

Elastic Seismic Response Coefficient

As shown in Fig. 9.12, the design response spectrum is composed of three segments demarcated by the periods of vibration:

$$
\begin{aligned}
T_m &= 0 \\
T_m &= T_0 = 0.2 T_S \\
T_m &= T_S = \frac{S_{D1}}{S_{DS}}
\end{aligned}
$$

The values of the elastic seismic response coefficient are determined using equations given in AASHTO Sec. 3.10.4.2, which are summarized in Table 9.11.

Table 9.11 Elastic Seismic Response Coefficient Equations

period, T_m	elastic seismic response coefficient, C_{sm}
$T_m \leq T_0$	$A_S + (S_{DS} - A_S)(T_m/T_0)$
$T_0 < T_m \leq T_s$	S_{DS}
$T_m > T_s$	S_{D1}/T_m

Example 9.36

Determine the reference periods used to define the shape of the response spectrum for the bridge of Ex. 9.31.

Solution

From Ex. 9.35, the adjusted response parameters are $S_{DS} = 1.45g$ and $S_{D1} = 0.78g$. From AASHTO Sec. 3.10.4.2, the reference periods are

$$T_S = \frac{S_{D1}}{S_{DS}} = \frac{0.78g}{1.45g}$$
$$= 0.538 \text{ sec}$$
$$T_0 = 0.2T_S = (0.2)(0.538 \text{ sec})$$
$$= 0.108 \text{ sec}$$

Seismic Performance Zone

The seismic performance zone (SPZ) is a function of the acceleration coefficient S_{D1} and is defined in AASHTO Sec. 3.10.6. The four categories are shown in Table 9.12; these determine the necessary requirements for selection of the design procedure, minimum support lengths, and substructure design details.

Table 9.12 Seismic Performance Zones

acceleration coefficient, S_{D1}	seismic zone
$S_{D1} \leq 0.15$	1
$0.15 < S_{D1} \leq 0.30$	2
$0.30 < S_{D1} \leq 0.50$	3
$0.50 < S_{D1}$	4

Example 9.37

Determine the seismic zone for the bridge of Ex. 9.31.

Solution

From AASHTO Sec. 3.10.6, for a value of the acceleration coefficient exceeding 0.50, the relevant seismic performance zone is

$$\text{SPZ} = 4$$

Selection of Analysis Procedure

In accordance with AASHTO Sec. 4.7.4.3, the analysis procedure selected depends on the seismic zone, importance category, and bridge regularity. This information is summarized in Table 9.13. A *regular bridge* is defined as having fewer than seven spans with no abrupt changes in weight, stiffness, or geometry. An *irregular bridge* does not satisfy the definition of a regular bridge, and in this type of structure, the higher modes of vibration significantly affect the seismic response. A detailed seismic analysis is not required for single-span bridges. Minimum support lengths are required, however, to accommodate the maximum inelastic displacement, in accordance with AASHTO Sec. 4.7.4.4.

Table 9.13 Selection of Analysis Procedure for Multispan Bridges

seismic zone	other bridges		essential bridges		critical bridges	
	regular	irregular	regular	irregular	regular	irregular
2	SM/UL	SM	SM/UL	MM	MM	MM
3	SM/UL	MM	MM	MM	MM	TH
4	SM/UL	MM	MM	MM	TH	TH

Note: UL = uniform load elastic method; SM = single-mode elastic method; MM = multimode elastic method; TH = time history method.

A seismic analysis is not required for bridges in seismic zone 1.

Example 9.38

Determine the required analysis procedure for the bridge of Ex. 9.31.

Solution

From Ex. 9.37, the seismic performance zone is 4. From AASHTO Sec. 4.7.4.3, for a regular bridge in seismic zone 4 with an importance category of "other," the required analysis procedure is UL or SM.

The Uniform Load Elastic Method

The uniform load elastic method is defined in AASHTO Sec. 4.7.4.3.2c as being suitable for regular bridges that respond principally in their fundamental mode. The method may be used for both transverse and longitudinal earthquake motions. The seven stages in the procedure are as follows.

1. Calculate the maximum lateral displacement, $v_{s(max)}$, due to a uniform unit load, p_o, as shown in Fig. 9.13 for a transverse load. The uniform load is resisted by the lateral stiffness of the superstructure and by the stiffness of the central column. The abutments are assumed to be rigid and to provide a pinned end restraint at each end of the superstructure. The maximum displacement and the corresponding reaction, V_o, in the column may be determined by the virtual work method.[15]

Figure 9.13 *Transverse Displacement Due to Unit Transverse Load*

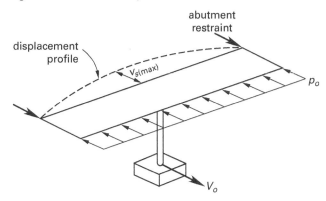

2. The bridge transverse stiffness is given by

$$K = \frac{p_o L}{v_{s(\max)}} \qquad \text{[AASHTO C4.7.4.3.2c-1]}$$

3. The total weight of the bridge superstructure and tributary substructure is

$$W = \int w(x)\,dx \qquad \text{[AASHTO C4.7.4.3.2c-2]}$$

4. The fundamental period of the bridge is given by

$$T_m = 2\pi\sqrt{\frac{W}{gK}} \qquad \text{[AASHTO C4.7.4.3.2c-3]}$$

5. The governing elastic seismic response coefficient, C_{sm}, is determined from AASHTO Sec. 3.10.4.2.

6. The uniform equivalent static seismic load is

$$p_e = \frac{C_{sm} W}{L} \qquad \text{[AASHTO C4.7.4.3.2c-4]}$$

7. Apply p_e to the bridge as shown in Fig. 9.14 and determine the member forces due to the seismic load.

Figure 9.14 *Equivalent Static Seismic Load Applied to Bridge*

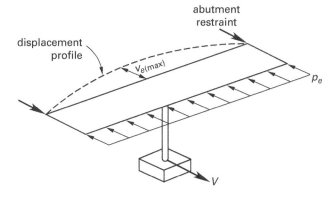

Example 9.39

Using the uniform load method, determine the elastic seismic design moment in the column, in the transverse direction, for the bridge of Ex. 9.31.

Solution

The stiffness of the column, fixed at the top and bottom, is given by

$$
\begin{aligned}
K_c &= \frac{12 E_c I_c}{h_c^3} \\
&= \frac{(12)\left(450{,}000\ \dfrac{\text{kips}}{\text{ft}^2}\right)(60\ \text{ft}^4)}{(30\ \text{ft})^3} \\
&= 12{,}000\ \text{kips/ft}
\end{aligned}
$$

The transverse reaction in the column due to a uniform unit transverse load, p_o, on the superstructure is

$$
\begin{aligned}
V_o &= v_{s(\max)} K_c \\
&= v_{s(\max)}(12{,}000\ \text{kips/ft})
\end{aligned}
$$

The maximum transverse displacement of the superstructure alone due to a uniform unit transverse load, p_o, is

$$
\begin{aligned}
\delta_p &= \frac{5 p_o L^4}{384 E_s I_s} \\
&= \frac{(5)\left(1.0\ \dfrac{\text{kip}}{\text{ft}}\right)(200\ \text{ft})^4}{(384)\left(450{,}000\ \dfrac{\text{kips}}{\text{ft}^2}\right)(4000\ \text{ft}^4)} \\
&= 0.0116\ \text{ft}
\end{aligned}
$$

The maximum transverse displacement of the superstructure due to the column reaction, V_o, is

$$
\begin{aligned}
\delta_v &= -\frac{V_o L^3}{48 E_s I_s} \\
&= -\frac{v_{s(\max)}\left(12{,}000\ \dfrac{\text{kips}}{\text{ft}}\right) L^3}{48 E_s I_s} \\
&= -\frac{v_{s(\max)}\left(12{,}000\ \dfrac{\text{kips}}{\text{ft}}\right)(200\ \text{ft})^3}{(48)\left(450{,}000\ \dfrac{\text{kips}}{\text{ft}^2}\right)(4000\ \text{ft}^4)} \\
&= -v_{s(\max)}(1.1111)
\end{aligned}
$$

The maximum transverse displacement of the superstructure due to p_o and V_o combined is

$$v_{s(\max)} = \delta_p + \delta_v$$
$$= 0.0116 \text{ ft} - v_{s(\max)}(1.1111)$$
$$= 0.00548 \text{ ft}$$

The bridge transverse stiffness is given by AASHTO Eq. C4.7.4.3.2c-1 as

$$K = \frac{p_o L}{v_{s(\max)}} = \frac{\left(1.0 \frac{\text{kip}}{\text{ft}}\right)(200 \text{ ft})}{(0.00548 \text{ ft})\left(12 \frac{\text{in}}{\text{ft}}\right)}$$
$$= 3041 \text{ kips/in} \quad (3040 \text{ kips/in})$$

The total weight of the bridge superstructure and of the tributary substructure is given by AASHTO Eq. C4.7.4.3.2c-2 as

$$W = \int w(x)\, dx = wL$$
$$= \left(10 \frac{\text{kips}}{\text{ft}}\right)(200 \text{ ft})$$
$$= 2000 \text{ kips}$$

The fundamental period of the bridge is given by AASHTO Eq. C4.7.4.3.2c-3 as

$$T_m = 2\pi\sqrt{\frac{W}{gK}} = 0.32\sqrt{\frac{W}{K}}$$
$$= 0.32 \frac{\text{sec}}{\text{in}^{-1}}\sqrt{\frac{2000 \text{ kips}}{3040 \frac{\text{kips}}{\text{in}}}}$$
$$= 0.26 \text{ sec} > T_0$$
$$< T_S$$

The elastic seismic response coefficient is given by AASHTO Eq. 3.10.4.2-4 as

$$C_{sm} = S_{DS}$$
$$= 1.45$$

The uniform equivalent static seismic load is given by AASHTO Eq. C4.7.4.3.2c-4 as

$$p_e = \frac{C_{sm}W}{L} = \frac{(1.45)(2000 \text{ kips})}{200 \text{ ft}}$$
$$= 14.5 \text{ kips/ft}$$

The maximum transverse displacement due to the equivalent seismic load is

$$v_{e(\max)} = \frac{p_e v_{s(\max)}}{p_o} = \frac{\left(14.5 \frac{\text{kips}}{\text{ft}}\right)(0.00548 \text{ ft})}{1.0 \frac{\text{kip}}{\text{ft}}}$$
$$= 0.079 \text{ ft}$$

The elastic transverse shear in the column is

$$V = v_{e(\max)} K_c$$
$$= (0.079)\left(12{,}000 \frac{\text{kips}}{\text{ft}}\right)$$
$$= 948 \text{ kips}$$

The elastic transverse moment in the column is

$$M = \frac{V h_c}{2} = \frac{(948 \text{ kips})(30 \text{ ft})}{2}$$
$$= 14{,}220 \text{ ft-kips}$$

The Single-Mode Elastic Method

The fundamental period and the equivalent static force are obtained by using the technique detailed in AASHTO Sec. 4.7.4.3.2b. The method may be used for both transverse and longitudinal earthquake motions. The six steps in the procedure are as follows.

step 1: Calculate the static displacements, $v_s(x)$, due to a uniform unit load, p_o, as shown in Fig. 9.15 for a longitudinal load. The uniform load is resisted by the lateral stiffness of the central column, with the abutments assumed to provide no restraint.

Figure 9.15 *Longitudinal Displacement Due to Unit Longitudinal Load*

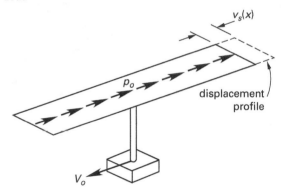

step 2: Calculate the factors α, β, and γ, which are given by

$$\alpha = \int v_s(x)\,dx \qquad \text{[AASHTO C4.7.4.3.2b-1]}$$

$$\beta = \int w(x)v_s(x)\,dx \qquad \text{[AASHTO C4.7.4.3.2b-2]}$$

$$\gamma = \int w(x)v_s^2(x)\,dx \qquad \text{[AASHTO C4.7.4.3.2b-3]}$$

The limits of the integrals extend over the whole length of the bridge.

step 3: The fundamental period is given by

$$T_m = 2\pi\sqrt{\frac{\gamma}{p_o g \alpha}} \qquad \text{[AASHTO C4.7.4.3.2b-4]}$$

step 4: The governing elastic seismic response coefficient, C_{sm}, is used to determine the elastic force in a member and is given by AASHTO Sec. 3.10.4.2, summarized in Table 9.11.

step 5: The equivalent static seismic load is

$$p_e(x) = \frac{\beta C_{sm} w(x) v_s(x)}{\gamma} \qquad \text{[AASHTO C4.7.4.3.2b-5]}$$

step 6: Apply $p_e(x)$ to the bridge as shown in Fig. 9.16 and determine the member forces due to the seismic load.

Figure 9.16 *Equivalent Static Seismic Load Applied to Bridge*

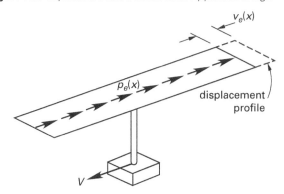

Example 9.40

Using the single-mode elastic method, determine the elastic seismic design moment in the column, in the longitudinal direction, for the bridge of Ex. 9.31.

Solution

The stiffness of the column, fixed at the top and bottom, is obtained from Ex. 9.39 as

$$K_c = 12{,}000 \text{ kips/ft}$$

Applying a uniform load of $p_o = 1.0$ kip/ft along the longitudinal axis of the bridge produces a longitudinal displacement of the superstructure of

$$v_s(x) = \frac{p_o L}{K_c} = \frac{\left(1.0\ \dfrac{\text{kip}}{\text{ft}}\right)(200 \text{ ft})}{12{,}000\ \dfrac{\text{kips}}{\text{ft}}}$$

$$= 0.0167 \text{ ft}$$

The factor α is given by AASHTO Eq. C4.7.4.3.2b-1 as

$$\alpha = \int v_s(x)\,dx$$

$$= (0.0167 \text{ ft})(200 \text{ ft})$$

$$= 3.333 \text{ ft}^2$$

The factor β is given by AASHTO Eq. C4.7.4.3.2b-2 as

$$\beta = \int w(x)v_s(x)\,dx$$

$$= \left(10\ \frac{\text{kips}}{\text{ft}}\right)\left(3.333 \text{ ft}^2\right)$$

$$= 33.333 \text{ ft-kips}$$

The factor γ is given by AASHTO Eq. C4.7.4.3.2b-3 as

$$\gamma = \int w(x)v_s^2(x)\,dx$$

$$= \left(10\ \frac{\text{kips}}{\text{ft}}\right)(0.0167 \text{ ft})^2(200 \text{ ft})$$

$$= 0.557 \text{ ft}^2\text{-kip}$$

The fundamental period is given by AASHTO Eq. C4.7.4.3.2b-4 as

$$T_m = 2\pi\sqrt{\frac{\gamma}{p_o g \alpha}}$$

$$= 2\pi\sqrt{\frac{0.557 \text{ ft}^2\text{-kip}}{\left(1.0\ \dfrac{\text{kip}}{\text{ft}}\right)\left(32.2\ \dfrac{\text{ft}}{\text{sec}^2}\right)(3.333 \text{ ft}^2)}}$$

$$= 0.45 \text{ sec} > T_0$$

$$< T_s$$

The elastic seismic response coefficient is given by AASHTO Eq. 3.10.4.2-4 as

$$C_{sm} = S_{DS}$$

$$= 1.45$$

The equivalent static seismic load is given by AASHTO Eq. C4.7.4.3.2b-5.

$$p_e(x) = \frac{\beta C_{sm} w(x) v_s(x)}{\gamma}$$

$$= \frac{(33.333 \text{ ft-kips})(1.45)\left(10 \, \frac{\text{kips}}{\text{ft}}\right)(0.0167 \text{ ft})}{0.557 \text{ ft}^2\text{-kip}}$$

$$= 14.49 \text{ kips/ft}$$

The longitudinal displacement due to the equivalent seismic load is

$$v_e(x) = \frac{p_e(x) v_s(x)}{p_0}$$

$$= \frac{\left(14.49 \, \frac{\text{kips}}{\text{ft}}\right)(0.0167 \text{ ft})}{1.0 \, \frac{\text{kip}}{\text{ft}}}$$

$$= 0.242 \text{ ft}$$

The elastic shear in the column in the longitudinal direction is

$$V = v_e(x) K_c$$

$$= (0.242 \text{ ft})\left(12{,}000 \, \frac{\text{kips}}{\text{ft}}\right)$$

$$= 2904 \text{ kips}$$

The elastic moment in the column in the longitudinal direction is

$$M = \frac{V h_c}{2} = \frac{(2904 \text{ kips})(30 \text{ ft})}{2}$$

$$= 43{,}560 \text{ ft-kips}$$

Response Modification Factor

The seismic design force for a member is determined by dividing the elastic force by the response modification factor, R. AASHTO Table 3.10.7.1-1 and Table 3.10.7.1-2 list the different structural systems and response modification factors. An abbreviated listing of response modification factors is provided in Table 9.14 and Table 9.15.

Table 9.14 *Response Modification Factors for Substructures*

substructure	importance category		
	critical	essential	other
wall-type pier: strong axis	1.5	1.5	2.0
single column	1.5	2.0	3.0
multiple column bents	1.5	3.5	5.0

Table 9.15 *Response Modification Factors for Connections*

connection	all importance categories
superstructure to abutment	0.8
superstructure to column or pier	1.0
column or pier to foundation	1.0

Example 9.41

Determine the seismic design moment in the column, in the longitudinal and transverse direction, for the bridge of Ex. 9.31.

Solution

From Ex. 9.32, the importance category is "other." The response modification factor for a single column is given in Table 9.14 as

$$R = 3$$

The reduced design moment in the column in the longitudinal direction is

$$M_R = \frac{M}{R}$$

$$= \frac{43{,}560 \text{ ft-kips}}{3}$$

$$= 14{,}520 \text{ ft-kips}$$

The reduced transverse design moment in the column in the transverse direction is

$$M_R = \frac{M}{R}$$

$$= \frac{14{,}220 \text{ ft-kips}}{3}$$

$$= 4740 \text{ ft-kips}$$

Combination of Orthogonal Seismic Forces

AASHTO Sec. 3.10.8 requires the combination of orthogonal seismic forces to account for the directional uncertainty of earthquake motions and the simultaneous occurrence of earthquake forces in two perpendicular horizontal directions. Two load combinations are specified as follows.

- *load case 1:* 100% of the forces due to a seismic event in the longitudinal direction plus 30% of the forces due to a seismic event in the transverse direction

- *load case 2:* 100% of the forces due to a seismic event in the transverse direction plus 30% of the forces due to a seismic event in the longitudinal direction

Example 9.42

Determine the resultant seismic design moment in the column, due to the longitudinal and transverse forces, for the bridge of Ex. 9.31.

Solution

From Ex. 9.41, load case 1 governs, and the longitudinal moment is

$$M_x = (1.0)(14{,}520 \text{ ft-kips})$$
$$= 14{,}520 \text{ ft-kips}$$

The corresponding transverse moment is

$$M_y = (0.3)(4740 \text{ ft-kips})$$
$$= 1422 \text{ ft-kips}$$

For a circular column, the maximum resultant moment is given by

$$M_R = \sqrt{M_x^2 + M_y^2}$$
$$= \sqrt{(14{,}520 \text{ ft-kips})^2 + (1422 \text{ ft-kips})^2}$$
$$= 14{,}589 \text{ ft-kips}$$

Minimum Seat-Width Requirements

In accordance with AASHTO Sec. 4.7.4.4, minimum support lengths are required at the expansion ends of all girders as shown in Fig. 9.17. For seismic zone 1, with $A_s \geq 0.05$, the minimum support length in inches is given by

$$N = (8 + 0.02L + 0.08H)(1 + 0.000125S^2)$$
[AASHTO 4.7.4.4-1]

Figure 9.17 *Minimum Seat-Width Requirements*

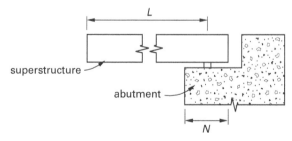

For seismic zone 1, with $A_s < 0.05$, the minimum support length is

$$N = (6 + 0.015L + 0.06H)(1 + 0.000125S^2)$$

For seismic zones 2, 3, and 4, the minimum support length is

$$N = (12 + 0.03L + 0.12H)(1 + 0.000125S^2)$$

The terms in these expressions are defined as follows.

L = length in feet of the bridge deck to the adjacent expansion joint or the end of the bridge deck

H = average height in feet of the columns

= 0 for a single-span bridge

S = angle of skew of the support in degrees measured from a line normal to the span

Example 9.43

Determine the minimum support length for the bridge of Ex. 9.31.

Solution

From Ex. 9.37, the seismic zone is 4.

From AASHTO Sec. 4.7.4.4, the minimum support length is given by

$$N = (12 + 0.03L + 0.12H)(1 + 0.000125S^2)$$
$$= \left(12 \text{ in} + \left(0.03 \, \frac{\text{in}}{\text{ft}}\right)(200 \text{ ft}) + \left(0.12 \, \frac{\text{in}}{\text{ft}}\right)(30 \text{ ft})\right)$$
$$\times \left(1 + (0.000125)(0.0)^2\right)$$
$$= 21.6 \text{ in}$$

7. REFERENCES

1. American Association of State Highway and Transportation Officials. *AASHTO LRFD Bridge Design Specifications*, Eighth ed. Washington, DC: American Association of State Highway and Transportation Officials, 2017.

2. American Institute of Steel Construction. *Moments, Shears, and Reactions: Continuous Highway Bridge Tables*. Chicago, IL: American Institute of Steel Construction, 1986.

3. Graudenz, Heinz. *Bending Moment Coefficients in Continuous Beams*. London: Pitman, 1964.

4. Portland Cement Association. *Influence Lines Drawn as Deflection Curves*. Skokie, IL: Portland Cement Association, 1948.

5. Williams, Alan. "The Determination of Influence Lines for Bridge Decks Monolithic with Their Piers." *Structural Engineer* 42, no. 5 (1964): 161–166.

6. Morice, Peter B. and G. Little. *The Analysis of Right Bridge Decks Subjected to Abnormal Loading.* London: Cement and Concrete Association, 1973.

7. West, Robert. *Recommendations on the Use of Grillage Analysis for Slab and Pseudo-Slab Bridge Decks.* London: Cement and Concrete Association, 1973.

8. Loo, Y. C. and A. R. Cusens. "A Refined Finite Strip Method for the Analysis of Orthotropic Plates." *Proceedings, Institution of Civil Engineers* 48, no. 1 (1971): 85–91.

9. Davis, J. D., I. J. Somerville, and O. C. Zienkiewicz. "Analysis of Various Types of Bridges by the Finite Element Method." *Proceedings of the Conference on Developments in Bridge Design and Construction, Cardiff,* (March 1971).

10. Westergaard, H. M. "Computation of Stresses in Bridge Slabs Due to Wheel Loads." *Public Roads* 11, no. 1 (1930): 1–23.

11. Pucher, Adolf. *Influence Surfaces of Elastic Plates.* New York, NY: Springer-Verlag, 1964.

12. American Concrete Institute. *Building Code Requirements for Structural Concrete and Commentary.* Farmington Hills, MI: American Concrete Institute, 2014.

13. Federal Highway Administration. *Seismic Design and Retrofit Manual for Highway Bridges.* Federal Highway Administration, 1987.

14. Imbsen, R. A. "Seismic Design of Bridges." *Boston Society of Engineers, Fall Lecture Series.* Boston Society of Engineers (1991).

15. Williams, Alan. *Structural Analysis in Theory and Practice.* Burlington, MA: Elsevier/International Codes Council, 2009.

10 Vertical and Other Forces Practice Problems

PRACTICE PROBLEMS

1. A uniform two-span continuous beam with a flexural rigidity of 40,000 ft^2-kips is shown.

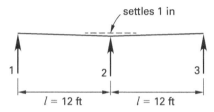

The central support settles 1 in. The bending moment produced at the location of the central support is most nearly

- (A)   69 ft-kips
- (B)   72 ft-kips
- (C)   74 ft-kips
- (D)   76 ft-kips

2. The cylindrical caisson shown is anchored in a soil with a density of 120 lbf/ft^3 and an angle of internal friction of 35°. The water table is 15 ft above the bottom of the caisson, which is sealed with a concrete plug.

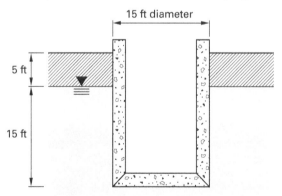

The force on the concrete plug is most nearly

- (A)   160 kips
- (B)   162 kips
- (C)   164 kips
- (D)   166 kips

3. The frame shown has a flexural rigidity of 40,000 ft^2-kips and a coefficient of linear expansion of 7×10^{-6} 1/°F for all members.

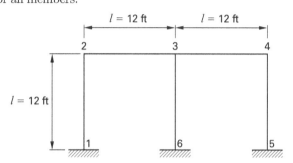

The frame is subjected to a temperature rise of 50°F. The bending moment produced at the bottom of column 12 is most nearly

- (A)   60 in-kips
- (B)   61 in-kips
- (C)   62 in-kips
- (D)   63 in-kips

4. The floor and roof framing layouts and elevation for a three-story office facility are shown. The construction consists of tilt-up concrete walls with panelized plywood roof and floors and pipe columns. The roof is nominally flat, and public access is not permitted. Rain, snow, and corridor loads may be neglected. The roof dead load is 15 lbf/ft², the floor dead load is 20 lbf/ft², and the column dead load is 40 lbf/ft.

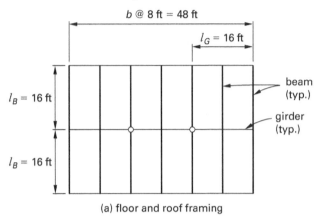

(a) floor and roof framing

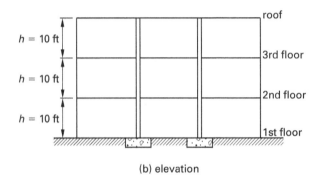

(b) elevation

The nominal force on the footing of an interior column due to dead load is most nearly

(A) 11 kips

(B) 13 kips

(C) 15 kips

(D) 17 kips

5. For the office facility in Prob. 4, the nominal force on the footing of an interior column due to roof live load is most nearly

(A) 2 kips

(B) 3 kips

(C) 4 kips

(D) 5 kips

6. For the office facility in Prob. 4, the nominal force on the footing of an interior column due to floor live load is most nearly

(A) 21 kips

(B) 22 kips

(C) 23 kips

(D) 24 kips

7. For the office facility in Prob. 4, the design load on the footing of an interior column using strength level load combinations is most nearly

(A) 57 kips

(B) 62 kips

(C) 64 kips

(D) 66 kips

8. For the office facility in Prob. 4, the design load on the footing of an interior column using allowable stress level load combinations is most nearly

(A) 36 kips

(B) 38 kips

(C) 40 kips

(D) 42 kips

SOLUTIONS

1. Remove the central support and apply a concentrated load, W, of 100 kips at the midpoint of the beam as shown.

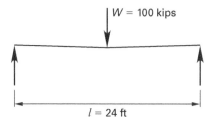

From AISC 360 Table 3-23, Part 7, the deflection produced at the midpoint of the beam is

$$
\begin{aligned}
\delta &= \frac{Wl^3}{48EI} \\
&= \frac{(100 \text{ kips})(24 \text{ ft})^3}{(48)(40{,}000 \text{ ft}^2\text{-kips})} \\
&= 0.72 \text{ ft}
\end{aligned}
$$

To produce a deflection of 1 in at the midpoint of the beam requires a concentrated load at the midpoint of the beam of

$$
\begin{aligned}
W' &= \frac{W(1 \text{ in})}{\delta} \\
&= \frac{(100 \text{ kips})(1 \text{ in})}{(0.72 \text{ ft})\left(12 \dfrac{\text{in}}{\text{ft}}\right)} \\
&= 11.57 \text{ kips}
\end{aligned}
$$

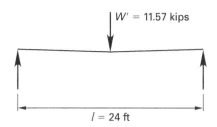

With a concentrated load, W', of 11.57 kips at the midpoint of the beam as shown, the bending moment at the midpoint of the beam is

$$
\begin{aligned}
M &= \frac{W'l}{4} \\
&= \frac{(11.57 \text{ kips})(24 \text{ ft})}{4} \\
&= 69.42 \text{ ft-kips} \quad (69 \text{ ft-kips})
\end{aligned}
$$

The answer is (A).

2. The density of freshwater is given by ASCE/SEI7 Table C3-2 as

$$
\gamma_W = 62 \text{ lbf/ft}^3
$$

The hydrostatic pressure at the bottom of the caisson is

$$
\begin{aligned}
p_W &= \gamma_W h \\
&= \left(62 \frac{\text{lbf}}{\text{ft}^3}\right)(15 \text{ ft}) \\
&= 930 \text{ lbf/ft}^2
\end{aligned}
$$

The area of the plug is

$$
\begin{aligned}
A &= \frac{\pi d^2}{4} \\
&= \frac{\pi (15 \text{ ft})^2}{4} \\
&= 176.63 \text{ ft}^2
\end{aligned}
$$

The force on the plug is

$$
\begin{aligned}
P &= p_W A \\
&= \frac{\left(930 \dfrac{\text{lbf}}{\text{ft}^2}\right)(176.63 \text{ ft}^2)}{1000 \dfrac{\text{lbf}}{\text{ft}}} \\
&= 164.27 \text{ kips} \quad (164 \text{ kips})
\end{aligned}
$$

The answer is (C).

3. Use the moment distribution method. Because of the symmetry of the structure and the loading,

- a sway distribution is unnecessary

- the equal extensions of the columns do not produce any lateral deformations and need not be calculated

- the central column remains vertical and undeformed (no moments in the central column)

- the extensions of beams 23 and 34 produce lateral displacements at the tops of columns 21 and 45 respectively, and the corresponding fixed-end moments must be distributed throughout the frame

- no rotation occurs at node 3, and beams 23 and 34 effectively have fixed ends at node 3

- distribution is required in only one-half of the frame, 123, with the ends of the beams fixed at node 3

The modified stiffness of the members is

$$
s_{21} = s_{23} = \frac{4EI}{l}
$$

The distribution factor is

$$d_{21} = \frac{s_{21}}{s_{21} + s_{23}}$$
$$= \frac{1}{2}$$
$$= d_{23}$$

Due to the temperature rise, beam 23 extends a distance of

$$\delta = l_{23}t\alpha$$
$$\doteq (12 \text{ ft})(50°F)\left(7 \times 10^{-6}\frac{1}{°F}\right)$$
$$= 0.0042 \text{ ft}$$

The initial fixed-end moments at each end of column 21 are

$$M_{F\,21} = \frac{6EI\delta}{l^2} \quad \text{[clockwise moments positive]}$$
$$= \frac{(6)(40{,}000 \text{ ft}^2\text{-kips})(0.0042 \text{ ft})\left(12\,\dfrac{\text{in}}{\text{ft}}\right)}{(12 \text{ ft})^2}$$
$$= 84 \text{ in-kips}$$

The distribution of moments is given in the table shown.

node	1	2		3
member	12	21	23	32
distribution factor	0	½	½	0
fixed-end moments	84	84		
distribution		−42	−42	
carryover	−21			−21
final moments	63	42	−42	−21

The final moments are as shown.

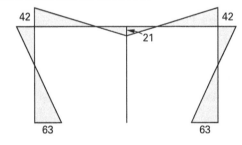

The bending moment produced at bottom of column 12 is

$$M_{12} = 63 \text{ in-kips}$$

The answer is (D).

4. The tributary area of the column is

$$A_T = l_B l_G$$
$$= (16 \text{ ft})(16 \text{ ft})$$
$$= 256 \text{ ft}^2$$

The distributed dead load from the roof is

$$w_{DR} = 15 \text{ lbf/ft}^2$$

The roof dead load applied to the column is

$$V_{DR} = w_{DR}A_T$$
$$= \frac{\left(15\,\dfrac{\text{lbf}}{\text{ft}^2}\right)(256 \text{ ft}^2)}{1000\,\dfrac{\text{lbf}}{\text{kip}}}$$
$$= 3.84 \text{ kips}$$

The distributed dead load from a floor is

$$w_{DF} = 20 \text{ lbf/ft}^2$$

The total floor dead load applied to the column by the second and third floors is

$$V_{DF} = 2w_{DF}A_T$$
$$= \frac{(2)\left(20\,\dfrac{\text{lbf}}{\text{ft}^2}\right)(256 \text{ ft}^2)}{1000\,\dfrac{\text{lbf}}{\text{kip}}}$$
$$= 10.24 \text{ kips}$$

The length of the column is

$$l_C = 3h$$
$$= (3)(10 \text{ ft})$$
$$= 30 \text{ ft}$$

The weight of the column is

$$V_{DC} = w_{DC}l_C$$
$$= \frac{\left(40\,\dfrac{\text{lbf}}{\text{ft}}\right)(30 \text{ ft})}{1000\,\dfrac{\text{lbf}}{\text{kip}}}$$
$$= 1.20 \text{ kips}$$

The total nominal dead load on the footing is

$$
\begin{aligned}
V_{DT} &= V_{DR} + V_{DF} + V_{DC} \\
&= 3.84 \text{ kips} + 10.24 \text{ kips} + 1.20 \text{ kips} \\
&= 15.28 \text{ kips} \quad (15 \text{ kips})
\end{aligned}
$$

The answer is (C).

5. From IBC Table 1607.1, the unreduced roof live load for a roof without public access is $L_o = 20 \text{ lbf/ft}^2$.

As calculated in the solution to Prob. 4, the tributary area for the column is

$$A_T = 256 \text{ ft}^2$$

The tributary area is greater than 200 ft^2 and less than 600 ft^2, so use IBC Eq. 16-28 to calculate the reduction factor R_1.

$$
\begin{aligned}
R_1 &= 1.2 - 0.001 A_T \\
&= 1.2 - \left(0.001 \ \frac{1}{\text{ft}^2} \right) (256 \text{ ft}^2) \\
&= 0.94
\end{aligned}
$$

For a flat roof, the rise per foot is

$$F = 0$$

This is less than 4, so use IBC Eq. 16-30 to calculate the reduction factor R_2.

$$R_2 = 1.0$$

From IBC Eq. 16-26, the reduced roof live load is

$$
\begin{aligned}
L_r &= L_o R_1 R_2 \\
&= \left(20 \ \frac{\text{lbf}}{\text{ft}^2} \right) (0.94)(1.0) \\
&= 18.8 \text{ lbf/ft}^2 \quad \text{[satisfactory]}
\end{aligned}
$$

This is greater than 12 lbf/ft^2 and less than 20 lbf/ft^2, so the minimum and maximum values do not govern. The nominal roof live load on the footing is

$$
\begin{aligned}
V_{LR} &= L_r A_T \\
&= \dfrac{\left(18.8 \ \dfrac{\text{lbf}}{\text{ft}^2} \right) (256 \text{ ft}^2)}{1000 \ \dfrac{\text{lbf}}{\text{kip}}} \\
&= 4.81 \text{ kips} \quad (5 \text{ kips})
\end{aligned}
$$

The answer is (D).

6. From IBC Table 1607.1, the reducible floor live load for office occupancy is

$$L_o = 50 \text{ lbf/ft}^2$$

From Table 1.1, the live load element factor for an interior column is

$$K_{LL} = 4$$

As calculated in the solution to Prob. 4, the tributary area for the column is

$$A_T = 256 \text{ ft}^2$$

Second floor to third floor

The influence area at the third floor is

$$
\begin{aligned}
A_I &= K_{LL} A_T \\
&= (4)(256 \text{ ft}^2) \\
&= 1024 \text{ ft}^2
\end{aligned}
$$

This is greater than 400 ft^2, so use IBC Eq. 16-23 to calculate the reduced live load at the third floor.

$$
\begin{aligned}
L &= L_o \left(0.25 + \frac{15}{\sqrt{K_{LL} A_T}} \right) \\
&= \left(50 \ \frac{\text{lbf}}{\text{ft}^2} \right) \left(0.25 + \frac{15}{\sqrt{1024}} \right) \\
&= 35.94 \text{ lbf/ft}^2 \quad \text{[satisfactory]}
\end{aligned}
$$

This is greater than $0.5 L_o = 25 \text{ lbf/ft}^2$, so this is satisfactory for a member supporting one floor.

In accordance with IBC Sec. 1607.5, an additional 15 lbf/ft^2 must be added to allow for the nonreducible weight of moveable partitions. The total live load intensity is

$$
\begin{aligned}
L &= 35.94 \ \frac{\text{lbf}}{\text{ft}^2} + 15 \ \frac{\text{lbf}}{\text{ft}^2} \\
&= 50.94 \text{ lbf/ft}^2
\end{aligned}
$$

The live load applied to the column at the third floor level is

$$
\begin{aligned}
W_3 &= L A_T \\
&= \dfrac{\left(50.94 \ \dfrac{\text{lbf}}{\text{ft}^2} \right) (256 \text{ ft}^2)}{1000 \ \dfrac{\text{lbf}}{\text{kip}}} \\
&= 13.04 \text{ kips}
\end{aligned}
$$

First floor to second floor

The column supports the floor live load from the second and third floors. The combined tributary area is

$$A_T = (2)(256 \text{ ft}^2)$$
$$= 512 \text{ ft}^2$$

The influence area at the second floor is

$$A_I = K_{LL}A_T$$
$$= (4)(512 \text{ ft}^2)$$
$$= 2048 \text{ ft}^2$$

This is greater than 400 ft^2, so use IBC Eq. 16-23 to calculate the reduced live load at the second floor.

$$L = L_o\left(0.25 + \frac{15}{\sqrt{K_{LL}A_T}}\right)$$
$$= \left(50 \frac{\text{lbf}}{\text{ft}^2}\right)\left(0.25 + \frac{15}{\sqrt{2048}}\right)$$
$$= 29.07 \text{ lbf/ft}^2 \quad \text{[satisfactory]}$$

This is greater than $0.4L_o = 20 \text{ lbf/ft}^2$, so this is satisfactory for a member supporting two floors.

In accordance with IBC Sec. 1607.5, an additional 15 lbf/ft^2 must be added to allow for the nonreducible weight of moveable partitions. The total live load intensity on both the second and third floor is

$$L = 29.07 \frac{\text{lbf}}{\text{ft}^2} + 15 \frac{\text{lbf}}{\text{ft}^2}$$
$$= 44.07 \text{ lbf/ft}^2$$

The total live load applied to the column at the second floor level is

$$W_2 = LA_T = \frac{\left(44.07 \frac{\text{lbf}}{\text{ft}^2}\right)(256 \text{ ft}^2)(2 \text{ floors})}{1000 \frac{\text{lbf}}{\text{kip}}}$$
$$= 22.56 \text{ kips}$$

The nominal force on the footing due to floor live load is

$$V_{LFT} = W_2$$
$$= 22.56 \text{ kips} \quad (23 \text{ kips})$$

Note that it is not necessary to calculate the load on the column at the third floor level. It is included here for completeness.

The answer is (C).

7. From IBC Eq. 16-2, for strength design, the design load on the column footing is

$$V_u = 1.2(D + F) + 1.6(L + H) + 0.5(L_r \text{ or } S \text{ or } R)$$
$$= 1.2V_{DT} + 1.6V_{LFT} + 0.5V_{LR}$$
$$= (1.2)(15.28 \text{ kips}) + (1.6)(22.56 \text{ kips}) + (0.5)(4.81 \text{ kips})$$
$$= 56.8 \text{ kips} \quad (57 \text{ kips})$$

The answer is (A).

8. From IBC Eq. 16-9, for allowable stress design, the design load on the column footing is

$$V_a = D + H + F + L$$
$$= V_{DT} + V_{LFT}$$
$$= 15.28 \text{ kips} + 22.56 \text{ kips}$$
$$= 37.8 \text{ kips} \quad (38 \text{ kips})$$

The answer is (B).

11 Reinforced Concrete Design Practice Problems

PRACTICE PROBLEMS

1. A reinforced concrete beam continuous over four spans and integral with columns at the ends is shown. The clear distance between supports is 15 ft and the beam supports a factored load of 10 kips/ft.

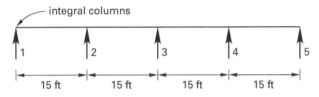

Design assumptions

- ACI Sec. 6.5.2 is applicable.

(a) The design shear force at the face of the first interior support is most nearly

- (A)   75 kips
- (B)   79 kips
- (C)   82 kips
- (D)   86 kips

(b) The design shear force at the face of the outer support is most nearly

- (A)   75 kips
- (B)   79 kips
- (C)   82 kips
- (D)   86 kips

(c) The design moment at the outer support is most nearly

- (A)   140 ft-kips
- (B)   160 ft-kips
- (C)   205 ft-kips
- (D)   230 ft-kips

(d) The design moment at the first interior support is most nearly

- (A)   140 ft-kips
- (B)   160 ft-kips
- (C)   205 ft-kips
- (D)   230 ft-kips

(e) The design moment at the center support is most nearly

- (A)   140 ft-kips
- (B)   160 ft-kips
- (C)   205 ft-kips
- (D)   230 ft-kips

(f) The design moment in the first span is most nearly

- (A)   140 ft-kips
- (B)   160 ft-kips
- (C)   205 ft-kips
- (D)   230 ft-kips

(g) The design moment in the second span is most nearly

- (A)   140 ft-kips
- (B)   160 ft-kips
- (C)   205 ft-kips
- (D)   230 ft-kips

2. A reinforced concrete beam of normal weight concrete with an effective depth of 20 in and a width of 12 in is reinforced with 3 in^2 of grade 60 reinforcement and has a concrete compressive strength of 3000 lbf/in^2. The maximum applied ultimate moment that the beam can support is most nearly

(A) 210 ft-kips

(B) 230 ft-kips

(C) 260 ft-kips

(D) 280 ft-kips

3. The reinforced concrete beam for Prob. 2 supports a factored shear force of 9 kips at the critical section. Is shear reinforcement required?

(A) Yes, it is required (26.3 kips $> 2V_u$).

(B) Yes, it is required (19.7 kips $< 2V_u$).

(C) No, it is not required (19.7 kips $> 2V_u$).

(D) No, it is not required (26.3 kips $< 2V_u$).

4. A short reinforced concrete column, 20 in square, is reinforced with ten no. 9 grade 60 bars and has a concrete strength of 4000 lbf/in^2. What is (a) the design axial load capacity and (b) the required spacing of the lateral ties?

(A) 700 kips; 20 in

(B) 1000 kips; 18 in

(C) 1020 kips; 18 in

(D) 1400 kips; 20 in

5. A simply supported reinforced concrete beam is reinforced with no. 9 grade 60 bars in bundles of three and has a concrete compressive strength of 4000 lbf/in^2. The reinforcement provided is 10% in excess of that required and has a clear cover equal to the equivalent diameter of the bundled bars and a clear spacing of twice the equivalent diameter. The required development length of an individual bar is most nearly

(A) 40 in

(B) 60 in

(C) 65 in

(D) 70 in

6. A reinforced concrete flat plate floor without beams has 18 in square columns at 20 ft centers in one direction and 24 ft centers in the other direction, and it supports a factored distributed load of 200 lbf/ft^2.

Design assumptions

• ACI Sec. 8.10.2 is applicable.

• $\alpha_{f1} = \alpha_{f2} = \beta_t = 0$

• $\dfrac{\alpha_{f1} l_2}{l_1} = 0$

• Consider moments in the direction of the shorter span only.

(a) The column strip moment at the end bay span is most nearly

(A) 35 ft-kips

(B) 40 ft-kips

(C) 65 ft-kips

(D) 110 ft-kips

(b) The column strip moment at the interior support of the end bay is most nearly

(A) 35 ft-kips

(B) 40 ft-kips

(C) 65 ft-kips

(D) 110 ft-kips

(c) The column strip moment in an interior bay span is most nearly

(A) 35 ft-kips

(B) 40 ft-kips

(C) 65 ft-kips

(D) 110 ft-kips

(d) The middle strip moment at the end bay span is most nearly

(A) 30 ft-kips

(B) 35 ft-kips

(C) 40 ft-kips

(D) 110 ft-kips

(e) The middle strip moment at the interior support of the end bay is most nearly

- (A)   30 ft-kips
- (B)   35 ft-kips
- (C)   40 ft-kips
- (D)   110 ft-kips

(f) The middle strip moment in an interior bay span is most nearly

- (A)   30 ft-kips
- (B)   35 ft-kips
- (C)   40 ft-kips
- (D)   110 ft-kips

SOLUTIONS

1. (a) The shear at the face of the first interior support is given by ACI Table 6.5.4 as

$$
\begin{aligned}
V_u &= \frac{1.15 w_u l_n}{2} \\
&= \frac{(1.15)\left(10 \; \dfrac{\text{kips}}{\text{ft}}\right)(15 \; \text{ft})}{2} \\
&= 86.3 \; \text{kips} \quad (86 \; \text{kips})
\end{aligned}
$$

The answer is (D).

(b) The shear at the face of all other supports is given by ACI Table 6.5.4 as

$$
V_u = \frac{w_u l_n}{2} = \frac{\left(10 \; \dfrac{\text{kips}}{\text{ft}}\right)(15 \; \text{ft})}{2} \\
= 75 \; \text{kips}
$$

The answer is (A).

(c) The bending moment is given by

$$
\begin{aligned}
M_u &= \tau w_u l_n^2 \\
&= \tau \left(10 \; \frac{\text{kips}}{\text{ft}}\right)(15 \; \text{ft})^2 \\
&= 2250\tau \; \text{ft-kips}
\end{aligned}
$$

Values of the bending moment coefficients, τ, and the bending moments, M_u, are

$\dfrac{\tau}{M_u}$	support 1	span 12	support 2	span 23	support 3
τ	1/16	1/14	1/10	1/16	1/11
M_u (ft-kips)	141	161	225	141	205

From the table, the design moment at the outer support is 141 ft-kips (140 ft-kips).

The answer is (A).

(d) From the table, the design moment at the first interior support is 225 ft-kips (230 ft-kips).

The answer is (D).

(e) From the table, design moment at the center support is 205 ft-kips.

The answer is (C).

(f) From the table, the design moment in the first span is 161 ft-kips (160 ft-kips).

The answer is (B).

(g) From the table, the design moment in the second span is 141 ft-kips (140 ft-kips).

The answer is (A).

2. The nominal moment of resistance is

$$M_n = A_s f_y d \left(1 - \frac{0.59 A_s f_y}{b_w d f'_c}\right)$$

$$= \frac{(3 \text{ in}^2)\left(60 \dfrac{\text{kips}}{\text{in}^2}\right)(20 \text{ in})}{12 \dfrac{\text{in}}{\text{ft}}}$$
$$\times \left(1 - \frac{(0.59)(3 \text{ in}^2)\left(60 \dfrac{\text{kips}}{\text{in}^2}\right)}{(12 \text{ in})(20 \text{ in})\left(3 \dfrac{\text{kips}}{\text{in}^2}\right)}\right)$$

$$= 255.8 \text{ ft-kips}$$

The reinforcement ratio of the beam is

$$\rho = \frac{A_s}{b_w d} = \frac{3 \text{ in}^2}{(12 \text{ in})(20 \text{ in})}$$
$$= 0.0125$$

The limiting reinforcement ratio for a tension-controlled section is

$$\rho_t = \frac{0.319 \beta_1 f'_c}{f_y} = \frac{(0.319)(0.85)\left(3 \dfrac{\text{kips}}{\text{in}^2}\right)}{60 \dfrac{\text{kips}}{\text{in}^2}}$$
$$= 0.0136 > \rho$$

The section is tension controlled, and the strength reduction factor is

$$\phi = 0.9$$

The maximum allowable ultimate moment is, then,

$$M_u = 0.9 M_n = (0.9)(255.8 \text{ ft-kips})$$
$$= 230.2 \text{ ft-kips} \quad (230 \text{ ft-kips})$$

The answer is (B).

3. The design shear capacity of the concrete section is, then,

$$\phi V_c = 2 \phi b_w d \lambda \sqrt{f'_c}$$

$$= \frac{(2)(0.75)(12 \text{ in})(20 \text{ in})(1.0)\sqrt{3000 \dfrac{\text{lbf}}{\text{in}^2}}}{1000 \dfrac{\text{lbf}}{\text{kip}}}$$

$$= 19.7 \text{ kips} > 2 V_u$$

Shear reinforcement is not required.

The answer is (C).

4. (a) The design axial load capacity is

$$\phi P_n = 0.80 \phi \left(0.85 f'_c (A_g - A_{st}) + A_{st} f_y\right)$$
$$= (0.80)(0.65)$$
$$\times \left(\begin{array}{l} (0.85)\left(4 \dfrac{\text{kips}}{\text{in}^2}\right)(400 \text{ in}^2 - 10 \text{ in}^2) \\[2mm] + (10 \text{ in}^2)\left(60 \dfrac{\text{kips}}{\text{in}^2}\right) \end{array} \right)$$
$$= 1002 \text{ kips} \quad (1000 \text{ kips})$$

(b) The minimum allowable tie size is

$$d_t = \text{no. 3 bar}$$

The maximum tie spacing must not be greater than

$$h = 20 \text{ in}$$
$$48 d_t = (48)(0.375 \text{ in})$$
$$= 18 \text{ in}$$
$$16 d_b = (16)(1.128 \text{ in})$$
$$= 18 \text{ in} \quad [\text{governs}]$$

The answer is (B).

5. The development length of a bar in a three-bar bundle is increased 20%.

The excess reinforcement factor is

$$E_{xr} = \frac{100}{110} = 0.91$$

The required development length is given by

$$l_d = 1.2 E_{xr} d_b \left[\frac{0.05 f_y \psi_t \psi_e}{\lambda \sqrt{f_c'}} \right]$$

$$= (1.2)(0.91)(1.13 \text{ in}) \left[\frac{(0.05)\left(60{,}000 \ \dfrac{\text{lbf}}{\text{in}^2}\right)}{(1.0)\sqrt{4000 \ \dfrac{\text{lbf}}{\text{in}^2}}} \right]$$

$$= 59 \text{ in} \quad (60 \text{ in})$$

The answer is (B).

6. The clear span is

$$l_n = l_1 - c_1 = 20 \text{ ft} - 1.5 \text{ ft}$$
$$= 18.5 \text{ ft} > 0.65 l_1 \quad \text{[satisfactory]}$$

The total factored static moment is

$$M_o = \frac{q_u l_2 l_n^2}{8}$$

$$= \frac{\left(0.2 \ \dfrac{\text{kip}}{\text{ft}^2}\right)(24 \text{ ft})(18.5 \text{ ft})^2}{8}$$

$$= 205 \text{ ft-kips}$$

The relevant coefficients, as given in the problem statement, are

$$\alpha_{f1} = \alpha_{f2} = \beta_t = 0$$
$$\frac{\alpha_{f1} l_2}{l_1} = 0$$

The total static moment is distributed as shown in the following table.

strip	coefficient/ moment	end span	interior support	interior span
full width	distribution coeff.	0.52	0.70	0.35
	moment (ft-kips)	107	144	72
column strip	distribution coeff.	0.60	0.75	0.60
	moment (ft-kips)	64	108	43
middle strip	distribution coeff.	0.40	0.25	0.40
	moment (ft-kips)	43	36	29

(a) From the table, the column strip moment in the end bay span is 64 ft-kips (65 ft-kips).

The answer is (C).

(b) From the table, the column strip moment at the interior support of the end bay is 108 ft-kips (110 ft-kips).

The answer is (D).

(c) From the table, the column strip moment in an interior bay span is 43 ft-kips (40 ft-kips).

The answer is (B).

(d) From the table, the middle strip moment at the end bay span is 43 ft-kips (40 ft-kips).

The answer is (C).

(e) From the table, the middle strip moment at the interior support of the end bay is 36 ft-kips (35 ft-kips).

The answer is (B).

(f) From the table, the middle strip moment in an interior bay span is 29 ft-kips (30 ft-kips).

The answer is (A).

12 Foundations and Retaining Structures Practice Problems

PRACTICE PROBLEMS

1. What is the factored net pressure on the footing shown?

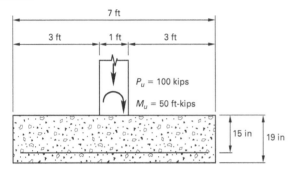

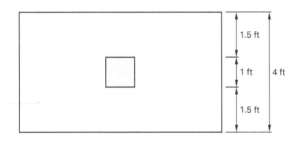

(A)   2.55 kips/ft^2 maximum, 1.02 kips/ft^2 minimum

(B)   5.10 kips/ft^2 maximum, 2.04 kips/ft^2 minimum

(C)   6.25 kips/ft^2 maximum, 0.89 kips/ft^2 minimum

(D)   9.67 kips/ft^2 maximum, -2.55 kips/ft^2 minimum

2. The rectangular footing of normal weight concrete of Prob. 1 has a concrete strength of 4000 lbf/in^2. Is the punching shear capacity adequate?

(A)   yes, 228 lbf/in$^2 > v_u$

(B)   yes, 190 lbf/in$^2 > v_u$

(C)   no, 53 lbf/in$^2 < v_u$

(D)   no, 60 lbf/in$^2 > v_u$

3. The flexural shear capacity for the rectangular footing of Prob. 1 is most nearly

(A)   68 kips

(B)   73 kips

(C)   82 kips

(D)   91 kips

4. The area of grade 60 reinforcement required in the direction of the applied moment for the rectangular footing of Prob. 1 is most nearly

(A)   1.50 in^2

(B)   1.55 in^2

(C)   1.65 in^2

(D)   1.70 in^2

5. For the retaining wall of Ex. 3.17, what is most nearly the minimum area of horizontal reinforcement required in the stem? Use no. 3 grade 60 bars.

(A)   0.3 in^2/ft

(B)   0.4 in^2/ft

(C)   0.7 in^2/ft

(D)   1.0 in^2/ft

SOLUTIONS

1. The equivalent eccentricity is

$$e = \frac{M_u}{P_u} = \frac{50 \text{ ft-kips}}{100 \text{ kips}} = 0.5 \text{ ft}$$

$$< \frac{L}{6} \quad [\text{within middle third}]$$

The net factored pressure on the footing is

$$q_u = \frac{P_u\left(1 \pm \frac{6e}{L}\right)}{BL}$$

$$= \frac{(100 \text{ kips})\left(1 \pm \frac{(6)(0.5 \text{ ft})}{7 \text{ ft}}\right)}{(4 \text{ ft})(7 \text{ ft})}$$

$$q_{u(\max)} = 5.10 \text{ kips/ft}^2$$

$$q_{u(\min)} = 2.04 \text{ kips/ft}^2$$

The answer is (B).

2. The length of the critical perimeter is

$$b_o = 4(c+d) = \frac{(4)(12 \text{ in} + 15 \text{ in})}{12 \frac{\text{in}}{\text{ft}}}$$

$$= 9 \text{ ft}$$

Shear caused by the axial load at the critical perimeter is

$$V_u = P_u - 0.5\left(q_{u(\max)} + q_{u(\min)}\right)\left(\frac{b_o}{4}\right)^2$$

$$= 100 \text{ kips} - (0.5)\left(5.10 \frac{\text{kips}}{\text{ft}^2} + 2.04 \frac{\text{kips}}{\text{ft}^2}\right)$$

$$\times (2.25 \text{ ft})^2$$

$$= 82 \text{ kips}$$

The polar moment of inertia of the critical perimeter, J_c, is given by the Portland Cement Association, *Notes on ACI 318-08*, Fig. 16-13, as

$$\frac{J_c}{y} = \frac{b_1 d(b_1 + 3b_2) + d^3}{3} \quad \begin{bmatrix} \text{for a footing with an} \\ \text{interior column} \end{bmatrix}$$

$$= \frac{(27 \text{ in})(15 \text{ in})(27 \text{ in} + (3)(27 \text{ in})) + (15 \text{ in})^3}{3}$$

$$= 15{,}705 \text{ in}^3$$

where:

$$y = (c+d)/2$$

b_1 = width of critical perimeter in the direction of M_u

$$= c + d$$

b_2 = width of critical perimeter perpendicular to b_1

$$= c + d$$

The fraction of the column moment transferred by shear is

$$\gamma_v = 1 - \gamma_f \quad [\text{ACI 8.4.4.2.2}]$$

$$= 1 - \frac{1}{1 + 0.67\sqrt{\dfrac{b_1}{b_2}}}$$

$$= 1 - \frac{1}{1.67}$$

$$= 0.40$$

From ACI Comm. R8.4.4.2.3, the combined shear stress due to the applied axial load and the column moment is

$$v_u = \frac{V_u}{db_o} + \frac{\gamma_v M_u y}{J_c}$$

$$= \frac{(82 \text{ kips})\left(1000 \frac{\text{lbf}}{\text{kip}}\right)}{(15 \text{ in})(108 \text{ in})}$$

$$+ \frac{(0.4)(600 \text{ in-kips})\left(1000 \frac{\text{lbf}}{\text{kip}}\right)}{15{,}705 \text{ in}^3}$$

$$= 66 \text{ lbf/in}^2$$

The ratio of the long side to the short side of the column is

$$\beta_c = \frac{c_2}{c_1} = \frac{12 \text{ in}}{12 \text{ in}}$$

$$= 1.00$$

$$< 2$$

The allowable shear stress for two-way action is given by ACI Table 22.6.5.2(a) as

$$\phi v_c = 4\phi\lambda\sqrt{f_c'}$$
$$\phi = \text{strength reduction factor}$$
$$= 0.75 \text{ from ACI Sec. } 9.3$$
$$\phi v_c = (4)(0.75)(1.0)\sqrt{4000\ \frac{\text{lbf}}{\text{in}^2}}$$
$$= 190\ \text{lbf/in}^2$$
$$> v_u \quad [\text{satisfactory}]$$

The answer is (B).

3. The distance of the critical section for flexural shear from the edge of the footing is

$$x = \frac{L}{2} - \frac{c}{2} - d = \frac{7\text{ ft}}{2} - \frac{1\text{ ft}}{2} - 1.25\text{ ft}$$
$$= 1.75\text{ ft}$$

The net factored pressure on the footing at this section is

$$q_{ux} = q_{u(\text{max})} - \frac{x(q_{u(\text{max})} - q_{u(\text{min})})}{L}$$
$$= 5.10\ \frac{\text{kips}}{\text{ft}^2} - \frac{(1.75\text{ ft})\left(5.10\ \frac{\text{kips}}{\text{ft}^2} - 2.04\ \frac{\text{kips}}{\text{ft}^2}\right)}{7\text{ ft}}$$
$$= 4.34\ \text{kips/ft}^2$$

The factored shear force at the critical section is

$$V_u = \frac{Bx(q_{u(\text{max})} + q_{ux})}{2}$$
$$= \frac{(4\text{ ft})(1.75\text{ ft})\left(5.10\ \frac{\text{kips}}{\text{ft}^2} + 4.34\ \frac{\text{kips}}{\text{ft}^2}\right)}{2}$$
$$= 33.04\ \text{kips}$$

The flexural shear capacity of the footing is given by ACI Eq. 22.5.5.1 as

$$\phi V_c = 2\phi B d\lambda\sqrt{f_c'}$$
$$= \frac{(2)(0.75)(48\text{ in})(15\text{ in})(1.0)\sqrt{4000\ \frac{\text{lbf}}{\text{in}^2}}}{1000\ \frac{\text{lbf}}{\text{kip}}}$$
$$= 68\ \text{kips}$$
$$> V_u \quad [\text{satisfactory}]$$

The answer is (A).

4. The net factored pressure on the footing at the face of the column is

$$q_{uc} = q_{u(\text{max})} - \left(\frac{L}{2} - \frac{c}{2}\right)\left(\frac{q_{u(\text{max})} - q_{u(\text{min})}}{L}\right)$$
$$= 5.10\ \frac{\text{kips}}{\text{ft}^2} - (3.5\text{ ft} - 0.5\text{ ft})$$
$$\times\left(\frac{5.10\ \frac{\text{kips}}{\text{ft}^2} - 2.04\ \frac{\text{kips}}{\text{ft}^2}}{7\text{ ft}}\right)$$
$$= 3.79\ \text{kips/ft}^2$$

The factored moment at the face of the column is

$$M_u = \frac{B\left(\dfrac{L}{2} - \dfrac{c}{2}\right)^2(2q_{u(\text{max})} + q_{uc})}{6}$$
$$= \frac{(4\text{ ft})(3\text{ ft})^2\left((2)\left(5.10\ \frac{\text{kips}}{\text{ft}^2}\right) + 3.79\ \frac{\text{kips}}{\text{ft}^2}\right)}{6}$$
$$= 83.94\ \text{ft-kips}$$

Assuming a tension-controlled section, the required reinforcement ratio is

$$\rho = \frac{0.85f_c'\left(1 - \sqrt{1 - \dfrac{M_u}{0.383bd^2f_c'}}\right)}{f_y}$$

$$= \frac{(0.85)\left(4\ \frac{\text{kips}}{\text{in}^2}\right)}{60\ \frac{\text{kips}}{\text{in}^2}}$$
$$\times\left(1 - \sqrt{1 - \frac{(83.94\text{ ft-kips})\left(12\ \frac{\text{in}}{\text{ft}}\right)\times\left(1000\ \frac{\text{lbf}}{\text{kip}}\right)}{(0.383)(48\text{ in})(15\text{ in})^2\times\left(4000\ \frac{\text{lbf}}{\text{in}^2}\right)}}\right)$$

$$= 0.0018$$
$$A_s = \rho B d$$
$$= (0.0018)(48\text{ in})(15\text{ in})$$
$$= 1.30\ \text{in}^2$$

The minimum reinforcement area is given by ACI Sec. 7.6.1.1 as

$$A_{s(\text{min})} = 0.0018Bh$$
$$= (0.0018)(48 \text{ in})(19 \text{ in})$$
$$= 1.64 \text{ in}^2 \quad (1.65 \text{ in}^2) \quad \text{[governs]}$$

The answer is (C).

5. From ACI Table 11.6.1, the required ratio of horizontal reinforcement in the stem is

$$\rho_{\text{hor}} = 0.0020$$
$$A_{sh} = \rho_{\text{hor}}bh$$
$$= (0.0020)(12 \text{ in})(18 \text{ in})$$
$$= 0.432 \text{ in}^2/\text{ft} \quad (0.4 \text{ in}^2/\text{ft})$$

The answer is (B).

13 Prestressed Concrete Design Practice Problems

PRACTICE PROBLEMS

1. The pretensioned beam of normal weight concrete shown is simply supported over a span of 20 ft and has a concrete strength at transfer of 4500 lbf/in². (See *Illustration for Prob. 1*.) What are the magnitude and the location of the initial prestressing force required to produce satisfactory stresses at midspan, immediately after transfer, without using auxiliary reinforcement?

(A) $P_i = 90$ kips, $e = 2.82$ in

(B) $P_i = 95$ kips, $e = 2.75$ in

(C) $P_i = 100$ kips, $e = 2.90$ in

(D) $P_i = 105$ kips, $e = 2.99$ in

2. The class U pretensioned beam in Prob. 1 has a long-term loss in prestress of 25% and a 28-day compressive strength of 6000 lbf/in²; normal cover is provided to the tendons. The maximum bending moment the beam can carry if all the superimposed load is sustained is most nearly

(A) 350,000 in-lbf

(B) 360,000 in-lbf

(C) 390,000 in-lbf

(D) 400,000 in-lbf

3. For the pretensioned beam in Prob. 1, the cracking moment strength is most nearly

(A) 380 in-kips

(B) 410 in-kips

(C) 430 in-kips

(D) 470 in-kips

4. The pretensioned beam in Prob. 1 is prestressed with low-relaxation tendons. The area of the low-relaxation prestressing tendons provided is 0.306 in² with a specified tensile strength of 270 kips/in² and a yield strength of 243 kips/in². The nominal flexural strength of the beam is most nearly

(A) 470 in-kips

(B) 530 in-kips

(C) 570 in-kips

(D) 600 in-kips

5. The pretensioned beam of Prob. 1 supports two concentrated loads each of 2.5 kips.

(a) The prestressing force required in the tendons to balance the applied loads is most nearly

(A) 40 kips

(B) 50 kips

(C) 60 kips

(D) 70 kips

(b) The stress in the beam caused by the prestressing force and the concentrated loads is most nearly

(A) 885 lbf/in²

(B) 890 lbf/in²

(C) 900 lbf/in²

(D) 930 lbf/in²

6. For the beam of Prob. 5, what are the stresses in the beam at midspan due to the loads, W, the prestressing force, and self-weight?

(A) $f_{be} = 570$ lbf/in²; $f_{te} = 1200$ lbf/in²

(B) $f_{be} = 585$ lbf/in²; $f_{te} = 1210$ lbf/in²

(C) $f_{be} = 590$ lbf/in²; $f_{te} = 1220$ lbf/in²

(D) $f_{be} = 610$ lbf/in²; $f_{te} = 1160$ lbf/in²

Illustration for Prob. 1

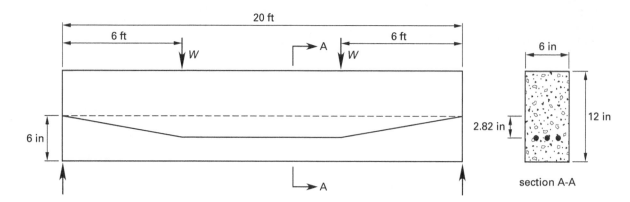

section A-A

SOLUTIONS

1. The properties of the concrete section are

$$A_g = 72 \text{ in}^2$$

$$S_t = S_b = 144 \text{ in}^3$$

At midspan, the self-weight moment is

$$M_G = \frac{w_c A_g l^2}{8}$$

$$= \frac{\left(150 \, \frac{\text{lbf}}{\text{ft}^3}\right)\left(\dfrac{72 \text{ in}^2}{\left(12 \, \frac{\text{in}}{\text{ft}}\right)^2}\right)(20 \text{ ft})^2\left(12 \, \frac{\text{in}}{\text{ft}}\right)}{8}$$

$$= 45{,}000 \text{ in-lbf}$$

At midspan, the permissible tensile stress in the top fiber without auxiliary reinforcement is given by ACI Table 24.5.3.2 as

$$f_{ti} = -3\sqrt{f'_{ci}} = -3\sqrt{4500 \, \frac{\text{lbf}}{\text{in}^2}}$$

$$= -201 \text{ lbf/in}^2$$

$$= \frac{P_i}{A_g} - \frac{P_i e}{S_t} + \frac{M_G}{S_t}$$

$$= \frac{P_i}{72 \text{ in}^2} - \frac{P_i e}{144 \text{ in}^3} + \frac{45{,}000 \text{ in-lbf}}{144 \text{ in}^3}$$

$$-514 \, \frac{\text{lbf}}{\text{in}^2} = \frac{P_i}{72 \text{ in}^2} - \frac{P_i e}{144 \text{ in}^3} \qquad \text{[Eq. 1]}$$

At midspan, the permissible compressive stress in the bottom fiber is given by ACI Table 24.5.3.1 as

$$f_{bi} = 0.6 f'_{ci} = 2700 \text{ lbf/in}^2$$

$$= \frac{P_i}{A_g} + \frac{P_i e}{S_b} - \frac{M_G}{S_b}$$

$$= \frac{P_i}{72 \text{ in}^2} + \frac{P_i e}{144 \text{ in}^3} - \frac{45{,}000 \text{ in-lbf}}{144 \text{ in}^3}$$

$$3013 \, \frac{\text{lbf}}{\text{in}^2} = \frac{P_i}{72 \text{ in}^2} + \frac{P_i e}{144 \text{ in}^3} \qquad \text{[Eq. 2]}$$

Solving Eq. [1] and Eq. [2] gives

$$P_i = 90 \text{ kips}$$

$$e = 2.82 \text{ in}$$

The answer is (A).

2. The permissible compressive stress at midspan, in the top fiber, due to the sustained load is

$$f_{te} = 0.45 f'_c$$

$$= (0.45)\left(6000 \, \frac{\text{lbf}}{\text{in}^2}\right)$$

$$= 2700 \text{ lbf/in}^2$$

$$= 0.75 P_i\left(\frac{1}{A_g} - \frac{e}{S_t}\right) + \frac{M_G}{S_t} + \frac{M_T}{S_t}$$

$$= (0.75)\left(-514 \, \frac{\text{lbf}}{\text{in}^2}\right) + 313 \, \frac{\text{lbf}}{\text{in}^2} + \frac{M_T}{144 \text{ in}^3}$$

$$M_T = 399{,}240 \text{ in-lbf}$$

The permissible tensile stress at midspan, in the bottom fiber, due to the total load is

$$f_{be} = -7.5\sqrt{f_c'}$$
$$= -7.5\sqrt{6000 \ \frac{\text{lbf}}{\text{in}^2}}$$
$$= -581 \ \text{lbf/in}^2$$
$$= 0.75 P_i\left(\frac{1}{A_g} + \frac{e}{S_b}\right) - \frac{M_G}{S_b} - \frac{M_T}{S_b}$$
$$= (0.75)\left(3013 \ \frac{\text{lbf}}{\text{in}^2}\right) - 313 \ \frac{\text{lbf}}{\text{in}^2} - \frac{M_T}{144 \ \text{in}^3}$$
$$M_T = 363{,}996 \ \text{in-lbf} \quad (360{,}000 \ \text{in-lbf}) \quad \text{[governs]}$$

The answer is (B).

3. The modulus of rupture is

$$f_r = 7.5\lambda\sqrt{f_c'}$$
$$= (7.5)(1.0)\sqrt{6000 \ \frac{\text{lbf}}{\text{in}^2}}$$
$$= 581 \ \text{lbf/in}^2$$

The cracking moment strength is

$$M_{cr} = S_b(P_e R_b + f_r)$$
$$= \frac{(144 \ \text{in}^3)\left((0.75)\left(3013 \ \frac{\text{lbf}}{\text{in}^2}\right) + 581 \ \frac{\text{lbf}}{\text{in}^2}\right)}{1000 \ \frac{\text{lbf}}{\text{kip}}}$$
$$= 409 \ \text{in-kips} \quad (410 \ \text{in-kips})$$

The answer is (B).

4. The relevant properties of the beam are

$$\gamma_p = 0.28 \quad \text{[for } f_{py}/f_{pu} \geq 0.9\text{]}$$
$$\rho_p = \frac{A_{ps}}{bd_p} = \frac{0.306 \ \text{in}^2}{(6 \ \text{in})(8.82 \ \text{in})}$$
$$= 0.00578$$
$$\beta_1 = 0.75 \quad \text{[from ACI Table 22.2.2.4.3]}$$

From ACI Eq. 20.3.2.3.1,

$$f_{ps} = f_{pu}\left(1 - \frac{\gamma_p \rho_p f_{pu}}{\beta_1 f_c'}\right)$$
$$= \left(270 \ \frac{\text{kips}}{\text{in}^2}\right)\left(1 - \frac{(0.28)(0.00578)\times\left(270 \ \frac{\text{kips}}{\text{in}^2}\right)}{(0.75)\left(6 \ \frac{\text{kips}}{\text{in}^2}\right)}\right)$$
$$= 244 \ \text{kips/in}^2$$

The depth of the stress block is given by

$$a = \frac{A_{ps}f_{ps}}{0.85 f_c' b} = \frac{(0.306 \ \text{in}^2)\left(244 \ \frac{\text{kips}}{\text{in}^2}\right)}{(0.85)\left(6 \ \frac{\text{kips}}{\text{in}^2}\right)(6 \ \text{in})}$$
$$= 2.44 \ \text{in}$$

The maximum depth of the stress block for a tension-controlled section is given by ACI Sec. R21.2.2 as

$$a_t = 0.375\beta_1 d_p = (0.375)(0.75)(8.82 \ \text{in})$$
$$= 2.48 \ \text{in}$$
$$> a$$

Therefore, the section is tension-controlled and $\phi = 0.9$.

The maximum nominal flexural strength is

$$M_n = (0.85 f_c' ab)\left(d_p - \frac{a}{2}\right)$$
$$= (0.85)\left(6 \ \frac{\text{kips}}{\text{in}^2}\right)(2.44 \ \text{in})(6 \ \text{in})$$
$$\times \left(8.82 \ \text{in} - \frac{2.44 \ \text{in}}{2}\right)$$
$$= 567 \ \text{in-kips} \quad (570 \ \text{in-kips})$$

The answer is (C).

5. The sag of the tendon is

$$g = 2.82 \text{ in}$$

(a) The prestressing force required in the tendons to balance the applied load exactly is

$$P = \frac{Wa}{g}$$
$$= \frac{(2.5 \text{ kips})(72 \text{ in})}{2.82 \text{ in}}$$
$$= 63.8 \text{ kips} \quad (60 \text{ kips})$$

The answer is (C).

(b) The uniform compressive stress throughout the beam is

$$f_c = \frac{P}{A_g} = \frac{63,800 \text{ lbf}}{72 \text{ in}^2}$$
$$= 886 \text{ lbf/in}^2 \quad (885 \text{ lbf/in}^2)$$

The answer is (A).

6. The moment produced at midspan by the beam self-weight is

$$M_G = \frac{w_G l^2}{8}$$
$$= \frac{\left(75 \dfrac{\text{lbf}}{\text{ft}}\right)(20 \text{ ft})^2 \left(12 \dfrac{\text{in}}{\text{ft}}\right)}{8}$$
$$= 45,000 \text{ in-lbf}$$

The resultant stresses in the beam at midspan are

$$f_{be} = f_c - \frac{M_G}{S}$$
$$= 886 \frac{\text{lbf}}{\text{in}^2} - \frac{45,000 \text{ in-lbf}}{144 \text{ in}^3}$$
$$= 573 \text{ lbf/in}^2 \quad (570 \text{ lbf/in}^2)$$
$$f_{te} = f_c + \frac{M_G}{S}$$
$$= 886 \frac{\text{lbf}}{\text{in}^2} + 313 \frac{\text{lbf}}{\text{in}^2}$$
$$= 1199 \text{ lbf/in}^2 \quad (1200 \text{ lbf/in}^2)$$

The answer is (A).

Structural Steel Design Practice Problems

PRACTICE PROBLEMS

(Answer options for ASD are given in parentheses.)

1. The pair of shear legs shown consists of two nonstandard steel tubes 13 ft long, of 3.5 in outside diameter and 3 in inside diameter, pinned together at the top and inclined to each other at an angle of 45°. The yield stress of the tubes is $F_y = 36$ kips/in². The legs are laterally braced at the top and are pinned at the base. The self-weight of the pipes is neglected. What is most nearly the maximum load the shear legs can lift? (For LRFD, use factored loads.)

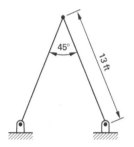

(A) 29 kips (19 kips)

(B) 31 kips (21 kips)

(C) 44 kips (30 kips)

(D) 58 kips (39 kips)

2. Both flanges of a W8 × 24 grade A36 steel section are each connected by six $\frac{3}{4}$ in diameter bolts to a steel bracket. A single row of three bolts is provided on each side of the beam web to both flanges, as shown.

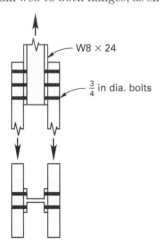

What is most nearly the capacity of the W section in direct tension?

(A) 180 kips (120 kips)

(B) 220 kips (150 kips)

(C) 230 kips (160 kips)

(D) 280 kips (190 kips)

3. A fixed-ended steel beam is shown, with the factored loads indicated. The distributed load shown includes the beam self-weight. Full lateral support is provided to the beam, which is of grade 50 steel. What is the lightest W section beam that can support the factored loads?

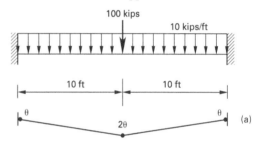

(A) W24 × 55

(B) W24 × 62

(C) W24 × 94

(D) W27 × 94

4. The simply supported composite beam shown consists of a 7½ in normal weight concrete slab cast on a W21 × 57 grade 50 steel beam. The beam forms part of a floor system with the beams spaced at 10 ft centers, and the concrete strength is 3000 lbf/in². The loads are indicated in the illustration; these include the weight of the concrete slab and the self-weight of the steel beam. (LRFD loads are factored.) Seventeen stud shear connectors of ¾ in diameter are provided between sections 1 and 2. Is the number of connectors adequate, and what is the available flexural strength?

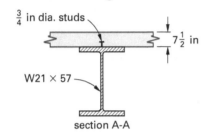

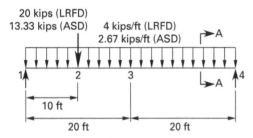

(A) No, $\phi M_n = 750$ ft-kips ($M_n/\Omega = 500$ ft-kips).

(B) No, $\phi M_n = 829$ ft-kips ($M_n/\Omega = 553$ ft-kips).

(C) Yes, $\phi M_n = 750$ ft-kips ($M_n/\Omega = 500$ ft-kips).

(D) Yes, $\phi M_n = 829$ ft-kips ($M_n/\Omega = 553$ ft-kips).

5. Both plates shown are of grade A36 steel, and the plates are connected with E70XX fillet welds. What is most nearly the available block shear capacity of the welded connection?

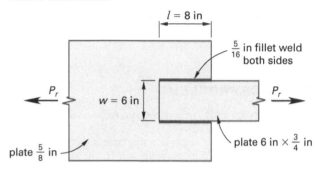

(A) 330 kips (220 kips)

(B) 360 kips (240 kips)

(C) 390 kips (260 kips)

(D) 420 kips (280 kips)

6. What is most nearly the available tensile capacity of the fillet welds in the welded connection shown in Prob. 5?

(A) 110 kips (70 kips)

(B) 180 kips (120 kips)

(C) 250 kips (160 kips)

(D) 330 kips (220 kips)

7. What is most nearly the available tensile capacity of the ¾ in plate in the welded connection shown in Prob. 5?

(A) 90 kips (60 kips)

(B) 110 kips (70 kips)

(C) 150 kips (100 kips)

(D) 230 kips (160 kips)

SOLUTIONS

1. The area of each pipe is

$$A = \pi(a^2 - b^2)$$
$$= \pi\big((1.75 \text{ in})^2 - (1.5 \text{ in})^2\big)$$
$$= 2.55 \text{ in}^2$$

The radius of gyration of each pipe is

$$r = \frac{\sqrt{a^2 + b^2}}{2} = \frac{\sqrt{(1.75 \text{ in})^2 + (1.5 \text{ in})^2}}{2} = 1.15 \text{ in}$$

The slenderness ratio of each pipe is given by

$$\frac{L_c}{r} = \frac{(1.0)(13 \text{ ft})\left(12 \dfrac{\text{in}}{\text{ft}}\right)}{1.15 \text{ in}}$$
$$= 135.7$$
$$< 200 \quad [\text{satisfactory}]$$

LRFD Method

From *AISC Manual* Table 4-14, the design axial stress is given by

$$\phi_c F_{cr} = 12.3 \text{ kips/in}^2$$

The design axial strength of each pipe is given by

$$\phi_c P_n = \phi_c F_{cr} A$$
$$= \left(12.3 \frac{\text{kips}}{\text{in}^2}\right)(2.55 \text{ in}^2)$$
$$= 31.37 \text{ kips}$$

The maximum factored load the pair of shear legs can lift is given by

$$P_u = 2\phi_c P_n \cos\phi$$
$$= (2)(31.37 \text{ kips})\cos 22.5°$$
$$= 58 \text{ kips}$$

The answer is (D).

ASD Method

From *AISC Manual* Table 4-14, the allowable axial stress is given by

$$\frac{F_{cr}}{\Omega_c} = 8.2 \text{ kips/in}^2$$

The allowable axial strength of each pipe is given by

$$\frac{P_n}{\Omega_c} = \frac{F_{cr} A}{\Omega_c}$$
$$= \left(8.2 \frac{\text{kips}}{\text{in}^2}\right)(2.55 \text{ in}^2)$$
$$= 20.91 \text{ kips}$$

The allowable load the pair of shear legs can lift is given by

$$P_a = 2\left(\frac{P_n}{\Omega_c}\right)\cos\phi$$
$$= (2)(20.91 \text{ kips})\cos 22.5°$$
$$= 38.64 \text{ kips} \quad (39 \text{ kips})$$

The answer is (D).

2. The relevant properties of the W8 × 24 are obtained from *AISC Manual* Table 1-1 and are

$$A_g = 7.08 \text{ in}^2$$
$$b_f = 6.5 \text{ in}^2$$
$$t_f = 0.40 \text{ in}$$
$$d = 7.93 \text{ in}$$

The hole diameter is

$$d_h = d_b + \frac{1}{8} \text{ in}$$
$$= \frac{3}{4} \text{ in} + \frac{1}{8} \text{ in}$$
$$= 0.875 \text{ in}$$

The ratio is

$$\frac{b_f}{d} = \frac{6.50 \text{ in}}{7.93 \text{ in}}$$
$$> 2/3$$

Three bolts are in line in the direction of stress, however, and hence from AISC 360 Table D3.1, Case 7,

$$U = 0.90$$

The net area is given by

$$A_n = A_g - 4d_h t_f$$
$$= 7.08 \text{ in}^2 - (4)(0.875 \text{ in})(0.40 \text{ in})$$
$$= 5.68 \text{ in}^2$$

LRFD Method

The design axial strength based on the gross section is

$$\phi_t P_n = 0.9 F_y A_g$$
$$= (0.9)\left(36 \ \frac{\text{kips}}{\text{in}^2}\right)(7.08 \ \text{in}^2)$$
$$= 229 \ \text{kips}$$

The design axial strength based on the net section is

$$\phi_t P_n = 0.75 U F_u A_n$$
$$= (0.75)(0.90)\left(58 \ \frac{\text{kips}}{\text{in}^2}\right)(5.68 \ \text{in}^2)$$
$$= 222 \ \text{kips} \quad (220 \ \text{kips}) \quad \text{[governs]}$$

The answer is (B).

ASD Method

The allowable axial strength based on the gross section is

$$\frac{P_n}{\Omega_t} = \frac{F_y A_g}{1.67}$$
$$= \frac{\left(36 \ \frac{\text{kips}}{\text{in}^2}\right)(7.08 \ \text{in}^2)}{1.67}$$
$$= 152.62 \ \text{kips}$$

The allowable axial strength based on the net section is

$$\frac{P_n}{\Omega_t} = \frac{U F_u A_n}{2.00}$$
$$= \frac{(0.90)\left(58 \ \frac{\text{kips}}{\text{in}^2}\right)(5.68 \ \text{in}^2)}{2.00}$$
$$= 148 \ \text{kips} \quad (150 \ \text{kips}) \quad \text{[governs]}$$

The answer is (B).

3. From the collapse mechanism shown in part (a) of the illustration, the required plastic moment of resistance is given by the equation

$$4M_p = (100 \ \text{kips})(10 \ \text{ft}) + \frac{(200 \ \text{kips})(10 \ \text{ft})}{2}$$
$$M_p = \frac{2000 \ \text{ft-kips}}{4}$$
$$= 500 \ \text{ft-kips}$$

From *AISC Manual* Table 3-2, a W24 × 55 has

$$\phi M_p = 503 \ \text{ft-kips}$$
$$> 500 \ \text{ft-kips} \quad \text{[satisfactory]}$$

The answer is (A).

4. *LRFD Method*

The factored bending moment at section 2 due to distributed load is

$$M_w = \frac{0.75 w L^2}{8}$$
$$= \frac{(0.75)\left(4 \ \frac{\text{kips}}{\text{ft}}\right)(40 \ \text{ft})^2}{8}$$
$$= 600 \ \text{ft-kips}$$

Due to point load,

$$M_W = \frac{Wab}{L}$$
$$= \frac{(20 \ \text{kips})(10 \ \text{ft})(30 \ \text{ft})}{40 \ \text{ft}}$$
$$= 150 \ \text{ft-kips}$$

The total moment, M_2, is

$$M_w = M_w + M_W$$
$$= 600 \ \text{ft-kips} + 150 \ \text{ft-kips}$$
$$= 750 \ \text{ft-kips}$$

The effective width of the concrete slab is the lesser of

- $$s = (10 \ \text{ft})\left(12 \ \frac{\text{in}}{\text{ft}}\right)$$
$$= 120 \ \text{in}$$

- $$\frac{L}{4} = \frac{(40 \ \text{ft})\left(12 \ \frac{\text{in}}{\text{ft}}\right)}{4}$$
$$= 120 \ \text{in} \quad \text{[governs]}$$

The nominal shear strength of a $\frac{3}{4}$ in diameter stud shear connector in 3000 lbf/in^2 concrete is obtained from *AISC Manual* Table 3-21 as

$$Q_n = 21.0 \ \text{kips}$$

The total horizontal shear transferred between sections 1 and 2 is given by

$$\begin{aligned} V_h &= \sum Q_n \\ &= nQ_n \\ &= (17 \text{ studs})(21.0 \text{ kips}) \\ &= 357 \text{ kips} \\ &< F_y A_s \end{aligned}$$

For this value of ΣQ_n, the plastic neutral axis lies below the top of the W21 × 57 at a distance, obtained from *AISC Manual* Table 3-19, of

$$Y_1 = 1.84 \text{ in}$$

The depth of the concrete stress block is given by

$$\begin{aligned} a &= \frac{\sum Q_n}{0.85 f_c' b} \\ &= \frac{357 \text{ kips}}{(0.85)\left(3 \dfrac{\text{kips}}{\text{in}^2}\right)(120 \text{ in})} \\ &= 1.17 \text{ in} \end{aligned}$$

The distance between the top of the steel beam and the centroid of the concrete slab force is

$$\begin{aligned} Y_2 &= Y_{\text{con}} - \frac{a}{2} \\ &= 7.5 \text{ in} - \frac{1.17 \text{ in}}{2} \\ &= 6.92 \text{ in} \end{aligned}$$

For these values of Y_1 and Y_2, the design flexural strength at section 2 is obtained from *AISC Manual* Table 3-19 as

$$\begin{aligned} \phi M_n &= 829 \text{ ft-kips} \\ &> 750 \text{ ft-kips} \quad \text{[satisfactory]} \end{aligned}$$

The shear connectors provided are adequate.

The answer is (D).

ASD Method

The factored bending moment at section 2 due to distributed load is

$$\begin{aligned} M_w &= \frac{0.75 w L^2}{8} = \frac{(0.75)\left(2.67 \dfrac{\text{kips}}{\text{ft}}\right)(40 \text{ ft})^2}{8} \\ &= 400 \text{ ft-kips} \end{aligned}$$

Due to point load,

$$\begin{aligned} M_W &= \frac{Wab}{L} = \frac{(13.33 \text{ kips})(10 \text{ ft})(30 \text{ ft})}{40 \text{ ft}} \\ &= 100 \text{ ft-kips} \end{aligned}$$

The total moment, M_2, is

$$\begin{aligned} M_w &= M_w + M_W \\ &= 400 \text{ ft-kips} + 100 \text{ ft-kips} \\ &= 500 \text{ ft-kips} \end{aligned}$$

The effective width of the concrete slab is the lesser of

- $$\begin{aligned} s &= (10 \text{ ft})\left(12 \frac{\text{in}}{\text{ft}}\right) \\ &= 120 \text{ in} \end{aligned}$$

- $$\begin{aligned} \frac{L}{4} &= \frac{(40 \text{ ft})\left(12 \dfrac{\text{in}}{\text{ft}}\right)}{4} \\ &= 120 \text{ in} \quad \text{[governs]} \end{aligned}$$

The nominal shear strength of a $\frac{3}{4}$ in diameter stud shear connector in 3000 lbf/in^2 concrete is obtained from *AISC Manual* Table 3-21 as

$$Q_n = 21.0 \text{ kips}$$

The total horizontal shear transferred between sections 1 and 2 is given by

$$\begin{aligned} V_h &= \sum Q_n \\ &= nQ_n \\ &= (17 \text{ studs})(21.0 \text{ kips}) \\ &= 357 \text{ kips} \\ &< F_y A_s \end{aligned}$$

For this value of ΣQ_n, the plastic neutral axis lies below the top of the W21 × 57 at a distance, obtained from *AISC Manual* Table 3-19, of

$$Y_1 = 1.84 \text{ in}$$

The depth of the concrete stress block is given by

$$\begin{aligned} a &= \frac{\sum Q_n}{0.85 f_c' b} \\ &= \frac{357 \text{ kips}}{(0.85)\left(3 \dfrac{\text{kips}}{\text{in}^2}\right)(120 \text{ in})} \\ &= 1.17 \text{ in} \end{aligned}$$

The distance between the top of the steel beam and the centroid of the concrete slab force is

$$Y_2 = Y_{con} - \frac{a}{2}$$
$$= 7.5 \text{ in} - \frac{1.17 \text{ in}}{2}$$
$$= 6.92 \text{ in}$$

For these values of Y_1 and Y_2, the allowable flexural strength at section 2 is obtained from *AISC Manual* Table 3-19 as

$$\frac{M_n}{\Omega} = 553 \text{ ft-kips}$$
$$> 500 \text{ ft-kips} \quad [\text{satisfactory}]$$

The shear connectors provided are adequate.

The answer is (D).

5. From the illustration, the gross shear area is

$$A_{gv} = (2)(8 \text{ in})(0.625 \text{ in})$$
$$= 10 \text{ in}^2$$
$$= A_{nv}$$

From the illustration, the net tension area is

$$A_{nt} = (6 \text{ in})(0.625 \text{ in})$$
$$= 3.75 \text{ in}^2$$
$$= A_{gt}$$

The tensile stress is uniform and the reduction coefficient is

$$U_{bs} = 1.0$$

The rupture strength in tension is given by

$$U_{bs}F_uA_{nt} = (1.0)\left(58 \frac{\text{kips}}{\text{in}^2}\right)(3.75 \text{ in}^2)$$
$$= 218 \text{ kips}$$

The yield strength in shear is given by

$$0.6F_yA_{gv} = (0.6)\left(36 \frac{\text{kips}}{\text{in}^2}\right)(10 \text{ in}^2)$$
$$= 216 \text{ kips}$$
$$< 0.6F_uA_{nv}$$

LRFD Method

Shear yielding governs and the resistance to block shear is given by AISC 360 Eq. J4-5 as

$$\phi R_n = \phi(0.6F_yA_{gv} + U_{bs}F_uA_{nt})$$
$$= (0.75)(216 \text{ kips} + 218 \text{ kips})$$
$$= 326 \text{ kips} \quad (330 \text{ kips})$$

The answer is (A).

ASD Method

Shear yielding governs and the allowable strength for block shear is given by AISC 360 Eq. J4-5 as

$$\frac{R_n}{\Omega} = \frac{0.6F_yA_{gv} + U_{bs}F_uA_{nt}}{\Omega}$$
$$= \frac{216 \text{ kips} + 218 \text{ kips}}{2.00}$$
$$= 217 \text{ kips} \quad (220 \text{ kips})$$

The answer is (A).

6. The total length of longitudinally loaded weld is

$$l_{wl} = (2)(8 \text{ in})$$
$$= 16 \text{ in}$$

For a linear weld group with $\theta = 0°$,

$$F_w = 0.60F_{EXX}$$
$$R_n = F_wA_w$$

LRFD Method

The design shear capacity of a $\frac{5}{16}$ in fillet weld is

$$Q_w = q_uD$$
$$= \left(1.39 \frac{\text{kips}}{\text{in}} \text{ per } \frac{1}{16} \text{ in}\right)(5 \text{ sixteenths})$$
$$= 6.95 \text{ kips/in}$$

Applying AISC 360 Eq. J2-4, the design strength of the weld is

$$\phi R_n = \phi R_{wl} = l_{wl}Q_w$$
$$= (16 \text{ in})\left(6.95 \frac{\text{kips}}{\text{in}}\right)$$
$$= 111 \text{ kips} \quad (110 \text{ kips})$$

The answer is (A).

ASD Method

The allowable shear capacity of a $\frac{5}{16}$ in fillet weld is

$$
\begin{aligned}
Q_w &= q_u D \\
&= \left(0.928 \ \frac{\text{kips}}{\text{in}} \ \text{per} \ \frac{1}{16} \ \text{in} \right)(5 \ \text{sixteenths}) \\
&= 4.64 \ \text{kips/in}
\end{aligned}
$$

Applying AISC 360 Eq. J2-4, the allowable strength of the weld is

$$
\begin{aligned}
\frac{R_n}{\Omega} = \frac{R_{wl}}{\Omega} &= l_{wl} Q_w \\
&= (16 \ \text{in}) \left(4.64 \ \frac{\text{kips}}{\text{in}} \right) \\
&= 74 \ \text{kips} \quad (70 \ \text{kips})
\end{aligned}
$$

The answer is (A).

7. The width of the $\frac{3}{4}$ in plate is

$$
w = 6 \ \text{in}
$$

From AISC 360 Table D3.1, Case 4, the length of the connection is

$$
\begin{aligned}
l &= \frac{(l_1 + l_2)}{2} \\
&= \frac{(8 \ \text{in} + 8 \ \text{in})}{2} \\
&= 8 \ \text{in}
\end{aligned}
$$

From AISC 360 Table D3.1, Case 4, the shear lag factor is

$$
U = \left(\frac{3l^2}{3l^2 + w^2} \right) \left(1 - \frac{\bar{x}}{l} \right)
$$

For a plate, this reduces to

$$
\begin{aligned}
U &= \left(\frac{3l^2}{3l^2 + w^2} \right) \\
&= \frac{(3)(8 \ \text{in})^2}{(3)(8 \ \text{in})^2 + (6 \ \text{in})^2} \\
&= 0.84
\end{aligned}
$$

The gross area of the $\frac{3}{4}$ in plate is given by

$$
\begin{aligned}
A_g &= wt \\
&= (6 \ \text{in})(0.75 \ \text{in}) \\
&= 4.5 \ \text{in}^2 \\
&= A_n
\end{aligned}
$$

The effective net area is given by AISC 360 Eq. D3-1 as

$$
\begin{aligned}
A_e &= A_n U \\
&= (4.5 \ \text{in}^2)(0.84) \\
&= 3.78 \ \text{in}^2
\end{aligned}
$$

LRFD Method

The corresponding design tensile capacity for tensile rupture is given by AISC 360 Sec. D2 as

$$
\begin{aligned}
\phi_t P_n &= \phi_t F_u A_e \\
&= (0.75) \left(58 \ \frac{\text{kips}}{\text{in}^2} \right)(3.78 \ \text{in}^2) \\
&= 164 \ \text{kips}
\end{aligned}
$$

The design tensile capacity for yielding of the gross section is given by AISC 360 Sec. D2 as

$$
\begin{aligned}
\phi_t P_n &= \phi_t F_y A_g \\
&= (0.9) \left(36 \ \frac{\text{kips}}{\text{in}^2} \right)(4.5 \ \text{in}^2) \\
&= 146 \ \text{kips} \quad (150 \ \text{kips}) \quad [\text{governs}]
\end{aligned}
$$

The answer is (C).

ASD Method

The corresponding allowable tensile capacity for tensile rupture is given by AISC 360 Sec. D2 as

$$
\begin{aligned}
\frac{P_n}{\Omega_t} &= \frac{F_u A_e}{\Omega_t} \\
&= \frac{\left(58 \ \frac{\text{kips}}{\text{in}^2} \right)(3.78 \ \text{in}^2)}{2.00} \\
&= 110 \ \text{kips}
\end{aligned}
$$

The allowable tensile capacity for yielding of the gross section is given by AISC 360 Sec. D2 as

$$
\begin{aligned}
\frac{P_n}{\Omega_t} &= \frac{F_y A_g}{\Omega_t} \\
&= \frac{\left(36 \ \frac{\text{kips}}{\text{in}^2} \right)(4.5 \ \text{in}^2)}{1.67} \\
&= 97 \ \text{kips} \quad (100 \ \text{kips}) \quad [\text{governs}]
\end{aligned}
$$

The answer is (C).

Wood Design Practice Problems

PRACTICE PROBLEMS

1. The select structural 3×10 Douglas fir-larch rafter shown is notched over a supporting 3 in wall. Based on the bearing stress in the rafter, what is most nearly the maximum available reaction at the support caused by snow loading? (ASD options are shown first. LRFD options are given in parentheses.)

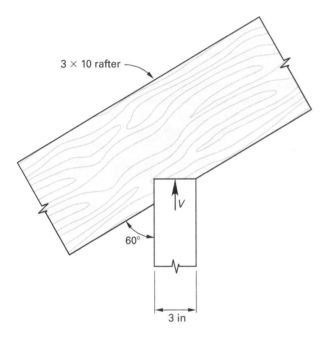

3 × 10 rafter

V

60°

3 in

(A) 5600 lbf (8300 lbf)

(B) 6300 lbf (9100 lbf)

(C) 6800 lbf (9900 lbf)

(D) 7500 lbf (10,700 lbf)

2. The select structural 4×10 Douglas fir-larch ledger shown supports a dead plus floor live load of 225 lbf/ft (ASD) or 390 lbf/ft (LRFD). Based on the $\frac{3}{4}$ in bolt design value in the ledger, what is most nearly the maximum allowable bolt spacing?

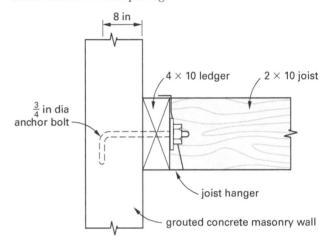

8 in

$\frac{3}{4}$ in dia anchor bolt

4 × 10 ledger

2 × 10 joist

joist hanger

grouted concrete masonry wall

(A) 2.8 ft

(B) 3.2 ft

(C) 3.6 ft

(D) 4.0 ft

3. The floor system in an office building consists of select structural $2 \times$ Douglas fir-larch joists at 16 in centers with $^{19}/_{32}$ plywood sheathing. Each joist supports a dead load of $w_D = 33.33$ lbf/ft plus floor live load of $w_L = 66.67$ lbf/ft over a span of $L = 16$ ft. Acceptable deflection due to live load is $\Delta_{ST} = L/360$ and acceptable deflection due to total load is $\Delta_T = L/240$. The depth of joist necessary to give acceptable stresses and deflection is most nearly

(A) 8 in

(B) 10 in

(C) 12 in

(D) 14 in

4. A select structural 6×6 Douglas fir-larch column is subjected to axial load due to dead plus floor live load. The column is 10 ft high and may be considered pin ended. What is most nearly the maximum load that may be applied? (ASD options are shown first. LRFD options are given in parentheses.)

(A) 18,300 lbf (31,300 lbf)

(B) 19,300 lbf (32,300 lbf)

(C) 21,300 lbf (34,300 lbf)

(D) 22,300 lbf (35,300 lbf)

SOLUTIONS

1. The reference design value for compressive bearing parallel to grain is tabulated in NDS Supp. Table 4A and is

$$F_c = 1700 \text{ lbf/in}^2$$

The applicable adjustment factors for compressive bearing parallel to grain are

$$C_t = 1.0$$
$$C_M = 1.0$$
$$C_F = 1.0$$
$$C_i = 1.0$$

The reference design value for compression perpendicular to grain is tabulated in NDS Supp. Table 4A and is

$$F_{c\perp} = 625 \text{ lbf/in}^2$$
$$C_M = 1.0, \quad C_t = 1.0, \quad C_i = 1.0$$

The bearing area factor for compression perpendicular to grain is specified in NDS Sec. 3.10.4 as

$$\begin{aligned}
C_b &= \frac{l_b + 0.375}{l_b} \\
&= \frac{3 \text{ in} + 0.375}{3 \text{ in}} \\
&= 1.125
\end{aligned}$$

ASD Method

From Table 6.5, the load duration factor for snow is

$$C_D = 1.15$$

The adjusted compressive bearing design value parallel to grain is

$$\begin{aligned}
F_c^* &= F_c C_t C_M C_F C_i C_D \\
&= \left(1700 \ \frac{\text{lbf}}{\text{in}^2}\right)(1.0)(1.0)(1.0)(1.0)(1.15) \\
&= 1955 \text{ lbf/in}^2
\end{aligned}$$

The adjusted compression design value perpendicular to grain is

$$\begin{aligned}
F_{c\perp}' &= F_{c\perp} C_b C_M C_t C_i \quad [C_D \text{ is not applicable}] \\
&= \left(625 \ \frac{\text{lbf}}{\text{in}^2}\right)(1.125)(1.0)(1.0)(1.0) \\
&= 703 \text{ lbf/in}^2
\end{aligned}$$

The allowable bearing design value at an angle θ to the grain is given by NDS Sec. 3.10.3 as

$$F_{\theta}' = \frac{F_c^* F_{c\perp}'}{F_c^* \sin^2\theta + F_{c\perp}' \cos^2\theta}$$

$$= \frac{\left(1955\ \dfrac{\text{lbf}}{\text{in}^2}\right)\left(703\ \dfrac{\text{lbf}}{\text{in}^2}\right)}{\left(1955\ \dfrac{\text{lbf}}{\text{in}^2}\right)(\sin^2 60°) + \left(703\ \dfrac{\text{lbf}}{\text{in}^2}\right)(\cos^2 60°)}$$

$$= 837\ \text{lbf/in}^2$$

The allowable reaction at the support is

$$V = F_{\theta}' b l_b = \left(837\ \frac{\text{lbf}}{\text{in}^2}\right)(2.5\ \text{in})(3\ \text{in})$$

$$= 6278\ \text{lbf} \quad (6300\ \text{lbf})$$

The answer is (B).

LRFD Method

From Table 6.4, the time effect factor for snow load is

$$\lambda = 0.8$$

The format conversion factor given in Table 6.2 is

$$K_F = 1.67 \quad [\text{for } F_{c\perp}]$$

$$K_F = 2.40 \quad [\text{for } F_c]$$

The resistance factor given in Table 6.3 is

$$\phi = 0.90 \quad [\text{for } F_{c\perp}]$$
$$\phi = 0.90 \quad [\text{for } F_c]$$

The adjusted factored compressive bearing design value parallel to grain is

$$F_c^* = F_c C_M C_t C_i C_F K_F \lambda \phi$$

$$= \left(1700\ \frac{\text{lbf}}{\text{in}^2}\right)(1.0)(1.0)(1.0)(1.0)$$

$$\times (2.40)(0.8)(0.90)$$

$$= 2938\ \text{lbf/in}^2$$

The adjusted factored compression design value perpendicular to grain is

$$F_{c\perp}' = F_{c\perp} C_b C_t C_i C_F K_F \phi \quad [\lambda \text{ is not applicable}]$$

$$= \left(625\ \frac{\text{lbf}}{\text{in}^2}\right)(1.125)(1.0)(1.0)(1.0)$$

$$\times (1.67)(0.90)$$

$$= 1057\ \text{lbf/in}^2$$

From NDS Sec. 3.10.3, the strength level bearing design value at an angle θ to the grain is

$$F_{\theta}' = \frac{F_c^* F_{c\perp}'}{F_c^* \sin^2\theta + F_{c\perp}' \cos^2\theta}$$

$$= \frac{\left(2938\ \dfrac{\text{lbf}}{\text{in}^2}\right)\left(1057\ \dfrac{\text{lbf}}{\text{in}^2}\right)}{\left(2938\ \dfrac{\text{lbf}}{\text{in}^2}\right)\sin^2 60° + \left(1057\ \dfrac{\text{lbf}}{\text{in}^2}\right)\cos^2 60°}$$

$$= 1210\ \text{lbf/in}^2$$

The available strength level reaction at the support is

$$V = F_{\theta}' b l_b$$

$$= \left(1210\ \frac{\text{lbf}}{\text{in}^2}\right)(2.5\ \text{in})(3\ \text{in})$$

$$= 9075\ \text{lbf} \quad (9100\ \text{lbf})$$

The answer is (B).

2. The nominal $\frac{3}{4}$ in diameter bolt design value for single shear perpendicular to grain into concrete is tabulated in NDS Table 12E as

$$Z_{\perp} = 900\ \text{lbf}$$
$$C_M = 1.0,\ \ C_t = 1.0,\ \ C_g = 1.0$$

From NDS Sec. 12.5.1, the geometry factor for the bolts is

$$C_{\Delta} = 1.0$$

ASD Method

From Table 6.5, the load duration factor is

$$C_D = 1.00$$

The allowable lateral design value is

$$Z_{\perp}' = Z_{\perp} C_{\Delta} C_M C_g C_t C_D$$

$$= (900\ \text{lbf})(1.0)(1.0)(1.0)(1.0)(1.00)$$

$$= 900\ \text{lbf}$$

The maximum allowable bolt spacing is

$$s = \frac{900 \text{ lbf}}{225 \frac{\text{lbf}}{\text{ft}}}$$
$$= 4.0 \text{ ft}$$

The answer is (D).

LRFD Method

From Table 6.4, the time effect factor for dead load plus floor live load is

$$\lambda = 0.8$$

The format conversion factor given in NDS Table 11.3.1 is

$$K_F = 3.32$$

The resistance factor given in NDS Table 11.3.1 is

$$\phi = 0.65$$

The adjusted factored lateral design value is

$$Z'_\perp = Z_\perp C_M C_g C_\Delta C_t K_F \lambda \phi$$
$$= (900 \text{ lbf})(1.0)(1.0)(1.0)(1.0)(3.32)(0.8)(0.65)$$
$$= 1554 \text{ lbf}$$

The available maximum bolt spacing is

$$s = \frac{1554 \text{ lbf}}{390 \frac{\text{lbf}}{\text{ft}}}$$
$$= 3.98 \text{ ft} \quad (4.0 \text{ ft})$$

The answer is (D).

3. The basic design values for bending and the modulus of elasticity are tabulated in NDS Supp. Table 4A and are

$$F_b = 1500 \text{ lbf/in}^2$$
$$E = 1.9 \times 10^6 \text{ lbf/in}^2$$
$$C_M = 1.0, \ C_L = 1.0, \ C_t = 1.0, \ C_i = 1.0$$

The applicable adjustment factors for bending stress are as follows.

C_F = size factor from NDS Supp. Table 4A,
   assuming a 10 in joist
   = 1.1
C_r = repetitive member factor from NDS Supp.
   Table 4A
   = 1.15

ASD Method

Applying IBC Eq. 16-9, the applied ASD load is

$$w = D + L$$
$$= 33.33 \frac{\text{lbf}}{\text{ft}} + 66.67 \frac{\text{lbf}}{\text{ft}}$$
$$= 100 \text{ lbf/ft}$$

From Table 6.5, the load duration factor is

$$C_D = 1.00$$

The adjusted bending stress is

$$F'_b = F_b C_D C_F C_r$$
$$= \left(1500 \frac{\text{lbf}}{\text{in}^2}\right)(1.00)(1.1)(1.15)$$
$$= 1898 \text{ lbf/in}^2$$

The applied moment on the joist is

$$M = \frac{wL^2}{8}$$
$$= \frac{\left(100 \frac{\text{lbf}}{\text{ft}}\right)(16 \text{ ft})^2 \left(12 \frac{\text{in}}{\text{ft}}\right)}{8}$$
$$= 38{,}400 \text{ in-lbf}$$

The required section modulus is

$$S_{xx} = \frac{M}{F'_b} = \frac{38{,}400 \text{ in-lbf}}{1898 \frac{\text{lbf}}{\text{in}^2}}$$
$$= 20.23 \text{ in}^3$$

Therefore, a 2×10 is adequate for acceptable stresses ($S_{xx} = 21.39 \text{ in}^3$).

LRFD Method

From Table 6.4, the time effect factor for dead load plus floor load is

$$\lambda = 0.8$$

The format conversion factor for bending given in Table 6.2 is

$$K_F = 2.54$$

The resistance factor given in Table 6.3 is

$$\phi = 0.85$$

Applying IBC Eq. 16-2, the applied LRFD load is

$$w = 1.2D + 1.6L$$
$$= (1.2)\left(33.33 \ \frac{\text{lbf}}{\text{ft}}\right) + (1.6)\left(66.67 \ \frac{\text{lbf}}{\text{ft}}\right)$$
$$= 146.67 \ \text{lbf/ft}$$

The adjusted factored bending stress is

$$F_b' = F_b C_F C_r C_M C_t C_L C_i K_F \lambda \phi$$
$$= \left(1500 \ \frac{\text{lbf}}{\text{in}^2}\right)(1.1)(1.15)(1.0)(1.0)$$
$$\quad \times (1.0)(1.0)(2.54)(0.8)(0.85)$$
$$= 3277 \ \text{lbf/in}^2$$

The applied moment on the joint is

$$M = \frac{wL^2}{8}$$
$$= \frac{\left(146.67 \ \frac{\text{lbf}}{\text{ft}}\right)(16 \ \text{ft})^2\left(12 \ \frac{\text{in}}{\text{ft}}\right)}{8}$$
$$= 56{,}321 \ \text{in-lbf}$$

The required section modulus is

$$S_{xx} = \frac{M}{F_b'} = \frac{56{,}321 \ \text{in-lbf}}{3277 \ \frac{\text{lbf}}{\text{in}^2}}$$
$$= 17.19 \ \text{in}^3$$

A 2×10 is adequate for acceptable stresses ($S_{xx} = 21.39 \ \text{in}^3$).

ASD and LRFD: Check the Deflection

The adjusted modulus of elasticity is

$$E' = E$$
$$= 1.9 \times 10^6 \ \text{lbf/in}^2$$

The floor live load is

$$w_L = 66.67 \ \text{lbf/ft}$$

The floor dead load is

$$w_D = 33.33 \ \text{lbf/ft}$$

The required live load deflection is

$$\Delta_{ST} = \frac{L}{360} = \frac{(16 \ \text{ft})\left(12 \ \frac{\text{in}}{\text{ft}}\right)}{360}$$
$$= 0.53 \ \text{in}$$

The corresponding required moment of inertia is

$$I_{xx} = \frac{5 w_L L^4}{384 E \Delta_{ST}}$$
$$= \frac{(5)\left(66.67 \ \frac{\text{lbf}}{\text{ft}}\right)(16 \ \text{ft})^4\left(12 \ \frac{\text{in}}{\text{ft}}\right)^3}{(384)\left(1.9 \times 10^6 \ \frac{\text{lbf}}{\text{in}^2}\right)(0.53 \ \text{in})}$$
$$= 97.63 \ \text{in}^4$$

Therefore, a 2×10 is acceptable for live load deflection ($I_{xx} = 98.9 \ \text{in}^4$).

The required deflection for total load is given as

$$\Delta_T = \frac{L}{240} = \frac{(16 \ \text{ft})\left(12 \ \frac{\text{in}}{\text{ft}}\right)}{240}$$
$$= 0.80 \ \text{in}$$

For long-term loads, NDS Sec. 3.5.2 specifies a creep factor, K_{cr}, of 1.5 for seasoned lumber. Therefore, to determine the total deflection, Δ_T, the applicable equivalent total load is

$$w_T = w_L + K_{cr} w_D$$
$$= 66.67 \ \frac{\text{lbf}}{\text{ft}} + (1.5)\left(33.33 \ \frac{\text{lbf}}{\text{ft}}\right)$$
$$= 116.67 \ \text{lbf/ft}$$

The corresponding required moment of inertia is

$$I_{xx} = \frac{5 w_T L^4}{384 E \Delta_T}$$
$$= \frac{(5)\left(116.67 \ \frac{\text{lbf}}{\text{ft}}\right)(16 \ \text{ft})^4\left(12 \ \frac{\text{in}}{\text{ft}}\right)^3}{(384)\left(1.9 \times 10^6 \ \frac{\text{lbf}}{\text{in}^2}\right)(0.80 \ \text{in})}$$
$$= 113.18 \ \text{in}^4 \quad \text{[governs]}$$

Therefore, a 2×12 is necessary to control deflection ($I_{xx} = 178 \ \text{in}^4$).

The answer is (C).

4. The reference design values for compression and modulus of elasticity are tabulated in NDS Supp. Table 4D and are

$$F_c = 1150 \text{ lbf/in}^2$$
$$E_{min} = 0.58 \times 10^6 \text{ lbf/in}^2$$
$$C_M = 1.0, \quad C_t = 1.0, \quad C_i = 1.0$$

The applicable adjustment factor for compression is

$$C_F = \text{size factor from NDS Supp. Table 4D}$$
$$= 1.0$$

The slenderness ratio is

$$\frac{K_e l}{d} = \frac{(1.0)(10 \text{ ft})\left(12 \dfrac{\text{in}}{\text{ft}}\right)}{5.5 \text{ in}}$$
$$= 21.82$$

ASD Method

The adjusted modulus of elasticity is

$$E'_{min} = E_{min} C_M C_t C_i$$
$$= \left(0.58 \times 10^6 \frac{\text{lbf}}{\text{in}^2}\right)(1.0)(1.0)(1.0)$$
$$= 0.58 \times 10^6 \text{ lbf/in}^2$$

From Table 6.5, the load duration factor for dead load and floor load is

$$C_D = 1.00$$

The reference compression design value multiplied by all applicable adjustment factors except C_P is given by

$$F_c^* = F_c C_F C_D C_f C_M C_i$$
$$= \left(1150 \frac{\text{lbf}}{\text{in}^2}\right)(1.0)(1.00)(1.0)(1.0)(1.0)$$
$$= 1150 \text{ lbf/in}^2$$

The critical buckling design value is

$$F_{cE} = \frac{0.822 E'_{min}}{\left(\dfrac{l_e}{d}\right)^2}$$
$$= \frac{(0.822)\left(0.58 \times 10^6 \dfrac{\text{lbf}}{\text{in}^2}\right)}{\left(\dfrac{(10 \text{ ft})\left(12 \dfrac{\text{in}}{\text{ft}}\right)}{5.5 \text{ in}}\right)^2}$$
$$= 1002 \text{ lbf/in}^2$$

The ratio of F_{cE} to F_c^* is

$$F' = \frac{F_{cE}}{F_c^*} = \frac{1002 \dfrac{\text{lbf}}{\text{in}^2}}{1150 \dfrac{\text{lbf}}{\text{in}^2}}$$
$$= 0.87$$

The column parameter for sawn lumber is obtained from NDS Sec. 3.7.1.5 as

$$c = 0.8$$

The column stability factor specified by NDS Sec. 3.7.1 is

$$C_P = \frac{1.0 + F'}{2c} - \sqrt{\left(\frac{1.0 + F'}{2c}\right)^2 - \frac{F'}{c}}$$
$$= \frac{1.0 + 0.87}{(2)(0.8)} - \sqrt{\left(\frac{1.0 + 0.87}{(2)(0.8)}\right)^2 - \frac{0.87}{0.8}}$$
$$= 0.641$$

The allowable compression design value parallel to grain is

$$F_c' = F_c C_D C_M C_t C_F C_i C_P$$
$$= \left(1150 \frac{\text{lbf}}{\text{in}^2}\right)(1.00)(1.0)(1.0)(1.0)(1.0)(0.641)$$
$$= 737 \text{ lbf/in}^2$$

The allowable load on the column is given by

$$P = F_c'A$$
$$= \left(737 \ \frac{\text{lbf}}{\text{in}^2}\right)(30.25 \ \text{in}^2)$$
$$= 22{,}294 \ \text{lbf} \quad (22{,}300 \ \text{lbf})$$

The answer is (D).

LRFD Method

Using values from Table 6.2 and Table 6.3, the adjusted factored modulus of elasticity is

$$E_{\text{min}}' = E_{\text{min}}C_MC_tC_iK_F\phi_s$$
$$= \left(0.58 \times 10^6 \ \frac{\text{lbf}}{\text{in}^2}\right)(1.0)(1.0)(1.0)(1.76)(0.85)$$
$$= 0.87 \times 10^6 \ \text{lbf/in}^2$$

From Table 6.4, the time effect factor for dead load and floor load is

$$\lambda = 0.8$$

From Table 6.2, the format conversion factor for compression is $K_F = 2.40$

From Table 6.3, the resistance factor for compression is $\phi = 0.90$

The reference compression design value multiplied by all applicable adjustment factors except C_P is given by

$$F_c^* = F_cC_FC_tC_MC_iK_F\lambda\phi_c$$
$$= \left(1150 \ \frac{\text{lbf}}{\text{in}^2}\right)(1.0)(1.0)(1.0)(1.0)$$
$$\times (2.40)(0.8)(0.90)$$
$$= 1987 \ \text{lbf/in}^2$$

The critical buckling design value is

$$F_{cE} = \frac{0.822E_{\text{min}}'}{\left(\dfrac{l_e}{d}\right)^2}$$
$$= \frac{(0.822)\left(0.87 \times 10^6 \ \dfrac{\text{lbf}}{\text{in}^2}\right)}{\left(\dfrac{(10 \ \text{ft})\left(12 \ \dfrac{\text{in}}{\text{ft}}\right)}{5.5 \ \text{in}}\right)^2}$$
$$= 1502 \ \text{lbf/in}^2$$

The ratio of F_{cE} to F_c^* is

$$F' = \frac{F_{cE}}{F_c^*} = \frac{1502 \ \dfrac{\text{lbf}}{\text{in}^2}}{1987 \ \dfrac{\text{lbf}}{\text{in}^2}}$$
$$= 0.756$$

The column parameter for sawn lumber is obtained from NDS Sec. 3.7.1.5 as

$$c = 0.8$$

The column stability factor specified by NDS Sec. 3.7.1 is

$$C_P = \frac{1.0 + F'}{2c} - \sqrt{\left(\frac{1.0 + F'}{2c}\right)^2 - \frac{F'}{c}}$$
$$= \frac{1.0 + 0.756}{(2)(0.8)} - \sqrt{\left(\frac{1.0 + 0.756}{(2)(0.8)}\right)^2 - \frac{0.756}{0.8}}$$
$$= 0.588$$

The adjusted factored compression design value parallel to grain is

$$F_c' = F_cC_FC_tC_MC_iC_PK_F\lambda\phi_c$$
$$= \left(1150 \ \frac{\text{lbf}}{\text{in}^2}\right)(1.0)(1.0)(1.0)(1.0)(0.588)$$
$$\times (2.40)(0.8)(0.90)$$
$$= 1168 \ \text{lbf/in}^2$$

The strength level load on the column is given by

$$P = F_c'A$$
$$= \left(1168 \ \frac{\text{lbf}}{\text{in}^2}\right)(30.25 \ \text{in}^2)$$
$$= 35{,}332 \ \text{lbf} \quad (35{,}300 \ \text{lbf})$$

The answer is (D).

16 Reinforced Masonry Design Practice Problems

PRACTICE PROBLEMS

(Answer options for SD are given in parentheses.)

1. The nominal 8 in solid grouted concrete block masonry bearing wall shown in the illustration has a specified strength of 1500 lbf/in² and a modulus of elasticity of 1,000,000 lbf/in². The wall supports an axial load, including its own weight, as shown in the illustration. The wall, has a height of 15 ft, and may be considered pinned at the top and bottom. Ignore accidental eccentricity. The wall is not part of the lateral-force resisting system. Determine if the wall is adequate. What is most nearly the minimum required vertical reinforcement in the wall? The wall is assigned to seismic design category C.

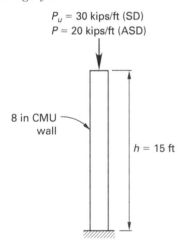

$P_u = 30$ kips/ft (SD)
$P = 20$ kips/ft (ASD)

8 in CMU wall

$h = 15$ ft

(A)   no. 4 at 48 in on center

(B)   no. 4 at 72 in on center

(C)   no. 4 at 96 in on center

(D)   no. 4 at 120 in on center

2. The nominal 8 in solid grouted concrete block masonry retaining wall shown in the illustration has a specified strength of 1500 lbf/in² and a modulus of elasticity of 1,000,000 lbf/in². The reinforcement consists of no. 4 grade 60 bars at 16 in centers. The wall retains a soil with an equivalent fluid pressure of $q = 30$ lbf/ft²/ft, and the self-weight of the wall may be neglected. Is the wall adequate?

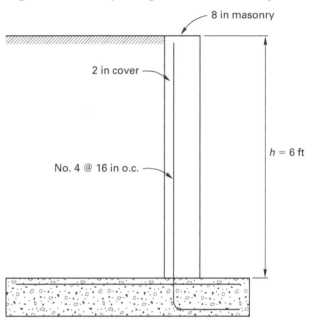

8 in masonry

2 in cover

No. 4 @ 16 in o.c.

$h = 6$ ft

(A)   No, $f_b = 272$ lbf/in², $f_s = 25{,}000$ lbf/in².

(No, $\phi M_n = 4100$ ft-lbf.)

(B)   No, $f_b = 300$ lbf/in², $f_s = 18{,}000$ lbf/in².

(No, $\phi M_n = 3500$ ft-lbf.)

(C)   Yes, $f_b = 300$ lbf/in², $f_s = 24{,}000$ lbf/in².

(Yes, $\phi M_n = 4000$ ft-lbf.)

(D)   Yes, $f_b = 272$ lbf/in², $f_s = 17{,}880$ lbf/in².

(Yes, $\phi M_n = 3400$ ft-lbf.)

3. The 8 in solid grouted concrete block masonry beam shown in the illustration may be considered simply supported over an effective span of 15 ft. The masonry has a compressive strength of 1500 lbf/in² and a modulus of elasticity of 1,000,000 lbf/in², and the reinforcement consists of two no. 7 grade 60 bars. The effective depth is 36 in, the overall depth is 40 in, and the beam is laterally braced at both supports. Is the beam adequate in flexure to support a uniformly distributed load, including its own weight, of 2500 lbf/ft?

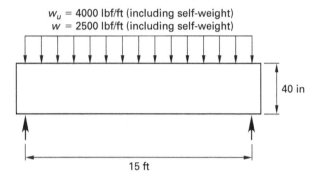

(A)   No, $f_b = 250$ lbf/in², $f_s = 20,150$ lbf/in².

(No, $\phi M_n = 200$ ft-kips.)

(B)   No, $f_b = 425$ lbf/in², $f_s = 16,870$ lbf/in².

(No, $\phi M_n = 150$ ft-kips.)

(C)   Yes, $f_b = 500$ lbf/in², $f_s = 22,470$ lbf/in².

(Yes, $\phi M_n = 170$ ft-kips.)

(D)   Yes, $f_b = 250$ lbf/in², $f_s = 15,975$ lbf/in².

(Yes, $\phi M_n = 130$ ft-kips.)

4. For the masonry beam of Prob. 3, the shear reinforcement required at each support is most nearly

(A)   0.030 in²/ft

(B)   0.034 in²/ft

(C)   0.038 in²/ft

(D)   0.042 in²/ft

SOLUTIONS

1. For axial loading, considering a 1 ft length of wall, the relevant parameters of the wall are

$$b = 12 \text{ in}$$
$$t = \text{nominal depth of wall}$$
$$= 7.63 \text{ in}$$
$$h = \text{effective column height}$$
$$= 15 \text{ ft}$$
$$A_n = \text{effective column area}$$
$$= bt$$
$$= (12 \text{ in})(7.63 \text{ in})$$
$$= 91.5 \text{ in}^2$$
$$P = 20 \text{ kips}$$

The radius of gyration of the wall is

$$r = \sqrt{\frac{I_n}{A_n}}$$
$$= 0.289t$$
$$= (0.289)(7.63 \text{ in})$$
$$= 2.21 \text{ in}$$

The slenderness ratio of the wall is

$$\frac{h}{r} = \frac{(15 \text{ ft})\left(12 \frac{\text{in}}{\text{ft}}\right)}{2.21 \text{ in}}$$
$$= 81.4$$
$$< 99$$

ASD Method

Ignoring vertical reinforcement in conformity with TMS 402 Sec. 8.3.2(d), the allowable wall load is given by TMS 402 Eq. 8-13 as

$$P_a = (0.25 f'_m A_n)\left[1.0 - \left(\frac{h}{140r}\right)^2\right]$$
$$= (0.25)\left(\frac{1500 \frac{\text{lbf}}{\text{in}^2}}{1000 \frac{\text{lbf}}{\text{kip}}}\right)(91.5 \text{ in}^2)\left[1.0 - \left(\frac{81.4}{140}\right)^2\right]$$
$$= 22.7 \text{ kips}$$
$$> P \quad [\text{satisfactory}]$$

The wall is adequate for axial loading, and in accordance with TMS 402 Sec. 7.4.3.1, the necessary minimum vertical reinforcement is no. 4 bars at 120 in centers. The required reinforcement details are shown in the illustration.

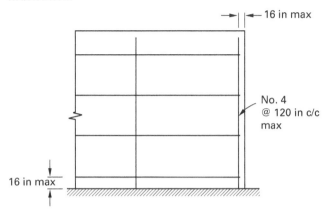

The answer is (D).

SD Method

The slenderness ratio of the wall is

$$\frac{h}{r} = 81.4$$
$$< 99 \quad \text{[TMS 402 Eq. 9-11 is applicable]}$$

The nominal axial strength, given in TMS 402 Eq. 9-11, is

$$P_n = 0.80\left(0.80A_n f'_m\left(1 - \left(\frac{h}{140r}\right)^2\right)\right)$$

$$= (0.80)(0.80)(91.5 \text{ in}^2)\left(\frac{1500 \dfrac{\text{lbf}}{\text{in}^2}}{1000 \dfrac{\text{lbf}}{\text{kip}}}\right)$$

$$\times \left(1 - \left(\frac{81.4}{140}\right)^2\right)$$

$$= 58.1 \text{ kips}$$

Ignoring vertical reinforcement and considering the wall unreinforced, then in accordance with TMS 402 Sec. 9.1.4.3 the design wall strength is

$$\phi P_n = (0.6)(58.1 \text{ kips})$$
$$= 34.9 \text{ kips}$$
$$> P \quad \text{[satisfactory]}$$

The wall is adequate for axial loading and, in accordance with TMS 402 Sec. 7.4.3.1, the necessary minimum vertical reinforcement requires no. 4 bars at 120 in centers. The required reinforcement details are shown.

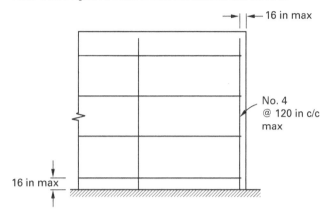

The answer is (D).

2. The allowable stresses, in accordance with TMS 402 Sec. 8.3.2, Sec. 8.3.3, and Sec. 8.3.4.2.2 are

$$F_b = 0.45f'_m$$
$$= (0.45)\left(1500 \frac{\text{lbf}}{\text{in}^2}\right)$$
$$= 675 \text{ lbf/in}^2$$
$$F_s = 32{,}000 \text{ lbf/in}^2$$

At the base of the wall, the bending moment produced in a 1 ft length of wall by the backfill is

$$M = \frac{qh^3}{6}$$
$$= \frac{\left(30 \dfrac{\text{lbf}}{\text{ft}^2}\right)(6 \text{ ft})^3}{6}$$
$$= 1080 \text{ ft-lbf}$$

With vertical reinforcement consisting of no. 4 bars at 16 in centers located with 2 in cover to the earth face, the relevant parameters of the wall are

$$b_w = 12 \text{ in}$$
$$d = 7.63 \text{ in} - 2 \text{ in} - 0.25 \text{ in}$$
$$= 5.38 \text{ in}$$
$$A_s = 0.15 \text{ in}^2$$
$$n = 29$$
$$\rho = \frac{A_s}{b_w d}$$
$$= \frac{0.15 \text{ in}^2}{(12 \text{ in})(5.38 \text{ in})}$$
$$= 0.00232$$
$$\rho n = (0.00232)(29)$$
$$= 0.0674$$

ASD Method

From App. 2.B, the stresses, in accordance with TMS 402 Sec. 8.3.2, are

$$k = \sqrt{2\rho n + (\rho n)^2} - \rho n$$
$$= \sqrt{(2)(0.0674) + (0.0674)^2} - 0.0674$$
$$= 0.306$$
$$j = 1 - \frac{k}{3} = 1 - \frac{0.306}{3}$$
$$= 0.898$$
$$f_b = \frac{2M}{jk b_w d^2}$$
$$= \frac{(2)(1080 \text{ ft-lbf})\left(12 \frac{\text{in}}{\text{ft}}\right)}{(0.898)(0.306)(12 \text{ in})(5.38 \text{ in})^2}$$
$$= 272 \text{ lbf/in}^2$$
$$< F_b \quad [\text{satisfactory}]$$

$$f_s = \frac{M}{jd A_s}$$
$$= \frac{(1080 \text{ ft-lbf})\left(12 \frac{\text{in}}{\text{ft}}\right)}{(0.898)(5.38 \text{ in})(0.15 \text{ in}^2)}$$
$$= 17{,}880 \text{ lbf/in}^2$$
$$< F_s \quad [\text{satisfactory}]$$

The wall is adequate.

The answer is (D).

SD Method

The factored moment at the base of the wall is given by IBC Eq. 16-2 as

$$M_u = 1.6M$$
$$= (1.6)(1080 \text{ ft-lbf})$$
$$= 1728 \text{ ft-lbf}$$

The stress block depth is

$$a = \frac{A_s f_y}{0.80 b_w f'_m}$$
$$= \frac{(0.15 \text{ in}^2)\left(60 \frac{\text{kips}}{\text{in}^2}\right)\left(1000 \frac{\text{lbf}}{\text{kips}}\right)}{(0.80)(12 \text{ in})\left(1500 \frac{\text{lbf}}{\text{in}^2}\right)}$$
$$= 0.63 \text{ in}$$

The nominal strength is

$$M_n = A_s f_y \left(d - \frac{a}{2}\right)$$
$$= \frac{(0.15 \text{ in}^2)\left(60 \frac{\text{kips}}{\text{in}^2}\right)\left(5.38 \text{ in} - \frac{0.63 \text{ in}}{2}\right)\times \left(1000 \frac{\text{lbf}}{\text{kip}}\right)}{12 \frac{\text{in}}{\text{ft}}}$$
$$= 3799 \text{ ft-lbf}$$

The design strength is

$$\phi M_n = (0.9)(3799 \text{ ft-lbf})$$
$$= 3419 \text{ ft-kips} \quad (3400 \text{ ft-lbf})$$
$$> M_u \quad [\text{satisfactory}]$$

The wall is adequate.

The answer is (D).

3. The allowable stresses, in accordance with TMS 402 Sec. 8.3.2 and Sec. 8.3.4.2.2, are

$$F_b = 0.45 f'_m$$
$$= (0.45)\left(1500 \frac{\text{lbf}}{\text{in}^2}\right)$$
$$= 675 \text{ lbf/in}^2$$
$$F_s = 32{,}000 \text{ lbf/in}^2$$

The relevant parameters of the beam are

$$E_m = 1,000,000 \text{ lbf/in}^2$$
$$E_s = 29,000,000 \text{ lbf/in}^2$$
$$b = 7.63 \text{ in}$$
$$d = 36 \text{ in}$$
$$l = 12 \text{ ft}$$
$$A_s = 1.20 \text{ in}^2$$

$$\frac{l}{b} = \frac{(12 \text{ ft})\left(12 \frac{\text{in}}{\text{ft}}\right)}{7.63 \text{ in}}$$
$$= 18.9$$
$$< 32 \quad \begin{bmatrix} \text{satisfies TMS 402} \\ \text{Sec. 5.2.1.2(a)} \end{bmatrix}$$

$$\frac{120b^2}{d} = \frac{(120)(7.63 \text{ in})^2}{36 \text{ in}}$$
$$= 194 \text{ in}$$
$$> l \quad [\text{satisfies TMS 402 Sec. 5.2.1.2(b)}]$$

$$n = \frac{E_s}{E_m}$$
$$= \frac{29,000,000 \dfrac{\text{lbf}}{\text{in}^2}}{1,000,000 \dfrac{\text{lbf}}{\text{in}^2}}$$
$$= 29$$

$$\rho = \frac{A_s}{bd}$$
$$= \frac{1.20 \text{ in}^2}{(7.63 \text{ in})(36 \text{ in})}$$
$$= 0.00437$$
$$\rho n = (0.00437)(29)$$
$$= 0.127$$

ASD Method

At midspan, the bending moment produced by the distributed load is

$$M = \frac{wl^2}{8} = \frac{\left(2500 \dfrac{\text{lbf}}{\text{ft}}\right)(15 \text{ ft})^2}{(8)\left(1000 \dfrac{\text{lbf}}{\text{kip}}\right)}$$
$$= 70.3 \text{ ft-kips}$$

From App. 2.B, the beam stresses, in accordance with TMS 402 Sec. 8.3, are

$$k = \sqrt{2\rho n + (\rho n)^2} - \rho n$$
$$= \sqrt{(2)(0.127) + (0.127)^2} - 0.127$$
$$= 0.393$$
$$j = 1 - \frac{k}{3} = 1 - \frac{0.393}{3}$$
$$= 0.869$$
$$f_b = \frac{2M}{jkbd^2}$$
$$= \frac{(2)(70.3 \text{ ft-kips})\left(12 \dfrac{\text{in}}{\text{ft}}\right)\left(1000 \dfrac{\text{lbf}}{\text{kip}}\right)}{(0.869)(0.393)(7.63 \text{ in})(36 \text{ in})^2}$$
$$= 500 \text{ lbf/in}^2$$
$$< F_b \quad [\text{satisfactory}]$$
$$f_s = \frac{M}{jdA_s}$$
$$= \frac{(70.3 \text{ ft-kips})\left(12 \dfrac{\text{in}}{\text{ft}}\right)\left(1000 \dfrac{\text{lbf}}{\text{kip}}\right)}{(0.869)(36 \text{ in})(1.20 \text{ in}^2)}$$
$$= 22,472 \text{ lbf/in}^2 \quad (22,470 \text{ lbf/in}^2)$$
$$< F_s \quad [\text{satisfactory}]$$

The beam is adequate in flexure.

The answer is (C).

SD Method

The factored moment produced by the distributed load at midspan is

$$M_u = \frac{w_u l^2}{8} = \frac{\left(4000 \dfrac{\text{lbf}}{\text{ft}}\right)(15 \text{ ft})^2}{(8)\left(1000 \dfrac{\text{lbf}}{\text{kip}}\right)}$$
$$= 112.50 \text{ ft-kips}$$

The stress block depth is

$$a = \frac{A_s f_y}{0.80 b f'_m}$$
$$= \frac{(1.20 \text{ in}^2)\left(60 \dfrac{\text{kips}}{\text{in}^2}\right)\left(1000 \dfrac{\text{lbf}}{\text{kip}}\right)}{(0.80)(7.63 \text{ in})\left(1500 \dfrac{\text{lbf}}{\text{in}^2}\right)}$$
$$= 7.86 \text{ in}$$

Practice Problems

The nominal strength is

$$M_n = A_s f_y \left(d - \frac{a}{2} \right)$$

$$= \frac{(1.20 \text{ in}^2) \left(60 \dfrac{\text{kips}}{\text{in}^2} \right) \left(36 \text{ in} - \dfrac{7.86 \text{ in}}{2} \right)}{12 \dfrac{\text{in}}{\text{ft}}}$$

$$= 192 \text{ ft-kips}$$

The design strength is

$$\phi M_n = (0.9)(192 \text{ ft-kips})$$

$$= 173 \text{ ft-kips} \quad (170 \text{ ft-kips})$$

$$> M_u \quad [\text{satisfactory}]$$

The beam is adequte in flexure.

The answer is (C).

4. *ASD Method*

For a *solid* grouted beam, the maximum permitted shear stress, assuming $M/(Vd) = 1$, is given by TMS 402 Eq. 8-24 as

$$F_v = 2\sqrt{f'_m}$$

$$= 2 \sqrt{1500 \dfrac{\text{lbf}}{\text{in}^2}}$$

$$= 77.5 \text{ lbf/in}^2$$

The shear force at a distance of $d/2$ from each support is given by

$$V = \frac{w(l - d)}{2}$$

$$= \frac{\left(2500 \dfrac{\text{lbf}}{\text{ft}} \right) \left(15 \text{ ft} - \left(\dfrac{36 \text{ in}}{12 \dfrac{\text{in}}{\text{ft}}} \right) \right)}{(2) \left(1000 \dfrac{\text{lbf}}{\text{kip}} \right)}$$

$$= 15.0 \text{ kips}$$

The shear stress at a distance of $d/2$ from each support is given by TMS 402 Eq. 8-21 as

$$f_v = \frac{V}{A_{nv}}$$

$$= \frac{(15.0 \text{ kips}) \left(1000 \dfrac{\text{lbf}}{\text{kip}} \right)}{(7.63 \text{ in})(40 \text{ in})}$$

$$= 49.2 \text{ lbf/in}^2$$

$$< 77.5 \text{ lbf/in}^2 \quad \begin{bmatrix} \text{satisfies TMS 402} \\ \text{Sec. 8.3.5.1.2} \end{bmatrix}$$

The allowable shear stress in a beam without shear reinforcement is given by TMS 402 Eq. 8-26. Since $P = 0 \text{ lbf/in}^2$,

$$F_{vm} = \frac{1}{2} \left(\left(4.0 - 1.75 \left(\frac{M}{Vd} \right) \right) \sqrt{f'_m} \right) + 0.25 \left(\frac{P}{A_n} \right)$$

$$= \left(\frac{1}{2} \right) \left((4.0 - (1.75)(1.0)) \sqrt{1500 \dfrac{\text{lbf}}{\text{in}^2}} \right) + 0 \dfrac{\text{lbf}}{\text{in}^2}$$

$$= 43.6 \text{ lbf/in}^2$$

$$< f_v = 49.2 \text{ lbf/in}^2 \quad \begin{bmatrix} \text{shear reinforcement} \\ \text{is required} \end{bmatrix}$$

The shear stress required from shear reinforcement is given by TMS 402 Eq. 8-21 as

$$F_{vs} = f_v - F_{vm}$$

$$= 49.2 \dfrac{\text{lbf}}{\text{in}^2} - 43.6 \dfrac{\text{lbf}}{\text{in}^2}$$

$$= 5.6 \text{ lbf/in}^2$$

The area of shear reinforcement required per foot is given by TMS 402 Eq. 8-27 as

$$A_v = \frac{2 F_{vs} A_{nv} s}{F_s d_v}$$

$$= \frac{(2) \left(5.6 \dfrac{\text{lbf}}{\text{in}^2} \right) (7.63 \text{ in})(40 \text{ in}) \left(12 \dfrac{\text{in}}{\text{ft}} \right)}{\left(32{,}000 \dfrac{\text{lbf}}{\text{in}^2} \right) (40 \text{ in})}$$

$$= 0.03205 \text{ in}^2/\text{ft} \quad (0.034 \text{ in}^2/\text{ft})$$

The answer is (B).

SD Method

The factored shear force at a distance, $d/2$, from each support is

$$V_u = \frac{w_u(l-d)}{2}$$

$$= \frac{\left(4000 \ \frac{\text{lbf}}{\text{ft}}\right)\left(15 \ \text{ft} - \left(\frac{36 \ \text{in}}{12 \ \frac{\text{in}}{\text{ft}}}\right)\right)}{(2)\left(1000 \ \frac{\text{lbf}}{\text{kip}}\right)}$$

$$= 24 \ \text{kips}$$

The maximum nominal shear capacity permitted, assuming $M_u/(V_u d_v) = 1.0$, is limited by TMS 402 Eq. 9-19 to

$$V_n \leq 4A_{nv}\sqrt{f'_m} = \frac{(4)(7.63 \ \text{in})(40 \ \text{in})\sqrt{1500 \ \frac{\text{lbf}}{\text{in}^2}}}{1000 \ \frac{\text{lbf}}{\text{kip}}}$$

$$= 47.28 \ \text{kips}$$

The maximum design shear capacity permitted is

$$\phi V_n = (0.8)(47.28 \ \text{kips})$$
$$= 37.82 \ \text{kips}$$
$$> V_u = 24 \ \text{kips} \quad [\text{satisfactory}]$$

The nominal shear capacity of the beam without shear reinforcement is given by TMS 402 Eq. 9-20. Since $P_u = 0 \ \text{lbf}$,

$$V_{nm} = \left(4.0 - 1.75\left(\frac{M_u}{V_u d_v}\right)\right)A_{nv}\sqrt{f'_m} + 0.25P_u$$

$$= \frac{\big(4.0 - (1.75)(1.0)\big)(7.63 \ \text{in})(40 \ \text{in})}{1000 \ \frac{\text{lbf}}{\text{kip}}}$$
$$\quad \times \sqrt{1500 \ \frac{\text{lbf}}{\text{in}^2}} + 0 \ \text{lbf}$$

$$= 26.60 \ \text{kips}$$

$$\phi V_{nm} = (0.8)(26.60 \ \text{kips})$$
$$= 21.3 \ \text{kips}$$
$$< V_u = 24 \ \text{kips} \quad \begin{bmatrix} \text{shear reinforcement} \\ \text{is required} \end{bmatrix}$$

The design shear capacity required from shear reinforcement is given by TMS 402 Eq. 9-17 as

$$\phi V_{ns} = V_u - \phi V_{nm}$$
$$= 24 \ \text{kips} - 21.3 \ \text{kips}$$
$$= 2.7 \ \text{kips}$$

The area of shear reinforcement required per foot is given by TMS 402 Eq. 9–21 as

$$A_v = \frac{2\phi V_{ns}s}{\phi F_y d_v}$$

$$= \frac{(2)(2.7 \ \text{kips})\left(12 \ \frac{\text{in}}{\text{ft}}\right)}{(0.8)\left(60 \ \frac{\text{kips}}{\text{in}^2}\right)(40 \ \text{in})}$$

$$= 0.034 \ \text{in}^2/\text{ft}$$

The answer is (B).

17 Lateral Forces Practice Problems

PRACTICE PROBLEMS

Lateral Force-Resisting Systems

1. A dual system with steel moment frames and steel special concentrically braced frames is subjected to a seismic force of 280 kips. The relative stiffnesses of the braced frame and the moment frame are 100:40. What is most nearly the design lateral force for each frame?

(A) 200 kips (braced frame); 80 kips (moment frame)

(B) 210 kips (braced frame); 90 kips (moment frame)

(C) 220 kips (braced frame); 100 kips (moment frame)

(D) 230 kips (braced frame); 110 kips (moment frame)

2. A dual system with steel moment frames and steel special concentrically braced frames is subjected to a seismic force of 280 kips. The relative stiffnesses of the braced frame and the moment frame are 100:15. What is most nearly the design lateral force for each frame?

(A) 230 kips (braced frame); 60 kips (moment frame)

(B) 240 kips (braced frame); 70 kips (moment frame)

(C) 250 kips (braced frame); 80 kips (moment frame)

(D) 260 kips (braced frame); 90 kips (moment frame)

3. The single-story building shown has a rigid roof diaphragm that is acted on by an east-west force of 40 kips. The building's center of gravity is located at its center.

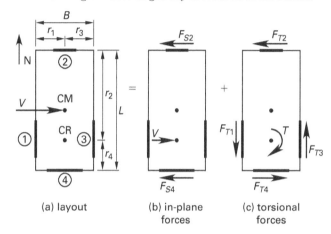

(a) layout   (b) in-plane forces   (c) torsional forces

The following parameters apply.

- $L = 80$ ft
- $B = 40$ ft
- $R_4 = 3R$
- $R_1 = R_2 = R_3 = 1R$
- $e_y = 20$ ft
- allow for accidental eccentricity
- $r_1 = r_3 = 20$ ft
- $r_4 = 20$ ft
- $r_2 = 60$ ft
- $J = (5600 \text{ ft}^2)R$
- neglect the amplification factor

The force produced in wall 2 is most nearly

(A) 17 kips

(B) 18 kips

(C) 19 kips

(D) 20 kips

Seismic Design

4. A one-story industrial building is assigned to seismic design category D and is located in an area with a 0.2 sec acceleration coefficient of $S_{DS} = 1.0$ and a 1.0 sec acceleration coefficient of $S_{D1} = 0.7$. Details are shown in the illustration. The weight of the wood roof is 15 lbf/ft², and the weight of the masonry walls is 75 lbf/ft². The walls are considered hinged at their base. The roof sheathing is ¹⁵/₃₂ in Structural I grade plywood and is blocked. For north-south seismic loads, what is most nearly the service level unit shear along the diaphragm boundaries?

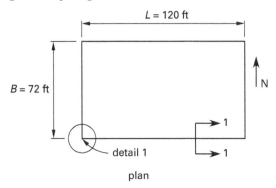

plan

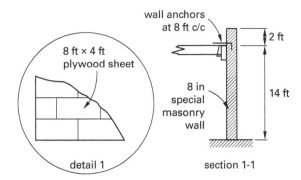

(A) 285 lbf/ft

(B) 320 lbf/ft

(C) 380 lbf/ft

(D) 410 lbf/ft

5. For the building of Prob. 4, wall anchors are provided to the masonry wall at 8 ft centers. What is most nearly the strength level design force in each anchor?

(A) 4000 lbf

(B) 4200 lbf

(C) 4400 lbf

(D) 4600 lbf

Wind Design

6. The office building shown is located in a suburban area, which is subjected to a wind speed of $V = 100$ mi/hr. Wind flows normal to the 60 ft long side of the building. The building's risk category is II. The building is enclosed, has flexible roof and floor diaphragms, and it is not sensitive to dynamic effects, nor is it located on a site at which channeling or buffeting occur. What are most nearly the wind loads (in lbf/ft²) at the center of the windward wall for the main wind force-resisting system? Use the analytical envelope method.

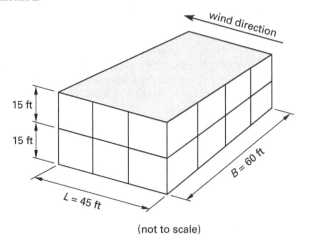

(not to scale)

(A) 9 lbf/ft² (internal suction); 3 lbf/ft² (internal pressure)

(B) 10 lbf/ft² (internal suction); 4 lbf/ft² (internal pressure)

(C) 11 lbf/ft² (internal suction); 5 lbf/ft² (internal pressure)

(D) 12 lbf/ft² (internal suction); 6 lbf/ft² (internal pressure)

7. The roof framing of the building in Prob. 6 consists of joists spaced at 5 ft centers that span 15 ft parallel to the long side of the building. Wind flows normal to the 60 ft long side of the building. What is most nearly the design wind load that acts on a roof beam in zone 2? Use the envelope design method.

(A) 72 lbf/ft (down); 140 lbf/ft (up)

(B) 76 lbf/ft (down); 145 lbf/ft (up)

(C) 80 lbf/ft (down); 150 lbf/ft (up)

(D) 84 lbf/ft (down); 155 lbf/ft (up)

SOLUTIONS

1. The frames will resist applied force in proportion to their relative stiffnesses. Find the design force for the braced frame.

$$F = \frac{(100)(280 \text{ kips})}{100 + 40}$$
$$= 200 \text{ kips}$$

Find the design force for the moment frame.

$$F = 280 \text{ kips} - 200 \text{ kips}$$
$$= 80 \text{ kips}$$

ASCE/SEI7 Table 12.2-1 states that a moment frame must resist a minimum of 25% of the applied lateral force. Therefore, the minimum force for the moment frame is

$$F = (0.25)(280 \text{ kips})$$
$$= 70 \text{ kips} < 80 \text{ kips} \quad [\text{satisfactory}]$$

The answer is (A).

2. In accordance with ASCE/SEI7 Table 12.2-1, the moment frame in a dual system must be designed for 25% of the applied lateral force. Therefore, the moment frame must be designed for the force

$$F = (0.25)(280 \text{ kips})$$
$$= 70 \text{ kips}$$

The braced frame will resist the applied force proportional to its stiffness. Therefore, the braced frame must be designed for the force

$$F = \frac{(100)(280 \text{ kips})}{100 + 15}$$
$$= 243 \text{ kips} \quad (240 \text{ kips})$$

The answer is (B).

3. Accidental eccentricity, in accordance with ASCE/SEI7 Sec. 12.8.4.2, is

$$e_a = \pm 0.05 L$$
$$= \pm (0.05)(80 \text{ ft})$$
$$= \pm 4 \text{ ft}$$

An accidental displacement of the center of mass to the north gives a maximum eccentricity of

$$e = e_y + e_a$$
$$= 20 \text{ ft} + 4.0 \text{ ft}$$
$$= 24 \text{ ft}$$

The maximum eccentricity governs for the force in wall 2 since the torsional force and the in-plane force act in the same direction and are additive. The maximum torsional moment acting about the center of rigidity is

$$T = Ve$$
$$= (40 \text{ kips})(24 \text{ ft})$$
$$= 960 \text{ ft-kips}$$

The torsional shear force in wall 2 is

$$F_{T2} = \frac{T r_2 R_2}{J}$$
$$= \frac{(960 \text{ ft-kips})(60 \text{ ft})(1R)}{(5600 \text{ ft}^2)R}$$
$$= 10.29 \text{ kips}$$

The in-plane shear force in wall 2 is

$$F_{S2} = \frac{V R_2}{R_2 + R_4}$$
$$= \frac{(40 \text{ kips})1R}{1R + 3R}$$
$$= 10 \text{ kips}$$

The total force in wall 2 is

$$F_2 = F_{S2} + F_{T2}$$
$$= 10 \text{ kips} + 10.29 \text{ kips}$$
$$= 20.29 \text{ kips} \quad (20 \text{ kips})$$

The answer is (D).

4. The relevant dead load tributary to the roof diaphragm in the north-south direction is due to the north and south wall and the roof dead load and is obtained as

$$w_r = \left(15 \frac{\text{lbf}}{\text{ft}^2}\right)(72 \text{ ft})$$
$$= 1080 \text{ lbf/ft}$$

$$w_{\text{N+S walls}} = \frac{(2 \text{ walls})\left(75 \frac{\text{lbf}}{\text{ft}^2}\right)(16 \text{ ft})^2}{(2)(14 \text{ ft})}$$
$$= 1371 \text{ lbf/ft}$$

The total dead load tributary to the roof diaphragm is

$$w_{px} = \frac{\left(1080 \frac{\text{lbf}}{\text{ft}} + 1371 \frac{\text{lbf}}{\text{ft}}\right)(120 \text{ ft})}{1000 \frac{\text{lbf}}{\text{kip}}}$$
$$= 294 \text{ kips}$$

Practice Problems

For a standard occupancy structure, the importance factor is

$$I_e = 1.0$$

For a bearing wall structure with special reinforced masonry shear walls, the value of the response modification factor is obtained from Table 8.6 as

$$R = 5.0$$

For this type of structure, the maximum value of the seismic response coefficient controls, and the value of the seismic response coefficient is given by ASCE/SEI7 Sec. 12.8.1.1 as

$$
\begin{aligned}
C_s &= \frac{S_{DS}I_e}{R} \\
&= \frac{(1.0)(1.0)}{5} \\
&= 0.2
\end{aligned}
$$

In accordance with ASCE/SEI7 Sec. 12.10.1.1, the force acting on the roof diaphragm of a one-story building may be taken as

$$
\begin{aligned}
F_{px} &= C_s w_{px} \\
&= 0.2 w_{px}
\end{aligned}
$$

The minimum allowable force acting on the roof diaphragm is given by ASCE/SEI7 Sec. 12.10.1.1 as

$$
\begin{aligned}
F_{px(\min)} &= 0.2 S_{DS} I_e w_{px} \\
&= (0.2)(1.0)(1.0) w_{px} \\
&= 0.2 w_{px}
\end{aligned}
$$

The force on the diaphragm is given by

$$
\begin{aligned}
F_{px} &= (0.2)(294 \text{ kips}) \\
&= 59 \text{ kips}
\end{aligned}
$$

This value of F_{px} is at the strength level, and the equivalent service level value for design of the diaphragm using allowable stress design is given by ASCE/SEI7 Sec. 2.4.1 as

$$
\begin{aligned}
F'_{px} &= 0.7 F_{px} \\
&= (0.7)(59 \text{ kips}) \\
&= 41 \text{ kips}
\end{aligned}
$$

The service level design unit shear along the diaphragm boundary is

$$
\begin{aligned}
q &= \frac{F'_{px}}{2B} \\
&= \frac{(41 \text{ kips})\left(1000 \dfrac{\text{lbf}}{\text{kip}}\right)}{(2)(72 \text{ ft})} \\
&= 285 \text{ lbf/ft}
\end{aligned}
$$

The required nail spacing is obtained from SDPWS Table 4.2A with a case 1 plywood layout applicable, all edges blocked, and 2 in framing. Using both $^{15}/_{32}$ in Structural I grade plywood and 10d nails with $1\frac{1}{2}$ in penetration, a nail spacing of 6 in at the diaphragm boundaries, and 6 in at all other panel edges gives an allowable unit shear in accordance with SDPWS Sec. 4.2.3 of

$$
q_a = \frac{640 \dfrac{\text{lbf}}{\text{ft}}}{2} = 320 \text{ lbf/ft}
$$

$$> 285 \text{ lbf/ft} \quad [\text{satisfactory}]$$

The answer is (A).

5. The relevant weight of the element tributary to the wall anchors is obtained from Prob. 4 as

$$
\begin{aligned}
W_w &= \frac{1371 \dfrac{\text{lbf}}{\text{ft}}}{2} \\
&= 685.5 \text{ lbf/ft}
\end{aligned}
$$

The weight of wall tributary to each anchor is

$$
\begin{aligned}
W_p &= W_w s \\
&= \left(685.5 \frac{\text{lbf}}{\text{ft}}\right)(8 \text{ ft}) \\
&= 5484 \text{ lbf}
\end{aligned}
$$

From the illustration in Prob. 4, the span of the flexible diaphragm is $L_f = 120$ ft.

The amplification factor for diaphragm flexibility is

$$
\begin{aligned}
k_a &= 1.0 + \frac{L_f}{100} \\
&= 1.0 + \frac{120 \text{ ft}}{100 \text{ ft}} \\
&= 2.2 \quad (2.0 \text{ maximum})
\end{aligned}
$$

For seismic design category D, the seismic lateral force on an anchor is given by ASCE/SEI7 Eq. 12.11-1 as

$$F_p = 0.4 S_{DS} k_a I_e W_p$$
$$= (0.4)(1.0)(2.0)(1.0)(5484 \text{ lbf})$$
$$= 4387 \text{ lbf} \quad [\text{governs}]$$

The minimum permissible force on one anchor is

$$F_p = 0.2 k_a I_e W_p$$
$$= (0.2)(2.0)(1.0)(5484 \text{ lbf})$$
$$= 2194 \text{ lbf}$$
$$< 4387 \text{ lbf}$$

The required seismic design force for the anchors is

$$F_p = 4387 \text{ lbf} \quad (4400 \text{ lbf})$$

The answer is (C).

6. For the two-story building with flexible diaphragms shown in the illustration, the illustration in ASCE/SEI7 Fig. 28.3-1 Note 5 specifies that torsional load cases may be neglected.

For a suburban location, the exposure is category B and the wind speed, V, is given as 100 mi/hr. The relevant parameters are

K_h = velocity pressure exposure coefficient

    = 0.7, from Table 8.17 for a height of 30 ft

        for cladding and components

K_{zt} = topographic factor

    = 1.0, from ASCE/SEI7 Fig. 26.8-1

K_d = wind directionality factor

    = 0.85, from ASCE/SEI7 Table 26.6-1

K_e = ground elevation factor

    = 1.0, as permitted by ASCE/SEI7 Sec. 26.9

The velocity pressure, q_h, at a roof height of 30 ft above the ground is given by ASCE/SEI7 Eq. 26.10.1 as

$$q_h = 0.00256 K_h K_{zt} K_d K_e V^2$$
$$= (0.00256)(0.7)(1.0)(0.85)(1.0)\left(100 \, \frac{\text{mi}}{\text{hr}}\right)^2$$
$$= 15.23 \text{ lbf/ft}^2$$

The product of the internal pressure coefficient and the gust effect factor for an enclosed building is obtained from Table 8.14 as

$$(GC_{pi}) = \pm 0.18$$

For wind acting transversely, load case A is applicable, and the product of the external pressure coefficient and the gust effect factor acting on zone 1 for an enclosed building is obtained from Table 8.15 as

$$(GC_{pf}) = 0.40$$

The design wind pressure acting on zone 1 is obtained from ASCE/SEI7 Eq. 28.3-1 as

$$p = q_h \big((GC_p) - (GC_{pi})\big)$$
$$= \left(15.23 \, \frac{\text{lbf}}{\text{ft}^2}\right)(0.4 + 0.18)$$
$$= 8.83 \, \frac{\text{lbf}}{\text{ft}^2} \quad (9 \text{ lbf/ft}^2) \quad [\text{for internal suction}]$$
$$= \left(15.23 \, \frac{\text{lbf}}{\text{ft}^2}\right)(0.4 - 0.18)$$
$$= 3.35 \text{ lbf/ft}^2 \quad (3 \text{ lbf/ft}^2) \quad [\text{for internal pressure}]$$

The answer is (A).

7. The effective tributary width of a roof joist is defined in ASCE/SEI7 Sec. 26.2 as the larger of

$$b_e = \text{joist spacing}$$
$$= 5 \text{ ft}$$

Or,

$$b_e = \frac{l}{3}$$
$$= \frac{15 \text{ ft}}{3}$$
$$= 5 \text{ ft} \quad [\text{governs}]$$

The effective wind area attributed to the roof joist is then

$$A = b_e l$$
$$= (5 \text{ ft})(15 \text{ ft})$$
$$= 75 \text{ ft}^2$$

For a suburban location, the exposure is category B and the wind speed, V, is given as 100 mi/hr. The relevant parameters are

K_h = velocity pressure exposure coefficient
  = 0.7, from Table 8.17 for a height of 30 ft
     for cladding and components
K_{zt} = topographic factor
  = 1.0, from ASCE/SEI7 Fig. 26.8-1
K_d = wind directionality factor
  = 0.85, from ASCE/SEI7 Table 26.6-1
K_e = ground elevation factor
  = 1.0, as permitted by ASCE/SEI7 Sec. 26.9

The velocity pressure, q_h, at a roof height of 30 ft above the ground is given by ASCE/SEI7 Eq. 26.10.1 as

$$q_h = 0.00256 K_h K_{zt} K_d K_e V^2$$
$$= (0.00256)(0.7)(1.0)(0.85)(1.0)\left(100\ \frac{\text{mi}}{\text{hr}}\right)^2$$
$$= 15.23\ \text{lbf/ft}^2$$

The product of the internal pressure coefficient and the gust effect factor for an enclosed building is obtained from Table 8.14 as

$$(GC_{pi}) = \pm 0.18$$

For an effective wind area of $A = 75$ ft^2, the negative external pressure coefficient for zone 2 is obtained from ASCE/SEI7 Table 30.3-2A as

$$(GC_p) = -1.8$$

The negative design wind pressure on a roof joist for zone 2 is obtained from ASCE/SEI7 Eq. 30.3-1 as

$$p = q_h\big((GC_p) - (GC_{pi})\big)$$
$$= \left(15.23\ \frac{\text{lbf}}{\text{ft}^2}\right)(-1.8 - 0.18)$$
$$= -30.16\ \text{lbf/ft}^2$$

The upward load on the roof joist over zone 2 is

$$w = ps$$
$$= \left(-30.16\ \frac{\text{lbf}}{\text{ft}^2}\right)(5\ \text{ft})$$
$$= -151\ \text{lbf/ft} \quad (150\ \text{lbf/ft})$$

For an effective wind area of $A = 75$ ft^2, the positive external pressure coefficient for zone 2 is obtained from ASCE/SEI7 Table 30.3-2A as

$$(GC_p) = 0.2$$

The positive design wind pressure on a roof joist for zone 2 is obtained from ASCE/SEI7 Eq. 30.3-1 as

$$p = q_h\big((GC_p) + (GC_{pi})\big)$$
$$= \left(15.23\ \frac{\text{lbf}}{\text{ft}^2}\right)(0.2 + 0.18)$$
$$= 5.75\ \text{lbf/ft}^2$$

In accordance with ASCE/SEI7 Sec. 30.2.2, the minimum pressure allowed is $p = 16$ lbf/ft^2.

The downward load on the roof joist over zone 2 is

$$w = ps$$
$$= \left(16\ \frac{\text{lbf}}{\text{ft}^2}\right)(5\ \text{ft})$$
$$= 80\ \text{lbf/ft}$$

The answer is (C).

18 Bridge Design Practice Problems

PRACTICE PROBLEMS

1. The reinforced concrete T-beam bridge shown in the illustration is simply supported over a span of 40 ft. The deck has an overall width of 39 ft with five supporting beams, and has three 12 ft design lanes. The ratio of the modulus of elasticity of the beam and the deck slab is $n = 1.0$. The superimposed dead load on an interior beam due to surfacing is 0.25 kip/ft, and due to parapets is 0.2 kip/ft. For an interior beam, what is most nearly the bending moment produced by the permanent loads?

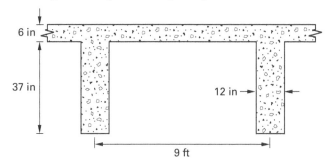

(A) 200 ft-kips

(B) 250 ft-kips

(C) 300 ft-kips

(D) 325 ft-kips

2. For the reinforced concrete T-beam bridge of Prob. 1, the maximum bending moment produced in an interior beam by the design lane load in combination with the design tandem is most nearly

(A) 470 ft-kips

(B) 600 ft-kips

(C) 630 ft-kips

(D) 730 ft-kips

For Prob. 3 through Prob. 5, assume $M_C = 250$ ft-kips, $M_W = 50$ ft-kips, and $M_L = 627$ ft-kips.

3. For the reinforced concrete T-beam bridge of Prob. 1, the strength I limit state factored moment for design of an interior beam is most nearly

(A) 1420 ft-kips

(B) 1485 ft-kips

(C) 1500 ft-kips

(D) 1660 ft-kips

4. For an interior beam of the reinforced concrete T-beam bridge of Prob. 1, the concrete strength is 4000 lbf/in^2, and the reinforcement consists of nine no. 9 grade 60 bars. The stress block depth is most nearly

(A) 1.25 in

(B) 1.35 in

(C) 1.45 in

(D) 1.55 in

5. For an interior beam of the reinforced concrete T-beam bridge of Prob. 1, are the fatigue stress limits satisfactory? Ignore the effects of the 8 kip axle and assume a value of $g = 0.65$ for the load distribution factor to the interior beam.

(A) No, $(\Delta F)_{TH} = 20.25$ kips/in^2.

(B) No, $(\Delta F)_{TH} = 21.83$ kips/in^2.

(C) Yes, $(\Delta F)_{TH} = 20.25$ kips/in^2.

(D) Yes, $(\Delta F)_{TH} = 21.83$ kips/in^2.

SOLUTIONS

1. The dead load acting on one beam due to the weight of the parapets, the weight of the deck slab, and the self-weight of the beam is

$$w_C = 0.20 \ \frac{\text{kip}}{\text{ft}}$$
$$+ \left(0.15 \ \frac{\text{kip}}{\text{ft}^3} \right) \left(\frac{(0.5 \ \text{ft})(39 \ \text{ft})}{5 \ \text{beams}} + (1.0 \ \text{ft})(3.08 \ \text{ft}) \right)$$
$$= 1.25 \ \text{kips/ft}$$

The bending moment produced in an interior beam at the center of the span by the parapets, deck slab, and beam self-weight is

$$M_C = \frac{w_C L^2}{8}$$
$$= \frac{\left(1.25 \ \dfrac{\text{kips}}{\text{ft}} \right)(40 \ \text{ft})^2}{8}$$
$$= 250 \ \text{ft-kips}$$

The bending moment produced in an interior beam at the center of the span by the surfacing is

$$M_W = \frac{w_W L^2}{8}$$
$$= \frac{\left(0.25 \ \dfrac{\text{kip}}{\text{ft}} \right)(40 \ \text{ft})^2}{8}$$
$$= 50 \ \text{ft-kips}$$

The total dead load bending moment produced in an interior beam at the center of the span is

$$M_D = M_C + M_W$$
$$= 250 \ \text{ft-kips} + 50 \ \text{ft-kips}$$
$$= 300 \ \text{ft-kips}$$

The answer is (C).

2. The bending moment produced at the center of the span by the design lane load is

$$M_{LL} = \frac{w_{LL} L^2}{8}$$
$$= \frac{\left(0.64 \ \dfrac{\text{kip}}{\text{ft}} \right)(40 \ \text{ft})^2}{8}$$
$$= 128 \ \text{ft-kips}$$

As shown in the illustration, the maximum moment due to the design tandem is produced under the lead axle of the design tandem when it is located 1 ft beyond the center of the span, and is given by

$$M_{DT} = \frac{(25 \ \text{kips})(19 \ \text{ft})(21 \ \text{ft} + 17 \ \text{ft})}{40 \ \text{ft}}$$
$$= 451 \ \text{ft-kips}$$

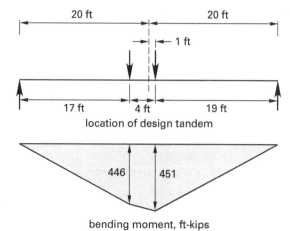

location of design tandem

bending moment, ft-kips

The dynamic load allowance for the span moment is given by AASHTO Table 3.6.2.1-1 as

$$IM = 33\%$$

It is applied to the static axle loads of the design tandem and the dynamic factor to be applied is

$$I = 1 + \frac{IM}{100}$$
$$= 1 + \frac{33}{100}$$
$$= 1.33$$

The moment caused by the design tandem, including the dynamic load allowance, is

$$M_T = (1.33)(451 \ \text{ft-kips})$$
$$= 600 \ \text{ft-kips}$$

The combined moment produced by the design lane load and the design tandem, including the dynamic load allowance, is

$$M_{LL+T} = 128 \ \text{ft-kips} + 600 \ \text{ft-kips}$$
$$= 728 \ \text{ft-kips}$$

The ratio of the modulus of elasticity of the beam and the deck slab is given as

$$n = \frac{E_B}{E_D} = 1.0$$

The moment of inertia of the beam is

$$I = \frac{bd^3}{12}$$
$$= \frac{(12 \text{ in})(37 \text{ in})^3}{12}$$
$$= 50{,}653 \text{ in}^4$$

The area of the beam is

$$A = bd$$
$$= (12 \text{ in})(37 \text{ in})$$
$$= 444 \text{ in}^2$$

The distance between the centers of gravity of the beam and the deck slab is

$$e_g = \frac{t_s + d}{2}$$
$$= \frac{6 \text{ in} + 37 \text{ in}}{2}$$
$$= 21.5 \text{ in}$$

The longitudinal stiffness parameter of the deck is defined by AASHTO Eq. 4.6.2.2.1-1 as

$$K_g = n(I + A e_g^2)$$
$$= (1.0)\left(50{,}653 \text{ in}^4 + (444 \text{ in}^2)(21.5 \text{ in})^2\right)$$
$$= 255{,}892 \text{ in}^4 \quad \begin{bmatrix} \text{complies with AASHTO} \\ \text{Table 4.6.2.2.2b-1} \end{bmatrix}$$
$$> 10{,}000 \text{ in}^4$$
$$< 7{,}000{,}000 \text{ in}^4$$

The beam spacing is

$$S = 9 \text{ ft} \quad [\text{complies with AASHTO Table 4.6.2.2.2b-1}]$$
$$> 3.5 \text{ ft}$$
$$< 16.0 \text{ ft}$$

The deck slab thickness is

$$t_s = 6 \text{ in} \quad [\text{complies with AASHTO Table 4.6.2.2.2b-1}]$$
$$> 4.5 \text{ in}$$
$$< 12.0 \text{ in}$$

The superstructure span is

$$L = 40 \text{ ft} \quad [\text{complies with AASHTO Table 4.6.2.2.2b-1}]$$
$$> 20 \text{ ft}$$
$$< 240 \text{ ft}$$

The number of beams in the deck is

$$N_b = 5 \quad [\text{complies with AASHTO Table 4.6.2.2.2b-1}]$$

Therefore, the configuration of the deck is in full conformity with the requirements of AASHTO Table 4.6.2.2.2b-1.

With one lane loaded, AASHTO Table 4.6.2.2.2b-1 gives the distribution factor for moment as

$$g_1 = 0.06 + \left(\frac{S}{14}\right)^{0.4}\left(\frac{S}{L}\right)^{0.3}\left(\frac{K_g}{12.0 L t_s^3}\right)^{0.1}$$
$$= 0.06 + \left(\frac{9 \text{ ft}}{14}\right)^{0.4}\left(\frac{9 \text{ ft}}{40 \text{ ft}}\right)^{0.3}$$
$$\times \left(\frac{255{,}892 \text{ in}^4}{(12.0)(40 \text{ ft})(6 \text{ in})^3}\right)^{0.1}$$
$$= 0.646$$

With two lanes loaded, as shown in the illustration, AASHTO Table 4.6.2.2.2b-1 gives the distribution factor for moment as

$$g_m = 0.075 + \left(\frac{S}{9.5}\right)^{0.6}\left(\frac{S}{L}\right)^{0.2}\left(\frac{K_g}{12.0 L t_s^3}\right)^{0.1}$$
$$= 0.075 + \left(\frac{9 \text{ ft}}{9.5}\right)^{0.6}\left(\frac{9 \text{ ft}}{40 \text{ ft}}\right)^{0.2}$$
$$\times \left(\frac{255{,}892 \text{ in}^4}{(12.0)(40 \text{ ft})(6 \text{ in})^3}\right)^{0.1}$$
$$= 0.861 \quad [\text{governs}]$$

The live load moment for the design of an interior beam is

$$M_L = g_m M_{L+T} = (0.861)(728 \text{ ft-kips})$$
$$= 627 \text{ ft-kips} \quad (630 \text{ ft-kips})$$

The answer is (C).

3. The relevant service level moments are

M_C = moment produced by the parapets, deck slab, and beam self-weight
   = 250 ft-kips

M_W = moment produced by the wearing surface
   = 50 ft-kips

M_L = moment produced by the design lane load and the design tandem, including the dynamic load allowance
   = 627 ft-kips

The factored design moment for the strength I limit state is given by AASHTO Eq. 3.4.1-1 and AASHTO Table 3.4.1-1 and Table 3.4.1-2 as

$$
\begin{aligned}
M_u &= \eta_i(\gamma_p M_C + \gamma_p M_W + \gamma_L M_L) \\
&= (1.0)(1.25 M_C + 1.5 M_W + 1.75 M_L) \\
&= (1.25)(250 \text{ ft-kips}) + (1.5)(50 \text{ ft-kips}) \\
&\quad + (1.75)(627 \text{ ft-kips}) \\
&= 1485 \text{ ft-kips}
\end{aligned}
$$

The answer is (B).

4. The effective compression flange width is given by AASHTO Sec. 4.6.2.6.1 as the minimum of

$$
\begin{aligned}
b &= S \\
&= (9 \text{ ft})\left(12 \; \frac{\text{in}}{\text{ft}}\right) \\
&= 108 \text{ in}
\end{aligned}
$$

The height of the centroid of the tensile reinforcement is

$$
\begin{aligned}
\bar{c} &= 2 \text{ in} + (2.5)(1.125 \text{ in}) \\
&= 4.81 \text{ in}
\end{aligned}
$$

The effective depth is

$$
\begin{aligned}
d &= h - \bar{c} \\
&= 43 \text{ in} - 4.81 \text{ in} \\
&= 38.19 \text{ in}
\end{aligned}
$$

Assuming that the stress block lies within the flange, the required tension reinforcement is determined from the principles of AASHTO Sec. 5.7. The design moment factor is

$$
\begin{aligned}
K_u &= \frac{M_u}{b_w d^2} \\
&= \frac{(1485 \text{ ft-kips})\left(12 \; \frac{\text{in}}{\text{ft}}\right)\left(1000 \; \frac{\text{lbf}}{\text{kip}}\right)}{(108 \text{ in})(38.19 \text{ in})^2} \\
&= 113 \text{ lbf/in}^2 \\
\frac{K_u}{f_c'} &= \frac{113 \; \dfrac{\text{lbf}}{\text{in}^2}}{4000 \; \dfrac{\text{lbf}}{\text{in}^2}} \\
&= 0.0283
\end{aligned}
$$

From App. 2.A, the corresponding tension reinforcement index is

$$
\begin{aligned}
\omega &= 0.032 \\
&< 0.319\beta_1 \\
&= (0.319)(0.85) \\
&= 0.271
\end{aligned}
$$

Therefore, the section is tension controlled, and $\phi = 0.90$.

The required reinforcement ratio is

$$
\begin{aligned}
\rho &= \frac{\omega f_c'}{f_y} \\
&= \frac{(0.032)\left(4000 \; \dfrac{\text{lbf}}{\text{in}^2}\right)}{60{,}000 \; \dfrac{\text{lbf}}{\text{in}^2}} \\
&= 0.00213
\end{aligned}
$$

The reinforcement area required is

$$
\begin{aligned}
A_s &= \rho b d \\
&= (0.00213)(108 \text{ in})(38.19 \text{ in}) \\
&= 8.79 \text{ in}^2
\end{aligned}
$$

For nine no. 9 bars as shown in the illustration, the reinforcement area provided is

$$
\begin{aligned}
A_s &= 9 \text{ in}^2 \\
&> 8.79 \text{ in}^2 \quad [\text{satisfactory}]
\end{aligned}
$$

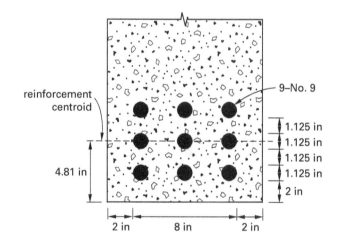

The stress block depth is

$$a = \frac{A_s f_y}{0.85 b f_c'}$$

$$= \frac{(9 \text{ in}^2)\left(60{,}000 \ \dfrac{\text{lbf}}{\text{in}^2}\right)}{(0.85)(108 \text{ in})\left(4000 \ \dfrac{\text{lbf}}{\text{in}^2}\right)}$$

$$= 1.47 \text{ in} \quad (1.45 \text{ in})$$

$$< h_f \quad \begin{bmatrix} \text{The stress block is contained} \\ \text{within the flange.} \end{bmatrix}$$

The answer is (C).

5. The lever arm for elastic design is conservatively obtained from Prob. 4 as

$$l_a = d - \frac{h_f}{2}$$

$$= 38.19 \text{ in} - \frac{6 \text{ in}}{2}$$

$$= 35.19 \text{ in}$$

Fatigue limits are determined using 150% of the stress produced by the design truck plus dynamic load allowance. The distance between the 32 kip axles of the design truck is fixed at 30 ft. Ignoring the effect of the 8 kip lead axle, the maximum moment is developed at the location of the 32 kip axle when this is positioned at the center of the span. The bending moment produced in an interior beam is

$$M_{DT} = \frac{gWL}{4}$$

$$= \frac{(0.65)(32 \text{ kips})(40 \text{ ft})}{4}$$

$$= 208 \text{ ft-kips}$$

The dynamic load allowance for the span moment is given by AASHTO Table 3.6.2.1-1 as

$$IM = 15\%$$

This is applied to the static axle loads of the design truck and the dynamic factor to be applied is

$$I = 1 + \frac{IM}{100}$$

$$= 1 + \frac{15}{100}$$

$$= 1.15$$

150% of the moment caused by the design truck, including the dynamic load allowance, is

$$M_f = (1.5)(1.15) M_{DT}$$

$$= (1.5)(1.15)(208 \text{ ft-kips})$$

$$= 359 \text{ ft-kips}$$

This is the maximum moment range producing fatigue.

The corresponding maximum factored stress range is

$$\gamma(\Delta f) = \frac{M_f}{l_a A_s}$$

$$= \frac{(359 \text{ ft-kips})\left(12 \ \dfrac{\text{in}}{\text{ft}}\right)}{(35.19 \text{ in})(9 \text{ in}^2)}$$

$$= 13.60 \text{ kips/in}^2$$

The permanent dead load moment produced in an interior beam at the center of the span is derived in Prob. 1 as

$$M_D = 300 \text{ ft-kips}$$

This is the minimum moment at the center of the span.

The corresponding minimum stress is

$$f_{\min} = \frac{M_D}{l_a A_s}$$

$$= \frac{(300 \text{ ft-kips})\left(12 \ \dfrac{\text{in}}{\text{ft}}\right)}{(35.19 \text{ in})(9 \text{ in}^2)}$$

$$= 11.37 \text{ kips/in}^2$$

The constant-amplitude fatigue threshold is given by AASHTO Eq. 5.5.3.2-1 as

$$(\Delta F)_{\text{TH}} = 26 - 22\left(\frac{f_{\min}}{f_y}\right)$$

$$= 26 - (22)\left(\frac{11.37 \ \dfrac{\text{kips}}{\text{in}^2}}{60 \ \dfrac{\text{kips}}{\text{in}^2}}\right)$$

$$= 21.83 \text{ kips/in}^2$$

$$> \gamma(\Delta f) \quad [\text{satisfactory}]$$

The answer is (D).

Appendices

Appendices

APPENDIX 2.A
Values of $M_u/f_c'bd^2$ for a Tension-Controlled Section

ω	0.000	0.001	0.002	0.003	0.004	0.005	0.006	0.007	0.008	0.009
0	0.0000	0.0009	0.0018	0.0027	0.0036	0.0045	0.0054	0.0063	0.0072	0.0081
0.01	0.0089	0.0098	0.0107	0.0116	0.0125	0.0134	0.0143	0.0151	0.0160	0.0169
0.02	0.0178	0.0187	0.0195	0.0204	0.0213	0.0222	0.0230	0.0239	0.0248	0.0257
0.03	0.0265	0.0274	0.0283	0.0291	0.0300	0.0309	0.0317	0.0326	0.0334	0.0343
0.04	0.0352	0.0360	0.0369	0.0377	0.0386	0.0394	0.0403	0.0411	0.0420	0.0428
0.05	0.0437	0.0445	0.0454	0.0462	0.0471	0.0479	0.0487	0.0496	0.0504	0.0513
0.06	0.0521	0.0529	0.0538	0.0546	0.0554	0.0563	0.0571	0.0579	0.0588	0.0596
0.07	0.0604	0.0612	0.0621	0.0629	0.0637	0.0645	0.0653	0.0662	0.0670	0.0678
0.08	0.0686	0.0694	0.0702	0.0711	0.0719	0.0727	0.0735	0.0743	0.0751	0.0759
0.09	0.0767	0.0775	0.0783	0.0791	0.0799	0.0807	0.0815	0.0823	0.0831	0.0839
0.10	0.0847	0.0855	0.0863	0.0871	0.0879	0.0887	0.0895	0.0902	0.0910	0.0918
0.11	0.0926	0.0934	0.0942	0.0949	0.0957	0.0965	0.0973	0.0981	0.0988	0.0996
0.12	0.1004	0.1011	0.1019	0.1027	0.1035	0.1042	0.1050	0.1058	0.1065	0.1073
0.13	0.1081	0.1088	0.1096	0.1103	0.1111	0.1119	0.1126	0.1134	0.1141	0.1149
0.14	0.1156	0.1164	0.1171	0.1179	0.1186	0.1194	0.1201	0.1209	0.1216	0.1223
0.15	0.1231	0.1238	0.1246	0.1253	0.1260	0.1268	0.1275	0.1283	0.1290	0.1297
0.16	0.1304	0.1312	0.1319	0.1326	0.1334	0.1341	0.1348	0.1355	0.1363	0.1370
0.17	0.1377	0.1384	0.1391	0.1399	0.1406	0.1413	0.1420	0.1427	0.1434	0.1441
0.18	0.1448	0.1456	0.1463	0.1470	0.1477	0.1484	0.1491	0.1498	0.1505	0.1512
0.19	0.1519	0.1526	0.1533	0.1540	0.1547	0.1554	0.1561	0.1568	0.1574	0.1581
0.20	0.1588	0.1595	0.1602	0.1609	0.1616	0.1623	0.1629	0.1636	0.1643	0.1650
0.21	0.1657	0.1663	0.1670	0.1677	0.1684	0.1690	0.1697	0.1704	0.1710	0.1717
0.22	0.1724	0.1730	0.1737	0.1744	0.1750	0.1757	0.1764	0.1770	0.1777	0.1783
0.23	0.1790	0.1797	0.1803	0.1810	0.1816	0.1823	0.1829	0.1836	0.1842	0.1849
0.24	0.1855	0.1862	0.1868	0.1874	0.1881	0.1887	0.1894	0.1900	0.1906	0.1913
0.25	0.1919	0.1925	0.1932	0.1938	0.1944	0.1951	0.1957	0.1963	0.1970	0.1976
0.26	0.1982	0.1988	0.1995	0.2001	0.2007	0.2013	0.2019	0.2026	0.2032	0.2038
0.27	0.2044	0.2050	0.2056	0.2062	0.2069	0.2075	0.2081	0.2087	0.2093	0.2099
0.28	0.2105	0.2111	0.2117	0.2123	0.2129	0.2135	0.2141	0.2147	0.2153	0.2159
0.29	0.2165	0.2171	0.2177	0.2183	0.2188	0.2194	0.2200	0.2206	0.2212	0.2218
0.30	0.2224	0.2229	0.2235	0.2241	0.2247	0.2253	0.2258	0.2264	0.2270	0.2276
0.31	0.2281	0.2287	0.2293	0.2298	0.2304	0.2310	0.2315	0.2321	0.2327	0.2332
0.32	0.2338	0.2344	0.2349	0.2355	0.2360	0.2366	0.2371	0.2377	0.2382	0.2388
0.33	0.2393	0.2399	0.2404	0.2410	0.2415	0.2421	0.2426	0.2432	0.2437	0.2443
0.34	0.2448	0.2453	0.2459	0.2464	0.2470	0.2475	0.2480	0.2486	0.2491	0.2496
0.35	0.2501	0.2507	0.2512	0.2517	0.2523	0.2528	0.2533	0.2538	0.2543	0.2549
0.36	0.2554	0.2559	0.2564	0.2569	0.2575	0.2580	0.2585	0.2590	0.2595	0.2600
0.37	0.2605	0.2610	0.2615	0.2620	0.2625	0.2631	0.2636	0.2641	0.2646	0.2651
0.38	0.2656	0.2661	0.2665	0.2670	0.2675	0.2680	0.2685	0.2690	0.2695	0.2700
0.39	0.2705	0.2710	0.2715	0.2719	0.2724	0.2729	0.2734	0.2739	0.2743	0.2748

Appendices

APPENDIX 2.B
Values of the Neutral Axis Depth Factor, k

ρn	0.000	0.001	0.002	0.003	0.004	0.005	0.006	0.007	0.008	0.009
0	0.0000	0.0437	0.0613	0.0745	0.0855	0.0951	0.1037	0.1115	0.1187	0.1255
0.01	0.1318	0.1377	0.1434	0.1488	0.1539	0.1589	0.1636	0.1682	0.1726	0.1769
0.02	0.1810	0.1850	0.1889	0.1927	0.1964	0.2000	0.2035	0.2069	0.2103	0.2136
0.03	0.2168	0.2199	0.2230	0.2260	0.2290	0.2319	0.2347	0.2375	0.2403	0.2430
0.04	0.2457	0.2483	0.2509	0.2534	0.2559	0.2584	0.2608	0.2632	0.2655	0.2679
0.05	0.2702	0.2724	0.2747	0.2769	0.2790	0.2812	0.2833	0.2854	0.2875	0.2895
0.06	0.2916	0.2936	0.2956	0.2975	0.2995	0.3014	0.3033	0.3051	0.3070	0.3088
0.07	0.3107	0.3125	0.3142	0.3160	0.3178	0.3195	0.3212	0.3229	0.3246	0.3263
0.08	0.3279	0.3296	0.3312	0.3328	0.3344	0.3360	0.3376	0.3391	0.3407	0.3422
0.09	0.3437	0.3452	0.3467	0.3482	0.3497	0.3511	0.3526	0.3540	0.3554	0.3569
0.10	0.3583	0.3597	0.3610	0.3624	0.3638	0.3651	0.3665	0.3678	0.3691	0.3705
0.11	0.3718	0.3731	0.3744	0.3756	0.3769	0.3782	0.3794	0.3807	0.3819	0.3832
0.12	0.3844	0.3856	0.3868	0.3880	0.3892	0.3904	0.3916	0.3927	0.3939	0.3951
0.13	0.3962	0.3974	0.3985	0.3996	0.4007	0.4019	0.4030	0.4041	0.4052	0.4063
0.14	0.4074	0.4084	0.4095	0.4106	0.4116	0.4127	0.4137	0.4148	0.4158	0.4169
0.15	0.4179	0.4189	0.4199	0.4209	0.4219	0.4229	0.4239	0.4249	0.4259	0.4269
0.16	0.4279	0.4288	0.4298	0.4308	0.4317	0.4327	0.4336	0.4346	0.4355	0.4364
0.17	0.4374	0.4383	0.4392	0.4401	0.4410	0.4419	0.4429	0.4437	0.4446	0.4455
0.18	0.4464	0.4473	0.4482	0.4491	0.4499	0.4508	0.4516	0.4525	0.4534	0.4542
0.19	0.4551	0.4559	0.4567	0.4576	0.4584	0.4592	0.4601	0.4609	0.4617	0.4625
0.20	0.4633	0.4641	0.4649	0.4657	0.4665	0.4673	0.4681	0.4689	0.4697	0.4705
0.21	0.4712	0.4720	0.4728	0.4736	0.4743	0.4751	0.4758	0.4766	0.4774	0.4781
0.22	0.4789	0.4796	0.4803	0.4811	0.4818	0.4825	0.4833	0.4840	0.4847	0.4855
0.23	0.4862	0.4869	0.4876	0.4883	0.4890	0.4897	0.4904	0.4911	0.4918	0.4925
0.24	0.4932	0.4939	0.4946	0.4953	0.4960	0.4966	0.4973	0.4980	0.4987	0.4993
0.25	0.5000	0.5007	0.5013	0.5020	0.5026	0.5033	0.5040	0.5046	0.5053	0.5059
0.26	0.5066	0.5072	0.5078	0.5085	0.5091	0.5097	0.5104	0.5110	0.5116	0.5123
0.27	0.5129	0.5135	0.5141	0.5147	0.5154	0.5160	0.5166	0.5172	0.5178	0.5184
0.28	0.5190	0.5196	0.5202	0.5208	0.5214	0.5220	0.5226	0.5232	0.5238	0.5243
0.29	0.5249	0.5255	0.5261	0.5267	0.5272	0.5278	0.5284	0.5290	0.5295	0.5301
0.30	0.5307	0.5312	0.5318	0.5323	0.5329	0.5335	0.5340	0.5346	0.5351	0.5357
0.31	0.5362	0.5368	0.5373	0.5379	0.5384	0.5389	0.5395	0.5400	0.5406	0.5411
0.32	0.5416	0.5422	0.5427	0.5432	0.5437	0.5443	0.5448	0.5453	0.5458	0.5464
0.33	0.5469	0.5474	0.5479	0.5484	0.5489	0.5494	0.5499	0.5505	0.5510	0.5515
0.34	0.5520	0.5525	0.5530	0.5535	0.5540	0.5545	0.5550	0.5554	0.5559	0.5564
0.35	0.5569	0.5574	0.5579	0.5584	0.5589	0.5593	0.5598	0.5603	0.5608	0.5613
0.36	0.5617	0.5622	0.5627	0.5632	0.5636	0.5641	0.5646	0.5650	0.5655	0.5660
0.37	0.5664	0.5669	0.5674	0.5678	0.5683	0.5687	0.5692	0.5696	0.5701	0.5705
0.38	0.5710	0.5714	0.5719	0.5723	0.5728	0.5732	0.5737	0.5741	0.5746	0.5750
0.39	0.5755	0.5759	0.5763	0.5768	0.5772	0.5776	0.5781	0.5785	0.5789	0.5794

Appendices

APPENDIX 2.C
Interaction Diagram: Tied Circular Column
$(f_c' = 4 \text{ kips/in}^2, f_y = 60 \text{ kips/in}^2, \gamma = 0.60)$

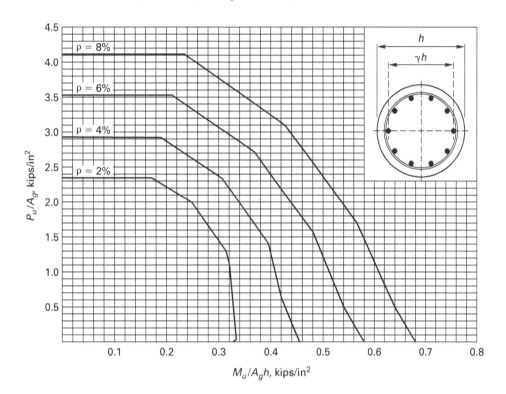

APPENDIX 2.D
Interaction Diagram: Tied Circular Column
$(f'_c = 4 \text{ kips/in}^2,\ f_y = 60 \text{ kips/in}^2,\ \gamma = 0.75)$

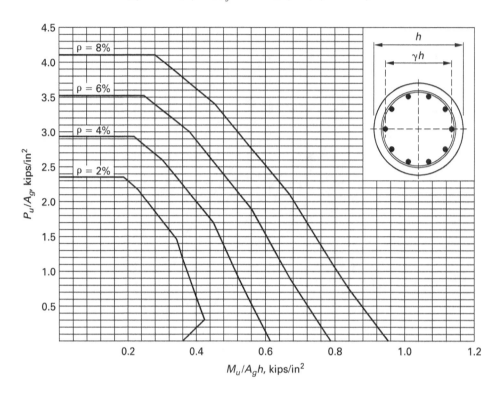

APPENDIX 2.E
Interaction Diagram: Tied Circular Column
($f_c' = 4$ kips/in², $f_y = 60$ kips/in², $\gamma = 0.90$)

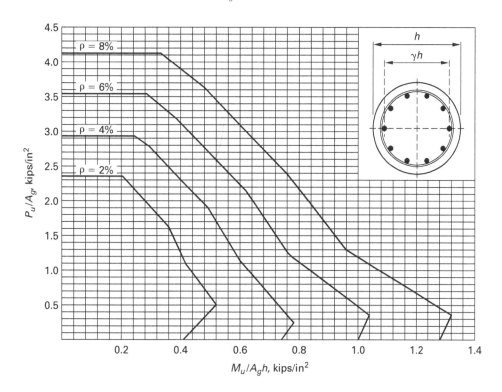

APPENDIX 2.F
Interaction Diagram: Tied Square Column
$(f_c' = 4 \text{ kips/in}^2,\ f_y = 60 \text{ kips/in}^2,\ \gamma = 0.60)$

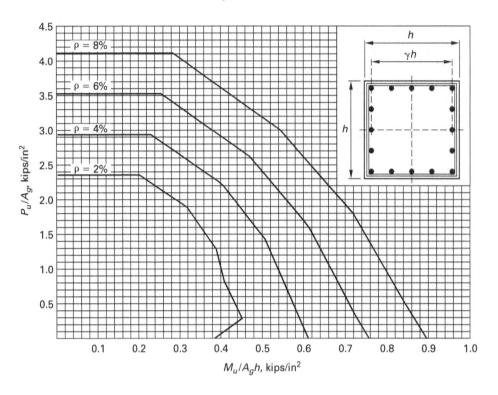

APPENDIX 2.G
Interaction Diagram: Tied Square Column
$(f_c' = 4 \text{ kips/in}^2,\ f_y = 60 \text{ kips/in}^2,\ \gamma = 0.75)$

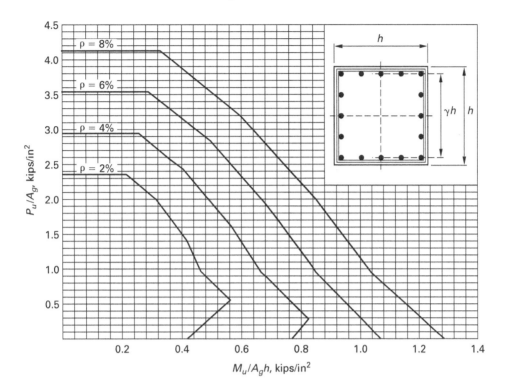

APPENDIX 2.H
Interaction Diagram: Tied Square Column
($f_c' = 4$ kips/in^2, $f_y = 60$ kips/in^2, $\gamma = 0.90$)

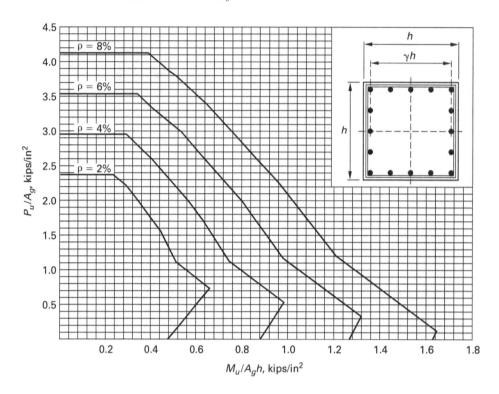

Index

INDEX - N

INDEX - P

INDEX - R